Hymenoptera: Biology, Taxonomy and Integrated Management

Hymenoptera: Biology, Taxonomy and Integrated Management

Collection Editors

Mar Ferrer-Suay

Andrew Polaszek

MDPI

Basel • Beijing • Wuhan • Barcelona • Belgrade • Novi Sad • Cluj • Manchester

Collection Editors

Mar Ferrer-Suay
Departament Zoologia
Universitat de València
Valencia
Spain

Andrew Polaszek
Department of Life Sciences
Natural History Museum
London
UK

Editorial Office
MDPI AG
Grosspeteranlage 5
4052 Basel, Switzerland

This is a reprint of the Topical Collection, published open access by the journal *Insects* (ISSN 2075-4450), freely accessible at: https://www.mdpi.com/journal/insects/topical_collections/Hymenoptera_Biology.

For citation purposes, cite each article independently as indicated on the article page online and as indicated below:

Lastname, A.A.; Lastname, B.B. Article Title. *Journal Name* **Year**, *Volume Number*, Page Range.

ISBN 978-3-7258-5411-0 (Hbk)
ISBN 978-3-7258-5412-7 (PDF)
https://doi.org/10.3390/books978-3-7258-5412-7

Contents

About the Editors

Mar Ferrer-Suay

Mar Ferrer-Suay is an entomologist specializing in the taxonomy of Hymenoptera Parasitica, with a focus on integrative approaches combining morphology, molecular data, and trophic web analysis. She is a leading global expert on the hyperparasitoid subfamily Charipinae, which influences primary parasitoid populations and, indirectly, pest dynamics in agricultural systems. Her research aims to define species boundaries, identify molecular markers, and clarify phylogenetic relationships, providing essential knowledge for integrated pest management (IPM). She has published 130 articles (58 in JCR journals) with 1,111 citations (h-index 17) and has presented at ~20 national and international conferences. She serves as editor for three peer-reviewed journals and is a member of the expert committee for the 2nd International Electronic Conference on Entomology. Dr. Ferrer-Suay supervises undergraduate, master's, and PhD research, has completed 500+ hours of teaching training, and actively contributes to teaching innovation and scientific outreach. She is Co-Director of the Master's Degree in Pest Control and Management and currently Vice-Dean of Quality and Innovation at the Faculty of Biological Sciences, University of Valencia.

Andrew Polaszek

Andrew Polaszek is a Principal Researcher in Entomology at the Natural History Museum, London. After studying Agricultural Entomology at the University of Newcastle-upon-Tyne, he completed a PhD at Imperial College, London. He started his professional career as an entomologist at the Commonwealth Institute of Entomology, focusing on parasitoid wasps. He has published more than 250 scientific papers, including 3 books (as editor). He was the lead author on a paper establishing the ZooBank register of scientific names of animals (Nature 437, 477 2005). He served as President of the International Society of Hymenopterists (2016–2018) and was Keeper of Entomology (Head of Department) at the Natural History Museum (2010–2012). Much of Andrew's work has been collaborative and of an applied nature, with successful classical biological control projects and input into the development of integrated pest management programmes, especially in the tropics. He is committed to public outreach, science education, and the facilitation of descriptive taxonomy using modern techniques. A current priority is the use of high-quality images to inform society at all levels about the importance of parasitoids and other wasps, as well as bees and ants, as ecosystem service providers.

Preface

As Collection Editors, we are delighted to present this Reprint, "Hymenoptera: Biology, Taxonomy and Integrated Management", dedicated to advancing knowledge of one of the most diverse insect taxa—Hymenoptera—and its pivotal role in ecological balance and agricultural sustainability. Hymenoptera, with their rich diversity and the ecological significance of parasitoids as biological control agents, remain a cornerstone of pest management strategies globally. While ongoing research continues to uncover new species, a vast number of undiscovered taxa persist, underscoring the urgency and value of this field.

This Reprint aims to bridge gaps in current entomological understanding by highlighting the critical interplay between taxonomy, biology, and applied management. It showcases cutting-edge work by researchers worldwide, from the description of new species that enrich our grasp of biodiversity to the application of molecular tools in resolving cryptic species complexes. By emphasizing accurate field identification as a foundational step, we underscore how such efforts inform behavioral and biological insights, directly enhancing the efficacy of integrated pest management (IPM) programs. Additionally, contributions on phylogeny and taxonomic revisions provide essential frameworks for understanding evolutionary relationships and refining conservation priorities.

Targeted at entomologists, ecologists, agricultural scientists, and biodiversity researchers, this collection brings together original research and reviews that address key themes: new species discovery, molecular systematics, cryptic diversity, phylogeny, and IPM innovation. We believe this Reprint will serve as a vital resource, inspiring further inquiry and collaboration to unlock the full potential of Hymenoptera in addressing global challenges like pest resistance and biodiversity loss.

We extend our gratitude to all contributors for their rigorous work and dedication, which have made this Reprint a testament to the vitality of Hymenoptera research.

Mar Ferrer-Suay and Andrew Polaszek
Collection Editors

insects MDPI

Article

Notes on the Genus *Aceratoneuromyia* Girault (Hymenoptera: Eulophidae) †

Ning Huangfu [1,2], Huan-Xi Cao [3,*] and Chao-Dong Zhu [1,2]

1 Key Laboratory of Zoological Systematics and Evolution, Institute of Zoology, Chinese Academy of Sciences, Beijing 100101, China; huangfuning@ioz.ac.cn (N.H.); zhucd@ioz.ac.cn (C.-D.Z.)
2 College of Biological Sciences, University of Chinese Academy of Sciences, Beijing 100049, China
3 National Animal Collection Resource Center, Institute of Zoology, Chinese Academy of Sciences, Beijing 100101, China
* Correspondence: caohuanxi@ioz.ac.cn
† Nomenclatural Acts: urn:lsid:zoobank.org:act:0AFC7A7A-FB71-4D87-8610-A00C1E9B337D; urn:lsid:zoobank.org:act:98AB6CD3-0518-47CD-B6D4-15568C2EB1F5; urn:lsid:zoobank.org:act:B6AB4126-0278-46FA-8BA7-7AEB3C7F9F4B.

Simple Summary: Fruit flies in the family Tephritidae are economically important pests of edible fruits, with some known hymenopteran parasitoids. Although *Aceratoneuromyia indica* (Silvestri) is one of the most common parasitoids and has been used in biological control of fruit flies, its establishment in biocontrol is impeded by the difficulties of proper identity. The genus *Aceratoneuromyia* Girault (Hymenoptera: Eulophidae) is studied here, and the generic name *Trjapitzinichus* Kostjukov and Kosheleva is proposed as a new synonym under *Aceratoneuromyia* as well. Based on morphology and DNA barcodes, *A. bilinis* Huangfu and Cao sp. nov., *A. carinata* Cao and Zhu sp. nov. and *A. trilinus* Cao and Zhu sp. nov. are described and illustrated. The well-known parasitoid of fruit flies, *A. indica*, is treated here with the diagnosis and illustrations. In addition, this study provided a morphological diagnosis for *Aceratoneuromyia* as well as a key to world species of this genus. We also briefly discussed the relationship of *Aceratoneuromyia* with other possibly close groups based on available morphological data. Together with DNA barcodes of *A. bilinis* sp. nov. and *A. indica* generated here, this study provided essential and useful information for the species identity of *Aceratoneuromyia*.

Abstract: Fruit flies in the family Tephritidae are well known as economically important pests of edible fruits and can often cause serious damage and losses to both agriculture and the economy. One of the common parasitoids of fruit flies, *Aceratoneuromyia indica* (Silvestri), has been used in biological programs. However, the biocontrol utilities of parasitoids are impeded by the difficulties of proper identification. Species of the genus *Aceratoneuromyia* Girault (Hymenoptera: Eulophidae), usually developed as parasitoids of fruit flies, are studied here. *Trjapitzinichus* Kostjukov and Kosheleva is proposed as a new synonym under *Aceratoneuromyia*. Three new species of *Aceratoneuromyia*, *A. bilinis* Huangfu and Cao sp. nov., *A. carinata* Cao and Zhu sp. nov., and *A. trilinus* Cao and Zhu sp. nov., are described and illustrated from China. *Aceratoneuromyia indica* is also treated here with diagnosis and illustrations. DNA barcodes of *A. bilinis* and *A. indica* and a key to the world species of *Aceratoneuromyia* are provided. This study provided important identification information of parasitoids with morphology and molecular evidence, which is useful for imperative needs regarding the identity of parasitoids attacking fruit flies.

Keywords: Chalcidoidea; new taxon; parasitoid; dipteran host; DNA barcoding

1. Introduction

Many members of dipteran flies are of great economic importance, and some of them are serious pests in agriculture. Among these pests, some species of the family

Tephritidae known as fruit flies attack mostly mature or maturing edible fruits and can make serious fruit infestation and cause significant economic losses, such as Oriental fruit fly, *Bactrocera dorsalis* (Hendel), and medfly, *Ceratitis capitata* (Wiedemann) [1,2]. There are a few projects on the fruit fly parasitoids [3–5], but the utilities of these parasitoids in the field release were possibly not well established as reported before [2,6]. *Aceratoneuromyia indica* (Silvestri), which is developed as a gregarious and koinobiont endoparasitoid, has been widely reported as a successful biological control agent against fruit flies worldwide [7,8]. It has been also recorded in China [1]. However, to some extent, the colony establishments of parasitoids used for field release are hindered by difficulties of identification and little knowledge of their biology.

Species of *Aceratoneuromyia* Girault have been known as primary gregarious endoparasitoids of Diptera, mainly Anthomyiidae and Tephritidae, or as hyperparasitoid (e.g., *A. granularis* Domenichini) of Lepidoptera (*Phragmites australis* (M.C. Day) through parasitic Sarcophagidae of Diptera (e.g., *Cleigastra apicalis* (Meigen)) [7,9]. This genus is one of several small genera in the subfamily Tetrastichinae (Hymenoptera: Eulophidae), with ten described species before the establishment of *Trjapitzinichus* Kostjukov and Kosheleva [7,9,10]. Graham [7] revised the European species of *Aceratoneuromyia* and provided keys to the females of seven species and males of five species. Kostjukov and Kosheleva [10] established a new genus *Trjapitzinichus* to transfer three of these 10 known species, *A. evanescens* (Ratzeburg), *A. lakica* Kostjukov and Gunasheva, and *A. polita* Graham, from *Aceratoneuromyia* to *Trjapitzinichus*, which are included as valid in the Universal Chalcidoidea Database (UCD) [9]. Another two species of *Trjapitzinichus* described based on only holotype by Kostjukov and Kosheleva [11] were not included in UCD. Therefore, seven species have been included in the genus *Aceratoneuromyia* and five species in *Trjapitzinichus* [9] before this study. However, the validity of *Trjapitzinichus* and its relationship with *Aceratoneuromyia* remains elusive based on morphology only, which requires further confirmation by molecular evidence from at least these two groups and other closely related taxa. Before that, this study considers *Trjapitzinichus* as a junior synonym under *Aceratoneuromyia*.

Species of *Aceratoneuromyia* are known mainly from the Old World [7]. Most species of *Aceratoneuromyia* were described or recorded from Palearctic (Europe), and three were described from Oriental (India, Japan), which is hypothesized to result from two possible aspects. On the one hand, *Aceratoneuromyia* species have a natural distribution in Europe; on the other hand, Europe has more taxonomists working on Chalcidoidea, increasing the possibilities of discovering species of *Aceratoneuromyia*. However, the European species *A. fimbriata* Graham has been recorded from the New World and *A. indica* also has been widely spread throughout the world as a biological control agent of fruit flies [8,9]. Despite their importance as a parasitic group and occupying an important niche in the ecosystem, it remains unclear about the phylogenetic relationships of *Aceratoneuromyia* with the other genera in Tetrastichinae. LaSalle [8] inferred that the similarities between *Aceratoneuromyia* and *Melittobia* are possibly due to convergence based on the available evidence at that time.

This study aims to describe three new species of *Aceratoneuromyia* and synonymize the name *Trjapitzinichus* with *Aceratoneuromyia* and discuss the possible relationships of *Aceratoneuromyia* with other possibly close genera in Tetrastichinae based on morphology. Although DNA barcode fragments of two species of *Aceratoneuromyia* (*A. bilinis* and *A. indica*) were generated in this study, by referring to descriptions and specimens of *Aceratoneuromyia* species, this study summarized the concept of *Aceratoneuromyia*. Clear classification and identification of these important parasitoids will facilitate their practical applications in agriculture as potential biocontrol agents and help to better understand interactions with their hosts and other organisms from evolutionary and ecological perspectives.

2. Materials and Methods

2.1. Insect Collection and Rearing

The live larvae of *Pelmatops ichneumoneus* were originally collected during field trips in Yunnan, China, 2020 and reared in the laboratory by feeding on its host plant *Rubus multibracteatus* Levl. et Vant. (Rosaceae), which was transplanted from the collection locality in Yunnan. Larvae and pupae of *Bactrocera dorsalis* were lab-reared and originally collected from field in Fujian, China. The host pupae were placed in plastic cups with a mesh cloth cover and moved to the laboratory of the Institute of Zoology, Chinese Academy of Sciences (IZCAS) to rear adults of fruit flies and parasitoid wasps. Emerged wasps were then preserved in 100% ethanol for further use after emergence.

2.2. Specimens Preparations

Specimens were examined with a Nikon SMZ 1500 stereomicroscope fitted with a 10 mm ocular grid having 100 divisions. Some ethanol-preserved specimens of parasitoid wasps were critical-point dried with a Leica EM CPD300 automated critical point dryer for morphological studies. Several critical-point dried specimens were dissected into head, mesosoma, and metasoma for scanning electron microscopy (SEM), which were subsequently sputter-coated with gold using a Leica EM SCD050 super cool sputter coater. Micrographs were produced using an FEI Quanta 450 environmental scanning electron microscope. Habitus pictures were taken with a Nikon D7000 digital camera connected to the Nikon SMZ 1500 stereomicroscope and then stacked in Helicon Focus software to generate single highly focused images. All images were processed with Adobe Photoshop CC 2019.

Morphological terminology follows Gibson [12]. Abbreviations used for morphological characters are as follows: F1–F3, funiculars 1–3; C1–C3, clavomeres 1–3; MLM, midlobe of mesoscutum; Gtn, gastral tergite number; POL, the shortest distance between the posterior ocelli; OOL, the shortest distance between an eye and posterior ocellus.

2.3. Molecular Analyses

Whole genomic DNA of 15 *Aceratoneuromyia* individuals (Table 1) was extracted using the DNeasy Blood & Tissue Kit (Qiagen GmbH, Hilden, Germany) following the manufacturer's instructions. Two primer pairs, LCO1490 (5′-GGTCA ACAAA TCATA AA-GAT ATTGG-3′) [13] and HCOout (5′-CCAGG TAAAA TTAAA ATATA AACTTC-3′) [14], C1-J-1514 (5′-TATCA ACCAA TCATA AAGAT ATTGG-3′) [13] and C1-N-2194 (5′-CCCGG TAAAA TTAAA ATATA AAC-3′) [15], were used to amplify COI (mitochondrial cytochrome c oxidase I) sequences. All polymerase chain reactions (PCR) were conducted using LongGene T20 Multi-Block Thermal Cycler (Hangzhou, China) using La Taq polymerase (Takara, Japan) under the following conditions: initial denaturation for 1 min at 94 °C; 4 cycles of 1 min at 94 °C, 1.5 min at 45 °C, and 1.5 min at 72 °C; 35 cycles of 2 min at 94 °C, 1.5 min at 58 °C, and 1 min at 72 °C s; followed by a final extension at 72 °C for 5 min. Sequencing was performed in a single direction. COI sequences were examined and aligned in BioEdit version 7.0.9.0 [16]. The COI alignment matrix after trimming was translated into the amino acids in MEGA7 [17] to check for stop codons. A phylogenetic tree by the Kimura 2-parameter (K2P) model with Neighbor-Joining (NJ) method was constructed in MEGA 7.0 with 1000 bootstrap replicates to calculate support values for nodes. Voucher specimens are deposited in IZCAS. The COI sequences generated in this study have been deposited in GenBank under accession numbers ON260937–ON260951.

Table 1. Information of specimens sequenced with GenBank accession number of COI.

Code	Species	Sex	Host	Primer Pair	Accession Number
CHX_ACE_833	*A. indica*	female	*Bactrocera dorsalis*	LCO1490 & HCOout	ON260937
CHX_ACE_834	*A. indica*	female	*Bactrocera dorsalis*	LCO1490 & HCOout	ON260938
CHX_ACE_835	*A. indica*	female	*Bactrocera dorsalis*	LCO1490 & HCOout	ON260939
CHX_ACE_836	*A. indica*	female	*Bactrocera dorsalis*	LCO1490 & HCOout	ON260940
CHX_ACE_837	*A. indica*	female	*Bactrocera dorsalis*	LCO1490 & HCOout	ON260941
CHX_ACE_838	*A. indica*	female	*Bactrocera dorsalis*	LCO1490 & HCOout	ON260942
CHX_ACE_839	*A. indica*	male	*Bactrocera dorsalis*	LCO1490 & HCOout	ON260943
CHX_ACE_840	*A. indica*	male	*Bactrocera dorsalis*	LCO1490 & HCOout	ON260944

Table 1. *Cont.*

Code	Species	Sex	Host	Primer Pair	Accession Number
CHX_ACE_841	*A. indica*	male	*Bactrocera dorsalis*	LCO1490 & HCOout	ON260945
CHX_ACE_842	*A. indica*	male	*Bactrocera dorsalis*	LCO1490 & HCOout	ON260946
CHX_ACE_843	*A. indica*	male	*Bactrocera dorsalis*	LCO1490 & HCOout	ON260947
CHX_ACE_844	*A. indica*	male	*Bactrocera dorsalis*	LCO1490 & HCOout	ON260948
CHX_ACE_788	*A. bilinis*	female	*Pelmatops ichneumoneus*	C1-J-1514 & C1-N-2194	ON260949
CHX_ACE_789	*A. bilinis*	female	*Pelmatops ichneumoneus*	C1-J-1514 & C1-N-2194	ON260950
CHX_ACE_790	*A. bilinis*	female	*Pelmatops ichneumoneus*	C1-J-1514 & C1-N-2194	ON260951

3. Results

During recent programs on the biology of Tephritidae in China, fresh materials of two species of *Aceratoneuromyia* were respectively collected and reared from the pupae of *Bactrocera dorsalis* (Hendel) and *Pelmatops ichneumoneus* (Westwood) in the family Tephritidae. The species reared from *Bactrocera dorsali* was identified as *A. indica*, and the other species reared from *Pelmatops ichneumoneus* was identified as a species new to science. By comparing with specimens deposited in the insect collection of IZCAS, two additional new species of *Aceratoneuromyia* were discovered as well. As a result, three new species were treated and illustrated in this study.

Fifteen COI sequences were successfully sequenced from 3 females of *Aceratoneuromyia bilinis*, and 6 females plus 6 males of *A. indica*. A COI matrix with a length of 657 base pairs was obtained based on these 15 COI sequences after alignment and trimming, which has no insertion or deletion. A NJ tree (Figure 1) by K2P distances was generated based on this COI alignment.

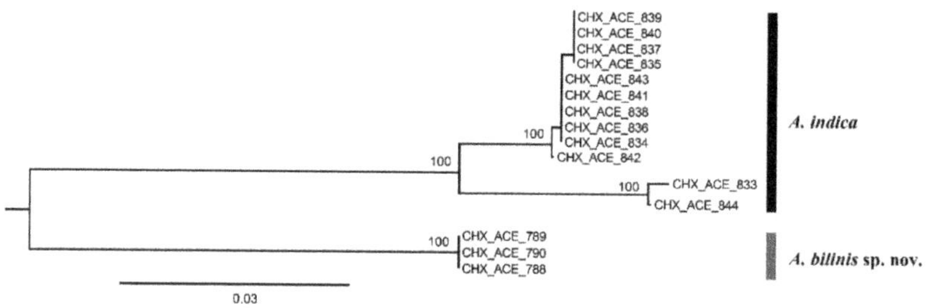

Figure 1. A NJ tree by K2P distances based on COI sequences from 15 *Aceratoneuromyia* specimens. The numbers above nodes represent bootstrap values.

3.1. *Morphological Diagnosis and Species Treatments of Aceratoneuromyia*

The genus *Aceratoneuromyia* Girault, 1917

Aceratoneuromyia Girault, 1917 [18]: 151. Type species: *Aceratoneuromyia australia* Girault, by original designation.

Trjapitzinichus Kostjukov & Kosheleva, 2006 [10]: 108. Type species: *Entedon evanescens* Ratzeburg, by original designation. Syn. nov.

Diagnosis. Antenna with very long apical seta of terminal spine (about 2× as long as terminal spine, sometimes broken off) and with long, scattered erect or semi-erect setae (e.g., Figures 3d and 9b) as well as male antenna, with each funicular having a short transverse row of long setae on the dorsal surface (e.g., Figures 5e and 9h); female antenna with funiculars usually quadrate to transverse (e.g., Figure 9h), rarely distinctly longer than broad; male antennal scape usually with (e.g., Figure 3d) and rarely without (e.g., A. evanescens) ventral plaque. Malar sulcus present, from weak to groove-like. Frontofacial sutures (e.g., Figures 3b and 9a,g) not or only slightly divergent dorsally, sometimes absent. Scutellum with (e.g., Figure 9a) or without (e.g., Figure 9d,i) submedian lines, if present, then weak and incomplete, with sublateral lines distinct and laterally carinate (e.g., Figure 9a); scutellum with anterior pair of setae situated in the posterior half and close to the posterior pair of setae (e.g., Figure 9a,d,i). Gaster usually slender, rarely rounded, with ovipositor sheaths not visible even ventrally (e.g., Figures 4e and 9f,k); spiracles of penultimate segment of gaster rather large (relatively smaller in males), usually visible in dorsal view, rarely directed laterally and not visible in dorsal view (e.g., Figures 4f and 9e,j); female gaster usually tending to be convex dorsally (e.g., Figures 2a and 8a). Hypopygium extending distinctly more than half length of the gaster, nearly to the apex (e.g., Figures 4e and 9f,k).

(a) **(b)**

Figure 2. *Aceratoneuromyia bilinis* sp. nov. (**a**) Habitus of female (paratype, IOZ(E)221423) in dorsolateral view; (**b**) Habitus of male (paratype, IOZ(E)221422) in dorsolateral view. Scale bar: 0.5 mm.

Remarks. The diagnosis of *Aceratoneuromyia* was listed here by referring to the three newly described species, descriptions of the other known species, as well as the diagnosis given by Bouček [19], Graham [7], LaSalle [8], and Ikeda [20]. Ikeda [20] has proposed a modified concept for *Aceratoneuromyia* with the addition of a new species from Japan and South Korea, *A. kamijoi* Ikeda, which disagrees in some characters with the concept previously given. This study follows those modifications to interpret the genus *Aceratoneuromyia*. Based on this genus concept of *Aceratoneuromyia* and re-descriptions of *Trjapitzinichus* by Kostjukov & Kosheleva [11], sometimes the placement of some intermediated species into one of them could be confusing. Furthermore, the separation of *Trjapitzinichus* from *Aceratoneuromyia* by Kostjukov & Kosheleva [10,11] obscures the differences between these two genera. It is hereby considered unreasonable to keep them as separate genera and *Trjapitzinichus* is synonymized with *Aceratoneuromyia* here.

3.1.1. *Aceratoneuromyia bilinis* Huangfu & Cao, sp. nov.

Diagnosis. Female antenna clavate, with F1 subquadrate, F2 and F3 transverse (Figure 3d), with scape not reaching anterior ocellus (Figure 2a). Frontofacial sutures narrowly separated from each other and then passed as an extremely short line that just reaching shallow scrobes in frontal view (Figures 3b and 5b), and divergent and visible dorsally, sometimes connecting the vertexal sutures (Figures 3d and 5b). Pronotum with a row of 4 long erect setae near posterior margin (Figures 3a and 5a). MLM with 3 adnotaular setae (Figures 3a and 5a). Scutellum with somewhat distinct but incomplete submedian lines (Figures 3a and 5a). Dorsellum large, about 2× as broad as high. Metasoma slender (Figure 2a,b, Figures 4d and 5c), as long as the combined length of mesosoma and head; in female, convex dorsally in female (Figure 2a), dorsolateral posterior margin of Gt_5 circled along outer margin of large spiracles of Gt_6 (Figure 4f). The longest cercal seta 2× length of the other 3 setae equal in length and weakly curved apically (Figures 4e and 5c).

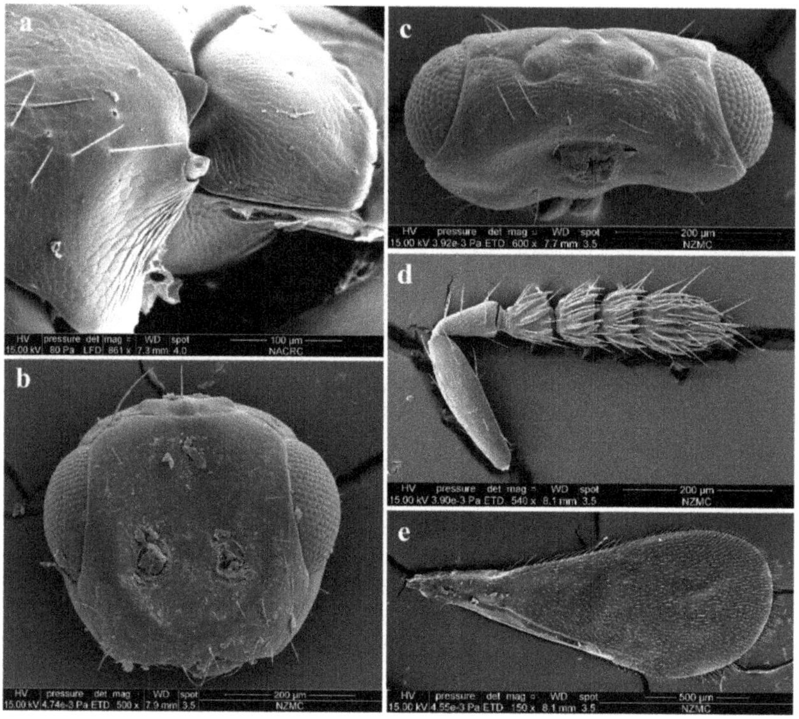

Figure 3. *Aceratoneuromyia bilinis* sp. nov., female. (**a**) Lateral spiracles of pronotum in dorsolateral view; (**b**) Face; (**c**) Head in dorsal view; (**d**) Antenna in lateral view; (**e**) Forewing.

Female (Figures 2a, 3 and 4). Body length 2.0–2.3 mm. Body black with a very weak dark coppery green tinge (Figure 2a), with tegula fuscous, with black setae. Brown antenna except scape and pedicel yellowish-brown ventrally (Figure 2a). Legs with fore and hind coxae dark brown, mid coxae brown, remainder parts testaceous except dark claws (Figure 2a). Wings hyaline, with brown veins.

Antenna (Figure 3d) with 3 funiculars and 3 clavomeres, setae of pedicel and flagellum standing out strongly; scape not reaching anterior ocellus, about 2.5× as long as pedicel; funicle proximally distinctly stouter than pedicel; pedicel distinctly longer than F1 (0.7:0.5); F1–F3 gradually decreasing in length (0.5:0.4:0.3) and increasing in width, F1 wider apically, length equal to apical width, F2 and especially F3 transverse; clava as long as F2 plus F3, clavomeres decreasing in length, C3 with a short but distinct terminal spine that has a distinctly long seta (about 2× length of terminal spine). Each flagellomere with longitudinal sensilla and apically with a circle of scattered, mushroom-shaped capitate peg sensilla; each flagellomere except C3 truncate apically.

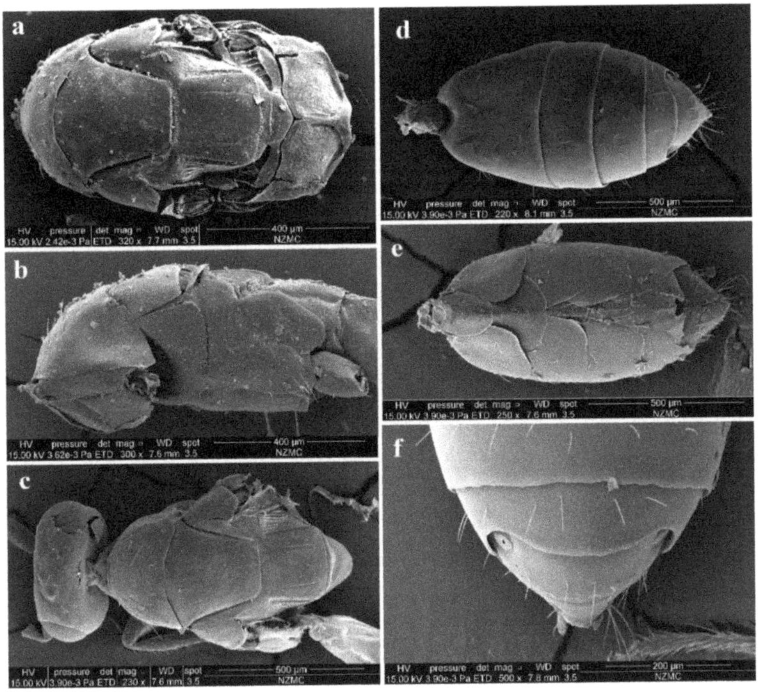

Figure 4. *Aceratoneuromyia bilinis* sp. nov., female. (**a**) Mesosoma in dorsal view; (**b**) Mesosoma in lateral view; (**c**) Head plus thorax in dorsal view; (**d**) Metasoma in dorsal view; (**e**) Metasoma in ventral view; (**f**) $Gt_{5–6}$ in dorsal view.

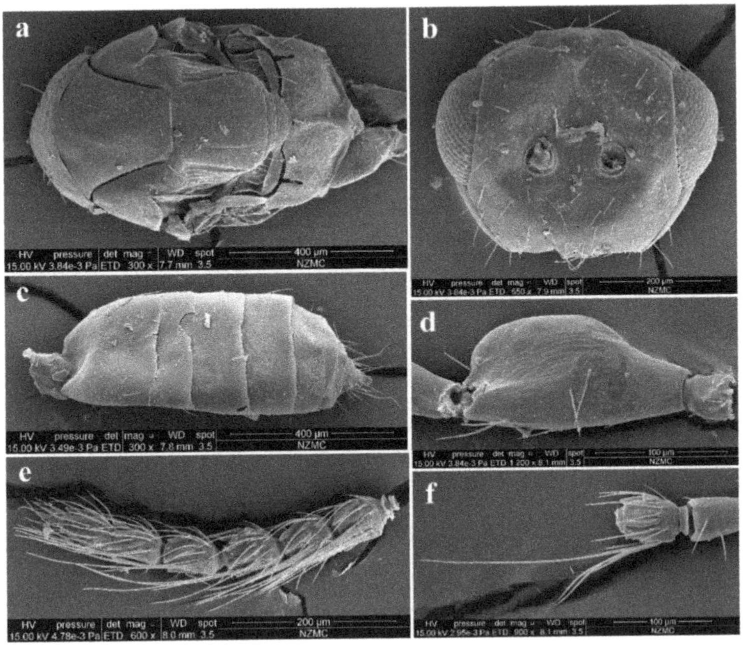

Figure 5. *Aceratoneuromyia bilinis* sp. nov., male. (**a**) Mesosoma in dorsal view; (**b**) Face; (**c**) Metasoma in dorsal view; (**d**) Scape of antenna; (**e**) Flagellum of antenna; (**f**) F1 with dorsal setae.

Head (Figure 3b,c) slightly broader than mesoscutum in dorsal view (4.0:3.5), about 3× as broad as long (4.0:1.3); vertex with long and erect setae, and a vertex suture at each side. OOL as long as POL (0.70:0.70). Ocelli arranged in a strong obtuse triangle. Head in frontal view about 1.14× as broad as high (4.0:3.5). Face with extremely fine delicately engraved reticulation, much weaker than that of thorax. Frons with a short and narrow Y-shaped frontofacial suture, with V-shaped diverging lines narrowly separated from each other, extending to vertex dorsally and sometimes connecting vertex suture, and with extremely short vertical line reaching scrobes (Figure 3b,c). Toruli inserted slightly above lower margin of eyes. Eyes with few of short and white hairs, diameter larger than malar space (2.0:1.0). Malar space 0.4–0.5× length of mouth opening, and malar sulcus almost straight and weakly curved in lateral view. Anterior margin of clypeus bilobed (Figure 3b).

Pronotum long, sloping and visible in dorsal view, with distinct and engraved reticulation, neck and collar not delimited, without posterior carina; with some setae at sides and 4 long setae near posterior margin (Figure 4a). Mesoscutum with extremely fine delicately engraved and longitudinal reticulation having areoles 2× or less as long as broad, notaular grooves deep; MLM with 2–3 adnotaular setae, without median line, with posterior margin straight (Figure 4a). Axillae strongly advanced, with engraved reticulation in the anterior half and with strongly raised striations (like carinae) in the posterior half (Figure 4a,c). Scutellum flattened in profile, slightly broader than long (2.5:2.1), sculptured rather more finely than mesoscutum; submedian lines present, distinctly nearer to sublateral lines than to each other, enclosing a space about 2.0–2.2× as long as broad; however, submedian lines are often incomplete, superficially indicated basally and apically, with two pairs of setae subequal in length, anterior pair situated well behind the middle and close to posterior pair, and posterior pair situated near posterior margin; scutellum with a sloping frenum (Figure 4a). Dorsellum about 1.86× as broad as high (1.3:0.7), about 0.54× length of propodeum (0.7:1.3), faintly engraved reticulate along outer margins (sculptures invisible under stereomicroscope), almost smooth; slightly incised in the middle of posterior margin; lateral panel of metanotum with more or less longitudinal but irregular carinae in the anterior 3/5 and smooth in the posterior 2/5 (Figure 4a). Propodeum raised-reticulate, with reticulation much coarser than that of scutellum (Figure 4a); with a strong medina carina having a short sulcus anteriorly, median carina distinctly raised, broadening caudad and then extending laterally; with rim of spiracle invisible, covered by the lobe of callus, while with a hole at the same position of spiracle and a broad sulcus from anterior margin to the spiracle position, with weak paraspiracular carina that is difficult to discern under certain lights of stereomicroscope; callus with raised reticulation as coarse as that of median area of propodeum, with 1 erect seta only. Lateral panel of pronotum weakly reticulate, prepectus, mesepimeron and mesepisternum nearly smooth; acropleuron smooth; metapleuron with raised reticulation like propodeum (Figure 4b).

Petiole distinct, nearly as long as broad, but partially hidden under certain angles (Figures 2a and 4d). Gaster (Figure 4d,e) slender, about 2× as long as broad and nearly as long as head plus mesosoma, based on the median length, the largest Gt_1 1.8× Gt_2, Gt_2 not clearly delimited from Gt_1, Gt_2 2/3 length of Gt_3, Gt_3 as long as Gt_4; sclerotized, convex dorsally, almost as hard as thorax and not easily collapsed; gastral tergites with superfacial engraved reticulation, almost polished under stereomicroscope even under SEM (Figure 4d); dorsolateral posterior margin of Gt_5 circled along outer rim of distinctly large spiracle of Gt_6, large spiracle visible and distinct in dorsal view (Figure 4f); Gt_7 with 4 cercal setae, the longest seta weakly curved apically and about 2× as long as the other 3 setae that are subequal in length (Figure 4d–f). Tip of hypopygium situated at 0.8× length of gaster measured from the base; anterior margin of hypopygium truncate, posterior margin bidentate (Figure 4e).

Legs with coxa, femora, and tibia-engraved reticulate; hind coxae compressed with dorsal carina on the apical 2/3 (distinct under certain lights). Fore wing (Figure 3e) densely setose, speculum small, somewhat closed below, not extending below marginal vein; costal cell 10–12× as long as wide, with 19 short setae on underside; submarginal vein with

5 setae on dorsal surface; marginal vein distinctly longer than submarginal vein (4.8:2.5); postmarginal vein absent.

Male (Figures 2b and 5). Body length about 2 mm (Figure 2b). Differs from female as follows. Antenna with 3 funiculars and 3 clavomeres (Figure 5e), with scape expanded and flattened, about 2.25× as long as broad, ventral plaque about 0.6× length of scape (Figure 5d); scape and pedicel brow, flagellum dark brown (Figure 2b); pedicel 1.6× as long as broad, about 0.5× length of scape and 1.4× length of F1; funicle nearly filiform, about as stout as pedicel; F1–F4 subequal in length, about 1.25× as broad as long; C1 and C2 subequal in length, C3 about 0.4× length of C2, spine 0.85× length of C3, with apical seta nearly 3× as long as the spine; clava slightly broader than funicle, about 2.6× as long as broad, shorter than F2 plus F3, F1–C2 with a short transverse row of dark and long setae on the dorsal surface (Figure 5e,f). Gaster less sclerotized, dorsolateral posterior margin of Gt_5 not circled, Gt_6 with spiracle much smaller, difficult to discern in dorsal view under lights of a stereomicroscope (Figure 5c).

Etymology. The name *bilinis* refers to the submedian lines on scutellum.

Type material. Holotype, female, CHINA, Yunnan, Yingjiang, 2020, coll. Yong Wang & Ning Huangfu, host: the larva-pupa of *Pelmatops ichneumoneus* (Westwood) (IZ-CAS, IOZ(E)221421). Paratypes: 10 females and 3 males, same data as holotype (IZCAS, IOZ(E)221422–IOZ(E)221434).

Host. This species, a gregarious endoparasitoid, parasitizes the larva of Tephritidae host *Pelmatops ichneumoneus* (Westwood) and emerges from the pupa of *P. ichneumoneus*.

Distribution. Oriental (China: Yunnan).

Remarks. Before synonymizing *Trjapitzinichus* with *Aceratoneuromyia*, like *A. kamijoi* Ikeda, *A. bilinis* is another intermediated species by having submedian lines although incomplete on scutellum, torulli slightly above the lower margins of eyes, and distinct malar sulcus. This confirms the necessary modifications for the concept of *Aceratoneuromyia* by Ikeda [20] and suggests that it is reasonable to synonymize *Trjapitzinichus* with *Aceratoneuromyia*.

3.1.2. *Aceratoneuromyia carinata* Cao & Zhu, sp. nov.

Diagnosis (female). Antenna with F1 longer than broad, F2–F3 subquadrate, with scape not reaching anterior ocellus. Frontofacial sutures broadly separated from each other and then passed as a line reaching deep scrobes in frontal view, and divergent and visible dorsally, connecting the vertexal sutures. Pronotum with a row of 7–8 long erect setae near posterior margin (Figure 7a). MLM with 3 adnotaular setae (Figure 7b). Scutellum with traces of incomplete submedian lines (Figures 6a and 7d). Frenum of scutellum large, almost as long as dorsellum (Figure 7b). Gaster slender, slightly longer than the combined length of mesosoma and head, somewhat flattened dorsally, and dorsolateral posterior margin of Gt_5 circled along outer margin of large spiracles of Gt_6 (Figures 6a,b and 7c). The longest cercal seta slightly longer than the other 2 setae equal in length, and very weakly curved apically (Figure 7c).

Female (Figures 6 and 7). Body length 1.5–1.7 mm. Body dark brown without metallic tinge (Figure 6a,b), with tegula yellow, with pale setae. Antenna brown, with scape and pedicel yellowish ventrally, and flagellum paler ventrally. Legs with coxa, femora and claw brown, with remainder parts as well as trochanter, base and tip of femora yellow (Figure 6a,b). Wings hyaline, with brown veins.

Antenna with 3 funiculars and 3 clavomeres, pale setae of pedicel and flagellum standing out strongly; scape not reaching anterior ocellus, nearly 3× as long as broad (1.7:0.6), about 3.4× as long as pedicel (1.7:0.5); funicle stouter than pedicel, pedicel as long as F1 (0.5:0.5); each funicular equal in length (0.5:0.5:0.5), F1 longer than broad (0.5:0.4), F2–F3 subquadrate (0.5:0.5); clava slightly shorter than F2 plus F3 (1.2:1.0), C2 as long as but broader than C1, C3 with a short but distinct terminal spine. Each flagellomere with longitudinal sensilla and apically with a circle of scattered, mushroom-shaped capitate peg sensilla; each flagellomere truncate apically, except C3.

Figure 6. *Aceratoneuromyia carinata* sp. nov., female. (**a**) Habitus of the holotype in dorsal view; (**b**) Habitus of the paratype (IOZ(E)221436) in lateral view. Scale bar: 0.5 mm.

Head as long as mesoscutum in dorsal view (4.0:4.0), about 3× as broad as long (4.0:1.3); vertex with long and erect setae, with vertexal suture at each side. OOL as long as POL (0.6:0.6) (Figure 7a). Ocelli arranged in a strong obtuse triangle (Figure 7a). Head in frontal view about 1.33× as broad as high (4.0:3.0). Face with extremely fine delicately engraved reticulation, much weaker than that of thorax. Frons with a Y-shaped frontofacial suture, with diverging lines broadly separated from each other, extending to vertex dorsally and connecting vertex suture. Toruli inserted slightly below the lower margin of eyes, scrobes distinct and deep. Eyes with sparsely short and white hairs, diameter 2× as long as malar space (1.6:0.8). Malar space 0.5–0.6× length of mouth opening (0.8:1.5), and malar sulcus distinct and almost straight. Clypeus delimited laterally by pale line, with anterior margin bilobed.

Pronotum long, sloping and visible in dorsal view, distinctly engraved reticulate, neck and collar not delimited, without posterior carina; with some scattered setae and 6–8 long setae near posterior margin (Figure 7a). Mesoscutum with extremely fine delicately engraved and longitudinal reticulation, notaular grooves deep; MLM with 3 adnotaular setae, without median line, with posterior margin straight (Figure 7d). Axillae strongly advanced, with engraved reticulation anteriorly and with strongly raised striations (like carinae) in most posterior part (Figure 7d). Scutellum flattened in profile, distinctly broader than long (2.0:1.5), sculptured rather more finely than mesoscutum; with traces of incomplete submedian lines, superficially indicated basally and apically, distinctly nearer to sublateral lines than to each other, enclosing a space about 1.5× as long as broad (1.5:1.0); with two pairs of setae subequal in length, anterior pair situated well behind the middle and close to posterior pair, and posterior pair situated near posterior margin; scutellum apically with a moderately long frenum, as high as dorsellum (0.4:0.4), and not well differentiated from dorsellum (Figure 7a,d). Dorsellum about 2.75× as broad as high (1.1:0.4), about 0.54× median length of propodeum (0.4:1.3), faintly engraved-reticulate; slightly incised in the middle of posterior margin; lateral panel of metanotum with more or less longitudinal carinae in the anterior half and smooth in the posterior half (Figure 7b,d). Propodeum with distinct scale-like reticulation (Figure 7b,g); with a strong medina carina slightly raised, broadening caudad and then extending laterally; with rim of spiracle visible, propodeal spiracle separated by its diameter from metanotum; callus with raised reticulation more distinct than that of median area of propodeum, with 2 long erect setae and 2 short setae at post-lateral corner. Lateral panel of pronotum weakly reticulate, prepectus, mesepimeron and mesepisternum nearly smooth; acropleuron smooth; metapleuron with raised reticulation like propodeum.

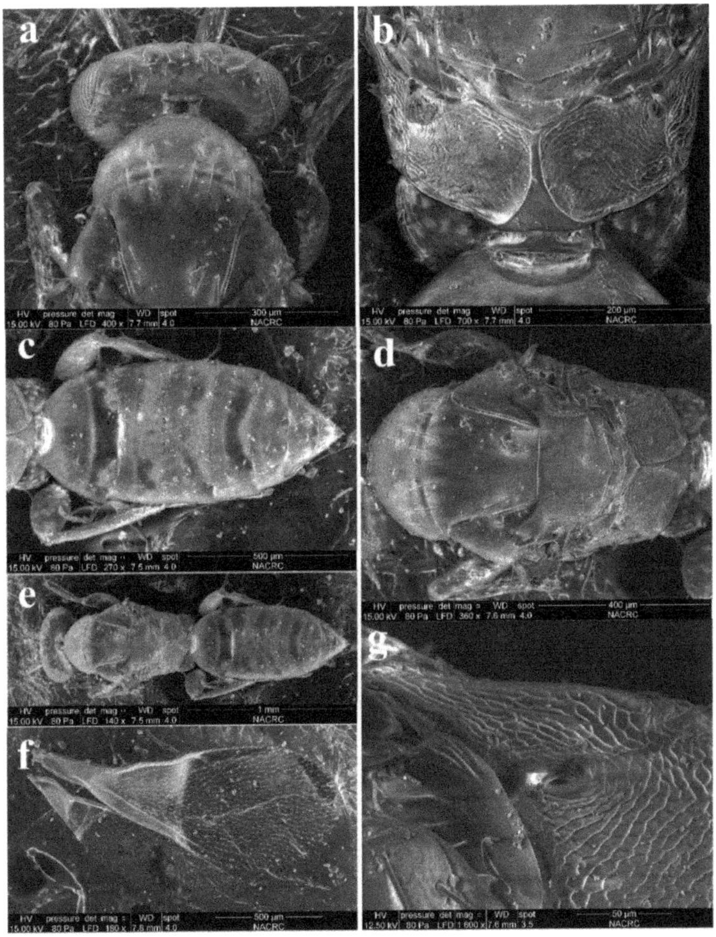

Figure 7. *Aceratoneuromyia carinata* sp. nov., female. (**a**) Head and pronotum in dorsal view; (**b**) Propodeum; (**c**) Metasoma in dorsal view; (**d**) Mesosoma in dorsal view; (**e**) Habitus in dorsal view; (**f**) Forewing; (**g**) Propodeal spiracle.

Petiole transverse, hidden under a stereomicroscope (Figure 7b,c). Gaster (Figure 7c) slender, more than 2× as long as broad (8.2:3.8) and slightly longer than head plus mesosoma (8.2:8.0), based on the median length, the largest Gt_1 about 1.83× Gt_2 (2.2:1.2), Gt_2 as long as Gt_3 and Gt_4 (1.2:1.2:1.2); alutaceous, with engraved reticulation; dorsolateral posterior margin of Gt_5 circled along outer rim of distinctly large spiracle of Gt_6, large spiracle visible and distinct in dorsal view; Gt_7 with 3 cercal setae, the longest seta weakly curved apically and slightly longer than the other 2 setae that are subequal in length. Tip of hypopygium nearly reaching the apex of gaster, situated at 7/8 length of gaster measured from the base; anterior margin of hypopygium truncate, posterior margin bidentate.

Legs with coxa, femora, and tibia-engraved reticulate; hind coxae compressed with distinct and complete dorsal carina. Fore wing (Figure 7f) densely setose, speculum small, closed below, not extending below marginal vein; costal cell 10–12× as long as wide, with 7–8 short setae on underside; submarginal vein with 4–5 setae on dorsal surface; marginal vein distinctly longer than submarginal vein (4.0:3.2); postmarginal vein absent.

Male. Unknown.

Etymology. The name *carinata* refers to the complete dorsal carina on the hind coxa.

Type material. Holotype, female, CHINA, Jiangsu, Nanjing, Nanjing Forestry University, V.1964, coll. Tian-Lin Chen, host: *Pegomyia phyllostachys* Fan (IZCAS, IOZ(E)221435). Paratypes: 3 females, same data as holotype (IZCAS, IOZ(E)221436–IOZ(E)221438).

Host. This species was recorded from *Pegomyia phyllostachys* Fan (Diptera: Anthomyiidae), possibly from the host pupa as an endoparasitoid.

Distribution. Oriental (China: Jiangsu).

Remarks. *Aceratoneuromyia carinata,* along with the other two species, *A. claridgei* Graham and *A. granularis* Domenichini, has pronotum with a row of more than 4 long setae near posterior margin, while other species usually have a row of 4 long setae. However, *A. carinata* is readily differentiated from the other two species by the presence of submedian lines on the scutellum which seems unclear under SEM because old specimens were not sputter-coated with gold. Additionally, *A. carinata* has hind coxa with a complete dorsal carina and ventrally paler antenna.

3.1.3. *Aceratoneuromyia indica* (Silvestri, 1910)

Syntomosphyrum indicum Silvestri, 1910 [21]: 230–244. Syntypes, USNM (not examined).

Melittobia indicum (Silvestri, 1910), Kurdjumov, 1913 [22]: 243–256.

Aceratoneuromyia australia Girault, 1917 [18]: 151. Syntypes, USNM (not examined). Synonymized by Gahan, 1938 [23]: 221.

Aceratoneuromyia indica (Silvestri, 1910), Gahan, 1938 [23]: 227.

Diagnosis. Female (Figure 8a) antenna clavate, with F1 somewhat as long as broad, with F2 and F3 somewhat more broad than long (Figure 9b), with scape not reaching anterior ocellus. Male scape with white ventral plaque about 0.8× length of scape (Figures 8b and 9h). Frontofacial sutures broadly separated from each other and then passed as a line that reaching shallow scrobes in frontal view (Figure 9a,g), invisible dorsally because easily collapsed head. Pronotum with a row of 4 long erect setae near posterior margin (Figure 9d,i). MLM with 2 adnotaular setae subequal in length (Figure 9d,i). Scutellum without submedian lines (Figure 9d,i). Dorsellum distinct, about 2× as broad as high (Figure 9d,i). Gaster shorter than (Figure 9e) or as long as (Figure 9j) mesosoma, convex dorsally (Figure 9e,j). The longest cercal seta about 1.8× length of the other 3 setae equal in length and distinctly curved apically (Figure 9f,k).

Material examined. 10 females 8 males, CHINA, Fujian, Fuzhou, 2.II.2022, lab-reared from the pupae of *Bactrocera dorsalis* (Hendel), coll. Qing-E Ji & Ning Huangfu (IZCAS).

Host. This species can widely parasitize the larva of fruit flies in the family Tephritidae of Diptera [9].

Distribution. *Aceratoneuromyia indica* was originally described in India and now has been thought to be widely introduced as a biological control agent to all zoological regions [7–9]. However, currently, it is difficult to define its original distribution pattern based on occurrence records only.

(a)

(b)

Figure 8. *Aceratoneuromyia indica* (Silvestri). (**a**) Habitus of female in lateral view; (**b**) Habitus of male in lateral view. Scale bar: 0.5 mm.

3.1.4. *Aceratoneuromyia trilinus* Cao & Zhu, sp. nov.

Diagnosis. Antenna with F1–F2 longer than broad, F3 subquadrate, with scape somewhat reaching anterior ocellus. Frontofacial sutures extremely short, almost absent, invisible in dorsal view; scrobes very shallow. Pronotum with a row of 4 long erect setae near posterior margin (Figure 11b). MLM with 2 adnotaular setae, with a trace of median line posteriorly (Figure 11b). Scutellum with incomplete submedian lines as well as a trace of median line in the anterior half (Figure 11b,e). Dorsellum very short, about 3× as broad as high (Figure 11b,c). Gaster slender, longer than the combined length of mesosoma and head (Figure 10a), dorsolateral posterior margin of Gt_5 circled along outer margin of large spiracles of Gt_6 (Figure 11a). The longest cercal seta 1.5× length of the other 3 setae equal in length and weakly curved apically.

Female. Body length 1.8–2.2 mm. Body dark brown without metallic tinge (Figure 10a,b), with pale setae. Antenna brown, with scape and pedicel paler dorsally and brownish yellow ventrally. Legs yellow, except for brown coxa and claw (Figure 10a). Wings hyaline, with brown veins.

Antenna with 3 funiculars and 3 clavomeres (Figure 11h), pale setae of pedicel and flagellum standing out strongly; scape somewhat reaching anterior ocellus, about 3× as long as pedicel (1.5:0.5), 3.78× as long as broad (1.5:0.4); funicle as stout as pedicel; pedicel, F1 and F2 equal in length (0.6:0.6:0.6) and longer than broad (0.6:0.5), F3 subquadrate (0.5:0.5); clava slightly shorter than F2 plus F3 (1.0:1.1), C1 longer than broad (0.7:0.5) and longer than F3 (0.7:0.5), C3 short, nearly as long as terminal spine. Each flagellomere with longitudinal sensilla and apically with a circle of scattered, mushroom-shaped capitate peg sensilla; each flagellomere except C3 truncate apically.

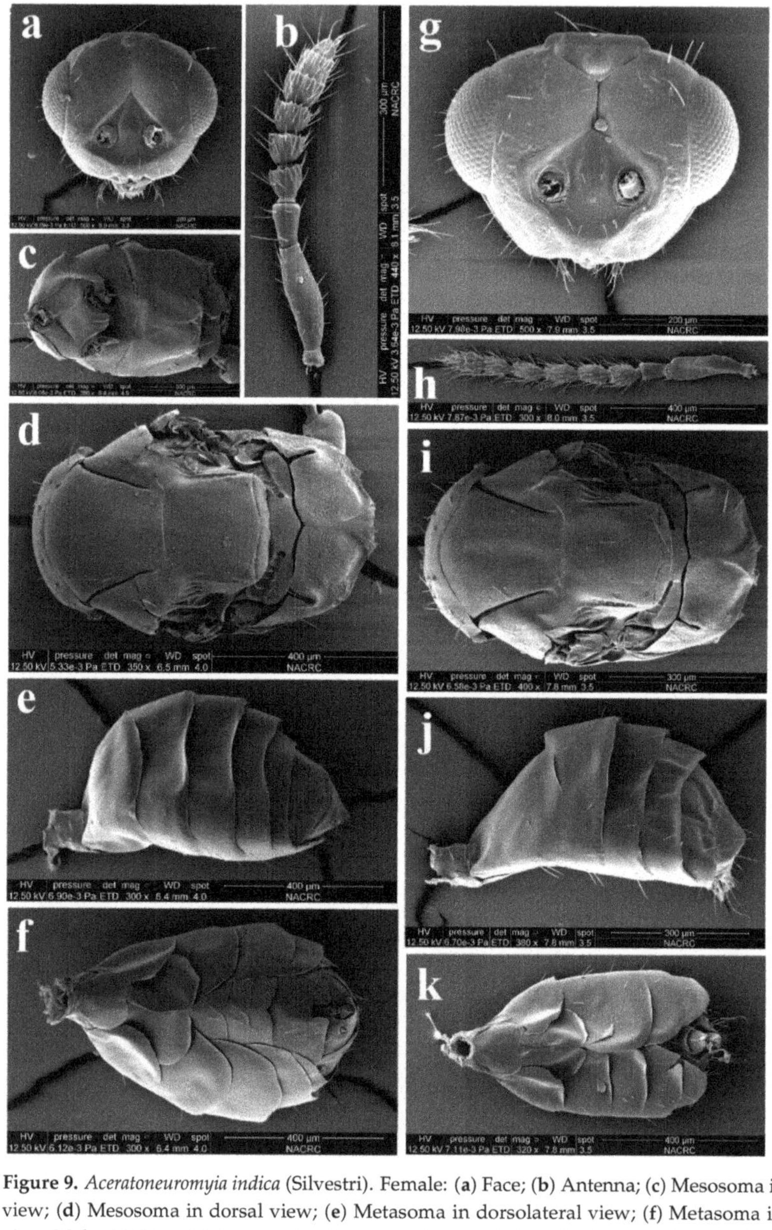

Figure 9. *Aceratoneuromyia indica* (Silvestri). Female: (**a**) Face; (**b**) Antenna; (**c**) Mesosoma in ventral view; (**d**) Mesosoma in dorsal view; (**e**) Metasoma in dorsolateral view; (**f**) Metasoma in ventral view; Male: (**g**) Face; (**h**) Antenna; (**i**) Mesosoma in dorsal view; (**j**) Metasoma in dorsolateral view; (**k**) Metasoma in ventral view.

(a) **(b)**

Figure 10. *Trjapitzinichus trilinus* sp. nov., female. (**a**) Habitus of the holotype in lateral view; (**b**) Habitus of the paratype (IOZ(E)221439) in dorsal view. Scale bar: 0.5 mm.

Head nearly as broad as mesoscutum in dorsal view (4.0:4.2), about 3.08× as broad as median length (4.0:1.3); vertex with long and erect setae, with weak vertexal suture at each side. OOL as long as POL (0.70:0.70). Ocelli arranged in a strong obtuse triangle. Head in frontal view, about 1.25× as broad as high (4.0:3.2). Face with extremely fine delicately engraved reticulation, much weaker than that of thorax. Frons with extremely short frontofacial suture, almost absent. Toruli inserted slightly above the lower margin of eyes. Eyes with sparsely short and white hairs, almost bare, diameter 1.8× as long as malar space (1.8:1.0). Malar space 0.5× length of mouth opening (1.0:2.0), with malar sulcus distinct and almost straight. Anterior margin of clypeus bilobed (Figure 3h). Pronotum long, sloping and visible in dorsal view, with distinct and engraved reticulation, neck and collar not delimited, without posterior carina; with some scattered setae and 4 long setae near posterior margin (Figure 11b); with spiracle on strong and distinct conical tubercle laterally (Figure 11d). Mesoscutum with extremely fine delicately engraved and longitudinal reticulation, notaular grooves deep; MLM with 2 adnotaular setae, with posterior margin straight (Figure 11b). Axillae strongly advanced, with engraved reticulation in about the anterior half and with strongly raised striations (like carinae) in about the posterior half (Figure 11c,d). Scutellum more or less flattened in profile, distinctly broader than long (2.5:2.0), sculptured rather more finely than mesoscutum; with traces of incomplete submedian lines, superficially indicated basally and apically, distinctly nearer to sublateral lines than to each other, enclosing a space about 2.3–2.5× as long as broad (2.5:1.0), as well as with a trace of median line in the anterior half; with two pairs of setae subequal in length, anterior pair situated well behind the middle and close to posterior pair, and posterior pair situated near posterior margin; scutellum with an extremely short frenum (Figure 11b,e). Dorsellum about 3× as broad as high (1.5:0.5), about 0.5× length of propodeum (0.5:1.0), faintly engraved-reticulate; slightly incised in the middle of posterior margin; lateral panel of metanotum with more or less longitudinal but irregular carinae in the anterior 4/5 and smooth in the posterior 1/5 (Figure 11c). Propodeum raised-reticulate, with reticulation slightly coarser than that of scutellum (Figure 11c,f); with a strong medina carina distinctly raised, broadening caudad and then extending laterally; with rim of spiracle visible, propodeal spiracle separated by 0.5–0.8 its diameter from metanotum; callus with raised reticulation as coarse as that of median area of propodeum, with 3 long erect setae and 1 long seta at post-lateral corner, not at the same plane with median area because of elevated median area (Figure 11f). Lateral panel of pronotum weakly reticultate, prepectus, mesepimeron and mesepisternum nearly smooth; acropleuron smooth; metapleuron with raised reticulation like propodeum.

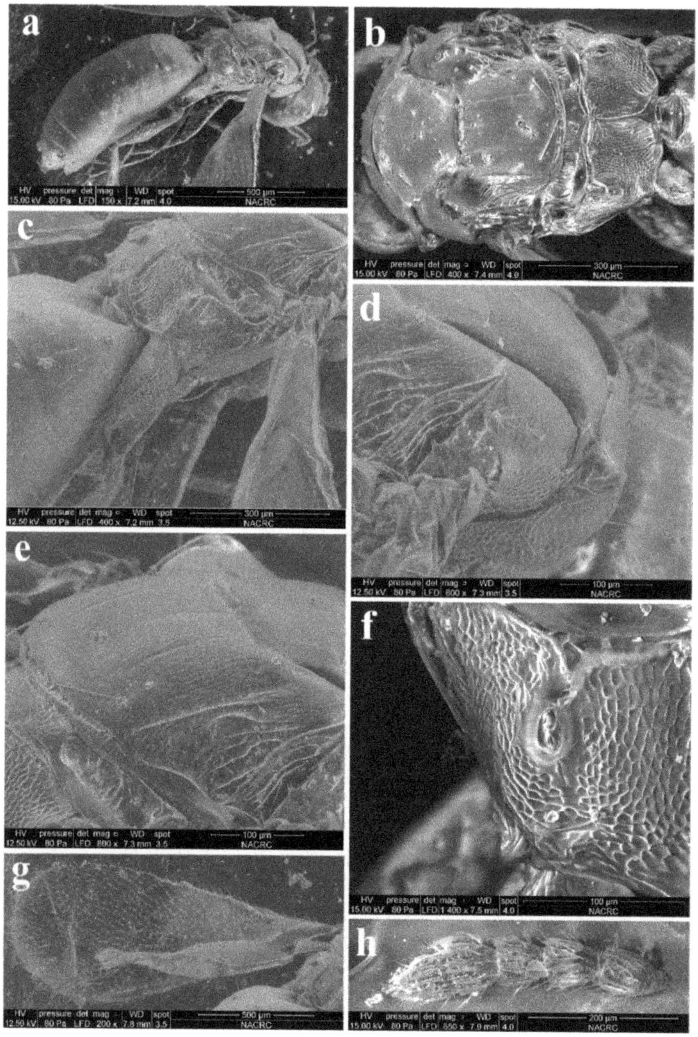

Figure 11. *Trjapitzinichus trilinus* sp. nov., female. (**a**) Habitus in lateral view; (**b**) Mesosoma in dorsal view; (**c**) Propodeum in lateral view; (**d**) Conical tubercle with spiracle in dorsolateral view; (**e**) scutellum in dorsolateral view; (**f**) Propodeal spiracle in dorsal view; (**g**) Forewing; (**h**) Flagellum in lateral view.

Petiole transverse (1.0:0.2) (Figure 11c). Gaster slender, about 2.64× as long as broad (9.5:3.6) and slightly longer than head plus mesosoma (9.5:7.5) (Figure 10a), based on the median length, the largest Gt_1 1.25× Gt_2 (2.5:2.0), Gt_2, Gt_3 and G_4 equal in length; Gt_1

smooth, the other tergites with superfacial engraved reticulation; dorsolateral posterior margin of Gt_5 circled along outer rim of distinctly large spiracle of Gt_6, large spiracle visible and distinct in dorsal view (Figure 11a); Gt_7 with 4 cercal setae, the longest seta weakly curved apically and about 1.5× as long as the other 3 setae that are subequal in length. Tip of hypopygium almost reaching the apex; anterior margin of hypopygium truncate, posterior margin bidentate.

Legs with coxa (Figure 11c), femora, and tibia engraved-reticulate; hind coxae compressed with complete dorsal carina. Fore wing (Figure 11g) densely setose, speculum small, closed below, not extending below marginal vein; costal cell 10–12× as long as wide, with 11 short setae on underside; submarginal vein with 4 setae on dorsal surface; marginal vein distinctly longer than submarginal vein; postmarginal vein absent.

Male. Unknown.

Etymology. The name *trilinus* refers to a median line and two submedian lines on the scutellum.

Type material. Holotype, female, CHINA, Liaoning, Liaoyang, Miaobu, 15.VIII.1979, host: the pupa of *Stilprotia salicis* (Linnaeus), coll. Hua Wei (IZCAS, IOZ(E)221441); Paratypes: 2 females, CHINA, Hebei, Pingquan, VIII.1978, host: the pupa of *Dendrolimus superans* (Butler) (IZCAS, IOZ(E)221439–IOZ(E)221440).

Hosts. Based on the label information of the above type material, two females were reared from the pupa of *Dendrolimus superans* (Butler) (Lepidoptera: Lasiocampidae), and one female was reared from the pupa of *Stilpnotia candida* Staudinger (Lepidoptera: Lymantriidae). The possibility that *A. trilinus* may be a hyperparasitoid of lepidopteran moths through other parasitoids is suggested by the host records of *A. evanescens* summarized in Noyes [9].

Distribution. Palaearctic (China: Hebei, Liaoning).

Remarks. *Aceratoneuromyia trilinus* is uniquely differentiated from the other known species in this genus by the trace of a median line on the scutellum.

3.2. Key to Species of *Aceratoneuromyia*

1. Pronotum subconical with distinct spiracles on conical tubercles; thorax weakly arched, propodeal slope as 15°–25° (e.g., Figures 10a and 11a).....2 (previously under *Trjapitzinichus*)

- Pronotum without conical tubercles and with indistinct spiracles; flattened thorax, mesoscutum, scutellum and propodeum almost at the same plane (e.g., Figures 2 and 8).......7

2(1) Pronotum with 2 long setae near its posterior margin; scutellum 2.15× as broad as long and with one pair of setae; propodeal spiracle separated by about 1.5× its diameter from metanotum; length of longest seta on vertex 1.3× maximum diameter of posterior ocellus; eye bare. Russia......................*A. lakicus* (Kostjukov & Gunasheva)

- Pronotum with 4 long setae near posterior margin; scutellum 1.3–1.4× as broad as long, with two pairs of setae; propodeal spiracle separated by less than its diameter from metanotum; eye with short, sparse setae..3

3(2) Antennal scape somewhat longer than an eye, reaching anterior ocellus; pronotum bare, except for a row of 4 long setae near its posterior margin and sometimes 2–3 short ones at sides; propodeal callus polished, virtually smooth; metapleuron shiny but with weak alutaceous sculpture. Europe........................*A. politus* (Graham)

- Antennal scape at most as long as an eye, not reaching anterior ocellus; pronotum with scattered short setae except for a row of setae near its posterior margin; propodeal callus and metapleuron distinctly alutaceous, sometimes with slightly raised reticulation..4

4(3) Scutellum with a trace of median line anteriorly (Figure 11e)...*A. trilinus* Cao & Zhu

- Scutellum without median line..5

5(4) Antenna with each funicular longer than broad; 3–4 tarsal segments dark; costal cell 15× as long as broad. North Korea...............................*A. leleji* Kostjukov & Kosheleva

- Antenna at least with F3 distinctly or slightly transverse; tarsal segments testaceous; costal cell at most 12× as long as broad........6

6(5) Eye 1.15× as long as broad; spine of clava 0.7× length of C3; pronotum 0.2× length of mesoscutum; MLM 1.45× as broad as long; propodeal spiracle separated by 0.25 its length from metanotum; tegula black; small species with body length 1.2 mm. Russia........*A. sugonjaevi* Kostjukov & Kosheleva

- Eye 1.3× as long as broad; spine of clava 0.5× length of C3; pronotum 0.3× length of mesoscutum; MLM about 1.3× as broad as long; propodeal spiracles separated by about 0.3 its length from metanotum; tegula fuscous; body length 1.4–1.6 mm. Germany........*A. evanescens* (Ratzeburg)

7(1) Females........8

- Males........16

8(7) Scutellum with submedian lines or with traces of submedian lines (e.g., Figure 3a).......9

- Scutellum without submedian lines (e.g., Figure 9d)........12

9(8) Scutellum only with traces of submedian lines (e.g., *Figure 9* in Ikeda [20]); torulli situated below the lower margin of eyes; frontofacial sutures distinct, scrobal grooves deep (e.g., *Figure 1* in Ikeda [20])........10

- Scutellum with somewhat distinct submedian lines although incomplete; torulus situated above the lower margin of eyes; frontofacial sutures absent or indistinct, scrobal grooves shallow (e.g., Figure 3b)........11

10(9) Vertex without vertexal sutures; pronotum with 4 long setae near posterior margin; propodeum with anterior part of median carina having narrow groove or small fovea; outer rim of spiracle partially covered by the raised lobe of callus; hind coxa without dorsal carina. Europe........*A. atherigonae* Ferriere

- Vertex with vertexal sutures; pronotum with 7–8 long setae near posterior margin; propodeum with anterior part of median carina without narrow groove or fovea; outer rim of spiracle not covered by the raised lobe of callus; hind coxa with complete dorsal carina (Figure 7b). China........*A. carinata* Cao & Zhu

11(9) Antenna with each funicular longer than broad; frontofacial sutures absent (*Figure 17* in Ikeda [20]); propodeal callus with 3–5 setae; spiracles of penultimate gastral tergite situated on lateral sides of gaster, invisible in dorsal view, Gt_5 with posterior margin straight (*Figure 23* in Ikeda [20]). Japan, South Korea........*A. kamijoi* Ikedai

- Antenna with F2 and F3 transverse (Figure 3d); frontofacial sutures short and narrowed separated from each other and divergent on the vertex (Figure 3b,c); propodeal callus with 1 seta; spiracles of penultimate gastral tergite situated on dorsolateral sides of gaster, distinct and visible in dorsal view, Gt_5 with posterior margin emarginated at dorsolateral corners (Figure 4f). China........*A. bilinis* Huangfu & Cao

12(8) Pronotum with a row of 8–12 long setae near posterior margin........13

- Pronotum with a row of 4 long setae near posterior margin........14

13(12) Pronotum with a row of 10–12 long setae near posterior margin; MLM with 3 adnotaular setae on each side; Larger species, 1.4–1.7 mm. UK........*A. claridgei* Graham

- Pronotum with a row of 8–10 long setae near posterior margin. MLM with 2–3 adnotaular setae on each side. Smaller species, length 1.1–1.4 mm. Europe...*A. granularis* Domenichini

14(12) F1 and F3 longer than broad, F2 subquadrate. Forewing with costal cell 21× as long as broad. Propodeam distinctly longer than scutellum. India........*A. wayanadensis* Narendran and Santhosh

- Antenna with each funicular transverse. Forewing with costal cell less than 14× as long as broad. Propodeam distinctly shorter than scutellum..15

15(14) Anterior adnotaular seta shorter than the posterior; tibiae infuscate with base and tip testaceous; forewing sparsely pilose all over, with costal cell more than 12× as long as broad. Czechoslovakia, USA..*A. filbriata* Graham

- Adnotaular setae equal in length; tibiae completely testaceous (Figure 8a); forewing tending to be thickly pilose all over, with costal cell less than 10× as long as broad. Worldwide...*A. indica* (Silvestri)

16(7) Antennal scape with ventral plaque only about 0.4 its length, situated in the upper half...17

- Antennal scape with ventral plaque 0.55–0.6× its length, extending somewhat into the lower half...18

17(16) Anterior adnotaular seta shorter than the posterior; tibiae infuscate with base and tips testaceous. Forewing sparsely pilose all over. Small species, length about 1 mm..*A. filbriata* Graham

- Adnotaular setae equal in length; tibiae completely testaceous (Figure 8b); forewing tending to be thickly pilose all over. Species tending to be larger (Figure 8b)...*A. indica* (Silvestri)

18(16) Scutellum with submedian lines or with traces of submedian lines............................19

- Scutellum without submedian lines..21

19(18) Scutellum only with traces of submedian lines; torulli situated below the lower margin of eyes; vertex without vertexal sutures; frontofacial sutures distinct, scrobal grooves deep...*A. atherigonae* Ferriere

- Scutellum with distinct submedian lines; torulus situated above the lower margin of eyes; vertex with vertexal sutures; frontofacial sutures absent or indistinct, scrobal grooves shallow..20

20(19) Frontofacial sutures absent; propodeal callus with 3–5 setae.............*A. kamijoi* Ikedai

- Frontofacial sutures short and narrowed separated from each other and divergent on the vertex (Figure 5b); propodeal callus with 1 seta (Figure 5a).......*A. bilinis* Huangfu & Cao

21(18) Pronotum with a row of 10–12 long setae near posterior margin; MLM with 3 adnotaular setae on each side; Larger species, 1.4–1.7 mm. UK.......................*A. claridgei* Graham

- Pronotum with a row of 8–10 long setae near posterior margin. MLM with 2–3 adnotaular setae on each side. Smaller species, length 1.1–1.4 mm. Europe...*A. granularis* Domenichini

4. Discussion

Kostjukov and Kosheleva [10] established the genus *Trjapitzinichus* to include three known species originally placed in *Aceratoneuromyia* and only summarized some differences with *Aceratoneuromyia* without argumentation. Kostjukov and Kosheleva [11] provided re-descriptions for *Trjapitzinichus*, and they did not mention and discuss the variations in some of those differences between *Aceratoneuromyia* and *Trjapitzinichus*. However, some intermediate species (e.g., *A. kamijoi* and *A. bilinis*) show some characteristics shared by both two genera, such as distinct malar sulcus, submedian lines of scutellum, toruli above the lower margin of eyes, and lateral projections on pronotum. These variations obscure the borderline between these two genera and bring difficulties in separating them. Those three species assigned to *Trjapitzinichus* are possibly just aberrant members of *Aceratoneuromyia* despite their lateral conical tubercles on the pronotum. Therefore, the generic name *Trjapitzinichus* is treated as a junior synonym of *Aceratoneuromyia* here although generic delimitations of most Tetrastichinae genera require confirmation from molecular evidence.

The phylogenetic relationships between *Aceratoneuromyia* and the other Tetrastichinae genera remain unclear. Gahan [23] synonymized *Aceratoneuromyia* with *Melittobia*, which were subsequently considered to be two distinct genera by other Eulophidae taxonomists [7,8,24,25]. Nevertheless, the affinity between *Aceratoneuromyia* and *Melittobia* was not supported by courtship behavior [26] and morphology which was briefly discussed by Cao et al. [27]. Although species of both *Tachinobia* (one of three genera in the *Melittobia* complex) and *Aceratoneuromyia* are gregarious parasitoids of some Diptera hosts, between which have no intersection, the former mostly in Tachinidae and the latter mostly in Tephritidae [8,28]. In addition, the *Melittobia* complex shows notable sexual morphological dimorphism, such as in the antenna, wings and eyes (males of *Koucoreckia* unknown) [27], whereas *Aceratoneuromyia* exhibits sexual difference restricted to only the antenna (e.g., Figure 9b,h). Therefore, the similarity between *Aceratoneuromyia* and the *Melittobia* complex is likely to be due to convergence, which requires further confirmation by comprehensive molecular evidence. Currently, it is still extremely hard to discuss the systematic position of *Aceratoneuromyia* as little is known about the phylogeny of Tetrastichinae. In contrast, *Aceratoneuromyia* is hypothetically more close to *Aprostocetus* than to *Tetrastichus*, with these two genera regarded as the two main lineages of Tetrastichinae.

Author Contributions: H.-X.C. conceived and designed the study. N.H. collected and reared the Tephritidae hosts and parasitoid wasps. N.H. and H.-X.C. prepared the material, and generalized and analyzed the data. Writing—original draft preparation, N.H.; writing—review and editing, H.-X.C. and C.-D.Z.; visualization, N.H. and H.-X.C.; supervision, C.-D.Z.; project administration, C.-D.Z.; funding acquisition, C.-D.Z. All authors have read and agreed to the published version of the manuscript.

Funding: This research was funded by the National Science Fund for Distinguished Young Scholars to C.-D.Z. from the National Natural Science Foundation of China., grant number 31625024. H.-X.C. is partially supported by the National Science and Technology Fundamental Resources Investigation Program of China, grant number 2019FY101800.

Institutional Review Board Statement: Not applicable.

Informed Consent Statement: Not applicable.

Data Availability Statement: DNA sequence data are available on GenBank under accession numbers (ON260937–ON260951). Other data of this research were deposited in the Institute of Zoology, Chinese Academy of Sciences, Beijing, China.

Acknowledgments: We thank John La Salle for sharing his knowledge and experience in taxonomy to guide and encourage H.-X.C. to work on Eulophidae taxonomy. We thank Qing-E Ji (Fujian Agriculture and Forestry University) for providing the parasitized hosts of *Aceratoneuromyia indica* and Kui-Yan Zhang (IZCAS) for providing help in SEM. We thank the two anonymous reviewers for their comments on the manuscript. In particular, we thank the one reviewer who raised concerns about the proximity of *Trjapitzinichus* and *Aceratoneuromyia* when we treated *Trjapitzinichus* as a valid genus in the first version submitted.

Conflicts of Interest: The authors declare no conflict of interest. The funders had no role in the design of the study; in the collection, analyses, or interpretation of data; in the writing of the manuscript, or in the decision to publish the results.

References

1. Zhang, Y.P.; Li, D.S.; Zhao, Y.C.; Huang, S.H.; Zhang, B.X. *Aceratoneuromyia indica* (Silvestri), a New Recorded Species of Parasitoid for *Bactrocera dorsalis*(Hendel)in China and Its Parasitic Efficiency. *Chin. J. Biol. Control* **2009**, *25*, 106–111, (In Chinese with English abstract).
2. Ovruski, S.M.; Schliserman, P. Biological control of tephritid fruit flies in Argentina: Historical review, current status, and future trends for developing a parasitoid mass-release program. *Insects* **2012**, *3*, 870–888. [CrossRef] [PubMed]
3. Wharton, R.A.; Yoder, M.J. Parasitoids of Fruit-Infesting Tephritidae. 2018. Available online: http://paroffit.org (accessed on 19 April 2022).
4. Pourhaji, A.; Lotfalizadeh, H.; Farshbaf-Pourabad, R.; Gharali, B. Parasitic wasps (Hymenoptera: Chalcidoidea) associated with flower-heads of Asteraceae in Iran. *Biologia* **2020**, *75*, 547–565. [CrossRef]
5. Amini, A.; Sadeghi, H.; Lotfalizadeh, H.; Notton, D. Parasitoids (Hymenoptera: Pteromalidae, Diapriidae) of *Carpomya vesuviana* Costa (Diptera: Tephritidae) in South Khorasan province of Iran. *Biharean Biol.* **2014**, *8*, 122–123.

6. Ovruski, S.M.; Schliserman, P.; Olga, R.; Peñaloza, C.; Oroño, L.E.; Colin, C. The establishment of *Aceratoneuromyia indica* (Hymenoptera: Eulophidae) in three biogeographical regions of Argentina. *Fla. Entomol.* **2006**, *89*, 270–273. [CrossRef]
7. Graham, M.W.R.d.V. A reclassification of the European Tetrastichinae (Hymenoptera: Eulophidae): Revision of the remaiming genera. *Mem. Amer. Ent. Inst.* **1991**, *49*, 1–322.
8. LaSalle, J. North american genera of tetrastichinae (Hymenoptera: Eulophidae). *J. Nat. Hist.* **1994**, *28*, 109–236. [CrossRef]
9. Noyes, J. Universal Chalcidoidea Database. 2019. Available online: http://www.nhm.ac.uk/chalcidoids (accessed on 17 April 2022).
10. Kostjukov, V.V.; Kosheleva, O.V. Taxonomic status of the species described in the genus *Tetrastichus* Haliday from the USSR and description of *Trjapitzinichus* gen. nov. In *Biologicheskaya Zashchita Rasteniy—osnova stabilizatsii agrozkosistem 4*; Nadykta, V.D., Ismailov, V.Y., Sugonyaev, E.S., Levaschova, G.I., Eds.; Krasnodar, Russia, 2006; pp. 103–111. Available online: https://www.nhm.ac.uk/resources/research-curation/projects/chalcidoids/pdf_X/KostjuKo2006.pdf (accessed on 17 April 2022). (In Russian)
11. Kostjukov, V.V.; Kosheleva, O.V. Two new species of *Trjapitzinichus* Kostjukov et Kosheleva, 2006 (Hymenoptera: Eulophidae: Tetrastichinae) from the North Korea and Russia, with redescription of the genus. Евразиатский энтомологический журнал **2016**, *15*, 84–89. (In Russian)
12. Gibson, G.A.P. Morphology and terminology. In *Annotated Keys to the Genera of Nearctic Chalcidoidea (Hymenoptera)*; Gibson, G.A.P., Huber, J.T., Woolley, J.B., Eds.; NRC Research Press: Ottawa, ON, Canada, 1997; pp. 279–284.
13. Folmer, O.; Black, M.; Hoeh, W.; Lutz, R.; Vrijenhoek, R. DNA primers for amplification of mitochondrial cytochrome c oxidase subunit I from diverse metazoan invertebrates. *Mol. Mar. Biol. Biotechnol.* **1994**, *3*, 294–299.
14. Carpenter, J.M. Towards simultaneous analysis of morphological and molecular data in Hymenoptera. *Zool. Scr.* **1999**, *28*, 251–260. [CrossRef]
15. Simon, C.; Frati, F.; Beckenbach, A.; Crespi, B.; Liu, H.; Flook, P. Evolution, weighting, and phylogenetic utility of mitochondrial gene sequences and a compilation of conserved polymerase chain reaction primers. *Ann. Entomol. Soc. Am.* **1994**, *87*, 651–701. [CrossRef]
16. Hall, T. BioEdit: A user-friendly biological sequence alignment editor and analysis program for Windows 95/98/NT. *Nucleic Acids Symp. Ser.* **1999**, *41*, 95–98.
17. Kumar, S.; Stecher, G.; Tamura, K. MEGA7: Molecular evolutionary genetics analysis version 7.0 for bigger datasets. *Mol. Biol. Evol.* **2016**, *33*, 1870–1874. [CrossRef] [PubMed]
18. Girault, A.A. New Australian Chaleid-flies (Hymenoptera, Chalcididae). *Insecutor Inscitiae Menstruus* **1917**, *5*, 151.
19. Bouček, Z. *Australasian Chalcidoidea (Hymenoptera). A Biosystematic Revision of Genera of Fourteen Families, with a Reclassification of Species*; CAB International: Wallingford, UK, 1988.
20. Ikeda, E. An unusual new species of *Aceratoneuromyia* Girault (Hymenoptera, Eulophidae), with redescription of *A. therigonae* Ferrière. *Insecta Matsumurana New Ser.* **1999**, *55*, 1–11.
21. Silvestri, F. Introduzione in Italia d'un Imenottero indiano per combattere la mosca delle arance. *Boll. Lab. Zool. Gen. Agr. Portici* **1910**, *4*, 228–245.
22. Kurdjumov, N.V. Notes on Tetrastichini (Hymenoptera, Chalcidoidea). *Rus. Entomol. Obozr.* **1913**, *13*, 243–256.
23. Gahan, A.B. Notes on some genera and species of Chalcidoidea (Hymenoptera). *Proc. Entomol. Soc. Wash.* **1938**, *40*, 209–227.
24. Gradwell, G.R. The selection of a neotype for *Melittobia hawaiiensis* Perkins and re-erection of the genus *Aceratoneuromyia* Girault (Hym., Eulophidae). *Entomol. Mon. Mag.* **1959**, *94*, 277–278.
25. Sugonjaev, E.S.; Kostjukov, V.V. A note on the genus *Aceratoneuromyia* Girault (Hymenoptera, Eulophidae). *Zool. Zhurnal* **1976**, *55*, 1110–1112.
26. Bosch, H.A.J.i.d.; Assem, J.v.d. The taxonomic position of *Aceratoneuromyia granulans* Domenichini (Hymenoptera: Eulophidae) as judged by characteristics of its courtship behaviour. *Syst. Entomol.* **1986**, *11*, 19–23. [CrossRef]
27. Cao, H.X.; La Salle, J.; Fornoff, F.; Guo, P.F.; Zhu, C.D. Notes on *Kocourekia* Bouček (Hymenoptera: Eulophidae: Tetrastichinae) with description of a new species from China. *Zootaxa* **2017**, *4317*, 391–400. [CrossRef]
28. Bouček, Z. Descriptions of *Tachinobia* gen. n. and three new species of Tetrastichinae (Hymenoptera: Eulophidae), with a tentative key to genera. *Bull. Entomol. Res.* **1977**, *67*, 17–30. [CrossRef]

insects

MDPI

Article

First Comprehensive Analysis of Both Mitochondrial Characteristics and Mitogenome-Based Phylogenetics in the Subfamily Eumeninae (Hymenoptera: Vespidae)

Li Luo [1], James M. Carpenter [2], Bin Chen [1] and Tingjing Li [1,*]

[1] Chongqing Key Laboratory of Vector Insects, Chongqing Key Laboratory of Animal Biology, Institute of Entomology and Molecular Biology, College of Life Science, Chongqing Normal University, Chongqing 401331, China; lioly19941112@126.com (L.L.); bin.chen@cqnu.edu.cn (B.C.)

[2] Division of Invertebrate Zoology, American Museum of Natural History, Central Park West at 79th Street, New York, NY 10024, USA; carpente@amnh.org

* Correspondence: ltjing1979@hotmail.com; Tel.: +86-159-2297-9264

Simple Summary: The subfamily Eumeninae comprises more than 3900 described species and eumenine mitochondrial analyses are almost absent. In order to provide further evidence toward understanding the relationships within the subfamily, the characteristics of 54 eumenine mitogenomes were comparatively analyzed, among which 52 mitogenomes are newly annotated. Meanwhile, using both Maximum-likelihood (ML) and Bayesian inference (BI), comprehensive phylogenetic relationship in the subfamily were investigated based on two mitochondrial datasets.

Abstract: The subfamily Eumeninae plays a significant role in the biological control of agricultural pests. However, the characteristics of eumenine mitogenomes that are important molecular markers for phylogenetics are not clearly revealed. Here, 52 eumenine mitogenomes are newly sequenced and annotated, and the phylogenetic relationships of the subfamily are comprehensively analyzed based on 87 vespid mitogenomes. Through the comparative analysis of the 54 eumenine mitogenomes, the gene compositions of about one half of the 54 species match with ancestral insect mitogenome, and remaining others contain two *trnM* which are highly similar, with 51.86% (*Eumenes tripunctatus*) to 90.65% (*Pseumenes nigripectus*) sequence identities, which is unique among the reported mitogenomes of the family Vespidae. Moreover, the translocation *trnL1* upstream of *nad1* is a common rearrangement event in all eumenine mitogenomes. The results of phylogenetic analyses support the paraphyly of the subfamily Eumeninae and the tribe Odynerini, respectively, and the monophyly of the tribe Eumenini, and verify that the tribe Zethini is a valid subfamily Zethinae. In this study, the relationships between some genera such as *Allorhynchium* and *Pararrhynchium* or the taxonomic status of the subgenera such as *Eremodynerus* and *Dirhynchium* are found to be confusing and there should be further inquiry with more samples.

Keywords: Eumeninae; gene rearrangement; mitogenomes; phylogenetic; Vespidae

1. Introduction

The subfamily Eumeninae containing more than 3900 described species is the biggest of the family Vespidae (Hymenoptera). It is the primary lineage of the Vespidae [1], which plays a significant role in the biological control of agricultural pests because of its cosmopolitan predation of the larvae of Lepidoptera (e.g., Geometridae, Tortricidae) and Coleoptera (e.g., Chrysomelidae and Curculionidae) [2,3]. Most of them are solitary wasps, using mud to partition the cells, while some are primitively social, burrowing in the soil or wood, many species in Zethini construct their nests by exploiting masticated and salivated plant material such as *Zethus*, *Ischnocoelia*, *Elimus*, *Discoelius*, and *Protodiscoelius* using vegetable matter for cell partition, and *Psiliglossa* and *Raphiglossa* using pith [4,5]. Additionally, the high morphological diversity and complexity of Eumeninae leads to some difficulties in its classification, and some other difficulties may be attributed to its troubled taxonomic history [6]. Hence, the early classifications of this subfamily underwent a radical transformation. Based on the morphology of the mouthparts and the general shape of the metasoma, Latreille began the generic classification of Eumeninae and divided the current species of Eumeninae into three genera: *Eumenes*, *Odynerus*, and *Synagris* [7]. Later, de Saussure divided 'Euméniens' into Anomaloptéres, Euptéres, and Mischoptéres, after that he separated *Zethus*, *Calligaster*, and *Discoelius* from the rest of his section 'Euptéres' of the 'Eumhiens' as the group 'Zethites' [8,9]. Thereafter, Richards proposed a classification for the "Eumenidae" with three subfamilies: Raphiglossinae, Discoeliinae (=Zethinae), and Eumeninae [10], while Carpenter did not recognize the names Zethinae and Raphiglossinae after investigating the relationships among the subfamilies of the Vespidae with a cladistic treatment [11]. Recently, Hermes et al. corroborated the monophyly of Eumeninae and proposed three tribes of this subfamily, namely Zethini, Eumenini, and Odynerini, which was the most comprehensive classification of the Eumeninae based on morphology [12]. Whether the tribe Zethini should be upgraded to the subfamily Zethinae and the subfamily Eumeninae is monophyletic are worthy of further exploration. Additionally, with the continuous enrichment of molecular data, controversies over the classifications of Eumeninae appear in the disagreement between morphological and molecular data.

In the published research on phylogenetic relationships in the subfamily Eumeninae, some molecular data have been utilized. The first molecular study showed that the solitary Eumeninae was a sister taxon to the Polistinae + Vespinae cluster leveraged on nuclear 28S rDNA and mitochondrial 16S rDNA of 12 species from the family Vespidae including 3 eumenine species [13]. Thereafter, based on the analysis of four nuclear gene fragments (18S rDNA, 28S rDNA, *abdominal-A*, and *RNA polymerase II*) from 27 Vespidae species (containing 11 eumenine species), Hines et al. supported the division of Eumeninae into two separate monophyletic clades: "Zethinae" and "Eumeninae" [5]. Neither of the two studies clarified the generic relationships of Eumeninae because only a few species of Eumeninae were contained. Later, with a total of 49 transcriptomes of vespid wasps (containing 40 eumenine species), Bank et al. suggested the subfamily Eumeninae was paraphyletic and the "Zethini" were divided into two clades: Raphiglossinae and Zethinae [14]. Then, Piekarski et al. also suggested that "Zethini" should be a valid subfamily Zethinae [15]. So far, several nuclear gene fragments, mitochondrial fragments, and transcriptomes have not completely resolved the phylogenetic relationships of Eumeninae due to insufficient and unrepresentative sampling of taxa. Thus, to understand the evolution of the various biologies exhibited by Eumeninae, robust investigations of phylogenetic relationships are still needed.

The mitogenome is a widely accepted molecular marker used in phylogenetic studies due to its maternal inheritance as well as the higher rate of nucleotide substitution compared with nuclear DNA [16,17]. To date, two mitogenomes of Eumeninae have been published, which is insufficient to explore phylogenetic relationships [18]. In China, which spans two faunal regions (Palearctic and Oriental Regions), there is a total of more than 310 known species and subspecies in 58 genera of the subfamily Eumeninae [19–30], which constitutes a quarter of the total genera of the world. To clarify the phylogenetic relationships within Eumeninae, especially the placement of Zethini, 52 new mitogenomes of 33 genera of Eumeninae from China are obtained and analyzed by combining 35 published vespid mitogenomes in this study. Given that gene rearrangements are very informative for phylogenetic analysis and are exhibited extensively in some insect orders [17], these eumenine mitogenomes are compared with the gene order of ancestral mitogenome to investigate their distinctive rearrangement models. Additionally, the characters of the eumenine mitogenomes are compared with other vespids, in order to identify any unique character to support the classification of Eumeninae.

2. Materials and Methods

2.1. Sample Collection and DNA Extraction

The 52 species of the subfamily Eumeninae, which were firstly identified at least to genus level by taxonomic specialists, were collected and stored in 95% ethanol at −20 °C in Chongqing Normal University, Chongqing, China (Table 1). Thereafter, total DNA was isolated from the muscle tissues of the thorax using the DNeasy DNA Extraction kit (Qiagen) and according to its instructions. Finally, we followed the manufacturer's protocol of the Qubit dsDNA high-sensitivity kit (Invitrogen) to determine the DNA concentration for each sample.

Table 1. GenBank accession numbers for mitogenomes of Eumeninae newly sequenced and annotated in this study.

Species	bp	Gene Number	Accession No.
Allodynerus delphinalis	16,932	38	ON024142
Allodynerus mandschuricus	17,449	38	ON012816
Allorhynchium chinense	16,909	38	MK051021
Allorhynchium argentatum	17,972	38	MK051022
Allorhynchium radiatumus	17,349	38	ON055163
Ancistrocerus renimaculus	15,614	36	ON045342
Ancistrocerus tussaci	15,679	35	ON012815
Antepipona sp.	19,040	38	ON012817
Antepipona ovalis	17,514	36	ON012818
Anterhynchium abdominale	16,488	36	MK051029
Anterhynchium coracinum	16512	36	MK051028
Anterhynchium flavomarginatum	15,196	35	MK051026
Anterhynchium mellyi	18,692	38	ON012812
Apodynerus protuberantus	17,943	36	ON045341
Calligaster cyanoptera	16,316	38	ON012814
Delta pyriforme pyriforme	14,883	35	ON076029
Delta campaniforme esuriens	16,126	36	ON055486
Discoelius zonalis	15,435	38	ON076025
Eumenes buddha	16,048	37	ON076024
Eumenes tripunctatus	15,702	36	ON045343
Eumenes pomiformis	16,520	38	ON076031
Ectopioglossa sublaevis	16,203	36	ON045340
Ectopioglossa sanban	16,454	37	ON012813
Euodynerus dantici	17,493	37	ON076022
Euodynerus nipanicus	22,088	38	ON076021
Jucancistrocerus atrofasciatus	18,848	38	ON045348
Jucancistrocerus angustifrons	19,867	38	ON012819
Katamenes sichelii sichelii	14,807	37	ON076027
Labus pusillus	17,409	36	ON076026
Labus angularis	15,227	35	ON076030
Leptochilus sp.	14,241	36	ON045339
Orancistrocerus drewseni drewseni	17,636	38	ON045338
Oreumenes decorates	15,563	37	ON076028
Paralepromenes sp.	16,805	37	ON045337
Parancistrocerus samarensis	17,773	38	ON076023
Pararrhynchium striatum	20,403	38	ON045347
Pararrhynchium septemfasciatus	18,003	36	ON055487
Pareumenes quadrispinosus acutus	17,426	38	ON076020
Pseudepipona kozhevnikovi	15,650	35	ON076019
Pseudepipona przewalskyi	20,281	37	ON024141
Pseudozumia indosinensis	16,376	35	ON045335
Pseumenes nigripectus	17,773	38	ON045336
Pseumenes depressus	16,677	38	ON045346
Rhynchium quinquecinctum murotai	16,317	36	MK051030
Rhynchium brunneum brunneum	23,251	38	MK051031
Rhynchium brunneum ceylonicum	23,122	38	MK051032
Stenodynerus frauenfeldi	17,252	36	ON045334
Stenodynerus chinensis	17,194	38	ON045345
Subancistrocerus camicrus	18,035	38	ON045344
Symmorphus ambotretus	17,280	38	ON076018
Symmorphus lucens	17,865	38	ON076017
Zethus dolosus	16,306	38	ON076016

2.2. Whole-Genome Sequencing and Assembling

The Illumina TruSeq library, containing an average size of 350 bp, was sequenced using the Illumina Hiseq 2500 platform at BerryGenomics (Beijing, China). Then, high-quality reads (after deletion of low-quality reads) were used in de novo assembly with IDBA-UD by using the NGS QC Toolkit [31,32]. *COX1* and *srRNA* were amplified by standard PCR reactions and were used to identify mitogenome assemblies with at least 98% similarity sequences in BLAST [33,34]. Finally, the accuracy of the assembly was investigated by mapping clean reads onto the obtained mitochondrial scaffold in each library using Geneious 10.1.3 (http://www.geneious.com/. Accessed date: 12 January 2022), which allowed for up to 2% mismatches, a maximum gap size of 3 bp, and a minimum overlap of 100 bp.

2.3. Mitogenome Annotation and Sequence Analysis

Annotation of the assembled mitochondrial sequences was identified using Clustal X 1.8 with homologous sequences against the publicly available Eumeninae mitogenomes [35]. Unrecognized tRNA genes were found by use of tRNA scan-SE version 2.0.2 and secondary structure modeling was completed using ARWEN version 1.2 [36,37]. The nucleotide composition, AT content, GC-skew, and the Relative Synonymous Codon Usage (RSCU) were calculated in MEGA X [38]. Effective Number of Codons (ENC) and GC of silent 3rd codon posit (GC3s) were computed in Codon W 1.4 and non-synonymous (*Ka*) and synonymous (*Ks*) substitution ratio (*Ka/Ks*) of PCGs were calculated in DnaSP 5.0 [39,40]. Then, the gene arrangement events were detected in CREx [41].

2.4. Phylogenetic Analyses

A total of 87 Vespidae mitogenomes containing 52 newly sequenced eumenine mitogenomes and 35 species of the subfamilies Eumeninae, Stenogastrinae (three species), Polistinae (19 species), and Vespinae (11 species) downloaded from GenBank were selected as ingroups, and four species from Apoidea (*Hylaeus dilatatus, Andrena cineraria, Megachile sculpturalis*, and *Apis cerana*) were selected as outgroups (Table S1). In total, 13 PCGs and 2 rRNAs were extracted by PhyloSuite v 1.2.2 [42]. The individual alignments of PCGs were performed using the L-INS-i strategy of the MAFFT algorithm executed in the TranslatorX online platform, and rRNA genes were aligned individually using the G-INS-i strategy implemented in MAFFT version 7.205 [43,44]. GBlocks v.0.91b was used to remove all ambiguously aligned sites from 13 PCGs and two rRNAs [45]. After that, MEGA X was used to check and correct all the alignments [38].

Phylogenetic trees were inferred from two sets of data: (1) PCGR: 13 PCGs and 2 rRNAs; (2) PCG: 13 PCGs. Before the construction of trees, PartitionFinder version 2.1.1 [46] was used to simultaneously choose the best partition schemes and substitution models for each matrix with the Akaike Information Criterion (AIC) and greedy search algorithm (Table S2). A Bayesian inference (BI) tree was constructed in MrBayes v.3.2.7, approximately 10,000,000 generations were conducted for the matrix, with the average deviation of split frequencies below 0.01 which suggests that runs reach convergence and were sampled every 1000 generations with a burn-in of 25% [47]. Maximum likelihood (ML) was constructed on the PhyML online web server (http://www.atgc-montpellier.fr/phyml/. Accessed date: 12 January 2022) and the node support values were evaluated via a bootstrap test with 100 replicates [48]. In addition, for (maximum parsimony) MP, the matrix was analyzed through the use of Winclada slaving TNT [49,50]. New technology search algorithms were used with the default settings, except ratchet 200 iterations, with up:down perturbation 8:4; hits to minimum length 25. Bootstrapping was via traditional search, with 100 replicates.

3. Results and Discussion

3.1. Mitochondrial Genome Organization

We obtained 52 complete or partial mitogenomes, which were deposited in GenBank (Table 1). Most of them include 13 protein-coding genes (PCGs), 2 rRNA genes (rRNAs), a control region, and 22 or 23 tRNA genes (tRNAs), with a size from 14,241 (*Leptochilus* sp.) to 23,251 bp (*Rhynchium brunneum brunneum*) and some of the entire A+T rich regions as well as three tRNA genes (*trnI, trnQ* and *trnM*) were unable to be amplified in 23 mitogenomes (Figure S1). The composition of 29 complete mitogenomes are significantly biased toward adenine and thymine, with high A+T content from 78.6% (*Subancistrocerus camicrus*) to 84.7% (*Eumenes pomiformis*) which is similar to other hymenopteran mitogenomes [51] and the AT skews are from −0.09 (*Pararrhynchium striatum*) to 0.19 (*Antepipona* sp.) (Table S3).

Typically, the mitogenomes of metazoan animals are double-strand circular DNA composed of 37 genes including 13 PCGs, 22 tRNAs, 2 rRNAs, and a control region, and most genes are located on the J-strand (major strand), the remaining being on the N-strand [52]. In this study, some mitogenomes of Eumeninae generally match that of the inferred mitogenomes except for some *trnM* duplications (Figure 1).

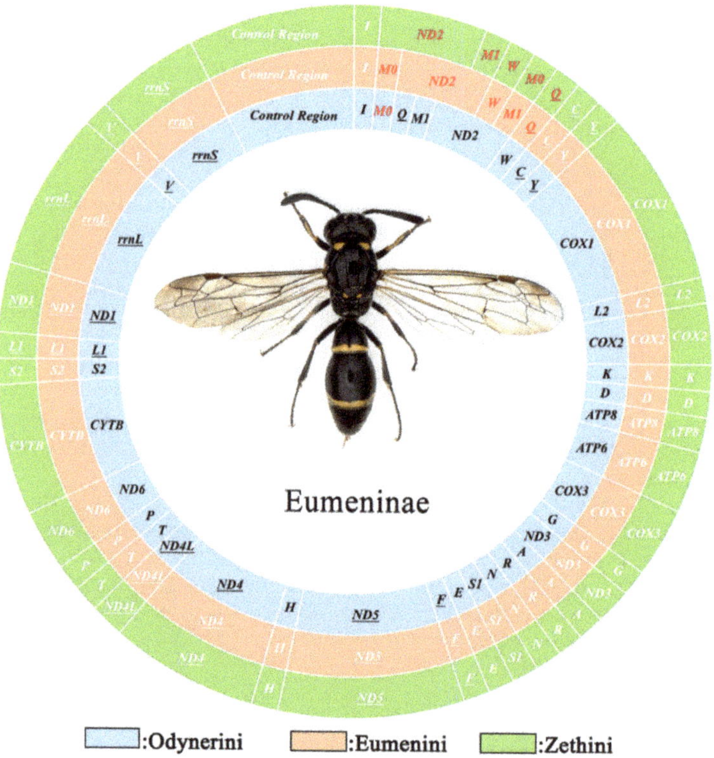

Figure 1. The mitochondrial genomes of Eumeninae. Circles of different colors indicate different tribes of Eumeninae. The red gene means its position is inconsistent with the ancestor insect.

Compared with other Vespidae, there are 26 in total in the 52 newly assembled mitogenomes of Eumeninae containing two *trnM* genes which are highly similar, with 51.86% (*Eumenes tripunctatus*) to 90.65% (*Pseumenes nigripectus*) sequence identities (Figure S2). The substitutions between *trnM0* and *trnM1* are identified in the Amino Acid acceptor (AA) arm, TψC (T) arm, Variable (V) loop, Anticodon (AC) arm, and the dihydorouridine (DHU) arm (Figure 2). The positions of *trnM0* and *trnM1* are different: some are connected and others are separated by *trnQ* and *trnW*. The duplication event is unique in the subfamily Eumeninae among the reported mitogenomes of Vespidae; meanwhile, it was reported in the mitogenomes of both *Ibalia leucospoides* (Hymenoptera: Cynipoidea) containing three *trnM* with 92–97% sequence identities and the genus *Pachycephus* (Hymenoptera: Cephidae) [18,53,54]. Moreover, there is another duplication of *trnL2* within the Eumeninae such as three regions of noncoding DNA containing four copies of *trnL2* in *Abispa ephippium* [55]. According to the existing reports, the duplication of tRNA is common in Hymenoptera; for instance, the copies of *trnD*, *trnA*, and *trnE* in the family Cephidae (Hymenoptera) and Trigonalyoidea (Hymenoptera), respectively [18,53]. Therefore, within the family Vespidae, the duplication of *trnM* may be one of the features to indicate whether a species belongs to the subfamily Eumeninae.

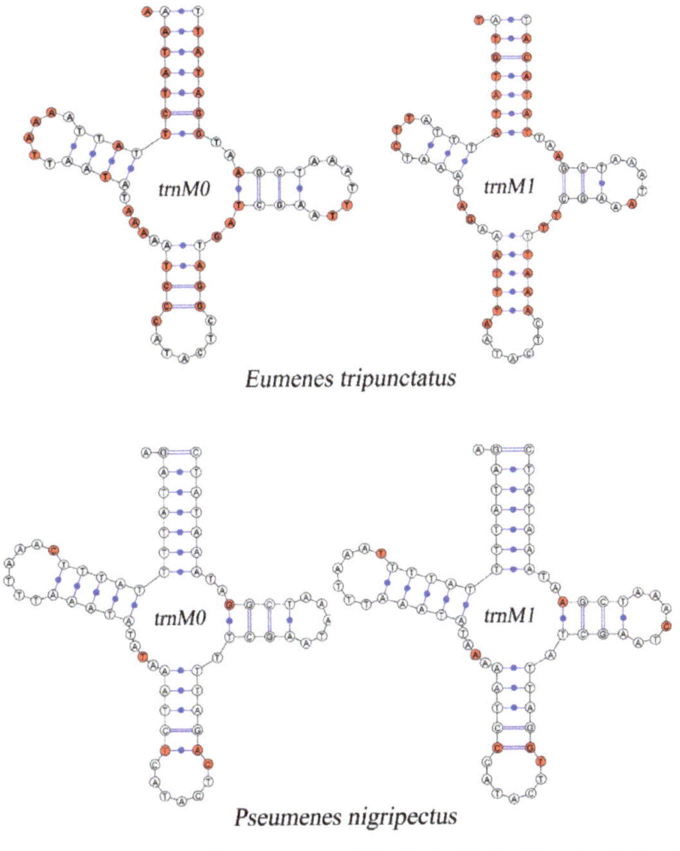

Figure 2. Inferred secondary structures of duplicated *trnM*. The substitutions in *trnM0* and *trnM1* compared with each other are indicated by red color.

3.2. Protein-Coding Genes and Codon Usage Patterns

All the PCGs start with the typical ATA, ATG, or ATT codons and stop with the complete TAA or TAG or truncate TA- or T – termination codons. The composition of PCGs is significantly biased toward adenine and thymine, with high A+T content from 75.9% to 84.4%, and the AT skews are always negative from −0.16 to −0.095 (Figure 3A). The A+T content of PCGs in other subfamilies of Vespidae was computed, showing that the value of A+T content in Stenogastrinae is higher than in three other subfamilies and in Vespinae it is the minimum (Figure 3B).

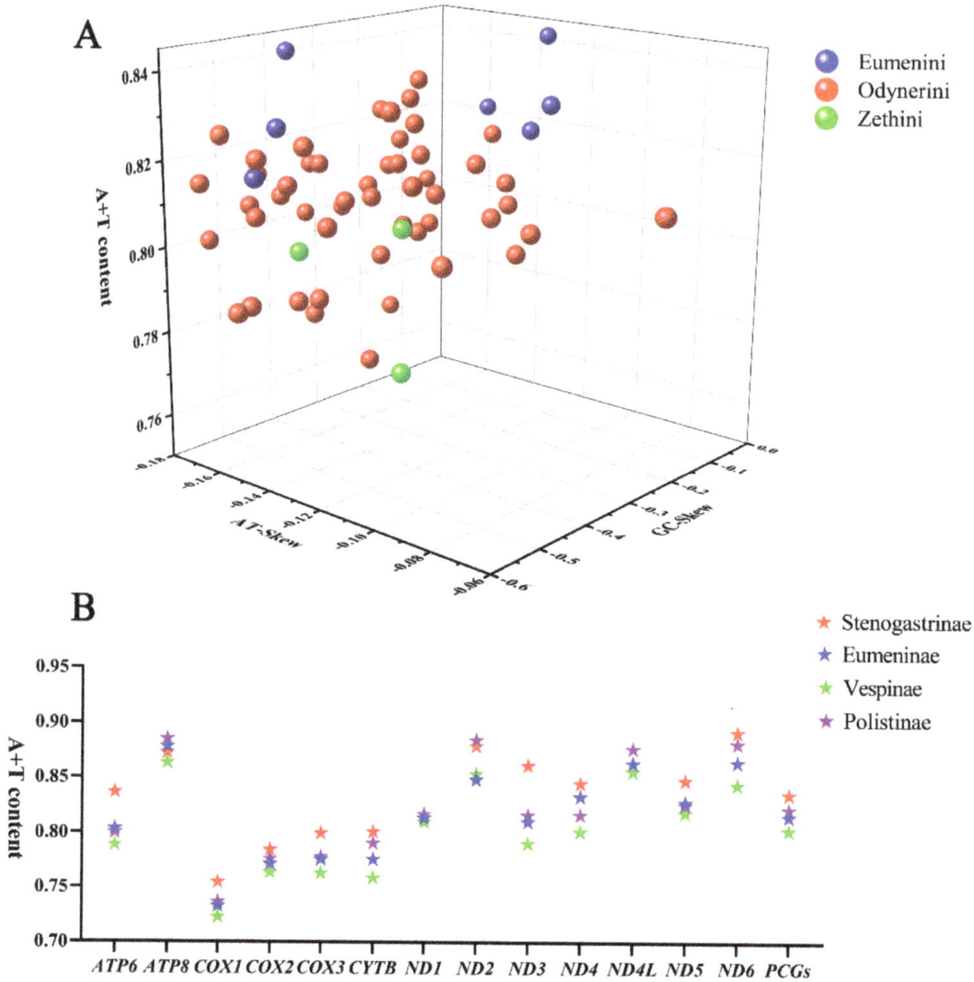

Figure 3. A+T content, AT-Skew, and GC-Skew of PCG in vespid mitogenomes. (**A**) The A+T content, AT-Skew, and GC-Skew of PCG in Eumeninae; (**B**) the A+T content of PCG in four subfamilies of Vespidae.

The Relative Synonymous Codon Usage (RSCU) values of codons such as UUA, GUU which ended with A or U, are all greater than 1.3 and those ending with G or C are all less than 1 (Figure 4). The RSCU value can directly reflect the frequency of codon usage: the RSCU value equivalent to 1 indicates that the codon has no preference, or the RSCU value greater than 1 illustrates that the frequency of the codon is relatively higher [56,57]. As a result, the optimal codons among PCGs of eumenine mitogenomes are codons ending with A or U, and accordingly, the third position of the codon in PCGs is significantly biased toward adenine and thymine with 90.5% A+T content. Additionally, the optimal codons of eumenine mitogenomes are consistent with those of Vespidae which frequently used UUU, UUA, AUU, and AUA, and among them, UUA (Leu2) is the one with the highest RSCU value. In addition, both UAA and AGA are the stop codons of the eumenine mitogenomes, of which UAA with the higher RSCU is the eumenine preferred codon.

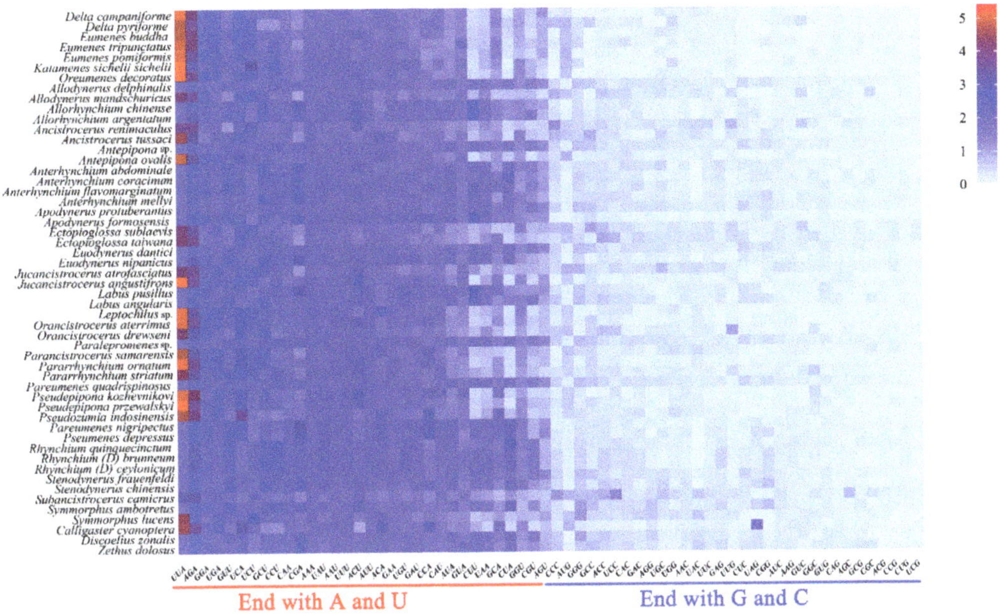

Figure 4. Relative synonymous codon usage (RSCU) of the mitogenomes of Eumeninae.

The synonymous codon usage bias is influenced by mutation pressure and natural selection, and an effective number of codons (ENC) standard curve can indicate that the determinant of codon preference is mutation pressure or natural selection [58]. Our result shows that all the points lie under the standard curve, which indicates that the codon usage bias is influenced by selection pressure (Figure S3). *Ka/Ks* is the ratio of the number of nonsynonymous substitutions per nonsynonymous site (*Ka*) to the number of synonymous substitutions per synonymous site (*Ks*), which could indicate something about the selective forces acting on the protein [59]. Thus, we computed the *Ka/Ks* value of PCGs from eumenine mitogenomes, and the result shows that all the *Ka/Ks* of PCGs except *ND4L* are less than 1, which indicates that only *ND4L* is under a positive selection and evolves rapidly, and other PCGs are under a purifying selection. Moreover, the lowest *Ka/Ks* value of *COX1* (0.11) indicates that it is conservative under environmental selection pressure and suitable for molecular barcoding (Figure 5).

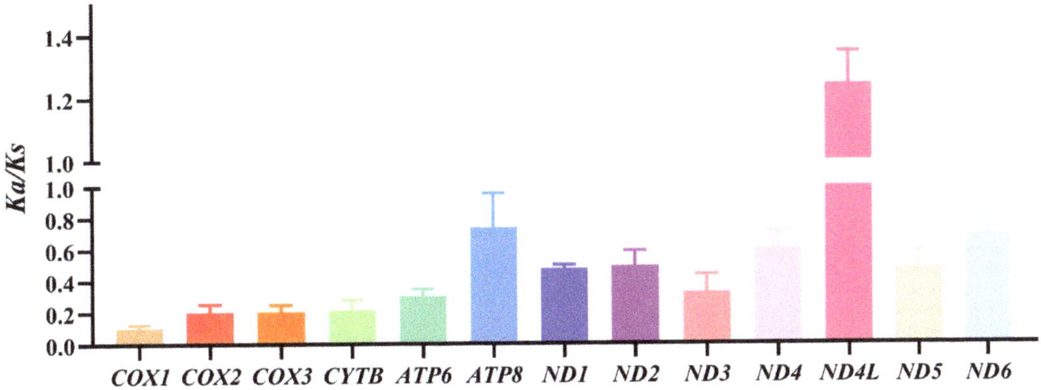

Figure 5. The *Ka/Ks* values of the subfamily Eumeninae are based on each PCG.

3.3. Gene Arrangement

Mitogenomes are usually stable in composition and gene arrangement is relatively conservative; therefore, recombination rarely occurs in the evolutionary history of insects [52,60]. As more and more mitogenomes of insects are reported, the rates of mitogenome rearrangement in Hymenoptera are accelerated [53,61]. The subfamily Eumeninae is the primary lineage of the Vespidae, and its gene rearrangement events are still poorly studied. Some eumenine mitogenomes contain a duplication of *trnM* and the positions of the two *trnM* are different, which means that the different mechanisms occurred in the gene rearrangement of Eumeninae. We investigated more rearrangement events of 54 eumenine mitogenomes and found that all eumenine mitogenomes contain a translocation *trnL1* upstream of *nad1* (Figure 6). Because gene duplications are not allowed in CREx, the rearrangement events in cluster *trnQ-trnM-ND2-trnW* are inferred as three patterns compared with the ancestral mitochondrial gene order (Figure S4): the tandem duplication of *trnM* occurs in all three clusters of Eumeninae, and then the distinct recombination occurs in the three clusters, respectively. In the tribe Zethini, the recombination occurs in *trnQ-trnM0* and *trnM1-ND2* after *trnM* duplicated to *trnM0-trnM1*, and subsequently, it occurs between *trnM0-trnQ* and *ND2-trnM1-trnW*, and there is another rearrangement type: from the ancestral order *trnQ-trnM-ND2-trnW* to *ND2-trnW-trnM0-trnM1* in *Calligaster cyanoptera*. In the tribe Odynerini, the recombination between *trnM1* and *trnQ-trnM0* occurs in most species, and the recombination *trnQ-trnM0* after *trnM* duplicated to *trnM0-trnM1* occurs in *Allodynerus delphinalis* and *Allodynerus mandschuricus*. In the same way, the recombination occurs in *trnQ-trnM1* after *trnM* duplicated to *trnM0-trnM1*, and then the recombination occurs between *trnM1 trnQ* and *trnM0-ND2-trnW* in most species of the tribe Eumenini. As mentioned above, the three tribes of the subfamily Eumeninae possess their distinctive rearrangement pattern. Our results provided additional evidence that the majority of mitogenome rearrangements occur in tRNAs in hymenopteran insects and also showed that the gene block *trnI-trnQ-trnM-ND2* may be the hot spot of rearrangement in Hymenoptera because the rearrangement events of this block are found in many hymenopteran lineages, such as the rearrangement events of the gene block *CR-trnI-trnQ-trnM-ND2-trnW-trnC-trnY* in all the Icheumonoid lineages and rearrangement of *CR-trnI-trnQ-trnM* in the mitochondrial genome of *Allantus luctifer* [62,63].

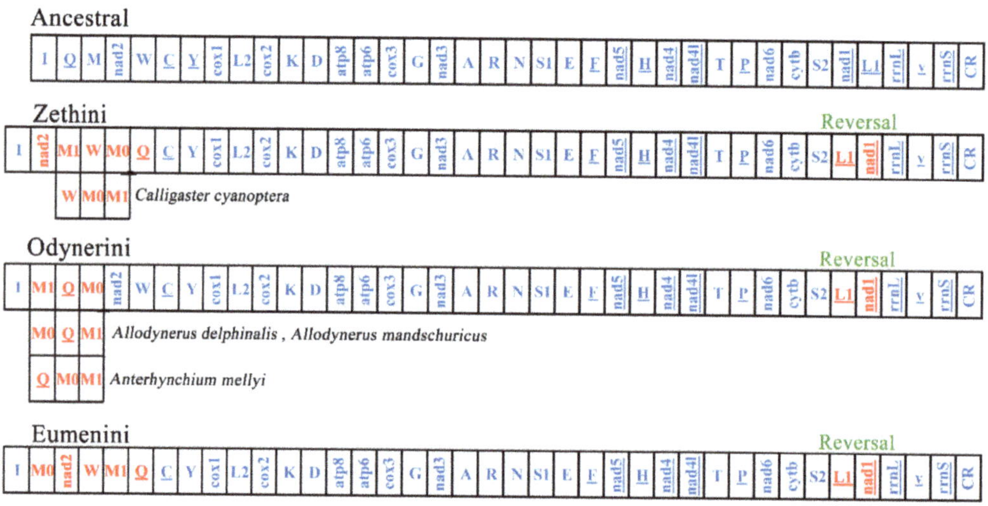

Figure 6. The rearrangement event in three tribes of the subfamily Eumeninae. The red genes represent its' position changes compared with ancestral mitogenomes.

3.4. Phylogenetic Relationship of Vespidae

The results of substitution saturation show that Iss > Iss.c and p = 0.0000 within both PCG and PCGR (Table S4), which indicates that the sequences are not saturated and can be used for phylogenetic analysis. In this study, phylogenetic analyses of two concatenated nucleotides (PCG and PCGR) were conducted, both representing four subfamilies (Stenogastrinae, Eumeninae, Polistinae, and Vespinae) of Vespidae and the outgroup (*Hylaeus dilatatus*, *Andrena cineraria*, *Megachile sculpturalis*, and *Apis cerana*). The two concatenated nucleotides were subjected to Bayesian inference (BI) and maximum likelihood (ML) analyses, resulting in four trees where the positions of the four subfamilies are congruent (Figure 7). In these trees, the phylogenetic relationships of the Vespidae are as follows: Stenogastrinae + ("Eumeninae" + (Zethini + (Polistinae + Vespinae))). According to our results, the subfamilies Stenogastrinae, Polistinae, and Vespinae are undoubtedly monophyletic, but nevertheless, the subfamily Eumeninae excluding Zethini is monophyletic. Additionally, Stenogastrinae is a sister lineage to other subfamilies of Vespidae with high bootstrap support values (BS) and Bayesian posterior probabilities values (PP) (BS = 100, PP = 1), which is consistent with some recent phylogenomic studies [64,65]. The study reveals that the tribe Zethini of the subfamily Eumeninae is an independent branch and more closely related to Polistinae and Vespinae, which is similar to previous studies [5,13,15]. As the tree shows in Figure 7, the position of the Zethini ("Zethinae") is between solitary Eumeninae and eusocial Vespinae + Polistinae. The genus *Calligaster* and the subgenus *Zethoides* of genus *Zethus* in Zethini ("Zethinae") have been cited as exemplifying the critical evolutionary stages of subsocial and communal behavior which connects solitary and eusocial wasps because it is reported that some species of both *Zethus* and *Calligaster* construct their nests with plant material rather than the typical eumenine nest construction with mud [66,67].

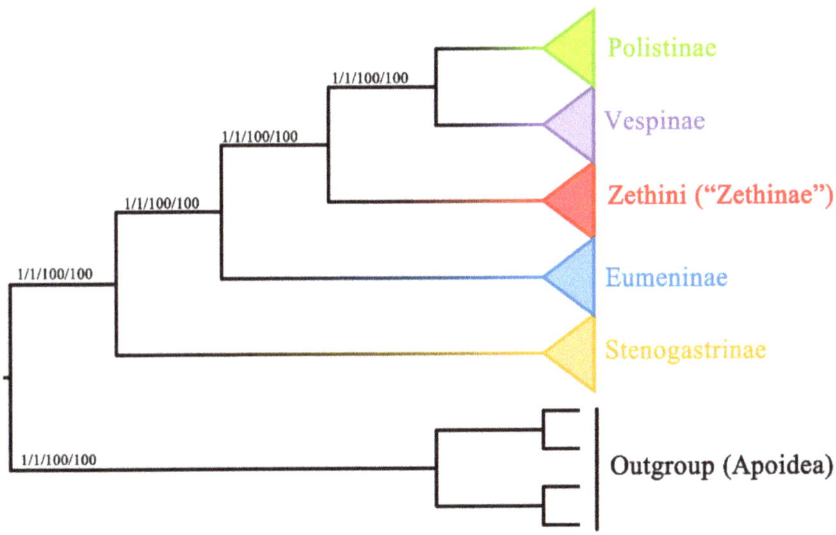

Figure 7. Phylogenetic trees of the Vespidae inferred from PCG and PCGR by ML and BI. Each node shows the Bayesian posterior probabilities (PP)/maximum likelihood bootstrap support (BS) values.

3.5. Phylogenetic Relationship within Eumeninae

The ingroup relationships of the family Vespidae are congruent in the obtained trees with the same methods, respectively, and the notable difference between the obtained tree topologies with ML and BI methods is that *Pseudozumia indosinensis* belongs to clade VIII in ML trees with low bootstrap support value (BS = 39, 38), while in BI trees it belongs to clade IX with high Bayesian posterior probability (PP = 1,1) (Figure S5). That the BS of a branch is lower than 50 means that the relationship has not been supported. In order to verify the accuracy of the obtained trees, a MP analysis was also performed, and the results were (of course) similar to the BI trees (Figure S6). Therefore, with the high Bayesian posterior probabilities, it is more likely that *Pseudozumia indosinensis* belongs to clade IX in BI trees. Of course, it may be because the differing placement in the ML trees is an artifact of bootstrap values below 50. The placement of the genus *Pseudozumia* in BI trees is also consistent with the result of Piekarski et al. In their research, a maximum-likelihood tree of Vespidae inferred from 235 selected loci obtained also shows the genus *Pseudozumia* as a sister group to *Orancistrocerus* which belongs to clade IX in this study [15]. Furthermore, the illustration in Figure 8 is identical to the results from analyzing the data of PCG and PCGR with the BI method. Both ML and BI reveal that the tribe Eumenini is monophyletic and the tribe Odynerini is paraphyletic containing 10 clades. The results also consistently indicate that the clade II to clade X is the sister group to the tribe Eumenini.

Within the tribe Eumenini, the sister relationship of (*Oreumenes* + *Delta*) + (*Katamenes* + *Eumenes*) is strongly supported by all datasets in this study (BS = 100; PP = 1). Hermes et al. found that the genera *Oreumenes, Delta, Katamenes,* and *Eumenes* all belong to their clade 3 of Eumenini, and the genus *Eumenes* was recovered as a sister to the remaining taxa of Eumenini [6]. The differences in phylogenetic relationships of the four genera between our study and Hermes et al. might be attributed to our limited generic sample, which is not enough to clarify the comprehensive relationships of all genera of the tribe Eumenini. Therefore, to clearly understand the phylogenetic relationships within the tribe Eumenini, more data are needed.

The tribe Odynerini is the biggest one within the subfamily Eumeninae [6]. Here, we investigated 24 genera of Odynerini to illustrate their phylogenetic relationships. The results show that Odynerini comprising 10 major clades (I-X) is paraphyletic, and clade I (the genus *Abispa*) is a sister group to all remaining Eumeninae. Bank et al. reported that the genus *Alastor* (clade A) is inferred as a sister lineage to all remaining Eumeninae based on transcriptomes of 49 vespid wasps [14]. Our study does not contain any species in the genus *Alastor*, and Bank *et al.*'s study did not contain any species in the genus *Abispa*, whereas that of Piekarskis et al. containing both *Abispa* and *Alastor*, is consistent with the standpoint of Bank et al. [14,15]. In clade II, the genus *Leptochilus* is inferred as a sister lineage to the genus *Labus*. In succession, the genus *Symmorphus* is an independent clade III and sister group to clades IV-X. In clades IV-X, there is a sister-group relationship between clades IV-V and VI-X. Within clades IV-V, clade IV is a sister group to clade V, while *Jucancistrocerus* (*Jucancistrocerus*) *angustifrons* and *Jucancistrocerus* (*Eremodynerus*) *atrofasciatus* are located at clades IV and V, respectively, which may support subgenera *Eremodynerus* being a valid genus [68]. Of course, more morphological evidence of more species should be investigated to confirm our results in further research. Within clades VI-X, clade VI is a sister group to clades VII-X and is composed of *Antepipona* + (*Paralepromenes* + *Apodynerus*). Then, clade VII is a sister group to clades VIII-X, composed of *Pararrhynchium* and *Allorhynchium*, while *Pararrhynchium striatum* is located within the genus *Allorhynchium*. The misidentification of *Pararrhynchium striatum* is eliminated by the examination of specimens, so *Pararrhynchium striatum* should be transferred to the genus *Allorhynchium*, or these two genera are synonymized. Likewise, there is a sister relationship between clade VIII and clades IX-X, and the phylogenetic relationships of clade VIII are as follows: (*Pareumenes* + (*Ectopioglossa* + *Pseumenes*)). Finally, within these two clades IX and X, they are sisters to each other, and the phylogenetic relationships of clade IX are as follows: (*Pseudozumia* + (*Euodynerus* + *Orancistrocerus*)), and of clade X they are (*Anterhynchium* (*Anterhynchium*) + (*Anterhynchium* (*Dirhynchium*) + *Rhynchium*)). As the results show, with high Bayesian posterior probabilities (PP = 1), the subgenus *Dirhynchium* of *Anterhynchium* is more closely related to the genus *Rhynchium* than the nominate subgenus *Anterhynchium*, which means the subgenus *Dirhynchium* should be upgraded to a valid genus. Again, further investigation is needed.

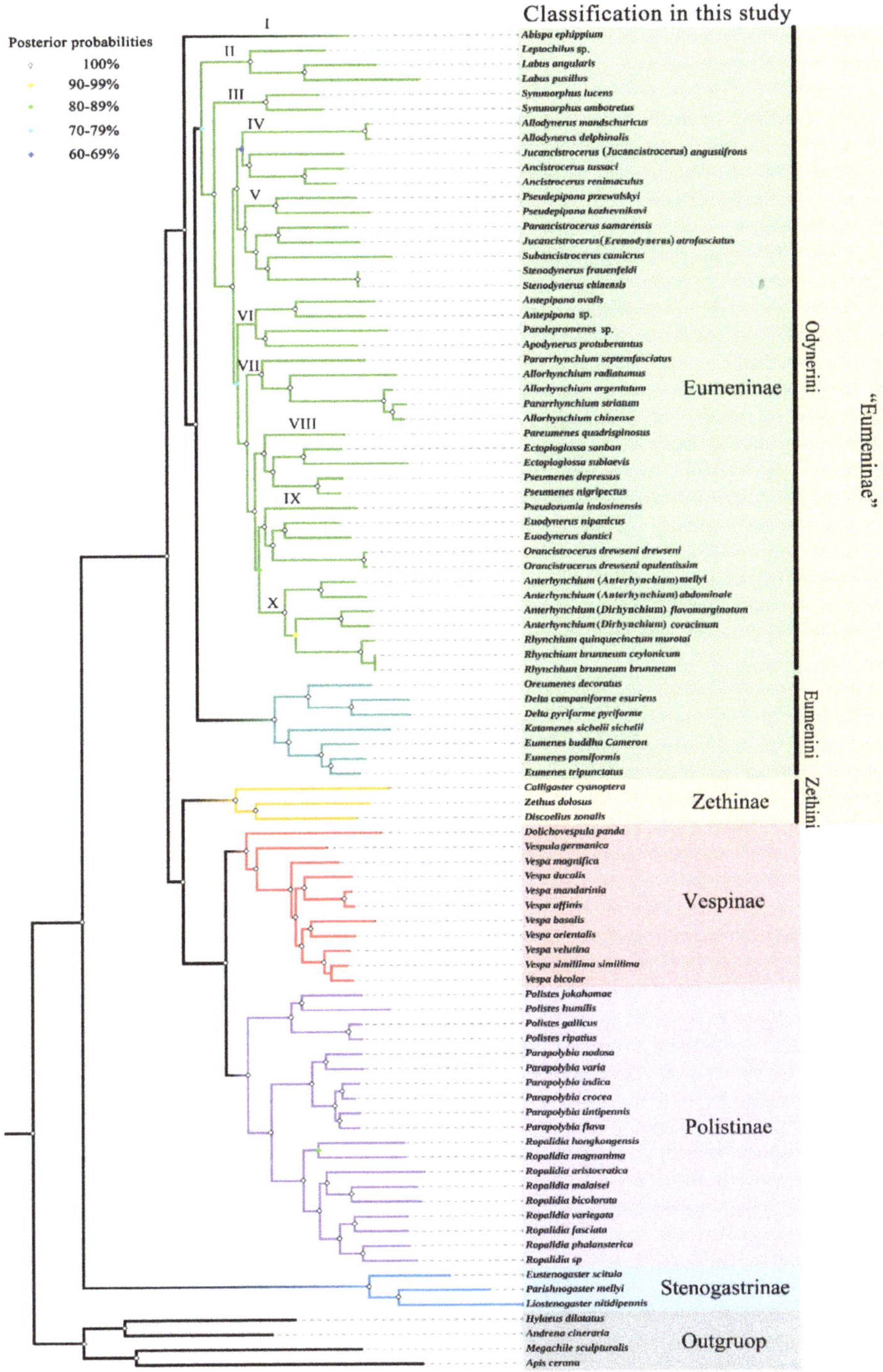

Figure 8. Phylogenetic tree of Vespidae inferred from PCG and PCGR by BI. Each node shows the Bayesian posterior probability (PP) values.

4. Conclusions

To sum up, the mitogenomes of Eumeninae are commonly found to contain two *trnM*, which differs remarkably from the gene orders of other Vespidae. This study based on mitogenomes further supports previously proposed relationships among Vespidae [5,14,68], especially the placement of the tribe Zethini and some genera of the subfamily Eumeninae, indicating that the tribe Zethini should be raised to Zethinae and that the tribe Eumenini is monophyletic and Odynerini is paraphyletic. Meanwhile, some issues have not been clearly resolved in this study. First, stable generic morphological characters are needed to support these two subgenera *Eremodynerus* and *Dirhynchium* as valid genera. Additionally, although *Pararrhynchium striatum* is proposed to be moved to *Allorhynchium*, it is possible that the relationship between these two genera is confused, which requires more species sampling and morphological characteristics to elucidate. Second, considering that only one limited mitogenome in some genera of Eumeninae, such as *Abispa*, *Apodynerus*, *Leptochilus*, *Parancistrocerus*, *Paraleptomenes*, *Pareumenes*, *Pseudozumia*, and *Subancistrocerus*, is presented in our analyses, the taxonomic status of these genera may be unstable and uncertain. In the end, the relationships of these taxa in this study still need to be verified by morphological and biological information. Therefore, to further advance the research on the systematic relationships of the subfamily Eumeninae, more taxon sampling and information about the morphological characteristics, molecular data, and biological behaviors are needed.

Supplementary Materials: The following supporting information can be downloaded at: https://www.mdpi.com/article/10.3390/insects13060529/s1, Supplementary Figure S1. The gene order of eumenine mitogenomes used in this study; Supplementary Figure S2. Inferred secondary structures of duplicated *trnM*. The substitutions in *trnM0* and *trnM1* compared with each other are indicated by red color; Supplementary Figure S3. ENC-GC_{3S} plot of total PCGs in 54 eumenine mitogenomes, the black curve shows the relationship between ENC values and GC_{3S} under random codon usage assumption; Supplementary Figure S4. The hypothesized pathway of the translocated inversion derived by recombination and duplication in three tribes of the subfamily Eumeninae. The red genes represent their positions; Supplementary Figure S5. Phylogenetic trees of Vespidae were inferred from PCG and PCGR by ML. Each node shows the bootstrap support values; Supplementary Figure S6. Phylogenetic trees of Vespidae. (A): Phylogenetic tree of Vespidae inferred from PCG by MP. (B): Phylogenetic tree of Vespidae inferred from PCGR by MP. Each nod shows the bootstrap support values; Supplementary Table S1. Mitochondrial genomes used for phylogenetic analysis in this study; Supplementary Table S2. The best partitioning scheme selected by PartitionFinder for different data matrices; Supplementary Table S3. Base composition, total length (bp), and AT-skew of complete Eumeninae mitogenomes; Supplementary Table S4. Substitution saturation test results.

Author Contributions: L.L. completed all the analyses and wrote the manuscript. J.M.C. provided data analysis, reviewed the manuscript, and approved the final version to be published. B.C. contributed design of the work, gave important comments on this study, and reviewed the manuscript. T.L. is responsible for the implementation of the entire project and revising the manuscript. All authors have read and agreed to the published version of the manuscript.

Funding: This research was funded by the National Natural Science Foundation of China (Nos: 31772490, 31372247, 31000976).

Institutional Review Board Statement: Not applicable.

Informed Consent Statement: Not applicable.

Data Availability Statement: Mitochondrial genome sequences are accessible on GenBank and accession numbers are contained within Table 1.

Acknowledgments: We are extremely grateful to Song Fan of China Agricultural University for his important comments on this study and to Shu-Lin He of Chongqing Normal University for his help with the analysis method of this study.

Conflicts of Interest: The authors declare no conflict of interest.

References

1. Cowan, D.P. *The Social Biology of Wasps*; Cornell University Press: Ithaca, NY, USA, 1991; pp. 33–73.
2. Jennings, D.T.; Houseweart, M.W. Predation by Eumenid Wasps (Hymenoptera: Eumenidae) on Spruce Budworm (Lepidoptera: Tortricidae) and Other Lepidopterous Larvae in Spruce-Fir Forests of Maine. *Ann. Entomol. Soc. Am.* **1984**, *49*, 39–45. [CrossRef]
3. Pannure, A.; Belavadi, V.V.; Carpenter, J.M. Taxonomic studies on potter wasps (Hymenoptera: Vespidae: Eumeninae) of south India. *Zootaxa* **2016**, *4171*, 1–50. [CrossRef] [PubMed]
4. Bohart, R.M.; Stange, L.A. *A Revision of the Genus Zethus in the Western Hemisphere (Hymenoptera, Eumenidae)*; University of California Press: Oakland, CA, USA, 1965; Volume 40, pp. 1–208.
5. Hines, H.M.; Hunt, J.H.; O'Connor, T.K.; Gillespie, J.J.; Cameron, S.A. Multigene phylogeny reveals eusociality evolved twice in vespid wasps. *Proc. Natl. Acad. Sci. USA* **2007**, *104*, 3295–3299. [CrossRef] [PubMed]
6. Carpenter, J.M.; Cumming, J.M. A character analysis of the North American potter wasps (Hymenoptera: Vespidae; Eumeninae). *J. Nat. Hist.* **1985**, *19*, 877–916. [CrossRef]
7. Latreille, P.A. *Histoire Naturelle, Générale et Particulière des Crustacés et des Insects: Ouvrage Faisant Suite aux Oeuvres de Leclerc de Buffon, et Partie du Cours Complet D'historie Naturelle Rédigé par C.S.*; Sonnini, F., Ed.; Dufart: Paris, France, 1802. [CrossRef]
8. de Saussure, H.F. *Etudes sur la Famille des Vespides 1. Monographie des Guepes Solitaires ou de la Tribe des Eumeniens*; Masson, V., Cherbuliez, J., Eds.; Victor Masson: Paris, France, 1852–1853; pp. 1–286.
9. de Saussure, H.F. *Etudes sur la Famille des Vespides 3. La Monographie des Masariens et un Supplement a la Monographie des Eumeniens*; Masson, V., Cherbuliez, J., Eds.; Victor Masson: Paris, France, 1854–1856; pp. 1–352.
10. Richards, O.W. *A Revisional Study of the Masarid Wasps (Hymenoptera, Vespoidea)*; British Museum (Natural History): London, UK, 1962; pp. 1–294.
11. Carpenter, J.M. The phylogenetic relationships and natural classification of the Vespoidea (Hymenptera). *Syst. Entomol.* **1982**, *7*, 11–38. [CrossRef]
12. Hermes, M.G.; Melo, G.; Carpenter, J.M. The higher-level phylogenetic relationships of the Eumeninae (Insecta, Hymenoptera, Vespidae), with emphasis onEumenessensu lato. *Cladistics* **2013**, *30*, 453–484. [CrossRef]
13. Schmitz, J.; Moritz, R. Molecular Phylogeny of Vespidae (Hymenoptera) and the Evolution of Sociality in Wasps. *Mol. Phylogenetics Evol.* **1998**, *9*, 183–191. [CrossRef]
14. Bank, S.; Sann, M.; Mayer, C.; Meusemann, K.; Donath, A.; Podsiadlowski, L.; Kozlov, A.; Petersen, M.; Krogmann, L.; Meier, R.; et al. Transcriptome and target DNA enrichment sequence data provide new insights into the phylogeny of vespid wasps (Hymenoptera: Aculeata: Vespidae). *Mol. Phylogenetics Evol.* **2017**, *116*, 213–226. [CrossRef]
15. Piekarski, P.K.; Carpenter, J.M.; Lemmon, A.R.; Lemmon, E.M.; Sharanowski, B.J. Phylogenomic Evidence Overturns Current Conceptions of Social Evolution in Wasps (Vespidae). *Mol. Biol. Evol.* **2018**, *35*, 2097–2109. [CrossRef]
16. Heraty, J.; Ronquist, F.; Carpenter, J.M.; Hawks, D.; Schulmeister, S.; Dowling, A.P.; Murray, D.; Munro, J.; Wheeler, W.C.; Schiff, N.; et al. Evolution of the hymenopteran megaradiation. *Mol. Phylogenetics Evol.* **2011**, *60*, 73–88. [CrossRef]
17. Wang, Y.; Cao, J.-J.; Li, W.-H. Complete Mitochondrial Genome of Suwallia teleckojensis (Plecoptera: Chloroperlidae) and Implications for the Higher Phylogeny of Stoneflies. *Int. J. Mol. Sci.* **2018**, *19*, 680. [CrossRef]
18. Mao, M.; Gibson, T.; Dowton, M. Higher-level phylogeny of the Hymenoptera inferred from mitochondrial genomes. *Mol. Phylogenetics Evol.* **2014**, *84*, 34–43. [CrossRef] [PubMed]
19. Li, T.-J.; Chen, B. Descriptions of four new species of *Pararrhynchium* de Saussure (Hymenoptera: Vespidae: Eumeninae) from China, with one newly recorded species and a key to Chinese species. *Orient. Insects* **2018**, *52*, 175–189. [CrossRef]
20. Li, T.-J.; Barthélémy, C.; Carpenter, J.M. The Eumeninae (Hymenoptera, Vespidae) of Hong Kong (China), with description of two new species, two new synonymies and a key to the known taxa. *J. Hymenopt. Res.* **2019**, *72*, 127–176. [CrossRef]
21. Tan, J.-L.; Carpenter, J.M.; Van Achterberg, C. An illustrated key to the genera of Eumeninae from China, with a checklist of species (Hymenoptera, Vespidae). *ZooKeys* **2018**, *740*, 109–149. [CrossRef] [PubMed]
22. Tan, J.-L.; Carpenter, J.M.; Van Achterberg, C. Most northern Oriental distribution of Zethus Fabricius (Hymenoptera, Vespidae, Eumeninae), with a new species from China. *J. Hymenopt. Res.* **2018**, *62*, 1–13. [CrossRef]
23. Wang, H.-C.; Chen, B.; Li, T.-J. A new species and a new record of the genus Discoelius Latreille, 1809 (Hymenoptera: Vespidae: Eumeninae) from China. *Zootaxa* **2019**, *4686*, 297–300. [CrossRef]
24. Wang, H.-C.; Chen, B.; Li, T.-J. Three new species of the genus Zethus Fabricius, 1804 (Hymenoptera, Vespidae, Eumeninae) from China, with an updated key to the Oriental species. *J. Hymenopt. Res.* **2019**, *71*, 209–224. [CrossRef]
25. Wang, H.-C.; Chen, B.; Li, T.-J. Taxonomy of the genus Ectopioglossa Perkins from China, with two new species and an updated key to the Oriental species (Hymenoptera: Vespidae: Eumeninae). *J. Asia-Pacific Entomol.* **2020**, *23*, 253–259. [CrossRef]
26. Zhang, X.; Chen, B.; Li, T.-J. Taxonomy of the genus Epsilon from China, with a new species and an updated key to the Oriental species (Hymenoptera, Vespidae, Eumeninae). *ZooKeys* **2020**, *910*, 131–142. [CrossRef]
27. Zhang, X.; Chen, B.; Li, T.-J. A taxonomic revision of Allodynerus Blüthgen (Hymenoptera: Vespidae: Eumeninae) from China. *Zootaxa* **2020**, *4750*, 545–559. [CrossRef] [PubMed]
28. Bai, Y.; Chen, B.; Li, T.-J. A new species and two new records of the genus Pseudepipona de Saussure, 1856 (Hymenoptera, Vespidae, Eumeninae) from China, with a key to the Chinese species. *J. Hymenopt. Res.* **2021**, *82*, 285–304. [CrossRef]
29. Bai, Y.; Chen, B.; Li, T.-J. Two newly recorded genera Malayepipona Giordani Soika and Megaodynerus Gusenleitner, with eight new species from China (Hymenoptera, Vespidae, Eumeninae). *Zootaxa* **2021**, *5060*, 371–391. [CrossRef]

30. Li, T.-J.; Bai, Y.; Chen, B. A revision of the genus Jucancistrocerus Blüthgen, 1938 from China, with review of three related genera (Hymenoptera: Vespidae: Eumeninae). *Zootaxa* **2022**, *5105*, 401–420. [CrossRef] [PubMed]
31. Peng, Y.; Leung, H.C.M.; Yiu, S.M.; Chin, F.Y.L. IBDA-UD: A de novo assembler for single-cell and metagenomic sequencing data with highly uneven depth. *Bioinformatics* **2012**, *28*, 1420–1428. [CrossRef] [PubMed]
32. Patel, R.K.; Mukesh, J.; Liu, Z. NGS QC Toolkit: A Toolkit for Quality Control of Next Generation Sequencing Data. *PLoS ONE* **2012**, *7*, e30619. [CrossRef] [PubMed]
33. Simon, C.; Buckley, T.R.; Frati, F.; Stewart, J.B.; Beckenbach, A.T. Incorporating Molecular Evolution into Phylogenetic Analysis, and a New Compilation of Conserved Polymerase Chain Reaction Primers for Animal Mitochondrial DNA. *Annu. Rev. Ecol. Evol. Syst.* **2006**, *37*, 545–579. [CrossRef]
34. Altschup, S.; Gish, W.; Miller, W.; Myers, E.; Lipman, D. Basic local alignment search tool. *J. Mol. Biol.* **1990**, *215*, 403–410. [CrossRef]
35. Thompson, J.D.; Gibson, T.J.; Plewniak, F.; Jeanmougin, F.; Higgins, D.G. The CLUSTAL_X windows interface: Flexible strategies for multiple sequence alignment aided by quality analysis tools. *Nucleic Acids Res.* **1997**, *25*, 4876–4882. [CrossRef]
36. Lowe, T.M.; Chan, P.P. tRNAscan-SE On-line: Search and contextual analysis of transfer RNA genes. *Nucleic Acids Res.* **2016**, *44*, W54–W57. [CrossRef]
37. Laslett, D.; Canbäck, B. ARWEN: A program to detect tRNA genes in metazoan mitochondrial nucleotide sequences. *Bioinformatics* **2008**, *24*, 172–175. [CrossRef] [PubMed]
38. Kumar, S.; Stecher, G.; Li, M.; Knyaz, C.; Tamura, K. MEGA X: Molecular Evolutionary Genetics Analysis across Computing Platforms. *Mol. Biol. Evol.* **2018**, *35*, 1547–1549. [CrossRef] [PubMed]
39. Cai, M.-S.; Cheng, A.-C.; Wang, M.-S.; Zhao, L.-C. Characterization of Synonymous Codon Usage Bias in the Duck Plague Virus UL35 Gene. *Karger* **2009**, *52*, 266–278. [CrossRef]
40. Librado, P.; Rozas, J. DnaSP v5: A software for comprehensive analysis of DNA polymorphism data. *Bioinformatics* **2009**, *25*, 1451–1452. [CrossRef] [PubMed]
41. Bernt, M.; Merkle, D.; Ramsch, D.; Fritzsch, G.; Perseke, M.; Bernhard, D.; Schlegel, M.; Stadler, P.F.; Middendorf, M. CREx: Inferring genomic rearrangements based on common intervals. *Bioinformatics* **2007**, *23*, 2957–2958. [CrossRef]
42. Zhang, D.; Gao, F.; Jakovlićc, I.; Zou, H.; Zhang, J.; Li, W.X.; Wang, G.T. PhyloSuite: An integrated and scalable desktop platform for streamlined molecular sequence data management and evolutionary phylogenetics studies. *Mol. Ecol. Resour.* **2020**, *20*, 348–355. [CrossRef]
43. Abascal, F.; Zardoya, R.; Telford, M.J. TranslatorX: Multiple alignment of nucleotide sequences guided by amino acid translations. *Nucleic Acids Res.* **2010**, *38*, W7–W13. [CrossRef]
44. Katoh, K.; Standley, D.M. MAFFT Multiple Sequence Alignment Software Version 7: Improvements in Performance and Usability. *Mol. Biol. Evol.* **2013**, *30*, 772–780. [CrossRef]
45. Talavera, G.; Castresana, J. Improvement of Phylogenies after Removing Divergent and Ambiguously Aligned Blocks from Protein Sequence Alignments. *Syst. Biol.* **2007**, *56*, 564–577. [CrossRef]
46. Lanfear, R.; Calcott, B.; Ho, S.Y.; Guindon, S. PartitionFinder: Combined Selection of Partitioning Schemes and Substitution Models for Phylogenetic Analyses. *Mol. Biol. Evol.* **2012**, *29*, 1695–1701. [CrossRef]
47. Ronquist, F.; Teslenko, M.; Mark, P.; Ayres, D.; Darling, A.; Hohna, S.; Larget, B.; Liu, L.; Suchard, M.; Huelsenbeck, J. MrBayes 3.2: Efcient Bayesian phyloge netic inference and model choice across a large model space. *Syst. Biol.* **2015**, *61*, 539–542. [CrossRef] [PubMed]
48. Guindon, S.; Dufayard, J.-F.; Lefort, V.; Anisimova, M.; Hordijk, W.; Gascuel, O. New Algorithms and Methods to Estimate Maximum-Likelihood Phylogenies: Assessing the Performance of PhyML 3.0. *Syst. Biol.* **2010**, *59*, 307–321. [CrossRef]
49. Goloboff, P.A.; Catalano, S.A. TNT version 1.5, including a full implementation of phylogenetic morphometrics. *Cladistics* **2016**, *32*, 221–238. [CrossRef] [PubMed]
50. Nixon, K.C. *Winclada*, version 1.98; Published by the author: Ithaca, NY, USA, 2015.
51. Tang, P.; Zhu, J.-C.; Zheng, B.-Y.; Wei, S.-J.; Sharkey, M.; Chen, X.-X. Mitochondrial phylogenomics of the Hymenoptera. *Mol. Phylogenetics Evol.* **2018**, *131*, 8–18. [CrossRef]
52. Boore, J.L. Animal mitochondrial genomes. *Nucleic Acids Res.* **1999**, *27*, 1767–1780. [CrossRef] [PubMed]
53. Mao, M.; Valerio, A.; Austin, A.D.; Dowton, M.; Johnson, N.F. The first mitochondrial genome for the wasp superfamily Platygastroidea: The egg parasitoid *Trissolcus basalis*. *Genome* **2012**, *55*, 194–204. [CrossRef]
54. Korkmaz, E.M.; Dogan, O.; Durel, B.S.; Budak, M.; Basibuyuk, H.H. Mitogenome organization and evolutionary history of the subfamily Cephinae (Hymenoptera: Cephidae). *Syst. Entomol.* **2018**, *43*, 606–618. [CrossRef]
55. Cameron, S.L.; Dowton, M.; Castro, L.R. Mitochondrial genome organization and phylogeny of two vespid wasps. *Genome* **2008**, *51*, 800–808. [CrossRef] [PubMed]
56. Sharp, P.M.; Li, W.-H. An evolutionary perspective on synonymous codon usage in unicellular organisms. *J. Mol. Evol.* **1986**, *24*, 28–38. [CrossRef]
57. Yu, T.; Li, J.; Yang, Y.; Qi, L. Codon usage patterns and adaptive evolution of marine unicellular cyanobacteria Synechococcus and Prochlorococcus. *Mol. Phylogenetics Evol.* **2012**, *61*, 206–213. [CrossRef]
58. Wei, L.; He, J.; Jia, X.; Qi, Q.; Liang, Z.; Zheng, H. Analysis of codon usage bias of mitochondrial genome in Bombyx moriand its relation to evolution. *BMC Evol. Biol.* **2014**, *14*, 262. [CrossRef] [PubMed]

59. Hurst, L.D. The Ka/Ks ratio: Diagnosing the form of sequence evolution. *Trends Genet.* **2002**, *18*, 486. [CrossRef]
60. Barr, C.M.; Neiman, M.; Taylor, D.R. Inheritance and recombination of mitochondrial genomes in plants, fungi and animals. *New Phytol.* **2005**, *168*, 39–50. [CrossRef] [PubMed]
61. Dowton, M.; Austin, A.D. Evolutionary dynamics of a mitochondrial rearrangement "hot spot" in the Hymenoptera. *Mol. Biol. Evol.* **1999**, *16*, 298–309. [CrossRef] [PubMed]
62. Feng, Z.B.; Wu, Y.F.; Yang, C.; Gu, X.H.; Wilson, J.J.; Li, H.; Cai, W.Z.; Yang, H.L.; Song, F. Evolution of tRNA gene rearrangement in the mitochondrial genome of ichneumonoid wasps (Hymenoptera: Ichneumonoidea). *Int. J. Biol. Macromol.* **2020**, *164*, 540–547. [CrossRef]
63. Wei, S.-J.; Niu, F.-F.; Du, B.-Z. Rearrangement oftrnQ-trnMin the mitochondrial genome ofAllantus luctifer(Smith) (Hymenoptera: Tenthredinidae). *Mitochondrial DNA* **2014**, *27*, 856–858. [CrossRef]
64. Peters, R.S.; Krogmann, L.; Mayer, C.; Donath, A.; Niehuis, O. Evolutionary History of the Hymenoptera. *Curr. Biol.* **2017**, *27*, 1013–1018. [CrossRef]
65. Huang, P.; Carpenter, J.M.; Chen, B.; Li, T.-J. The first divergence time estimation of the subfamily Stenogastrinae (Hymenoptera: Vespidae) based on mitochondrial phylogenomics. *Int. J. Biol. Macromol.* **2019**, *137*, 767–773. [CrossRef]
66. Ducke, A. Uber Phylogenie und Klassifikation der socialen Vespiden. *Zool. Jahr. Abt. Syst. Geogr. Biol. Tiere* **1914**, *36*, 303–330.
67. Rau, P. The Jungle Bees and Wasps of Barro Colorado Island. *Ann. Entomol. Soc. Am.* **1933**, *2*, 376. [CrossRef]
68. Blüthgen, P. Beiträge zur Kenntnis der paläarktischen und einiger athiopischer Faltenwespen (*Hym. Vespidae*). *Veröffentlichungen Aus Dem Dtsch. Kolon. Und Übersee Mus. Bremen* **1939**, *2*, 233–267.

insects MDPI

Article

Revision of the World Species of *Megaphragma* Timberlake (Hymenoptera: Trichogrammatidae)†

Andrew Polaszek [1], Lucian Fusu [2,*], Gennaro Viggiani [3], Andie Hall [4], Paul Hanson [5] and Alexey A. Polilov [6]

1 Department of Life Sciences, Natural History Museum, Cromwell Road, London SW7 5BD, UK; a.polaszek@nhm.ac.uk
2 Research Group in Invertebrate Diversity and Phylogenetics, Faculty of Biology, Alexandru Ioan Cuza University, Bd. Carol I, Nr. 11, 700506 Iasi, Romania
3 Laboratorio di Lotta biologica, Dipartimento di Agraria, Università degli Studi di Napoli "Federico II", Via Università 100, 80055 Portici, NA, Italy; genviggi@unina.it
4 Core Research Laboratories, Natural History Museum, Cromwell Road, London SW7 5BD, UK; a.hall@nhm.ac.uk
5 Escuela de Biología, Universidad de Costa Rica, San Pedro de Montes de Oca, San Jose 11501-2060, Costa Rica; phanson91@gmail.com
6 Department of Entomology, Faculty of Biology, Lomonosov Moscow State University, 119234 Moscow, Russia; polilov@gmail.com
* Correspondence: lucfusu@hotmail.com

† urn:lsid:zoobank.org:pub:39858D98-8117-407B-ACF2-FBBFE1B231EB.

Simple Summary: Parasitoid wasps of the genus *Megaphragma* are some of the smallest known insects, being as small as some unicellular protozoans. Their life history is not known in great detail, but all species with known biology are parasitoids of thrips eggs (Thysanoptera) and as such, they are potential biological control agents of these pests. At the current state of knowledge of the genus, it is impossible to identify with confidence most of the *Megaphragma* species (original descriptions lack essential details or illustrations; molecular markers are available for very few species; many species are still undescribed while others were described multiple times). We provide the first revision of the genus that includes the formal descriptions and naming of 22 species and a key to all 32 valid species.

Abstract: *Megaphragma* species are important models for basic organismal research, and many are potential biological control agents. We present the first extensive revision of species of the genus *Megaphragma* based on morphological and molecular data. Our revision includes all previously described species, 6 of which are synonymized, and 22 of which are described here as new. We also provide the first key to all species of the genus and reconstruct their phylogeny based on 28S and CO1 molecular markers. The following species are synonymized with *M. longiciliatum* Subba Rao: *M. aligarhensis* Yousuf and Shafee **syn. nov.**; *M. amalphitanum* Viggiani **syn. nov.**; *M. decochaetum* Lin **syn. nov.**; *M. magniclava* Yousuf and Shafee **syn. nov.**; *M. shimalianum* Hayat **syn. nov.** *M. anomalifuniculi* Yuan and Lou **syn. nov.** is synonymized with *M. polychaetum* Lin. The following species are described as new: *M. antecessor* Polaszek and Fusu **sp. nov.**; *M. breviclavum* Polaszek and Fusu **sp. nov.**; *M. chienleei* Polaszek and Fusu **sp. nov.**; *M. cockerilli* Polaszek and Fusu **sp. nov.**; *M. digitatum* Polaszek and Fusu **sp. nov.**; *M. fanenitrakely* Polaszek and Fusu **sp. nov.**; *M. funiculatum* Fusu, Polaszek, and Viggiani **sp. nov.**; *M. giraulti* Viggiani, Fusu, and Polaszek **sp. nov.**; *M. hansoni* Polaszek, Fusu, and Viggiani **sp. nov.**; *M. kinuthiae* Polaszek, Fusu, and Viggiani **sp. nov.**; *M. liui* Polaszek and Fusu **sp. nov.**; *M. momookherjeeae* Polaszek and Fusu **sp. nov.**; *M. nowickii* Polaszek, Fusu, and Viggiani **sp. nov.**; *M. noyesi* Polaszek and Fusu **sp. nov.**; *M. pintoi* Viggiani **sp. nov.**; *M. polilovi* Polaszek, Fusu, and Viggiani **sp. nov.**; *M. rivelloi* Viggiani **sp. nov.**; *M. tamoi* Polaszek, Fusu, and Viggiani **sp. nov.**; *M. tridens* Fusu, and Polaszek **sp. nov.**; *M. uniclavum* Polaszek and Fusu **sp. nov.**; *M. vanlentereni* Polaszek and Fusu **sp. nov.**; *M. viggianii* Fusu, Polaszek, and Polilov **sp. nov**.

Keywords: species delimitation; egg parasitoid; Chalcidoidea; integrative taxonomy; molecular phylogeny; DNA barcoding; mini-barcode

1. Introduction

Trichogrammatidae (Hymenoptera: Chalcidoidea) is a family of egg parasitoids that consists of approximately 100 genera and 1000 species [1–3]. The genus *Megaphragma* (tribe Oligositini) currently contains 15 species, all of which are egg parasitoids of Thysanoptera [1]. It includes some of the smallest insects: most species of this genus have a body length of only 0.16–0.3 mm, which is about the same size as a larger unicellular organism such as *Paramecium* [4]. Huber and Noyes [5] provided a review of the body size limit in insects with only three genera of Mymaridae having species smaller than the smallest *Megaphragma*. The genus has been recorded on all continents except Antarctica; though most of the species are confined to the tropics and warmer temperate regions. According to the available data, all species of this genus are egg parasitoids of thrips, but the biology of most species remains unknown. The type of the genus, *M. mymaripenne* Timberlake, was examined by Viggiani [6], who gave details on several previously unused features, in particular on the sculpture present on tergites of the metasoma. The same author described several new species of *Megaphragma* [7,8]. Lin [9] described five new species from China, and Hayat [10] revised the Indian species. The *Megaphragma* of Argentina were studied by Viggiani et al. [11]. Detailed biological data are available for *M. mymaripenne*, *M. polilovi* **sp. nov.** (under the name *M. mymaripenne*), and *M. longiciliatum* Subba Rao (under the name *M. amalphitanum* Viggiani) [12–14].

Due to their extremely small body size, *Megaphragma* species have become model organisms for studying the miniaturization of insects [15,16] and solving neurobiological problems [17]. The general anatomy and anatomical features associated with miniaturization have been described [18]; the structure of the eye [19], antenna [20,21], and leg structures used for grooming [22], and peculiar features of the genome [23–25] have been studied. Anucleate neurons have been found in three species of *Megaphragma* [18,26] and the unique phenomenon of lysis of the bodies and nuclei of cells at the pupal stage of development has been described [4]. Analysis of the connectome of *Megaphragma* [27,28] and reconstruction of sensory organs at the cellular and subcellular level [28] are currently underway.

In addition to their value for basic research, *Megaphragma* species are potential biological agents for the control of thrips, many of which are important agricultural pests [29,30].

Most *Megaphragma* species descriptions are very brief, and genetic markers are available only for one named species [23,25,31] and three unnamed species [32,33]. Identification keys to species are available only for a few regions and include only selected species. Many specimens cannot therefore be identified. Thus, the lack of a revision makes it extremely difficult to work with these wasps, which are important for basic and applied research.

2. Materials and Methods

2.1. Specimens and Depository Abbreviations

Specimens on slides, mostly type material, were received or deposited in the following institutions: Alexandru Ioan Cuza University of Iaşi, Romania, Lucian Fusu collection (**AICF**); Aligarh Muslim University, Aligarh, India (**AMU**: Mohammad Hayat); Australian National Insect Collection, Canberra, Australia (**ANIC**: Nicole Fisher); Canadian National Collection of Insects, Ottawa, Canada (**CNCI**: John Huber); CIRAD/UMR CBGP, Montferrier-sur-Lez, France (**CIRAD**: Gerard Delvare); Università di Napoli "Federico II", Dipartimento di Agraria, Collezione di Entomologia, Portici, Italia (**DACE**: Gennaro Viggiani); Department of Zoology, Plant Protection College, Fujian Agricultural and Forestry University Fuzhou, Fujian, China (**FAFU**: Naiquan Lin); International Institute of Tropical Agriculture, Cotonou, Benin (**IITA**: Georg Goergen); Museo de Zoología, Universidad de Costa Rica (**MZUCR**). Natural History Museum of Oman (**NHMO**); Natural History Museum, London, UK (**NHMUK**: Natalie Dale-Skey); Musée Royal de l'Afrique Centrale, Tervuren, Belgium (**MRAC**: Eliane de Coninck); Plant and Food Research New Zealand (formerly **DSIR**: Jocelyn Berry); University of California, Riverside, USA (**UCRC**: Serguei Triapitsyn). Additional material was received for identification from several institutions, and a number of recent collections by the authors and Dr John Noyes (NHMUK) contributed substantial material to this revision.

2.2. Morphology

All material was examined on microscope slides for morphological characters using an Olympus BX63 microscope with Nomarski differential interference contrast (DIC) with 40× and 100× objectives. Since the lysis buffer used for DNA extraction (see below) contains proteinase K, there is no need to clear the body with KOH as usually performed before slide mounting. Instead, specimens were extracted from the lysis buffer with an adjustable volume pipette (0.5 to 10 µL) set at 1–2 µL to reduce liquid loss and transferred to distilled water to remove unwanted reagents. Afterward, they were dehydrated using a series of ethanol solutions of increasing concentration and cleared in clove oil as described by Noyes [34]. Afterward, some specimens were mounted laterally in Canada balsam while others were dissected and wings, antennae, head, and body were mounted separately under different coverslips following [34]. However, mounting the abdomen dorsal side up has the disadvantage of making the setae on the sides of the tergites very difficult to see. Where possible, the setae on the disc of the fore wing were counted on the upper and lower surfaces. Body colour was observed on both card-mounted specimens and on slide-mounted specimens in which the generally unremarkable body pigmentation remains preserved.

Selected specimens were dried using a critical point drier and examined with an electron microscope as described in Polilov [26].

Morphological terminology (see Figure 1) broadly follows Pinto [2], while terminology of the antennal sensilla follows Diakova et al. [20], albeit with abbreviations that follow the descriptions; e.g., "**UST**" for "uniporous sensilla trichodea" instead of Diakova et al. [20], who use "TS-UP".

The following abbreviations are used for morphological terms (see Figure 1b): **ASC** = aporous sensilla chaetica; **C1** = 1st (basal) clavomere; **C2** = 2nd (central or apical) clavomere; **C3** = 3rd (apical) clavomere; **MPS** = multiporous placoid sensilla; **MT** = microtrichia (referred to as "aporous sensilla trichodea, type 1" by Diakova et al. [20]); **SB** = sensilla basiconica; **SS** = sensilla styloconica; **T1, T2** etc. = metasomal tergite 1, 2 etc.; **UST** = uniporous sensilla trichodea.

In species with a single discal fore wing seta, its length is important: "short" = shorter than or equal to the distance between the 2 proximal wing fringe setae (i.e., those closest to the seta); long = longer than the distance between the 3 proximal wing fringe setae (see Figure 1c).

A peculiar type of metafemoral spine with a unique shape, structure, and position is present in all species of the *ghesquierei*-group. Probably non-homologous metafemoral spines are present in other species groups.

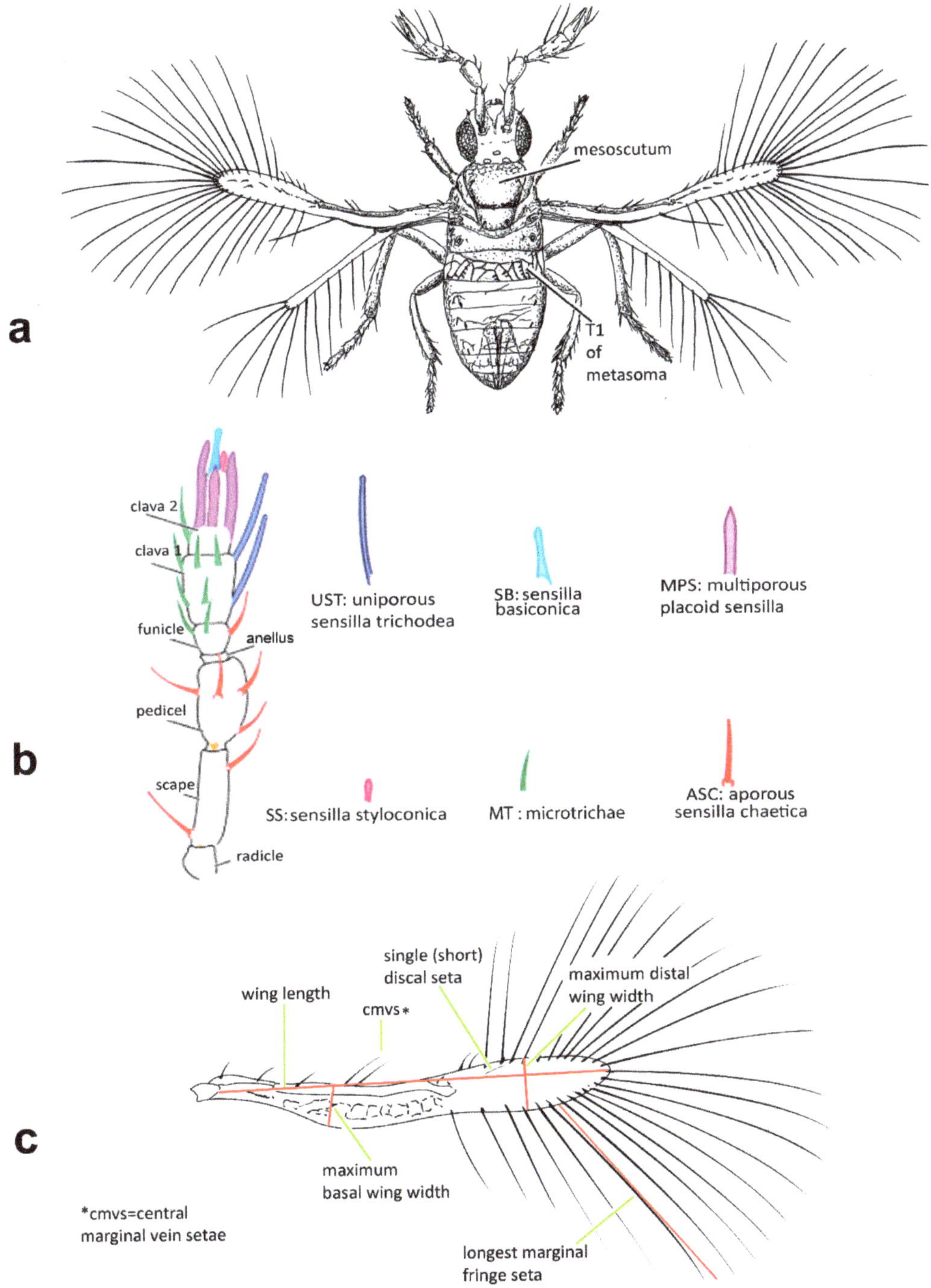

Figure 1. Line drawings of *Megaphragma* species: (**a**) *M. mymaripenne*, female habitus; (**b**) *Megaphragma* sp., stylized antenna; (**c**) *Megaphragma* sp. *ghesquierei*-group, fore wing.

2.3. Molecular Methods

DNA was extracted using a Qiagen DNeasy Blood and Tissue Kit with modifications as described in Cruaud et al. [35]; specimens were lysed whole for 6–8 h without grinding, then frozen at −80 °C overnight and thawed at room temperature before addition of buffer AL. After about the first 100 extractions the freezing stage was omitted as it appeared not to increase DNA yield significantly.

The primer pair D23F (5′-GAGAGTTCAAGAGTACGTG-3′) [36] and 28Sb also known as D3B (5′-TCGGAAGGAACCAGCTACTA-3′) [37,38] was used to amplify an approximately 850 bp fragment from the 5′ end of the nuclear ribosomal 28S gene spanning the D2–D3 region. In the instances where there was no detectable PCR product, we performed a second PCR using 1 μL of the primary PCR product and the semi-nested primer pair D23F combined with the newly designed reverse primer 28Sbsn (5′-GATGGTTCGATTAGTCTTTCG-3′), which amplified an approximately 800 bp fragment of the 28S rDNA.

The CO1 gene was amplified using the standard primer pair LCO1480 and HCO2198 [39], which amplifies the DNA barcode region for animals [40,41]. In case of failed reactions, we used a pair of internal primers from Fusu and Polaszek [42] that amplify the standard barcode region from two overlapping fragments (mini-barcodes, [43]): a modified LCO1490, named LCO1490M (5′-CAACAAATCATAAAGATATTGG-3′), pairs with MChaR1 (5′-CCYGTTCCAAYAAATATTCT-3′), and MChaF1 (5′-CCTCGAATAAATAATATAAGATT-3′) pairs with HCO2198.

The PCR conditions were as described in Fusu and Polaszek [42] except the standard barcode region was amplified at 42 °C.

All PCR products were checked by gel electrophoresis in 1% agarose gels, cleaned using AxyPrep PCR clean-up beads as per manufacturer's instructions, then sequenced bi-directionally using BigDye terminator reaction mix v.3.1 in a 3730xl DNA analyser (Applied Biosystems) at the NHMUK sequencing facility.

The forward and reverse sequences were assembled and corrected using the Staden Package v.1.7.0 [44]. The resulting sequences were aligned in Mega v.7.0.14 [45] with the Clustal W program [46] for the CO1 gene; the 28S gene was aligned with the MAFFT web server [47] using the E-INS-i algorithm, a gap opening penalty of 2, leave gappy regions option activated, and UPGMA as a tree-building method for the guide tree. The CO1 sequences were also translated to amino acids to detect eventual stop codons that indicate NUMTs. The two alignments were first used in single-gene phylogenetic analyses in RAxML-NG v.1.0.0 [48] to detect eventual long branches and misplaced sequences (that might indicate pseudogenes or contaminants) that are to be checked/removed prior to the concatenation of the two datasets. A phylogenetic analysis of the concatenated but unpartitioned dataset using a simple substitution model (K2P) was also conducted in RAxML-NG since over-parameterization of the substitution and partition models might be a problem in a maximum likelihood framework [49], especially when using a comparatively small alignment. For the partitioned analyses, data blocks were delimited in Mesquite v.3.10 [50], CO1 being divided by codon position and 28S was treated as one block. The best partitioning scheme and substitution models were selected using PartitionFinder2 v.2.1.1 [51], with branch lengths proportionally linked and the search option set to all.

Partitioned analyses were run in RAxML-NG [48], which is maximum likelihood (ML) based, and MrBayes v.3.2.7 [52], which is based on Bayesian inference (BI) with the following substitution models as indicated by PartitionFinder2: GTR+G for 28S, HKY+I+G, TIM+I+G, and TIM+G for the 1st, 2nd, and 3rd codon positions of CO1, respectively. For MrBayes, we substituted TIM with GTR, since the former model is not available in this program. In MrBayes two parallel analyses, each with four chains, were run for 10^7 MCMC generations, with trees and lnLs sampled every 100 generations; all estimated parameters were unlinked among partitions except for branch lengths; convergence of all parameters and estimated sample size values (ESS) above 200 were assessed by examining the trace files in Tracer v1.7.1 [53]. Support for the maximum likelihood analysis was estimated with rapid bootstrapping (number of replicates determined by the autoMRE criterion [54]).

Bootstrap percentages (BP) over 85% were considered as strong support and BP smaller than 65% as weak. Posterior probabilities (PP) over 0.95 were considered as strong support and those below 0.90 as weak. The trees were imported and modified in FigTree v1.4.4 [55] and Adobe Illustrator.

All sequences were uploaded to GenBank (accession numbers ON555486–ON555643 for 28S and ON557406–ON557518 for CO1). Since a part of the DNA extractions did not yield PCR products, the presence of a DNA code after the label data of a specimen does not necessarily mean that it has an associated DNA sequence. A complete list of specimens with associated DNA sequences and their repository is provided in Appendix A.

2.4. New Species Left Undescribed

We have identified several species that are clearly new based either on their DNA sequences or morphology (or both), but are not described herein for one or more of the following reasons:

- Species known from males only. Within (e.g.) the *ghesquierei*-group, several new species have been identified (at least 7 or 8—see Figure 2), which are known only from males. Since in most cases females are essential for species recognition (e.g., antennal structure, ovipositor length), we have refrained from describing these species here.
- Incomplete specimens. In several instances, new species are indicated by both morphology and DNA sequences, but a crucial morphological character is missing, most often the antennae. These specimens and their sequences have been curated pending the discovery of fresh, complete specimens.
- Poorly-mounted specimens. In a few cases, slide-mounted specimens not represented by DNA sequences appear to be very likely new species. In many cases, the material is simply not in good enough condition for the designation of a holotype to represent the species.

Clearly, there is overlap and gradation between the above categories, and we have used our discretion when deciding whether or not to describe specimens. In all cases, information as to our opinion of species status is included on the specimens.

2.5. A Note on Figures Supplementing the Descriptions

While all new species are fully described, in many cases there are aspects of the morphology that differ so little between species that images of these structures would be superfluous. In these cases, "cf Figure" is used, where the reader is referred to a figure that to all intents and purposes can serve to illustrate the species while actually depicting a different one. This is especially true for many species of the *ghesquierei*-group, where several species are morphologically indistinguishable, and to a lesser extent for *M. mymaripenne*, *M. noyesi* **sp. nov.**, and *M. polilovi* **sp. nov.** in the *mymaripenne*-group. In every case of extreme morphological similarity, robust molecular data are available to support separate species status. In addition, where possible, illustrations were made from holotypes. Under each photograph, we mention whether it is that of a holotype, neotype, or paratype.

3. Results

3.1. Phylogenetic Analyses

We obtained DNA sequences for a total of 170 *Megaphragma* specimens (158 sequences for 28S and 113 sequences for CO1) (Appendix A). The 28S alignment was 1068 bp in length, while the CO1 alignment was 652 bp, though only shorter sequences (DNA mini-barcodes) were obtained for some species/specimens.

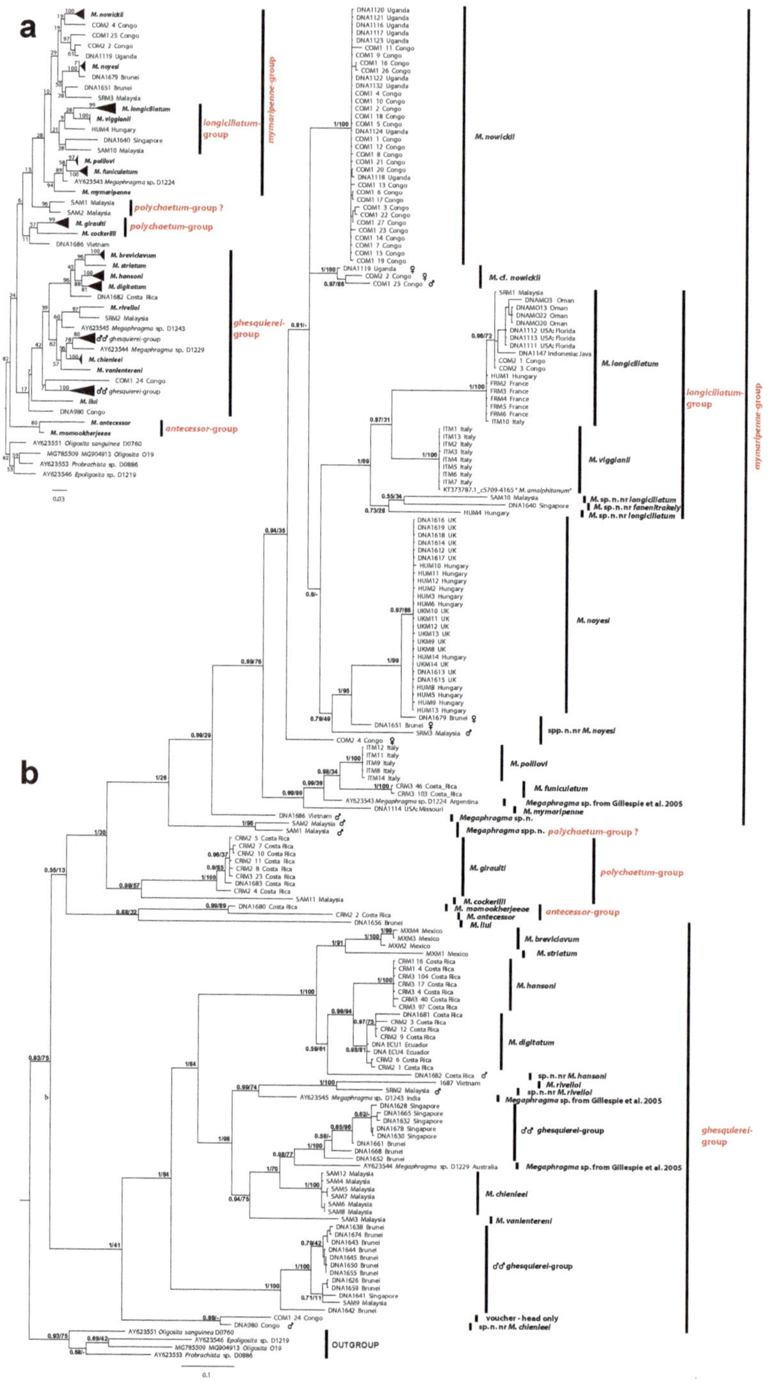

Figure 2. Combined CO1 and 28S sequences analysis of 174 *Megaphragma* specimens (170 from this study, 4 from GenBank) with species delineated and species-groups indicated: (**a**) unpartitioned ML analysis (bootstrap values indicated at nodes); (**b**) partitioned BI analysis (both posterior probabilities and bootstrap values indicated at nodes).

The phylogenetic trees from the single-locus analyses are in general agreement though, for example, the position of *M. antecessor* **sp. nov.** differs drastically between the two, while some species are present in one data matrix but not the other (Supplementary Figures S2 and S3). They are also in general agreement regarding the clustering of specimens into putative species: all species that are distinct based on CO1 are also distinct based on 28S. An exception is two specimens of *M. digitatum* **sp. nov.** that are very divergent on CO1, but almost identical on 28S. Both the ML and BI trees from the combined and partitioned dataset show similar topologies, with minor differences; hence, in Figure 2b on the BI tree, both posterior probabilities and bootstrap support were plotted at the nodes. The species that were included in the analysis are split into two major groups: one consisting of mostly species of the *mymaripenne*-group (species groups are discussed below), the other of species of the *ghesquierei*-group. Both have high posterior probabilities (0.99 and 1, respectively) but low bootstrap support, indicating strong support based on a low number of characters. Two other small groups are formed by two species each in the *antecessor*- and *polychaetum*-groups. The *antecessor*-group is sister to *M. liui* **sp. nov.** in both of these trees, while two unnamed species of the *polychaetum*-group (SAM1 and SAM2) are basal to the *mymaripenne*-group instead of clustering with *M. giraulti* **sp. nov.** and *M. cockerilli* **sp. nov.** (the other two species of the group). The two species of the *longiciliatum*-group, though forming a monophyletic group with a posterior probability of 1, and a medium bootstrap support of 69%, render the *mymaripenne*-group paraphyletic.

The two trees based on the combined analysis of both genes have a major difference from the unpartitioned analysis that used a simple substitution model and not the best fit model (Figure 2a); in this latter tree, *M. liui* is recovered in a basal position in the *ghesquierei*-group, where it belongs based on its morphology. Another major difference between the partitioned analyses and the unpartitioned one is the position of the *antecessor*-group that is not sister to the *ghesquierei*-group in the first analyses (Figure 2b) as would be expected by its morphology, while it is retrieved as basal to all other *Megaphragma* in the second tree (Figure 2a).

Even very short DNA sequences are sufficient to place a specimen, though in some cases this is also the explanation for the unstable and likely erroneous position in the phylogenies for some species. For example, HUM9 is correctly placed in the cluster with other *M. noyesi* based on a 296 bp CO1 sequence and the same is true for *M. momookherjeeae* **sp. nov.** retrieved as sister to *M. antecessor* based on a 394 bp CO1 sequence (the morphology of both species places them in the *antecessor*-group). *Megaphragma antecessor*, *M. liui*, and *M. momookherjeeae* that have their positions on the trees drastically altered depending on the analysis (partitioned versus unpartitioned) are represented by short sequences: 519 bp for 28S and 366 for CO1, 344 bp for 28S and 370 for CO1, and 394 bp for CO1, respectively.

3.2. Taxonomy

***Megaphragma* Timberlake, 1924**

Megaphragma Timberlake, 1924. *Proc. Haw. Entomol. Soc.* 5: 412–414. Type species: *Megaphragma mymaripenne* Timberlake, by original designation.

Sethosiella Kryger, 1932. *Bulletin de la Société Royale d'Egypte* 16: 38–39. Type species: *Sethosiella priesneri* Kryger, by original designation. Synonymy by Ghesquière 1939, p. 36.

Paramegaphragma Lin, 1992. *Entomotaxonomia* 14(2): 133–135, 138. Type species: *Paramegaphragma stenopterum* Lin, by original designation. Synonymy by Delvare 1993, p. 151.

Diagnosis. Female (Figure 1a). Body rather compact, extremely small, length 0.16–0.3 mm. Antenna (Figure 1b) inserted at mid level of the internal orbital line, with short radicle, scape usually elongate, pedicel, anellus, single funicle segment present or absent, clava one -, two -, or three-segmented. Antennal formula: 1 (scape), 1 (pedicel), (1) (anellus), 1 (funicle), 2 (clava); or 1,1,(1),1,1 or 1,1,(1),0,3. The antenna is counted as four- or five-segmented, since the anellus is not counted among the antennomeres. Claval segment

1 without multiporous placoid sensilla. Mandible with two small teeth. Maxillary palp very small and labial palp vestigial. Eye black unless otherwise stated. Mesosoma rather high, usually shorter than metasoma. Pronotum very short; mid lobe of mesoscutum not much longer than wide, either smooth or with polygonal or striate sculpture; one pair of adnotaular setae. Scutellum shorter than mid lobe of mesoscutum, with a pair of setae. Metanotum short; propodeum slightly longer than metanotum, or, in the middle, even longer, with a well-developed central area (disc) that may bear crenulae. Propodeal spiracle placed in an oval groove, and near the internal margin with two very small setae. Fore wing (Figure 1c) extremely narrow compared with other Trichogrammatidae genera, 5.3–10× as long as maximum discal width, with short submarginal vein; costal cell and parastigma not distinct; marginal vein very long, with one short seta at the base and with one or two setae centrally, which when paired may be of similar or very different lengths; stigmal vein very short with one or two short setae on the stigma; disc with one or a few setae in one or two rows or glabrous (when there is one seta it is located on the dorsal surface of the wing, when discal setae are more numerous they are located on both dorsal and ventral surfaces of the wing, cf Figure 22d,e). Hind wing without discal fringe on front margin. Legs robust, often with striate sculpture on coxae, also on femora and tibiae. T7 and T8, respectively, without spiracle and cercus.

Male: As female, but often with postanellar antennomeres shaped differently. Genitalia tubular, very simple and usually small.

Relationship. The closest relatives of *Megaphragma* appear to be *Prestwichia* Lubbock and *Sinepalpigramma* Viggiani and Pinto [56]. Unfortunately, sequences for neither of these genera were available for comparison. We have used an *Epoligosita* Girault, two *Oligosita* Walker, and a *Probrachista* Viggiani species as outgroups. These Oligositinae genera are close phylogenetically to *Megaphragma* according to a previous molecular study [33]. Species-group relationships are discussed below.

Distribution: Cosmopolitan.

Hosts and biology. The known species of *Megaphragma* are all egg parasitoids of Thysanoptera (Supplementary Figure S4) [57–59]. Biological data are available only for a few species, e.g., *M. mymaripenne, M. longiciliatum* (as *M. amalphitanum*) [12–14], and are given below where available. It is interesting that at the same locality there may be more than one species of *Megaphragma,* even in Europe. *Megaphragma viggianii* and *M. polilovi* were found in Italy at the same locality and on the same host, while in a single sample from near Barkás Lake in Hungary, there are three species (*M. longiciliatum, M. noyesi,* and the undescribed species represented by the specimen HUM4, close to *M. longiciliatum* but distinct genetically).

Species groups in *Megaphragma*

On the basis of present knowledge, the following species group are proposed in *Megaphragma*:

M. mymaripenne-group: antenna with a single funicle segment that is longer than wide (this feature also shared by *polychaetum*- and *longiciliatum*- groups); **T1** with longitudinal and/or transverse cells, with some denticles laterally within the cells (Figures 17f, 18f and 20b); **T2–T4** each with a pair of short setae.

Included species: *M. funiculatum* Polaszek and Fusu **sp. nov.**, *M. mymaripenne* Timberlake, *M. nowickii* Polaszek, Fusu, and Viggiani **sp. nov.**, *M. noyesi* Polaszek and Fusu **sp. nov.**, *M. polilovi* Polaszek, Fusu, and Viggiani **sp. nov**.

M. longiciliatum-group: same as *mymaripenne*-group, but without cells on **T1**. According to the phylogenetic analysis, the group appears to be derived from within the *mymaripenne*-group having lost the denticulate cells on **T1**.

Included species: *M. longiciliatum* Subba Rao, *M. fanenitrakely* Polaszek and Fusu **sp. nov.**, *M. priesneri* Kryger, *M. viggianii* Polaszek, Fusu, and Polilov **sp. nov.** The species of the *macrostigmum*-group (*M. caribea* and *M. macrostigmum*) characterized by a four-segmented antenna, might be derived species within this group.

M. polychaetum-group: antenna with a long, cylindrical, funicle segment; spatulate sensilla at the end of each clava segment, and a robust terminal sensillum on **C2**; fore wing disc with more than seven setae, often arranged in two rows. Male antenna is particularly distinctive, with an elongate **C1**, short **C2** usually with very long sensilla.

Included species: *M. cockerilli* Polaszek and Fusu **sp. nov.**, *M. giraulti* **sp. nov.**, *M. polychaetum* Lin, *M. kinuthiae* Polaszek, Fusu, and Viggiani **sp. nov.** Our molecular analysis also includes two males of this group, representing two undescribed species (vouchers SAM1 and SAM2, NHMUK). They have the antennal structure characteristic for males of the group, but our analyses recover them basal to the *mymaripenne*- and *longiciliatum*-groups instead of clustering them with the other two species of the *polychaetum*-group.

M. ghesquierei-group: antenna without funicle segment and with clava three-segmented, because the funicle is fused with the clava along an oblique suture. Fore wing disc with one seta on the dorsal surface or no setae. Propodeum characteristically produced centrally, almost always with a row of crenulae. Metafemur with a robust spine close to the connection with the tibia. Because of the intergradation in the structure of the antenna between the *ghesquierei* and other groups, we do not currently consider *Paramegaphragma* as a valid genus for the species in the *ghesquierei*- plus *stenopterum*- groups. It is possible that future analyses, especially including multigene or reduced genome representation data, may lead to the reinstatement of *Paramegaphragma* Lin as a valid genus. The two species formerly assigned to *Paramegaphragma* by Lin [9], *M. stenopterum* and *M. macrostigmum*, are not closely related and clearly belong to different species-groups (*stenopterum*-group and *macrostigmum*-group, respectively), though on morphological grounds *stenopterum*-group is clearly related to *ghesquierei*-group or even integral part of it. This is another reason for not recognizing *Paramegaphragma*.

Included species: *M. breviclavum* **sp. nov.**, *M. chienleei* **sp. nov.**, *M. deflectum* Lin, *M. digitatum* **sp. nov.**, *M. ghesquierei* Ghesquière, *M. hansoni* **sp. nov.**, *M. liui* **sp. nov.**, *M. pintoi* Viggiani **sp. nov.**, *M. rivelloi* **sp. nov.**, *M. striatum* Viggiani, *M. tamoi* Polaszek, Fusu, and Viggiani **sp. nov.**, *M. tridens* Fusu and Polaszek **sp. nov.**, *M. vanlentereni* Polaszek and Fusu **sp. nov**.

M. stenopterum-group: same as *M. ghesquierei* but with clava two-segmented. The antennal structure is very suggestive of the *ghesquierei*-group, given the similarity between the apparent **C1** of the *stenopterum*-group and that of the *ghesquierei*-group; i.e., it is actually a funicle completely fused to the clava. In the *antecessor*-group, the funicle is distinct albeit transverse and anneliform. Pending further evidence, we consider the *stenopterum*-group as possibly nested within the *ghesquierei*-group. *Megaphragma macrostigmum* and *M. caribea* (*macrostigmum*-group) were considered by previous authors to belong in a group with *M. stenopterum* [60], and *M. macrostigmum* with *M. stenopterum* were both originally included by Lin [9] in his genus *Paramegaphragma*. However, the former two species lack any of the obvious apomorphies of the *ghesquierei*-group except for the apparently lost funicle. Members of the *macrostigmum*-group are otherwise similar in the structure of the fore wing and sculpture of the mesoscutum to the species in the *longiciliatum*-, *mymaripenne*-, and *polychaetum*- groups and are probably not related to the *ghesquierei*- and *stenopterum*-groups. Our molecular analysis did not include *M. stenopterum*, the only member of this species group.

Included species: *M. stenopterum* (Lin).

M. antecessor-group: antenna with a transverse funicle segment not much larger than the anellus, and clava one- or two-segmented. Metatibia with a characteristic row of setae (Figures 11d and 21j). The structure of the antenna seems intermediate between that characteristic of the *longiciliatum*- and *mymaripenne*- groups and that of the *ghesquierei*-group. In the latter species group, the antenna is apparently without a funicle, as the funicle is completely fused with the clava by an oblique suture and, hence, the clava appears three-segmented. Our phylogenetic analysis shows that *M. antecessor* and *M. momookherjeeae*, while resembling the *ghesquierei* species group in many features (including fore wing structure and the robust spine on metatibia), appear outside it, and basal to all remaining

Megaphragma except members of *ghesquierei*-group (partitioned analyses) or the most basal species group of all *Megaphragma* (unpartitioned analysis).

Included species: *M. antecessor* Polaszek and Fusu **sp. nov.**, *M. momookherjeeae* Polaszek and Fusu **sp. nov.**, *M. uniclavum* Polaszek and Fusu **sp. nov**.

M. macrostigmum-group: as explained above, *M. macrostigmum* and *M. caribea*, while undoubtedly very closely related to each other, appear to have no connection with the *ghesquierei*-group (our molecular analysis did not include either of these two species). The antenna has the clava two-segmented and no funicle as in the *stenopterum*-group; the fore wing structure, however, is not similar to the *ghesquierei*-group but suggestive of the *longiciliatum*-group, especially *M. priesneri*; metafemur without the robust spine characteristic for the *antecessor-*, *ghesquierei-*, and *stenopterum-* groups.

Included species: *M. caribea* Delvare, *M. macrostigmum* (Lin).

Previously described species

***Megaphragma caribea* Delvare** (Figure 3a–c, Figure 12d–f and Figure 19a–c)

Megaphragma caribea Delvare, 1993. *Revue fr. Ent. (n.s.)* 15(4): 149–151.

Diagnosis. Female: Antenna four-segmented (excluding anellus), without funicle, clava two-segmented (Figure 3a and Figure 12d); **C1** with 16 **MT**, 2 long **UST**, 1 **SS**; **C2** with 4 **MPS**, 2 **MT**, 1 SB, 1 prominent apical **SS** (Figure 3a and Figure 12d).

Mid lobe of mesoscutum with large, but shallow polygonal cells (Figures 12e and 19c). Propodeum with central area short. Fore wing 5.5× as long as maximum discal width, marginal vein about twice length of submarginal vein, with two rather long setae (of equal length) present at midpoint of marginal vein (Figure 3c). Metasoma with lines of microspines evident on **T2–T6** (cf Figure 19a). Ovipositor 1.7× as long as mesotibia.

Body yellow, with the following slightly darkened: occiput, meso- and meta-coxae, apices of meso- and metafemora. Metasoma with pale brown transverse bands.

Male: Antenna with **C1** longer than in female (Figure 19b).

Material examined. Paratypes: **GUADELOUPE**: Vieux Habitants, 17.XI.1988, coll. J. Etienne, ex eggs *Selenothrips rubrocinctus* (Giard) on *Psidium guajava*. Slides n. 8002.4, 8002.6, 8002.9 (2♀, 1♂, NHMUK).

Non-types: **COLOMBIA**: Cartagena, i.2015, with *Heliothrips haemorrhoidalis* and *Selenothrips rubrocinctus* on *Terminalia catappa*, coll. A.A. Polilov (1♂, AICF).

Species-group placement: *M. macrostigmum*-group—possibly a subgroup of the *M. longiciliatum*-group.

Distribution: Colombia, Guadeloupe.

Host: *Selenothrips rubrocinctus* (Giard).

DNA data: no DNA sequences.

Comments: The species was described in detail by the author. *Megaphragma caribea* is clearly close to *M. macrostigmum* (Lin). At present, their discrimination is based on the absence of long **UST** on the basal clava (**C1**) of the antenna of the latter species (Figure 6a). Since the original description did not indicate whether the species-group name *caribea* is a noun or an adjective, following Art. 31.2.2. of ICZN, we treat it as a noun and do not make a gender agreement.

***Megaphragma deflectum* Lin** (Figure 3d–f)

Megaphragma deflectum Lin, 1992. *Entomotaxonomia* 14(2): 130–131.

Diagnosis. Female: Antenna (Figure 3e) without funicle, clava three-segmented, with **C1** and **C2** almost fused; **C1** with 1 **UST**; **C2** with 1 **UST**, and ≥4 **MT**; **C3** with ≥2 **MPS**, 1 **MT**, and 1 SB.

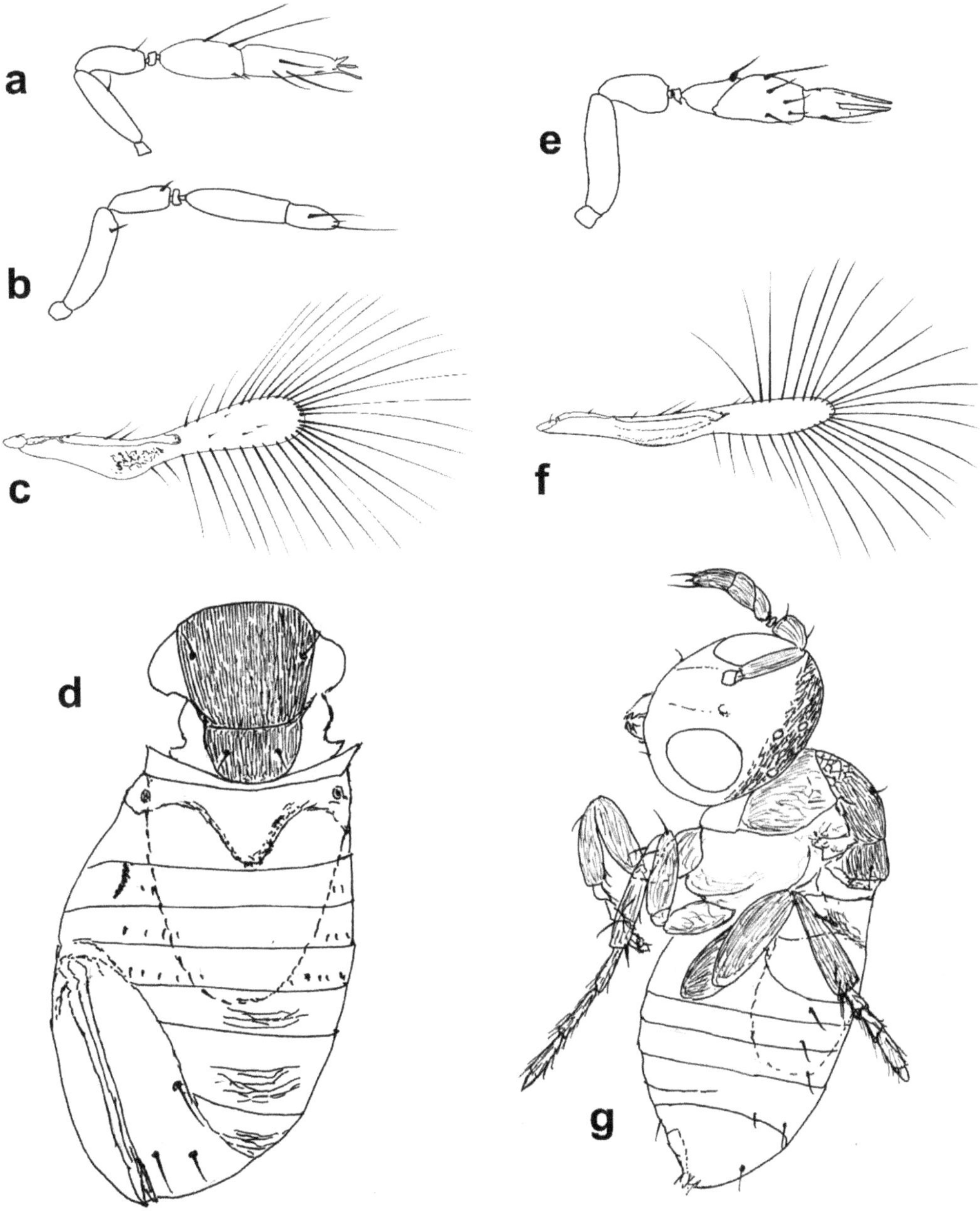

Figure 3. Line drawings of *Megaphragma* species: (**a**) *M. caribea*, female antenna (Paratype); (**b**) *M. caribea*, male antenna (Paratype); (**c**) *M. caribea*, female fore wing (Paratype); (**d**) *M. deflectum*, female body dorsal (Holotype); (**e**) *M. deflectum*, female antenna (Holotype); (**f**) *M. deflectum*, female fore wing (Holotype); (**g**) *M. ghesquierei*, male body lateral (Paratype).

Mid lobe of mesoscutum (Figure 3d) with longitudinal striate sculpture extending to scutellum; propodeum (Figure 3d) with central area extended posteriorly, crenulae absent. Fore wing (Figure 3f) 9× as long as maximum width; longest fringe seta 5× as long as maximum discal width. Fore wing disc without setae. Marginal vein with two long subequal setae centrally. Stigmal vein not enlarged, with two sensilla apically. Middle tibia with one large spine basally; metafemur with spine. **T1** with elongate cells laterally, 2–3× as long as wide; **T2–T4** without setae laterally. Ovipositor 1.7× as long as mesotibia.

Body largely brown, the following paler: legs except coxae and metafemur. Antenna with pedicel pale; scape, **C1–C3** darker. Fore wing strongly infuscate basally; stigmal and marginal veins brown; marginal vein very dark centrally.

Male: As female but **C3** with fewer **MPS** and with ASC apically.

Material examined. Holotype ♀ (FAFU). **CHINA**: Wuyishan, Fujian, 19.x.1987, Wang Jiashe col.

Paratype: **CHINA**: Fuzhou, Fujian, 8.v.1987, N.Q. Lin col. (1♀, FAFU).

Non-type: **CHINA**: Wuyishan, Fujian, 10.x.1987, Wang Jiashe col. (1♂, FAFU).

Species-group placement: *M. ghesquierei*-group.

Distribution: China.

DNA data: no DNA sequences.

Comments: This species was correctly considered allied to *M. ghesquierei* Ghesquière mostly due to features of the antenna (Figure 3e), fore wing (Figure 3f), and other characteristics of the body, but *M. deflectum* can be distinguished easily from that species by the sculpture of the mid lobe of the mesoscutum and the central area of propodeum (Figure 3d). The male "allotype" (paratype) of *M. deflectum* is actually a male of *M. rivelloi* **sp. nov.** (see below).

***Megaphragma ghesquierei* Ghesquière** (Figures 3g and 4a,b)

Megaphragma ghesquierei Ghesquière, 1939. *Rev. Zool. Bot. Afr.* 33(1): 36–38.

Diagnosis. Female: Antenna without funicle, clava three-segmented, **C1** distal margin transverse (cf Figure 3g). **C1** with ≥1 **MT**; **C2** with ≥2 **MT**, ≥2 **UST**; **C3** with ≥2 **MPS**, 1 **MT**, and 1 **SB**.

Mid lobe of mesoscutum anteriorly with reticulate sculpture, remainder with longitudinal striation continuing onto scutellum (cf Figure 3g). Propodeum with a large subtriangular central area. Fore wing 7× as long as maximum width (Figure 4a); the disc pointed distally, without setae. Metasoma with tergites with some short transverse striation centrally, and each with a pair of lateral setae (Figure 4b).

Body dark brown, with the following paler: frons and occiput, scutellum and propodeum, tarsi. Metasoma with tergites and sternites appearing as dark bands (in the slide-mounted types). Fore wing basally strongly infuscate with a dark marginal vein.

Male: Similar to female in all aspects of morphology except genitalia characters.

Material examined. Holotype ♀ (MRAC). **D. R. CONGO**: Rutshuru, i.1938, ex eggs of *Panchaetothrips noxius* Priesner on *Coffea arabica*.

Paratypes: **D. R. CONGO**: 1♂, on slide with holotype; 3♀ on one slide, with data as holotype except "Neotopotype" in Ghesquière's writing (MRAC).

Non-type: 1♂, labeled type in the Nowicki collection, no other data (DACE).

Species-group placement: *M. ghesquierei*-group.

Distribution: D. R. Congo.

Host: *Panchaetothrips noxius* Priesner

DNA data: no DNA sequences.

Comments: The species is rather easily recognizable by the combination of features of the antenna, mid lobe of mesoscutum, propodeum, fore wing, and metasomal tergites.

The species was intended to be described by Nowicki, but was published by Ghesquière [61] (p. 36) because Nowicki's manuscript on several African Trichogrammatidae never reached the journal Revue de zoologie et de botanique Africaines in Tervuren where Ghesquière was working. Ghesquière [61] (p. 37) gives the date of collection as "XII.1937", but, as given above, the holotype is labeled: "I.1938".

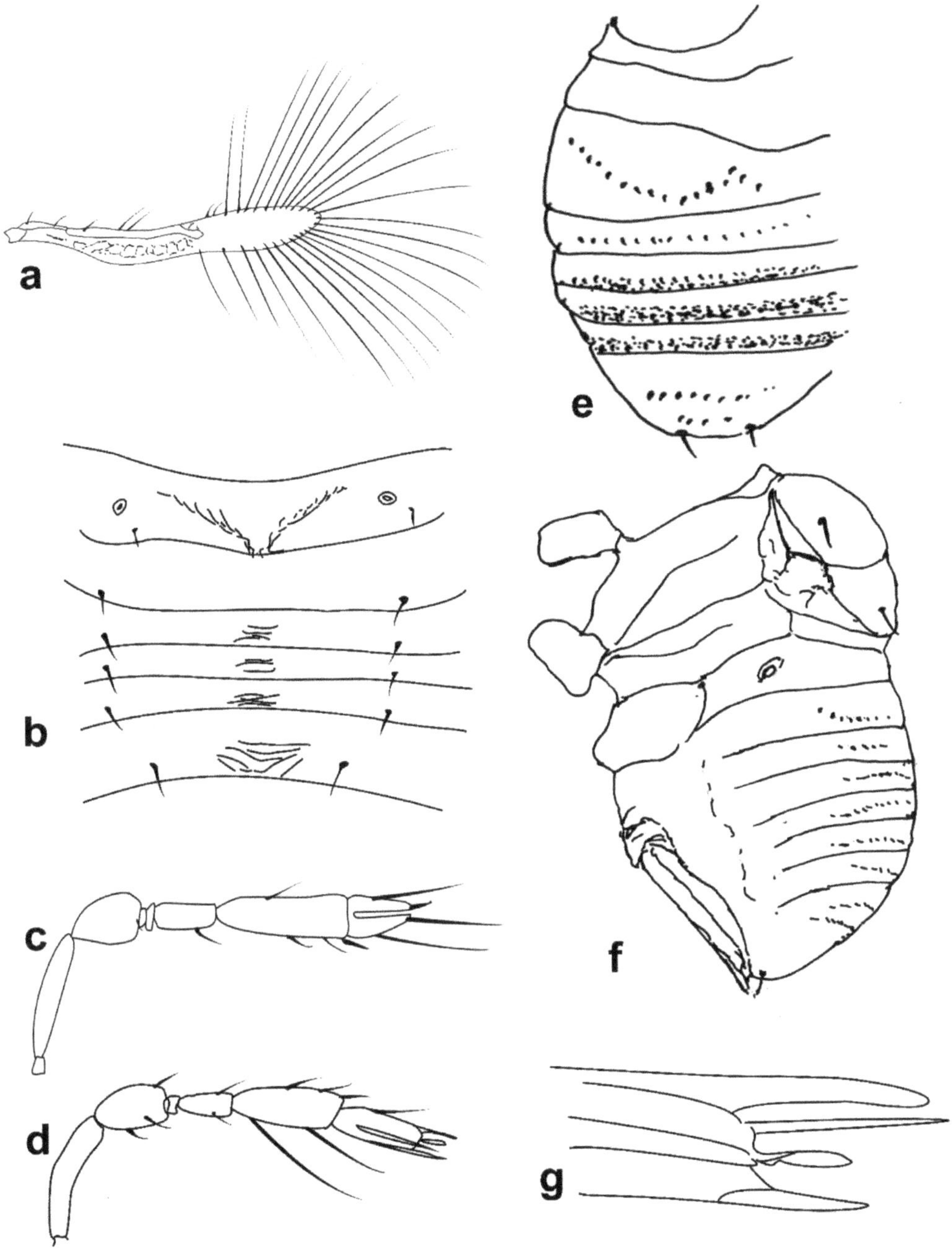

Figure 4. Line drawings of *Megaphragma* species: (**a**) *M. ghesquierei*, female fore wing (Holotype); (**b**) *M. ghesquierei*, male propodeum and tergites (Paratype); (**c**) *M. giraulti*, male antenna (Paratype); (**d**) *M. giraulti*, female antenna (Holotype); (**e**) *M. giraulti*, male metasoma (Paratype); (**f**) *M. giraulti*, female lateral meso- and metasoma (Holotype); (**g**) *M. giraulti*, female antenna, detail (Holotype).

***Megaphragma longiciliatum* Subba Rao** (Figure 5e–h and Figure 16e)

Megaphragma longiciliatum Subba Rao, 1969. *Proc. R. Ent. Soc. Lond. (B)* 38(7–8): 114.
Megaphragma aligarhensis Yousuf and Shafee, 1988. *Indian J. syst. Ent.* 4(2) [1987]: 114. **Syn. nov.**
Megaphragma amalphitanum Viggiani in Viggiani and Bernardo, 1997. *Boll. Zool. Agr. Bach. Ser. II* 29(1): 51–55. **Syn. nov.**
Megaphragma magniclava Yousuf and Shafee, 1988. *Indian J. syst. Ent.* 4(2) [1987]: 115–116. **Syn. nov.**
Megaphragma decochaetum Lin, 1992. *Entomotaxonomia* 14(2): 131–132. **Syn. nov.**
Megaphragma shimalianum Hayat, 2009. *Oriental Insects* 43: 212–213. **Syn. nov.**

Diagnosis. Female: Antenna (Figure 5e and Figure 16e) with clava two-segmented. Funicle with ≥2 **MT**; **C2** with ≥6 **MT**, 1 **UST**; **C3** with ≥3 **MPS**, and 1 **UST**.

Mid lobe of mesoscutum anteriorly with reticulate sculpture. Propodeum with a very short central area. Fore wing 8× as long as wide (Figure 5g). Metasoma (Figure 5h) without subpolygonal sculpture on tergites, but with some ridges, **T2–T4** each with a pair of long setae. Ovipositor 1.1× as long as mesotibia.

Body brown to dark brown, with the following paler: antenna, legs. Metasoma with tergites and sternites appearing as dark bands (in the slide-mounted types). Fore wing completely hyaline.

Male: Similar to female in most characters except genitalia; antennal funicle slightly more elongate than in female, and clava darker than remainder of antenna. **C2** without long **UST**; **C3** shorter than in female (Figure 5f).

Material examined. Holotype ♀ *M. longiciliatum* (NHMUK). **INDIA**: Bangalore, Avati, ex. *Frankliniella lilivora* Takahashi on *Polyanthes tuberosa*, x.1968, V. P. Rao. Paratypes: 13♀ 1♂, same data as holotype (NHMUK).

Holotype ♀ *M. aligarhensis* (AMU). **INDIA**: Aligarh, IX.1985, M. Yousuf.

Holotype ♀ *M. amalphitanum* (DACE). **ITALY**: Vietri sul mare (SA), x.1994, coll. G. Viggiani, ex egg of *Heliothrips haemorrhoidalis* on *Viburnum tinus*. Paratypes: 36♀, 32♂, mostly obtained from the same host collected in the same holotype locality (DACE).

Holotype ♀ *M. decochaetum* (FAFU). **CHINA**: Fuzhou, Fujian, 30.vi.1987, coll. Lin. Paratype: CHINA: Fuzhou, Fujian, 30.vi.1987, coll. Lin (1♂, FAFU).

Holotype ♀ *M. magniclava* (AMU). **INDIA**: Aligarh, 25.x.1985, M. Yousuf.

Paratypes *M. shimalianum*. **INDIA**: Uttar Pradesh, Mainpuri Malau, slide XIV 1, 2, 4, 6.ix.2007, F. R. Khan col. (3♀, 1♂, AMU); Firozabad, Nagla Prabhu, slide IX, 4.ix.2007, F. R. Khan col. (12♀, AMU).

Non-types: **ARGENTINA**: INTA Oliveros Santa Fe, v.2004, ex *Caliothrips phaseoli*, A. M. Molinari col. (2♀, 7♂, DACE, AICF); San Miguel de Tucuman, x–xi.2006, ex Thysanoptera eggs on corn, E. Luft col. (2♂, DACE, AICF); Salta Prov., Aguas Blancas, Routa 19, 22.72° S, 64.40° W, 447 m, 23.iii.2003, swp rainforest along Bolivia border, J. Munro 003-03-23-01 (1♀, UCRC); Salta Prov. Rosario de la Frontera (grounds of Hotel Termas), 25.84° S, 64.93° W, 447 m, 20.iii.2003, sweeping, J. Munro 003-03-20-10 (1♀, UCRC); Salta Prov., RN81, 66 km E. jct RP 24, 23.24° S, 63.40° W, 260 m, 24.iii. 2003, swp Dry Chaco, J. Munro 003-03-24-01 (6♀, AICF, UCRC). **AUSTRALIA**: WA, Margaret R, Warner Glen Rd, Stone Cottages, 34°04.44′ S, 115°08.14′ E, eucalyptus forest, YPT, 15–16.xi.2002, George, Owen, Hawks, Munro PEET02-010P (1♀, UCRC). **CHINA**: 24.v.1987 and 26.v.1987, coll. Lin, identified as *M. decochaetum* (1♀, 1♂, FAFU). **D. R. CONGO**: Province Orientale, Yangambi Biosphere Reserve 0°45.822′ N 24°30.285′ E, 15.v.2012, screen sweep primary forest, A. Polaszek col. BMNH 2012-88, DNA: COM 2.1 and COM 2.3 (2♀, AICF, NHMUK). **FRANCE**: Dept Gironde, St Colombe (nr Castillon-la-Bataille), Pitray, 1.viii.2000, S. Bessart, M. van Helden (3♀, UCRC); Dordogne, 3.5 km E Issigeac, 44°43′ N 0°38′ E, 100 m, 31.vii.2013, J.S. Noyes col. NHM(Ent.) 2013-144, DNA: FRM2 to FRM6 (4♀, 1♂, AICF, NHMUK). **HUNGARY**: Őrség Nemzeti Park, Barkás Lake, 46°52′ N 16°26′ E, 268 m, 28.vi.2010, screen-sweep, J.S. Noyes col., BMNH(Ent) 2010-63, DNA: HUM1 (1♂, NHMUK). **INDIA**: Uttar Pradesh, New

Delhi, IARI, 220 m, 28°37′51″ N 77°09′50″ E, 5–7.xi.2003, pan trap, J. Heraty col. (1♀ 4♂, UCRC); Karnataka, W of Mudigere, 850–912 m, 13°07′05″ N 75°30′20″ E, 24.xi.2003, sweep evergreen forest, J. Heraty col. (3♀, UCRC). **INDONESIA**: W Java, Gunung Halimun NP, Tea-Forest Junction, 1066 m, 6°41′07″ S 106°31′16″ E, 17.ix.2015, screen-sweep, A. Polaszek col., DNA1147 (1♀, NHMUK). **ITALY**: Vietri sul mare, Benincasa, 40°40′ N 44°20′ E, 17.vii.2013, ex *Heliothrips haemorrhoidalis* on *Viburnum tinus*, G. Viggiani, DNA: ITM10 (1♀, NHMUK). **MALAYSIA**: Sarawak, Mentawai 4°14′ N 114°52′ E, ix.2011, screen sweep, A. Polaszek col., DNA: SRM1 (1♀, NHMUK). **OMAN**: Hajar Mts, screen-sweep, 20.i.2017 A. Polaszek col., DNA: MO3, MO13, MO20, MO22 (MO13 was destroyed during the DNA extraction) (2♀ 1♂, NHMO). **PAPUA NEW GUINEA**: Central Province, 15km SE Port Moresby, 1.i.1986, screen-sweep eucalyptus grassland, G. Gordh col. 86-01-01-1 (1♀, ANIC). **UNITED ARAB EMIRATES**: Abu Dhabi Emirate, Al Ain, Al Khabisi garden, 24°13.521N 55°41.95E, 25–30.iii.2019, yellow pan trap, A. Polaszek, B. Howarth col. (1♀, NHMUK). **USA**: Florida, Lake Seminole Park, Seminole, 27°50–51′ N 82°46′ W 9.vii.2015, sweep, Z. Lahey col., DNA1111, 1112, 1113 (3 specimens, NHMUK, currently misplaced).

Species-group placement: *M. longiciliatum*-group.

Distribution: Argentina, Australia, China, D. R. Congo, France, India, Indonesia (Java), Italy, Malaysia (Borneo, Sarawak), Oman, Papua New Guinea, Portugal [62] (as *M. amalphitanum*), UAE, and USA.

Hosts: *Caliothrips phaseoli* (Hood) (Argentina); *Frankliniella lilivora* Takahashi (India); *Heliothrips haemorrhoidalis* (Bouché) (Italy). The record from Argentina "ex Thysanoptera eggs on corn" could be from *Frankliniella williamsi* Hood.

DNA data: CO1: 8 sequences from 4 countries: D. R. Congo, France, Hungary, Malaysia (Sarawak); 28S: 18 sequences from 8 countries: D. R. Congo, France, Hungary, Indonesia (Java), Italy, Malaysia (Sarawak), Oman, USA.

Comments: *Megaphragma longiciliatum* is the most widely distributed *Megaphragma* species; hence the large number of synonyms. We have examined 150 specimens from 14 countries and have DNA sequences for 18 specimens from 8 very widely distributed countries. We have carefully assessed morphological variation within the specimens examined, and consider that it encompasses the morphological characteristics of the type material of the species synonymized above [7,9,10,63,64].

The holotype of *M. longiciliatum* is in extremely poor condition. The mountant, presumably gum chloral, has turned black. It is to be hoped that in a few years' time, the holotype will be destroyed completely, and one of the paratypes, all of which are still in excellent condition, can be designated a neotype. Unfortunately, there is no current provision under the Code to legitimately replace a holotype specimen that has deteriorated irremediably.

***Megaphragma macrostigmum* (Lin)** (Figure 6a–d)

Paramegaphragma macrostigmum Lin, 1992. *Entomotaxonomia* 14(2): 135–136.

Megaphragma macrostigmum: Delvare, 1993. *Revue fr. Ent. (n.s.)* 15(4): 151.

Diagnosis. Female: Antenna without funicle and clava two-segmented (Figure 6a); **C1** with two short **MT**; **C2** with one **MPS**, one **SB**, and one **UST**.

Mid lobe of mesoscutum with some large, but shallow polygonal cells. Propodeum with a very short central area. Fore wing 5.3× as long as maximum discal width, with two rather long setae in the middle of marginal vein (Figure 6c); disc with 4–5 setae not in a row. Metasoma with a line of microspines evident on **T2–T6** (cf Figure 6d). Ovipositor 2.1× as long as mesotibia. The main features of the antenna and fore wing are illustrated in Figure 6a,c.

Body uniformly pale brown, fore wing slightly to moderately infuscate below marginal vein.

Male: Similar to female in most characters except genitalia; antennal funicle slightly shorter than in female, **C1** longer, **C2** with long **UST** (Figure 6b).

Material examined. Holotype ♂ (FAFU). **CHINA**: Fuzhou, Fujian, 31.viii.1987, N.Q. Lin col.

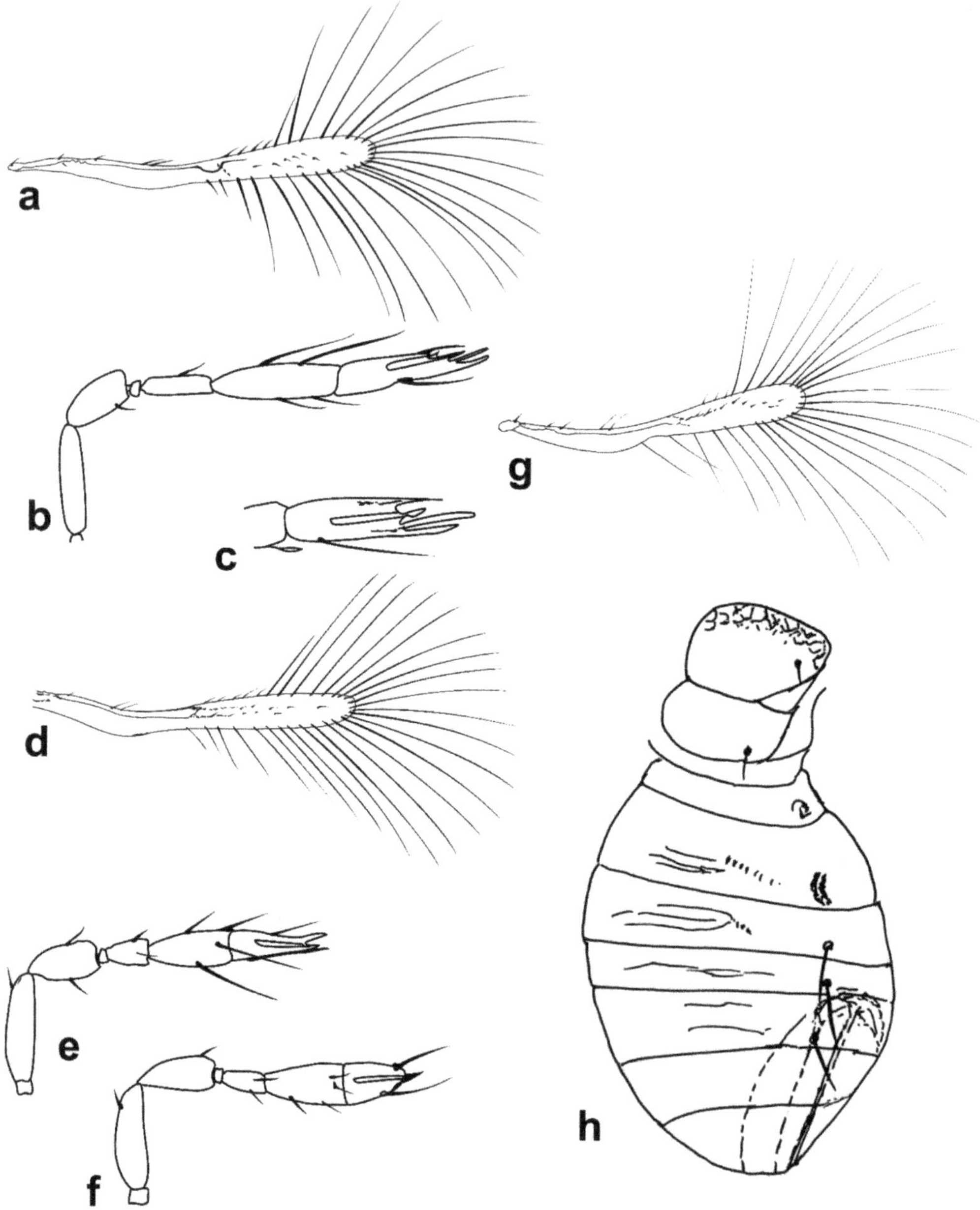

Figure 5. Line drawings of *Megaphragma* species: (**a**) *M. giraulti*, female fore wing (Holotype); (**b**) *M. kinuthiae*, female antenna; (**c**) *M. kinuthiae*, female antenna (detail); (**d**) *M. kinuthiae*, female fore wing; (**e**) *M. longiciliatum*, female antenna; (**f**) *M. longiciliatum*, male antenna; (**g**) *M. longiciliatum*, female fore wing; (**h**) *M. longiciliatum*, female dorsal meso-and metasoma.

Paratype: **CHINA**: Guangzhou, 3.xi.1985 N.Q. Lin col. (1♀, FAFU)

Non-type: **CHINA**: Guangzhou, 30.x.1985, N.Q. Lin col. (1♂, FAFU).

Species-group placement: *M. macrostigmum* group—possibly a subgroup of the *M. longiciliatum*-group.

Distribution: China.

DNA data: no DNA sequences.

Comments: This species is very similar to *M. caribea*; at present, the only difference from the latter species appears to be the absence of long sensilla on **C2**.

***Megaphragma mymaripenne* Timberlake** (Figure 1a, Figure 6e–i and Figure 17d)

Megaphragma mymaripenne Timberlake, 1924. *Proc. Haw. Entomol. Soc.* 5: 414–415.

Megaphragma mymaripenne: Viggiani, 1997. *Boll. Lab. Ent. agr. Filippo Silvestri* 53: 117–122.

Diagnosis. Female: Antenna (Figure 6h and Figure 17d) with clava two-segmented, funicle with ≥4 **MT**; **C1** trapezoid in lateral view with length 1.5× maximum width or less (longer in dorsal or ventral view), with 2 **UST**, but without linear sensilla (**MPS**), ≥9 **MT**; **C2** with ≥3 **MPS**, 1 **MT**, 1 **SB**, and 1 **SS** (Figure 6i).

Mid lobe of mesoscutum anteriorly with subpolygonal sculpture, but often appearing smooth in slide-mounts. Propodeum with a very short central area. Fore wing (Figure 6e) 9–10× as long as wide, marginal vein with two long setae in the middle, setae on disc more or less regularly in a row of 10–15 setae, and longest fringe seta 5–6× as long as maximum disc width. **T1** with sculpture represented by a combination of transverse and longitudinal cells, lateral ones twice as long as wide; sides of some cells with denticles present. The subsequent tergites show rather variable sculpture, differing from the pattern on the first tergum. **T2–T4** each with a pair of very short setae.

Body uniformly pale brown, scutellum paler than mesoscutum. Legs pale, wings hyaline. Clava slightly darker than the remainder of the antenna.

Male (hitherto undescribed): same as female but antenna slender, with funicle twice as long as wide and **C1** about 1.7× as long as **C2**. **T1** with sculpture not as complete as in the female. Genitalia simple, tubular, 4.5× as long as wide (cf Figure 7f).

Material examined. Holotype ♀ (USNM). **USA**: Hawaii, Mountain View, i.1920, C.E. Pemberton col.

Paratype: same data and on the same slide with holotype (1♀, USNM).

Non-types: **ARGENTINA**: San Miguel de Tucuman, x–xi.2006, ex Thysanoptera eggs on corn, E. Luft col. (5♀, DACE); ix.2006, from corn, E. Vinla col. (3♂, DACE); Salta Prov., Rosario de la Frontera, 25.83° S 64.88° W, 745 m, 20.iii.2003, sweep forest, J. Munro 003-03-20-01 (1♀, UCRC); La Rioja Prov., Chuquis, 28°53′40″ S 67°00′31″ W, 1575 m, 17.iii.2003, sweep acacia scrub, J. Munro 003-03-17-05 (1♀, UCRC); Salta Prov., Orán, rd to San Andres along Rio Blanca, 23.11° S, 64.52° W, 535 m, 23.iii.2003, sweep scrub and ginger, J. Munro 003-03-23-02 (2♀, UCRC). **BRAZIL**: Santa Catarina, Nova Teutonia, 17.x.1949, F. Plaumann col. BM 1957-341 (1♀, NHMUK). **COSTA RICA**: Limón, Hitoy-Cerere Reserve, 9°40′ N 83°02′ W, 100 m, 24–26.ii.2008, J.S. Noyes col. NHM(E)2010-21AQ (1♀, NHMUK). **DOMINICAN REPUBLIC**: San Cristobal, S. Cristobal Manomatuey, 20 km NW valley, 500 m, 23.iii.1991, L. Masner col. (1♀, UCRC). **ECUADOR**: 1♀, Galapagos Is., Sta Cruz, Bellavista 2 mi N, 360 m, guava thicket, v–vii.1985, S. and J. Peck col. (CNCI). **GUADELOUPE**: Petit Borg, Domaine Duclos, 28.ii.1989, with *Solenothrips rubrocinctus* and *Heliothrips haemorrhoidalis* on *Inga ingoides*, J. Etienne col. (3♀, NHMUK). **ISRAEL**: Bet Dagan, ix.1996, ex *Heliothrips haemorrhoidalis*, M. Wysoki col. (7♀, DACE, AICF). **MEXICO**: Chiapas, 6.2 miles N Berriozabal, premontane rain forest, 9.viii.1990, 4000′ J.B. Woolley col. (3♀, 2♂, UCRC). **USA**: California, Orange Co., Irvine, 13.vi.1990, ex *Heliothrips haemorrhoidalis* on avocado, H.G. Johnson (1♀, NHMUK); California, Orange Co., South Coast Field Station, El Toro, ex *Heliothrips haemorrhoidalis* on avocado, H.G. Johnson (2♀, 1♂, UCRC); California, Orange Co., 10.ix.1989, ex *Heliothrips haemorrhoidalis* on avocado N. Hessein col. (1♀, DACE); Virginia, Montgomery Co., 8 km NW Blacksburg, 19–30.vi.1987, **MT**, rural, 1000 m, BRC HYM. TEAM (1♀, CNCI); California, San Diego Co., Valley Center, Weslilac Rd, Playa Grove, on avocado, H.G. Johnson (7♀, UCRC);

Missouri, Parkville, 39°12′17″ N, 94°40′38″ W, 5.vii.2015, swept, Z. Lahey col., DNA: 1114 (1♀, NHMUK—currently misplaced).

Species-group placement: *M. mymaripenne*-group.

Distribution: Argentina, Brazil, Chile, Costa Rica, Dominican Republic, Ecuador, Guadeloupe, Israel, Mexico, USA, and Venezuela.

Hosts: *Megaphragma mymaripenne* is a solitary egg endoparasitoid of several species of Panchaetothripinae (Thripidae). The most common host is the widespread *Heliothrips haemorrhoidalis*. The populations recorded in the USA [12] are represented mainly by females. The population reared in Argentina from maize and identified as *M. mymaripenne* [11] differs from the known populations of the species: the reared specimens from maize appear to be normally bisexual.

DNA data: 28S: 1 sequence, Missouri (USA).

Comments: This species was described in detail by Timberlake [65], and additional features were given by Viggiani [6]. *Megaphragma mymaripenne* is extremely difficult to distinguish morphologically from the closely related species *M. polilovi*, and even from the more distantly related species *M. noyesi*, with which it has been previously confused. They differ, however, in the length and shape of **C1**, length of the scape and colour of the radicle, and length of the ovipositor, respectively, as outlined in the key. Without the molecular data, these subtle differences would have been overlooked or treated as intraspecific variability. The correlation between the molecular clades and morphological characters indicates, however, that there are three species involved.

Records from Israel are the only Old-World records for this species; previous records of *M. mymaripenne*, e.g., from Italy [13,14,66], turned out to be misidentifications of the new species *M. polilovi*.

***Megaphragma polychaetum* Lin** (Figure 8a–c)

Megaphragma polychaetum Lin, 1992. *Entomotaxonomia* 14(2): 132–133.

Megaphragma anomalifuniculi Yuan et Lou in Yuan et al., 1997. *Journal of Northeast Normal University* 4: 62–63. **Syn. nov**.

Diagnosis. *Female*: Antenna long and narrow (Figure 8a) with pedicel shorter than the subcylindrical funicle, which has two **MT**. Clava two-segmented, **C1** twice as long as funicle, with two **MT** and two long **UST**; **C2** with two **MPS**, two **MT**, and a terminal basiconic sensillum (**SB**) slightly shorter than half **C2** length.

Mid lobe of mesoscutum with subpolygonal sculpture. Propodeum with a very short central area (Figure 8c). Fore wing (Figure 8b) 8–9× as long as wide, with two short setae in the middle of the marginal vein, and a disc with two distinct rows of 6–8 setae (Figure 8b). Tergites of metasoma without sculpture, but with some short and strong setae. The ovipositor is 1.1× as long as the mesotibia.

Head (including antenna), metasoma, meso-, and metacoxae are very dark. Remainder of body, including legs, pale brown. Fore wing strongly infuscate basally.

Male: Unknown.

Material examined. Paratypes: **CHINA**: Wuyishan, Fujian, 30.vii.1987, Wang Jiashe col. (3♀, FAFU); 10.vii.1987, 14-051, 14-052 (2♀, FAFU).

Species-group placement: *M. polychaetum*-group.

Distribution: China.

DNA data: no DNA sequences. DNA sequences are very likely to be close to those of *M. cockerilli* **sp. nov.** (see below).

Comments: The type material of *M. anomalifuniculi* was not available to the authors. According to the illustration given by Yuan and Lou [67], *M. anomalifuniculi* appears to be similar, if not identical, to *M. polychaetum* Lin. The features concerning the funicular segment appear to derive from a preparation artifact.

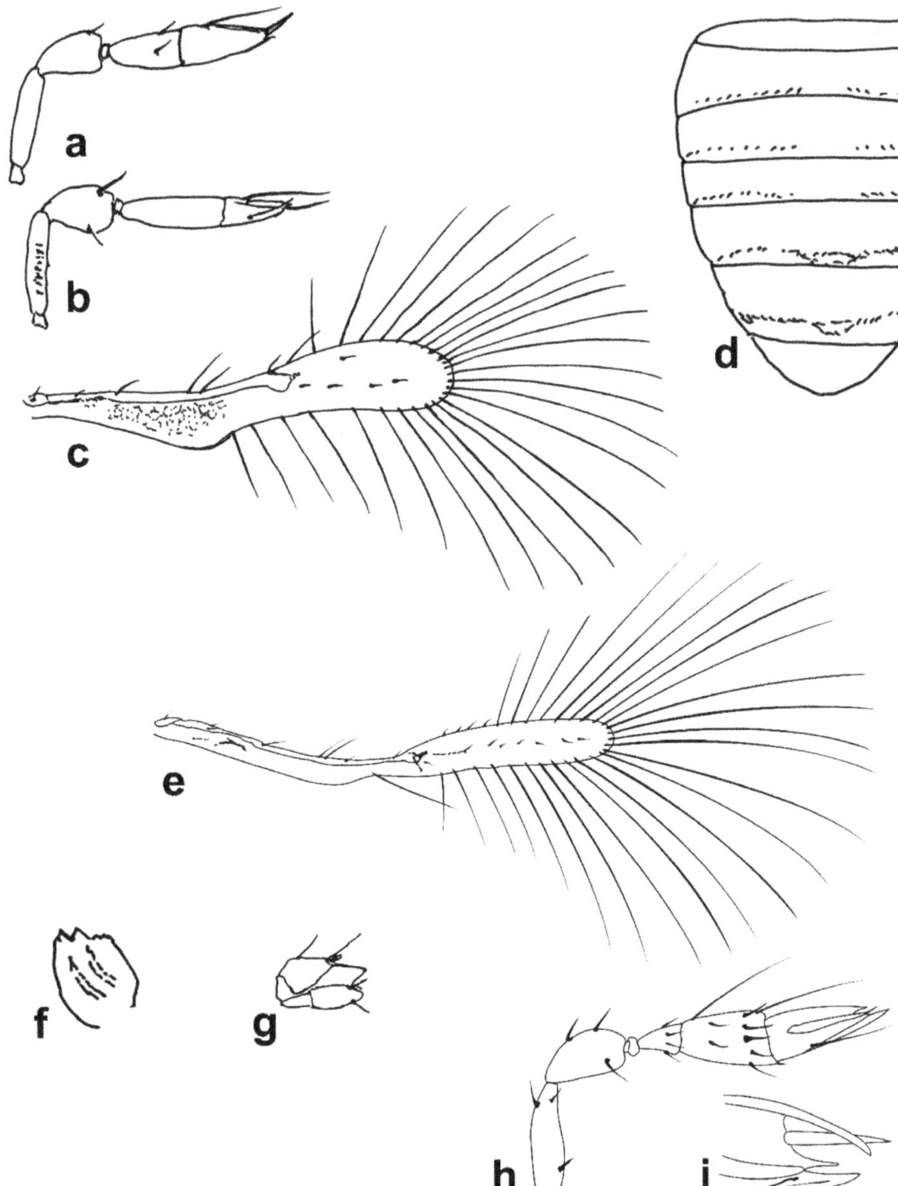

Figure 6. Line drawings of *Megaphragma* species: (**a**) *M. macrostigmum*, female antenna (Paratype); (**b**) *M. macrostigmum*, male antenna (Holotype); (**c**) *M. macrostigmum*, female fore wing (Paratype); (**d**) *M. macrostigmum*, male dorsal metasoma (Holotype); (**e**) *M. mymaripenne*, female fore wing (Holotype); (**f**) *M. mymaripenne*, mandible (Holotype); (**g**) *M. mymaripenne*, maxillary palp (Holotype); (**h**) *M. mymaripenne*, female antenna (Holotype); (**i**) *M. mymaripenne*, female antenna (detail of apex) (Holotype).

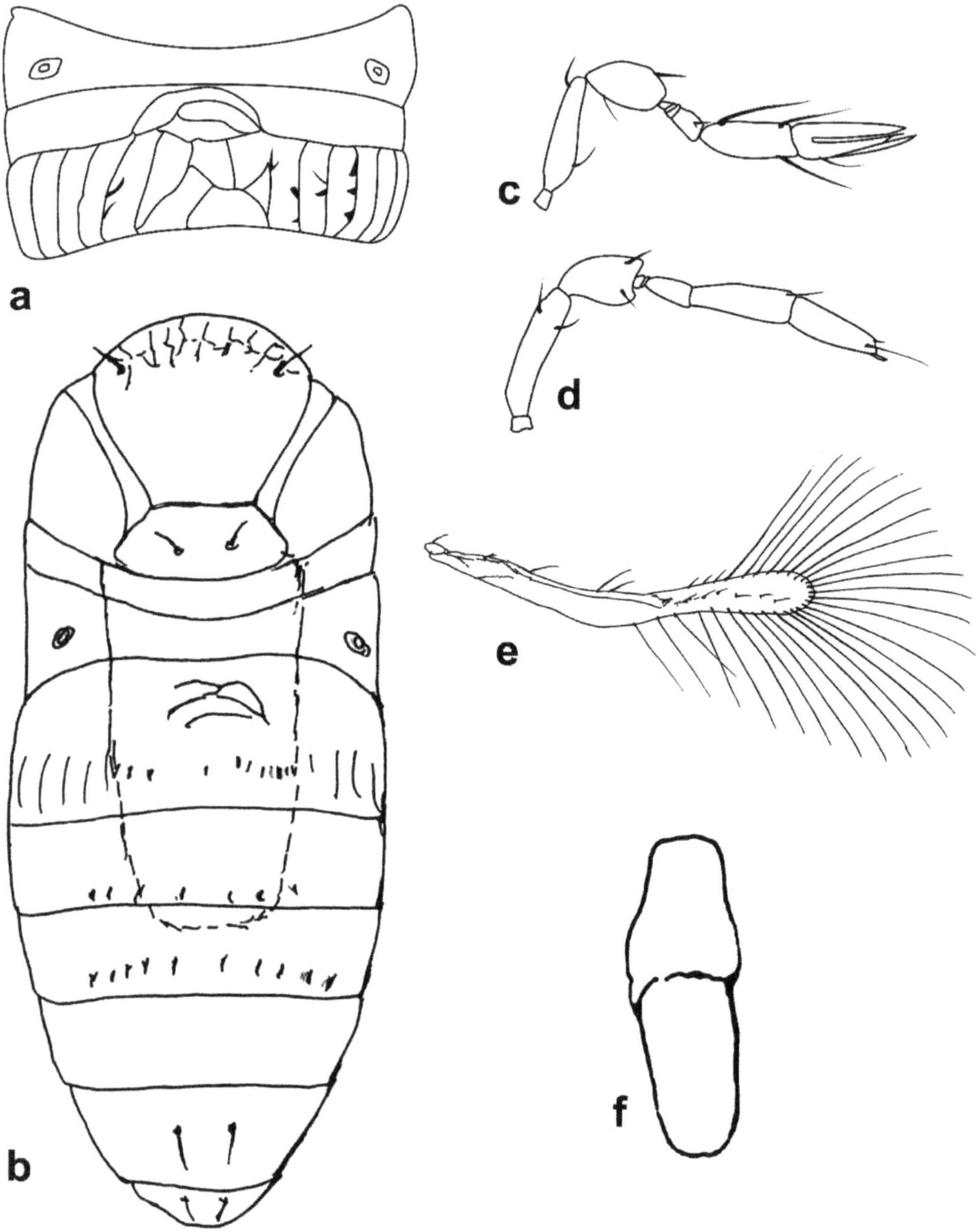

Figure 7. Line drawings of *Megaphragma* species: (**a**) *M. nowickii*, female propodeum and **T1** (Holotype); (**b**) *M. nowickii*, male meso- and metasoma (Paratype); (**c**) *M. nowickii*, female antenna (Holotype); (**d**) *M. nowickii*, male antenna (Paratype); (**e**) *M. nowickii*, female fore wing (Holotype); (**f**) *M. nowickii*, male aedeagus (Paratype).

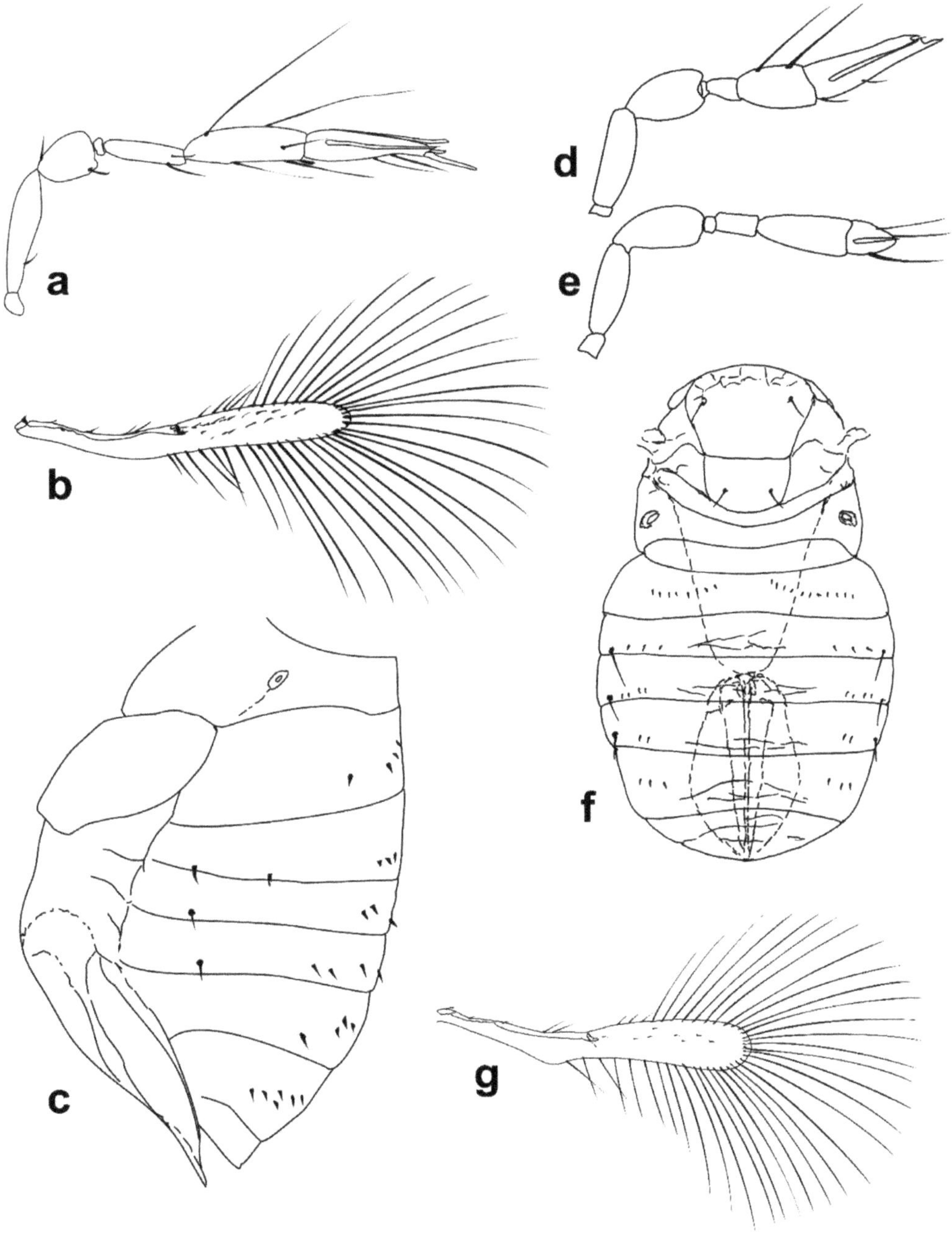

Figure 8. Line drawings of *Megaphragma* species: (**a**) *M. polychaetum*, female antenna (Paratype); (**b**) *M. polychaetum*, female fore wing (Paratype); (**c**) *M. polychaetum*, female propodeum and metasoma, lateral view (Paratype); (**d**) *M. priesneri*, female antenna (Neotype); (**e**) *M. priesneri*, male antenna (non-type); (**f**) *M. priesneri*, female dorsal meso- and metasoma (Neotype); (**g**) *M. priesneri*, female fore wing (Neotype).

Megaphragma priesneri (Kryger) (Figure 8d–g and Figure 22e)

Sethosiella priesneri Kryger, 1932. *Bulletin de la Société Royale d'Egypte* 16: 40.
Megaphragma priesneri: Ghesquière, 1839. *Rev. Zool. Bot. Afr.* 33(1): 38.

Diagnosis. Female: Antenna (Figure 8d) with pedicel slightly shorter than scape, funicle as long as half pedicel. Clava two-segmented with two long **UST** on **C1** of female; **C2** with one **MPS** and two **MT**.

Mid lobe of mesoscutum (Figure 8f) anteriorly with subpolygonal sculpture; propodeum with a very short central area. Fore wing (Figure 8g) 7× as long as wide, with maximum distal width less than 2× width measured at apex of marginal vein (Figure 8g); maximum fringe seta length 4× maximum discal width; setae on ventral disc surface short, penultimate one not reaching to the base of the distal (Figure 22e). **T1** without sculpture, but with a row of microspines; **T2–T4** each with a pair of setae, shorter than their corresponding tergites. Ovipositor 1.1× mesotibia.

Entire head and body are very dark. Legs and antenna paler. Wings hyaline.

Male: Similar to female in most characters except genitalia. Antenna with funicle and **C1** more elongate than in female, without **UST** on **C1**; **C2** much shorter than in female (Figure 8e).

Material examined. Neotype ♀ (NHMUK), here designated. **EGYPT**: Tanta, 30.11.30, vine leaves with *Retithrips*.

Non-types: **ISRAEL**: Higwe Yisrael, xi.1996, M. Wysoki coll., ex eggs *Retithrips syriacus* on *Vitis vinifera* (15♀, 3♂, NHMUK, DACE, AICF).

Species-group placement: *M. longiciliatum*-group.

Distribution: Egypt, Israel.

Host: *Retithrips syriacus* (Mayet).

DNA data: no DNA sequences.

Comments: Following extensive inquiries over the decades since 1990 in Egypt and Denmark, the holotype (and indeed the remainder of the type series of four specimens) appears to be lost. A specimen with data almost identical to the holotype is in the NHMUK, but has aberrant antennae. Nevertheless, we here designate that specimen as neotype, given that the data are very similar to those of the original type [68] (only the collection date differs by less than a month). Furthermore, all of the remaining morphology accords perfectly with the original description. Unfortunately, extensive efforts to collect fresh specimens in both Egypt and Israel failed.

The neotype designation for *M. priesneri* (Kryger) satisfies the provisions of Article 75.3 of the International Code of Zoological Nomenclature by: (1) clarifying the taxonomic identity of the species in its accepted modern concept (Article 75.3.1); (2) defining the combination of features of the sculpture of the mesoscutum and **T1**, propodeal structure and wing proportions as diagnostic for the species (Article 75.3.2); (3) providing data and description sufficient to ensure recognition of the specimen designated (Article 75.3.3); (4) giving reasons (no references available heretofore) for believing that the original type material is lost (Article 75.3.4); (5) selecting a neotype specimen consistent with the original description of the species and that was collected not long (less than 1 month) after the original description (specimen in this case) and, as such, represents the type species (Article 75.3.5); (6) choosing a neotype from the originally cited type locality, Tanta, Egypt (Article 75.3.6); and (7) recording that the neotype is the property of a recognized scientific institution, NHMUK in London (Article 75.3.7).

Megaphragma stenopterum (Lin) (Figure 9e–h)

Paramegaphragma stenopterum Lin, 1992. *Entomotaxonomia* 14(2): 134–135.
Megaphragma stenopterum: Delvare, 1993. *Revue fr. Ent. (n.s.)* 15(4): 151.

Diagnosis. Female: Antenna (Figure 9e) without funicle, clava two-segmented, and **C2** twice as long as **C1**. **C1** with two **MT**; **C2** with one **MPS**, two **MT**, one **SB**, and one **UST**.

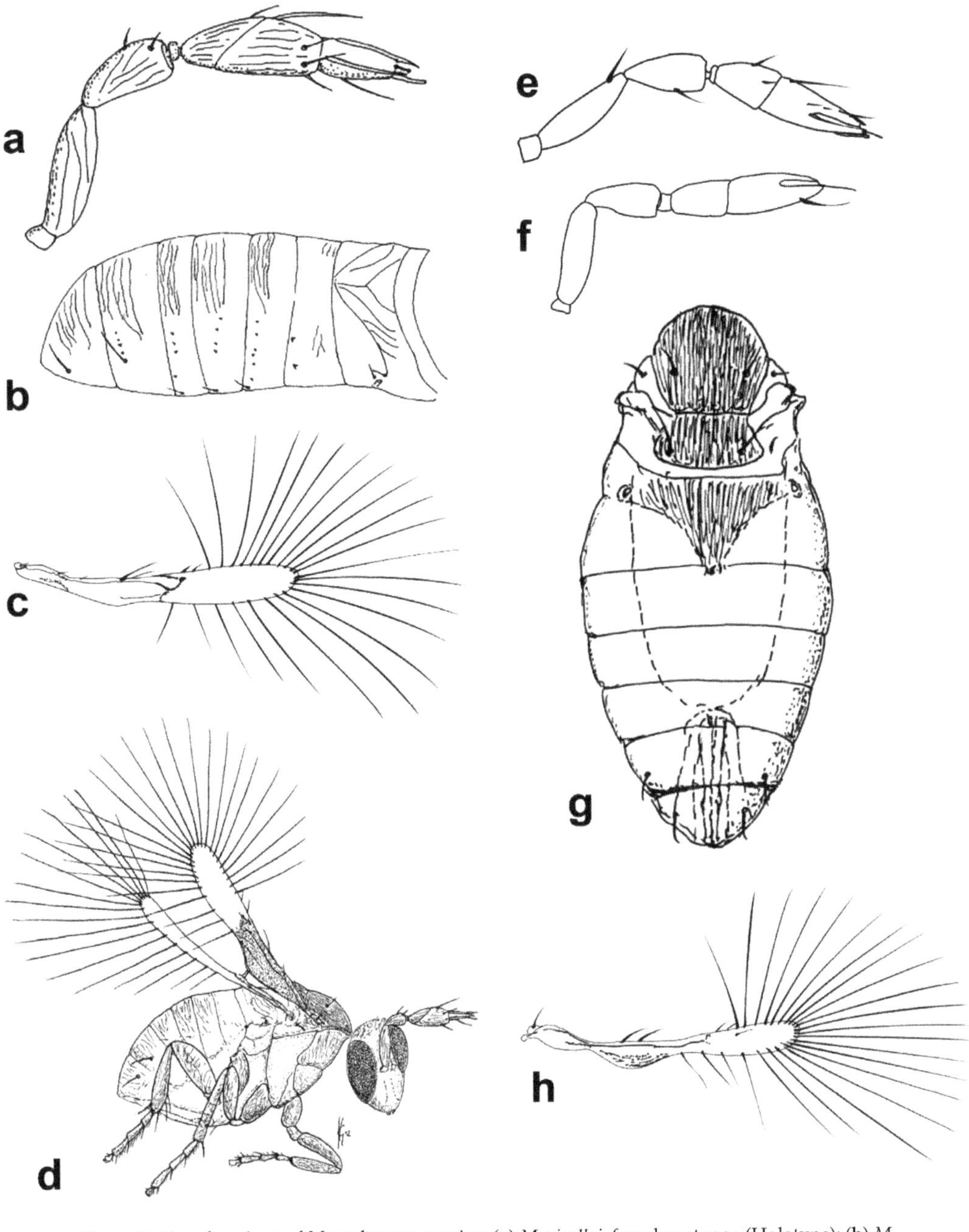

Figure 9. Line drawings of *Megaphragma* species: (**a**) *M. rivelloi*, female antenna (Holotype); (**b**) *M. rivelloi*, female metasoma (Holotype); (**c**) *M. rivelloi*, female fore wing (Holotype); (**d**) *M. rivelloi*, female habitus (Holotype); (**e**) *M. stenopterum*, female antenna (Paratype); (**f**) *M. stenopterum*, male antenna (Paratype); (**g**) *M. stenopterum*, female dorsal meso- and metasoma (Paratype); (**h**) *M. stenopterum*, female fore wing (Paratype).

Mid lobe of mesoscutum, scutellum, and central area of propodeum longitudinally striate (Figure 9g). Fore wing (Figure 9h) 9× as long as wide, and longest fringe seta 7× as long as maximum discal width, with two long central setae on the marginal vein, one long discal seta; hind margin sinuate. Tergites of metasoma without sculpture or crenulae (Figure 9g).

Body brown, the head darker brown. Fore wing basally strongly infuscate.

Male: Almost no discernible differences from female except genitalia characters. Even the antennae are very similar (Figure 9e,f).

Material examined. Paratype: **CHINA**: Fuzhou, Fujian, 20.xii.1987, N.Q. Lin col. (1♀, FAFU).

Non-types: **CHINA**: Fuzhou, Fujian, 6.x.1987, N.Q. Lin col. (1♀, FAFU); Fuzhou, Fujian, 20.vi.1987, N.Q. Lin col. (1♂, FAU).

Species-group placement: *M. stenopterum*-group.

Distribution: China.

DNA data: no DNA sequences.

Comments: This species, described in detail by the author, has the unique combination of a single seta on the fore wing and four-segmented antenna without any apparent funicle. A transverse, anelliform funicle is present in *M. uniclavum*, the only other species with a four-segmented antenna *and* a single seta on the fore wing.

There are differences between the Chinese text and the English text of the original description concerning the collecting dates of the type series. The examined paratype is mentioned in the English part but not in the Chinese part.

***Megaphragma striatum* Viggiani** (Figure 10a–d and Figure 21b,c)

Megaphragma striatum Viggiani, 1997. *Boll. Lab. Ent. Agr. Filippo Silvestri* 53: 119–120.

Diagnosis. Female: Antenna (Figure 10a and Figure 21b) without funicle, clava three-segmented, with **C1** having a transverse distal margin. **C1** with ≥1 **MT**; **C2** ≥8 **MT**, and 2 **UST**; **C3** with 2 **MPS**, ≥2 **MT**, and 1 **SB**.

Mid lobe of mesoscutum and scutellum longitudinally striate (Figure 10d). Propodeum with a pronounced subtriangular central area (Figure 21c). Fore wing 8× as long as wide, with one long central seta on the marginal vein, one discal seta, longest fringe seta 4–5× as long as maximum discal width. Metasoma with a row of crenulae on **T2** (Figure 21c).

Head and metasoma very dark, mesoscutum brown, the remainder, including legs and antenna, paler. Fore wing infuscate basally.

Male: Similar to female in most characters except genitalia.

Material examined. Holotype ♀ (DACE). **MEXICO**: Chiapas, Ocozocoautla, El Aquacero, 1800–2200′, 8.8.1990, coll. JB Woolley.

Paratypes: 2♀, 2♂, same data as holotype (UCRC).

Non-types: **MEXICO**: Tamaulipas, Alta Cima (nr Goméz Farias), 23°01′ N 99°09′ W, 2.xi.2009, screen-sweep A. Polaszek col., DNA: MXM1 (1♀, NHMUK); 1♀, 1♂, same data as holotype, but not mentioned in the original description (CNCI, UCRC).

Species-group placement: *M. ghesquierei*-group.

Distribution: Argentina, Belize, Costa Rica, Mexico.

DNA data: CO1: one sequence; 28S: one sequence (both Mexico).

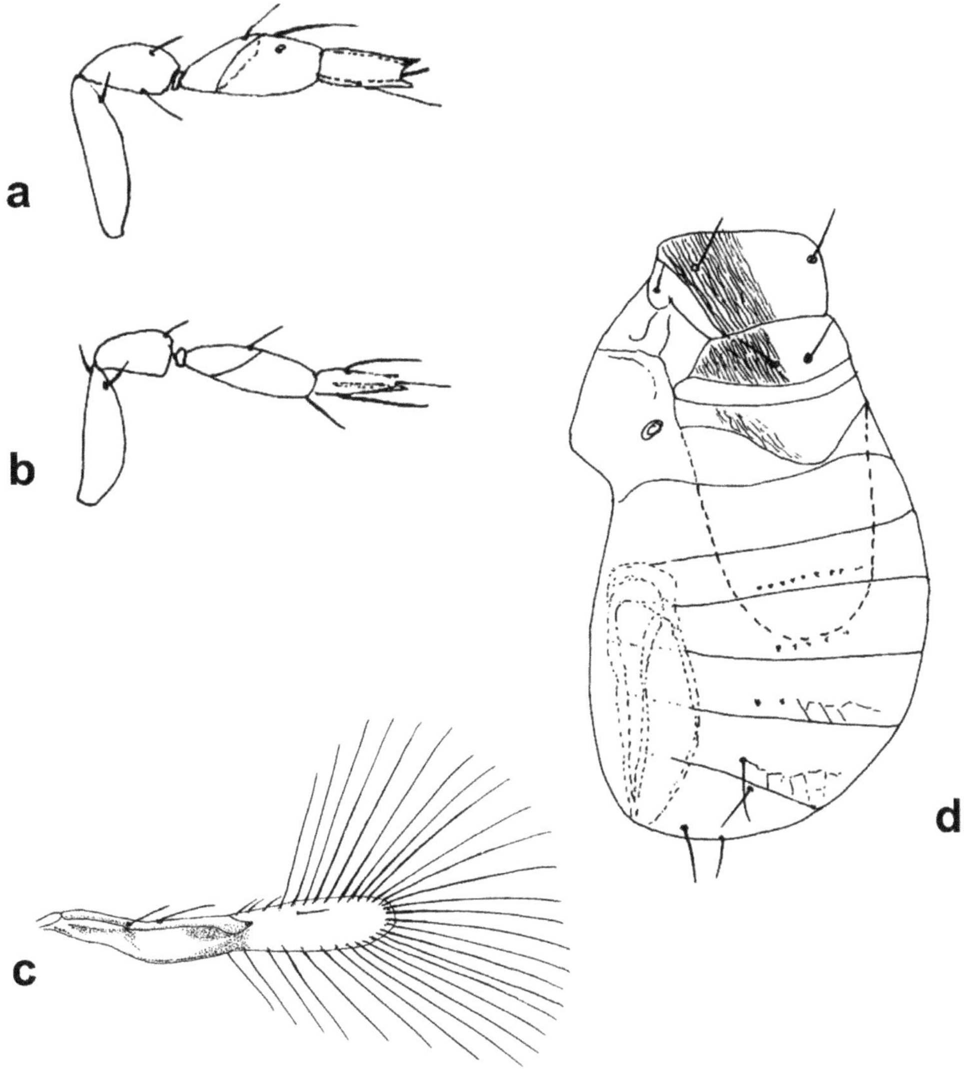

Figure 10. Line drawings of *Megaphragma* species: (**a**) *M. striatum*, female antenna (Holotype); (**b**) *M. striatum*, male antenna (Paratype); (**c**) *M. striatum*, female fore wing (Holotype); (**d**) *M. striatum*, female dorsal meso- and metasoma (Holotype).

Descriptions of new species

Megaphragma antecessor **Polaszek and Fusu sp. nov.** (Figure 11a–d)

urn:lsid:zoobank.org:act:A68D7B18-ADE4-4EDC-BB04-32A77E063469

Description. Female: Head (Figure 11c) with toruli vertical, in contact with each other medially; area below toruli with fine longitudinal sculpture, 1 min seta present laterally on each side. Antenna (Figure 11a) five-segmented (excluding anellus), transverse funicle present; clava two-segmented but these almost completely fused. **C1** with ≥6 **MT**, 3 **UST**; **C2** with 3 **MPS** and 2 **UST**; **SB** not detected but presumably present. Base of **C2** with one (apparent) **SS**.

Mid lobe of mesoscutum (Figure 11d) with fine longitudinal striation; vertical/ventral anterior mid lobe of mesoscutum with coarse, reticulate sculpture (Figure 11d); propodeum with subtriangular area centrally (Figure 11d) with 3–4 large crenulae; propodeum with hind margin arcuate. Fore wing (Figure 11b) 8.5× as long as maximum width, maximum distal width is 92× the maximum basal width; disc with a single short seta, and longest fringe seta 6.5× maximum discal width. Marginal vein with four setae, the second (from the wing base) robust and blunt; central setae equal in length. Campaniform sensillum present below second seta, a line joining the sensillum to the fourth seta. Stigmal vein with a row of three campaniform sensilla apically. Mesotibia with two large spines basally; metafemur with spine; metatibia with a row of fine, blunt setae extending almost the entire inner length, increasing abruptly in length at the distal tibia (exact length not visible in Figure 11d since setae positioned almost vertically; a similar row of setae is found in *M. momookherjeeae* and *M. uniclavum*, Figure 21j). **T1** with smooth area centrally, flanked by two or three longitudinal grooves and a longitudinal cell laterally, extending for the length of the tergum; **T1–T4** with very long setae laterally, each longer than its tergum; **T2** with a curved row of 6–8 spicules on each side. Ovipositor 1.6× as long as mesotibia.

Body brown. Occiput and face entirely brown, vertex paler. Antenna pale brown, pedicel paler. Mesosoma with the following brown: mid lobe of mesoscutum centrally, side lobes, axillae, propodeum laterally; remainder of mesosoma pale. Entire metasoma brown, except **T1** centrally pale. Fore wing distinctly infuscate basally, below, and including marginal vein.

Male: Unknown.

Material examined. Holotype ♀ (deposited in NHMUK). **COSTA RICA**: Puntarenas, Est. Biol. Monteverde, 10°19′ N 83°49′ W, 1540–1890 m, 26.ii.2007, J.S. Noyes BMNH(E) 2010-21, DNA: CRM2.2.

Species group placement: *M. antecessor*-group. *Megaphragma antecessor* clusters in different DNA sequence analyses with *M. momookherjeeae* **sp. nov.** (CO1 and combined analysis, strong support) or as basal to all other *Megaphragma* (28S, weak support). The shapes of the fore wing, propodeum, and to some extent the antenna, are strongly suggestive of the *ghesquierei*-group; presumably unique aspects of both DNA sequences prevent it from clustering with the species in that group.

Distribution: Costa Rica

DNA data: CO1: one sequence; 28S: one sequence.

Etymology: From the Latin *antecessor* (predecessor, precursor), in reference to the basal position of this species in the phylogenetic analyses. Noun in apposition.

Figure 11. Photographs of *Megaphragma* species: (**a**) *M. antecessor*, female antenna (Holotype); (**b**) *M. antecessor*, female fore wing (Holotype); (**c**) *M. antecessor*, female head (Holotype); (**d**) *M. antecessor*, female dorsal meso- and metasoma (Holotype). Scale bars 20 μm.

***Megaphragma breviclavum* Polaszek and Fusu sp. nov.** (Figure 12a–c)

urn:lsid:zoobank.org:act:FF10048F-0F8B-4B17-AF74-6E7C504C2ED4

Description. Female: Antenna (Figure 12a) five-segmented (excluding anellus); funicle absent; hence, clava three-segmented, with **C1** and **C2** almost fused. **C1** without **UST**; **C2** with 2 **UST**, and abundant **MT**; **C3** with 2–3 **MPS**, **SB**, and **SS**.

Mid lobe of mesoscutum (Figure 12b) with longitudinal striate sculpture extending to scutellum; propodeum (Figure 12b) with central area extended posteriorly, crenulae absent. Fore wing (Figure 12c) 7.5× as long as maximum width; the disc with a single long seta; longest fringe seta 4.5× as long as maximum discal width. Marginal vein with one long seta centrally, extending to apex of marginal vein. Stigmal vein moderately enlarged, with four sensilla apically. Mesotibiae with one large spine basally; metafemora with spine. **T1** with elongate cells laterally, 2–3× as long as wide; **T2–T4** without setae laterally; **T5** with long setae laterally. Ovipositor 1.7× as long as mesotibia.

Body largely brown, the following paler: legs except coxae and metafemur. Antenna with pedicel pale; scape and **C1–C3** darker. Fore wing strongly infuscate basally; stigmal and marginal vein brown; marginal vein very dark centrally.

Male: **C1** and **C2** with scattered **SS**; **C2** with 2–3 **MT** apically; **C3** with long apical and ventral **UST**. Colour and morphology largely as in female.

Material examined. Holotype ♀ (deposited in NHMUK). **MEXICO**: Tamaulipas, Alta Cima (nr Gómez Farias), 23°01′ N 99°09′ W, 2.ii.2009, A. Polaszek col. NHM(E) 2010-21, DNA: MXM2.

Paratypes: 1♀, 1♂ with same data as holotype, DNA: MXM3 and MXM4 (NHMUK).

Species-group placement: *M. ghesquierei*-group.

Distribution: Mexico.

DNA data: CO1: three sequences; 28S: three sequences.

Etymology: A noun in apposition referring to the comparatively short clava.

***Megaphragma chienleei* Polaszek and Fusu sp. nov.** (Figure 12g,h and Figure 13a–c)

urn:lsid:zoobank.org:act:0ECA37D4-69CF-412F-B081-5F140B3EBA1D

Description. Female: Antenna (Figure 12g) five-segmented (excluding anellus); funicle absent (though anellus extremely large); hence, clava three-segmented, with **C1** and **C2** strongly overlapping, almost fused; **C1** with 1 elongate, apical **UST**; **C2** with ≥10 **MT**; and **C3** with 3 very long **UST**.

Mid lobe of mesoscutum (cf Figure 13b) smooth with some irregular longitudinal striate sculpture; propodeum (cf Figure 13a) elongate and curved centrally and posteriorly, crenulae present. **T1** without elongate cells laterally; **T2–T4** with short setae laterally. Ovipositor 2× as long as mesotibia. Mesotibia with one large spine basally; metafemur with spine. Fore wing (cf Figure 13c) 8× as long as maximum width, maximum distal width equal to maximum basal width; discal setae absent, longest fringe seta 4.7× as long as maximum discal width. Marginal vein with two long setae centrally, approximately equal in length. Stigmal vein moderately enlarged, with three sensilla apically.

Head and body uniformly very pale brown. **C3** darker than remainder of antenna (Figure 12g). Fore wing basally infuscate (cf Figure 13c).

Male: Characteristics as for female (Figure 13a–c) (except antenna and genitalia); although, metasoma darker than in female. Antenna as in Figure 12h, with a much shorter **C3** compared to the female.

Material examined. Holotype ♀ (deposited in AICF). **MALAYSIA**: Sabah (Borneo), Danum Valley, 05°01′ N 117°49′ E, 16.ix.2012, fogged tree, T. Cockerill, DNA: SAM12.

Paratypes: **MALAYSIA**: Sabah (Borneo), Maliau Basin Studies Centre, Belian Trail, 04°44′ N 116°58′ E, 20.ix.2012, screen-sweep, A. Polaszek NHM(E) 2010-21, DNA: SAM4 to SAM8 (4♂, 1♀, AICF, NHMUK).

Species-group placement: *M. ghesquierei*-group.

Distribution: Malaysia (Borneo, Sabah).

DNA data: CO1: four sequences; 28S: six sequences.

Figure 12. Photographs of *Megaphragma* species: (**a**) *M. breviclavum*, female antenna (Holotype); (**b**) *M. breviclavum*, female dorsal meso- and metasoma, composite image (Holotype); (**c**) *M. breviclavum*, female fore wing (Holotype); (**d**) *M. caribea*, female antenna (flagellum only) (Paratype); (**e**) *M. caribea*, female dorsal mesosoma (Paratype); (**f**) *M. caribea*, female fore wing (Paratype); (**g**) *M. chienleei*, female antenna (Holotype); (**h**) *M. chienleei*, male antenna (Paratype). Scale bars 20 µm except 50 µm for b, c, and f.

Etymology: Named for pitcher-plant (*Nepenthes*) botanist and wildlife photographer Chien C. Lee (Sarawak, Malaysian Borneo).

Megaphragma cockerilli **Polaszek and Fusu sp. nov.** (Figure 13d–f)

urn:lsid:zoobank.org:act:09B45F53-180F-4204-9AC7-69DE9E352132

Description. Female: Antenna (Figure 13d) five-segmented (excluding anellus); pedicel as long as funicle; funicle 4× as long as wide; **C2** longer than **C1**. **C1** with two prominent dorsal **UST**, proximal **UST** almost as long as entire clava; **C2** with two **MT** and an **SB**, which is only slightly shorter than **C2**. **MPS** apparently absent.

Mid lobe of mesoscutum (Figure 13e) entirely with large, coarse reticulation; propodeum with a rhomboid, laterally arcuate central area, its hind margin truncate, with fine crenulae. **T1–T4** largely smooth, with scattered denticles and no setae laterally. **T5** and **T6** with a pair of long setae centrally. Ovipositor 1.1× as long as mesotibia. Mesotibia without spines basally; metafemur without spine; metatibia with a row of five spines within the distal inner half; a robust spine towards the apex of the outer surface. Fore wing (Figure 13f) 8.5× as long as maximum width, maximum distal width 1.4× maximum basal width; discal setae arranged in 3–4 rows, of 4–6 setae per row, longest fringe seta 5× as long as the maximum discal width. Marginal vein with two long setae centrally, of equal length. Stigmal vein with two sensilla apically.

Body largely dark brown, mesosoma paler laterally; antenna very dark brown. Fore wing infuscate basally.

Male: Unknown.

Material examined: Holotype ♀ (deposited in AICF): **MALAYSIA**: Sabah (Borneo), Danum Valley, 05°01′ N 117°49′ E, 16.ix.2012, fogged tree, T. Cockerill, DNA: SAM11.

Species-group placement: *polychaetum*-group. Very close to *M. polychaetum*, differing by the extremely elongate terminal sensillum basiconicum.

Distribution: Malaysia (Borneo, Sabah)

DNA data: CO1: one sequence; 28S: one sequence.

Etymology: Named for our colleague and friend, Dr Tim Cockerill, collector of this species (Falmouth University, UK).

Megaphragma digitatum **Polaszek and Fusu sp. nov.** (Figure 14a–c)

urn:lsid:zoobank.org:act:136D58D7-A1FC-4D46-8F63-A7780E4D0871

Description. Female: Antenna (Figure 14a) five-segmented (excluding anellus); funicle absent; hence, clava three-segmented, with **C1** and **C2** almost fused; **C1** with ≥4 **MT**, without **UST**; **C2** with 2 **UST** and abundant **MT**; **C3** with 2–3 **MPS**, **SB**, and **SS**.

Mid lobe of mesoscutum (Figure 14b) with longitudinal striate sculpture extending to scutellum; propodeum with central area extending posteriorly, crenulae present; **T1** with one elongate cell or groove laterally, 2–3× as long as wide; **T2–T4** without setae laterally; **T5** with long setae laterally. Ovipositor 1.5× as long as mesotibia. Mesotibia with one large spine basally; metafemur with spine. Fore wing (Figure 14c) 8.5× as long as maximum width, maximum distal width equal to maximum basal width; the disc with a single long seta; longest fringe seta 5× as long as maximum discal width. Marginal vein with two setae centrally; proximal seta 5–7× as long as distal seta, extending to the end of the marginal vein (in Figure 14c, the distal seta is barely visible in the space between the proximal one and the marginal vein). Stigmal vein moderately enlarged, with two sensilla apically.

Body largely brown, the following paler: legs except coxae and metafemora. Antenna with pedicel pale; scape and **C1–C3** darker. Fore wing strongly infuscate basally; stigmal and marginal vein brown; marginal vein very dark centrally.

Male: Largely as in female. **C1** and **C2** with scattered **SS**; **C2** with 2–3 **MT** apically; **C3** with long apical and ventral **UST**.

Material examined. Holotype ♀ (deposited in NHMUK). **COSTA RICA**: Puntarenas, Est. Biol. Monteverde, 10°19′ N 83°49′ W, 1540–1890 m, 26.ii.2007, J.S. Noyes BMNH(E) 2010-21, DNA: CRM2.1.

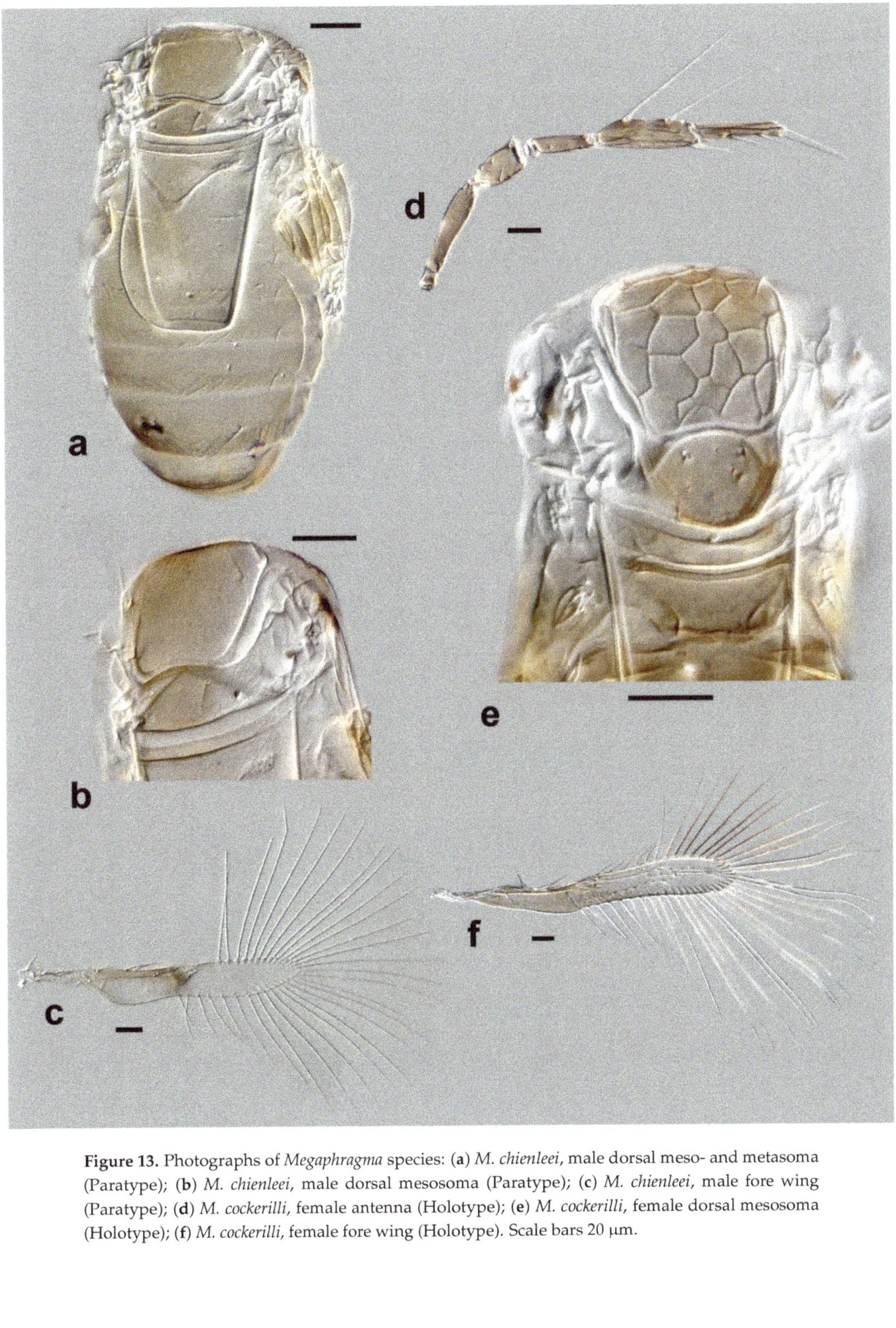

Figure 13. Photographs of *Megaphragma* species: (**a**) *M. chienleei*, male dorsal meso- and metasoma (Paratype); (**b**) *M. chienleei*, male dorsal mesosoma (Paratype); (**c**) *M. chienleei*, male fore wing (Paratype); (**d**) *M. cockerilli*, female antenna (Holotype); (**e**) *M. cockerilli*, female dorsal mesosoma (Holotype); (**f**) *M. cockerilli*, female fore wing (Holotype). Scale bars 20 μm.

Paratypes: **COSTA RICA**: same data as holotype except DNA: CRM2.3, 2.6, 2.9, 2.12 (2♂, 2♀, AICF, MZUCR); Cartago, 12.5 km S Turrialba Rancho Naturalista, 1000 m, 9°50″ N 83°34″ W, 12-14.ii.2017, J.S. Noyes BMNH(E) 2017-39, DNA1681 (1♀, NHMUK). **ECUADOR**: Km 26.5 road Dura–Tambo, Estación Experimental Litoral Sur, INIAP, 21.xi.2017, ex *Chaetanaphothrips signipennis* on *Musa paradisiaca*, M. Arias col., DNA: ECU3 (3♀, 1♂ AICF, NHMUK, UCRC); same data except DNA: ECU4 (3♀ 2♂, AICF, NHMUK, UCRC); same data but ex *Frankliniella parvula*, DNA: ECU1 (2♀, NHMUK).

Non-type: same data as holotype (1♀, without wings, NHMUK).

Species-group placement: *ghesquierei*-group.

Distribution: Costa Rica, Ecuador.

Hosts: *Chaetanaphothrips signipennis* (Bagnall); *Frankliniella parvula* Hood.

DNA data: CO1: two sequences (Costa Rica); 28S: eight sequences (six Costa Rica, two Ecuador).

Etymology: The species name refers to the digitate **C3**.

***Megaphragma fanenitrakely* Polaszek and Fusu sp. nov.** (Figure 14d–f)

urn:lsid:zoobank.org:act:4DE1AD1D-C800-48C7-898B-24353F5355F0

Description. Female: Antenna (Figure 14d) five-segmented (excluding anellus), with pedicel slightly longer than funicle (12:8); clava two-segmented, **C1** with two **UST**; one **SB** at the apex of **C1** and **C2**; apex of **C2** (Figure 14d) also with two elongate **MPS** and a long **SB**.

Mesoscutum with mid lobe (Figure 14e) entirely with coarse, reticulate sculpture; metanotum and propodeum medially short. Metasoma with a row of microspines on each segment. **T1** without cells. Ovipositor 1.1× as long as mesotibia. Mesotibia without spines basally. Metafemur without prominent spine. Fore wing (Figure 14f) 9× as long as wide, maximum distal width 1.5× maximum basal width; the disc with 10 setae is irregularly arranged in 1–2 rows, and the longest fringe seta 6× maximum discal width. Marginal vein with two long setae centrally, of equal length. Stigmal vein moderately enlarged, with three sensilla apically.

Body largely pale brown, mesosoma paler laterally; antenna brown. Wings hyaline.

Male: Unknown.

Material examined. Holotype ♀ (deposited in NHMUK). **MADAGASCAR**: Nosy Komba, 13°27′45″ S 48°20′18″ E, 460 m, 22.vi.2015, screen-sweep, A. Polaszek col. BMNH(E)2015-122.

Paratypes: **MADAGASCAR**: Nosy Komba, closed canopy forest, 13°27′11″ S 48°20′4″ E, 170 m, 19.vi.2015, yellow pan trap, A. Polaszek col. BMNH(E)2015-122 (2♀, NHMUK).

Species-group placement: *longiciliatum*-group.

Distribution: Madagascar.

DNA data: no DNA sequences.

Etymology: A noun in apposition; "fanenitra kely" = "tiny wasp" (Malagasy).

***Megaphragma funiculatum* Fusu, Polaszek, and Viggiani sp. nov.** (Figure 14g,h and Figure 15a,b)

urn:lsid:zoobank.org:act:B6273DD4-14D8-4C91-9057-489883DA0DDE

Description. Female: Antenna (Figure 14g) five-segmented (excluding anellus), pedicel twice as long as funicle, the latter trapezoid, and slightly longer than wide; **C1** slightly shorter than **C2**; **C1** with 2 dorsal **UST**; **C2** with ≥3 **MPS** 1 **SB** and a short **SS**.

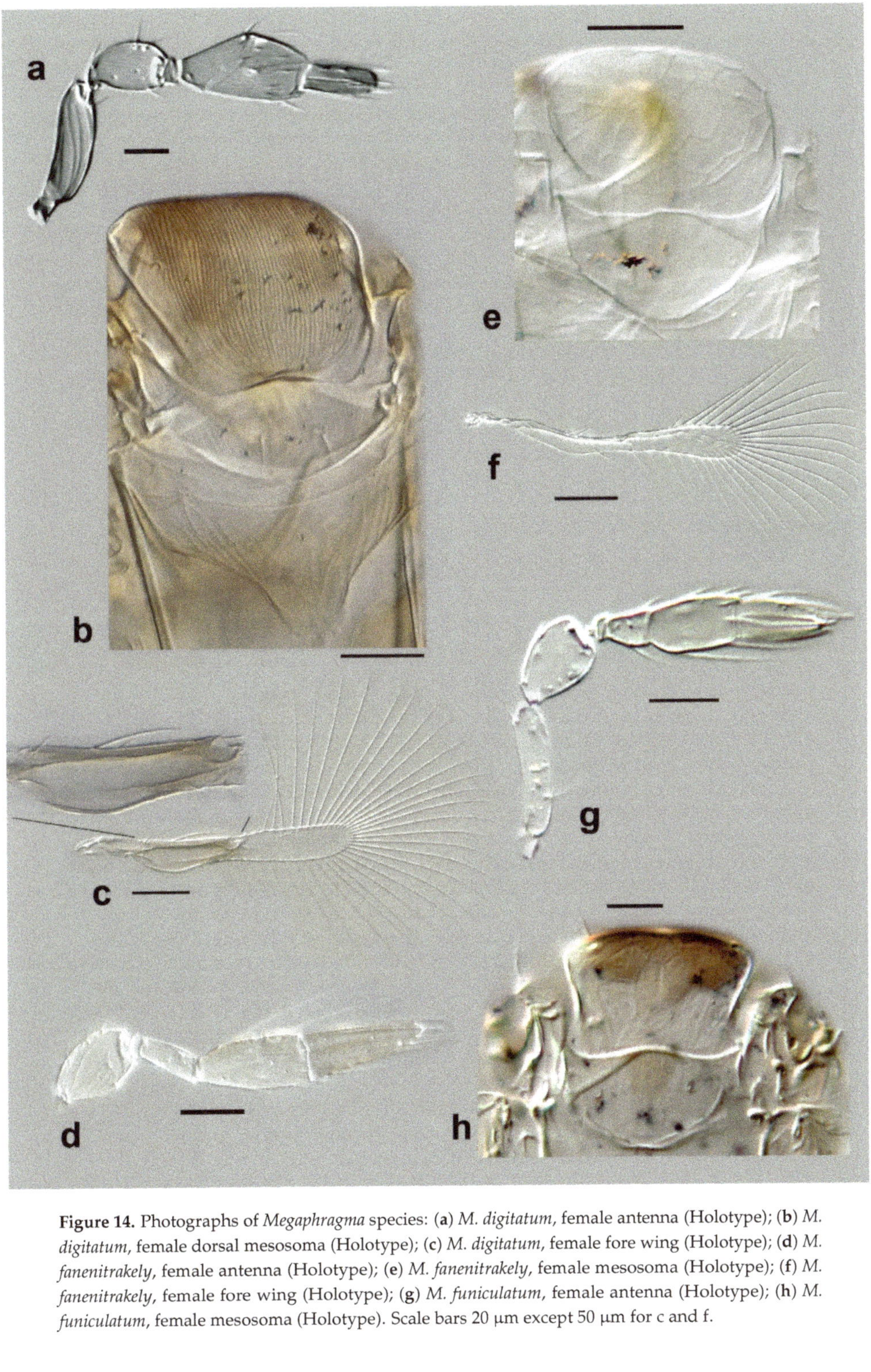

Figure 14. Photographs of *Megaphragma* species: (**a**) *M. digitatum*, female antenna (Holotype); (**b**) *M. digitatum*, female dorsal mesosoma (Holotype); (**c**) *M. digitatum*, female fore wing (Holotype); (**d**) *M. fanenitrakely*, female antenna (Holotype); (**e**) *M. fanenitrakely*, female mesosoma (Holotype); (**f**) *M. fanenitrakely*, female fore wing (Holotype); (**g**) *M. funiculatum*, female antenna (Holotype); (**h**) *M. funiculatum*, female mesosoma (Holotype). Scale bars 20 μm except 50 μm for c and f.

Mid lobe of mesoscutum smooth, with weakly impressed large subpolygonal sculpture (not visible in paratype which has been strongly macerated). Metanotum and propodeum relatively long centrally, each about half the length of scutellum. Propodeum with short central area, without crenulae. **T1** (Figure 15a) sculpture with cells converging centrally, lateral cells 2–3× as long as wide, without denticles (some denticles present on innermost cells). **T2–T4** without long setae laterally, each with similar sculpture comprising a central irregular oval cell and elongate lateral cells; those on **T3** and **T4** divided medially. Ovipositor 1.3× as long as mesotibia. Mesotibia with one large spine basally; metafemur without spine. Fore wing (Figure 15b) 10× as long as wide, maximum distal width 1.2× maximum basal width; the disc with a single irregular row of five setae; longest fringe seta 5.5× maximum discal width. Marginal vein with two long setae centrally.

Head, including antenna, pale; mesosoma largely pale, anterior half of mesoscutal mid lobe brown; metasoma entirely brown, **T1** darker than the remainder. Wings hyaline.

Male: Unknown.

Material examined. Holotype ♀ (deposited in NHMUK). **COSTA RICA**: Limón, Hitoy-Cerere Reserve, 100 m, 9°40′ N 83°02′ W, 24–26.ii.2008, J.S. Noyes col. NHM(E) 2010-21AQ, DNA: CRM 3.103.

Paratypes: same data as holotype, DNA: CRM 3.46 (1♀, AICF), CRM 3.100 (1♀, MZUCR).

Species-group placement: *mymaripenne*-group.

Distribution: Costa Rica.

DNA data: 28S: two sequences.

Etymology: Named for the distinctive funicle.

***Megaphragma giraulti* Viggiani, Fusu, and Polaszek sp. nov.** (Figure 4c–g, Figures 5a and 15c)

urn:lsid:zoobank.org:act:9F02CA78-49F3-49D3-A4B8-DC4E3A3033EF

Megaphragma sp.: Huber and Noyes, 2013. *J. Hymenopt. Res.* 32: 37, Figs 49–51.

Description. Female: Antenna (Figures 4d and 15c) five-segmented (excluding anellus), with pedicel slightly longer than funicle (12:8); clava two-segmented, **C1** 1.5× as **C2**, with ≥10 **MT** and 2 **UST**; one **SB** at the apex of **C1** and **C2**; apex of **C2** (Figure 4g) also with 2 elongate **MPS**, a long **SB** and **UST**.

Mid lobe of mesoscutum and scutellum without apparent sculpture; metanotum and propodeum medially short. Metasoma with a row of microspines on each segment. **T1** without cells (Figure 4f). Ovipositor 1.1× as long as mesotibia. Mesotibia without spines basally. Metafemur without prominent spine. Fore wing (Figure 5a) 9× as long as wide, maximum distal width 1.5× maximum basal width; disc with 10 setae irregularly arranged in 1–2 rows; fringe with longest seta 6× maximum discal width. Marginal vein with two long setae centrally, of equal length. Stigmal vein moderately enlarged, with three sensilla apically.

Body brown/yellow. Mesosoma largely pale, but mid lobe of mesoscutum brown anteriorly. Scape and pedicel pale, **C1–C3** brown. Fore wing slightly infuscate basally.

Male: As female, but antenna (Figure 4c) with **C1** approximately 2× **C2**. Metasoma (Figure 4e) with a row of microspines on each segment.

Material examined. Holotype ♀ (deposited in NHMUK). **COSTA RICA**: Puntarenas, Est. Biol. Monteverde, 10°19′ N 83°49′ W, 1540–1890 m, 26.ii.2007, J.S. Noyes BMNH(E) 2010-21, DNA: CRM2.4.

Paratypes: **COSTA RICA**: same data as holotype, DNA: CRM2.5 (1♀, AICF), CRM2.7, 2.8, 2.10, 2.11 (4♀, DACE, MZUCR, NHMUK); Limón, Hitoy-Cerere Reserve, 9°40′ N 83°02′ W, 100 m, 24–26.ii.2008, J.S. Noyes NHM(E) 2010-21AQ, DNA: CRM 3.23/B11, CRM 3.105, B9, E1 (2♀, 2♂, NHMUK); Cartago, 12.5 km S Turrialba, Rancho Naturalista, 1000 m, 9°50″ N 83°34″ W, 12–14.ii.2017, J.S. Noyes BMNH(E) 2017-39, DNA1683 (1♀, NHMUK); Heredia, La Selva Biol. Sta., 10°26′ N 84°01′ W, 75 m, 27–28.ii.2003, J. S. Noyes (1♂, DACE).

Non-types: **ARGENTINA**: Salta Prov. Orán, road to San Andres along Rio Blanca, 399 m, 23.09° S 63.37° W, 23.iii.2003, J. Munro 003-03-23-02 (1♀, UCRC). **USA**: Northampton, 7 km S Jackson, 23.ix–14.xi.1987, **MT**, Bald Cypress Swamp, BRC Hym Team (1♀, UCRC).

Species-group placement: *polychaetum*-group. The male antenna is very distinctive in the group, and *M. giraulti* male antenna agrees very well with several other species that definitely belong to the *polychaetum*-group (but without sequence data to back up this assertion).

Distribution: Argentina, Costa Rica, USA.

DNA data: CO1: four sequences; 28S: eight sequences (all Costa Rica).

Etymology: The species is named for A.A. Girault for his pioneering studies on the Trichogrammatidae.

***Megaphragma hansoni* Polaszek, Fusu, and Viggiani sp. nov.** (Figure 15d–f)

urn:lsid:zoobank.org:act:980B8787-BB40-4589-91D4-FB4262F4BA0F

Description. Female: Antenna (Figure 15d) five-segmented (excluding anellus); funicle absent; hence clava three-segmented, with **C1** and **C2** almost fused; **C1** with 1–2 **MT**; **C2** with some **MT** and 2 **UST**; **C3** with 2–3 **MT**, 2 **MPS**, and prominent **SB** and **SS**.

Mid lobe of mesoscutum (Figure 15e) with longitudinal striate sculpture extending to scutellum; propodeum (Figure 15e) with central area extended posteriorly, crenulae present; **T1** with elongate cells laterally, 2–3× as long as wide; **T2–T4** without setae laterally; **T5** with long setae laterally. Ovipositor 2× as long as mesotibia. Mesotibia with one large spine basally; metafemur with spine. Fore wing (Figure 15f) 7× as long as maximum width, maximum distal width equal to maximum basal width; disc with a single short seta; longest fringe seta 5× as long as maximum discal width. Marginal vein with one long seta centrally, extending almost to the end of the marginal vein; a minute additional seta next to it. Stigmal vein moderately enlarged, with four sensilla apically.

Body largely brown, the following paler: most of mesosoma except anterior half of mid lobe of mesoscutum, anterior half of **T1** and antenna. Fore wing slightly infuscate basally; stigmal and marginal vein pale brown.

Male: Largely as in female. **C1** and **C2** with scattered **SS**; **C2** with 2–3 **MT**, 1 apically; **C3** with long apical and ventral **UST**.

Material examined. Holotype ♀ (deposited in NHMUK). **COSTA RICA**: Limón, Hitoy-Cerere Reserve, 9°40′ N 83°02′ W, 100 m, 24–26.ii.2008, J.S. Noyes NHM(E) 2010-21AQ, DNA: CRM 3.101 (but no associated DNA sequence).

Paratypes: **COSTA RICA**: same data as holotype except DNA: CRM3.4, 3.17/B5, 3.40/D4, 3.97, 3.99, 3.104 (CRM3.40 lost after DNA extraction) (1♀, 4♂, AICF, NHMUK); Puntarenas, Est. Biol. La Gamba, 8°42′ N 83°12′ W, 150 m, 13–14.ii.2008, J.S. Noyes BMNH(E) 2010-21AQ, DNA: CRM1.1, 1.2, 1.4, 1.7, 1.12-1.19 (12♂, DACE, MZUCR, NHMUK, UCRC).

Species-group placement: *ghesquierei*-group.

Distribution: Costa Rica.

DNA data: 28S: seven sequences.

Etymology: Named for our colleague and co-author on this paper, Professor Paul Hanson, University of Costa Rica, San José.

***Megaphragma kinuthiae* Polaszek, Fusu, and Viggiani sp. nov.** (Figure 5b–d)

urn:lsid:zoobank.org:act:E6685BED-8C4E-46AB-9E70-9C0FBF6FCF6A

Description. Female: Antenna (Figure 5b) five-segmented (excluding anellus); pedicel as long as funicle; funicle 3× as long as wide; **C1** slightly longer or as long as **C2** with two dorsal **UST**; three elongate **MPS** extending beyond clava tip.

Figure 15. Photographs of *Megaphragma* species: (**a**) *M. funiculatum*, female metasoma (Holotype); (**b**) *M. funiculatum*, female fore wing (Holotype); (**c**) *M. giraulti*, female antenna (Holotype); (**d**) *M. hansoni*, female antenna (Holotype); (**e**) *M. hansoni*, female dorsal meso- and metasoma (Holotype); (**f**) *M. hansoni*, female fore wing (Holotype). Scale bars 20 μm.

Head with toruli separated by about their own width. Mid lobe of mesoscutum smooth; propodeum with straight hind margin, without crenulae; **T1** non-reticulate; **T2–T4** with short setae laterally. Mesotibia with two large spines basally; metafemur with spine; metatibia with a row of fine, blunt setae extending almost their entire inner length, increasing abruptly in length distally. Fore wing (Figure 5d) about 9× as long as maximum width, maximum distal width 1.3× maximum basal width; discal setae arranged in two rows, each with 4–5 setae, longest fringe seta 10× as long as maximum discal width. Marginal vein with two long setae centrally. Stigmal vein with two sensilla apically. Ovipositor 1.1× as long as mesotibia.

Body entirely yellow, anterior mesosoma brown, posterior metasoma slightly darker. Wings hyaline.

Male: Unknown.

Material examined. Holotype ♀ (deposited in NHMUK). **KENYA**: Meru, vi. 1965, ID. No. 2851, CIE 233, BM 196. T. F. Crowe, ex tea leaves.

Paratypes: same data as holotype (6♀, NHMUK). All specimens are on the same slide; the holotype is circled in red.

Species-group placement: *polychaetum*-group. The species appears closest to *M. giraulti* based on morphology.

Distribution: Kenya.

Host: Not identified, but possibly *Scirtothrips dorsalis* (Hood), a species common on tea in Kenya.

DNA data: no DNA sequences.

Etymology: Named for our colleague and friend Dr Wanja Kinuthia, National Museums of Kenya, Nairobi.

Megaphragma liui **Polaszek and Fusu sp. nov.** (Figure 16a–d)

urn:lsid:zoobank.org:act:35AD8001-C03B-4711-ADF5-22AFB681F184

Description. Female: Antenna (Figure 16a) five-segmented (excluding anellus), **C1** and **C2** strongly overlapping; **C3** elongate, more than half the length of **C1** and **C2**; 1 min **UST** on **C1**, two long **UST** on **C2**; three elongate **MPS** extending beyond clava tip.

Head with toruli very close together, separated by about one-third their own width. Mid lobe of mesoscutum (Figure 16b) with fine longitudinal striations, but also with distinct large reticulate cells; propodeum medially with strongly produced hind margin, with two crenulae (Figure 16b). **T1** smooth centrally, but with 8–10 elongate cells laterally (Figure 16b); **T2–T4** with short setae laterally, lateral cells present. Mesotibia without large spines basally, but a robust spine present at the apex of mesofemur; metafemur with spine; metatibia with a group of fine, sharp setae on inner surface apically. Metacoxa and metafemur (Figure 16c) with distinct longitudinal sculpture ventrally, contrasting with transverse sculpture dorsally. Fore wing (Figure 16d) 7× as long as maximum width, maximum distal width 1× maximum basal width; disc distally pointed, without setae (but one wing with a possible indication of a minute seta); longest fringe seta 4× as long as maximum discal width. Marginal vein with two setae centrally, the proximal one very robust, about 1.5× as long as distal. Stigmal vein with one elongate sensillum apically. Ovipositor 1.9× as long as mesotibia.

Body entirely brown, mesosoma pale posteriorly, **T1** with pale areas laterally. **C1** very dark, pedicel paler than the remainder of the antenna. Fore wing strongly infuscate basally. Legs dark, tarsi pale.

Male: Unknown.

Material examined. Holotype ♀ (deposited in UCRC). **BRUNEI**: Temburong Dist., Bukit Patoi trail, 41–290 m, 4°45′21″ N 115°10′30″ E, 4 July 2010, swp dipterocarp forest, J. Mottern M10-065, DNA1656.

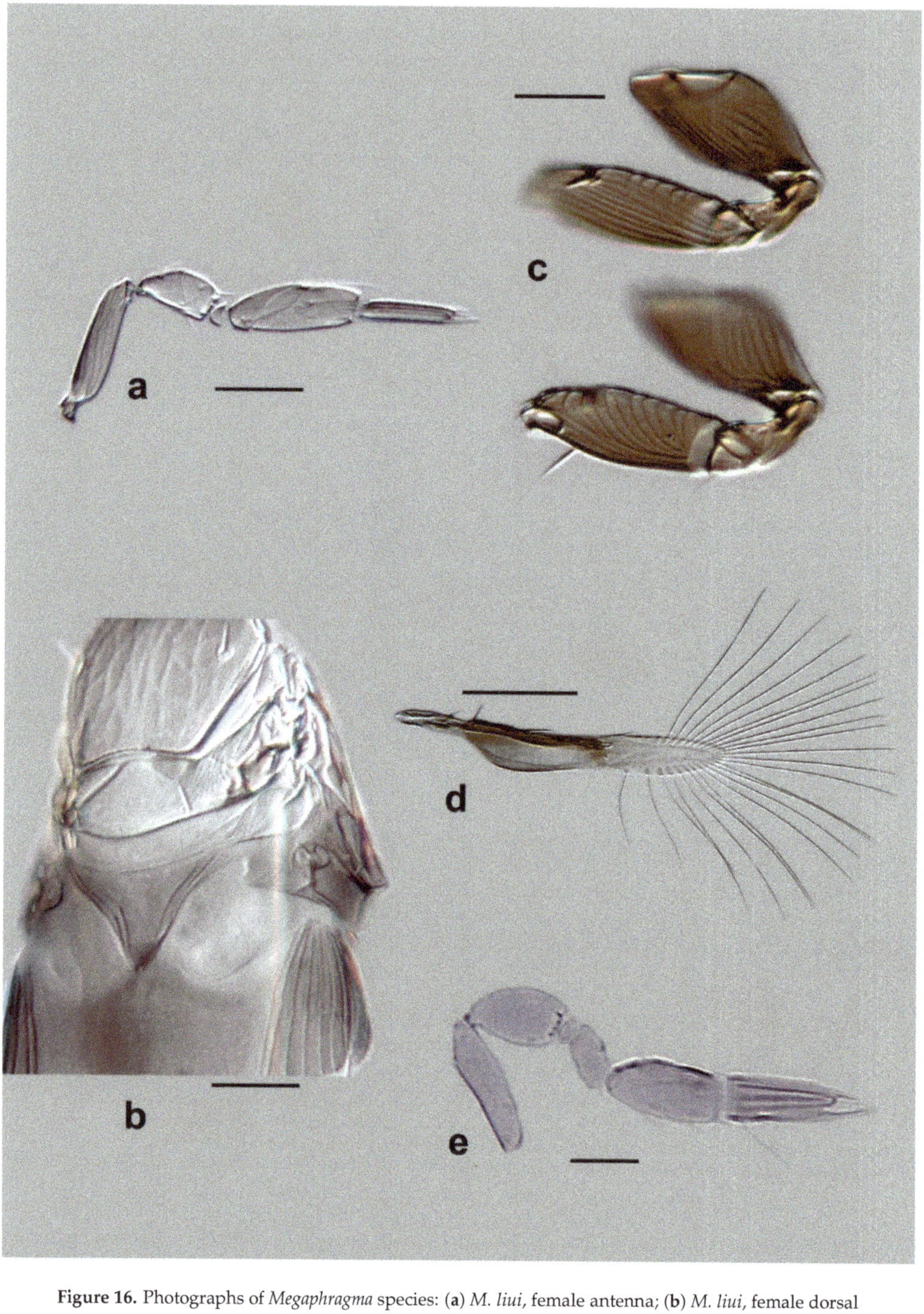

Figure 16. Photographs of *Megaphragma* species: (**a**) *M. liui*, female antenna; (**b**) *M. liui*, female dorsal meso- and metasoma (Holotype); (**c**) *M. liui*, female metacoxa and metafemur (Holotype); (**d**) *M. liui*, female fore wing (Holotype); (**e**) *M. longiciliatum*, female antenna (Holotype). Scale bars 20 μm except 50 μm for d.

Species-group placement: *ghesquierei*-group. In the concatenated and partitioned analysis, this species is not included in the group (Figure 2b), though in an unpartitioned analysis with a simple substitution model it is one of the most basal species of the *ghesquierei*-group (Figure 2a). It is also retrieved as part of the group in the tree based on the 28S sequences alone, where it is sister to *M. rivelloi* but on a very long branch (Supplementary Figure S3). Hence, morphology, and partly molecular analyses, indicate that our inclusion of the species is correct.

Distribution: Brunei.

DNA data: CO1: one sequence; 28S: one sequence.

Etymology: Named for our colleague and friend Prof. Shu-sheng Liu, Zhejiang University, Hangzhou, China.

***Megaphragma momookherjeeae* Polaszek and Fusu sp. nov.** (Figure 17a–c)

urn:lsid:zoobank.org:act:CC6535BA-271C-44CB-9697-CC048ED070E0

Description. Female: Antenna (Figure 17a) five-segmented (excluding anellus); transverse funicle present, clava two-segmented; **C1** longer than **C2; C1** with ≥5 **MT**, 2 **UST**, and with fine, longitudinal striation; **C2** with basal **SS**, 2 **UST**, and ≥4 **MPS**.

Mid lobe of mesoscutum smooth; propodeum without distinct central area. **T1–T4** largely smooth, **T2–T4** with long setae laterally. Ovipositor exserted, exceptionally long for the genus, more than 3× as long as mesotibia (Figure 17b). Mesotibia with a very robust spine basally, 0.4× tibial length; metafemur without spine; metatibia with a row of about 17 spines along almost the entire inner length, and 4 robust spines toward the apex of the outer surface. Fore wing (Figure 17c) 9× as long as maximum width; maximum distal width 1× maximum basal width; disc with a single, minute seta; longest fringe seta 6.5× as long as maximum discal width. Marginal vein apparently with three long setae centrally, of equal length. Stigmal vein with three sensilla apically.

Body largely dark brown; scutellum, propodeum, and lateral mesosoma paler (Figure 17b); antenna pale brown, **C2** darker. Fore wing basally infuscate.

Male: Unknown.

Material examined. Holotype ♀ (deposited in NHMUK). **COSTA RICA**: Cartago, 12.5 km S. Turrialba, Rancho Naturalista, 1000 m, 9°50″ N 83°34″ W, 12–14.ii.2017, J.S. Noyes BMNH(E) 2017-39, DNA1680.

Species-group placement: *M. antecessor*-group. Resembling also the *ghesquierei*-group in some features (e.g., fore wing with one seta), but clearly not clustering with that group in any molecular analyses. It is recovered as sister to *M. antecessor*, and the two clustered together as basal to all species-groups except the *ghesquierei*-group (partitioned analysis), or as the most basal species group of *Megaphragma* (unpartitioned analysis).

Distribution: Costa Rica.

DNA data: CO1: one sequence.

Etymology: Named for Mo Mookherjee, a friend of the first author (AP).

***Megaphragma nowickii* Polaszek, Fusu, and Viggiani sp. nov.** (Figures 7a–f and 17e–g)

urn:lsid:zoobank.org:act:126687C2-A7AE-4FC2-810E-11DE6E6D6784

Description. Female: Antenna (Figures 7c and 17g) five-segmented (excluding anellus); pedicel twice as long as funicle; funicle twice as long as wide; **C1** slightly shorter or as long as **C2; C1** with ≥7 **MT**; 2 **UST**; 3 elongate **MPS** extending beyond clava apex.

Mid lobe of mesoscutum smooth, anteriorly with subpolygonal sculpture (cf Figure 7b). Metanotum and propodeum narrow centrally, the latter without an extension or crenulae. **T1** sculpture (Figures 7a and 17f) with cells converging centrally, lateral cells 3× as long as wide, each with 3–5 inward-pointing denticles; **T2–T4** without long setae laterally, all with coarse reticulate sculpture becoming lateral distally. Ovipositor 1.7× as long as mesotibia. Mesotibia with two large spines basally; metafemur with spine; metatibiae with a row of fine, blunt setae extending almost their entire inner length, increasing abruptly in length distally. Fore wing (Figure 7e) 9.5× as long as wide, maximum distal width 1.1× maximum basal width; disc with a single row of six setae; longest fringe seta 4.5× maximum discal width. Marginal vein with two long setae centrally.

Body entirely brown, metasoma slightly darker posteriorly. Pleural parts of mesosoma and hind legs except for tarsi lighter. Wings hyaline.

Male: Largely as in female. Antenna with **C1** slightly longer than **C2**; **T1** (Figure 7b) with an incomplete pattern of cells. Aedeagus as in Figure 7f.

Material examined. Holotype ♀ (deposited in NHMUK). **D. R. CONGO**: Province Orientale, Yangambi Biosphere Reserve, 15.v.2012, N 0°48.837′ E 24°30.287′, screen sweep, primary forest, A. Polaszek col, BMNH(E) 2012-88, DNA: COM1.16.

Paratypes: **D. R. CONGO**: same data as holotype except DNA: COM1.1–1.15, 1.17–1.23, 1.26, 1.27 (2♀, 22♂, AICF, DACE, IITA, NHMUK). **UGANDA**: Mabira Forest, N0°23′22″ E33°00′22″, 1250 m, 1.iii.2015, A. Polaszek, screen sweep, DNA1132/F11 (1♀, NHMUK); Mabira Forest, N0°23′22″ E33°00′22″, 1250 m, 1.iii.2015 A. Polaszek, screen sweep, DNA1116–1118, 1120–1124 (1♀, 7♂, NHMUK).

Non-types: **BENIN**: Dept. Zou, Zogbodomey, Massi, 18.xii.1989, ex egg *Megalurothrips sjostedti* on *Pueraria*, M. Tamo col. 275 (1♂, 1♀, NHMUK, IITA).

Species-group placement: *mymaripenne*-group.

Distribution: Benin, D. R. Congo, Uganda.

Host: *Megalurothrips sjostedti* (Trybom).

DNA data: CO1: 25 sequences from D. R. Congo; 28S: 30 sequences from 2 countries: D. R. Congo (21), Uganda (9).

Etymology: Named for S. Nowicki for his outstanding contribution to the knowledge of the Trichogrammatidae.

***Megaphragma noyesi* Polaszek and Fusu sp. nov.** (Figure 18a)

urn:lsid:zoobank.org:act:63048F49-FCC2-49FF-9E2C-C0552BC1FBF4

Description. Female: Antenna (Figure 18a) five-segmented (excluding anellus); pedicel almost twice as long as funicle; funicle slightly longer than wide; **C1** shorter than **C2**; **C1** with two dorsal **UST**; **C2** with three elongate **MPS** extending beyond clava apex.

Mid lobe of mesoscutum and scutellum smooth. Metanotum and propodeum narrow centrally, the latter without an extension or crenulae.

Mesotibia without large spines basally; metafemur without spine; metatibia with a row of robust spines extending along the inner surface of distal half.

Fore wing 9× as long as wide, maximum distal width 1.3× maximum basal width; disc with a single row of six setae, and longest fringe seta 5× maximum discal width. Marginal vein with two long setae centrally, subequal in length (cf Figure 18g). **T1** sculpture with cells converging centrally, six lateral cells present, 2–3× as long as wide, each with 1–3 inward-pointing denticles; **T2–T4** with lateral cells indicated, with lateral setae not detected (cf Figure 18f). Ovipositor as long as mesotibia. Body largely brown, mesosoma largely pale. **T1** very dark brown in contrast to rest of body. Legs pale. Wings hyaline.

Male: Largely as in female. Antenna with **C1** longer than **C2**; **T1** with an incomplete pattern of cells. Genitalia tubular as in other species, 3× as long as wide.

Material examined. Holotype ♀ (deposited in NHMUK). **UK**: England, Surrey, Coulsdon Common, Happy Valley, 51°17′ N 0°07′ W, 168 m, viii.2013, J.S. Noyes BMNH(E) 2013-, DNA: UKM14.

Paratypes: **UK**: same data as holotype except DNA: UKM8–13 (UKM9 has DNA sequences but was lost during extraction) (4♀, 1♂, AICF, DACE, NHMUK); East Sussex, Brede High Wood, TQ79432018/19, 30.viii.2019–20.ix.2019, D. Binns col., DNA1612–1619 (1612, 1614 and 1618 have DNA sequences but were lost during extraction) (10♀, 3♂, NHMUK, UCRC). **HUNGARY**: Őrség Nemzeti Park, Barkás Lake, 46°52′ N 16°26′ E, 268 m, 28.vi.2010, J.S. Noyes screen sweeping, BMNH(Ent) 2010-63, DNA: HUM2 (1♀, NHMUK) (HUM3 has DNA sequences but was lost during extraction), HUM5 (1♀, AICF); 4–5 km SW Kőszeg, Meszes Völgy, 47°22′ N 16°31′ E, 431 m, 26.vi.2010, screen sweeping, J.S. Noyes BMNH(Ent) 2010-63, DNA: HUM6 (1♂, AICF); Őrség National Park, Lugosi Valley, 46°54′ N 16°27′ E, 231 m, 28.vi.2010, J.S. Noyes, DNA: HUM 8–14, BMNH(Ent) 2010-63 (3♀ 4♂, NHMUK).

Figure 17. Photographs of *Megaphragma* species: (**a**) *M. momookherjeeae*, female antenna (Holotype); (**b**) *M. momookherjeeae*, female lateral meso-and metasoma (Holotype); (**c**) *M. momookherjeeae*, female fore wing (Holotype); (**d**) *M. mymaripenne*, female antenna (Holotype); (**e**) *M. nowickii*, female dorsal meso- and metasoma (Holotype); (**f**) *M. nowickii*, female dorsal base of metasoma (Holotype); (**g**) *M. nowickii*, female antenna (Holotype). Scale bars 20 μm except 100 μm for c.

Non-type: **CZECH REPUBLIC**: Moravia, Vranov, River Dyje, ss riparian forest, 13.viii.1991, L. Masner (1♀, UCRC).

Species-group placement: *mymaripenne*-group.

Distribution: UK (England); Czech Republic, Hungary.

DNA data: CO1: 21 sequences from 2 countries; 28S: 23 sequences from 2 countries: Hungary, UK.

Etymology: Named for Dr John Noyes of the Natural History Museum, London, for his outstanding contribution to our knowledge of Chalcidoidea. As a chalcid collector, John is unmatched so far, and a major proportion of the material in this study was collected by him.

***Megaphragma pintoi* Viggiani sp. nov.** (Figure 18b–d)

urn:lsid:zoobank.org:act:9C3F4C6B-6290-4DB5-B4E2-6080C11F6C91

Description. Female: Antenna (Figure 18b) five-segmented (excluding anellus); pedicel slightly longer than funicle; funicle 3× as long as wide; **C1** slightly longer or as long as **C2**; **C1** with two **UST**; three **MPS** extending beyond **C2** apex.

Mid lobe of mesoscutum (Figure 18c) smooth, with 5–6 deep striae anteriorly; propodeum with straight hind margin, without crenulae. **T1** non-reticulate; **T2–T4** with short setae laterally. Ovipositor 1.1× as long as the mesotibia. Metafemur without spine but a prominent seta present. Fore wing (Figure 18d) 10× as long as maximum width, and maximum distal width 1.4× maximum basal width; discal setae arranged in 2–3 rows, each with 3–5 setae, longest fringe seta 5.5× as long as maximum discal width. Marginal vein with two long setae centrally. Stigmal vein with two sensilla apically.

Body entirely pale brown (eyes deep purple); legs and antenna pale. Wings hyaline.

Male: Unknown.

Material examined. Holotype ♀ (deposited in PFRNZ). **NEW ZEALAND**: Auckland, Mt Albert, Science Center, 17.v.1997, ex *Heliothrips haemorrhoidalis* ex acmena leaves, P. Stevens.

Non-type: **COLOMBIA**: Supata, 5°04′07″ N 74°16′48″ W, 1800 m, 31.12.2018, sweep, A.A. Polilov col. (1♀, AICF).

Species-group placement: *polychaetum*-group.

Distribution: Colombia, New Zealand.

DNA data: no DNA sequences.

Etymology: Named for Emeritus Professor John Pinto, formerly of University of California, Riverside, in recognition of his monumental contribution to our understanding of Trichogrammatidae.

Comments: Known so far from only two specimens to date; this species has an apparently extraordinary distribution, being known from Colombia and New Zealand. It seems very likely that it will turn up elsewhere and is probably another cosmopolitan *Megaphragma* species.

***Megaphragma polilovi* Polaszek, Fusu, and Viggiani sp. nov.** (Figure 18e–g, Figures 19d and 20a,b)

urn:lsid:zoobank.org:act:7A812908-FC9A-4F84-9A1C-1699DC922132

Megaphragma mymaripenne: Viggiani and Bernardo, 1998. *Boll. Zool. agr. Bach. Ser. II* 29: 51–55; Bernardo and Viggiani, 2003. *Boll. Lab. Entomol. agr. Filippo Silvestri* 58 [2002]: 77–85; Polilov, 2012. *Arthropod Struct. Dev.* 41(1): 29–34; Makarova et al., 2015. *Arthropod Struct. Dev.* 44(1): 21–32; Polilov, 2016. *At the Size Limit—Effects of Miniaturization in Insects*; Polilov, 2017. *PLoS ONE* 12(5): e0175566; Diakova et al., 2018. *PeerJ* 6: e6005 (misidentifications).

Description. Female: Antenna (Figures 18e and 20a) five-segmented (excluding anellus); pedicel almost twice as long as funicle; funicle slightly longer than wide, with 3 **MT** and 1 ASC; **C1** shorter than **C2**; **C1** with 17 **MT**, **SS**, and 2 **UST**; **C2** with **SB**, 2 **MT**, **SS**, and 4 elongate **MPS** extending beyond clava apex.

Mid lobe of mesoscutum and scutellum smooth (Figures 18f and 19d). Propodeum with a distinct central area with lateral boundaries in line with those of mesoscutum and scutellum (Figure 19d); two lateral lobes present behind propodeum central area (Figure 19d); propodeum without crenulae. **T1** sculpture (Figures 18f and 20b) with cells converging centrally, about six lateral cells, 2× as long as wide, mesal cells each with 2–3 inward-pointing denticles. **T2–T4** with 2–3 lateral cells, with short setae laterally. Ovipositor 1.3× as long as mesotibia. Mesotibia with two large spines basally; metafemur without spine; metatibia with a row of fine, blunt setae extending almost its entire inner length, increasing abruptly in length distally. Fore wing (Figure 18g) 9.5× as long as wide, maximum distal width 1.4× maximum basal width; disc with a single row of six setae, longest seta of fringe 5.8× maximum disc width. Marginal vein with two long setae centrally.

Antenna (Figure 18e) with radicle brown, very dark compared to the remainder of the antenna; remainder of body largely brown, mesosoma largely pale, mid lobe of mesoscutum brown anteriorly, an indistinct brown spot on the scutellum. Legs pale. Wings hyaline.

Male: Largely as in female. Antenna with **C1** slightly longer than **C2**; **T1** with an incomplete pattern of cells. Genitalia tubular as in other species, 3× as long as wide.

Material examined. Holotype ♀ (deposited in DACE). **ITALY**: Vietri sul Mare, Benincasa, 40°40′ N 14°44′ E, 17.vii.2013, G. Viggiani ex *Heliothrips haemorrhoidalis* on *Viburnum tinus*, DNA: ITM9.

Paratypes: **ITALY**: same data as holotype, DNA: ITM8, 11, 12, 14 (4♀, AICF, DACE, NHMUK).

Species group placement: *mymaripenne*-group.

Distribution: Italy.

Host: *Heliothrips haemorrhoidalis*. Males are very rare, and reproduction is normally thelytokous [14] (as *M. mymaripenne*).

DNA data: CO1: five sequences; 28S: five sequences; all from Italy.

Etymology: Named for our colleague Alexey Polilov, co-author of this paper, for his outstanding contribution to our knowledge of the Trichogrammatidae and miniaturization in insects.

Comments: Found in Italy at the same locality and on the same host as *M. viggianii*.

***Megaphragma rivelloi* Viggiani sp. nov.** (Figures 9a–d and 21a)

urn:lsid:zoobank.org:act:4436D2E2-590F-4329-A65B-AB562F0E04ED

Megaphragma sp. Viggiani, 2002. *Boll. Zool. Agr. Bach. Ser. II* 34: 449–450.

Description. Female: Antenna (Figures 9a and 21a) five-segmented (excluding anellus), pedicel approximately two-thirds the length of **C1** and **C2**; **C3** one-quarter shorter than **C2**; **C2** with a single **UST**; **C3** with **SB**, two **MPS**, **MT**, and apical **SB**.

Mid lobe of mesoscutum and scutellum with longitudinally striate sculpture; propodeum (Figure 9b) with a large subtriangular and crenulated central area. Metasomal tergites without reticulation but with transverse striations, laterally with a row of 2–7 microspines. Ovipositor approximately 1.3× as long as mesotibia. Fore wing (Figure 9c) 6–7× as long as wide; disc without setae; longest fringe seta 4–4.5× as long as the maximum discal width.

Body yellow with brown mostly on mesosoma; fore wing infuscate behind the venation.

Male: Largely as in female except for genitalia.

Material examined. Holotype ♀ (deposited in DACE): **ITALY**: Basilicata, Rivello (PZ), vii.2002, yellow sticky traps in a vineyard.

Figure 18. Photographs of *Megaphragma* species: (**a**) *M. noyesi*, female antenna (Holotype); (**b**) *M. pintoi*, female antenna (Holotype); (**c**) *M. pintoi*, female mesoscutum (Holotype); (**d**) *M. pintoi*, female fore wing (Holotype); (**e**) *M. polilovi*, female antenna (Holotype); (**f**) *M. polilovi*, female dorsal mesosoma (part) and metasoma (Holotype); (**g**) *M. polilovi*, female fore wing (Holotype). Scale bars 20 μm except 50 μm for a, d, and g.

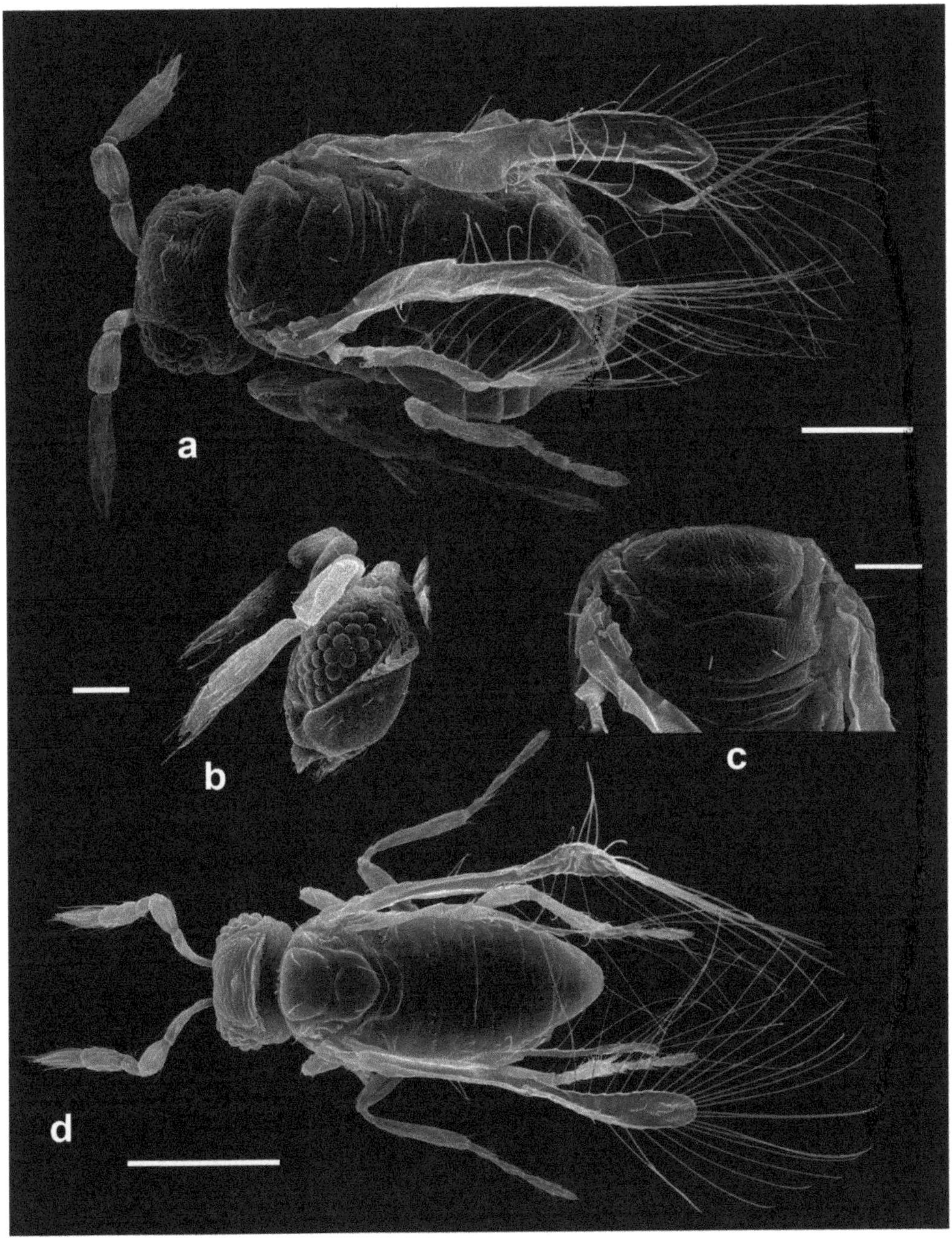

Figure 19. SEM micrographs of *Megaphragma* species (non-types): (**a**) *M. caribea*, male habitus; (**b**) *M. caribea*, male head and antennae; (**c**) *M. caribea*, male mesoscutum; (**d**) *M. polilovi*, female dorsal habitus. Scale bars 50 µm for a, 20 µm for b and c, and 100 µm for d.

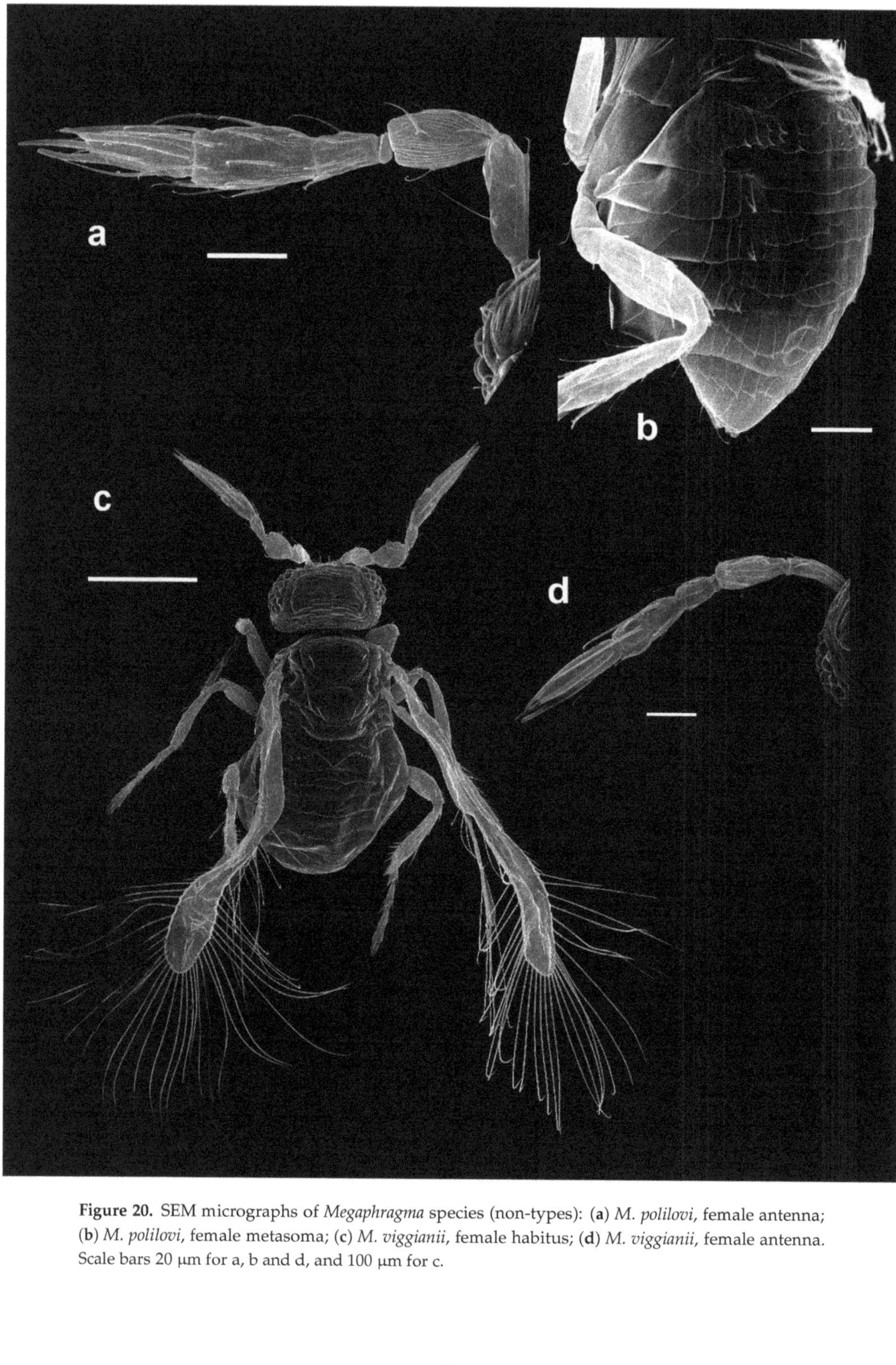

Figure 20. SEM micrographs of *Megaphragma* species (non-types): (**a**) *M. polilovi*, female antenna; (**b**) *M. polilovi*, female metasoma; (**c**) *M. viggianii*, female habitus; (**d**) *M. viggianii*, female antenna. Scale bars 20 μm for a, b and d, and 100 μm for c.

Paratypes: **ITALY**: same data as holotype (19♀, DACE, NHMUK). **INDIA**: UP, New Delhi, IARI; 220 m 28°37′51″ N, 77°09′50″ E, xi.5.2003, pan trap, J. Heraty col. H03-106 (1♀, UCRC). **INDONESIA**: W Java, Gunung Halimun NP, Tea-Forest Junction, 1066 m, 6°41′07″ S 106°31′16″ E, screen-sweep, 17.ix.2015, A. Polaszek, DNA1148 (1♀, NHMUK). **JAPAN**: Tokyo area, ?1984, Takagi col., ex *Scirtothrips dorsalis* on tea, A. Loomans leg. (5♀, AICF, NHMUK). **UK**: Surrey, Coulsdon Common, 25.viii.2002, J.S. Noyes, screen sweep (1♂, NHMUK). **VIETNAM**: Cat Tien NP, sweeping, N11°24′45″ E107°25′23″, 21.xi.2018, leg. A.A. Polilov, DNA1687 (1♀, AICF); Cat Tien NP, sweeping, N11°24′45″ E107°25′23″, 25.xi.2016, leg. A.A. Polilov (3♀, 2♂, AICF, UCRC).

Non-type: **CHINA**: 1♂ misidentified as *M. deflectum*, Wuyishan, Fujian, 19.x.1987 Wang Jiashe col. (FAU).

Species group placement: *ghesquierei*-group.

Distribution: China, India, Indonesia, Italy, Japan, UK, Vietnam.

Host: *Scirtothrips dorsalis* Hood.

DNA data: 28S: 1 sequence (Vietnam).

Etymology: Named for the ancient village of Rivello in Italy, where this extremely widespread species was first discovered. As noted above under *M. deflectum*, one paratype (male "allotype") of that species is in fact *M. rivelloi*.

Megaphragma tamoi **Polaszek, Fusu, and Viggiani sp. nov.** (Figure 21d,e)

urn:lsid:zoobank.org:act:8F82A348-AFC5-4618-A0C8-416867AC501C

Description. *Female*: Antenna (Figure 21d) five-segmented (excluding anellus); funicle absent; hence, clava three-segmented, with **C1** and **C2** almost fused; **C1** without **UST**; **C2** with one elongate **UST**, reaching more than half the length of **C3**; **C3** with **MPS**, **SB**, and **SS**.

Mid lobe of mesoscutum with longitudinal striate sculpture extending to scutellum (cf Figure 14b); propodeum with central area extended posteriorly, crenulae present (cf Figure 15e). **T2–T4** without setae laterally. Ovipositor 1.7× as long as mesotibia. Mesotibia with one large spine basally; metafemur with spine (cf Figure 16c, upper). Fore wing 6× as long as the maximum width, maximum distal width 1.2× maximum basal width; disc with a single long seta (Figure 21e), longest fringe seta 3× maximum discal width. Marginal vein with two setae centrally, equal in length, extending to the end of the marginal vein. Stigmal vein moderately enlarged, with two sensilla apically.

Largely brown, the following paler: legs except coxae and metafemur. Pedicel pale; scape, **C1–C3** darker. Fore wing strongly infuscate basally; stigmal and marginal vein brown; marginal vein very dark centrally.

Male: Largely as in female. **C1** and **C2** with scattered **SS**; **C2** with 2–3 **MT** apically; **C3** with long apical and ventral **UST**. **C3** is darker than preceding segments.

Material examined. Holotype ♀ (deposited NHMUK). **BENIN**: Agbotagon, 6°49′ N 2°12′ E, 22.iv.1990, M. Tamo 351.

Paratypes: **BENIN**: same data as holotype (8♀, DACE, IITA); Mono Province, 7.x.1988, M. Tamo, D-Vac on cowpea *Vigna unguiculata* (1♀, NHMUK); Mono Province, 25.ii.1988, M. Tamo, emergence cage *Megalurothrips sjostedti* on *Vigna unguiculata* (1♀, IITA); Dept Zou, Zogbodomey, 22.i.1990, ex egg *Megalurothrips sjostedti*, M. Tamo col. 275, 342 (2♀, IITA); Cotonou, IITA Station, 16.i.1990, ex egg *Megalurothrips sjostedti* on *Pueraria*, M. Tamo col. 336 (1♀, NHMUK).

Species group placement: *ghesquierei*-group.

Distribution: Benin.

Host: *Megalurothrips sjostedti*.

DNA data: no DNA sequences.

Etymology: Named for our colleague and friend Manu Tamo (IITA, Benin), collector of many *Megaphragma* specimens.

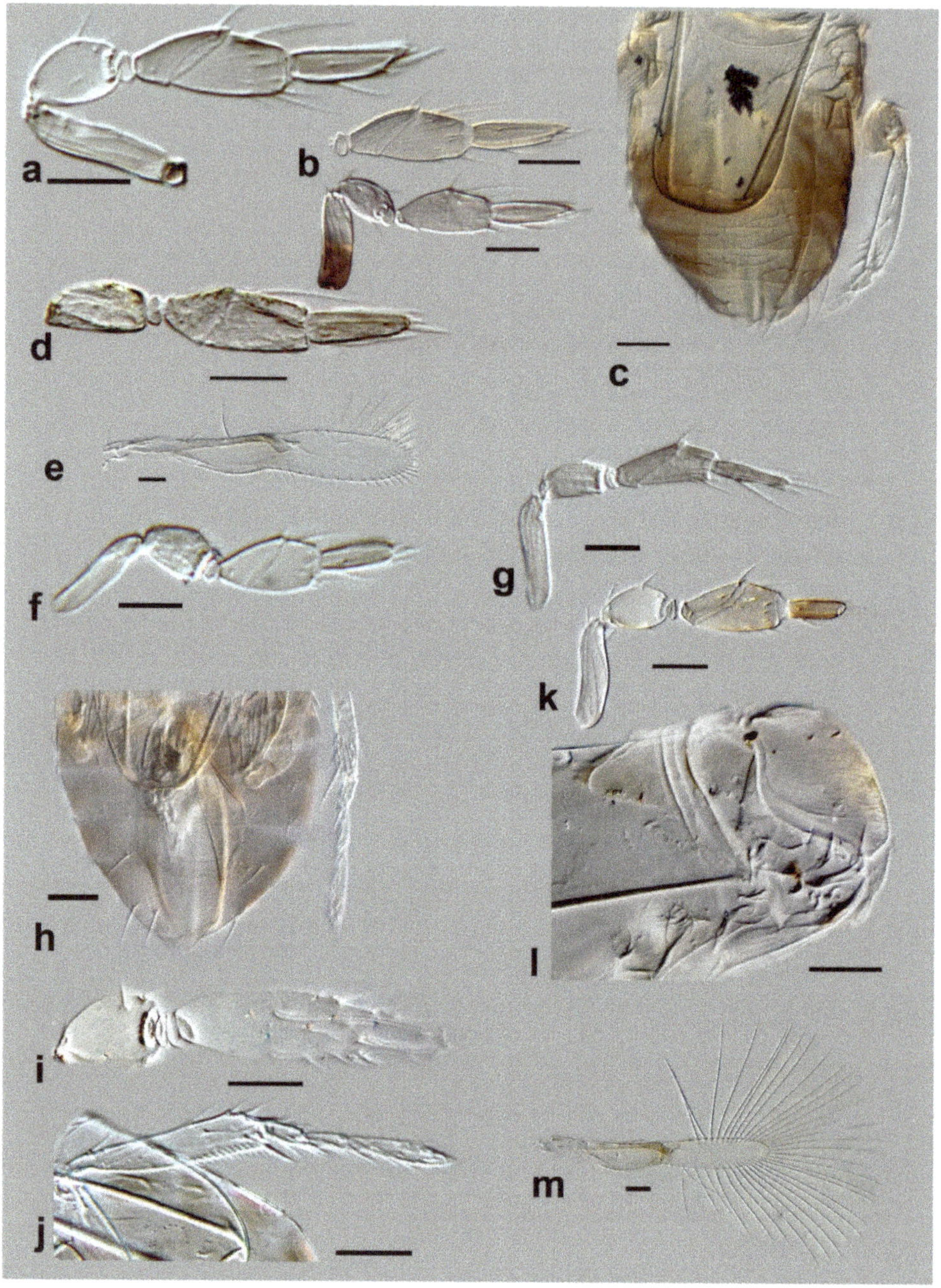

Figure 21. Photographs of *Megaphragma* species: (**a**) *M. rivelloi*, female antenna (Holotype); (**b**) *M. striatum*, female antenna (MXM1 above and Paratype below); (**c**) *M. striatum*, female metasoma (MXM1); (**d**) *M. tamoi*, female antenna (Holotype); (**e**) *M. tamoi*, female fore wing (Holotype); (**f**) *M. tridens*, female antenna (Holotype); (**g**) *M. tridens*, male antenna (Paratype); (**h**) *M. tridens*, female metasoma (Holotype); (**i**) *M. uniclavum*, female antenna (Holotype); (**j**) *M. uniclavum*, female metatibia (Holotype); (**k**) *M. vanlentereni*, female antenna (Holotype); (**l**) *M. vanlentereni*, female mesosoma (Holotype); (**m**) *M. vanlentereni*, female fore wing (Holotype). Scale bars 20 μm.

Megaphragma tridens Fusu and Polaszek sp. nov. (Figure 21f–h)

urn:lsid:zoobank.org:act:C77515FA-E1EA-4C74-B7DA-CE244A2BD05F

Description. Female: Antenna (Figure 21f) five-segmented (excluding anellus); funicle absent; hence, clava three-segmented, with **C1** and **C2** almost fused; **C1** apparently without **UST**; **C2** with 1 prominent **UST**, abundant **MT**, and 1 apical **MPS**; **C3** with **MT**, 2-3 **UST**, **SB**, and **SS**.

Mid lobe of mesoscutum with longitudinal striate sculpture (cf Figure 14b); propodeum with central area extended posteriorly, crenulae absent (cf Figure 21l). **T1** with elongate cells laterally, 2–3× as long as wide (cf Figure 12b); **T2–T4** with a short, robust seta near lateral margin; **T5** centrally with subparallel striations and with long setae laterally (Figure 21h). Ovipositor 1.7× as long as the mesotibia. Mesotibia with one large spine basally; metafemur with spine. Fore wing 8.5× as long as the maximum width, maximum distal width equal to maximum basal width; disc with a single long seta (cf Figure 12c), longest fringe seta 3.5× as long as maximum discal width. Marginal vein with one long seta centrally, extending to the end of the marginal vein. Stigmal vein moderately enlarged, with four sensilla apically.

Body largely brown, the following paler: legs except coxae and metafemur. Pedicel pale; scape, **C1–C3** darker. Fore wing slightly infuscate basally; stigmal and marginal vein brown.

Male: Largely as in female. Antennal clava (Figure 21g) with **C1** and **C2** with scattered **SS**; **C2** with 2-3 **MT** apically; **C3** with long apical and ventral **UST**.

Material examined. Holotype ♀ (deposited in NHMUK). **COSTA RICA**: Puntarenas, La Gamba Biol. Sta., 150 m, 8°42′ N 83°12′ W, 13–14.ii.2008, J.S. Noyes BMNH(E) 2010-21AQ, DNA: CRM1.10.

Paratypes: **COSTA RICA**: same data as holotype, DNA: CRM1.5, 1.8 (1♀, 1♂, AICF, MZUCR); Limón, Hitoy-Cerere Reserve, 9°40′ N 83°02′ W, 100 m, 24–26.ii.2008, J.S. Noyes NHM(E) 2010-21AQ, DNA: CRM 3.41, 3.102 (1♀ 1♂, AICF, NHMUK).

Species-group placement: *ghesquierei*-group.

Distribution: Costa Rica.

DNA data: no DNA sequences.

Etymology: From the Latin word *tridens*, in reference to the three apical sensilla on **C3**, resembling a longer central and two shorter lateral teeth of a trident (from Latin *tri*=three and *dens*=tooth). Noun in apposition.

Megaphragma uniclavum Polaszek and Fusu sp. nov. (Figure 21i,j)

urn:lsid:zoobank.org:act:1F77A0D1-B9C0-46D2-A553-3176AAF76708

Description. Female: Antenna (Figure 21i) four-segmented (excluding anellus); transverse funicle present; clava one-segmented (unique so far for the genus, though *antecessor* approaches this condition); 2 **UST**; 3–4 long **MPS** extending beyond half the clava.

Head with few features discernible due to mounting position, and extensively obscured by eye pigment. Mid lobe of mesoscutum with fine longitudinal striation (cf Figure 14b); vertical/ventral anterior mid lobe of mesoscutum with coarse, reticulate sculpture (cf Figure 3g); propodeum elongate centrally, longitudinally striate with three large crenulae. **T2–T4** with very long setae laterally, each longer than its tergum; **T3** and **T4** with large irregular cells laterally. Mesotibia with two large spines basally; metafemur with spine; metatibia with a row of fine, elongate setae extending almost their entire inner length (Figure 21j). Fore wing 5× as long as maximum width; longest fringe seta 6.7× as long as maximum discal width, maximum distal width versus maximum basal width is unclear; disc (cf Figure 11b) with single short seta. Other details of disc and venation are unclear. Stigmal vein with a row of three sensilla apically. Ovipositor 2× as long as the mesotibia.

Body uniformly pale. Distal metasoma darker. Wings hyaline.

Male: Unknown.

Material examined. Holotype ♀ (deposited in NHMUK). **COSTA RICA**: Heredia, La Selva BS., 10°26′ N 84°01′ W, 75 m, 28–29.ii.2008, J.S. Noyes BMNH(E) 2010-21 AQ.

Species-group placement: *antecessor*-group. *Megaphragma uniclavum* is so far unique for the genus having a single claval segment. Despite the very unusual fore-wing structure, it is clearly affiliated with *M. antecessor* (similar mesoscutal sculpture and row of setae on metatibia).

Distribution: Costa Rica.

DNA data: no DNA sequences.

Etymology: This species is so far the only *Megaphragma* known with a single claval segment; hence, the name *uniclavum*. Noun in apposition.

***Megaphragma vanlentereni* Polaszek and Fusu sp. nov.** (Figure 21k–m)

urn:lsid:zoobank.org:act:E76B71C9-D7A7-4005-A7E1-66EC26A146BB

Description. Female: Antenna (Figure 21k) five-segmented (excluding anellus); funicle absent; hence, clava three-segmented, with **C1** and **C2** almost fused; **C1** with 2 **UST**; **C3** with three **MPS** extending beyond apex of clava.

Mid lobe of mesoscutum (Figure 21l) with irregular longitudinal striate sculpture; propodeum (Figure 21l) elongate, curved centrally and posteriorly; crenulae present. **T1** without cells; **T2–T4** without setae laterally. Ovipositor 1.7× as long as mesotibia. Mesotibia with two large spines basally. Fore wing (Figure 21m) 6.5× as long as maximum width, maximum distal width 1.1× maximum basal width; disc with one short seta, longest fringe seta 4× as long as maximum discal width. Marginal vein with two long setae centrally, of equal length. Stigmal vein moderately enlarged, with three sensilla apically.

Body largely pale with dorsal mesosoma, including propodeum, pale brown. Scape and pedicel pale, **C1** and **C2** brown, **C3** very dark brown in contrast. Fore wing infuscate basally; stigmal and marginal vein distally brown.

Male: Unknown.

Material examined: Holotype ♀ (deposited in NHMUK). **MALAYSIA**, Sabah, Maliau Basin Studies Centre, Knowledge Trail, 04°44′ N 116°58′ E, 22.ix.2012, A. Polaszek, screen-sweep, NHM(E) 2010-21, DNA: SAM3.

Species group placement: *ghesquierei*-group.

Distribution: Malaysia (Borneo, Sabah).

DNA data: CO1: one sequence; 28S: one sequence (both Malaysia, Sabah).

Etymology: Named for our colleague and friend Joop van Lenteren, a pioneer of biocontrol, especially in greenhouses.

***Megaphragma viggianii* Fusu, Polaszek, and Polilov sp. nov.** (Figure 20c,d and Figure 22a–d)

urn:lsid:zoobank.org:act:05F51567-04A0-45D8-91F7-787A041BEB08

Megaphragma amalphitanum: Nedoluzhko et al. 2016. *Mitochondrial DNA Part A* 27(6): 4526–4527; Polilov, 2016. *At the Size Limit—Effects of Miniaturization in Insects*; Nedoluzhko et al., 2017. *Genom. Data* 11: 87–88; Polilov, 2017. *PLoS ONE* 12(5): e0175566; Prokhortchouk et al., 2017. *Mosc. Univ. Biol. Sci. Bull.* 72(1): 30–32; Diakova et al., 2018. *PeerJ* 6: e6005; Sharko et al., 2019. *PLoS ONE* 14(12): e0226485; Polilov et al., 2021. *Sci. Rep.* 11(1): 4717; Boudinot et al., 2020. *J. Zool.* 313(2): 99–113; Diakova and Polilov, 2021. *J. Hymenopt. Res.* 84: 69–73 (misidentifications).

Description. Female: Antenna (Figures 20d and 22a) five-segmented (excluding anellus); pedicel 2× as long as funicle; funicle slightly longer than wide; **C2** longer than **C1**; **C1** with 1 ASC, 2 **MT**; **C2** with 13 **MT**, 2 **UST**; **C3** with 2 **MPS**, 1 **SS**, and 3 **UST** extending beyond clava tip.

Mid lobe of mesoscutum (Figure 22b) posteriorly smooth, anteriorly with large, coarse reticulation; propodeum with broad, truncate hind margin, with two widely separated lobes distally and laterally, without crenulae. **T1** with a central "V" composed of minute denticles, a row of coarser denticles laterally (Figure 20c), and one elongate cell laterally, about 2× as long as wide. Metasoma dorsally with rows of denticles laterally on **T2** and **T3**; **T2–T4** with moderately long setae laterally. Ovipositor as long as the mesotibia.

Mesotibia without spines basally, a single seta present; metafemur without spine but with three robust setae; metatibia with a row of five or six fine setae extending along half its inner length. Fore wing (Figure 22c) 6× as long as maximum width; maximum distal width 1.5× maximum basal width and more than 2× width measured at the apex of the marginal vein; longest fringe seta 4× as long as the maximum discal width; disc with setae in one or two rows of 4–6 setae, and setae on ventral surface long, the penultimate one reaching to the base of the distal (Figure 22d). Marginal vein with two setae centrally, the proximal approximately 2× the length of the distal. Stigmal vein has four sensilla apically.

Head and metasoma dark brown; central/posterior mesoscutal mid lobe and lateral scutellum paler. Remainder of body, including legs, pale. Flagellum darker than remainder; wings hyaline.

Male: Antenna with pedicel almost as long as scape; funicle with two small setae; clava two-segmented; **C1** >2× **C2**; **C1** with three setae; **C2** with three flagelliform setae; two multiporous plate sensilla and a short terminal process present. Metasoma dorsally with rows of denticles laterally on **T1–T3**. One individual with funicle apparently fused with **C1**.

Material examined: Holotype ♀ (deposited in DACE). **ITALY**: Naples, Massa Lubrense, S. Agata sui Due Golfi, 24.x.2012, 40°36′ N 14°20′ E, 114 m, ex *Heliothrips haemorrhoidalis* in leaves of *Viburnum tinus* (G. Viggiani), DNA: ITM1.

Paratypes: **ITALY**: same data as holotype except DNA: ITM2 (1♂), ITM3 (1♂), ITM4 (1♂), ITM5 (1♂), ITM6 (1♀), ITM7 (1♀) (AICF, NHMUK); same data but without DNA extraction codes (2♀ NHMUK); Vietri sul mare, Benincasa, 40°40′ N 44°20′ E, 17.vii.2013, ex *Heliothrips haemorrhoidalis* on *Viburnum tinus* (G. Viggiani), DNA: ITM13 (1♀, NHMUK); Liguria, Santa Margherita, iv.2015, A. Polilov, ex *Heliothrips haemorrhoidalis* on *Viburnum tinus*, DNA1095, DNA1096 (2♀, NHMUK). **GREECE**: Koutsoupia, 39.81N 22.80E, 22.viii.2016, Leg. A.A. Polilov, ex eggs *Heliothrips haemorrhoidalis* on *Arbutus* sp. (2♀, 1♂, AICF).

Species-group placement: *longiciliatum*-group.

Distribution: Greece, Italy.

Host: *Heliothrips haemorrhoidalis*.

DNA data: CO1: eight sequences; 28S: eight sequences.

Etymology: Named for Professor Gennaro Viggiani, specialist of microhymenoptera and biocontrol, and instigator and co-author of this paper.

Comments: Found in Italy at the same locality and on the same host as *M. polilovi*. According to the CO1 sequence of the mitochondrial genome deposited on GenBank by Nedoluzhko et al. [23], their species is not *M. amalphitanum* (=*M. longiciliatum*) but *M. viggianii*. Our molecular analysis retrieves the two as sister species. Both have the mid lobe of the mesoscutum reticulate anteriorly, and long lateral setae on **T2–T4**, though these are longer in the former. In *M. longiciliatum* **T2–T4** have very long setae, at least as long as the tergite (Figure 5h), whereas in *M. viggianii* **T2–T4** have shorter setae, much shorter than the tergite (cf Figure 8f).

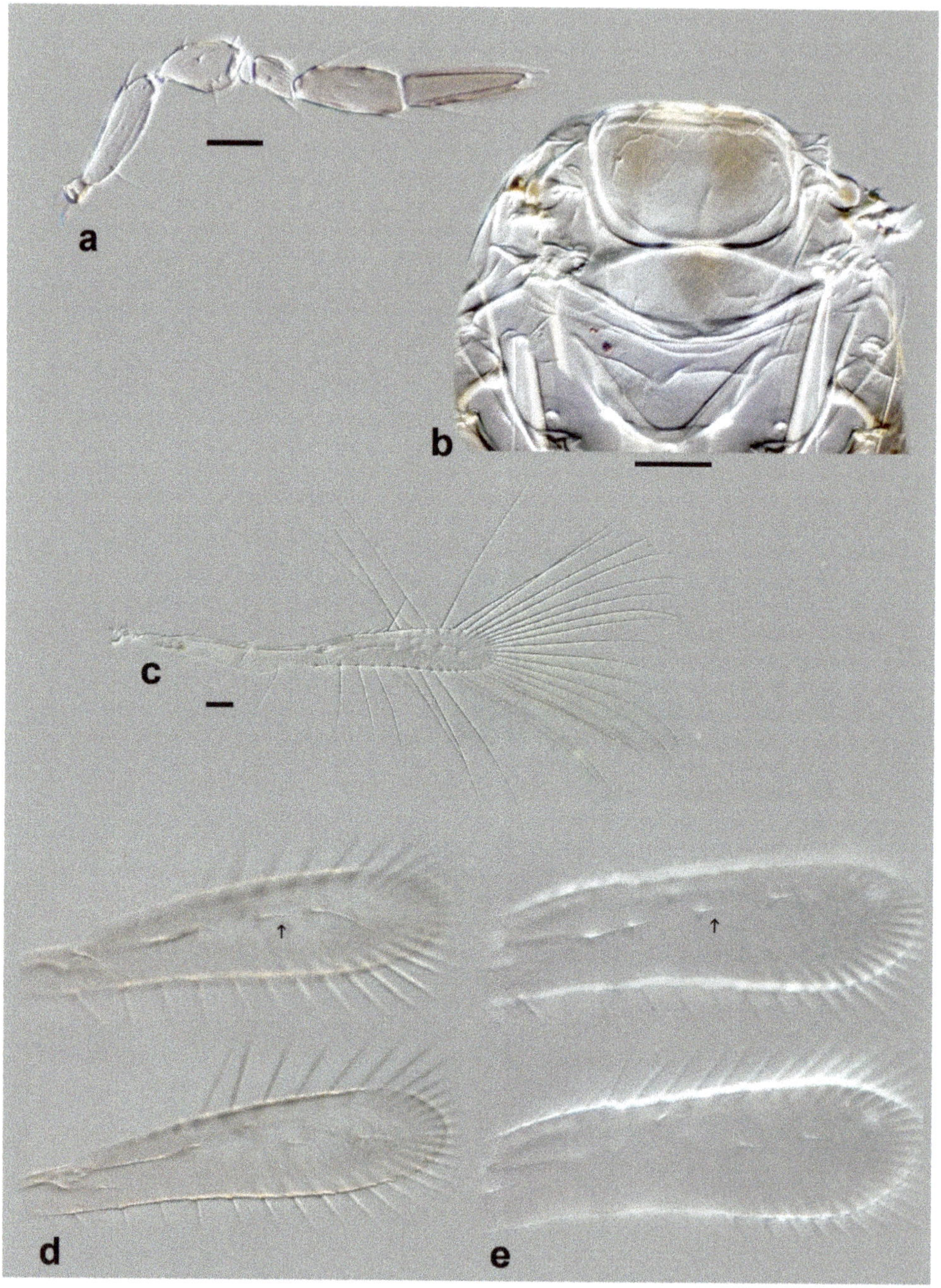

Figure 22. Photographs of *Megaphragma* species: (**a**) *M. viggianii*, female antenna (Holotype); (**b**) *M. viggianii*, female mesosoma (Holotype); (**c**) *M. viggianii*, female fore wing (Holotype); (**d**) *M. viggianii*, female ventral (above) and dorsal (below) wing surface, (Holotype); (**e**) *M. priesneri*, female ventral (above) and dorsal (below) wing surface (Neotype). Arrows point at the penultimate seta on ventral surface. Scale bars 20 μm.

Key to species of *Megaphragma*

1 Fore wing with a single discal seta (Figures 9h and 10c) or without any discal setae (Figures 3f and 4a) ... 2

- Fore wing with at least one line of 3 or more setae (Figures 3c and 5d) ... 17

2 Fore wing without discal seta (Figure 3f) ... 3

- Fore wing with one discal seta (Figure 9h) ... 6

3 Mid lobe of mesoscutum entirely with regular longitudinally striate sculpture (Figure 3d) ... 4

- Mid lobe of mesoscutum with reticulate sculpture anteriorly (Figure 3g), or with irregular longitudinal sculpture (Figure 13b) ... 5

4 Female: ovipositor more than 1.5× as long as mesotibia. Male: **C3** with terminal sensillum less than 2× length **C3** ... ***deflectum***

- Female: ovipositor less than 1.5× as long as mesotibia. Male: **C3** with terminal sensillum more than 2× length **C3** ... ***rivelloi***

5 Mid lobe of mesoscutum with reticulate sculpture anteriorly (Figure 3g) ... ***ghesquierei***

- Mid lobe of mesoscutum with irregular longitudinal sculpture (Figure 13b) ... ***chienleei***

6 Fore wing with discal seta short, ≤distance between 2 proximal wing fringe setae (Figure 11b) ... 7

- Fore wing with discal seta long, ≥distance between 3 proximal wing fringe setae (Figure 10c) ... 12

7 Female: ovipositor strongly exserted, more than 3× mesotibia (Figure 17b) ... ***momookherjeeae***

- Ovipositor slightly or not exserted, less than 2.5× mesotibia ... 8

8 Clava 1-segmented (Figure 21i). Marginal vein with both central setae minute ... ***uniclavum***

- Clava 2- or 3-segmented (if 2-segmented, **C1** partially fused with **C2**). Marginal vein with at least one of the central setae elongate ... 9

9 Marginal vein with 2 central setae, the distal one minute, less than 0.2× as long as the proximal seta (Figure 15f) ... ***hansoni***

- Marginal vein with 2 central setae, either subequal in length (Figure 11b), or the distal one about 0.5× as long as the proximal seta (Figure 16d) ... 10

10 Funicle present, transverse (Figure 11a); clava 2-segmented ... ***antecessor***

- Funicle absent; clava 3-segmented ... 11

11 Mid lobe of mesoscutum entirely with evident regular longitudinally striate sculpture while reticulate cells hardly visible (Figure 21l) ... ***vanlentereni***

- Mid lobe of mesoscutum entirely with fine regular longitudinally striate sculpture, surface divided into obvious reticulate cells (Figure 16b) ... ***liui***

12 Marginal vein with one central seta (Figure 12c) ... 13

- Marginal vein with two central setae ... 15

13 **T2–T4** of metasoma each with a robust seta near lateral margin; **T5** centrally with subparallel striations (Figure 21h). Male: **C3** elongate, length more than 3× basal width, with very long sensilla (Figure 21g) ... ***tridens***

- **T2–T4** of metasoma without a seta near lateral margin; **T5** centrally with reticulate sculpture (Figures 12b and 21c). Male: **C3** elongate or short, with less prominent sensilla (Figure 10b) ... 14

14 **T2–T5** each with rows of prominent denticles laterally; **T3** and **T4** centrally with reticulate sculpture (Figure 21c). Female: **C3** long (Figure 21b). Male: **C3** long, length more than 2× basal width (Figure 10b) ... ***striatum***

- **T2–T5** with denticles reduced to faint protrusions, especially on **T3** and **T4**; **T3** and **T4** centrally with sculpture consisting of faint striae (Figure 12b). Female: **C3** short (Figure 12a). Male: **C3** short, length less than 2× basal width ... ***breviclavum***

15 Clava 2-segmented ... ***stenopterum***

- Clava 3-segmented ... 16

16 Marginal vein proximal central seta 2–3× length of distal central seta (Figure 21e) . . . . ***tamoi***

- Marginal vein proximal central seta about 7× longer than distal seta (Figure 14c) . . . . ***digitatum***

17 Antenna 4-segmented (excluding anellus) **18**

- Antenna 5-segmented (excluding anellus) **19**

18 Female: **C1** with 2 very short **UST**, much shorter than **C1** (Figure 6a). Male: **C1** 2× as long as maximum width . . . ***macrostigmum***

- Female: **C1** with 2 long **UST**, each as long as **C1** (Figure 3a). Male: **C1** 3× as long as maximum width . . . . ***caribea***

19 **T1** with characteristic sculpture consisting of distinct cells with denticulate margins (Figures 1a, 17f and 18f) **20**

- **T1** various, but not as above; elongate cells may be present on sides of **T1** (Figures 12b and 16b) **24**

20 Female: **T2** sculpture with many closed cells (Figure 17f); ovipositor 1.7× mesotibia . . . . ***nowickii***

- Female: **T2** sculpture without closed cells (occasionally a single cell in noyesi); ovipositor 1.3× mesotibia or shorter **21**

21 Mid lobe of mesoscutum reticulate with comparatively large cells (Figure 14h) . . . . ***funiculatum***

- Mid lobe of mesoscutum smooth, or appearing longitudinally striate **22**

22 Female: **C1** trapezoid in lateral view, length 1.5× maximum width or less. Male: funicle trapezoid, slightly longer than wide . . . . ***mymaripenne***

- Female: **C1** parallel sided and elongate, 2× as long as wide or more. Male: funicle elongate, approximately 2× as long as wide (male unknown in *M. polilovi*) **23**

23 Scape 4× maximum width; radicle concolorous with scape, both pale (Figure 18a); ovipositor length 1× mesotibia . . . . ***noyesi***

- Scape 6× maximum width; radicle darker compared to scape (Figure 18e); ovipositor length 1.3× mesotibia . . . . ***polilovi***

24 Mid lobe of mesoscutum coarsely reticulate over most of its surface (Figures 13e and 14e) **25**

- Mid lobe of mesoscutum reticulate only anteriorly (Figure 3g), or sculpture different (smooth Figure 4f, striate Figure 3d) . . . **27**

25 Fore wing with base hyaline (not infuscate; Figure 14f) . . . . ***fanenitrakely***

- Fore wing with base infuscate (Figure 13f) **26**

26 **C2** with apical **SB** only slightly shorter than **C2** (Figure 13d) (male unknown) . . . . ***cockerilli***

- **C2** with apical **SB** less than $\frac{1}{2}$ length of **C2** (Figure 8a) . . . . ***polychaetum***

27 **T2–T4** with very long setae laterally (Figure 5h), at least as long as the tergite **28**

- **T2–T4** with short setae laterally (Figure 8f), much shorter than the tergite **29**

28 **T1** with long lateral setae. Female: funicle with long, robust **UST** (Figure 15c) . . . . ***giraulti***

- **T1** without long lateral setae (Figure 5h). Female: funicle without long, robust ventral **UST** (Figures 5e and 16e) . . . ***longiciliatum***

29 Mid lobe of mesoscutum reticulate anteriorly (Figures 8f and 22b) **30**

- Mid lobe of mesoscutum with different sculpture (Figure 18c), or smooth **31**

30 Fore wing with maximum distal width <2× width measured at apex of marginal vein (Figure 8g); setae on ventral fore wing disc short, penultimate one not reaching to the base of the distal (Figure 22e) . . . . ***priesneri***

- Fore wing with maximum distal width >2× width measured at apex of marginal vein (Figure 22c); setae on ventral fore wing disc long, penultimate one reaching to the base of the distal (Figure 22d) . . . . ***viggianii***

31 Mid lobe of mesoscutum smooth. Female: proximal **UST** attached close to the mid point of **C1**, shorter than **C1** (Figure 5b) . . . . ***kinuthiae***

- Mid lobe of mesoscutum with anterior striae (Figure 18c). Female: proximal **UST** attached close to the base of **C1**, longer than **C1** (Figure 18b) . . . . ***pintoi***

4. Discussion

The present study is the culmination of more than 10 years of intensive collecting and examination of several thousand *Megaphragma* specimens from all over the world, including type material of all but one of the previously described species. Without the molecular dimension, our conclusions would have been very different. For example, the separation of *M. noyesi* and *M. polilovi* from *M. mymaripenne* would not have been possible, and these species have been confused in the past. Within the *ghesquierei*-group, morphological differences between species that are very distinct based on DNA, are completely undetectable in many cases. Some of this is no doubt due to the limitations of light microscopy, even when using techniques such as Nomarski differential interference contrast (DIC), coupled with focus stacking. Morphological evolution at species level is apparent for several structures, most notably the antenna, but also the setae and spines on the middle and hind legs, the fore wing, the structure of the propodeum, and that of metasomal terga showing variation in microsculpture and chaetotaxy (of diagnostic value). Species can be relatively easily grouped based on features of these morphological characters, and the *ghesquierei*-group, in particular, can be defined based on features of three characters: fusion of the funicle with the clava, development of the central propodeum, and metafemoral spine. The *mymaripenne*-group is less easily defined, with the loss of the otherwise characteristic sculpture of **T1** having occurred in the *longiciliatum* subgroup. The shape of the fore wing, and its discal and marginal ciliation appear to be critical in reflecting species evolution in this genus. Future studies should carefully assess the setation of the upper and underside of the fore wing, something that is very difficult once the specimen has been slide-mounted. No doubt future studies, including more scanning electron microscope imaging, will reveal additional patterns of morphological variation in the genus.

Perhaps the most surprising discovery of this study is the extraordinary distribution of some species. *Megaphragma longiciliatum*, under our new and broader definition, is found from Southeast Asia to Northwest Europe, as well as in North America, the Congo, and the Middle East. In this respect, as well as in terms of their physical size, the *Megaphragma* species parallel some Protozoa. The phrase "everything is everywhere, but the environment selects" [69], originally applied to Protozoa, certainly seems to apply to several *Megaphragma* species. Previous theories attempting to explain ubiquitous distributions of particular species of organisms have attributed this to their large population sizes, rather than to any inherent properties of such groups [70]. This argument appears to be so entirely back to front (i.e., "some species are cosmopolitan because they have huge populations") that it can be easily dismissed. It is precisely the inherent property of minuteness, among other attributes discussed below, that is the main reason for these species having cosmopolitan distributions. Minuteness is directly related to dispersive ability, which can be largely passive for minute organisms; although, *Megaphragma* are known to be good at directional flight [28]. Our study, despite being very patchy in terms of the sample sizes of most species, suggests a mixed pattern of dispersal and distribution, with ubiquitous species as well as apparently endemic ones, as shown previously for about 200 Protozoa species [71]. Minuteness also directly affects the relationship between the species and its immediate microhabitat. We can assume that for an organism whose adults are around one-quarter of a millimeter, and whose developmental stages are entirely within a closed environment (the thrips egg), the macroecological factors of climate and temperature are less important, at least for some species. Hence, e.g., *M. rivelloi* appears to be as suited to the humid rainforests of SE Asia as it is to the much drier countryside of Southern England, and the same must be true for other species. Humidity, and especially avoidance of desiccation, are critical for the survival of minute terrestrial organisms, and *Megaphragma* species are known to have very thin cuticle [18] (see also Supplementary Figure S1). Even if air masses could transport minute hymenopterans quickly across the globe, desiccation would be a major impediment [72]. However, given that for most of their adult life *Megaphragma* species are likely to be in close proximity to living plant tissue, and hence access to moisture, it is probable that these immediate microclimatic conditions override the macroecological conditions already mentioned, and hence some of the remarkable latitudinal distributions of, e.g., *M. longiciliatum*, *M. pintoi*, and *M. rivelloi*. As well as both direct and passive colonization of new geographical areas, human movement of plant material containing thrips eggs, and *Megaphragma* life stages is inevitable, and has certainly contributed to a large extent to the cosmopolitan distribution of some species. Adaptation to, and dependence on, microhabitat conditions, especially humidity, have undoubtedly also contributed to their human-mediated distribution through the movement of plant material. Another explanation for species of egg parasitoids having extremely wide distributions has been their assumed defencelessness, and the relative uniformity of their hosts [73]. Whilst the former appears to be generally true, the latter clearly is not. Parasitoids that parasitize eggs are in some cases able to additionally parasitize Lepidoptera larvae, braconid cocoons, and even act as hyperparasitoids of other egg parasitoids. This extreme range of hosts has been reliably documented in *Centrodora darwini* (Girault) [74]. Other egg parasitoids appear to be extremely specific in which species they will either attack and/or develop successfully on, and this is particularly true of many species in the hyperdiverse genus *Telenomus* (e.g., [75]). Thus, a huge range of levels of host species specificity exists across egg parasitoids. In several taxa, eggs are entirely free of any parasitoids (aphids, whiteflies, and scale insects), while their (relatively) close relatives (leafhoppers and their relatives) are very heavily parasitized [76]. Clearly, the ability to evade or resist being parasitized at the egg stage is highly heterogeneous across the insects. In the case of *Megaphragma*, it is difficult to tell how host specific they are; in the few cases where there are rearing records from more than one host, they belong to different genera of the same family. Most likely as in other groups of parasitoid wasps, members of *Megaphragma* are a mix of generalists and specialist species [77].

As stated above, in several lineages *Megaphragma* has diversified to produce numerous cryptic species, many of which appear indistinguishable morphologically, at least using the techniques employed in this study. All future studies of these and similar organisms must rely to an extent on DNA data, and it may be that species will be described solely on differences in DNA where these differences can be demonstrated to be reasonably accurate proxies for biological species distinctions.

Supplementary Materials: The following supporting information can be downloaded at: https://www.mdpi.com/article/10.3390/insects13060561/s1, Figure S1: Confocal laser micrograph of *M. longiciliatum* (female from Oman). Blue areas are the least sclerotized and yellow the most strongly sclerotized (Photo A. P.); Figure S2: Phylogenetic tree of CO1 in *Megaphragma*, unpartitioned analysis in RAxML-NG; Figure S3. Phylogenetic tree of 28S in *Megaphragma*, unpartitioned analysis in RAxML-NG; Figure S4: a *Heliothrips haemorrhoidalis*, the host of *Megaphragma* sp., b eggs of *Heliothrips haemorrhoidalis* with exit hole of *Megaphragma* sp., c pupa of *Megaphragma* sp. in egg of *Heliothrips haemorrhoidalis* (Photo G. Viggiani).

Author Contributions: Conceptualization, A.P., L.F. and G.V.; methodology, A.P., L.F., A.H. and A.A.P.; validation, A.P. and L.F.; formal analysis, A.P. and L.F.; investigation, A.P., L.F., G.V., A.H., P.H. and A.A.P.; resources, A.P., A.A.P., P.H. and G.V.; data curation, A.P. and L.F.; writing—original draft preparation, A.P., L.F., G.V. and A.A.P.; writing—review and editing, A.P., L.F., G.V., A.H., P.H. and A.A.P.; visualization, A.P., L.F., G.V. and A.A.P.; project administration, A.P., P.H. and G.V.; funding acquisition, A.P., G.V. and A.A.P. All authors have read and agreed to the published version of the manuscript.

Funding: This research was funded by the UK Royal Society under International Joint Project with grant number JP100330/2010 awarded to A.P. and G.V.; A.A.P. gratefully acknowledges financial support from the Russian Foundation for Basic Research (project no. 20-04-00642).

Institutional Review Board Statement: Not applicable.

Informed Consent Statement: Not applicable.

Data Availability Statement: All sequences were deposited in GenBank. Voucher specimens are deposited in the collections listed under Specimens and depository abbreviations. All data and materials will be publicly available.

Acknowledgments: Claire Griffin at the NHM sequencing facility carried out all Sanger sequencing. Kiara van der Kroon-Polaszek enhanced many of the line drawings, all of which were originally drawn by Gennaro Viggiani. John Huber (Canadian National Collection of Insects, Arachnids and Nematodes, Agriculture and Agri-Food Canada), Zachary Lahey (Department of Evolution, Ecology, and Organismal Biology, The Ohio State University), and an anonymous reviewer are acknowledged for their constructive suggestions that greatly improved the manuscript.

Conflicts of Interest: The authors declare no conflict of interest.

Appendix A

Details of the specimens used for the molecular analyses are provided in Table A1.

Table A1. Voucher specimens, their repository, and accession numbers for the 28S and CO1 sequences.

Nr	Voucher	Species	Country	Type Status	Repository	28S		CO1	
						bp + gaps	Accession nr	bp	Accession nr
1	**COM1 1**	*M. nowickii*	D. R. Congo	Paratype	DACE	1095 (212 indels)	ON555486	295	ON557406
2	**COM1 2**	*M. nowickii*	D. R. Congo	Paratype	DACE	1095 (211 indels)	ON555494	652	ON557417
3	**COM1 3**	*M. nowickii*	D. R. Congo	Paratype	DACE	1095 (211 indels)	ON555502	652	ON557426
4	**COM1 4**	*M. nowickii*	D. R. Congo	Paratype	DACE	1000 (211 indels)	ON555503	652	ON557427
5	**COM1 5**	*M. nowickii*	D. R. Congo	Paratype	DACE	1092 (211 indels)	ON555504	652	ON557428
6	**COM1 6**	*M. nowickii*	D. R. Congo	Paratype	DACE	1085 (211 indels)	ON555505	652	ON557429
7	**COM1 7**	*M. nowickii*	D. R. Congo	Paratype	NHMUK	-	-	592	ON557430
8	**COM1 8**	*M. nowickii*	D. R. Congo	Paratype	NHMUK	1087 (212 indels)	ON555506	652	ON557431
9	**COM1 9**	*M. nowickii*	D. R. Congo	Paratype	NHMUK	1095 (211 indels)	ON555507	366	ON557432
10	**COM1 10**	*M. nowickii*	D. R. Congo	Paratype	NHMUK	1087 (211 indels)	ON555487	652	ON557407
11	**COM1 11**	*M. nowickii*	D. R. Congo	Paratype	NHMUK	414 (121 indels)	ON555488	652	ON557408
12	**COM1 12**	*M. nowickii*	D. R. Congo	Paratype	AICF	1093 (212 indels)	ON555489	652	ON557409
13	**COM1 13**	*M. nowickii*	D. R. Congo	Paratype	DACE	1048 (212 indels)	ON555490	652	ON557410
14	**COM1 14**	*M. nowickii*	D. R. Congo	Paratype	DACE	1049 (209 indels)	ON555491	652	ON557411
15	**COM1 15**	*M. nowickii*	D. R. Congo	Paratype	DACE	-	-	652	ON557412
16	**COM1 16**	*M. nowickii*	D. R. Congo	**Holotype**	NHMUK	1106 (211 indels)	ON555492	652	ON557413
17	**COM1 17**	*M. nowickii*	D. R. Congo	Paratype	DACE	-	-	652	ON557414
18	**COM1 18**	*M. nowickii*	D. R. Congo	Paratype	DACE	1106 (211 indels)	ON555493	652	ON557415
19	**COM1 19**	*M. nowickii*	D. R. Congo	Paratype	DACE	-	-	652	ON557416
20	**COM1 20**	*M. nowickii*	D. R. Congo	Paratype	DACE	1106 (212 indels)	ON555495	652	ON557418
21	**COM1 21**	*M. nowickii*	D. R. Congo	Paratype	AICF	1106 (212 indels)	ON555496	652	ON557419
22	**COM1 22**	*M. nowickii*	D. R. Congo	Paratype	IITA	1106 (208 indels)	ON555497	652	ON557420
23	**COM1 23**	*M. nowickii*	D. R. Congo	Paratype	IITA	1106 (209 indels)	ON555498	522	ON557421
24	**COM1 24**	*Megaphragma* sp.	D. R. Congo	NA	NHMUK	-	-	369	ON557422
25	**COM1 25**	*Megaphragma* sp.	D. R. Congo	NA	NHMUK	684 (245 indels)	ON555499	369	ON557423
26	**COM1 26**	*M. nowickii*	D. R. Congo	Paratype	IITA	1106 (211 indels)	ON555500	652	ON557424
27	**COM1 27**	*M. nowickii*	D. R. Congo	Paratype	IITA	1106 (211 indels)	ON555501	652	ON557425
28	**COM2 1**	*M. longiciliatum*	D. R. Congo	NA	AICF	918 (206 indels)	ON555508	652	ON557433
29	**COM2 2**	*Megaphragma* sp.	D. R. Congo	NA	NHMUK	1045 (239 indels)	ON555509	391	ON557434

Table A1. *Cont.*

Nr	Voucher	Species	Country	Type Status	Repository	28S		CO1	
						bp + gaps	Accession nr	bp	Accession nr
30	**COM2 3**	*M. longiciliatum*	D. R. Congo	NA	NHMUK	896 (206 indels)	ON555510	-	-
31	**COM2 4**	*Megaphragma* sp.	D. R. Congo	NA	NHMUK	680 (124 indels)	ON555511	652	ON557435
32	**CRM1 4**	*M. hansoni*	Costa Rica	Paratype	NHMUK	534 (117 indels)	ON555513	-	-
33	**CRM1 16**	*M. hansoni*	Costa Rica	Paratype	NHMUK	1106 (255 indels)	ON555512	-	-
34	**CRM2 1**	*M. digitatum*	Costa Rica	**Holotype**	NHMUK	1062 (264 indels)	ON555514	-	-
35	**CRM2 2**	*M. antecessor*	Costa Rica	**Holotype**	NHMUK	647 (123 indels)	ON555518	366	ON557437
36	**CRM2 3**	*M. digitatum*	Costa Rica	Paratype	MZUCR	1106 (265 indels)	ON555519	652	ON557438
37	**CRM2 4**	*M. giraulti*	Costa Rica	**Holotype**	NHMUK	679 (158 indels)	ON555520	-	-
38	**CRM2 5**	*M. giraulti*	Costa Rica	Paratype	AICF	689 (159 indels)	ON555521	291	ON557439
39	**CRM2 6**	*M. digitatum*	Costa Rica	Paratype	MZUCR	1036 (264 indels)	ON555522	-	-
40	**CRM2 7**	*M. giraulti*	Costa Rica	Paratype	DACE	1048 (269 indels)	ON555523	291	ON557440
41	**CRM2 8**	*M. giraulti*	Costa Rica	Paratype	MZUCR	679 (158 indels)	ON555524	-	-
42	**CRM2 9**	*M. digitatum*	Costa Rica	Paratype	MZUCR	679 (151 indels)	ON555525	-	-
43	**CRM2 10**	*M. giraulti*	Costa Rica	Paratype	NHMUK	646 (158 indels)	ON555515	296	ON557436
44	**CRM2 11**	*M. giraulti*	Costa Rica	Paratype	MZUCR	490 (156 indels)	ON555516	-	-
45	**CRM2 12**	*M. digitatum*	Costa Rica	Paratype	AICF	646 (151 indels)	ON555517	-	-
46	**CRM3 4**	*M. hansoni*	Costa Rica	Paratype	NHMUK	938 (250 indels)	ON555530	-	-
47	**CRM3 17**	*M. hansoni*	Costa Rica	Paratype	NHMUK	1106 (255 indels)	ON555528	-	-
48	**CRM3 23**	*M. giraulti*	Costa Rica	Paratype	NHMUK	1043 (269 indels)	ON555529	-	-
49	**CRM3 40**	*M. hansoni*	Costa Rica	NA	lost	691 (374 indels)	ON555531	-	-
50	**CRM3 46**	*M. funiculatum*	Costa Rica	Paratype	AICF	1101 (233 indels)	ON555532	-	-
51	**CRM3 97**	*M. hansoni*	Costa Rica	Paratype	NHMUK	473 (137 indels)	ON555533	-	-
52	**CRM3 103**	*M. funiculatum*	Costa Rica	**Holotype**	NHMUK	761 (210 indels)	ON555526	-	-
53	**CRM3 104**	*M. hansoni*	Costa Rica	Paratype	NHMUK	1025 (255 indels)	ON555527	-	-
54	**DNA ECU1**	*M. digitatum*	Ecuador	Paratype	NHMUK	679 (152 indels)	ON555534	-	-
55	**DNA ECU4**	*M. digitatum*	Ecuador	Paratype	AICF	849 (321 indels)	ON555535	-	-
56	**DNA980**	*Megaphragma* sp.	D. R. Congo	NA	NHMUK	1106 (261 indels)	ON555584	-	-
57	**DNA1111**	*M. longiciliatum*	USA	NA	NHMUK	1013 (225 indels)	ON555536	-	-
58	**DNA1112**	*M. longiciliatum*	USA	NA	NHMUK	1030 (225 indels)	ON555537	-	-
59	**DNA1113**	*M. longiciliatum*	USA	NA	NHMUK	912 (266 indels)	ON555538	-	-
60	**DNA1114**	*M. mymaripenne*	USA	NA	NHMUK	1056 (228 indels)	ON555539	-	-
61	**DNA1116**	*M. nowickii*	Uganda	Paratype	NHMUK	1085 (211 indels)	ON555540	-	-
62	**DNA1117**	*M. nowickii*	Uganda	Paratype	NHMUK	1085 (211 indels)	ON555541	-	-
63	**DNA1118**	*M. nowickii*	Uganda	Paratype	NHMUK	1085 (212 indels)	ON555542	-	-
64	**DNA1119**	*Megaphragma* sp.	Uganda	NA	NHMUK	1031 (238 indels)	ON555543	-	-
65	**DNA1120**	*M. nowickii*	Uganda	Paratype	NHMUK	473 (93 indels)	ON555544	-	-
66	**DNA1121**	*M. nowickii*	Uganda	Paratype	NHMUK	806 (181 indels)	ON555545	-	-
67	**DNA1122**	*M. nowickii*	Uganda	Paratype	NHMUK	1052 (211 indels)	ON555546	-	-
68	**DNA1123**	*M. nowickii*	Uganda	Paratype	NHMUK	1085 (211 indels)	ON555547	-	-
69	**DNA1124**	*M. nowickii*	Uganda	Paratype	NHMUK	1085 (212 indels)	ON555548	-	-
70	**DNA1132**	*M. nowickii*	Uganda	Paratype	NHMUK	1085 (209 indels)	ON555549	-	-
71	**DNA1147**	*M. longiciliatum*	Indonesia	NA	NHMUK	1097 (202 indels)	ON555550	-	-
72	**DNA1612**	*M. noyesi*	UK	NA	lost	1067 (213 indels)	ON555551	-	-
73	**DNA1613**	*M. noyesi*	UK	Paratype	NHMUK	1106 (213 indels)	ON555552	607	ON557441

Table A1. *Cont.*

Nr	Voucher	Species	Country	Type Status	Repository	28S		CO1	
						bp + gaps	Accession nr	bp	Accession nr
74	**DNA1614**	*M. noyesi*	UK	NA	lost	1106 (213 indels)	ON555553	-	-
75	**DNA1615**	*M. noyesi*	UK	Paratype	NHMUK	908 (189 indels)	ON555554	607	ON557442
76	**DNA1616**	*M. noyesi*	UK	Paratype	NHMUK	1106 (213 indels)	ON555555	607	ON557443
77	**DNA1617**	*M. noyesi*	UK	Paratype	UCRC	1106 (213 indels)	ON555556	-	-
78	**DNA1618**	*M. noyesi*	UK	NA	lost	1065 (213 indels)	ON555557	-	-
79	**DNA1619**	*M. noyesi*	UK	Paratype	UCRC	1106 (213 indels)	ON555558	607	ON557444
80	**DNA1626**	*Megaphragma* sp.	Brunei	NA	NHMUK	1067 (271 indels)	ON555559	-	-
81	**DNA1628**	*Megaphragma* sp.	Singapore	NA	NHMUK	1096 (266 indels)	ON555560	394	ON557445
82	**DNA1630**	*Megaphragma* sp.	Singapore	NA	NHMUK	-	-	394	ON557446
83	**DNA1632**	*Megaphragma* sp.	Singapore	NA	NHMUK	-	-	394	ON557447
84	**DNA1638**	*Megaphragma* sp.	Brunei	NA	NHMUK	1071 (265 indels)	ON555561	-	-
85	**DNA1640**	*Megaphragma* sp.	Singapore	NA	NHMUK	1046 (156 indels)	ON555562	394	ON557448
86	**DNA1641**	*Megaphragma* sp.	Singapore	NA	NHMUK	1106 (269 indels)	ON555563	394	ON557449
87	**DNA1642**	*Megaphragma* sp.	Brunei	NA	NHMUK	1106 (269 indels)	ON555564	394	ON557450
88	**DNA1643**	*Megaphragma* sp.	Brunei	NA	NHMUK	975 (264 indels)	ON555565	394	ON557451
89	**DNA1644**	*Megaphragma* sp.	Brunei	NA	NHMUK	1106 (271 indels)	ON555566	394	ON557452
90	**DNA1645**	*Megaphragma* sp.	Brunei	NA	NHMUK	1042 (268 indels)	ON555567	394	ON557453
91	**DNA1650**	*Megaphragma* sp.	Brunei	NA	NHMUK	1072 (270 indels)	ON555568	-	-
92	**DNA1651**	*Megaphragma* sp.	Brunei	NA	NHMUK	1048 (222 indels)	ON555569	394	ON557454
93	**DNA1652**	*Megaphragma* sp.	Brunei	NA	NHMUK	1106 (267 indels)	ON555570	394	ON557455
94	**DNA1655**	*Megaphragma* sp.	Brunei	NA	NHMUK	1000 (270 indels)	ON555571	394	ON557456
95	**DNA1656**	*M. liui*	Brunei	**Holotype**	UCRC	900 (556 indels)	ON555572	370	ON557457
96	**DNA1659**	*Megaphragma* sp.	Brunei	NA	NHMUK	1038 (270 indels)	ON555573	-	-
97	**DNA1661**	*Megaphragma* sp.	Brunei	NA	NHMUK	1106 (266 indels)	ON555574	-	-
98	**DNA1665**	*Megaphragma* sp.	Singapore	NA	NHMUK	1106 (266 indels)	ON555575	394	ON557458
99	**DNA1668**	*Megaphragma* sp.	Brunei	NA	NHMUK	1106 (267 indels)	ON555576	394	ON557459
100	**DNA1674**	*Megaphragma* sp.	Brunei	NA	NHMUK	1106 (265 indels)	ON555577	-	-
101	**DNA1678**	*Megaphragma* sp.	Singapore	NA	NHMUK	1095 (266 indels)	ON555578	394	ON557460
102	**DNA1679**	*Megaphragma* sp.	Brunei	NA	NHMUK	1106 (218 indels)	ON555579	-	-
103	**DNA1680**	*M. momookherjeeae*	Costa Rica	**Holotype**	NHMUK	-	-	394	ON557461
104	**DNA1681**	*M. digitatum*	Costa Rica	Paratype	NHMUK	1106 (265 indels)	ON555580	367	ON557462
105	**DNA1682**	*Megaphragma* sp.	Costa Rica	NA	NHMUK	932 (257 indels)	ON555581	-	-
106	**DNA1683**	*M. giraulti*	Costa Rica	Paratype	NHMUK	1055 (270 indels)	ON555582	394	ON557463
107	**DNA1686**	*Megaphragma* sp.	Vietnam	NA	NHMUK	-	-	652	ON557464
108	**DNA1687**	*M. rivelloi*	Vietnam	Paratype	AICF	1106 (259 indels)	ON555583	-	-
109	**DNAMO3**	*M. longiciliatum*	Oman	NA	NHMO	420 (102 indels)	ON555588	-	-
110	**DNAMO13**	*M. longiciliatum*	Oman	NA	NHMO	1106 (224 indels)	ON555585	-	-
111	**DNAMO20**	*M. longiciliatum*	Oman	NA	NHMO	1106 (224 indels)	ON555586	-	-
112	**DNAMO22**	*M. longiciliatum*	Oman	NA	NHMO	1106 (224 indels)	ON555587	-	-
113	**FRM2**	*M. longiciliatum*	France	NA	NHMUK	1106 (213 indels)	ON555589	652	ON557465
114	**FRM3**	*M. longiciliatum*	France	NA	NHMUK	1106 (213 indels)	ON555590	652	ON557466
115	**FRM4**	*M. longiciliatum*	France	NA	AICF	1106 (213 indels)	ON555591	652	ON557467
116	**FRM5**	*M. longiciliatum*	France	NA	NHMUK	1106 (213 indels)	ON555592	652	ON557468
117	**FRM6**	*M. longiciliatum*	France	NA	NHMUK	1106 (213 indels)	ON555593	652	ON557469

Table A1. *Cont.*

Nr	Voucher	Species	Country	Type Status	Repository	28S		CO1	
						bp + gaps	Accession nr	bp	Accession nr
118	**HUM1**	*M. longiciliatum*	Hungary	NA	NHMUK	679 (122 indels)	ON555594	652	ON557470
119	**HUM2**	*M. noyesi*	Hungary	Paratype	NHMUK	1060 (213 indels)	ON555599	650	ON557476
120	**HUM3**	*M. noyesi*	Hungary	NA	lost	1061 (213 indels)	ON555600	652	ON557477
121	**HUM4**	*Megaphragma* sp.	Hungary	NA	NHMUK	1010 (216 indels)	ON555601	652	ON557478
122	**HUM5**	*M. noyesi*	Hungary	Paratype	AICF	-	-	652	ON557479
123	**HUM6**	*M. noyesi*	Hungary	Paratype	AICF	1060 (213 indels)	ON555602	614	ON557480
124	**HUM8**	*M. noyesi*	Hungary	Paratype	NHMUK	641 (152 indels)	ON555603	652	ON557481
125	**HUM9**	*M. noyesi*	Hungary	Paratype	NHMUK	-	-	296	ON557482
126	**HUM10**	*M. noyesi*	Hungary	Paratype	NHMUK	679 (125 indels)	ON555595	652	ON557471
127	**HUM11**	*M. noyesi*	Hungary	Paratype	NHMUK	649 (125 indels)	ON555596	603	ON557472
128	**HUM12**	*M. noyesi*	Hungary	Paratype	NHMUK	1040 (213 indels)	ON555597	295	ON557473
129	**HUM13**	*M. noyesi*	Hungary	Paratype	NHMUK	-	-	574	ON557474
130	**HUM14**	*M. noyesi*	Hungary	Paratype	NHMUK	763 (192 indels)	ON555598	643	ON557475
131	**ITM1**	*M. viggianii*	Italy	**Holotype**	DACE	1106 (219 indels)	ON555604	652	ON557483
132	**ITM2**	*M. viggianii*	Italy	Paratype	NHMUK	1106 (219 indels)	ON555610	652	ON557488
133	**ITM3**	*M. viggianii*	Italy	Paratype	NHMUK	934 (217 indels)	ON555611	631	ON557489
134	**ITM4**	*M. viggianii*	Italy	Paratype	NHMUK	1106 (219 indels)	ON555612	652	ON557490
135	**ITM5**	*M. viggianii*	Italy	Paratype	AICF	1050 (219 indels)	ON555613	652	ON557491
136	**ITM6**	*M. viggianii*	Italy	Paratype	AICF	1106 (219 indels)	ON555614	652	ON557492
137	**ITM7**	*M. viggianii*	Italy	Paratype	NHMUK	1047 (219 indels)	ON555615	296	ON557493
138	**ITM8**	*M. polilovi*	Italy	Paratype	DACE	1080 (208 indels)	ON555616	641	ON557494
139	**ITM9**	*M. polilovi*	Italy	**Holotype**	DACE	1088 (208 indels)	ON555617	641	ON557495
140	**ITM10**	*M. longiciliatum*	Italy	NA	NHMUK	1081 (213 indels)	ON555605	-	-
141	**ITM11**	*M. polilovi*	Italy	Paratype	AICF	1080 (208 indels)	ON555606	641	ON557484
142	**ITM12**	*M. polilovi*	Italy	Paratype	NHMUK	1081 (208 indels)	ON555607	652	ON557485
143	**ITM13**	*M. viggianii*	Italy	Paratype	NHMUK	1081 (219 indels)	ON555608	641	ON557486
144	**ITM14**	*M. polilovi*	Italy	Paratype	NHMUK	1080 (208 indels)	ON555609	641	ON557487
145	**MXM1**	*M. striatum*	Mexico	NA	NHMUK	995 (297 indels)	ON555618	652	ON557496
146	**MXM2**	*M. breviclavum*	Mexico	**Holotype**	NHMUK	1010 (298 indels)	ON555619	652	ON557497
147	**MXM3**	*M. breviclavum*	Mexico	Paratype	NHMUK	878 (514 indels)	ON555620	652	ON557498
148	**MXM4**	*M. breviclavum*	Mexico	Paratype	NHMUK	1000 (298 indels)	ON555621	575	ON557499
149	**SAM1**	*Megaphragma* sp.	Malaysia	NA	NHMUK	1106 (269 indels)	ON555622	296	ON557500
150	**SAM2**	*Megaphragma* sp.	Malaysia	NA	NHMUK	1065 (248 indels)	ON555626	356	ON557503
151	**SAM3**	*M. vanlentereni*	Malaysia	**Holotype**	NHMUK	1017 (263 indels)	ON555627	652	ON557504
152	**SAM4**	*M. chienleei*	Malaysia	Paratype	NHMUK	1106 (268 indels)	ON555628	-	-
153	**SAM5**	*M. chienleei*	Malaysia	Paratype	NHMUK	1106 (268 indels)	ON555629	652	ON557505
154	**SAM6**	*M. chienleei*	Malaysia	Paratype	AICF	1106 (268 indels)	ON555630	652	ON557506
155	**SAM7**	*M. chienleei*	Malaysia	Paratype	NHMUK	1106 (268 indels)	ON555631	652	ON557507
156	**SAM8**	*M. chienleei*	Malaysia	Paratype	NHMUK	1106 (268 indels)	ON555632	652	ON557508
157	**SAM9**	*Megaphragma* sp.	Malaysia	NA	NHMUK	1106 (271 indels)	ON555633	652	ON557509
158	**SAM10**	*Megaphragma* sp.	Malaysia	NA	NHMUK	1080 (225 indels)	ON555623	641	ON557501
159	**SAM11**	*M. cockerilli*	Malaysia	**Holotype**	AICF	1073 (222 indels)	ON555624	460	ON557502
160	**SAM12**	*M. chienleei*	Malaysia	**Holotype**	AICF	1074 (268 indels)	ON555625	-	-

Table A1. *Cont.*

Nr	Voucher	Species	Country	Type Status	Repository	28S bp + gaps	28S Accession nr	CO1 bp	CO1 Accession nr
161	**SRM1**	*M. longiciliatum*	Malaysia	NA	NHMUK	513 (287 indels)	ON555634	652	ON557510
162	**SRM2**	*Megaphragma* sp.	Malaysia	NA	NHMUK	1044 (248 indels)	ON555635	624	ON557511
163	**SRM3**	*Megaphragma* sp.	Malaysia	NA	NHMUK	679 (124 indels)	ON555636	601	ON557512
164	**UKM8**	*M. noyesi*	UK	Paratype	NHMUK	647 (125 indels)	ON555642	-	-
165	**UKM9**	*M. noyesi*	UK	Paratype	lost	1106 (213 indels)	ON555643	652	ON557518
166	**UKM10**	*M. noyesi*	UK	Paratype	AICF	1106 (213 indels)	ON555637	652	ON557513
167	**UKM11**	*M. noyesi*	UK	Paratype	NHMUK	1106 (213 indels)	ON555638	652	ON557514
168	**UKM12**	*M. noyesi*	UK	Paratype	DACE	1106 (213 indels)	ON555639	652	ON557515
169	**UKM13**	*M. noyesi*	UK	Paratype	DACE	1106 (213 indels)	ON555640	652	ON557516
170	**UKM14**	*M. noyesi*	UK	**Holotype**	NHMUK	1047 (213 indels)	ON555641	652	ON557517
171	-	"*M. amalphitanum*"	**GenBank**	NA	-	-	-	652	KT373787
172	**D1224**	*Megaphragma* sp.	**GenBank**	NA	-	1048 (230 indels)	AY623543	-	-
173	**D1229**	*Megaphragma* sp.	**GenBank**	NA	-	1106 (260 indels)	AY623544	-	-
174	**D1243**	*Megaphragma* sp.	**GenBank**	NA	-	1106 (240 indels)	AY623545	-	-
175	**O19**	*Oligosita* sp.	**GenBank**	NA	-	659 (195 indels)	MG785509	603	MG904913
176	**D1219**	*Epoligosita* sp.	**GenBank**	NA	-	1106 (301 indels)	AY623546	-	-
177	**D0760**	*Oligosita sanguinea*	**GenBank**	NA	-	1106 (291 indels)	AY623551	-	-
178	**D0886**	*Probrachista* sp.	**GenBank**	NA	-	1106 (288 indels)	AY623553	-	-

References

1. Noyes, J.S. Universal Chalcidoidea Database. Available online: https://www.nhm.ac.uk/our-science/data/chalcidoids/database/ (accessed on 15 October 2021).
2. Pinto, J.D. A review of the New World genera of Trichogrammatidae. *J. Hymenopt. Res.* **2006**, *15*, 38–163.
3. Viggiani, G.; Dipartimento di Agraria, Laboratorio di Lotta Biologica, Università degli Studi di Napoli "Federico II", Portici, Italy. 2021; *Unpublished data.*
4. Polilov, A.A. The smallest insects evolve anucleate neurons. *Arthropod Struct. Dev.* **2012**, *41*, 29–34. [CrossRef] [PubMed]
5. Huber, J.; Noyes, J. A new genus and species of fairyfly, *Tinkerbella nana* (Hymenoptera, Mymaridae), with comments on its sister genus *Kikiki*, and discussion on small size limits in arthropods. *J. Hymenopt. Res.* **2013**, *32*, 17–44. [CrossRef]
6. Viggiani, G. Notes on the type of *Megaphragma* Timberlake (Hymenoptera: Trichogrammatidae). *Boll. Lab. Ent. Agr. Filippo Silvestri* **1997**, *53*, 117–122.
7. Viggiani, G.; Bernardo, U. Two species of *Megaphragma* (Hymenoptera Trichogrammatidae), egg-parasitoids of *Heliothrips haemorrhoidalis* Bouché (Thysanoptera) in southern Italy, with description of a new species. *Boll. Zool. Agr. Bach. Ser. II* **1997**, *29*, 51–55.
8. Viggiani, G. A species of *Megaphragma* (Hymenoptera: Chalcidoidea: Trichogrammatidae), one of the smallest known insects, new for Italy. *Boll. Zool. Agr. Bach. Ser. II* **2002**, *34*, 449–452.
9. Lin, N.Q. Descriptions of five new species of *Megaphragma* and *Paramegaphragma* gen. nov. (Hymenoptera: Trichogrammatidae) from China. *Entomotaxonomia* **1992**, *14*, 129–138.
10. Hayat, M. Records and descriptions of Trichogrammatidae from India (Hymenoptera: Chalcidoidea). *Orient. Insects* **2009**, *43*, 201–227. [CrossRef]
11. Viggiani, G.; Luft Albarracin, E.; Virla, E. The species of *Megaphragma* Timberlake (Hymenoptera: Trichogrammatidae) from Argentina. *Zootaxa* **2009**, *2308*, 65–68. [CrossRef]
12. Hessein, N.A.; McMurtry, J.A. Observations on *Megaphragma mymaripenne* Timberlake (Hymenoptera: Trichogrammatidae), an egg parasite of *Heliothrips haemorrhoidalis* (Bouché) (Thysanoptera: Thripidae). *Pan-Pac. Entomol.* **1988**, *64*, 250–254.
13. Viggiani, G.; Bernardo, U. Distribuzione e attività di ooparassitoidi di *Heliothrips haemorrhoidalis* Bouché (Thysanoptera) su piante della macchia mediterranea. *Phytophaga* **1998**, *8*, 13–20.
14. Bernardo, U.; Viggiani, G. Biological data on *Megaphragma amalphitanum* Viggiani and *Megaphragma mymaripenne* Timberlake (Hymenoptera: Trichogrammatidae), egg-parasitoid of *H. haemorrhoidalis* (Bouché) (Thysanoptera: Thripidae) in southern Italy. *Boll. Lab. Ent. Agr. Filippo Silvestri* **2003**, *58*, 77–85.
15. Polilov, A.A. *At the Size Limit—Effects of Miniaturization in Insects*; Springer International Publishing: Cham, Switzerland, 2016; p. 325.

16. Minelli, A.; Fusco, G. No limits: Breaking constrains in insect miniaturization. *Arthropod Struct. Dev.* **2019**, *48*, 4–11. [CrossRef] [PubMed]
17. Makarova, A.A.; Polilov, A.A.; Chklovskii, D.B. Small brains for big science. *Curr. Opin. Neurobiol.* **2021**, *71*, 77–83. [CrossRef]
18. Polilov, A.A. Anatomy of adult *Megaphragma* (Hymenoptera: Trichogrammatidae), one of the smallest insects, and new insight into insect miniaturization. *PLoS ONE* **2017**, *12*, e0175566. [CrossRef]
19. Makarova, A.; Polilov, A.; Fischer, S. Comparative morphological analysis of compound eye miniaturization in minute hymenoptera. *Arthropod Struct. Dev.* **2015**, *44*, 21–32. [CrossRef]
20. Diakova, A.V.; Makarova, A.A.; Polilov, A.A. Between extreme simplification and ideal optimization: Antennal sensilla morphology of miniaturized *Megaphragma* wasps (Hymenoptera: Trichogrammatidae). *PeerJ* **2018**, *6*, e6005. [CrossRef]
21. Diakova, A.V.; Polilov, A.A. Porous or non-porous? The challenge of studying unusual placoid sensilla of *Megaphragma* wasps (Hymenoptera, Trichogrammatidae) with electron microscopy. *J. Hymenopt. Res.* **2021**, *84*, 69–73. [CrossRef]
22. Boudinot, B.E.; Beutel, R.G.; Gorb, S.N.; Polilov, A.A. Functional diversity of attachment and grooming leg structures is retained in all but the smallest insects. *J. Zool.* **2020**, *313*, 99–113. [CrossRef]
23. Nedoluzhko, A.V.; Sharko, F.S.; Boulygina, E.S.; Tsygankova, S.V.; Sokolov, A.S.; Mazur, A.M.; Polilov, A.A.; Prokhortchouk, E.B.; Skryabin, K.G. Mitochondrial genome of *Megaphragma amalphitanum* (Hymenoptera: Trichogrammatidae). *Mitochondrial DNA Part A* **2016**, *27*, 4526–4527. [CrossRef]
24. Nedoluzhko, A.V.; Sharko, F.S.; Tsygankova, S.V.; Boulygina, E.S.; Sokolov, A.S.; Rastorguev, S.M.; Kadnikov, V.V.; Mardanov, A.V.; Ravin, N.V.; Mazur, A.M.; et al. Metagenomic analysis of microbial community of a parasitoid wasp *Megaphragma amalphitanum*. *Genom. Data* **2017**, *11*, 87–88. [CrossRef] [PubMed]
25. Sharko, F.S.; Nedoluzhko, A.V.; Le, B.M.; Tsygankova, S.V.; Boulygina, E.S.; Rastorguev, S.M.; Sokolov, A.S.; Rodriguez, F.; Mazur, A.M.; Polilov, A.A.; et al. A partial genome assembly of the miniature parasitoid wasp, *Megaphragma amalphitanum*. *PLoS ONE* **2019**, *14*, e0226485. [CrossRef]
26. Polilov, A.A. First record of *Megaphragma* (Hymenoptera, Trichogrammatidae) in Columbia, and third animal species known to have anucleate neurons. *J. Hymenopt. Res.* **2017**, *60*, 181–185. [CrossRef]
27. Polilov, A.A.; Makarova, A.A.; Pang, S.; Shan Xu, C.; Hess, H. Protocol for preparation of heterogeneous biological samples for 3D electron microscopy: A case study for insects. *Sci. Rep.* **2021**, *11*, 4717. [CrossRef] [PubMed]
28. Polilov, A.A.; Department of Entomology, Faculty of Biology, Lomonosov Moscow State University, Moscow, Russia. 2021; *Unpublished data*.
29. Loomans, A.J.M.; van Lenteren, J.C. Biological control of thrips pests: A review on thrips parasitoids. *Wagening. Agric. Univ. Pap.* **1995**, *95*, 89–201.
30. Loomans, A.J.M. *Parasitoids as Biological Control Agents of Thrips Pests*; Wageningen University: Wageningen, The Netherlands, 2003; p. 215.
31. Prokhortchouk, E.B.; Nedoluzhko, A.V.; Sharko, F.S.; Tsygankova, S.V.; Boulygina, E.S.; Rastorguev, S.M.; Sokolov, A.S.; Mazur, A.M.; Polilov, A.A.; Skryabin, K.G. Genome sequencing and transcriptome assembly of the parasitoid wasp *Megaphragma amalphitanum* (Hymenoptera: Trichogrammatidae). *Mosc. Univ. Biol. Sci. Bull.* **2017**, *72*, 30–32. [CrossRef]
32. Gillespie, J.J.; Munro, J.B.; Heraty, J.M.; Yoder, M.J.; Owen, A.K.; Carmichael, A.E. A secondary structural model of the 28S rRNA expansion segments D2 and D3 for chalcidoid wasps (Hymenoptera: Chalcidoidea). *Mol. Biol. Evol.* **2005**, *22*, 1593–1608. [CrossRef]
33. Owen, A.K.; George, J.; Pinto, J.D.; Heraty, J.M. A molecular phylogeny of the Trichogrammatidae (Hymenoptera: Chalcidoidea), with an evaluation of the utility of their male genitalia for higher level classification. *Syst. Entomol.* **2007**, *32*, 227–251. [CrossRef]
34. Noyes, J.S. Collecting and preserving chalcid wasps (Hymenoptera: Chalcidoidea). *J. Nat. Hist.* **1982**, *16*, 315–334. [CrossRef]
35. Cruaud, A.; Nidelet, S.; Arnal, P.; Weber, A.; Fusu, L.; Gumovsky, A.; Huber, J.; Polaszek, A.; Rasplus, J.-Y. Optimized DNA extraction and library preparation for minute arthropods: Application to target enrichment in chalcid wasps used for biocontrol. *Mol. Ecol. Resour.* **2019**, *19*, 702–710. [CrossRef]
36. Park, J.K.; O'Foighil, D. Sphaeriid and corbiculid clams represent separate heterodont bivalve radiations into freshwater environments. *Mol. Phylogenet. Evol.* **2000**, *14*, 75–88. [CrossRef] [PubMed]
37. Whiting, M.F.; Carpenter, J.C.; Wheeler, Q.D.; Wheeler, W.C. The Strepsiptera problem: Phylogeny of the holometabolous insect orders inferred from 18S and 28S ribosomal DNA sequences and morphology. *Syst. Biol.* **1997**, *46*, 1–68. [CrossRef] [PubMed]
38. Nunn, G.B.; Theisen, B.F.; Christensen, B.; Arctander, P. Simplicity correlated size growth of the nuclear 28S ribosomal RNA D3 expansion segment in the crustacean order Isopoda. *J. Mol. Evol.* **1996**, *42*, 211–223. [CrossRef]
39. Folmer, O.; Black, M.; Hoeh, W.; Lutz, R.; Vrijenhoek, R. DNA primers for amplification of mitochondrial cytochrome c oxidase subunit I from diverse metazoan invertebrates. *Mol. Mar. Biol. Biotech.* **1994**, *3*, 294–299.
40. Hebert, P.D.; Cywinska, A.; Ball, S.L.; Dewaard, J.R. Biological identifications through DNA barcodes. *Proc. R. Soc. Lond. Ser. B Biol. Sci.* **2003**, *270*, 313–321. [CrossRef]
41. Dascălu, M.M.; Caba, F.G.; Fusu, L. DNA barcoding in Dorcadionini (Coleoptera, Cerambycidae) uncovers mitochondrial-morphological discordance and the hybridogenic origin of several subspecies. *Org. Divers. Evol.* **2022**, *22*, 205–229. [CrossRef]
42. Fusu, L.; Polaszek, A. Description, DNA barcoding and phylogenetic placement of a remarkable new species of *Eopelma* (Hymenoptera: Eupelmidae) from Borneo. *Zootaxa* **2017**, *4263*, 557–566. [CrossRef]
43. Meusnier, I.; Singer, G.A.; Landry, J.F.; Hickey, D.A.; Hebert, P.D.; Hajibabaei, M. A universal DNA mini-barcode for biodiversity analysis. *BMC Genom.* **2008**, *9*, 214. [CrossRef]

44. Bonfield, J.K.; Smith, K.F.; Staden, R. A new DNA sequence assembly program. *Nucleic Acids Res.* **1995**, *24*, 4992–4999. [CrossRef]
45. Kumar, S.; Stecher, G.; Tamura, K. MEGA7: Molecular evolutionary genetics analysis version 7.0 for bigger datasets. *Mol. Biol. Evol.* **2016**, *33*, 1870–1874. [CrossRef]
46. Thompson, J.D.; Higgins, D.G.; Gibson, T.J. CLUSTAL W: Improving the sensitivity of progressive multiple sequence alignment through sequence weighting, position-specific gap penalties and weight matrix choice. *Nucleic Acids Res.* **1994**, *22*, 4673–4680. [CrossRef]
47. Katoh, K.; Rozewicki, J.; Yamada, K.D. MAFFT online service: Multiple sequence alignment, interactive sequence choice and visualization. *Brief. Bioinform.* **2019**, *20*, 1160–1166. [CrossRef] [PubMed]
48. Kozlov, A.M.; Darriba, D.; Flouri, T.; Morel, B.; Stamatakis, A. RAxML-NG: A fast, scalable and user-friendly tool for maximum likelihood phylogenetic inference. *Bioinformatics* **2019**, *35*, 4453–4455. [CrossRef] [PubMed]
49. Fabreti, L.G.; Höhna, S. Bayesian inference of phylogeny is robust to substitution model over-parameterization. *bioRxiv* **2022**. [CrossRef]
50. Maddison, W.P.; Maddison, D.R. Mesquite: A Modular System for Evolutionary Analysis, Version 3.10. Available online: http://www.mesquiteproject.org (accessed on 27 June 2016).
51. Lanfear, R.; Frandsen, P.B.; Wright, A.M.; Senfeld, T.; Calcott, B. PartitionFinder 2: New methods for selecting partitioned models of evolution for molecular and morphological phylogenetic analyses. *Mol. Biol. Evol.* **2016**, *34*, 772–773. [CrossRef] [PubMed]
52. Ronquist, F.; Teslenko, M.; van der Mark, P.; Ayres, D.L.; Darling, A.; Höhna, S.; Larget, B.; Liu, L.; Suchard, M.A.; Huelsenbeck, J.P. MrBayes 3.2: Efficient Bayesian phylogenetic inference and model choice across a large model space. *Syst. Biol.* **2012**, *61*, 539–542. [CrossRef] [PubMed]
53. Rambaut, A.; Drummond, A.J.; Xie, D.; Baele, G.; Suchard, M.A. Posterior summarisation in Bayesian phylogenetics using Tracer 1.7. *Syst. Biol.* **2018**, *67*, 901–904. [CrossRef]
54. Pattengale, N.D.; Alipour, M.; Bininda-Emonds, O.R.; Moret, B.M.; Stamatakis, A. How many bootstrap replicates are necessary? *J. Comput. Biol.* **2010**, *17*, 337–354. [CrossRef]
55. Rambaut, A. FigTree v1.4.4. Available online: https://github.com/rambaut/figtree/releases (accessed on 22 October 2019).
56. Pinto, J.D.; Viggiani, G. A review of the genera of Oligositini (Hymenoptera: Trichogrammatidae) with a preliminary hypothesis of phylogenetic relationships. *J. Hym. Res.* **2004**, *13*, 269–294.
57. Doutt, R.L.; Viggiani, G. The classification of the Trichogrammatidae (Hym. Trichogrammatidae). In *Proceedings of the California Academy of Sciences*; 4th Series; California Academy of Science: San Francisco, CA, USA, 1968; Volume 35, pp. 477–586.
58. Viggiani, G.; Bernardo, U.; Jesu, R. Caratterizzazione di due specie di *Megaphragma* Timberlake (Hymenoptera: Trichogrammatidae), ooparassitoidi di *Heliothrips haemorrhoidalis* Bouché (Thysanoptera) in Campania. In Proceedings of the Atti XVIII Congresso Nazionale Italiano di Entomologia, Maratea, Italy, 21–26 June 1998; p. 70.
59. Shibao, M.; Hosomi, A.; Tanaka, H. Seasonal fluctuation in percentage parasitism of the yellow tea thrips, *Scirtothrips dorsalis* Hood (Thysanoptera: Thripidae) by an egg parasitoid of *Megaphragma* (Hymenoptera: Trichogrammatidae) on grapes. *Entomol. Sci.* **2000**, *3*, 611–613.
60. Delvare, G. Sur les *Megaphragma* de Guadeloupe avec la description d'une espèce nouvelle [Hymenoptera, Trichogrammatidae]. *Revue Fr. Ent. (n.s.)* **1993**, *15*, 149–152.
61. Ghesquière, J. Contributions à l'étude des Hyménoptères du Congo Belge. *Rev. Zool. Bot. Afr.* **1939**, *33*, 33–41.
62. Pintureau, B.; Lassabliere, F.; Khatchadourian, C.; Daumal, J. Parasitoïdes oophages et symbiotes de deux thrips Européens. *Ann. Soc. Entomol. Fr.* **1999**, *35*, 416–420.
63. Subba Rao, B.R. A new species of *Megaphragma* (Hymenoptera: Trichogrammatidae) from India. *P. Roy. Entomol. Soc. B* **1969**, *38*, 114–116. [CrossRef]
64. Yousuf, M.; Shafee, A. Taxonomy of Indian Trichogrammatidae (Hymenoptera: Chalcidoidea). *Indian J. Syst. Entomol.* **1988**, *4*, 55–200.
65. Timberlake, P.H. Descriptions of new chalcid-flies from Hawaii and Mexico (Hymenoptera). *Proc. Haw. Entomol. Soc.* **1924**, *5*, 395–417.
66. Viggiani, G. Notizie preliminari su un *Megaphragma* (Hymenoptera: Trichogrammatidae), ooparassitoide di *Heliothrips haemorrhoidalis* Bouché (Thysanoptera: Tripidae) in Italia. *Boll. Lab. Ent. Agr. Filippo Silvestri* **1994**, *49*, 203–206.
67. Yuan, J.; Cong, B.; Lou, J. Two new species a new record genus and a new record species of Trichogrammatidae in China (Hymenoptera: Chalcidoidea). *J. Northeast. Norm. Univ.* **1997**, *4*, 62–66.
68. Kryger, I.P. One new genus and species, and three new species of Trichogramminae from Egypt with remarks upon *Neocentrobia hirticornis*, *Alaptus minimus*, and *Trichogramma evanescens*. *Bull. Société R. d'Egypte* **1932**, *16*, 38–44.
69. Baas-Becking, L.G.M. *Geobiologie, of Inleading Tot de Milieukunde*; W.P. Van Stockum & Zoon: Den Haag, The Netherlands, 1934; p. 263.
70. Fenchel, T.; Finlay, B.F. The ubiquity of small species: Patterns of local and global diversity. *Bioscience* **2004**, *54*, 777–784. [CrossRef]
71. Bass, D.; Richards, T.A.; Matthai, L.; Marsh, V.; Cavalier-Smith, T. DNA evidence for global dispersal and probable endemicity of protozoa. *BMC Evol. Biol.* **2007**, *7*, 162. [CrossRef] [PubMed]
72. Popovici, O.A.; Masner, L.; Viciriuc, M.; Pintilioaie, A.; Notton, D.G.; Talamas, E. New distribution data for some charismatic tramp species of Platygastroidea (Hymenoptera). *Zootaxa* **2018**, *4370*, 1–22. [CrossRef] [PubMed]
73. Huber, J.T.; Thuróczy, C. Review of *Anaphes* Haliday (Hymenoptera: Mymaridae) with key to species in Europe and a world catalogue. *Zootaxa* **2018**, *4376*, 1–104. [CrossRef] [PubMed]

74. Polaszek, A. Egg parasitism in the Aphelinidae (Hymenoptera: Chalcidoidea) with special reference to *Centrodora* and *Encarsia* species. *B. Entomol. Res.* **1991**, *81*, 97–106. [CrossRef]
75. Polaszek, A.; Kimani, S.W. *Telenomus* species (Hymenoptera: Scelionidae) attacking eggs of pyralid pests (Lepidoptera) in Africa: A review and guide to identification. *B. Entomol. Res.* **1990**, *80*, 57–71. [CrossRef]
76. Fatouros, N.E.; Cusumano, A.; Bin, F.; Polaszek, A.; van Lenteren, J.C. How to escape from insect egg parasitoids: A review of potential factors explaining parasitoid absence across the Insecta. *Proc. R. Soc. Lond. Ser. B Biol. Sci.* **2020**, *287*, 20200344. [CrossRef]
77. Al khatib, F.; Cruaud, A.; Fusu, L.; Genson, G.; Rasplus, J.-Y.; Ris, N.; Delvare, G. Multilocus phylogeny and ecological differentiation of the "*Eupelmus urozonus* species group" (Hymenoptera, Eupelmidae) in the West-Palaearctic. *BMC Evol. Biol.* **2016**, *16*, 1–20 [CrossRef]

insects

MDPI

Article

New Pamphiliids with Varying Venations from Lower Cretaceous Yixian Formation of Northeast China (Hymenoptera, Pamphiliidae) †

Jialiang Zhuang [1,2], Alexandr P. Rasnitsyn [3,4], Chungkun Shih [2,5], Dong Ren [2] and Mei Wang [1,*]

1 Key Laboratory of Forest Protection of National Forestry and Grassland Administration, Ecology and Nature Conservation Institute, Chinese Academy of Forestry, Beijing 100091, China
2 College of Life Sciences, Capital Normal University, 105 Xisanhuanbeilu, Haidian District, Beijing 100048, China
3 Paleontological Institute, Russian Academy of Sciences, 123, Profsoyuznaya ul., 117647 Moscow, Russia
4 Department of Palaeontology, Natural History Museum, Cromwell Road, London SW7 5BD, UK
5 Department of Paleobiology, National Museum of Natural History, Smithsonian Institution, Washington, DC 20013, USA
* Correspondence: wangmeicnu@163.com

† This published work and the nomenclatural acts it contains have been registered in ZooBank, the online registration system for the ICZN (International Code of Zoological Nomenclature). The LSID (Life Science Identifier) for this publication are: urn:lsid:zoobank.org:pub:7636CD49-8B2E-4634-AC61-46C99144AB46; urn:lsid:zoobank.org:act:D88FB212-38B3-4D66-BDF8-F7A6B15B8E91; urn:lsid:zoobank.org:act:0DF9F9C7-F685-4A8E-A987-DB9F9206D685; urn:lsid:zoobank.org:act:AE9C9861-1A71-4906-AD19-84897B07D65A; urn:lsid:zoobank.org:act:C8A1475B-C8F7-4D96-A04A-4254778DF54E.

Simple Summary: The presence of various morphologies of the ap-Cu, an extra vein with different lengths in the forewing, is a special character in Pamphiliidae. In Juralydinae, this vein is usually absent or extremely short. Here, we describe one new genus and three new species of Juralydinae with varying lengths of the ap-Cu from the Lower Cretaceous Yixian Formation in Duolun County, Inner Mongolia, China, and revise the diagnostic characteristics of Juralydinae. The new specimens expand the diversity of Pamphiliidae in the Mesozoic and strengthen our knowledge of venation variation.

Abstract: One new genus and three new species of Pamphiliidae, *Dolicholyda obtusata* gen. et sp. nov., *Dolicholyda confluens* sp. nov., and *Dolicholyda angusta* sp. nov. are described and illustrated. All of them were collected from the Lower Cretaceous Yixian Formation in Duolun County, Inner Mongolia, China. The new genus is established based on the following characters: body surface without punctations; forewing with pterostigma lanceolate and sclerotized around the margins; angle between 1-M and 1-Cu nearly 90°; cell 1mcu long and obviously longer than length of pterostigma. In most cases, the ap-Cu is present, and its length varied. Additionally, we revise the diagnostic characteristics of Juralydinae based on the new specimens. New findings enhance our understanding of the wing venation characteristics of fossil pamphiliids and expand the diversity of Pamphiliidae in the Mesozoic.

Keywords: Juralydinae; Sandaogou; venation variation; fossils; taxonomy

1. Introduction

Pamphilioidea, one of the oldest and basal superfamilies of Hymenoptera, consist of four families: Mirolydidae, Xyelydidae, Pamphiliidae and Megalodontesidae [1–4]. Its earliest fossil records are dated to the Early Jurassic [5,6]. As one of the relict families of this superfamily, Pamphiliidae also have fossil records in the late Middle Jurassic to early Late Jurassic age [7–9]. Currently, two extant subfamilies, Pamphiliinae Cameron 1890 and Cephalciinae Benson 1945, and an extinct subfamily, Juralydinae Rasnitsyn 1977, are attributed to Pamphiliidae [10].

However, the extinct subfamily Juralydinae was not initially recognized after its establishment because of the lack of important information for identification and diagnosis [11,12]. Extant taxa of Pamphiliidae have been typically identified by tiny and rarely preserved characters, such as the configuration of the pre-apical spur in fore tibia, the inner tooth in the tarsal claw, the body surface colour, etc., and little attention was paid to wing venations [13,14]. However, the classifications of fossil Pamphiliidae have mainly focused on the structures often preserved, especially wing venation and antennae. The contradictions in the focal characters used for classification and the limited number of fossil specimens make it challenging to compare and classify fossil vs. extant pamphiliids. Therefore, some distinguishing features of Juralydinae still need to be deliberated. Jouault et al. [15] transferred *Tapholyda caplani* (Cockerell, 1933) (from Oligocene and Miocene) to Juralydinae and revised the diagnoses of Juralydinae. So far, the subfamily Juralydinae includes all the fossil species except *Acantholyda erythrocephala* (Linnaeus, 1758) (from Miocene), possibly *Acantholyda ribesalbesensis* Peñalver and Arillo, 2002 (from Miocene) and *Ulteramus republicensis* Archibald and Rasnitsyn, 2015 (from lower Eocene and its subfamily is undetermined) [16–19].

Herein, we describe three new species in a new genus *Dolicholyda* gen. nov. in Juralydinae, from the Lower Cretaceous Yixian Formation, and revise the diagnoses of Juralydinae again. In addition, we summarize and discuss wing venations and their variations for extant and extinct pamphiliids.

2. Materials and Methods

The fossil specimens of the new genus *Dolicholyda* gen. nov. are housed in the Key Laboratory of Insect Evolution and Environmental Changes, College of Life Sciences, Capital Normal University, Beijing, China (CNUB; Dong Ren, Curator). All of them were collected from the Lower Yixian Formation, Sandaogou, Nanyingpan Village, Duolun County, Inner Mongolia, China. The extant specimens of Pamphiliidae used for wing venation photos are deposited in the Natural History Museum of the Chinese Academy of Forestry.

The stratigraphic attribution of the Sandaogou locality has been controversial. It was assigned to the Sandaogou, Jianchang, Duolun, Jiufotang Formations successively, and eventually classified into the Lower Yixian Formation [20,21]. However, the precise age of Sandaogou in Duolun County has been unknown for a long time [22]. This area, as a member of the Yixian Formation, consists of two rock members: the lower one (K1y1) is a set of acidic volcanic-sedimentary rock strata, and the upper one (K1y2) is a set of trachytic volcanic rock strata. In 2019, the LA-ICP-MA zircon U-Pb ages of (132.2 ± 0.6) Ma and (132.8 ± 0.8) Ma were obtained in the rhyolite of the K1y1 rock member by Xue et al. [23]. Considering that the sedimentary rocks in this area exist in K1y1, we tentatively treat 132 Ma as the age of the fossil material in this study, which matches the period of the Jehol Biota (135–120 Ma) [24–27].

The specimens were examined and photographed under a Nikon SMZ 25 stereo microscope with an attached Nikon DS-Ri2 digital camera system, either dry or wetted with 95% ethanol. Wing venation terminology is adapted from Huber and Sharkey [28]. For wing venations, 1-Rs, 2-Rs, 3-Rs, and 4-Rs mean, respectively, the 1st, 2nd, 3rd, and 4th segments of Rs; 1-M and 2-M mean, respectively, the 1st and 2nd segments of M; 1-Cu, 2-Cu, and 3-Cu refer, respectively, to the 1st, 2nd, and 3rd segments of Cu. We treat the extra vein located at the turning of M+Cu as "ap-Cu", which follows Archibald and Rasnitsyn [19].

It is worth noting that wings of specimens are broken at the locations of the anal fold and the jugal fold, for example, the location near vein 1A in the hind wing of specimen CNU-HYM-ND2022105p/c, and vein 1A in both forewings of specimen CNU-HYM-ND2022108. We have deliberately reconstructed the wing venations with reference to other fossil specimens in the line drawing pictures.

3. Results

Systematic Palaeontology.

Order Hymenoptera Linnaeus, 1758.

Suborder Symphyta Gerstaecker, 1867.

Superfamily Pamphilioidea Cameron, 1890.

Family Pamphiliidae Cameron, 1890.

Subfamily Juralydinae Rasnitsyn, 1977.

Type genus. *Juralyda* Rasnitsyn, 1977.

Diagnosis (modified after Jouault et al., 2022, [15]). Antenna with first flagellomere nearly as long as next three flagellomeres and distinctly thicker than these (also in some Cephalciinae, e.g., *Caenolyda* Konow, 1897). Forewing with Sc2 if present/preserved meeting R before 1-Rs (also in all pamphiliid genera except *Ulteramus* Archibald and Rasnitsyn, 2016 [19]). 1-Rs about half as long as, or slightly longer than, 1-M (the only uniquely diagnostic feature of the subfamily). M+Cu angular with or without ap-Cu of various length at elbow (uniquely diagnostic feature of the family Pamphiliidae except that M+Cu is gently bent and lacking ap-Cu in *Tapholyda* Rasnitsyn, 1983, [15]). Cell 1mcu usually shorter than length of pterostigma (except *Dolicholyda* gen. nov.).

Included genera. *Juralyda* Rasnitsyn, 1977; *Scabolyda* Wang, Rasnitsyn, Shih and Ren, 2014; *Atocus* Scudder, 1892; *Tapholyda* Rasnitsyn, 1983 and *Dolicholyda* gen. nov.

Remarks. The present addition of *Dolicholyda* gen. nov. to Juralydinae affects considerably the diagnosis of the subfamily as outlined by Jouault et al. [15]. A long 1-Rs is left the only uniquely diagnostic feature of the subfamily, with the diagnostic meaning of others (given in quotation marks below) changed by this additional genus, or for other reasons as follows:

- 'First flagellomere is distinctly thicker than following flagellomeres, twice to three times as long as second': the same holds true for *Caenolyda* (Cephalciinae); not thicker and less long in *Atocus* [29].
- 'M+Cu gently bent, with ap-Cu extremely short when present': this is correct for *Tapholyda* only, suggesting a secondary loss of the M+Cu angulation in that genus.
- 'cell 1mcu shorter than length of pterostigma': incorrect for *Dolicholyda* gen. nov.
- 'cell 2r much wider than cell 2r-m': the feature is characteristic of the vast majority of Pamphiliidae.
- '3r-m separated from 2m-cu by a distance equal or subequal to its length': the feature is characteristic of the vast majority of Pamphiliidae.

Genus: *Dolicholyda* Zhuang, Rasnitsyn, Shih and Wang, gen. nov.

Etymology: The generic name is a combination of Greek "*dolich-*", meaning long and referring to the long and narrow cell 1mcu, and the generic name *Lyda* (often used in Pamphiliidae). Gender: feminine.

Type species. *Dolicholyda obtusata* Zhuang, Rasnitsyn, Shih and Wang, sp. nov.

Diagnosis. Body surface without punctations. Length-width ratio of flagellomeres less than 3:1 excluding first flagellomere. Forewing with pterostigma lanceolate, not sclerotized inside and sclerotized along margins; Sc developed and bifid, Sc2 merging into R anteriad 1-Rs; Rs+M at least 1.5 times as long as 2-M; angle between 1-M and 1-Cu nearly 90°; cell 1r nearly twice as long as wide; cell 1mcu rectangular, long and narrow, and obviously longer than length of pterostigma; 3r-m inclined toward the wing apex. In most cases, ap-Cu present and of variable length.

Remarks. *Dolicholyda* gen. nov. is assigned to Pamphiliidae based on the combination of forewing with Sc developed; M+Cu angularly bent; antenna with pedicel not inflated, first flagellomere mostly 2–3 times as long as the second. The new genus matches the main diagnostic character of Juralydinae, forewing with 1-Rs distinct and nearly half length of 1-M (short or absent in Pamphiliinae and Cephalciinae) (see Remarks to the subfamily Juralydinae above).

Dolicholyda gen. nov. differs from other Juralydinae in having pterostigma sclerotized only along sides (other genera have pterostigma fully sclerotized except *Juralyda*), cell 1r nearly twice as long as wide (other genera have the ratio less than 1.5 times except twice in *Tapholyda*) and cell 1mcu longer than pterostigma (other genera have 1mcu shorter, except nearly isometric in *Scabolyda latusa* Zhuang, Shih, Wang and Ren, 2022 and *S. tenuis* Zhuang, Shih, Wang and Ren, 2022, ([30], Figures 2F and 4C)). It differs from all genera but *Atocus* by 1-M vertical to 1-Cu (in *Tapholyda* these form an acute angle, in *Juralyda* and *Scabolyda*, an obtuse angle). It differs from *Juralyda* in 3r-m strongly oblique (almost vertical in *Juralyda*). In order to better distinguish the genera of Juralydinae, we establish the relevant key to the identification of different genera in Juralydinae.

Key to the genera of Juralydinae

1. Forewing with Sc not forked..2
Forewing with Sc developed and forked..3
2. Forewing with pterostigma only sclerotized along margins, 3r-m almost vertical.........
....*Juralyda*
Forewing with pterostigma completely sclerotized, 3r-m clearly oblique...............*Atocus*
3. Forewing with M+Cu smoothly bent medially....................................*Tapholyda*
Forewing with M+Cu angularly bent...4
4. Forewing with pterostigma only sclerotized along margins, cell 1mcu obviously longer than length of pterostigma...*Dolicholyda* gen. nov.
Forewing with pterostigma completely sclerotized, cell 1mcu subequal or shorter than length of pterostigma...*Scabolyda*

Dolicholyda obtusata Zhuang, Rasnitsyn, Shih and Wang, sp. nov.

Material. Holotype, CNU-HYM-ND2022105p/c (part and counterpart; Figures 1 and 2) and paratype, CNU-HYM-ND2022106p/c (part and counterpart; Figure 3).

Etymology. The species name is a Latin word and means obtuse, referring to the obtuse apical part of cell a in hind wing.

Locality. Nanyingpan Village, Sandaogou Township, Duolun County, Inner Mongolia, China. Yixian Formation, Lower Cretaceous.

Diagnosis. In addition to the generic diagnosis, forewing with ap-Cu short; pterostigma slender and ca. 4.7 × as long as wide, its basal part blunt; Rs+M ca. 1.7–1.8 × as long as 2-M; 2m-cu angular medially. Hind wing with cell a wide and blunt apically; vein 2A sharply curved in the end.

Description. Holotype. Female. Head, thorax, and abdomen dark brown. Antenna with scape dark brown except the apical part drab; pedicel and possibly first flagellomere drab, other flagellomeres dark brown (Figure 1E).

Head flat and square, width slightly longer than length; clypeal margin slightly wavy; mandibles large and sharp; eyes medium-length. Antenna with 15~16 segments and approximately twice as long as width of head; scape at least twice as long as pedicel, and 1.5 × as wide as pedicel; pedicel 0.3 × as long, and 1.2 × as wide as first flagellomere; first flagellomere not subsegmented and nearly as long as length of next three flagellomeres combined, second flagellomere 2.7 times as long as wide and slightly shorter than third. Remaining flagellomeres gradually shortening.

Thorax without punctations. Pronotum slightly wider than the head (the other structures of thorax indistinct). Legs partly preserved and coxae inverted trapeziform. Abdomen with eight segments preserved, as wide as the thorax. Ovipositor with valvula 1 and valvula 2 partly visible (Figure 2A).

Forewing with pterostigma long and slender, sclerotized along margins, its basal part blunt, nearly 4.7 × as long as wide (Figure 1G,H). Sc bifid, Sc2 entering R before 1-Rs; R bent near the base of 1-Rs. 1-Rs inclined toward wing apex and nearly half the length of 1-M; 2-Rs ca. 0.6 × as long as Rs+M; 3-Rs strongly curved; 4-Rs almost half the length of 1-M. 2r-m located distad to 2r-rs and 1.3 × as long as 2-Rs. 3r-m inclined toward wing apex and meeting 5-Rs nearly at 116°. M+Cu bent sharply; ap-Cu very short and nearly as wide as the width of vein M+Cu (Figure 1C). Rs+M 1.7 × as long as 1-M; angle between 1-M and 1-Cu of 85°. Cell 1mcu nearly twice as long as wide, with cu-a located distad to the middle of the cell. 2-M 0.6 times as long as Rs+M; 3-Cu nearly 2.3 times as long as 1m-cu;

2m-cu curved near the middle. The base of vein 2A+3A having a protrusion at the bend close to vein 1A (Figure 1C). Cell 2a almost 3.5 times as long as wide.

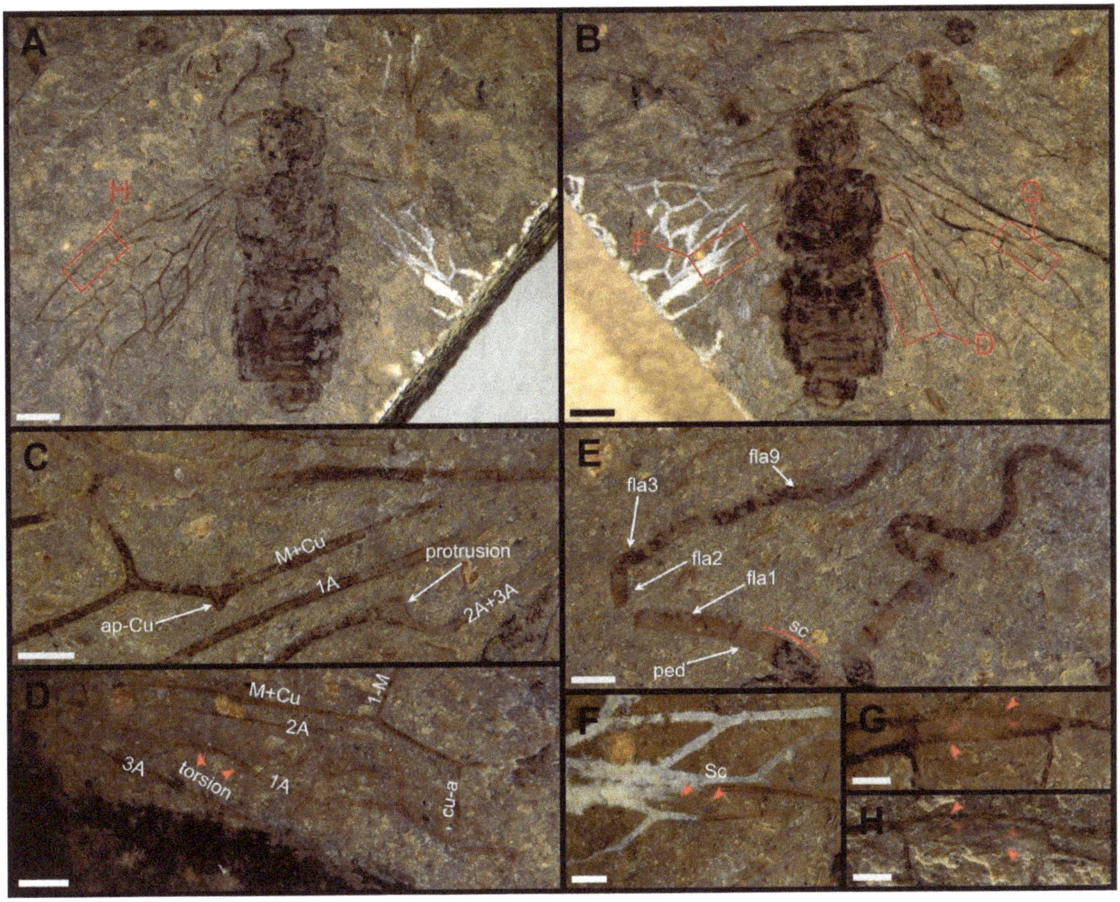

Figure 1. *Dolicholyda obtusata* gen. et sp. nov., Holotype (CNU-HYM-ND2022105p/c), female. (**A**), ventral view. (**B**), dorsal view. (**C**), ap-Cu and protrusion in forewing. (**D**), break and torsion in hind wing. (**E**), antennae. (**F**), Sc in hind wing. (**G**,**H**), pterostigma. Symbols: sc = scape; ped = pedicel; fla1, fla2, fla3, fla9 = 1st, 2nd, 3rd, and 9th flagellomere. Arrows in red mean the venations here are highlighted. Scale bars: 2 mm in (**A**,**B**); 0.5 mm in (**C**–**H**).

Hind wing with Sc present and partly preserved (Figure 1F). 1-Rs nearly as long as 1-M; 1r-m located distad to the base of 1-Rs, and nearly aligned with 1-M; 3r-m inclined toward the wing apex, slightly sigmoidal; m-cu bent, located distad to the middle of cell rm, and separated from 3r-m by a distance of almost 1.2 × its length; cu-a proximad to the middle of cell mcu and meeting 1-Cu at the angle of 117°. 1A nearly straight; 2A sharply bent apically (Figure 1D). Cell a blunt apically with apex width nearly equal to length of 1-M.

Paratype (Figure 3). Sex unknown. Body surface without punctations. Head subcircular. Thorax slightly wider than the width of head. Legs preserved with basal parts. Abdomen with eight segments visible. The structure of genitalia not preserved.

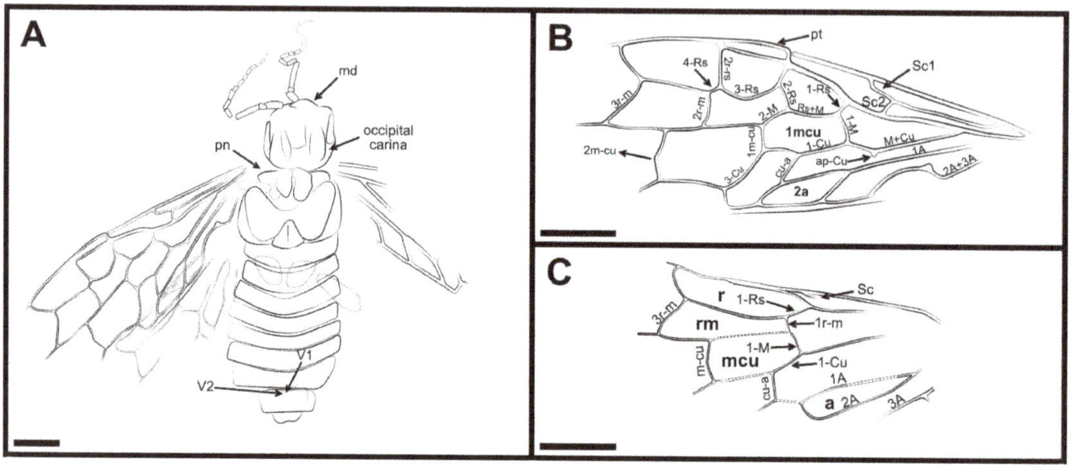

Figure 2. *Dolicholyda obtusata* gen. et sp. nov., (**A**), line drawing of ventral view. (**B**), line drawing of forewing. (**C**), Reconstruction of the hind wing venation (mirror flip for the anal field). Symbols: md = mandible; pn = pronotum; v1, v2 = 1st and 2nd valvifer; pt = pterostigma. Cell names are in bold. Scale bars: 2 mm in (**A–C**).

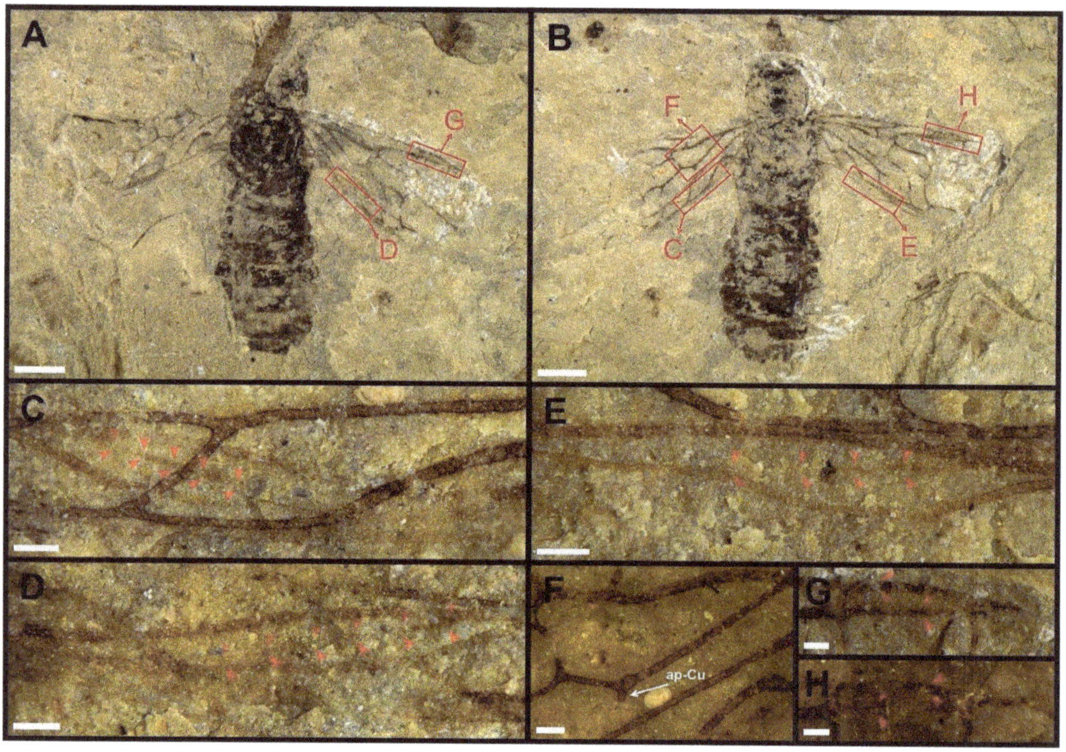

Figure 3. *Dolicholyda obtusata* gen. et sp. nov., Paratype (CNU-HYM-ND2022106p/c), sex unknown. (**A**), dorsal view. (**B**), ventral view. (**C–E**), Sc in hind wing. (**F**), ap-Cu in forewing under alcohol. (**G**), pterostigma. (**H**), pterostigma under alcohol. Arrows in red mean the venations here are highlighted. Scale bars: 2 mm in (**A**,**B**); 0.25 mm in (**C–H**).

Forewing with pterostigma slender and sclerotized along sides, nearly 4.7 × as long as wide (Figure 3G,H); Sc developed and forked, Sc1 entering C slightly distal to the base of 1-Rs. 1-Rs 0.4 times as long as 1-M and inclined toward wing apex. 2-Rs ca. 0.3 × as long as Rs+M. 4-Rs shorter than half of 1-M. 2r-m ca. 1.9 × as long as 2-Rs. Rs+M nearly 1.8 × as long as 2-M. Cell 1mcu longer than pterostigma and nearly twice as long as wide. Vein M+Cu bent, the extra stub ap-Cu extremely short (Figure 3F). 1-M meeting 1-Cu nearly at the angle of 94°. 3-Cu ca. 2.6 × as long as 1m-cu. Cell 2a nearly 3.3 × as long as wide.

In hind wing, cell c narrow in the basal part, slender fusiform. Sc developed and not forked, originating from the middle of vein R, and almost entering apex of vein C; its length occupying half of cell c (Figure 3C–E).

Dimensions (in mm). CNU-HYM-ND2022105p/c: body length (excluding antenna) 14.12; head width 3.23, length 2.95; the first flagellomere as preserved 1.16 in length; forewing length 11.81, width 4.86. CNU-HYM-ND2022106p/c: body length 12.73; head width 2.68, length 2.12; forewing length at least 8.11.

Dolicholyda confluens Zhuang, Rasnitsyn, Shih and Wang, sp. nov.

Material. Holotype, CNU-HYM-ND2022107p/c (part and counterpart; Figure 4).

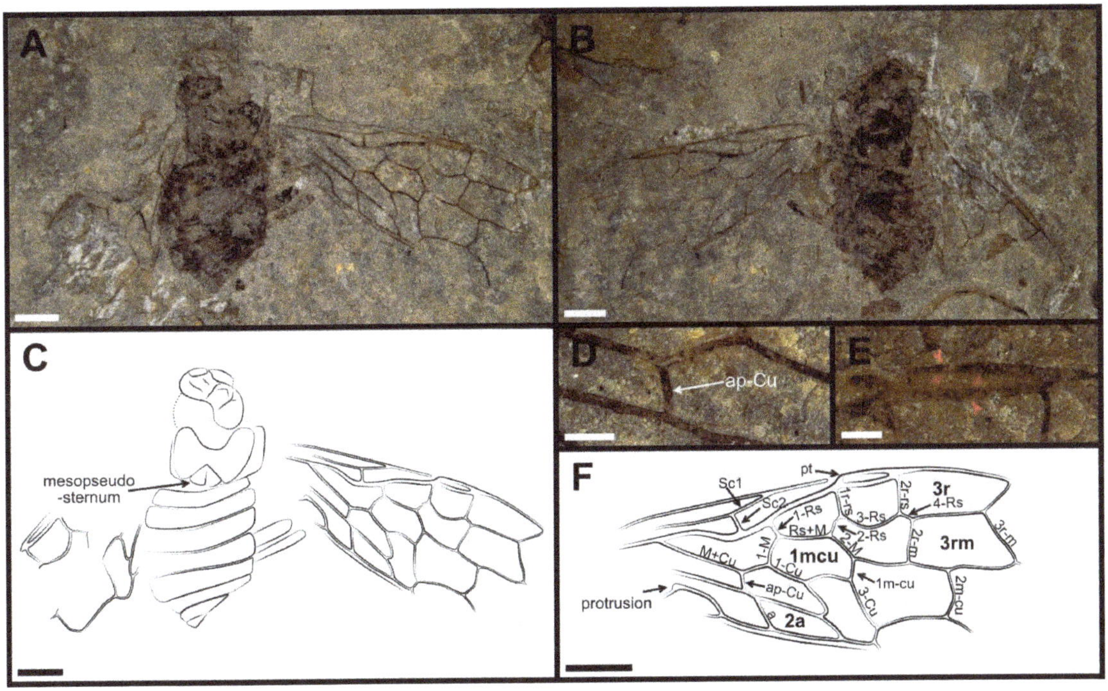

Figure 4. *Dolicholyda confluens* sp. nov., Holotype, (CNU-HYM-ND2022107p/c), sex unknown. (**A**), ventral view. (**B**), dorsal view. (**C**), line drawing of ventral view. (**D**), ap-Cu in forewing. (**E**), pterostigma. (**F**), line drawing of forewing. Arrows in red mean the venations here are highlighted. Cell names are in bold. Scale bars: 2 mm in (**A–C**,**F**); 0.5 mm in (**D**,**E**).

Etymology. The species epithet is a Latin word meaning confluence. It refers to the extremely long ap-Cu joining Cu and 1A in the forewing.

Locality. Nanyingpan Village, Sandaogou Township, Duolun County, Inner Mongolia, China. Yixian Formation, Lower Cretaceous.

Diagnosis. *Dolicholyda confluens* sp. nov. can be easily separated from all other pamphiliids by the extremely long ap-Cu of the forewing, which joins vein 1A in the end.

Additionally, the pterostigma is slender and ca. 4.8 × as long as wide; Rs+M ca. 1.5 × as long as 2-M; 2m-cu angular medially.

Description. Holotype (Figure 4). Sex unknown. The whole body dark brown except head light brown, and its surface lacking punctations (head and thorax damaged with distortion). Mesopseudosternum nearly triangular and large. Abdomen with eight segments, each segment short and wide. The widest segment 1.2 times as wide as thorax. Genitalia not preserved.

Forewing with pterostigma long and narrow, sclerotized around margins, nearly 4.8 × as long as wide (Figure 4E). Sc bifid, Sc2 joining R before the base of 1-Rs and obviously shorter than Sc1. R strongly bent at the confluence with Sc2. 1-Rs inclined toward wing apex, and ca. 0.6 × as long as 1-M; 3-Rs strongly arched; 4-Rs nearly half the length of 1-M. 2r-m parallel to 2r-rs and nearly twice as long as 2-Rs. 3r-m inclined toward wing apex. Cell 3rm as wide as cell 3r in its basal part. Rs+M 1.6 × as long as 1-M. Angle between 1-M and 1-Cu nearly of ca. 95°. M+Cu bending sharply distad to the middle part, ap-Cu stub extremely long and entering vein A anteriad crossvein a (Figure 4D). Rs+M ca. 1.5 × as long as 2-M. Cell 1mcu slightly longer than pterostigma and 1.7 × as long as wide. 3-Cu 2.7 × as long as 1m-cu; 2m-cu curved near its middle. Cell 2a almost 3.1 × as long as wide. Hind wing not preserved.

Dimensions (in mm). Holotype, CNU-HYM-ND2022107p/c: body length at least 11.87; head width 3.36; forewing length 13.00, width 5.60.

Dolicholyda angusta Zhuang, Rasnitsyn, Shih and Wang, sp. nov.

Material. Holotype, CNU-HYM-ND2022108 (Figure 5).

Etymology. The species name is a Latin word meaning narrow, referring to narrow cell a in hind wing.

Locality. Nanyingpan Village, Sandaogou Township, Duolun County, Inner Mongolia, China. Yixian Formation, Lower Cretaceous.

Diagnosis. In addition to the generic diagnosis, forewing with ap-Cu absent; pterostigma wide and ca. 3.5 × as long as wide, its basal part acute; Rs+M ca. 1.9 × as long as 2-M 2m-cu almost straight medially. Hind wing with cell a relatively narrow apically; vein 2A smoothly curved in the end.

Description. Holotype (Figure 5). Female. Body surface sepia and lacking punctations. Antennae partly preserved and dark brown.

Head rounded rectangular, transverse, with occipital carina apparently complete, with submedial branches toward occipital foramen, with structures delimited below occipital foramen possibly representing postgenae (lower part of postocciput) and lower tentorial bridge comparable to those in living Pamphiliidae ([31], Figure 136) except more narrow lower tentorial bridge (Figure 5E). Antennae obviously longer than the width of head, with at least sixteen segments; dark brown. Several terminal flagellomeres equal in lengths and widths, except apical flagellomere nearly twice as long as wide. Thorax poorly preserved. Legs short, poorly preserved. Abdomen with nine segments and the middle part inflated. Ovipositor very short, valvula 1 and valvula 2 widely separated basally and meeting only at the apex, valvula 3 located on both sides of valvula 2.

Forewing with pterostigma relatively wide, sclerotized at sides and nearly 3.5 × as long as wide, its basal part slightly acute. Sc bifid, Sc2 entering R anteriad base of 1-Rs. 1-Rs ca. 0.4 × as long as 1-M. 2-Rs nearly half length of Rs+M. 3-Rs curved, 4-Rs short and ca. 0.3 × as long as 1-M. 2r-m ca. 1.7 × as long as 2-Rs. 3r-m inclined toward apical part of wing. ap-Cu absent (Figure 5F). Angle between 1-M and 1-Cu nearly of 80°. Length proportions of vein 1-M: Rs+M: 1-Cu: 2-M = 1: 1.9: 1.7: 1; cu-a located slightly distad to the middle of cell 1mcu, 3-Cu ca. 2.5 × as long as 1m-cu. Cell 1mcu 2.2 × as long as wide and obviously longer than pterostigma. 2m-cu slightly curved. Cell 2a nearly 3.5 × as long as wide.

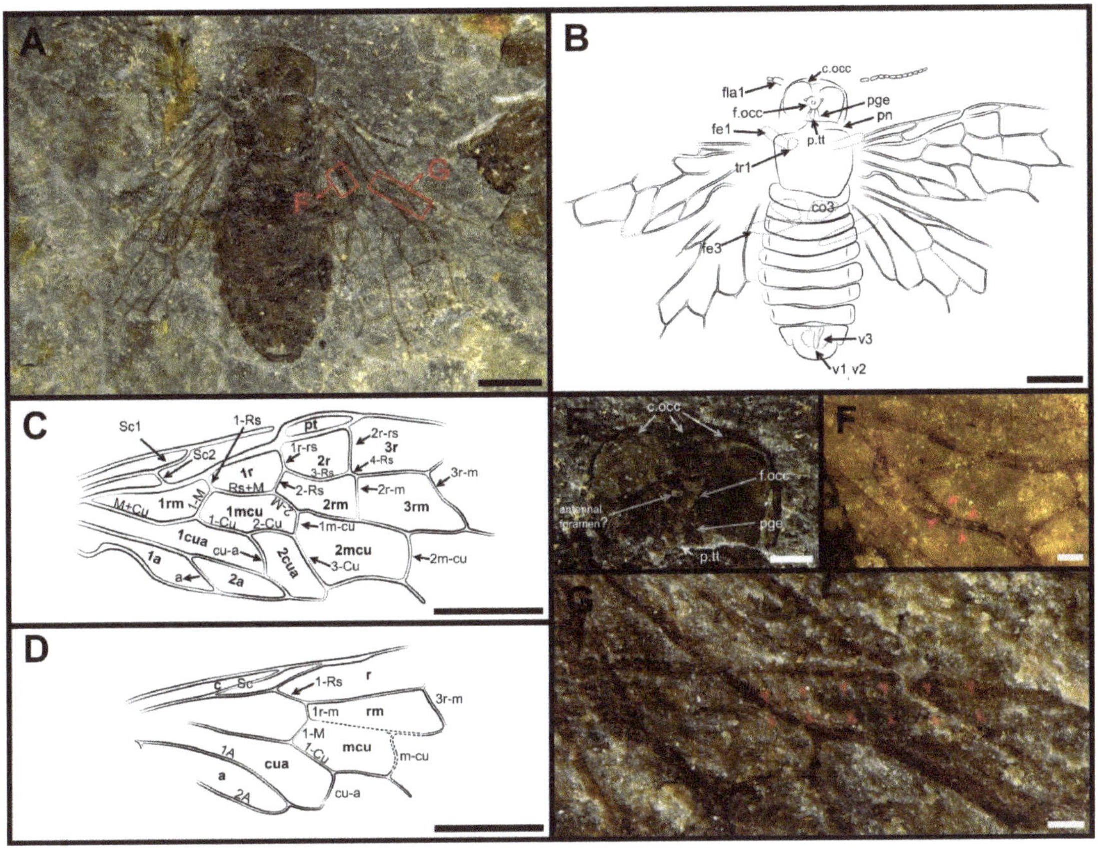

Figure 5. *Dolicholyda angusta* sp. nov., Holotype, (CNU-HYM-ND2022108), female. (**A**), dorsal view. (**B**), line drawing of dorsal view with forewing and hind wing artificially rotated from body and each other for better viewing. (**C**), line drawing of forewing after reconstruction (mirror flip for the anal field). (**D**), line drawing of hind wing. (**E**), head under alcohol. (**F**), vein M+Cu in forewing under alcohol. (**G**), Sc in hind wing. Symbols: fe1, fe3 = fore and hind femur; tr1 = fore trochanter; v3 = 3rd valvifer; co3 = hind coxa; c.occ = occipital carina; f.occ = occipital foramen; pge = postgena; p.tt = tentorial bridge. Cell names are in bold. Arrows in red mean the venations here are highlighted. Scale bars: 2 mm in (**A–D**); 0.5mm in (**E**); 0.25 mm in (**F**,**G**).

Hind wing with Sc developed, Sc extending out from the middle part of R and its length twice as long as 1-Rs (Figure 5G). 1-Rs ca. 2.5 × as long as 1r-m. 1-M nearly 1.9 × as long as 1r-m and angle between them of 144°; cu-a located at middle of cell mcu and angle between 1-Cu and cu-a nearly 100°. 1A gently curved throughout its length, 2A bent apically. Cell a narrow and its apical part acute.

Dimensions (in mm). Holotype: body length (excluding antenna) 9.95; head width 2.38, length 1.87; antenna at least 2.81 in length; forewing length 8.22, width 3.46.

Remarks. A small round structure clearly visible within the occipital foramen is difficult to interpret: it could be an antennal attachment orifice or, rather, simply a damage in the head cuticle.

In *Dolicholyda* gen. nov., we can easily distinguish species by the length of ap-Cu, which is absent in *D. angusta* sp. nov., short in *D. obtusata* sp. nov. and extremely long (even joins 1A) in *D. confluens* sp. nov. Leaving this feature aside, *D. obtusata* sp. nov. show much similarity with *D. confluens* sp. nov., especially on the characters of forewing with pterostigma slender and ca. 4.7 × as long as wide, its basal part blunt; 2m-cu angular medially. In contrast, in *D. angusta* sp. nov., pterostigma wide and ca. 3.5 × as long as wide, its basal part acute; 2m-cu almost straight medially. Additionally, *D. obtusata* sp. nov. also differs from *D. angusta* sp. nov. in hind wing with cell a blunt apically vs. acute in *D. angusta* sp. nov.

4. Discussion

The discovery of *Dolicholyda* gen. nov. in the Early Cretaceous Yixian Formation, NE China, sheds new light on the diversity and the characteristics of the subfamily Juralydinae and the family Pamphiliidae in general. The present study indicates wide morphological diversity of the subfamily, including the characters that were earlier treated as diagnostic features of the subfamily, such as length of cell 1mcu, presence or absence of angular bend of M+Cu, and the presence and length of ap-Cu at that bend. As a result, a long 1-Rs remains the only uniquely diagnostic character of Juralydinae.

The vein ap-Cu is present in most genera of extant Pamphiliidae while showing a wide diversity of variations in its length, except in *Acantholyda* with ap-Cu usually absent or extremely short [32]. However, among fossil pamphiliid taxa, most species lack ap-Cu or only possess an extremely short ap-Cu. Up to now, only *Atocus*, *Ulteramus* and our new genus *Dolicholyda* gen. nov. have ap-Cu (extremely short in *Atocus*). According to previous studies, the length change of ap-Cu is subject to intraspecific variation [18]. Shinohara [33] found that the length of ap-Cu was unstable in the same species of *Acantholyda*, sometimes absent or very short, and rarely long. But in general, variation is feeble. Our new genus also displayed varying lengths of ap-Cu in the Lower Cretaceous Pamphiliidae. Furthermore, *Dolicholyda confluens* has extremely long ap-Cu (Figure 4D,F), which is rare in fossils and there were no cases of ap-Cu joining 1A documented before. This unique case provides a new extreme value for the length variation of ap-Cu and indicates that ap-Cu absent or extremely short is not the stable character for Juralydinae.

Besides the stub of ap-Cu, there are also some bud-like protrusions at the bend of 2A+3A in the forewings of *D. obtusata* and *D. confluens* (Figures 1C and 4F). Similarly, some other records about wing variations in fossil Symphyta have been also reported in the Mesozoic sawflies (mainly referring to supernumerary veins there), such as *Chionoxyela nivea* Rasnitsyn, 1993; *Shartexyela mongolica* Rasnitsyn, 2008; *Xyelocerus diaphanous* Gao, Ren and Shih, 2009; *Hoplitolyda duolunica* Gao, Shih, Rasnitsyn and Ren, 2013; *Potrerilloxyela menendez* Lara, Rasnitsyn and Zavattieri, 2014; *Aethotoma aninomorpha* Gao, Shih, Engel and Ren, 2016; *Pamparaphilius khasurtensis* Kopylov and Rasnitsyn, 2017; *Trematothorax extravenosus* Kopylov and Rasnitsyn, 2017; and *Brachyoxyela conjunctiva* Dai, Shih, and Rasnitsyn, 2022 [34–42]. The variation of their wing veins looks irregular. Extra veins appear transversal, longitudinal, or annular. They may appear anywhere on the forewing or hind wing, located close to the wing base, middle or apex (Figure 6). Judging from the species referred to above, these abnormalities are more likely to occur in the wing apex (six out of eight of them in the apical part). To a certain extent, it reflects the instability of wing venations in Symphyta. Interestingly, the variations in *Chionoxyela nivea* and *Hoplitolyda duolunica* are located at the bend of M+Cu, which is similar to that of Pamphiliidae, except that it is inverted in direction (i.e., directed toward the anterior margin of the wing; Figure 6C,D).

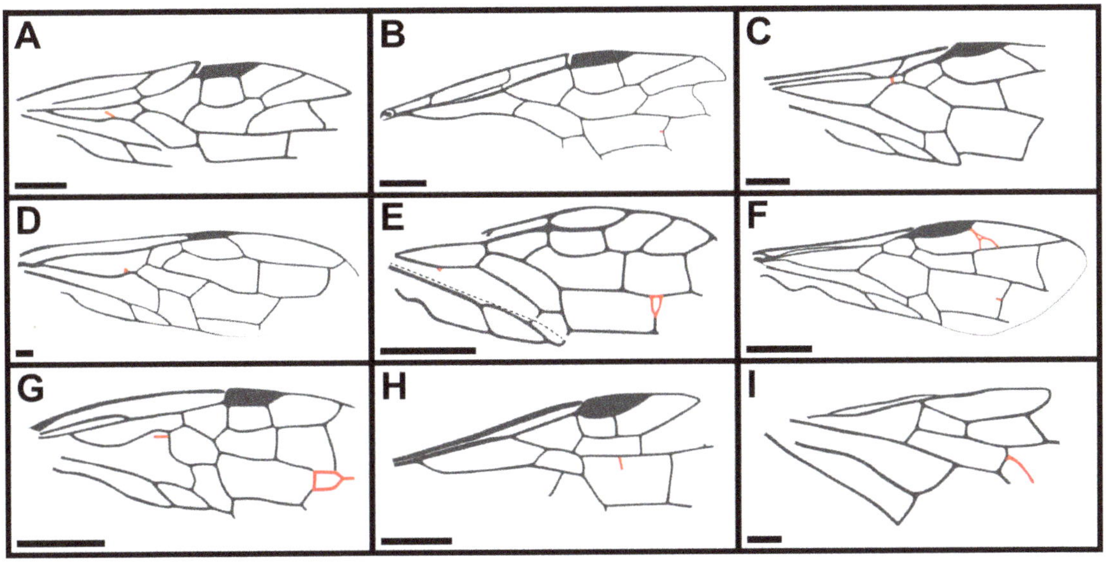

Figure 6. Several examples of wing venation variation in fossil Symphyta. (**A**), *Chionoxyela nivea* Rasnitsyn, 1993 (Xyelidae) ([34], Figure 12). (**B**), *Shartexyela mongolica* Rasnitsyn, 2008 (Xyelidae) ([35], Figure 1B). (**C**), *Xyelocerus diaphanous* Gao, Ren and Shih, 2009 (Xyelotomidae) ([36], Figure 2B). (**D**), *Hoplitolyda duolunica* Gao, Shih, Rasnitsyn and Ren, 2013 (Praesiricidae) ([37], Figure 3B). (**E**), *Potrerilloxyela menendez* Lara, Rasnitsyn and Zavattieri, 2014 (Xyelidae) ([38], Figure 4C). (**F**), *Aethotoma aninomorpha* Gao, Shih, Engel and Ren, 2016 (Xyelotomidae) ([39], Figure 2D). (**G**), *Pamparaphilius khasurtensis* Kopylov and Rasnitsyn, 2017 (Sepulcidae) ([40], Figure 1). (**H**), *Trematothorax extravenosus* Kopylov and Rasnitsyn, 2017 (Sepulcidae) ([41], Figure 5). (**I**), *Brachyoxyela conjunctiva* Dai, Shih and Rasnitsyn, 2022 (Xyelidae) ([42], Figure 3C). Scale bars: 2 mm in (**A–I**). Notes: specimens of *Hoplitolyda duolunica*, *Xyelocerus diaphanous* and *Aethotoma aninomorpha* have both wings preserved and only *Hoplitolyda duolunica* has symmetric venation variation. Other specimens have only a single wing preserved.

The presence of vein Sc in the hind wing is a feature inherited from a non-hymenopteran ancestor. According to the statistics of Vilhelmsen [1], only some of the primitive taxa currently have this feature, such as the Xyelidae and the Pamphiliidae. In more derived taxa, there are more records of Sc in the extinct fossil groups, such as Xyelydidae Rasnitsyn, 1968; Xyelotomidae Rasnitsyn, 1968; Siricidae Billberg, 1820 and Mirolydidae Wang, Rasnitsyn and Ren, 2017 [4,36,43]. Throughout the extant groups of Pamphiliidae, vein Sc is long and parallel to R (Figure 7A–D). Its anterior branch Sc1 joins C near the apex of cell c and the posterior branch Sc2 joins R at two-thirds of the length of vein R [32,44]. By contrast, in *Dolicholyda*, Sc is very short, not forked and slightly shorter than half length of cell c. It extends out from the middle part of R and enters C in its apical part (Figure 3C–E, Figures 5G and 7F). In *Scabolyda*, Sc is appressed to R except distally (a double vein is maybe present, Figure 7E). In general, there are some differences between the Sc of extant Pamphiliidae and fossil genera (*Dolicholyda* and *Scabolyda*). However, its diagnostic value is obscure, and we need more specimens to define its value as a diagnostic feature.

Figure 7. Sc in hind wings of Pamphiliidae. (**A**), *Acantholyda posticalis* Matsumura, 1912, ♀(Cephalciinae). (**B**), *Cephalcia yanqingensis* Xiao, 1987, ♀(Cephalciinae). (**C**), *Pamphilius ignymontiensis* Lacourt, 1973, ♂(Pamphiliinae). (**D**), *Neurotoma sibirica* Gussakovskij, 1935, ♀(Pamphiliinae). (**E**), *Scabolyda latusa* Zhuang, Shih, Wang and Ren, 2022, ♀(Juralydinae). (**F**), *Dolicholyda angusta* sp. nov., ♀(Juralydinae). Cell c are in bold. Arrows here mean the venations are highlighted. Scale bars: 0.5 mm in (**A**–**F**).

5. Conclusions

Dolicholyda obtusata gen. et sp. nov., *D. confluens* sp. nov., and *D. angusta* sp. nov. are described based on four fossil specimens from the Lower Cretaceous Yixian Formation in Duolun County, Inner Mongolia, China. The new species display varying lengths of ap-Cu in their forewings and an unforked Sc in the hind wings. We treat the long 1-RS as the only valid diagnostic character of Juralydinae based on the new material described herein. These new findings enhance our understanding of the wing venation characteristics of fossil pamphiliids and expand the diversity of Pamphiliidae in the Mesozoic.

Author Contributions: Conceptualization, J.Z. and M.W.; resources, D.R.; writing—original draft preparation, J.Z.; and M.W.; writing—review and editing, M.W., C.S. and A.P.R.; validation, C.S., D.R.

and M.W.; visualization, J.Z. and C.S.; supervision, M.W. and A.P.R.; project administration, M.W.; funding acquisition, M.W., A.P.R. and D.R. All authors have read and agreed to the published version of the manuscript.

Funding: In this work: M.W. was supported by the Fundamental Research Funds for the Central Non-profit Research Institute of the Chinese Academy of Forestry (Grant No. CAFYBB2019QB004) and the Youth Talent Support Program for Science & Technology Innovation of National Forestry and Grassland (2020132602). A.P.R. was partially supported by the Presidium of the Russian Academy of Sciences Program 'Origin of the Biosphere and Evolution of Geobiological Systems'. D.R. was supported by grants from the National Natural Science Foundation of China (42288201, 32020103006).

Institutional Review Board Statement: Not applicable.

Informed Consent Statement: Not applicable.

Data Availability Statement: All data from this study are available in this paper and the associated papers.

Acknowledgments: We would like to give our sincere gratitude to the editor and the two reviewers for providing us with valuable suggestions and comments to improve this manuscript. We are grateful to Liyang Dai, Jiajia Wang, Rui Dai (Capital Normal University, Beijing, China) for their helpful advice and Mengfei Liu (Chinese Academy of Forestry, Beijing, China) for the supply of photos about extant pamphiliids.

Conflicts of Interest: The authors declare no conflict of interest.

References

1. Vilhelmsen, L. Phylogeny and classification of the extant basal lineages of the Hymenoptera (Insecta). *Zool. J. Linn. Soc.* **2001**, *131*, 393–442. [CrossRef]
2. Ronquist, F.; Klopfstein, S.; Vilhelmsen, L.; Schulmeister, S.; Murray, D.L.; Rasnitsyn, A.P. A total-evidence approach to dating with fossils, applied to the early radiation of the Hymenoptera. *Syst. Biol.* **2012**, *61*, 973–999. [CrossRef] [PubMed]
3. Peters, R.; Krogmann, L.; Mayer, C.; Donath, A.; Gunkel, S.; Meusemann, K.; Kozlov, A.; Podsiadlowski, L.; Petersen, M.; Lanfear, R.; et al. Evolutionary history of the Hymenoptera. *Curr. Biol.* **2017**, *27*, 1013–1018. [CrossRef] [PubMed]
4. Wang, M.; Rasnitsyn, A.P.; Yang, Z.; Shih, C.K.; Wang, H.; Ren, D. Mirolydidae, a new family of Jurassic pamphilioid sawfly (Hymenoptera) highlighting mosaic evolution of lower Hymenoptera. *Sci. Rep.* **2017**, *7*, 43944. [CrossRef]
5. Rasnitsyn, A.P. Fossil Hymenoptera of the superfamily Pamphilioidea. *Paleontol. J.* **1983**, *2*, 56–70. (In Russian)
6. Rasnitsyn, A.P.; Zhang, H.C.; Wang, B. Bizarre fossil insects: Web-spinning sawflies of the genus *Ferganolyda* (Vespida, Pamphilioidea) from the Middle Jurassic of Daohugou, Inner Mongolia, China. *Paleontology* **2006**, *49*, 907–916. [CrossRef]
7. Wang, M.; Rasnitsyn, A.P.; Shih, C.K.; Ren, D. A new fossil genus in Pamphiliidae (Hymenoptera) from China. *Alcheringa* **2014**, *38*, 391–397. [CrossRef]
8. Huang, D.Y.; Cai, C.Y.; Fu, Y.Z.; Su, Y.T. The Middle- Late Jurassic Yanliao entomofauna. *Palaeoentomology* **2018**, *1*, 3–31. [CrossRef]
9. Lian, X.N.; Cai, C.Y.; Huang, D.Y. The early assemblage of Middle–Late Jurassic Yanliao biota: Checklist, bibliography and statistical analysis of described taxa from the Daohugou beds and coeval deposits. *Palaeoentomology* **2021**, *4*, 95–136. [CrossRef]
10. Rasnitsyn, A.P. New Hymenoptera from the Jurassic and Cretaceous of Asia. *Paleontol. J.* **1977**, *3*, 98–108. (In Russian)
11. Abe, M.; Smith, D.R. The Genus-Group Names of Symphyta (Hymenoptera) and Their Type Species. *Esakia* **1991**, *31*, 1–115. [CrossRef]
12. Taeger, A.; Blank, S.M.; Liston, A.D. World catalog of Symphyta (Hymenoptera). *Zootaxa* **2010**, *2580*, 1–1064. [CrossRef]
13. Benson, R.B. Classification of the Pamphiliidae (Hymenoptera Symphyta). In *Proceedings of the Royal Entomological Society of London*; Series B Taxonomy; Blackwell Publishing Ltd.: Oxford, UK, 1945; Volume 14, pp. 25–33.
14. Van Achterberg, C.; van Aartsen, B. The European Pamphiliidae (Hymenoptera: Symphyta), with special reference to The Netherlands. *Zool. Verh.* **1986**, *234*, 1–98.
15. Jouault, C.; Wei, M.C.; Niu, G.Y.; Nel, A. Revision of the Oligocene pamphiliid genus and species *Tapholyda caplani* (Cockerell, 1933). *Palaeoentomology* **2022**, *5*, 173–182. [CrossRef]
16. Riou, B. Descriptions de quelques insectes fossiles du Miocène supérieur de la Montagne d'Andance (Ardèche, France). *École Pra-Tique Hautes Études Biol. Evol. Insectes* **1999**, *11*, 123–133.
17. Penãlver, E.; Arillo, A. Primer registro fósil del género *Acantholyda* (Insecta: Hymenoptera: Pamphiliidae), Mioceno Inferior de Ribesalbes (España). *Rev. Esp. Paleontol.* **2002**, *17*, 73–81.
18. Nel, A. New and poorly known Cenozoic sawflies of France (Hymenoptera, Tenthredinoidea, Pamphilioidea). *Dtsch. Entomol. Z.* **2004**, *51*, 253–269. [CrossRef]
19. Archibald, S.B.; Rasnitsyn, A.P. New early Eocene Siricomorpha (Hymenoptera: Symphyta: Pamphiliidae, Siricidae, Cephidae) from the Okanagan Highlands, western North America. *Can. Entomol.* **2015**, *148*, 209–228. [CrossRef]

20. Bai, Z.; Xu, D.; Gu, D.; Tian, M.; Zhao, G.; Sun, K. *Five 1:50,000 Regional Geological Survey Reports in Duolun County, etc.*; China University of Geosciences Press: Hubei, China, 2000. (In Chinese)
21. Wan, X.Q.; Chen, P.J.; Wei, M.J. The Cretaceous system in China. *Acta. Geol. Sin.* **2007**, *81*, 957–983.
22. Jin, F.; Zhang, F.C.; Li, Z.H.; Zhang, J.Y.; Li, C.; Zhou, Z.H. On the horizon of *Protopteryx* and the early vertebrate fossil assemblages of the Jehol Biota. *Chin. Sci. Bull.* **2008**, *53*, 2820–2827. [CrossRef]
23. Xue, W.; Song, K.; Liu, X.; Peng, Y.; Shan, P.; Gao, L.; Zhi, C. Zircon U-Pb Chronology of Mesozoic Volcanic Rocks in Sandaogou Area, Duolun County, Inner Mongolia and Its Ore-Prospecting Significance. *Geol. Sci. Technol. Inf.* **2019**, *38*, 12. (In Chinese)
24. Ren, D.; Shih, C.K.; Gao, T.P.; Yao, Y.Z.; Wang, Y.J. *Rhythms of Insect Evolution: Evidence from the Jurassic and Cretaceous in Northern China*; Wiley Blackwell: New York, NY, USA, 2019; pp. 1–708.
25. Yang, S.H.; He, H.Y.; Jin, F.; Zhang, F.C.; Wu, Y.B.; Yu, Z.Q.; Li, Q.L.; Wang, M.; O'Connor, J.K.; Deng, C.; et al. The appearance and duration of the Jehol Biota: Constraint from SIMS U-Pb zircon dating for the Huajiying Formation in northern China. *Proc. Natl. Acad. Sci. USA* **2020**, *117*, 14299–14305. [CrossRef] [PubMed]
26. Gao, T.P.; Shih, C.K.; Ren, D. Behaviors and Interactions of Insects in Mid-Mesozoic Ecosystems of Northeastern China. *Annu. Rev. Entomol.* **2021**, *66*, 337–354. [CrossRef]
27. Xiao, L.F.; Labandeira, C.C.; Ren, D. Insect herbivory immediately before the eclipse of the gymnosperms: The Dawangzhangzi plant assemblage of Northeastern China. *Insect Sci.* **2022**, *29*, 1483–1520. [CrossRef]
28. Huber, T.H.; Sharkey, M.J. Structure. In *Hymenoptera of the World: An Identification Guide to Families*; Goulet, H., Huber, T.H., Eds.; Research Ottawa: Ottawa, ON, Canada, 1993; pp. 13–59.
29. Wang, M.; Rasnitsyn, A.P.; Li, H.; Shih, C.K.; Sharkey, J.M.; Ren, D. Phylogenetic analyses elucidate the inter-relationships of Pamphilioidea (Hymenoptera, Symphyta). *Cladistics* **2016**, *32*, 407–411. [CrossRef] [PubMed]
30. Zhuang, J.L.; Shih, C.K.; Wang, M.; Ren, D. Two New Fossil Sawflies of *Pamphiliidae* (Hymenoptera: Symphyta) from the Mesozoic of Northeastern China. *Insects* **2022**, *13*, 402. [CrossRef] [PubMed]
31. Rasnitsyn, A.P. *Origin and Evolution of Lower Hymenoptera*; Nauka: Moscow, Russia, 1969; Volume 123, pp. 1–196. (In Russian)
32. Jiang, J.G.; Wei, M.C. Two new species of *Onycholyda* Tacheuchi (Hymenoptera: Pamphiliidae) from China. *Entomotaxonomia* **2004**, *26*, 132–135. (In Chinese)
33. Shinohara, A. Conifer-feeding webspinning sawflies of the genus *Acantholyda* (Hymenoptera, Pamphiliidae) of Japan. *Species Divers.* **2001**, *6*, 23–63. [CrossRef]
34. Gromov, V.V.; Dmitriev, V.Y.; Zherikhin, V.V.; Lebedev, E.L.; Ponomarenko, A.G.; Rasnitsyn, A.P.; Sukatsheva, I.D. Cretaceous insects from Uljya River basin (West Okhot region). In *Mesozoic Insects and Ostracods from Asia*; Ponomarenko, A.G., Ed.; Nauka: Moscow, Russia, 1993; Volume 252, pp. 5–60. (In Russian)
35. Rasnitsyn, A.P. Hymenopterous insects (Insecta: Vespida) in the Upper Jurassic deposits of Shar Teg, SW Mongolia. *Russ. Entomol. J.* **2008**, *17*, 299–310.
36. Gao, T.P.; Ren, D.; Shih, C.K. The First Xyelotomidae (Hymenoptera) From the Middle Jurassic in China. *Ann. Entomol. Soc. Am.* **2009**, *102*, 588–596. [CrossRef]
37. Gao, T.P.; Shih, C.K.; Rasnitsyn, A.P.; Ren, D. *Hoplitolyda duolunica* gen. et sp. nov. (Insecta, Hymenoptera, Praesiricidae), the Hitherto Largest Sawfly from the Mesozoic of China. *PLoS ONE* **2013**, *8*, e62420. [CrossRef] [PubMed]
38. Lara, M.B.; Rasnitsyn, A.P.; Zavattieri, A.M. *Potereilloxyela menendezi* gen. et sp. nov. from the Late Triassic of Argentina: The oldest representative of Xyelidae (Hymenoptera: Symphyta) for Americas. *Paleontol. J.* **2014**, *48*, 182–190. (In Russian) [CrossRef]
39. Gao, T.P.; Shih, C.K.; Ren, D. A new xyelotomid (Hymenoptera) from the Middle Jurassic of China displaying enigmatic venational asymmetry. *BMC Evol. Biol.* **2016**, *16*, 155. [CrossRef] [PubMed]
40. Kopylov, D.S.; Rasnitsyn, A.P. New sepulcids (Hymenoptera: Sepulcidae) from the Lower Cretaceous of Asia. I. Parapamphiliinae and Xyelulinae. *Paleontol. J.* **2017**, *51*, 69–76. [CrossRef]
41. Kopylov, D.S.; Rasnitsyn, A.P. New sepulcids (Hymenoptera: Sepulcidae) from the Lower Cretaceous of Asia: II. Ghilarellinae and Trematothoracinae. *Paleontol. J.* **2017**, *51*, 291–303. [CrossRef]
42. Dai, L.Y.; Rasnitsyn, A.P.; Shih, C.K.; Wang, M.; Ren, D. New fossil Xyelidae (Insecta, Hymenoptera) from the Lower Cretaceous Yixian Formation of northeast China. *Cretac. Res.* **2022**, *137*, 105249. [CrossRef]
43. Rasnitsyn, A.P. New Mesozoic sawflies (Hymenoptera, Symphyta). In *Jurassic Insects of Karatau*; Rohdendorf, B.B., Ed.; Nauka: Moscow, Russia, 1968; pp. 190–236. (In Russian)
44. Wei, M.C. *Fauna of Insects from Xiaowutai Mountain*; Ren, G.D., Guo, S.B., Zhang, F., Eds.; Hebei University: Baoding, China, 2013; pp. 392–393. (In Chinese)

insects

MDPI

Article

The Association between the Distribution of Resin Beads and the Emergence of *Sirex noctilio* on Red Pine in North America

Hajar Faal [1,2,*] and Stephen A. Teale [1]

1 Department of Environmental and Forest Biology, College of Environmental Science and Forestry, State University of New York, Syracuse, NY 13210, USA
2 Forest Pest Methods Laboratory, USDA-APHIS-PPQ, 1398 W. Truck Rd., Buzzards Bay, Barnstable, MA 02542, USA
* Correspondence: hajar.faal@gmail.com

Simple Summary: The European woodwasp, *Sirex noctilio,* is an invasive pest in North America. The presence of resin beads as signs of female ovipositor activity has been effectively used in ground surveillance. Improving our understanding of the relationship between resin bead density and the number of emerging *S. noctilio,* its parasitoid, and native woodwasp strengthens integrated pest management programs. We also investigated the resin bead density associated with the height, diameter, or cardinal direction on red pines. Results indicated resin bead density was significantly associated with only the emergence of *S. noctilio,* not the emergence of its parasitoid, or native woodwasp. Resin beads were also most abundant on the north, east, and south sides of logs at 4.5 m above the ground. This study supports the use of resin beads as a first visible sign of *S. noctilio* colonization of host trees in North America.

Abstract: This study examined the relationships of the abundance and distribution of resin beads (signs of *Sirex noctilio* parent female ovipositor activity) with the abundance and distribution of emerging progeny of *S. noctilio, S. nigricornis* and their parasitoid *Ibalia leucospoides. S. noctilio* is native to Europe and is an invasive pest of pines in the Southern Hemisphere and North America; *S. nigricornis* is native to North America and is a secondary pest of dying pines. *I. leucospoides* is a parasitoid that has been widely deployed for biological control of *S. noctilio.* This study aimed to determine if the distribution of resin beads is associated with the height, diameter, or cardinal direction on red pines, *Pinus resinosa,* as well as the distribution of wood wasp and parasitoid emergence. Our results showed that among log sections taken at five heights, resin beads were most abundant on the north, east, and south sides of logs and mid log at 4.5 m above the ground. Emergence of *S. noctilio* was most abundant only from logs with more than five resin beads per square meter, while diameter and height were not contributing factor. None of variables evaluated (resin bead densities, height, and diameter) had significant effects on the emergence of *S. nigricornis* and *I. leucospoides.* These findings help clarify the biological significance of resin beads as indicators of *S. noctilio* colonization of host trees in North America.

Keywords: red pine; resin beads; cardinal direction; wood borers; *Sirex*; *Ibalia*

1. Introduction

The Eurasian woodwasp, *Sirex noctilio* (Hym.: Siricidae), is an invasive wood borer that is attracted to physiologically stressed pine trees of suppressed and intermediate crown position [1]. Physiological stresses intensify tissue permeability and result in enhanced monoterpene production [2], which increases the probability of *S. noctilio* attack. Monoterpene composition influences host selection by *S. noctilio,* and the amount and duration of monoterpene production are affected by tree resistance as well as the degree and duration of stress [3,4].

S. noctilio attack induces increased biosynthesis and accumulation of resin [5]. The resin fills the *S. noctilio* drill holes and egg galleries while a small amount flows to the bark surface forming a droplet which crystallizes into a resin bead [6–8]. However, resin pressure declines in stressed hosts [2], and *S. noctilio* attack intensifies the original stress and causes an additional drop in resin flow by injecting a phytotoxic venom (mucus) and a pathogenic fungus, *Amylostereum areolatum*, in their drill holes [9]. Due to the toxicity and quantity of venom, as well as the pathogenicity of the fungus, *S. noctilio* is the only siricid with the ability to overcome tree defenses and eventually kill the host tree [10].

Native *Sirex* species, wood boring beetles, and bark beetles commonly occur in the same host with *S. noctilio* [11–20]. *S. nigricornis*, a native siricid in North America, typically attacks dying pine trees [21], but may occur in trees attacked by *S. noctilio* because it exploits the fungal symbiont deposited by *S. noctilio* [20]. Subcortical interspecific competition can adversely influence biological control programs using the nematode *Deladenus siridicola*. For instance, *Ips* bark beetles often arrive first and promptly inoculate trees with blue stain fungi [15,16,22]. In addition, wood-boring beetle larvae directly compete with *S. noctilio* larvae for nutrition and space [23] or feed on *S. noctilio* larvae [24]. These arboreal pests may limit the access of *S. noctilio* to nutrition sources, depending on which species arrives first [23].

The parasitic wasp, *Ibalia leucospoides* (Hymenoptera: Ibaliidae), has been widley introduced as a biological control agent against *S. noctilio* in the Southern Hemisphere [25,26]. The extensive use of *Ibalia* as a parasitoid in *Sirex* management prompted many investigations into its biology and ecology [26–28]. *Ibalia* females use the oviposition holes of siricid woodwasps to access their host [1,29,30], but it is not known if, or to what extent, resin beads interfere with the parasitoid's access to late-stage eggs and early instar larval siricids.

Resin beads, as the first visible sign of *S. noctilio* attacks, are the most effective visual cue used in ground surveillance to locate infested host trees. The present study aims to improve our understanding of the relationship between resin bead density and distribution and the number of emerging *S. noctilio*. The objectives are to (i) examine whether the density of resin beads is associated with the cardinal directions, tree height, or tree diameter at different heights; (ii) compare the distribution of wood-boring wasps (*S. noctilio*, *S. nigricornis*, and *I. leucospoides*) emerging from trees with tree height, tree diameter, and the density of resin beads; and (iii) determine whether the emergence of siricids is influenced by the abundance of wood boring beetle galleries.

2. Materials and Methods

2.1. Distribution of Resin Beads

Twelve red pines, *Pinus resinosa*, with signs of *S. noctilio* infestation, were selected at Chimney Bluffs State Park (43°16′50″ N, 76°54′48″ W), 10 km nw of Wolcott, NY, USA in December 2017. Their crown class was observed and recorded, their height was measured using a clinometer and meter tape [31], and tree diameter was measured at breast height (DBH). Each tree was sampled at five 0.5 m long sections with midpoints at heights of 1.5, 3, 4.5, 6, and 7.5 m above ground. The circumference of each section was measured at the midpoint, and the density of resin beads was calculated on each section and recorded relative to cardinal direction. The density of resin beads in each cardinal direction quadrant were counted for each section. The sample sections were marked for subsequent dissection.

2.2. Distribution of Insects

On 26 March 2018, ten of the trees were felled and the marked sample sections were excised. The unsampled tree sections were discarded. Two trees were abandoned due to woodpecker damage. The 50 sampled sections were sealed with hot wax (Gulf Wax, Royal Oak Enterprises, LLC, Roswell, GA, USA). The log sections were individually placed in mesh bags and kept at room temperature. Beginning with the first emerging wasp, emergence was recorded daily for six weeks, and subsequently once weekly until the end of September 2018. Then, each bolt was split to a maximum thickness of 1 cm using an electric log splitter (Grizzly Industrial, Inc., Model H8170/H8171, China). All live and dead wood borer larvae and adults were recorded as well as the presence or absence of blue stain fungi. The number of unemerged wasps for each section was estimated using methods of Standley (2012). The galleries of wood boring beetle larvae were extensively coalesced and difficult to count, so only the presence or absence of galleries was recorded for wood boring beetles.

2.3. Data Analysis

2.3.1. Distribution of Resin Beads

The density of resin beads was calculated for each sampled section and cardinal direction. The trunk diameter (X) was categorized into three groups, 10–15, 15–20, or 20–25 cm in diameter. The distribution of resin beads as a function of tree height and diameter, and cardinal direction was evaluated using Generalized Linear Models (GLM) with a negative binomial distribution and the least squares means (LSMeans) statement in SAS 9.4.

2.3.2. Distribution of Insects

The distribution of *S. noctilio*, *S. nigricornis*, *I. leucospoides* emergences was evaluated as a function of height, trunk diameter, and resin bead density (fixed effects), while individual trees were treated as a random effect. Resin bead densities (Y) were categorized into four groups: 0, 0–5, 5–10, >10 resin beads per square meter. The effect of resin bead density on wood borer survival was first assessed using Generalized Linear Mixed Models (GLMM) with a Poisson distribution and logarithmic link function, but the Poisson regression was replaced with a negative binomial due to overdispersion and formal Akaike information criterion (AIC) model selection. The model coefficient (β) indicates a change in the independent variable. The expected response variable changes by a multiplicative factor of exp(β) for each unit change of an independent variable if the other independent variables in the model are held constant. GLM with a negative binomial distribution was applied to the significant response variables derived from GLMM to distinguish which level of the independent variable had a significant effect on the response variable. Degrees of freedom for the fixed effects F-test were adjusted for statistical dependence using Satterthwaite formulas.

3. Results

3.1. Distribution of Resin Beads

The sampled red pines had suppressed and intermediate crowns, heights of 9–12 m and DBH of 15.24–28 cm. The density of resin beads ranged from 0 to 23 per square meter of log surface (average = 6.75 ± 0.82/m^2). The density of resin beads was influenced by the cardinal direction (p = 0.04, X^2 = 4.09) (Figure 1), the tree height (p < 0.01, X^2 = 21.63) (Figure 2), but not trunk diameter (p = 0.77, X^2 = 0.09) (Figure 3).

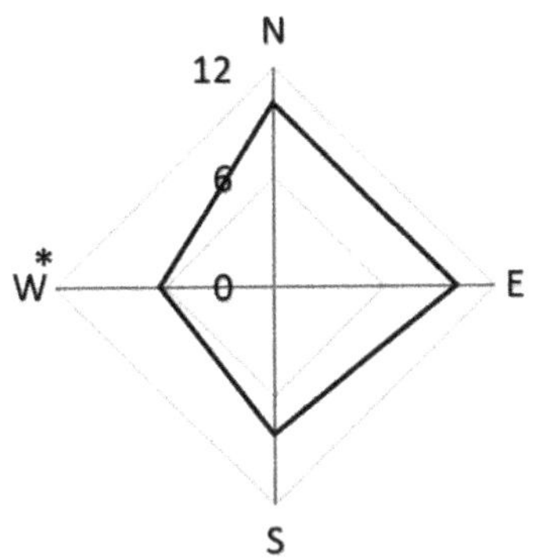

Figure 1. The density of resin beads per square meter on red pines, *Pinus resinosa*, by cardinal direction. The asterisk (*) indicates significantly lower resin bead density (p = 0.04).

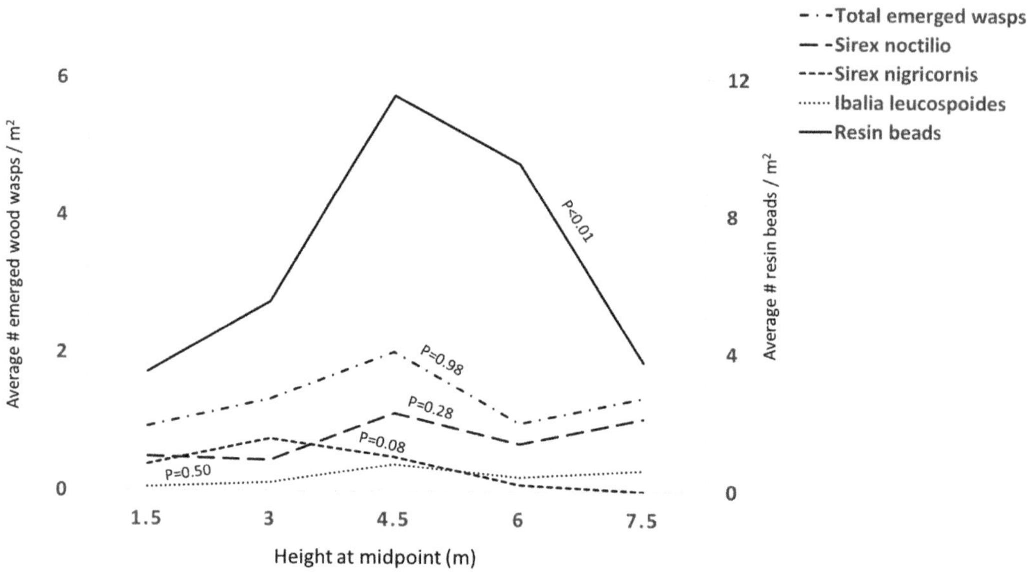

Figure 2. The density of resin beads and emerged wood wasps per square meter on log sections with heights of 1.5, 3, 4.5, 6, and 7.5 m above ground. The height was significant only for resin bead density ($X^2 = 21.63$, $p < 0.01$), not for the emergence of *Sirex noctilio* ($p = 0.28$), *S. nigricornis* ($p = 0.08$), and *Ibalia leucospoides* ($p = 0.50$).

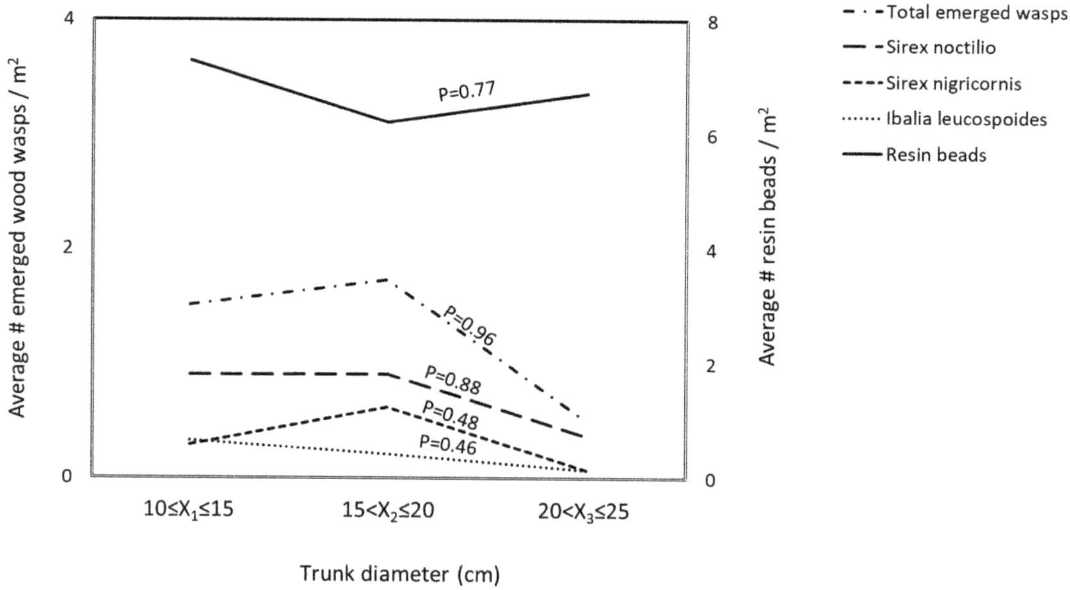

Figure 3. The density of resin beads and the average number of emerged wood wasps per square meter on log sections of red pines, *Pinus resinosa*, by diameter (X). Log diameter was not a significant factor influencing resin bead density ($p = 0.77$), nor the emergence of *Sirex noctilio* ($p = 0.88$), *S. nigricornis* ($p = 0.48$), *Ibalia leucospoides* ($p = 0.46$), and the total wasps ($p = 0.96$).

The majority of resin beads was observed on the north (10.1 resin beads/m^2), east (9.9 resin beads/m^2), and south (8.1 resin beads/m^2), but not west (6.3 resin beads/m^2) aspects of the log sections (Figure 1) and on the mid bole sections at 4.5 m (11.5 resin beads/m^2), followed by 6 m above the ground (9.5 resin beads/m^2) (Figure 2).

3.2. Distribution of Insects

S. noctilio and its parasitoid *I. leucospoides* simultaneously started to emerge on 18 May 2018 and continued emerging for four weeks, while emergence of *S. nigricornis* began three months later on 10 August and continued for two weeks. From fifty half-meter-long log sections with a cumulative surface area of 135.5 m^2, the total numbers of emerged *S. noctilio*, *S. nigricornis*, and *I. leucospoides* were 101 (average = 0.82 ± 0.2/m^2), 46 (average = 0.35 ± 0.1/m^2), and 26 (average = 0.22 ± 0.1/m^2), respectively. The average sex ratios (male to total) of *S. noctilio*, *S. nigricornis*, and *I. leucospoides* were 0.71 ± 0.07, 0.40 ± 0.12, and 0.62 ± 0.11, respectively.

The height of the log sections was not a significant variable for emerged *S. noctilio*, *S. nigricornis*, and *I. leucospoides* (Table 1, Figure 2). The log diameter did not affect the emergence of the three species (Table 1, Figure 3). A significant number of *S. noctilio* emerged from logs with more than five resin beads per square meter ($F = 5.30$, $p < 0.01$), however, the emergence of *S. nigricornis* and *I. leucospoides* was not affected by the density of resin beads (Table 1, Figure 4). None of the independent variables influced the total number of unemerged wasps (Table 1). There were no unemerged wasps in the log sections that were free of blue stain fungus and beetles.

Table 1. Results of Generalized Linear Mixed Model analysis for the wood borer wasps.

Variables	Emerged			Total Emerged Wasps	Total Unemerged Wasps
	Sirex noctilio	*Sirex nigricornis*	*Ibalia leucospoides*		
Resin beads					
β [1]	0.48	−0.07	0.48	0.35	0.19
F	4.05	0.04	2.21	3.41	0.20
P	0.05 *	0.84	0.15	0.07	0.66
Df [2]	1, 37	1, 37	1, 37	1, 37	1, 37
Height					
β	0.16	−0.40	0.14	0.02	−0.07
F	1.19	3.34	0.46	0.00	0.06
P	0.28	0.08	0.50	0.98	0.81
Df	1, 37	1, 37	1, 37	1, 37	1, 37
Diameter					
β	0.08	−0.53	−0.38	0.03	−0.37
F	0.02	0.49	0.54	0.00	0.30
P	0.88	0.48	0.46	0.96	0.59
Df	1, 37	1, 37	1, 37	1, 37	1, 37

[1] β = Negative binomial regression coefficients. The sign of the coefficient represents the direction of effect of that independent variable on the response variable. [2] Degrees of freedoms (Num Df and Den Df) are indicated, respectively. Asterisks denote significant variables: * $\alpha = 0.05$.

Wood borer beetles preferred thicker parts of the trunk and there were no wood borer galleries in sections from the 7.5 m height. Wood borer beetles shared 42% of the log space with siricids, solely utilized 18% of the log sections, while *S. noctilio* colonized 24%. The blue stain fungi colonized the entire bolt in 9% of the total examined log sections and partially colonized the remaining log sections.

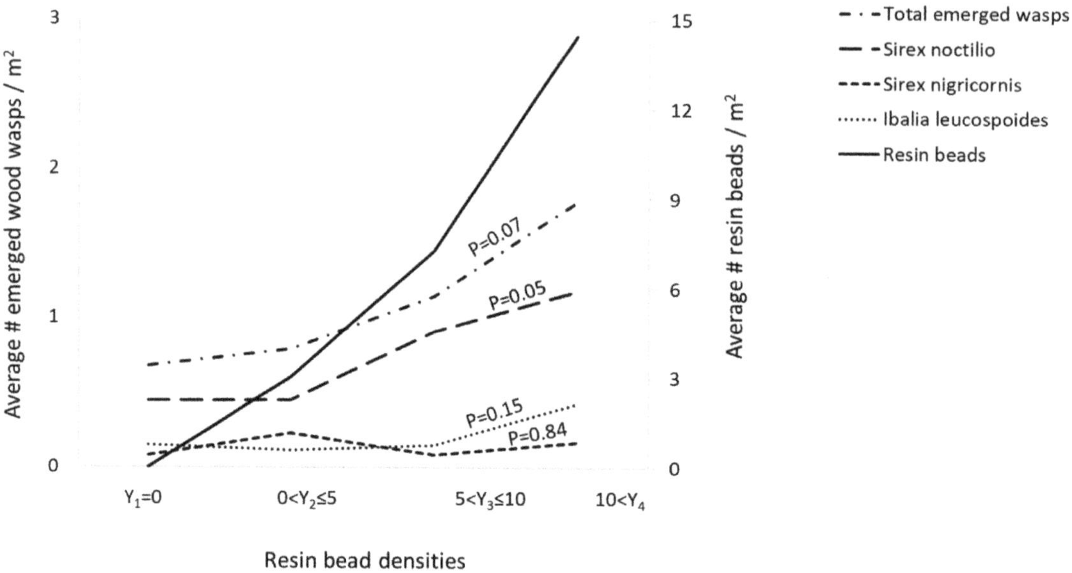

Figure 4. The mean number of emerged wood wasps per square meter collected from infested red pines, *Pinus resinosa*, versus the densities of resin beads (Y). The density of resin beads was significant only for the emergence of *Sirex noctilio* ($p = 0.05$), but not for *S. nigricornis* ($p = 0.84$), *Ibalia leucospoides* ($p = 0.15$), and total wasps ($p = 0.07$). A significant number of *S. noctilio* emerged from sections with higher resin bead densities.

4. Discussion

The distribution of resin beads was influenced by tree height and cardinal direction. The density of resin beads was higher on the north, east, and south sides and on mid bole log sections 4.5 m above the ground. The density of resin beads was the only variable significantly associated with the number of emerged *S. noctilio*.

The density of resin beads varies between dominant and suppressed trees, and is related to several factors, including tree resistance to *S. noctilio* attacks [6,32], tree physiology [2,33], tree health, and prevailing winds [34–38].

Resinosis is the initial and fundamental barrier for *S. noctilio* to overcome, as evidenced by its high mortality in early stages of development [17]. The degree of tree resistance to *S. noctilio* attack can determine the distribution of resin beads; for example, Madden [8] reported random patterns on susceptible trees and aggregated patterns on resistant trees. We did not observe any aggregated patterns on the trees studied, and we did not evaluate tree resistance. Additionally, host susceptibility is related to tree age or size at the time of the attack; for instance, *S. noctilio* females prefer smaller diameter trees [39]. Thompson et al. [24]) also showed that *S. noctilio* made fewer oviposition wounds in larger trees than smaller trees. In the current study, tree diameter was not a significant factor for the emergence of any wasps.

Phloem and cambium physiology influence the number of *S. noctilio* drills and eggs [33]. In *P. radiata,* the rate of respiration varies with tree size and at different heights within a tree [2]. The maximum respiration rate occurs in trees with a diameter of 15 cm and between 2.5 to 4 m above the ground. The phloem in this area of the stem exhibits minimal diurnal fluctuations and maximal respiration. Any physiological disturbance first occurs in that vulnerable area, causing a change in tissue permeability and release of volatile monoterpenes that attract *S. noctilio* [2]. In our study, more resin beads were observed at heights of 4.5 m, which is a range similar to that described by Ryan et al. [40].

In the current study, more *S. noctilio* emergence from red pine logs with more resin beads may relate to tree health. Red pine can be a prime host for *S. noctilio* in the absence of Scots pine in the northeastern U.S. [41,42]. Red pine, *P. resinosa*, has been extensively planted in the northern U.S. and eastern Canada because of its ability to grow on poor quality sites [43]. Variation in red pine growth rates is strongly influenced by soil characteristics due to relatively low genetic diversity [44,45]. Red pines grow poorly or die in highly compacted, water-logged, or unfavorable pH soils. Chimney Bluffs State Park along to Lake Ontario, Wolcott, NY, USA was a prime location for this research because the *S. noctilio* population has been established in this location since at least 2010 [46]. This stand is unmanaged, and in 2011 the stand density and basal area were 1200 $\pm$ 113 tree/ha and 0.4 $\pm$ 0.03 m^2/ha, respectively [46]. This suggests that the effect of poor-quality site may be a contributing factor in maintaining the population of *S. noctilio* in the northeastern U.S.

Insect activity is often more prevalent on the south facing side of host trees, where they exploit the extended exposure to sun in winter [47,48], or avoid bird predators [49] or parasitoids [50]. It is likely that wind drift and prevailing winds also influence insect activity, for instance bark beetles fly upwind towards pheromone sources located on the tree bole [34–38]. In this study, the minimum resin bead densities occurred on the west side of the trees studied, which may be related to the upwind orientation of insects toward semiochemical sources prior to landing [35]. Coutts [32] reported that *S. noctilio* mostly attacked the northeast side of host trees which is opposite the prevailing wind direction at that location. Thus, the orientation of insects to semiochemical sources may explain the radial distribution of resin beads.

The majority of subcortical insects spatially partition their habitats to maximize the exploitation of resources [51], for example bark beetles indirectly partition the tree by employing their symbiont, blue stain fungi, to colonize larger areas because they outcompete *A. areolatum* [15]. The failure of *A. areolatum* growth would directly obstruct the development of woodwasp larvae [52]. The blue stain fungi colonized the entire bolt in 9% of the total examined log sections and partially colonized the remaining log sections, and eventually impeded the development of other subcortical insects, either entirely or partially. This result is in agreement with Foelker et al. [53], who reported that a 10% increase in blue stain volume caused an 11.2% decrease in *S. noctilio* survival. Therefore, the competition between wood borers and bark beetles jeopardizes the survival and development of *S. noctilio*.

S. noctilio has a wide range of host plants with different resin canal densities [54]. Thus, describing factors that contribute to the success or failure of commericial hosts to produce resin beads in response to *S. noctilio* attacks may be fruitful topic to investigate. In the current study, we showed that the density of resin beads is a reliable indicator of the presence of *S. noctilio* red pines, *P. resinosa*, growing on a poor-quality site. By enhancing our understanding of the relationship between resin beads and the presence of *S. noctilio*, our results should improve surveillance of this important pest.

Author Contributions: Conceptualization, S.A.T. and H.F.; methodology, S.A.T. and H.F.; validation, S.A.T. and H.F.; formal analysis, H.F.; investigation, H.F.; resources, S.A.T.; data curation, S.A.T. and H.F.; writing—original draft preparation, H.F.; writing—review and editing, S.A.T. and H.F.; visualization, S.A.T. and H.F.; supervision, S.A.T.; project administration, S.A.T.; funding acquisition, H.F. All authors have read and agreed to the published version of the manuscript.

Funding: This research was funded by the Edna Bailey Sussman Foundation in the summer of 2018.

Data Availability Statement: The data presented in this study are openly available in Figures 1–4, and Table 1.

Acknowledgments: The authors thank Thomas Hughes for issuing a permit to access and fall infested trees at Chimney Bluffs State Park, Lianjun Zhang for his comments on the data analysis, Joseph Marascalchi for his field assistance, Julia Fairbanks for her field and lab assistance, and Terry Ettinger for granting access to the storage facility. A special thanks to Michael Marascalchi for his comments on the first draft of this manuscript.

Conflicts of Interest: The authors declare no conflict of interest. The funders had no role in the design of the study; in the collection, analyses, or interpretation of data; in the writing of the manuscript; or in the decision to publish the results.

References

1. Slippers, B.; Hurley, B.P.; Wingfield, M.J. *Sirex woodwasp*: A model for evolving management paradigms of invasive forest pests. *Annu. Rev. Entomol.* **2015**, *60*, 601–619. [CrossRef] [PubMed]
2. Madden, J.L. Physiological reactions of Pinus radiata to attack by woodwasp, *Sirex noctilio* F.(Hymenoptera: Siricidae). *Bull. Entomol. Res.* **1977**, *67*, 405–426. [CrossRef]
3. Böröczky, K.; Zylstra, K.E.; McCartney, N.B.; Mastro, V.C.; Tumlinson, J.H. Volatile profile differences and the associated *Sirex noctilio* activity in two host tree species in the northeastern United States. *J. Chem. Ecol.* **2012**, *38*, 213–221. [CrossRef] [PubMed]
4. Ferrenberg, S.; Langenhan, J.M.; Loskot, S.A.; Rozal, L.M.; Mitton, J.B. Resin monoterpene defenses decline within three widespread species of pine (Pinus) along a 1530-m elevational gradient. *Ecosphere* **2017**, *8*, e01975. [CrossRef]
5. Hudgins, J.W.; Christiansen, E.; Franceschi, V.R. Induction of anatomically based defense responses in stems of diverse conifers by methyl jasmonate: A phylogenetic perspective. *Tree Physiol.* **2004**, *24*, 251–264. [CrossRef]
6. Coutts, M.P.; Dolezal, J.E. Polyphenols and Resin in the Resistance Mechanisms of Pinus Radiata Attacked by the Wood Wasp, Sirex Noctilio, and Its Associated Fungus; Department of National Development, Forestry and Timber Bureau, Australia, 1966. Department of National Development, Forestry and Timber Bureau: Hawthorn, VIC, Australia.
7. Madden, J.L.; Coutts, M.P. The role of fungi in the biology and ecology of woodwasps (Hymenoptera: Siricidae). In *Insect–Fungus Symbiosis*; Allanheld, Osmun & Co.: Montclair, NJ, USA, 1979; pp. 165–174.
8. Madden, J.L. Oviposition behaviour of the woodwasp, *Sirex noctilio* F. *Aust. J. Zool.* **1974**, *22*, 341–351. [CrossRef]
9. Madden, J.L.; Irvine, C.J. The use of lure trees for the detection of Sirex noctilio in the field. *Aust. For.* **1971**, *35*, 164–166. [CrossRef]
10. Slippers, B.; De Groot, P.; Wingfield, M.J. The Sirex Woodwasp and its Fungal Symbiont: Research and Management of a Worldwide Invasive Pest. Springer: Dordrecht, The Netherlands, 2011.
11. Wermelinger, B.; Rigling, A.; Schneider Mathis, D.; Dobbertin, M. Assessing the role of bark- and wood-boring insects in the decline of Scots pine (*Pinus sylvestris*) in the Swiss Rhone valley. *Ecol. Entomol.* **2008**, *33*, 239–249. [CrossRef]
12. Hurley, B.P.; Slippers, B.; Wingfield, M.J.; Dyer, C.; Slippers, J. Perception and Knowledge of the Sirex Woodwasp and Other Forest Pests in South Africa. *Agric. For. Entomol.* **2012**, *14*, 306–316. [CrossRef]
13. Ryan, K.; de Groot, P.; Davis, C.; Smith, S.M. Effect of two bark beetle-vectored fungi on the on-host search and oviposition behavior of the introduced woodwasp *Sirex noctilio* (Hymenoptera: Siricidae) on *Pinus sylvestris* trees and logs. *J. Insect Behav.* **2012**, *25*, 453–466. [CrossRef]
14. Ryan, K.; de Groot, P.; Smith, S.M. Evidence of interaction between *Sirex noctilio* and other species inhabiting the bole of Pinus. *Agric. For. Entomol.* **2012**, *14*, 187–195. [CrossRef]
15. Yousuf, F.; Gurr, G.M.; Carnegie, A.J.; Bedding, R.A.; Bashford, R.; Gitau, C.W.; Nicol, H.I. The bark beetle, *Ips grandicollis*, disrupts biological control of the woodwasp, *Sirex noctilio*, via fungal symbiont interactions. *FEMS Microbiol. Ecol.* **2014**, *88*, 38–47. [CrossRef] [PubMed]
16. Yousuf, F.; Carnegie, A.J.; Bashford, R.; Bedding, R.A.; Nicol, H.I.; Gurr, G.M. Bark beetle (*Ips grandicollis*) disruption of woodwasp (*Sirex noctilio*) biocontrol: Direct and indirect mechanisms. *For. Ecol. Manag.* **2014**, *323*, 98–104. [CrossRef]
17. Haavik, L.J.; Dodds, K.J.; Allison, J.D. Do native insects and associated fungi limit non-native woodwasp, *Sirex noctilio*, survival in a newly invaded environment? *PLoS ONE* **2015**, *10*, e0138516. [CrossRef] [PubMed]
18. Bittner, T.D.; Hajek, A.E.; Liebherr, J.K. Associations among Serropalpus substriatus (Coleoptera: Melandryidae) and Sirex (Hymenoptera: Siricidae) Communities. *Great Lakes Entomol.* **2016**, *49*, 18–26.
19. Clarke, C.W.; Carnegie, A.J.; Yousuf, F.; Bashford, R.; Nicol, H.I.; Bedding, R.A.; Gurr, G.M. Minimizing the disruptive effect of *Ips grandicollis* (Coleoptera: Scolytinae) on biocontrol of *Sirex noctilio* (Hymenoptera: Siricidae). *For. Ecol. Manag.* **2016**, *381*, 134–143. [CrossRef]
20. Hajek, A.E.; Tobin, P.C.; Kroll, S.A.; Long, S.J. Symbionts mediate oviposition behaviour in invasive and native woodwasps. *Agric. For. Entomol.* **2018**, *20*, 442–450. [CrossRef]
21. Madden, J.L. Sirex in Australasia. In *Dynamics of Forest Insect Populations*; Plenum Pub. Corp.: New York, NY, USA, 1988; pp. 407–429.
22. Carnegie, A.J.; Bashford, R. Sirex woodwasp in Australia: Current management strategies, research and emerging issues. In *The Sirex woodwasp and Its Fungal Symbiont*; Springer: Dordrecht, The Netherlands, 2012; pp. 175–201.
23. Williams, D.W.; Zylstra, K.E.; Mastro, V.C. Ecological considerations in using Deladenus (=Beddingia) siricidicola for the biological control of Sirex noctilio in North America. In *The Sirex woodwasp and Its Fungal Symbiont*; Springer: Dordrecht, The Netherlands, 2012; pp. 135–148.
24. Thompson, B.M.; Bodart, J.; Gruner, D.S. Community resistance to an invasive forest insect–fungus mutualism. *Ecosphere* **2019**, *10*, e02609. [CrossRef]
25. Carnegie, A.J.; Eldridge, R.H.; Waterson, D.G. History and management of *Sirex woodwasp* in pine plantations in New South Wales, Australia. *N. Z. J. For. Sci.* **2005**, *35*, 3–24.

26. Fischbein, D.; Corley, J.C. Classical biological control of an invasive forest pest: A world perspective of the management of *Sirex noctilio* using the parasitoid *Ibalia leucospoides* (Hymenoptera: Ibaliidae). *Bull. Entomol. Res.* **2015**, *105*, 1–12. [CrossRef]
27. Faal, H.; Cha, D.H.; Hajek, A.E.; Teale, S.A. A double-edged sword: Amylostereum areolatum odors attract both *Sirex noctilio* (Hymenoptera: Siricidae) and its parasitoid, *Ibalia leucospoides. Fungal Ecol.* **2021**, *54*, 101108. [CrossRef]
28. Faal, H.; Silk, P.J.; Mayo, P.D.; Teale, S.A. Courtship behavior and identification of a sex pheromone in *Ibalia leucospoides* (Hymenoptera: Ibaliidae), a larval parasitoid of *Sirex noctilio* (Hymenoptera: Siricidae). *PeerJ* **2021**, *9*, e12266. [CrossRef] [PubMed]
29. Fernández-Arhex, V.; Corley, J.C. The functional response of *Ibalia leucospoides* (Hymenoptera: Ibaliidae), a parasitoid of *Sirex noctilio* (Hymenoptera: Siricidae). *Biocontrol Sci. Technol.* **2005**, *15*, 207–212. [CrossRef]
30. Hurley, B.P.; Fitza, K.N.E.; Wingfield, M.J.; Slippers, B. Sequence data reflect the introduction pathways of the *Sirex woodwasp* parasitoid, *Ibalia leucospoides* (Ibaliidae, Hymenoptera). *Agric. For. Entomol.* **2020**, *22*, 129–135. [CrossRef]
31. Bragg, D.C. An improved tree height measurement technique tested on mature southern pines. *South. J. Appl. For.* **2008**, *32*, 38–43. [CrossRef]
32. Coutts, M.P. Sirex noctilio and the Physiology of Pinus radiata: Some Studies of Interactions between the Insect, the Fungus and the Tree in Tasmania. Department of National Development, Forestry and Timber Bureau: Canberra, ACT, Austrilia, 1965.
33. Madden, J.L. Physiological aspects of host tree favourability for the woodwasp, *Sirex noctilio* F. In Proceedings of the Ecological Society of Australia; 1968; pp. 147–149.
34. Chapman, J.A. Field studies on attack flight and log selection by the ambrosia beetle *Trypodendron lineatum* (Oliv.)(Coleoptera: Scolytidae). *Can. Entomol.* **1962**, *94*, 74–92. [CrossRef]
35. Byers, J.A. Upwind flight orientation to pheromone in western pine beetle tested with rotating wind vane traps. *J. Chem. Ecol.* **1988**, *14*, 189–198. [CrossRef]
36. Safranyik, L.; Silversides, R.; McMullen, L.H.; Linton, D.A. An empirical approach to modeling the local dispersal of the mountain pine beetle (*Dendroctonus ponderosae* Hopk.) (Col., Scolytidae) in relation to sources of attraction, wind direction and speed. *J. Appl. Entomol.* **1989**, *108*, 498–511. [CrossRef]
37. Safranyik, L.; Linton, D.A.; Silversides, R.; McMullen, L.H. Dispersal of released mountain pine beetles under the canopy of a mature lodgepole pine stand. *J. Appl. Entomol.* **1992**, *113*, 441–450. [CrossRef]
38. Mahroof, R.M.; Edde, P.A.; Robertson, B.; Puckette, J.A.; Phillips, T.W. Dispersal of *Rhyzopertha dominica* (Coleoptera: Bostrichidae) in different habitats. *Environ. Entomol.* **2010**, *39*, 930–938. [CrossRef]
39. Madden, J.L. An analysis of an outbreak of the woodwasp, *Sirex noctilio* F. (Hymenoptera, Siricidae), in *Pinus radiata. Bull. Entomol. Res.* **1975**, *65*, 491–500. [CrossRef]
40. Ryan, K.; de Groot, P.; Smith, S.M.; Turgeon, J.J. Seasonal occurrence and spatial distribution of resinosis, a symptom of *Sirex noctilio* (Hymenoptera: Siricidae) injury, on boles of *Pinus sylvestris* (Pinaceae). *Can. Entomol.* **2013**, *145*, 117–122. [CrossRef]
41. Dodds, K.J.; de Groot, P.; Orwig, D.A. The impact of *Sirex noctilio* in *Pinus resinosa* and *Pinus sylvestris* stands in New York and Ontario. *Can. J. For. Res.* **2010**, *40*, 212–223. [CrossRef]
42. Dodds, K.J.; Cooke, R.R.; Gilmore, D.W. Silvicultural options to reduce pine susceptibility to attack by a newly detected invasive species, *Sirex noctilio. North. J. Appl. For.* **2007**, *24*, 165–167. [CrossRef]
43. Rudolf, P.O. Red pine (*Pinus resinosa* Ait.). *Silv. N. Am.* **1990**, *1*, 442–455.
44. McLaughlin, J.A.; Hsiang, T.; Hayden, G.H.; Greifenhagen, S. Abiotic and biotic factors used to assess decline risk in red pine (*Pinus resinosa* Ait.) plantations. *For. Chron.* **2011**, *87*, 99–115. [CrossRef]
45. Mosseler, A.; Egger, K.N.; Hughes, G.A. Low levels of genetic diversity in red pine confirmed by random amplified polymorphic DNA markers. *Can. J. For. Res.* **1992**, *22*, 1332–1337. [CrossRef]
46. Standley, C.R. Emergence Phenology and Ecological Interactions between the Exotic *Sirex noctilio*, Native Siricids, and a Shared Guild of Native Parasitoids. Master's Thesis, State University of New York College of Environmental Science and Forestry, Syracuse, NY, USA, 2012.
47. Albert, J.; Platek, M.; Cizek, L. Vertical stratification and microhabitat selection by the Great Capricorn Beetle (*Cerambyx cerdo*) (Coleoptera: Cerambycidae) in open-grown, veteran oaks. *Eur. J. Entomol.* **2012**, *109*, 553–559. [CrossRef]
48. Kašák, J.; Foit, J. Shortage of declining and damaged sun-exposed trees in European mountain forests limits saproxylic beetles: A case study on the endangered longhorn beetle *Ropalopus ungaricus* (Coleoptera: Cerambycidae). *J. Insect Conserv.* **2018**, *22*, 171–181. [CrossRef]
49. Bereczki, K.; Molnár, D.; Csóka, G.; Báldi, A. Factors affecting the bird predation of low density gypsy moth egg masses in three types of hardwood forests in southwest Hungary. *Bull. Insectol.* **2017**, *70*, 201–207.
50. Zargaran, M.R.; Safaralizadeh, M.H.; Pourmirza, A.A.; Valizadegan, O. Effect of cardinal directions on gall morphology and parasitization of the gall wasp, *Cynips quercusfolii. J. Insect Sci.* **2011**, *11*, 169. [CrossRef] [PubMed]
51. Schoeller, E.N. Characterization of the Subcortical Interactions between Larvae of the Southern Pine Sawyer Monochamus titillator (F.) and the Larvae of the Southern Pine Beetle Guild Using Molecular Gut Analyses. Master's Thesis, Louisiana State University, Baton Rouge, LA, USA, 2011.
52. Thompson, B.M.; Bodart, J.; McEwen, C.; Gruner, D.S. Adaptations for symbiont-mediated external digestion in *Sirex noctilio* (Hymenoptera: Siricidae). *Ann. Entomol. Soc. Am.* **2014**, *107*, 453–460. [CrossRef]

53. Foelker, C.J.; Standley, C.R.; Parry, D.; Fierke, M.K. Complex ecological relationships among an assemblage of indigenous hymenopteran parasitoids, the exotic European woodwasp (Sirex noctilio; Hymenoptera: Siricidae), and a native congener. *Can Entomol.* **2016**, *148*, 532–542. [CrossRef]
54. Bookwalter, J.D.; Riggins, J.J.; Dean, J.F.D.; Mastro, V.C.; Schimleck, L.R.; Sullivan, B.T.; Gandhi, K.J.K. Colonization and development of *Sirex noctilio* (Hymenoptera: Siricidae) in bolts of a native pine host and six species of pine grown in the southeastern United States. *J. Entomol. Sci.* **2019**, *54*, 1–18. [CrossRef]

insects

MDPI

Article

Chinese Species of the Genus *Pseudanaphes* Noyes & Valentine (Hymenoptera: Mymaridae) with Description of a New Species †

Zhulidezi Aishan [1,2,3], Huan-Xi Cao [4,*], Hong-Ying Hu [1,3] and Chao-Dong Zhu [2,5]

1 College of Life Science and Technology, Xinjiang University, Urumqi 830017, China
2 Key Laboratory of Zoological Systematics and Evolution, Institute of Zoology, Chinese Academy of Sciences, Beijing 100101, China
3 Xinjiang Key Laboratory of Biological Resources and Genetic Engineering, Urumqi 830017, China
4 National Animal Collection Resource Center, Institute of Zoology, Chinese Academy of Sciences, Beijing 100101, China
5 College of Life Sciences, University of Chinese Academy of Sciences (UCAS), Beijing 100049, China
* Correspondence: caohuanxi@ioz.ac.cn
† urn:lsid:zoobank.org:pub:8F85152C-5E7B-4E26-A328-70837ACE68F6.

Simple Summary: Parasitoids of the family Mymaridae (Hymenoptera: Chalcidoidea) are difficult to collect and study, in part because of their small body size of 0.2–1.5 mm. Currently, more than 1500 species have been described and recorded. However, molecular data are rarely generated and used in relation to these small parasitoids. The genus *Pseudanaphes* is one of the smaller genera, with only five known species, for which molecular data were not available until this study. In this study, a new species of *Pseudanaphes*, *P. yadongicus* Aishan & Cao sp. nov., is separated from the known species *P. zhaoi* Lin by combining morphological and molecular data, both collected in Tibet, China. The DNA barcode and illustrations of *P. zhaoi* are also provided simultaneously. The DNA barcodes, morphological diagnosis, and a key to the world species of this genus will facilitate studies on this genus.

Abstract: The fairyfly Mymaridae (Hymenoptera: Chalcidoidea) are widely distributed worldwide, but species of this family have rarely been collected and recorded from the Qinghai–Tibet Plateau. In this study, mymarids collected in Tibet, China, are identified based on morphology and molecular data. Two species of the genus *Pseudanaphes* Noyes & Valentine are treated and illustrated here, including a known species, *P. zhaoi* Lin, and a new species, *P. yadongicus* Aishan & Cao sp. nov. In addition, a key to the world species of *Pseudanaphes* (females) and DNA barcodes for *P. yadongicus* and *P. zhaoi* are provided.

Keywords: Chalcidoidea; new taxon; China; taxonomy; COI

1. Introduction

The genus *Pseudanaphes* Noyes & Valentine is one of several smaller genera in the family Mymaridae (Hymenoptera: Chalcidoidea), species of which are not commonly or numerously collected. This genus is characterized by relatively hairy fore wings and long marginal veins and was established by Noyes and Valentine (1989), who recorded an Australian species, *Pseudanaphes hirtus* Noyes & Valentine, from Titirangi, New Zealand [1]. After that, another species, *P. zhaoi* Lin, was discovered in the Palaearctic region (Shaanxi, China) [2]. Two Australian species, *P. lincolni* (Girault) and *P. particoxae* (Girault), were originally described briefly without illustrations based on a female specimen under the generic name *Polynemoidea* Girault [3,4]. New (1974) provided measurements and illustrations for these two species by examining the type specimens, and Lin (2007) transferred these two species to *Pseudanaphes* [5,6]. Later, Rehmat and Anis (2011) described another new species, *P. sikkimianus*, from India, based on only a single female specimen [7]. As a result, five species were known in the genus *Pseudanaphes* prior to this study. Nevertheless, the host information of *Pseudanaphes* is currently unknown, with only the type species of this genus, *P. hirtus*, being possibly associated with the eggs of weevils (Coleoptera: Curculionidae) [8].

Currently, the systematic relationship of *Pseudanaphes* with other mymarids remains unclear due to the lack of comprehensive molecular evidence and insufficient collections of this genus and possibly related groups. Usually, *Pseudanaphes* is placed in the *Anaphes* group, along with *Anaphes* Haliday, *Dorya* Noyes & Valentine, and *Erythmelus* Enock [6]. Among these genera, the species of *Pseudanaphes* are most similar to some species of *Anaphes* Haliday and can be distinguished from the latter by the rounded apex of the fore wing with a longer marginal vein and the absence of a longitudinal medina groove on the propodeum.

Species of the genus *Pseudanaphes* are rarely reported, either because they are not frequently discovered and collected in the field or because their identification is difficult. This genus also remains poorly studied in China. In this study, a new species from Tibet, China, is described, and diagnosis and illustrations are provided for a known Chinese species, *P. zhaoi* Lin. In addition, DNA barcodes of the new species and *P. zhaoi* are provided here, which are also reported for the first time for this genus.

2. Materials and Methods

2.1. Specimens Collection

All examined specimens of the two Chinese *Pseudanaphes* species were collected using Malaise traps and preserved in 99% ethanol at −20 °C until use. The holotype and one paratype of the newly described species were deposited in the Institute of Zoology, Chinese Academy of Sciences, Beijing, China (IZCAS), and the remaining paratypes and specimens of *P. zhaoi* were deposited in the Insect Collection of the College of Life Science and Technology, Urumqi, Xinjiang, China (ICXU).

2.2. Taxonomic Studies

Specimens were dissected and mounted on slides by using Canada balsam. All measurements were taken from slide-mounted specimens at 65×, 100×, 250×, or 400× magnification with a Leica compound microscope fitted with an eyepiece reticle. Body length, in micrometers (μm), was measured from the transverse trabecula to the metasomal apex, excluding the exserted part of the ovipositor. The remaining measurements are given either in micrometers or as ratios. Photographs were taken from slide-mounted specimens using a Nikon Ni-E system, and images were processed by Adobe Photoshop.

Abbreviations used are as follows: F1–6 = antennal funicular 1–6; POL = the shortest distance between posterior ocelli; OOL = the shortest distance between posterior ocellus and eye margin.

Terminology for morphological features follows Lin [6] and Rehmat and Anis [7].

2.3. DNA Extraction, Amplification, and Sequencing

Genomic DNA was extracted from 16 specimens (Table 1) using the whole body of the specimens and following the protocol of the DNeasy Blood & Tissue Kit (Qiagen, Germany). After DNA extraction, the voucher specimens were dissected and mounted on slides for further morphological examination. The voucher specimens were deposited in IZCAS.

The COI sequences were amplified using two primer pairs, Dicopus-COI-F (5′-ATCCT GGTTC ATTTT TAGGA AAT-3′) and Gonat-COI-R (5′-GCTCC NGCTA ATACW GGTAA TG-3′) and Alaptus-COI-F (5′-ATCCA GGTTC ATTTT TAGGA AAT-3′) and Gonat-COI-R (5′-GCTCC NGCTA ATACW GGTAA TG-3′), which were designed by this study based on published data of Mymaridae from the NCBI database. All PCR reactions were performed in a total volume of 30 μL, including 15 μL premix Taq polymerase (Takara, Japan), 1 μL of each forward and reverse primer, 3 μL DNA template, and 10 μL distilled water. PCR conditions followed Triapitsyn et al. [9]. Sequencing was performed in both directions.

Table 1. Information on sequenced specimens with GenBank accession number of COI.

Code	Species	Sex	Primer Pair	Accession Number
PS001	*P. yadongicus*	female	Dicopus-COI-F and Gonat-COI-R	ON834699
PS002	*P. yadongicus*	female	Alaptus-COI-F and Gonat-COI-R	ON834700
PS003	*P. yadongicus*	female	Dicopus-COI-F and Gonat-COI-R	ON834701
PS004	*P. yadongicus*	female	Dicopus-COI-F and Gonat-COI-R	ON834702
PS005	*P. yadongicus*	female	Dicopus-COI-F and Gonat-COI-R	ON834703
PS006	*P. yadongicus*	female	Dicopus-COI-F and Gonat-COI-R	ON834704
PS007	*P. yadongicus*	female	Alaptus-COI-F and Gonat-COI-R	ON834705
PS008	*P. zhaoi*	female	Dicopus-COI-F and Gonat-COI-R	ON834706
PS009	*P. zhaoi*	female	Dicopus-COI-F and Gonat-COI-R	ON834707
PS010	*P. zhaoi*	female	Dicopus-COI-F and Gonat-COI-R	ON834708
PS011	*P. zhaoi*	female	Dicopus-COI-F and Gonat-COI-R	ON834709
PS012	*P. zhaoi*	female	Dicopus-COI-F and Gonat-COI-R	ON834710
PS013	*P. zhaoi*	female	Dicopus-COI-F and Gonat-COI-R	ON834711
PS014	*P. zhaoi*	female	Alaptus-COI-F and Gonat-COI-R	ON834712
PS015	*P. zhaoi*	female	Alaptus-COI-F and Gonat-COI-R	ON834713
PS016	*P. zhaoi*	female	Alaptus-COI-F and Gonat-COI-R	ON834714

2.4. Sequence Analysis

All sequences were verified by the NCBI Nucleotide Blast tool. Sequences from both directions were assembled and edited in Sequencher version 4.5 (Gene Codes Corporation, Ann Arbor, MI, USA) and aligned in BioEdit version 7.0.9.0 [10] to create the COI matrix used for distance calculations. The COI matrix was translated into amino acids in MEGA7.0 [11] to check for stop codons. A neighbor-joining (NJ) tree based on the Kimura two-parameter (K2P) distances was constructed using MEGA7.0 with 1000 bootstrap replicates to generate support values for the nodes. The COI sequences generated in this study were deposited in GenBank under accession numbers ON834699–ON834714.

3. Results

The genus *Pseudanaphes*, with one described species and one new species from China, was treated in this study. The DNA barcodes (COI) of seven females of *P. yadongicus* and nine females of *P. zhaoi* were successfully obtained. After alignment and trimming, the COI matrix without insertions or deletions, with a length of 401 base pairs, was used to obtain a NJ tree (Figure 1) by K2P distances. Based on 401 bp COI sequences, these two species were separated with high bootstrap support of 100%. The interspecific divergence (K2P distance) between these two species varied from 11.7% to 12.5%. *Pseudanaphes yadongicus* displayed no intraspecific variation, and *P. zhaoi* showed the maximum intraspecific distance at 0.8%.

3.1. *Pseudanaphes* Noyes & Valentine, 1989

Pseudanaphes Noyes & Valentine, 1989 [1], pp. 47–48. Type species: *Pseudanaphes hirtus* Noyes & Valentine, by original designation; Lin et al. 2007 [6], pp. 15,18,47 (list of Australia species).

Diagnosis. The genus *Pseudanaphes* Noyes & Valentine can be characterized by the following characters: female antenna with six-segmented funicle, three-segmented clava; clava large and slightly shorter than the combined length of six funiculars, and with one row of campaniform sensilla; fore wing apex evenly rounded, with venation extending at least 0.4× wing length and with disc cilia dense and with a slight brownish tinge and curved, infuscate band below proximal half of marginal vein; four-segmented tarsi. In males, antenna with 11 flagellomeres; body color and size similar to females.

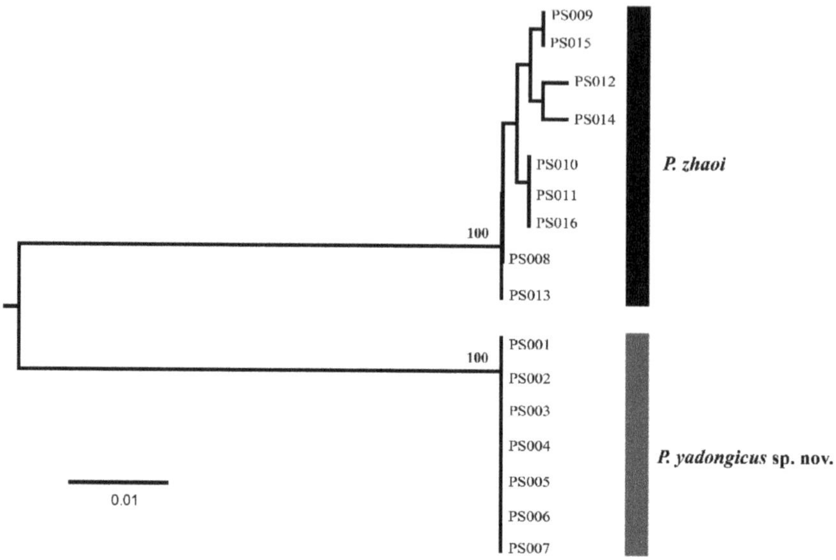

Figure 1. A NJ tree by K2P distances based on COI sequences from 16 *Pseudanaphes* specimens. The numbers above nodes indicate bootstrap values.

The *Anaphes* group consists of four genera, *Anaphes*, *Erythmelus*, *Dorya*, and *Pseudanaphes*. The genus *Pseudanaphes* can be distinguished from the other three genera by the combination of the following characters: large and three-segmented antenna clava; fore wing apex evenly rounded, with a curved, dark mark below parastigma and a longer venation that extends at least 0.4× wing length. These characters are also useful in distinguishing species of *Pseudanapahes* from the most similar genus, *Anaphes* Haliday. Compared to the females, the species of *Anaphes* have a one- or two-segmented clava and an asymmetrical fore wing, often with an oblique row of microtrichia extending from just below the apex of the stigmal vein towards the wing apex. Lin (2007) stated that, for *Pseudanapahes*, there is no longitudinal median groove on the propodeum [6]. In contrast, a median longitudinal groove was observed on the propodeum (Figures 2c and 3c) for the *Pseudanapahes* specimens examined in this study. Along with this character, the clava with one row of campaniform sensilla (Figure 3b) was suggested to be diagnostic for the genus *Pseudanapahes*.

3.1.1. *Pseudanaphes yadongicus* Aishan & Cao, sp. nov.

Diagnosis. Antenna with the first claval segment the longest, apical segment the shortest, the first and second segments of clava with two longitudinal sensilla on each, and the third segment of clava with three longitudinal sensilla. Fore wing disc densely setose with setae extending to base of submarginal vein. Ovipositor long and exserted beyond apex of metasoma; ovipositor length equal to length of metasoma.

Description. Female (holotype and paratypes). Body length 1450–1470 μm. Body color brown. Antenna brown, except scape, and pedicel yellowish brown. Fore wing hyaline, slightly infuscate below submarginal vein, and with a usual, curved, infuscate band below proximal half of marginal vein. Legs yellow brown, except tarsi yellow.

Head (Figure 2a) brown and smooth, in front view wider than height (56: 36); eye 0.69× height of head (0.68–0.70×); posterior ocellus very close to eye margin, POL: OOL = 4.2. Antenna (Figure 2b) with scape about 3.75× as long as wide; pedicel about 2.0× (1.95–2.00×) as long as wide, as long as F1; the combined length of F4–F6 slightly shorter than that of three preceding ones; F2 and F3 equal in length; F5 and F6 equal in length; F6 with one longitudinal sensillum; clava about 4.77× (4.77–4.85×) as long as wide, the first segment of clava the longest, apical segment of clava the shortest, the first and second segments of clava with two sensilla on each, the third segment of clava with three sensilla.

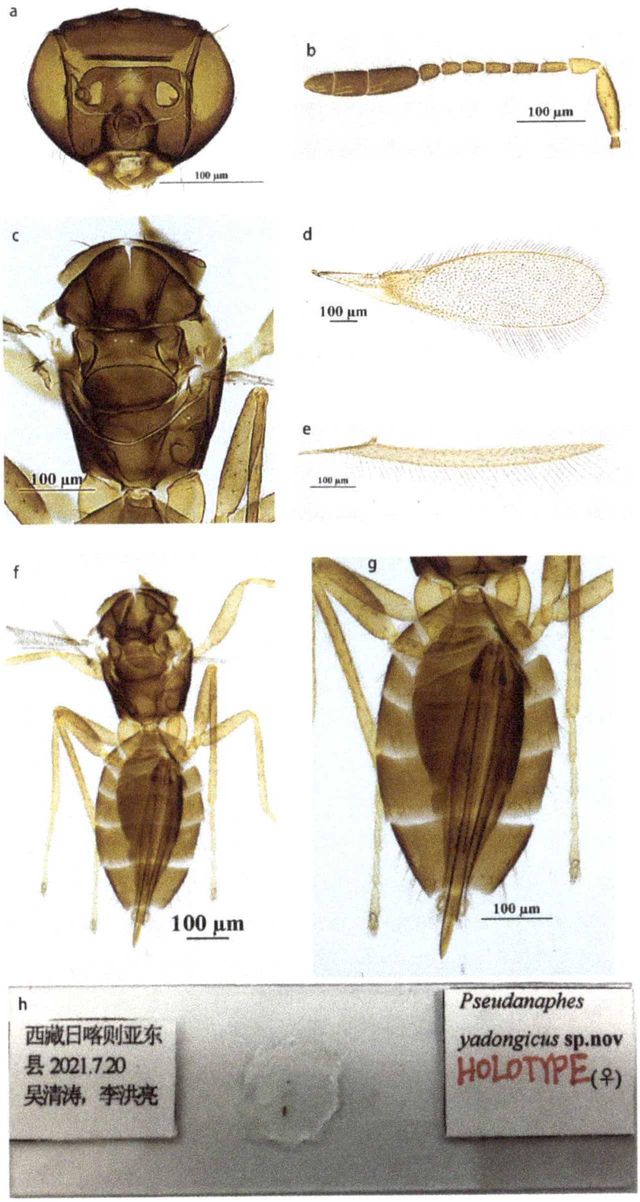

Figure 2. *Pseudanaphes yadongicus* sp. nov. female. (**a**) Head; (**b**) antenna; (**c**) mesosoma in dorsal view; (**d**) fore wing; (**e**) hind wing; (**f**) body in dorsal view; (**g**) metasoma in dorsal view; (**h**) holotype slide.

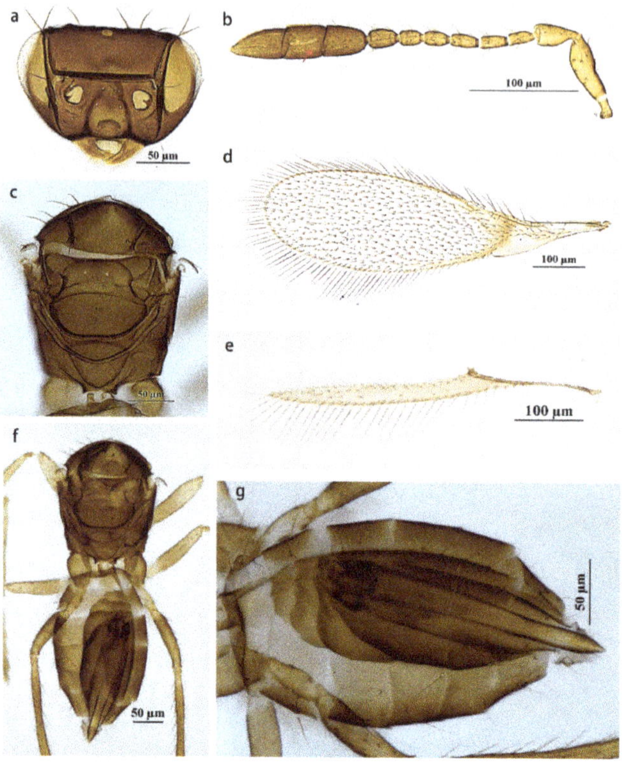

Figure 3. *Pseudanaphes zhaoi*. Female. (**a**) Head; (**b**) antenna; (**c**) mesosoma in dorsal view; (**d**) fore wing; (**e**) hind wing; (**f**) body in dorsal view; (**g**) metasoma in dorsal view.

Mesosoma (Figure 2c,f) smooth and about 0.69× length of metasoma. Lateral plate of pronotum with six setae on each, midlobe of mesoscutum with two setae, each side lobe with one, and each axilla with one seta; scutellum without setae; propodeum with a pair of setae on both sides. Fore wing (Figure 2d) about 3.81× as long as wide, disc with a slight brownish tinge and a curved, infuscate band below proximal half of marginal vein, and infuscate below submarginal vein; densely setose, with setae extending nearly to the wing base; longest marginal seta 0.49× (0.48–0.52×) greatest width of wing. Hind wing (Figure 1e) 21× as long as wide; longest marginal seta 2× greatest width of wing.

Metasoma (Figure 2g) smooth. Ovipositor originating from about the level of the second metasomal tergite and occupying about 0.99× length of metasoma, exserted beyond apex of metasoma by 0.13× (0.13–0.14×) its own length, 2.0× (1.9–2.0×) as long as mesotibia.

Measurements of the holotype (Figure 2h). Antenna: scape: 150: 40; pedicel: 60: 30; F1: 60; F2: 50; F3: 50; F4: 45; F5: 40; F6: 40; clava: 260. Mesosoma: 500; fore wing: 1640: 430; longest marginal seta: 210. Hind wing: 1260: 60; longest marginal seta: 20. Metasoma: 720; ovipositor: 715.

MALE. Unknown.

Etymology. The species is named after the Yadong county of Tibet, China, where the type material was collected.

Type material. Holotype, female, China, Tibet, Yadong County, 20 July 2021, Qing-Tao Wu and Hong-Liang Li (IZCAS, IOZ(E) 221465). Paratypes: one female, same data as the holotype (IZCAS, IOZ(E) 221466); five females, same data as the holotype (ICXU).

Biology. Unknown.

Distribution. China (Tibet).

Comments. Females of *P. yadongicus* can be recognized by the combination of characters described in the diagnosis. Accordingly, the females of this species resemble those of *P. lincolni* (Girault). In females, *P. yadongicus* differs from the latter species in antenna color, relative length of F1 and the pedicel, relative length of the three claval segments, and relative length of the extended ovipositor. Females of *P. yadongicus* have brown antenna with yellowish brown scape and pedicel, F1 and pedicel of subequal length, the longest F1 of the funicle, and the longest first segment of the clava (Figure 2b); they have an ovipositor notably exserted beyond the apex of the metasoma (Figure 2g), whereas, in comparison to *P. yadongicus*, females of *P. lincolni* have black antennae, F1 longer than pedicel, F2 and pedicel of subequal length, the longest F2 of the funicle, and the ovipositor not notably exserted beyond the apex of the metasoma as in the former species.

3.1.2. *Pseudanaphes zhaoi* Lin, 1997

Pseudanaphes zhaoi Lin 1997 [2], pp. 98–100. Type locality: Shaanxi, China.

Diagnosis. Antenna (Figure 3b): F5 the shortest funicular segment; the first and second claval segments subequal in length, apical segment longer than each of them; the first segment of clava without sensilla, the second segment of clava with two longitudinal sensilla, and the third segment of clava with three longitudinal sensilla. Mesosoma (Figure 3c) about 0.77× length of metasoma. Fore wing (Figure 3d) disc with a slight brownish tinge and a curved, infuscate band below proximal half of marginal vein and slightly infuscate below submarginal vein; densely setose, with setae reaching about half length of submarginal vein. Hind wing (Figure 3e) 16.92× as long as wide; longest marginal seta 3.08× greatest width of wing. Metasoma (Figure 3f): ovipositor (Figure 3g) not or barely exserted beyond the metasomal apex.

Material examined. Six females, China, Tibet, Yadong County, 20 July 2021, Q. Wu (ICXU); three females, Yadong County, 1 July 2020, Qing-Tao Wu and Hong-Liang Li (ICXU).

Distribution. China (Tibet, Shaanxi).

Biology. Unknown.

Comments. The specimens studied here differ from the holotype in the following characters: F6 with one longitudinal sensillum; fore wing 3.25–3.58× as long as wide, with longest marginal seta 0.42–0.52× greatest width of the fore wing; ovipositor 1.39–1.42× as long as the mesotibia. In addition, for one specimen of the material examined above, the apical two claval segments, F2 and F3, are coalesced (Figure 4a,b). These characteristics fall within the range of intraspecific variations.

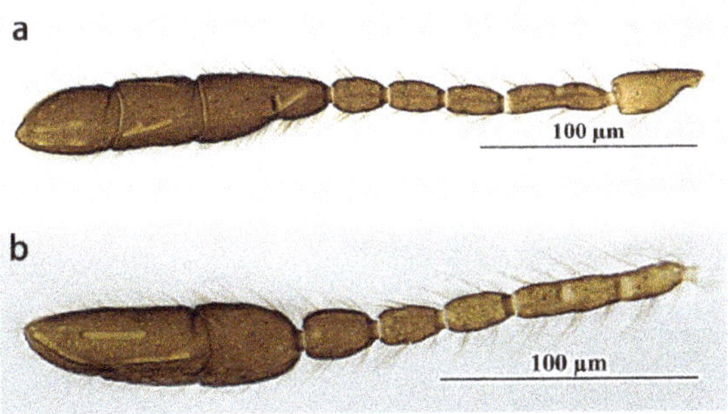

Figure 4. *Pseudanaphes zhaoi*. Female. (**a**), (**b**) Antenna (without scape and pedicel).

3.2. *Key to Species of Pseudanaphes (Female)*

urn:lsid:zoobank.org:act:7B041410-9F8A-4AB2-A693-BB83DA19F0FF

1. The first claval segment longer than the other two claval segments . . . 2

- The first claval segment shorter than at least one of the other two claval segments ... 3
2. Ovipositor notably exserted beyond apex of metasoma; clava at most 4.85× as long as wide, apical claval segment the shortest ... *P. yadongicus* Aishan & Cao, sp. nov.
- Ovipositor not exserted beyond apex of metasoma; clava 5.91× as long as wide, second claval segment the shortest ... *P. lincolni* (Girault)
3. Metasoma with anterior scutellum with reticulate sculpture near posterior margin ... P. hirtus Noyes & Valentine
- Metasoma smooth, without reticulate sculpture ... 4
4. Mesothorax golden; first segment of clava the shortest, clava 3.42× as long as wide ... *P. particoxae* (Girault)
- Mesothorax brown to dark brown; first segment of clava not the shortest, clava at least 3.9× as long as wide ... 5
5. F5 and F4 subequal in length, F4 without longitudinal sensilla; anterior scutellum without setae; fore wing 3.33× as long as wide ... *P. sikkimianus* Rehmat & Anis
- F5 distinctly shorter than F4, F4 with one longitudinal sensillum; anterior scutellum with one pair of setae; fore wing 2.85× as long as wide ... *P. zhaoi* Lin

4. Discussion

Almost all species of Mymaridae are important parasitoids of the eggs of a wide variety of insect pests [12]. Currently, more than 1500 species in 115 genera of Mymaridae are known worldwide [2,9]. Although a total of 31,839 COI sequences of Mymaridae have been recorded in GenBank (National Center for Biotechnology Information, NCBI), most of them were sequenced from unnamed species with a high repeat rate of 80%, and 673 of them were sequenced from only thirteen named species of *Anaphes* Haliday, *Anagrus* Haliday, *Cosmocomoidea* Howard, and *Ooctonus* Haliday [13–20]. In this study, COI primers for the genus *Pseudanaphes* and COI sequences of *Pseudanaphes* were developed and generated for the first time, which provide a reference for studies on the family Mymaridae. The results of molecular analysis support the species delimitation of these two *Pseudanaphes* species, *P. yadongicus* and *P. zhaoi*. In addition, it is suggested that the antennal clava with one row of campaniform sensilla could be used as a diagnosis for the genus *Pseudanaphes*. This will facilitate the identification and discovery of *Pseudanaphes* species or even other mymarids.

Author Contributions: Z.A. and H.-X.C. prepared the material and generalized and analyzed the data. Writing—original draft preparation, Z.A.; writing—review and editing, H.-X.C., H.-Y.H. and C.-D.Z.; supervision, C.-D.Z. and H.-Y.H.; project administration, C.-D.Z.; funding acquisition, Z.A. All authors have read and agreed to the published version of the manuscript.

Funding: This study is supported by the National Natural Science Foundation of China (32260123), the second Tibetan Plateau scientific expedition and research (STEP) program (2019QZKK05010605), and the PhD research startup foundation of Xinjiang University (62031224613).

Institutional Review Board Statement: Not applicable.

Informed Consent Statement: Not applicable.

Data Availability Statement: DNA sequence data are available on GenBank under accession numbers ON834699–ON834714. Other data of this research were deposited in the Institute of Zoology, Chinese Academy of Sciences, Beijing, China.

Acknowledgments: We thank Qing-Tao Wu, Hong-Liang Li, and Ze-Qing Nui (IZCAS) for collecting specimens from Tibet, China. We thank Mei Xiong (IZCAS) for assistance in designing the primers utilized in the COI amplifications. Editors and two anonymous reviewers provided helpful comments on earlier drafts of this manuscript.

Conflicts of Interest: The authors declare no conflict of interest. The funders had no role in the design of the study; in the collection, analyses, or interpretation of data; in the writing of the manuscript; or in the decision to publish the results.

References

1. Noyes, J.S.; Valentine, E.W. Mymaridae (Insecta: Hymenoptera)-introduction and review of genera. *Fauna N. Z.* **1989**, *17*, 47–48.
2. Lin, N.Q. A new species of *Pseudanaphes* (Hymenoptera: Mymaridae) from China. *Wuyi Sci. J.* **1997**, *13*, 98–100.
3. Girault, A.A. Australian Hymenoptera Chalcidoidea—II. Supplement. *Mem. Qld. Mus.* **1913**, *13*, 107–129.
4. Girault, A.A. New Trichogrammatidae and Mymaridae from Australia (Hym). *Rev. Entomol. 1938 9*, 382–396.
5. New, T.R. The relationships of the genus *Australomymar* (Hymenoptera: Mymaridea). *Entomophaga* **1974**, *19*, 229–235. [CrossRef]
6. Lin, N.Q.; Huber, J.T.; LaSalle, J. The Australian genera of Mymaridae (Hymenoptera: Chalcidoidea). *Zootaxa* **2007**, *1596*, 1–111. [CrossRef]
7. Rehmat, T.; Anis, S.B. First record of *Pseudanaphes* (Hymenoptera: Mymaridae) from India, with description of a new species. *Biosystematica* **2011**, *5*, 14–16.
8. Noyes, J. Universal Chalcidoidea Database. 2019. Available online: http://www.nhm.ac.uk/chalcidoids (accessed on 11 June 2022).
9. Triapitsyn, S.V.; Rugman-Jones, P.F.; Tretiakov, P.S.; Daane, K.M.; Wilson, H. Reassessment of molecular and morphological variation within the *Anagrus atomus* species complex (Hymenoptera: Mymaridae): Egg parasitoids of leafhoppers (Hemiptera: Cicadellidae) in Europe and North America. *J. Nat. Hist.* **2020**, *54*, 1735–1758. [CrossRef]
10. Hall, T. BioEdit: A user-friendly biological sequence alignment editor and analysis program for Windows 95/98/NT. *NucleicAcids Symp. Ser.* **1999**, *41*, 95–98.
11. Kumar, S.; Stecher, G.; Tamura, K. MEGA7: Molecular evolutionary genetics analysis version 7.0 for bigger datasets. *Mol. Biol. Evol.* **2016**, *33*, 1870–1874. [CrossRef] [PubMed]
12. Tuber, J.T. Systematics, biology, and hosts of the Mymaridae and Mymarommatidae (Insecta: Hymenoptera): 1758–1984. *Entomography* **1986**, *4*, 185–243.
13. Zhou, Q.S.; Xi, Y.Q.; Fang, Y.; Zhang, X.; Li, X.J.; Liu, C.L.; Niu, Z.Q.; Zhu, C.D.; Qiao, G.X.; Zhang, Y.Z. Application of DNA barcoding to the identification of hymenoptera parasitoids from the soybean aphid (*Aphis glycines*) in china. *Insect Sci.* **2014**, *21*, 363–373. [CrossRef] [PubMed]
14. Huber, J.T.; Read, J.D.; Triapitsyn, S.V. Illustrated key to genera and catalogue of Mymaridae (Hymenoptera) in America north of Mexico. *Zootaxa* **2020**, *4773*, 4773. [CrossRef] [PubMed]
15. Triapitsyn, S.V. Revised Taxonomy of the Common Northern European Fairyfly *Polynema* (*Doriclytus*) *Atratum* Haliday, 1833 (Hymenoptera: Mymaridae), with Fourteen New Synonymies. *Ann. Zool. Fenn.* **2021**, *58*, 87–107. [CrossRef]
16. Triapitsyn, S.V.; Koponen, M.; Vikberg, V.; Várkonyi, G. Taxonomy, annotated new records and a checklist of Mymaridae (Hymenoptera) of Finland, with description of a new species of *Eustochus*. *Acta Entomol. Musei Natl. Pragae* **2020**, *60*, 565–589. [CrossRef]
17. Triapitsyn, S.V.; Rugman-Jones, P.F.; Jeong, G.; Morse, J.G.; Stouthamer, R. Morphological and molecular differentiation of the *Anagrus epos* species complex (Hymenoptera: Mymaridae), egg parasitoids of leafhoppers Hemiptera: Cicadellidae) in North America. *Zootaxa* **2010**, *2428*, 1. [CrossRef]
18. Triapitsyn, S.V.; Rugman-Jones, P.F.; Tretiakov, P.S.; Shih, H.T.; Huang, S.H. New synonymies in the *Anagrus incarnatus* haliday 'species complex' (Hymenoptera: Mymaridae) including a common parasitoid of economically important planthopper (Hemiptera: Delphacidae) pests of rice in Asia. *J. Nat. Hist.* **2018**, *52*, 2795–2822. [CrossRef]
19. Munro, J.B.; Heraty, J.M.; Burks, R.A.; Hawks, D.; Mottern, J.; Cruaud, A.; Rasplus, J.-Y.; Jansta, P. A molecular phylogeny of the Chalcidoidea (Hymenoptera). *PLoS ONE* **2011**, *6*, e27023. [CrossRef] [PubMed]
20. Noort, S.V.; Triapitsyn, S.V. Revision of the *Polynema dikobraz* species-group with description of a remarkable new species from South Africa (Hymenoptera, Chalcidoidea, Mymaridae). *ZooKeys* **2018**, *783*, 67–84. [CrossRef] [PubMed]

insects

MDPI

Article

Descriptions of Three New Species of the Genus *Acerataspis* Uchida, 1934 (Hymenoptera, Ichneumonidae, Metopiinae), with an Illustrated Identification Key to Extant Species

Jing-Xian Liu [1], Alexey Reshchikov [2] and Hua-Yan Chen [3,*]

[1] Department of Entomology, College of Plant Protection, South China Agricultural University, Guangzhou 510642, China
[2] Institute of Eastern Himalaya Biodiversity Research, Dali University, Dali 671003, China
[3] Key Laboratory of Plant Resources Conservation and Sustainable Utilization, South China Botanical Garden, Chinese Academy of Sciences, Guangzhou 510650, China
* Correspondence: huayanc@scbg.ac.cn

Simple Summary: *Acerataspis* Uchida, 1934 is a genus of ichneumonid parasitoids distributed in the Oriental and Palaearctic regions. To date, only seven extant species have been recorded. The host of the genus is still not known. This paper reviews all valid extant species of the genus based on both morphological and molecular analyses and describes three new species.

Abstract: The Asian genus *Acerataspis* Uchida, 1934 is reviewed based on both morphology and DNA barcodes. Ten species are recognized in total, of which three species from Yunnan Province of China are described as new: *Acerataspis maliae* sp. nov., *A. seperata* sp. nov. and *A. similis* sp. nov. The male of *A. fukienensis* Chao, 1957 is described and illustrated for the first time. The genus is recorded from Thailand and Southeast Asia for the first time. An illustrated key to all known extant species is provided. With the supplement of DNA barcodes, a few diagnostic morphological characters are found useful in species identification.

Keywords: Darwin wasps; Ichneumonoidea; parasitoid; DNA barcodes; new species

1. Introduction

The genus *Acerataspis* Uchida 1934 is a small genus in the subfamily Metopiinae (Hymenoptera, Ichneumonidae), previously comprising seven extant species all distributed in Asia [1–6], and one fossil from the Late Eocene of Colorado, USA [7]. Most of the species (*A. cruralis* Chiu, 1962, *A. formosana* Cushman, 1937, *A. fukienensis* Chao, 1957, *A. fusiformis* Morley, 1913, *A. szechuanensis* Chao, 1962) are known from the Oriental part of China [3–5,8–11], and two species, *A. clavata* Uchida, 1934 and *A. sinensis* Michener, 1940, range from the Eastern Palaearctic to the Oriental region [4,12–15]. *Acerataspis fusiformis* was described from the Karen Hills in Eastern Myanmar [1] and was recently recorded from the Guangxi Province of China [11]. Townes mentioned a few undescribed specimens of this species deposited at the American Entomological Institute (AEI) from Indonesia and Philippines [2]. There are also a number of specimens of unidentified barcoded species originating from Chattogram, Bangladesh at the BOLD "(available at http://bins.boldsystems.org/index.php/Public_SearchTerms?query=Acerataspis[tax], accessed on 1 February 2023)" and Cameron Highlands, Malaysia [16].

To date, nothing is known about the biology of the genus. However, phylogenetic analysis of the subfamily with *Acerataspis* species included showed a close relationship with the genus *Metopius* Panzer, 1806 [16], which are known to be koinobiont, larval-pupal endoparasitoids of Lepidoptera, mainly Lasiocampidae, Erebidae, Notodontinae, Noctuidae, and Geometridae [6,17]. Members of both genera are coloured with black and yellow stripes, which probably indicates mimicry of aculeate wasps of the subfamily Eumeninae (Hymenoptera, Vespidae) [18].

In this study, we examined species of *Acerataspis* from Asia and reviewed it based on morphological and molecular analyses, with an illustrated key to species.

2. Materials and Methods

The specimens examined are deposited in the following institutions (curators in parentheses):

CAS: California Academy of Sciences, California, USA (Robert Zuparko).

DEI: Deutsches Entomologisches Institute, Eberswalde, Germany (Andreas Taeger).

HKU: Hong Kong University, Hong Kong, China (Benoit Guénard).

IOZ: Institute of Zoology, Chinese Academy of Science, Beijing, China (Kui-Yan Zhang).

MCSN: Museo Civico di Storia Naturale "Giacomo Doria", Genoa, Italy (Maria Tavano).

QSBG: Queen Sirikit Botanic Garden, Chiang Mai, Thailand (Wichai Srisuka).

SCAU: South China Agricultural University, Guangzhou, China (Jing-Xian Liu).

TARI: Taiwan Agricultural Research Institute, Taichung, China (Chi-Feng Lee).

Specimens were newly sampled for this study in Northern Thailand using Malaise traps during the "Tea Fauna" project (https://teafauna.com) in an old secondary *Decteracarpus* forest with *Camellia sinensis* var. *assamica* (Masters) Kitamura in the understory [19]. Specimens from Hong Kong were sampled using Malaise traps during long-term monitoring by Christophe Barthélémy [20]. Specimens from other parts of China, such as Guangdong, Yunnan and Hainan provinces, were either collected by Malaise trap or sweep net by Li Ma, Shi-Xiao Luo and Hua-Yan Chen.

Specimens were examined using a Zeiss (Stemi-508) stereomicroscope (Carl Zeiss AG, Jena, Germany). Photos were taken at SCAU using a Keyence VHX-5000 (Keyence Corporation, Osaka, Japan) and Leica S8AP0 stereomicroscope (Leica Camera AG, Wetzlar, Germany) mounted with a digital camera. Photos of the type specimen of *A. formosana* were taken at the DEI using a Leica DFC 495 digital camera (Leica Camera AG, Wetzlar, Germany) and M205 C microscope (Leica Camera AG, Wetzlar, Germany). Photos of the type specimen of *A. cruralis* were taken at TARI using a LEICA MC 190 HD (Leica Camera AG, Wetzlar, Germany). Composite images with an extended depth of field were created from stacks of images using Combine ZP software and arranged and partly enhanced using Ulead PhotoImpact X3.

Morphological terminology follows that of Broad et al. 2018 [17]. Abbreviations used in the text are as follows: POL = the shortest distance between the posterior ocelli; OD = the longest diameter of a posterior ocellus; OOL = the shortest distance between a posterior ocellus and compound eye. T1–T7 refers to the metasomal tergites 1–7.

To supplement molecular species identification of *Acerataspis* for future work, the "DNA barcoding" region—mainly for the Arthropod mitochondrial cytochrome oxidase subunit 1 (*CO1*)—was amplified from species with fresh available specimens. Genomic DNA was extracted from one leg of each specimen using the TIANamp Micro DNA Kit (Tiangen Biotech, Beijing, China), following the manufacturer's protocols. Polymerase chain reactions (PCRs) were performed using Tks Gflex™ DNA Polymerase (Takara) with the LCO1490/HCO2198 primer pair [21] and conducted in a T100™ Thermal Cycler (Bio-Rad). Thermocycling conditions were: an initial denaturing step at 94 °C for 5 min, followed by 35 cycles of 94 °C for 30 s, 50 °C for 30 s, 72 °C for 30 s and an additional extension at 72 °C for 5 min. Amplicons were directly sequenced in both directions with forward and reverse primers on an Applied Biosystems (ABI) 3730XL by Guangzhou Tianyi Huiyuan Gene Technology Co., Ltd. (Guangzhou, China). Chromatograms were assembled with Geneious 11.0.3. All sequences generated from this study are deposited in GenBank (for accession numbers see Table 1).

Sequences of *Acerataspis clavata* (Uchida, 1934) were downloaded from Genbank. Sequences were aligned by codons using MUSCLE implemented in Geneious. The p-distances within and between species were calculated in MEGA 11 [22]. The aligned sequences were then analyzed using RAxML as implemented in Geneious ver. 11.0.3 to generate a maximum likelihood (ML) tree to show the affinities among the studied species. Each codon position of the *CO1* sequence was treated as a different data partition, and the best-fit substitution model was determined using ModelFinder [23]. The best-fit model according to the Bayesian Information Criterion was GTR + Γ+I. Automatic bootstopping criterion was selected as the appropriate number of bootstraps; 300 replicates were run. A sequence of *Chorinaeus funebris* (Gravenhorst, 1829) (Ichneumonidae, Metopiinae) (KR931343) was downloaded from Genbank and used as an outgroup.

Table 1. List of sequenced taxa and accession numbers.

Code	Species	Sex	GanBank Accession No.
SCAU-3011979	*Acerataspis clavata* (Uchida, 1934)	female	OP525229
SCAU-3011980	*Acerataspis clavata* (Uchida, 1934)	male	OP525229
SCAU-3011983	*Acerataspis clavata* (Uchida, 1934)	male	OP525229
SCAU-3013718	*Acerataspis clavata* (Uchida, 1934)	female	OP525229
SCAU-3013714	*Acerataspis clavata* (Uchida, 1934)	male	OP525229
SCAU-3011049	*Acerataspis clavata* (Uchida, 1934)	male	OP525229
SCAU-3011050	*Acerataspis clavata* (Uchida, 1934)	female	OP525229
SCBG-E0000053	*Acerataspis clavata* (Uchida, 1934)	male	OP525229
SCBG-E0000055	*Acerataspis clavata* (Uchida, 1934)	male	OP525229
SCBG-E0000045	*Acerataspis fukienensis* Chao, 1957	male	OP525229
SCBG-E0000048	*Acerataspis fukienensis* Chao, 1957	female	OP525229
SCAU-3013715	*Acerataspis fusiformis* (Morley, 1913)	female	OP525229
SCAU-3013716	*Acerataspis fusiformis* (Morley, 1913)	male	OP525229
SCAU-3013717	*Acerataspis fusiformis* (Morley, 1913)	male	OP525229
SCAU-3011024	*Acerataspis fusiformis* (Morley, 1913)	male	OP525229
SCAU-3011028	*Acerataspis fusiformis* (Morley, 1913)	female	OP525229
SCBG-E0000043	*Acerataspis fusiformis* (Morley, 1913)	male	OP525229
SCBG-E0000042	*Acerataspis maliae* sp. n.	female	OP525229
SCBG-E0000047	*Acerataspis maliae* sp. n.	female	OP525229
SCAU-3011978	*Acerataspis separata* sp. n.	male	OP525229
SCAU-3011982	*Acerataspis separata* sp. n.	female	OP525229
SCBG-E0000044	*Acerataspis separata* sp. n.	male	OP525229
SCBG-E0000050	*Acerataspis separata* sp. n.	male	OP525229
SCBG-E0000051	*Acerataspis separata* sp. n.	male	OP525229
SCBG-E0000052	*Acerataspis separata* sp. n.	male	OP525229
SCBG-E0000054	*Acerataspis separata* sp. n.	male	OP525229
SCAU-3011981	*Acerataspis similis* sp. n.	female	OP525229

3. Results

3.1. Morphological Examination

In this study, 35 specimens representing ten morphospecies were examined from China, Myanmar, Thailand, and Russia, and ten morphospecies were recognized. Seven species were identified as previously described species and three are considered as new to science species as described below (Tables 1 and 2). *Acerataspis clavata* and *A. fusiformis* are recorded from Hong Kong and Thailand; these are the first documented records of the genus from these regions as well as Southeast Asia.

3.2. DNA Barcoding

The present study generated 27 *CO1* sequences, representing six morphospecies with 642–684 bps. When sequences from GenBank and BOLD databases were compared using BLAST, only the sequences of *Acerataspis clavata* (Uchida, 1934) received close matches with two sequences also labeled as *A. clavata* (Uchida, 1934), but only share a 92.7% similarity. Genetic distances of the 27 sequences are provided in Table S1. Intraspecific distances of the *CO1* sequences are generally less than 4%. Interspecific distances range between

9.7% and 16.5% (Table S2). Each species recovered on the tree is clearly separated from all neighboring species, as shown in Figure 1.

Table 2. List of species of *Acerataspis* (Bold species indicate newly described species in this study).

Species	Distribution
Acerataspis clavata (Uchida, 1934)	Oriental and Palaearctic
Acerataspis cruralis Chiu, 1962	Oriental
Acerataspis formosana Cushman, 1937	Oriental
Acerataspis fukienensis Chao, 1957	Oriental
Acerataspis fusiformis (Morley, 1913)	Oriental
Acerataspis maliae **sp. nov.**	Oriental
Acerataspis separata **sp. nov.**	Oriental
Acerataspis similis **sp. nov.**	Oriental
Acerataspis sinensis Michener, 1940	Oriental and Palaearctic
Acerataspis szechuanensis Chao, 1962	Oriental

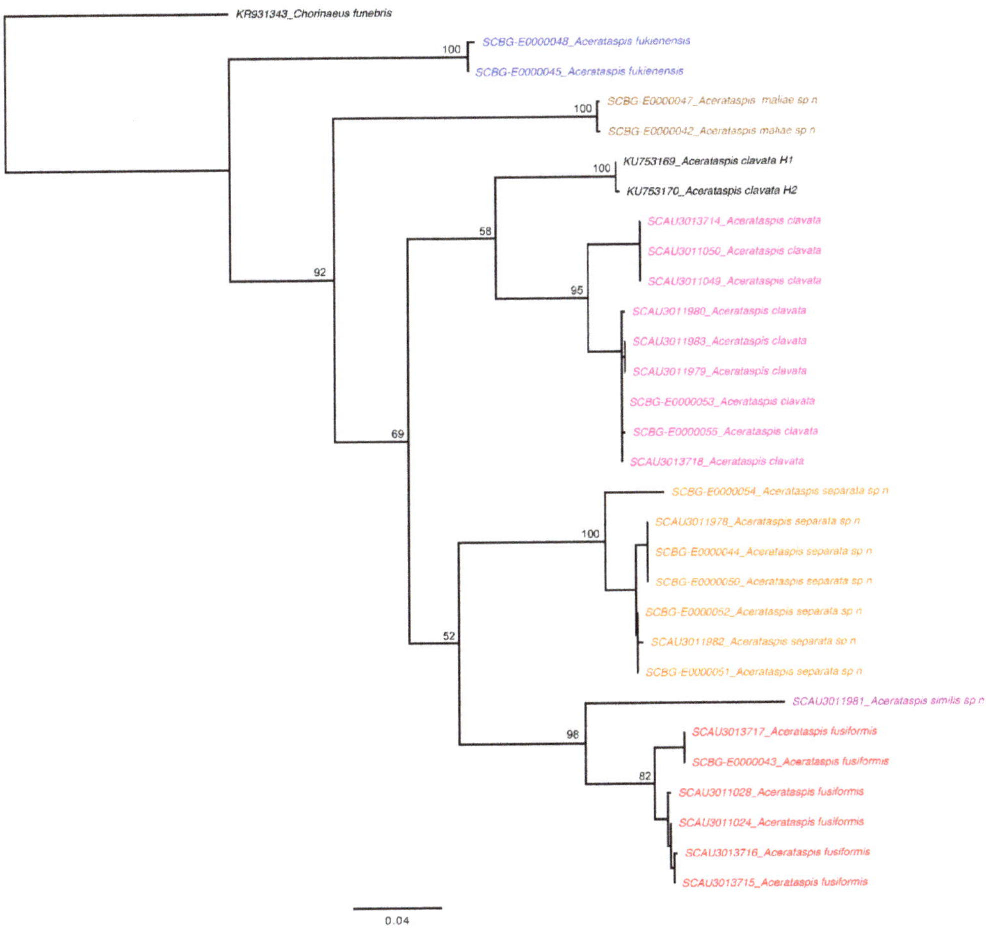

Figure 1. Maximum Likelihood tree demonstrating the clusters of *Acerataspis* species, based on *CO1*. The numbers at nodes show the boostrap values. The scale bar represents 0.04 substitutions per site.

3.3. *Systematics*

***Acerataspis* Uchida, 1934** (Figures 2 and 3).

Ceratapsis Uchida, 1934, 275 [24]. Preoccupied by *Cerataspis* Gray, 1847. Type: *Cerataspis clavata* Uchida. Original designation.

Acerataspis Uchida, 1934, 23 [12]. New name for *Cerataspis* Uchida, not Gray. Type species: *Acerataspis clavata* Uchida, by original designation.

Diagnosis. Fore wing length 7.0–10.0 mm. Face and clypeus evenly moderately convex (Figure 2A). Clypeal sulcus absent. Upper margin of face continued dorsally as an intra-antennal projection with a deep central groove. Temple very short, mostly flat (Figure 2B). Occipital carina complete, but not reaching mandibular base ventrally, nor meeting hypostomal carina (Figure 2C). Subocular sulcus absent. Mandible with upper tooth a little larger than lower tooth. Notaulus absent. Scutellum short, transverse, prolonged lateroposteriorly as a tooth (Figure 2D). Propodeum with areas, anterior transverse carina present. Sternaulus broadly impressed. Metapleuron with juxtacoxal carina complete or incomplete. Fore wing with areolet sessile or petiolate, receiving 2m-cu vein before middle (Figure 2E), hind wing with nervellus interrupted below middle. Middle tibia with a single slender spur in male, with two spurs in female. Hind tibia with two spurs. All claws pectinate (Figure 2F). Metasoma clavate, T1–T3 with a pair of lateromedian longitudinal carinae from base to apex (Figure 3A), T4 of some species also with a pair of short lateromedian longitudinal carinae on anterior base, T6 rounded and turned under (Figure 3C,D). Ovipositor short, thin, tapered to apex, without a dorsal subapical notch (Figure 3B).

Figure 2. *Acerataspis* spp (*A. clavata*: (**A**,**D–F**); *A. fusiformis*: (**B**,**C**)). (**A**) Head, frontal view; (**B**) head, dorsal view; (**C**) head, posterior lateral view; (**D**) mesosoma, dorsal view; (**E**) fore wing; (**F**) fore tarsal claw of female.

Comment. In previous studies of Metopiinae, the presence or absence of pectinae of tarsal claws were used to separate species in the genus *Hypsicera* [4,25]. However, this character was more or less overlooked or roughly described by authors in previous studies of *Aceratapsis*. Based on the materials of *Aceratapsis* from China, we found that the tarsal claw pectination of the female is a useful diagnostic character for species identification (although sometimes the pattern of pectination in the hind legs of some female specimens may vary between left hind leg and right hind leg). The proximal pecten of the claw is also very small and needs to be observed at high magnification.

Biology. No host records have been reported.

Distribution. Palaearctic and Oriental regions [6].

Figure 3. *Aceratapsis* spp (*A. clavata*: (**A**,**B**,**D**); *A. fusiformis*: (**C**)). (**A**) Metasoma, dorsal view (female); (**B**) metasoma, lateral view (female); (**C**) T4–T6 of male specimen, central view; (**D**) metasoma of male, lateral view.

Key to species of *Aceratapsis* (only *A. formosa*, *A. clavata*, *A. fukienensis*, *A. fusiformis*, *A. separate* sp. nov. known from both sexes)

1. Female (1a) . . . 2

-. Male (1aa) . . . 11

1a 1aa

2. Hind tibia entirely black or blackish (2a) . . . 3

-. Hind tibia with proximal half white or yellow (2aa) . . . 6

3. Fore wing with 1cu-a opposite M&RS (3a); fore tarsal claw with 3–4 large pectinae (3b); T4 with a pair of short lateromedian longitudinal carinae on anterior half (3c) . . . 4

-. Fore wing with 1cu-a postfurcal to M&RS (3aa); claw with six or more fine pectinae (3bb); T4 without caraine (3cc) or a single of very short median longitudinal carinae on anterior half (3dd) . . . 5

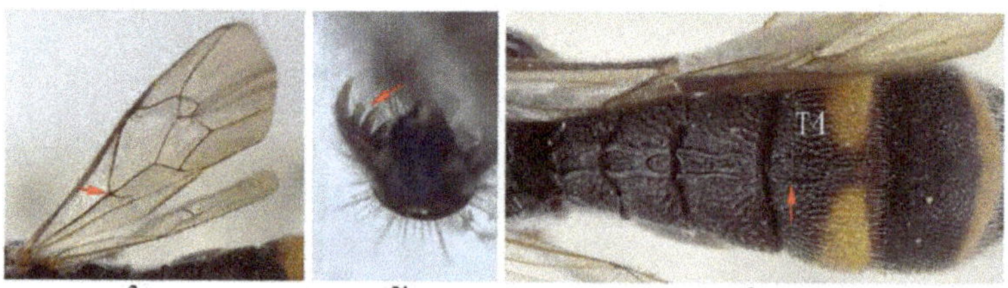

3a 3b 3c

3cc 3dd

4. Fore wing with areolet sessile (4a); T1–T3 with yellowish or whitish marks (4b); T4 with apical yellow band complete (4b) . . . *A. formosana*

-. Fore wing with areolet petiolate (4aa); T1–T3 black (4bb); T4 with apical yellow band interrupted dorsomedially (4cc) . . . *A. fukienensis*

4aa 4bb 4cc

5. Scutellum entirely black (5a); subtegular ridge blackish brown (5b); T2 punctate and without irregular longitudinal carina between the latero-median longitudinal carinae (5c); T4 without carina anteriorly (5d) . . . *A. sinensis*

-. Scutellum anteriorly black and posteriorly largely orange yellow (5aa); subtegular ridge entirely orange yellow (5bb); T2 with an irregular longitudinal carina between the latero-median longitudinal carinae (5cc); T4 with a very short and weakly defined carina anteriorly medially (5dd) . . . *A. szechuanensis*

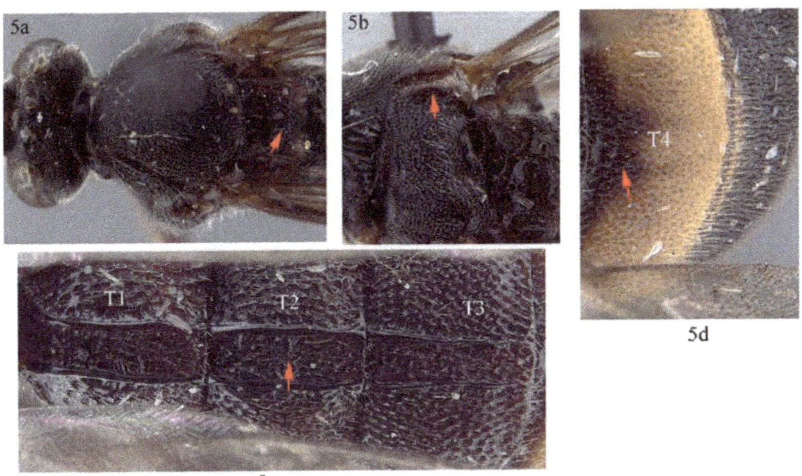

5d

5c

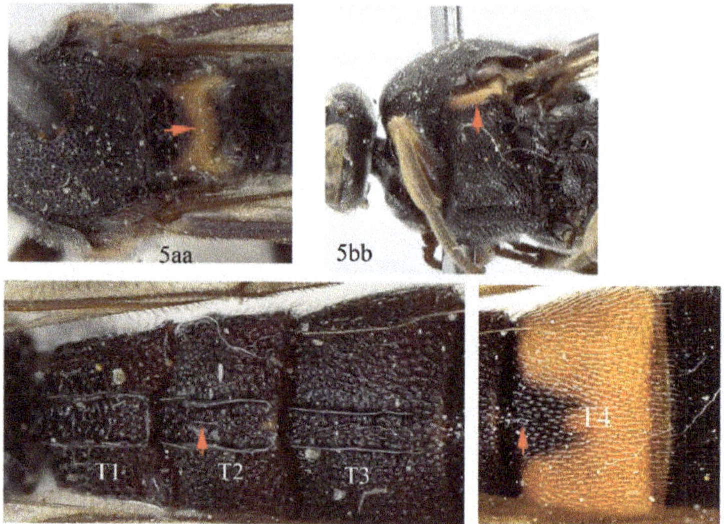

5aa 5bb 5cc 5dd

6. T4 with posterior yellow or whitish band dorso-medially interrupted (6a) ... *A. separata* sp. nov.

-. T4 with posterior yellow or whitish band complete (6aa) ... 7

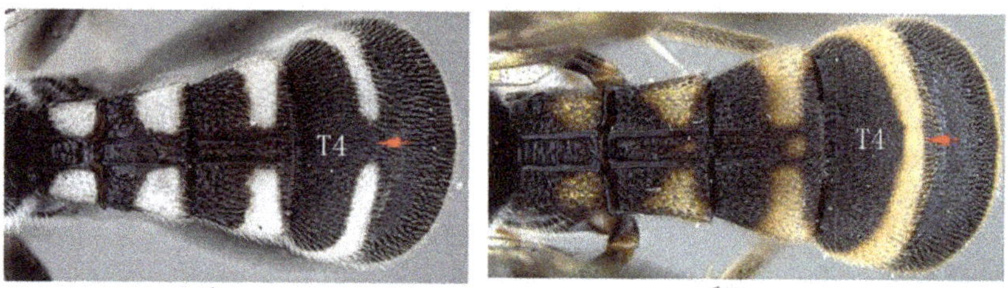

6a 6aa

7. Fore tarsal claw with 3–4 pectinae (7a); fore wing with 1cu-a opposite M&RS (7b) (except for *Acerataspis maliae* sp.nov., 1cu-a slightly postfurcal to M&RS) ... 8

-. Fore tarsal claw with eight pectinae (7aa); fore wing with 1cu-a postfurcal to M&RS (7bb) ... 10

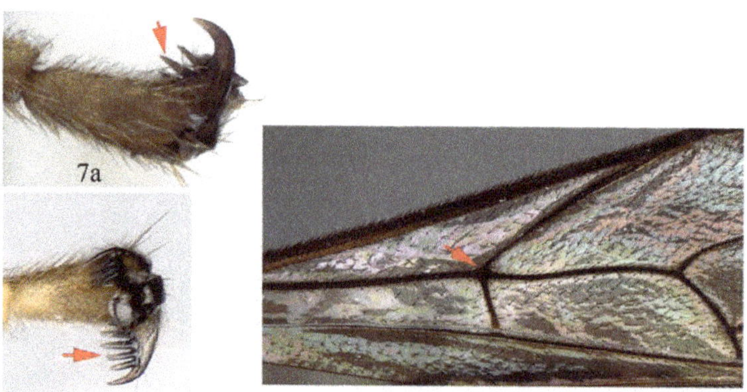

7a 7a 7b

7aa

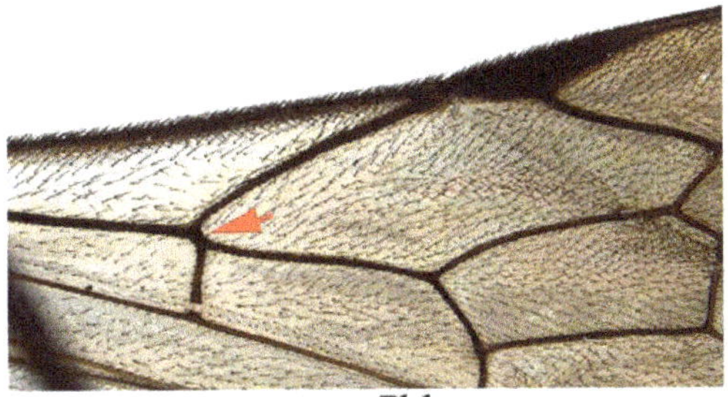

7bb

8. Hind tarsal claw with seven unequal pectinae (8a), fore wing with 1cu-a slightly postfurcal to M&RS (8b) ... *A. maliae* sp. nov.

-. Hind tarsal claw with three or four unequal pectinae (8aa), fore wing with 1cu-a opposite M&RS (8bb) ... 9

8a

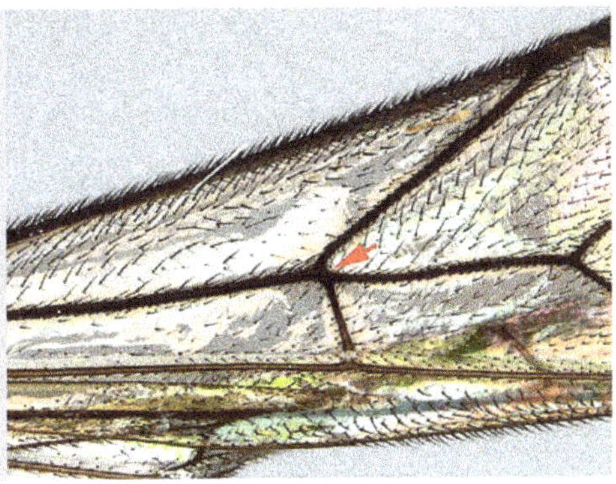

8b

8aa

8aa

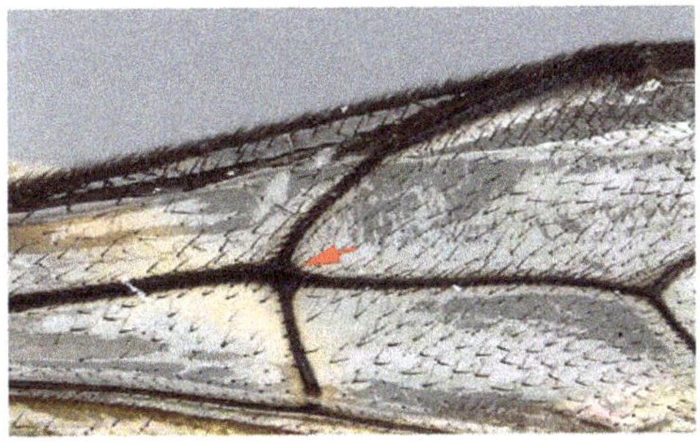

8bb

9. Hind tarsal claw with three pectinae (9a); metanotum orange yellow (9b) ... *A. fusiformis*

-. Hind tarsal claw with four pectinae (9aa); metanotum black (9bb) ... *A. similis* sp. nov.

9a

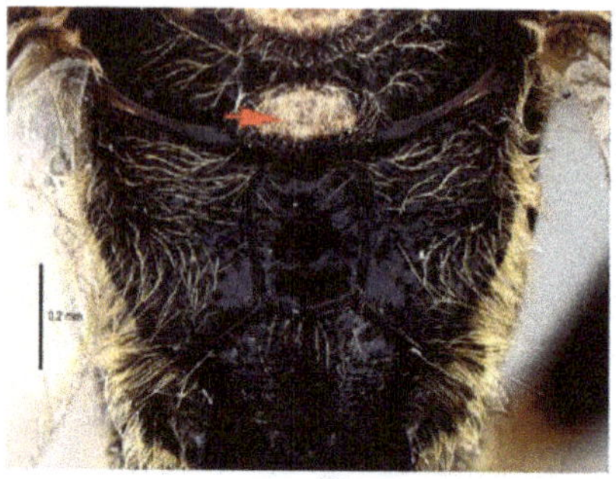

9b

9aa

9bb

10. T2 and T3 entirely ferruginous, posterior margin of T2 straight (10a); T4 largely ferruginous with the extreme base black (10b) ... *A. cruralis*

-. T2 and T3 anteriorly black and posteriorly yellowish brown, posterior margin of T2 convex (10aa); T4 black on anterior half and yellow on posterior half (10bb) ... *A. clavata*

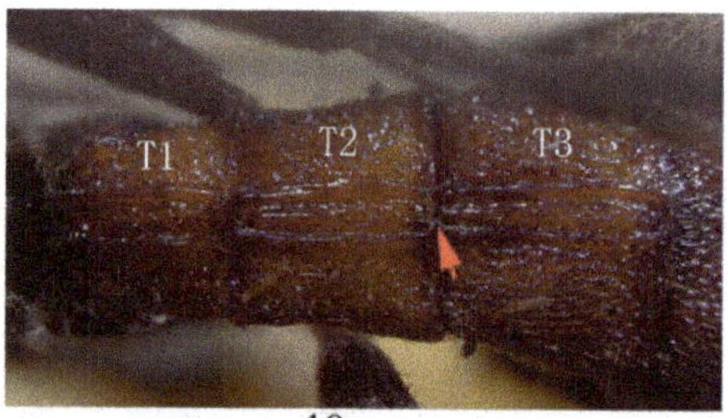

10a

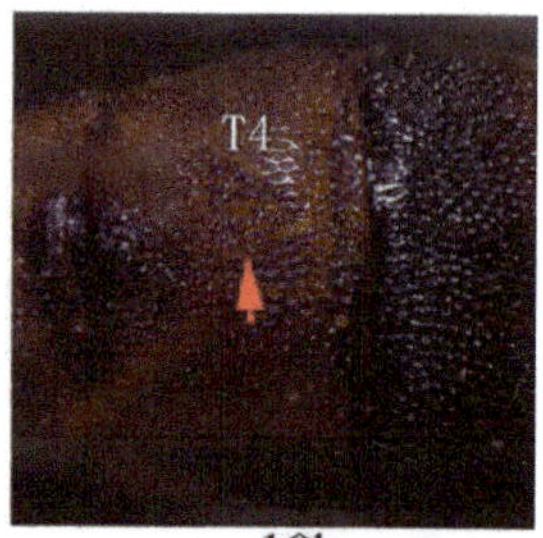

10b

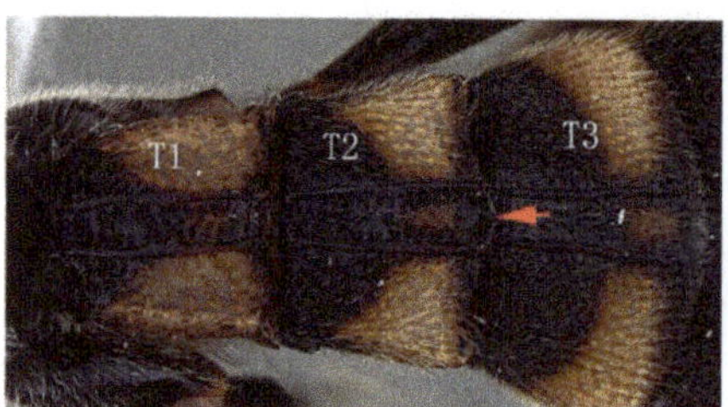

10aa

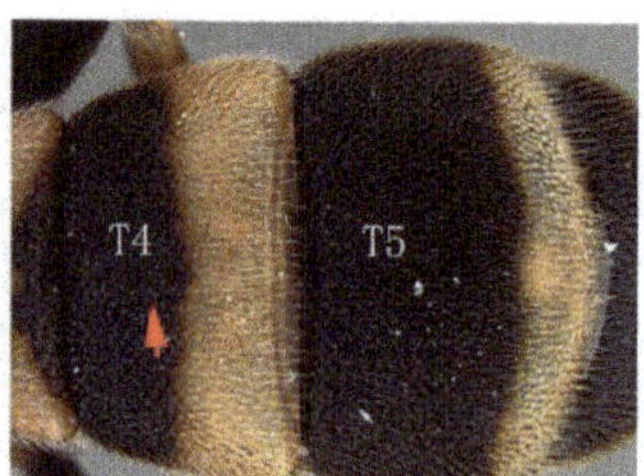

10bb

11. T4 with a pair of short lateromedian longitudinal carinae on anterior 0.45 of tergite (11a); hind tibia black (11b) ... 12

-. T4 without a pair of short lateromedian longitudinal carinae anteriorly (11aa); hind tibia proximally yellow or white (11bb) ... 13

11a

11b

11aa

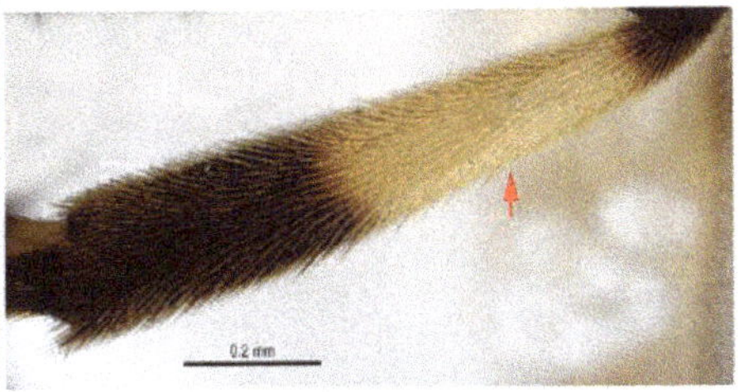

11bb

12. T1–T3 entirely black (12a) ... *A. fukienensis*

-. T1–T3 with fuscous or brownish marks ... *A. formosana*

12a

13. Fore wing with 1cu-a opposite M&RS (13a) ... 14

-. Fore wing with 1cu-a postfurcal to M&RS (13aa) ... 15

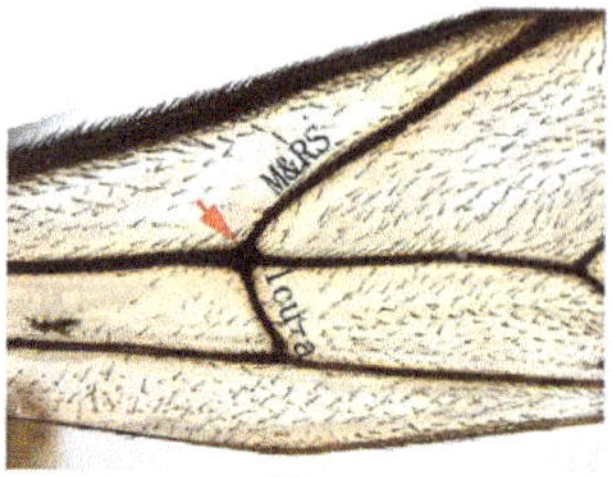

13a

13aa

14. Hind tarsal claw with 3–4 pectinae (14a); metanotum yellow (14b); area superomedia of propodeum longer than wide (14c) . . . *A. fusiformis*

-. Hind tarsal claw with six pectinae (14aa); metanotum black (14bb); area superomedia of propodeum wider than long (14cc) . . . *A. clavata*

14a

14b, 14c

14aa

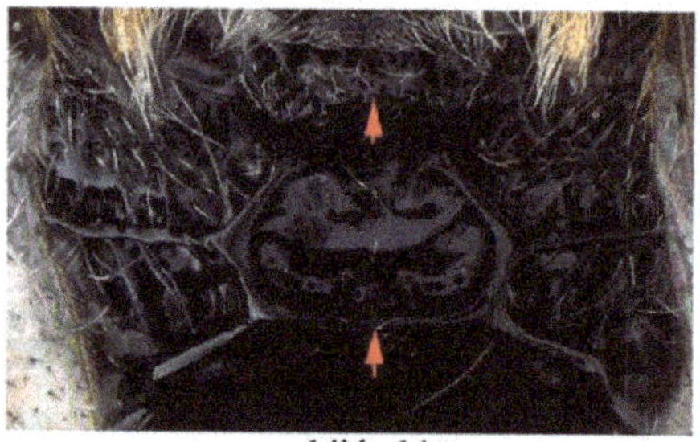

14bb, 14cc

15. Hind tarsal claw with 3–4 pectinae (15a); metanotum yellow (15b) . . . *A. fusiformis*

-. Hind tarsal claw with more than five pectinae (15aa), metanotum black (15bb) . . . 16

15a

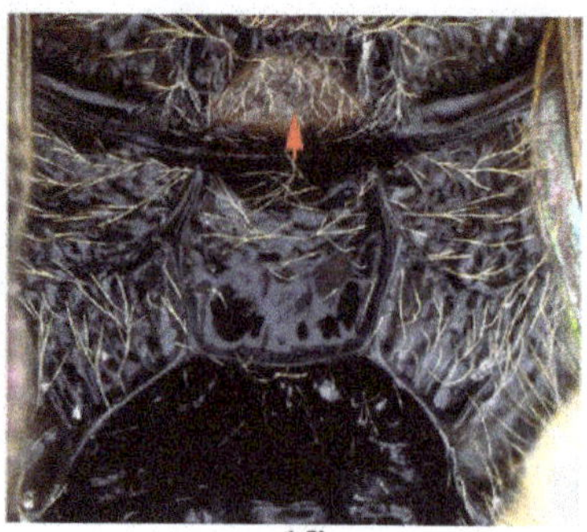

15b

15aa

15bb

16. T4 with posterior white band interrupted dorso-medially(16a); terminal flagellomere stout and 2.8 times as long as wide (16b) . . . *A. separata* sp. nov.

-. T4 with posterior white band complete (16aa); terminal flagellomere slender and 5.0–5.4 times as long as wide (16bb) . . . *A. clavata*

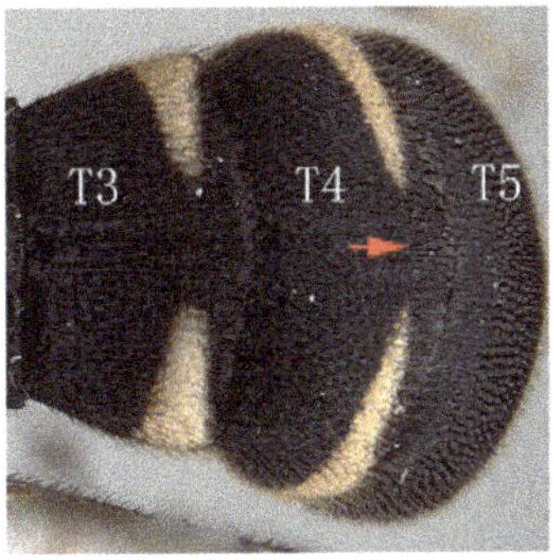

16a

16b

16aa

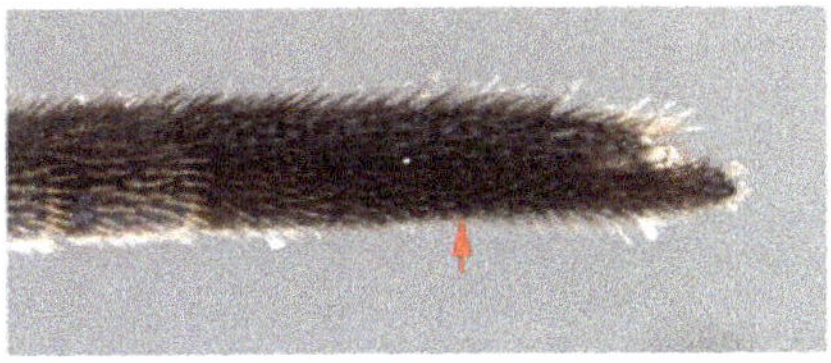

16bb

3.4. Species Descriptions

3.4.1. *Acerataspis clavata* (Uchida, 1934) (Figures 4–6)

Cerataspis clavata Uchida, 1934, Transactions of the Sapporo Natural History Society. 13(3): 275 [12]. Type, male, from Japan.

Materials examined. CHINA: 1♀1♂, Yunnan Province, Xishuangbanna Tropical Botanical Garden of the Chinese Academy of Science, 15.V–17.VI.2018, Malaise trap, 21°52′ N, 101°19′ E, alt. 543 m, Li Ma leg., SCAU-3011979, SCAU-3011983 (SCAU); 2♂, Yunnan, Xishuangbanna Botanical Garden, rubber plantation, 21°52′ N, 101°19′ E, alt. 543 m,

Malaise trap, 16.III–11.IV.2021, Li Ma leg., SCBG-E0000053, SCBG-E0000055 (SCAU); 1♂, Yunnan Province, Xishuangbanna Tropical Botany Garden of Chinese Academy of Science, Rain Forest, 17.V–18.VI.2018, Malaise trap, alt. 560 m, 21°52′ N, 101°19′ E, Li Ma leg., SCAU-3011980 (SCAU); 1♂, Guangdong Province, Shenzhen, Tanglangshan, LSX-653, GD8, 30.VI–31.VIII.2020, 22°34′ N, 113°58′E, Luo Shi-xiao leg., Malaise trap, SCAU-3011049 (SCAU); 1♀, Guangdong Province, Shenzhen, Fairylake Botanical Garden, 1.VIII–8.IX.2020, 22°34′ N, 114°10′ E, Luo Shi-xiao leg., SCAU-3011050 (SCAU); 1♂, Hong Kong, Ping Shan Chai, 22°29′8.30″ N 114°11′14.74″ E, alt. 140 m, Malaise trap (M348), 14–30.IV.2018, Christophe Barthélémy leg., SCAU-3013714 (HKU); RUSSIA: 1♀, Primorje, Vladivostok, Sedanka, VII.1920, René Malaise leg., NHRS-HEVA000020635 (NHRS); THAILAND: 1♀, Chiang Mai, Mae Taeng, Pa Pae, 19°14′30.6″ N, 98°39′14.1″ E, old secondary forest with *Camellia sinensis* var. *assamica*, Malaise trap (Data#1). 21.III–11.IV.2017, Monsoon Tea leg. SCAU-3013718 (QSBG).

Diagnosis. Fore wing (Figure 4) with 1cu-a postfurcal to M&RS, areolet sessile. Propodeum with anterior transverse carina joining area superomedia at middle. T1–T3 with a pair of median carinae. All tarsal claws (Figure 6A–C) of female and male densely pectinate, with at least six pectinae on each claw. Antenna with scape largely black, only ventrally yellow. Hind tibia with proximal 0.4 yellowish white. T1–T3 with apical yellow band (band sometimes interrupted medially), T4 with apical margin yellow, usually not interrupted dorsomedially (Figures 4 and 5A).

Variation. One specimen (specimen ID: SCAU-3013718) from Thailand with five pectinae on fore claw (Figure 6D) and apical yellow band of T4 dorsomedially interrupted (Figure 5B).

Distribution. Oriental and Eastern Palaearctic regions (newly recorded from Hong Kong and Thailand).

Differential diagnosis. This species is very similar to *A. fusiformis* (Morley, 1913) in colour pattern, it is slightly different from the latter in having all tarsal claws densely pectinate (female fore tarsal claw of *A. fusiformis* with only three pectinae), area superomedia of male laterally expanded (based on the materials from China, area superomedia weakly convergent posteriorly), fore wing with 1cu-a postfurcal to M&RS (fore wing with 1cu-a opposite M&RS of *A. fusiformis*), metanotum usually black (except one specimen from Yunnan slightly brown) (metascutellum distinctly yellow in *A. fusiformis*).

Comments. The specimen (ID: SCAU-3013718) from Thailand with five pectinae on fore tarsal claw and apical transverse band on T4 interrupted dorsomedially is an unusual variation within species, and it was determined based on DNA barcoding.

Figure 4. *Acerataspis clavata* (Uchida, 1934), female. (**A**) habitus, dorsal view; (**B**) head, frontal view (non-type specimen from China, specimen ID: SCAU-3011979). (Note: Facial colouration was altered after the DNA extraction experiment; the original colour of the species is yellow).

Figure 5. *Acerataspis clavata* (Uchida, 1934). (**A**) Male habitus, dorsal view (non-type specimen from China, specimen ID: SCAU-3011980); (**B**) metasoma of a female specimen from Thailand, dorsal view.

Figure 6. Tarsal claws of *Acerataspis clavata* (female). (**A**) Fore tarsal claw; (**B**) mid tarsal claw; (**C**) hind tarsal claw ((**A–C**), specimen ID: SCAU3011979, from China); (**D**) fore tarsal claw with five pectinae (specimen from Thailand).

3.4.2. *Acerataspis cruralis* Chiu, 1962 (Figures 7 and 8)

Acerataspis cruralis Chiu, 1962, 20: 5 [4]. Holotype, female, from Taiwan.

Materials examined. Holotype, 1♀, Labeled "Oiwake (Wushe), FORMOSA, 2. IX. 1940, Col. J. Sonan; Holotpye, type, S.C. Chiu 197; Holotype ♀, *Acerataspis cruralis*, DET.S.C. CHIU" (TARI).

Diagnosis. Propedeum with anterior transverse carina joining area superomedia at middle, area dentipara densely and finely punctate. Fore wing (Figure 8D) with 1cu-a postfurcal to M&RS; areolet sessile. T1 with a pair of carinae (Figure 8B); T2 with three longitudinal median carinae (Figure 8B), the median one obscure with apical half absent; T3 with three longitudinal carinae, the median one weak (Figure 8B); T4 strongly punctate, without carina (Figure 8E). Fore tarsal claw strongly pectinate, with 6–7 pectinae (Figure 8C). Hind leg black, proximal half of tibia yellow with proximal base black (Figure 7). Metanotum yellow. Metasoma ferruginous (Figure 7), with base of T1, base of T4, anterior half of T5, and T6 blackish to black.

Distribution. Oriental region [26].

Differential diagnosis. This species is very similar to *A. clavata*, especially in colour pattern, with differences between them as follows: *A. cruralis* with metascutellum yellow and T1 (except for the base black) to T4 (except for the extreme base black) mainly ferruginous, (*A. clavata* with metascutellum entirely black, T1–T4 black, apical margin with yellow marks or yellow band). Hind margin of T2 straight in *A. cruralis* (convex in its middle in *A. clavata*), fore claw weakly curved (Figure 8C) in *A. cruralis* (more strongly curved in *A. clavata*) (Figure 6A).

Figure 7. *Acerataspis cruralis* Chiu, habitus of holotype female.

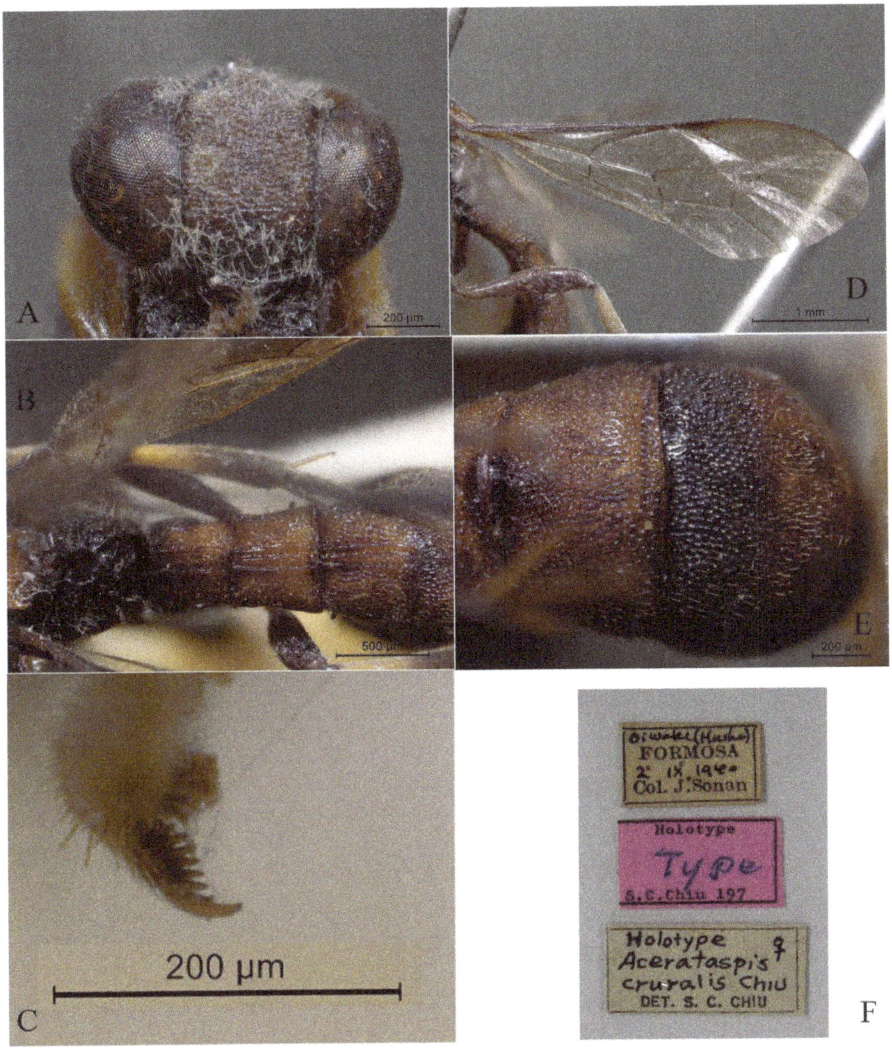

Figure 8. *Acerataspis cruralis* Chiu, holotype female. (**A**) Head, frontal view; (**B**) propodeum and T1–T3, dorsal view; (**C**) fore tarsal claw; (**D**) fore wing; (**E**) T4 and T5, dorsal view; (**F**) labels of holotype.

3.4.3. *Acerataspis formosana* Cushman, 1937 (Figures 9 and 10)

Acerataspis formosana Cushman, 1937, 4: 291 [8]. Type female.

Material examined. Holotype. 1♀, labelled "Suisharyo, Formosa, H.Sauter, X. 1911; Holotypus; *Acerataspis formosana*, Type, Cush; DEI-GISHym, 16611" (DEI). Figure posted on 21.iii.2019, 18:33, by Andreas Taeger. (Note: Type materials were examined based on the type photos provided by Andreas Taeger.)

Diagnosis. Propedeum with the area dentipara rugose. Fore tarsal claw (Figure 9H) with 2–3 large teeth. Fore wing with 1cu-a opposite to M&RS (Figure 10C). T1 and T2 with a pair of median longitudinal carinae (Figure 9D). T3 with median carinae distinct on anterior half, obscure and weakened from middle to apex; T4 anteriorly with a pair of short lateromedian longitudinal carinae (Figure 9C). Hind leg entirely black (Figure 10B,D).

Differential diagnosis. This species is similar to *A. clavata* in the colour pattern of metasoma, with differences between them as follows: *A. formosana* with hind tibia entirely black or blackish brown (hind tibia of *A. clavata* proximally with a white band), *A. formosana* with fore wing with 1cu-a vein opposite M&RS (fore wing with 1cu-a postfurcal to M&RS in *A. clavata*), T4 anteriorly with a pair of short lateromedian longitudinal carinae (Figure 9C) (T4 without a pair of short lateromedian longitudinal carinae in *A. clavata*).

Figure 9. *Acerataspis formosana* Cushman, female. (**A**) Head and mesosoma, lateral view; (**B**) head and mesosoma, dorsal view; (**C**) T3–T6, dorsal view; (**D**) scutellum, metanotum, propodeum and T1–T3, dorsal view; (**E**) head, dorsal view; (**F**) head, anterior view; (**G**) head, anterior ventral view; (**H**) fore tarsal claw; (**I**) ovipositor, lateral view; (**J**) labels of holotype. (Images: A. Taeger).

Figure 10. *Acerataspis formosana* Cushman, holotype, female. (**A**) Habitus, dorsal view; (**B**,**D**) habitus, lateral view; (**C**) wings; (**E**) labels. (Images: A. Taeger).

3.4.4. *Acerataspis fukienensis* Chao, 1957 (Figures 11–15)

Acerataspis fukienensis Chao, 1957, 7:105 [27]. Holotype female from China.

Materials examined. Holotype, 1♀, Label "HOLOTYPE; Shaowu, Fukien/China 1945.VIII-20/H.F. Chao Coll.; *Acerataspis fukienensis*/Chao, sp. nov./Holotype" (IOZ). **Other materials:** CHINA: 1♀1♂, Yunnan Province, Xishuangbanna, Menghai, Menghun Town, Area E1, rain forest, 21°44.944′ N 100°26.182′ E, 1677 m, 25.VI–15.VII.2021, Li Ma leg., Malaise trap, SCBG-E0000045, SCBG-E0000048 (SCAU).

Diagnosis. Area superomedia of propedeum medially with a transverse carina which divides the area into two parts: anterior transverse carina joining area superomedia at middle. Fore wing with 1cu-a opposite M&RS (Figure 12A), areolet petiolate, T4 with a pair of carinae on anterior half (Figure 12D).

Description. Female (holotype) (Figure 11). Body length 9.0 mm, antenna length 9.0 mm.

Head. Face and clypeus densely punctate, punctures on central area more or less transverse. Mesosoma. Pronotum with dorsal 0.3 strongly punctate, lower part polished with several transverse short wrinkles. Epomia strong. Mesoscutum strongly punctate. Scuto-scutellar groove with 3short longitudinal carinae. Scutellum irregularly and coarsely punctate. Propedeum with lateromedian longitudinal carinae nearly parallel; anterior transverse carina joining area superomedia at middle; area superomedia separated from area basalis by the median abscissa of anterior transverse carina; area petiolaris punctate. Mesopleuron densely and coarsely punctate, and setose. Metapleuron polished, with minute punctures, setose, juxtacoxal carina weak.

Wings. Fore wing (Figure 12A) with 1cu-a opposite M&RS; areolet petiolate, receiving 2m-cu before middle. Hind wing with nervellus interrupted at lower 0.4, distal abscissa of CU weakly pigmented.

Legs. Fore and hind tarsal claw with three pectinae (Figure 12B,C).

Metasoma. Coarsely and densely punctate. T1–T3 with a pair of paralleled lateromedian longitudinal carinae. T1 and T2 rugose punctate, with central area between median carinae irregularly rugose. Lateromedian longitudinal carinae of T2 connected by a transverse carina beyond middle. T3 with a weak carina between the median carinae, punctate, transverse, T4 with a pair of lateromedian longitudinal carinae on anterior half (Figure 12D).

Colour (Figure 11). Antenna reddish brown, with apical flagellomeres blackish brown; scape largely black with apex dorsally reddish brown, ventrally yellow; pedicel basally black, apically reddish brown. Face largely yellow, with lower 0.3 black. Mandible brown. Tegula blackish brown. Subtegular ridge reddish brown. Scutellum black with apex yellowish brown. Fore coxa blackish brown, trochanter and femur dark brown (with ventral side lighter), tibia and tarsus yellowish brown. Mid and hind legs blackish brown, tarsus dark brown, spurs yellowish white. T1–T3 black, T4 with apical medially interrupted yellow band, T5 with apical margin dorsally medially yellow. Hind leg entirely blackish. Wings hyaline, pterostigma and veins blackish brown.

Variation (n = 1): Scutellum black with yellowish-brown mark on apex to entirely black.

Male (Figures 13–15). Body length 7.4 mm, fore wing length 5.5 mm.

Head. Face rugose punctate (Figure 13B), 1.3 times as high as wide. Frons rugulose dorsally and transversely wrinkled above antennal sockets. Malar space 0.5 times as mandibular basal width. POL:OD:OOL = 7:7:6. Antenna with 47 flagellomeres, distal 12 flagellomeres with sensillar plates on dorsal side, apical segment of antenna 2.0 times as long as its basal width.

Mesosoma. Pronotum (Figure 14A) in lateral view dorsally and posteriorly strongly punctate, centrally anteriorly with a small glabrous patch, and with several transverse wrinkles on lower part. Mesoscutum (Figure 14B) strongly and coarsely punctate. Scuto-scutellar groove with 3–4 short carinae. Scutellum strongly and coarsely punctate, with lateral carinae reaching to the apex of scutellum. Metanotum slightly convex with an M-shaped wrinkle. Mesopleuron (Figure 14A) strongly convex above sternaulus, strongly rugose punctate; epicnemial carina reaching to upper 0.6 of mesopleuron. mesopleural furrow glabrous. Metapleuron slightly convex, centrally punctulate, juxtacoxal carina fine and complete. Propodeum short (Figure 14C), weakly convex, area externa and area dentipara strongly rugose; area superomedia rugulose, weakly arched at middle, 0.9 times as long as wide, receiving anterior transverse carina near middle; area lateralis rugose; lateral longitudinal carina with anterior abscissa absent, spiracle round.

Wings. Fore wing (Figure 14D) with 1cu-a opposite M&RS, areolet sessile, and receiving 2m-cu before middle, hind wing with Cu & cu-a vein weakly interrupted at lower 1/3, distal abscissa of CU weakly pigmented with proximal section spurious.

Legs. Tarsal claws (Figure 15A,B) strongly pectinate, fore claw with 10 pectinae, mid and hind claws with 7–9 pectinae. Middle tibia with two spurs (Figure 15C). Hind femur 5.0 times as long as maximum width, the longer spur of hind tibia 0.5 times as long as basitarsus.

Metasoma (Figure 14F). Strongly and coarsely punctate. T1–T4 with a pair of median longitudinal carinae. Median longitudinal carinae of T1 and T2 reaching to posterior margin of T1 and T2, and closed posteriorly, median longitudinal carinae of T3 reaching to posterior 0.8 of tergite and posteriorly open; median longitudinal carinae of T4 fine and reaching to posterior 0.6 of T4, and with longitudinal wrinkles in between; area of median longitudinal carinae of T1–T3 irregularly sculptured. T5 and T6 strongly and closely punctate.

Colour. Body mainly black (Figure 13A), covered with white setae. Face largely yellow with lower 0.25 black. Mandible black. Palpi black. Antenna ventrally yellowish brown and dorsally blackish brown. Scutellum yellowish orange. Tegula blackish brown, mediolaterally tinged with yellowish brown. Subtegular ridge orange yellow. Metasoma black with orange marks on T4 and T5, T4 posteriorly with a irregular transverse yellowish orange band (posterior margin of T4 black), T5 with a narrow orange transverse band posteriorly. Fore leg blackish brown, trochanter with apical margin brownish yellow; femur anteriorly brownish yellow (proximal base of femur blackish brown), posteriorly black with proximal apex brownish yellow. Mid and hind legs black, with spurs light blackish brown. Wings hyaline, veins and pterostigma blackish brown.

Distribution. Oriental region.

Comments. Chao [26] stated that the holotype of *Acerataspis fukienensis* was deposited in 'Fukien Agricultural College' (=Fujian Agricultural and Forestry University), but the type materials of this species are currently deposited in the Institute of Zoology, Chinese Academy of Science (Beijing).

Differential diagnosis. This species is similar to *A. sinensis* and *A. szechuanensis* Chao,1962, but can separated from these species by: fore wing with 1cu-a opposite M&RS (fore wing of *A. sinensis* and *A. szechuanensis* with 1cu-a distinctly postfurcal to M&RS); areolet petiolate (areolet of *A. sinensis* and *A. szechuanensis* sessile); fore and hind claws with three stout pectinae (with six fine pectinae in *A. sinensis* and *A. szechuanensis*); T4 with a pair of carinae on anterior half (T4 of *A. sinensis* without carina, of *A. szechuanensis* with a single short carina anteriorly); T4 with apical yellow narrow band medially interrupted (T4 of *A. sinensis* and *A. szechuanensis* with apical yellow band complete and largely broadly yellow); anterior transverse carina of propodeum joining area superomedia near middle (both *A. sinensis* and *A. szechuanensis* with anterior transverse carina joining area superomedia near posterior transverse carina).

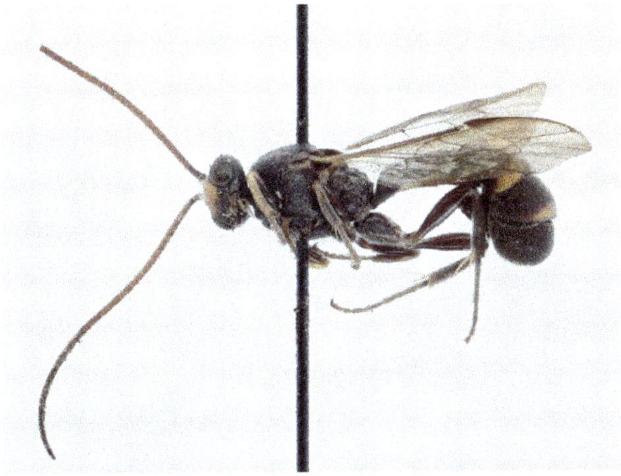

Figure 11. *Acerataspis fukienensis* Chao, holotype female, habitus, lateral. (Image: Huang Zhengzhong).

Figure 12. *Acerataspis fukienensis* Chao, holotype female. (**A**) Fore wing; (**B**) fore tarsal claw; (**C**) hind tarsal claw; (**D**) metasoma, dorsal view; (**E**) labels (Image: Huang Zhengzhong).

Figure 13. *Acerataspis fukienensis* Chao (non-type specimen ID: SCBG-E0000045), male. (**A**) habitus, lateral view; (**B**) head, frontal view.

Figure 14. *Acerataspis fukienensis* Chao (non-type specimen ID: SCBG-E0000045), male. (**A**) Head and mesosoma, lateral view; (**B**) head and mesosoma, dorsal view; (**C**) scutellum, metanotum and propodeum, dorsal view; (**D**) fore wing; (**E**) hind leg; (**F**) T1–T4, dorsal view.

Figure 15. *Acerataspis fukienensis* Chao (non-type specimen ID: (**A–C**), SCBG-E0000045, (**D–F**), SCBG-E0000048). (**A**) Fore claw of male; (**B**) hind tarsus of male; (**C**) middle tibial spurs of male; (**D**) fore tarsal claw of female; (**E**) hind tarsus of female.

3.4.5. *Acerataspis fusiformis* (Morley, 1913) (Figures 16–20)

Metopius fusiformis Morley, 1913: 268 [1]. Holotye female from Myanmar.

Materials examined. Holotype: 1♀, labelled "Type C.M; Carin Cheba 900–1100 m, L.Fea V XII-88; named by Claude Morley, *Metopius fusiformis*, ♀, ii.1913" (MCSN). Approximate coordinates 19°19′ N 96°48′ E (Fea 1888). Examined by photos provided by Maria Tavano (Museum of Natural History "Giacomo Doria" (Museo Civico di Storia Naturale Giacomo Doria) Genova, Italy). **Non-type materials:** CHINA: 1♀, Hong Kong, Ping Shan Chai, 22°29′8.30″ N 114°11′14.74″ E, alt. 140 m, Malaise trap (M335), 17–31.III.2018, Christophe Barthélémy leg., SCAU-3013715 (HKU); 1♂, Hong Kong, Ping Shan Chai, 22°29′8.30″ N 114°11′14.74″ E, alt. 140 m, Malaise trap (M338), 31.III–14.IV.2018, Christophe Barthélémy leg., SCAU-3013716 (HKU); 1♂, Hainan Province, Nankai Town, LSX-727, 19°4′20.77″ N, 109°22′28.2″ E, HN9, 30.IV–30.V.2020, Malaise trap, SCAU-3011024 (SCAU); 1♀, Hainan Province, Nankai Town, 19°4′43.32″ N,109°23′38.74″ E, HN10, 30.IV–30.V.2020, Malaise trap, Luo Shi-Xiao leg., SCAU-3011028 (SCAU); 1♂, Yunnan, Xishuangbanna Botanical Garden, rubber forest, 21°52′ N, 101°19′ E, alt. 543 m, 16.V–19.VI.2021, Malaise trap, Li Ma leg., SCBG-E0000043 (SCAU). THAILAND: 1♂, Chiang Mai, Mae Taeng, Pa Pae, 19°14′30.6″ N, 98°39′14.1″ E, old sec. forest with *Camellia sinensis* var. *assamica*, Malaise trap (Dara#1), 04.V–25.V.2017 (#5), Monsoon Tea leg. SCAU-3013717 (QSBG).

Diagnosis. Fore wing with 1cu-a opposite M&RS, areolet sessile (Figures 17 and 18). Propodeum with costa joining area superomedia at middle (Figure 19C). Fore tarsal claw of female with three pectinae (Figure 19F). T1–T3 with a pair of median longitudinal carinae, with a weak longitudinal wrinkle in between. T4 without median carina. Antenna with scape entirely yellow. Metanotum yellow. Mid femur black with proximal base and apex yellow. Hind tibia with basal half yellow.

Description. Female. Body length 8.0–11.0 mm, fore wing length 7.0–7.8 mm.

Head. Face rugose punctate (Figure 17B), 1.3 times as high as wide, antenna with 60 flagellomeres. Frons with minute punctures on dorsal half, smooth on lower half.

Mandible with upper tooth longer than lower tooth, outer margin of upper tooth weakly convex. Malar space 0.3 times width of mandibular base. POL:OD:OOL = 10:10:6.

Mesosoma (Figure 19A,B). Pronotum polished and transversely crenulate on lower half, dorsally densely punctate and setose. Epomia weak. Mesoscutum densely and coarsely punctate. Scutellum transverse, shallowly punctate, moderately densely setose. Propodeum (Figure 19C) with anterior transverse carina joining area superomedia on anterior 0.6, area superomedia slightly convergent posteriorly from the meeting point with anterior transverse carina; area externa punctate and setose; area dentipara with sparse punctures, laterally punctate along lateral longitudinal carina; area lateralis rugulose and densely setose; area postero-externa rugulose; spiracle oval and touching pleural carina. Mesopleuron punctate reticulate (Figure 19A), with speculum and posterior margin along mesepisternum glabrous and impunctate, mesopleuron strongly convex above sternaulus; epicnemial carina sharp and well defined on lower 0.4 (below the sternaulus) and obscure on upper 0.6 (the convex part of mesopleuron above sternaulus); sternaulus broadly impressed. Metapleuron sparsely punctulate, setose.

Wings. Fore wing with 1cu-a opposite to M&RS (Figure 19D), areolet sessile and receiving 2m-cu before middle. Hind wing with nervellus interrupted at lower 0.3, distal abscissa of CU weakly pigmented.

Legs. All tarsal claws with three pectinae (Figure 19F–H), fore tarsal claw with the middle pecten longer than the other two pectinae. Longer spur of hind tibia 0.67 times length of basitarsus. Hind femur 4.1 times its maximum width, 1.0 times the length of tibia.

Metasoma. Strongly punctate reticulate. T1–T3 with a pair of lateromedian longitudinal carinae, area between lateromedian longitudinal carinae rugose, or sometimes with irregular longitudinal wrinkles. T1 without dorsolateral carina. T3 with lateromedian longitudinal carinae reaching to posterior 0.88. T4 without lateromedian longitudinal carinae. Anterior lateral corner of T4–T6 around spiracle with a distinct longitudinal impression. Ovipositor 0.72 times as long as hind tibia. Ovipositor sheath slender, sparsely setose.

Colour (Figure 17A). Body black with yellow marks. Face with upper 0.6 yellow and lower 0.3 black. Mandible dark brown with extreme base black. Palpi whitish yellow. Scape yellow with laterally outer side black, pedicel dorsally black and ventrally yellow, flagellum dorsally blackish brown, ventrally fuscous on basal 0.6, and gradually becoming blackish brown on apical 0.3. Pronotum black with a yellow spot on posterior lateral corner. Tegula, parategula, scutellum, subtegular ridge and metanotum entirely yellow. Fore coxa dark brown with apex yellow, fore trochanter yellow, fore femur blackish on basal half and yellow on apical half, fore tibia and tarsus yellow; mid coxa blackish on basal 0.6 and yellow on apical 0.3, mid trochanter yellow, mid femur black with apical 0.2 yellow, mid tibia and tarsus yellow; hind coxa black, hind trochanter basally fuscous and apically yellow, hind femur black with extreme basal 0.1 whitish yellow, hind tibia with proximal 0.6 yellow (except the extreme base black) and apical 0.4 black, hind tarsus with 1–4 segments dark fuscous, 5th segment brownish yellow (Figure 19E). Wings hyaline, wing base yellow, veins and pterstigma blackish brown. T1 laterally largely yellow, basal 0.3 and area between lateromedian longitudinal carinae black, T2 and T3 black with a pair of moderately large yellow marks on posterior lateral corner (extreme apical margin of T2 and T3 black), T4 (except for extreme apical margin black) and T5 apically with a yellow band, T6 black. Ovipositor sheaths fuscous.

Male (Figure 18). Body length 8.6 mm, fore wing length 6.8 mm. Similar to female, except: propodeum shorter than female (Figure 20D); fore and middle tarsal claws densely pectinate, with 9–12 pectinae (Figure 20F), hind claw with 3–4 pectinae (Figure 20G). Scape dorsally black and ventrally yellow. T1–T3 posteriorly with a pair of separated yellow marks (Figure 20H,I). Fore wing with 1cu-a opposite M&RS.

Distribution. Oriental region (newly recorded from Hong Kong, Yunnan and Thailand).

Comments. The head and antenna of the type was separated into two parts. We reviewed historical literature of the field notes of Leonardo Fea, who collected the holotype specimen of *A. fusiformis* during his travels to the Karen Hills of Bristish Burma [27,28].

The Karen Hills, or Kayah-Karen Mountains, is a large highland area in eastern Myanmar stretching from Shan Mountain southward to Tenassurin Mountain, although Fea had defined the Karen Hills as a restricted area between two tributaries of the Sittaung River (Thauk Ye Kupt and Paunglaung rivers) and the Nam Pawn River, a tributary of the Salween River. We reconstructed his sampling sites using his field diaries and supplementary maps [27,28] and estimate the approximate type locality of *A. fusiformis* to be 19°19′ N 96°48′ E.

There are some variations between some female specimens (n = 1, specimen ID: SCAU-3013715), propodeum with area dentipara densely punctate, lateromedian longitudinal carinae subparallel, area petiolaris transversely wrinkled with punctures in between. Yellow marks on T1–T4 larger (Figure 17), T1 largely yellow except for black anterior base, T2 with yellow marks not interrupted at middle (Figure 17). The pecten number of hind tarsal claws is different between left leg and right leg in one specimen (n = 1, specimen ID: SCAU-3011028) from Hainan Province of China (three pectinae on left hind claw and four pectinae on the right one). Variations in male specimens as follows: in SCAU-3013716, T4 with apical yellow band interrupted at middle (Figures 18 and 20H); in No. SCAU-3013717, with 1cu-a postfurcal to M&RS) (Figure 20E).

Due to the brief original description of *A. fusiformis* and similar colour patterns of *A. fusiformis* and *A. clavata*, many specimens of *A. fusiformis* from different regions in previous studies [5,10,13,14,29–31] were incorrectly determined as *A. clavata* (Uchida, 1934). However, their *CO1* genetic distances between *A. fusiformis* and *A. clavata* were 11–12% (Table S2). Therefore, their previous identification should be checked by using the main differential diagnostic characters: fore claw with three visible pectinae, fore wing with 1cu-a opposite to M&RS, and scutellum yellow in *A. fusiformis*; while in *A. clavata*, tarsal claw strongly pectinated, fore wing with 1cu-a postfurcal to M&RS, and scutellum black.

It is necessary to mention both *A. fusiformis* and *A. clavata* were collected in the same sampling sites in Thailand and Hong Kong. In Northern Thailand, the sampling site was in an old secondary *Decteracarpus* forest with *Camellia sinensis* var. *assamica* in the understory. There are similar communities with *Camellia sinensis* var. *assamica* around the type locality of *A. fusiformis* on the Karen Hills at Leonardo Fea's sampling sites [32].

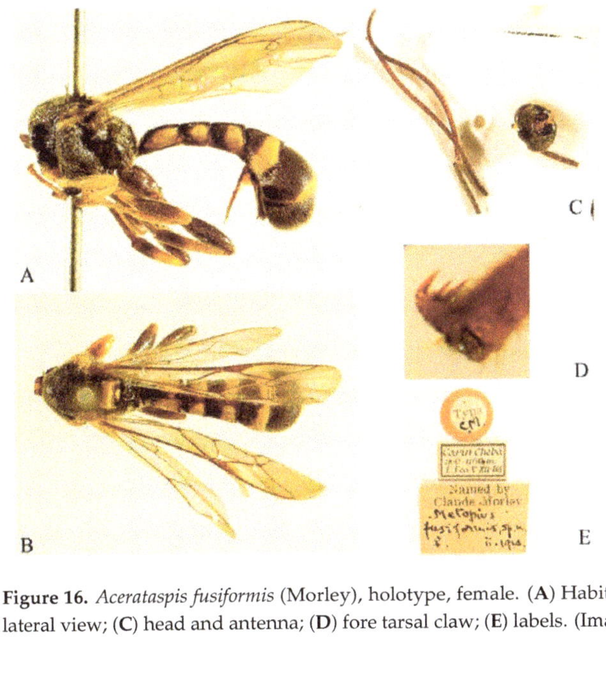

Figure 16. *Acerataspis fusiformis* (Morley), holotype, female. (**A**) Habitus, lateral view; (**B**) habitus, lateral view; (**C**) head and antenna; (**D**) fore tarsal claw; (**E**) labels. (Images: Maria Tavano).

Figure 17. *Acerataspis fusiformis* (Morley) (non-type, specimen ID: SCAU-3013715), female. (**A**) habitus, dorsal view; (**B**) head, frontal view. (Note: Facial colouration was altered after the DNA extraction experiment; the original colour of the species is yellow).

Figure 18. *Acerataspis fusiformis* (Morley) (non-type, specimen ID: SCAU-3013716), male, dorsal view.

Figure 19. *Acerataspis fusiformis* (Morley) (non-type, specimen ID: SCAU-3013715), female. (**A**) Head and anterior part of mesosoma, lateral view; (**B**) head and mesosoma, dorsal view; (**C**) metanotum and propodeum, dorsal view; (**D**) fore wing veins; (**E**) hind leg; (**F**) fore tarsal claw; (**G**) mid tarsal claw; (**H**) hind tarsal claw.

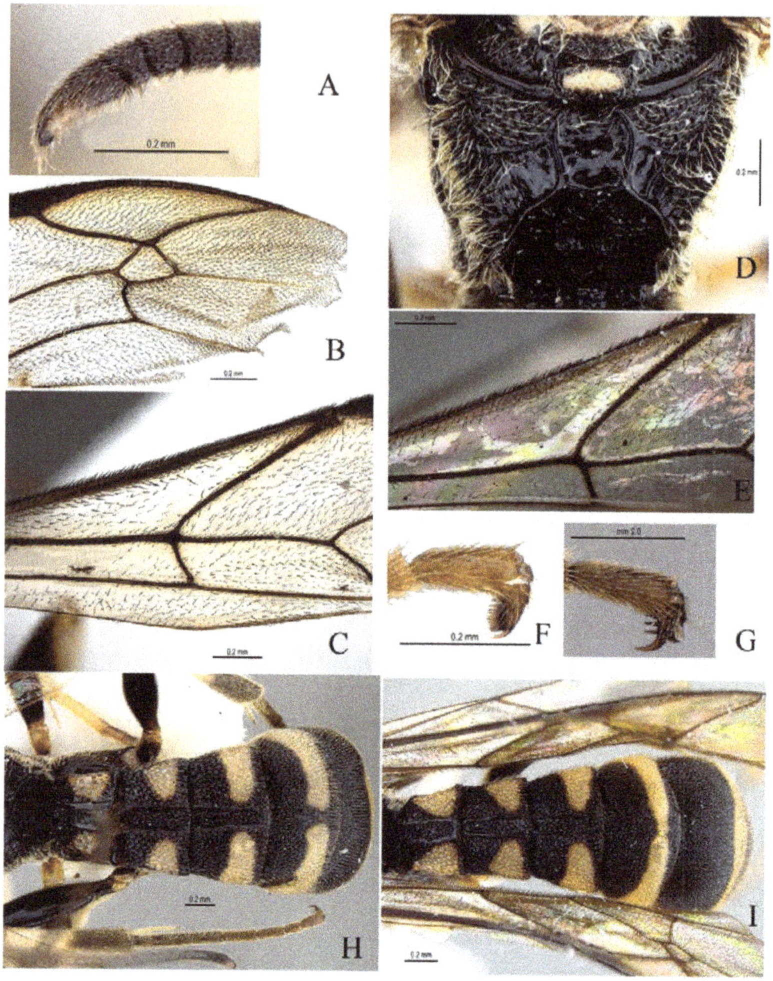

Figure 20. *Acerataspis fusiformis* (Morley) (non-type, specimen ID: SCAU-3013716, SCAU-3013717), male. (**A**) Apical part of flagellum; (**B**,**C**,**E**) fore wing veins ((**B**,**C**), SCAU-3013716, E, SCAU-3013717); (**D**) metanotum and propodeum, dorsal view; (**F**) fore tarsal claw; (**G**) hind tarsal claw; (**H**) metasoma with apical transverse band on T4 separated, dorsal view (SCAU-3013716); (**I**) metasoma with apical transverse band complete (SCAU-3013717).

3.4.6. *Acerataspis maliae* Liu & Chen sp. nov. (Figures 21–23)

Materials examined. Holotype, 1♀, CHINA: Yunnan, Xishuangbanna, Menghai, Menghun Town, alt. 1,621 m, Area E1, rainforest, Malaise trap, 21°44.944′ N 100°26.182′ E, 25.VI–15.VII.2021, Li Ma leg., SCBG-E0000047 (SCAU). Paratype. CHINA: 1♀, Yunnan Province, Xishuangbanna, Menghai, Bulangshan Village, alt. 1706 m, Area A2, forest, 21°45.061′ N 100°21.661′ E, 15.IX–15.X.2021, Malaise trap, Li Ma leg., SCBG-E0000042 (SCAU).

Diagnosis. Fore and mid tarsal claws with four pectinae, hind claw with seven pectinae; fore wing with 1cu-a slightly postfurcal to M&RS, areolet weakly sessile above. Scape yellow with a black stripe laterally (Figure 22A).

Description. Holotype. Female. Body length 7.7 mm, fore wing length 5.7 mm.

Head. Face strongly rugose punctate (Figure 21B), 1.3 times as high as wide. Malar space 0.44 times as long as the basal width of mandible. Frons dorsally densely punctulate, and glabrous above antennal sockets. POL:OD:OOL = 11:9:8. Antenna with 58 flagellomeres.

Mesosoma. Pronotum (Figure 22B) dorsal 0.3 strongly punctate, anteriorly centrally with a glabrous patch, and with several transverse short carinae on lower 0.3. Mesoscutum strongly punctate (Figure 22C), punctures on central area somewhat sparser than those on lateral and posterior areas. Scuto-scutellar groove with three short longitudinal carinae. Scutellum sparsely and coarsely punctate (Figure 22D). Epicnemium anteriorly crenulated. Epicnemial carina reaching to upper 0.5 of mesopleuron. Mesopleuron (Figure 22B) strongly convex above sternaulus, strongly punctate reticulate, mesopleural furrow weakly crenulated. Metapleuron (Figure 23F) sparely punctulate, area between punctures polished, juxtacoxal carina indicated by medially interrupted wrinkles. Propodeum (Figure 22D) with areas rugose punctate, area superomedia weakly wrinkled, polished, as long as its maximum width, and receiving anterior transverse carina at middle; lateral longitudinal carina with anterior abscissa weak; area lateralis rugulose. area postero-externa strongly rugose.

Wings. Fore wing (Figure 23D) with 1cu-a slightly postfurcal to M&RS, areolet slightly sessile above, receiving 2m-cu before middle. Hind wing with nervellus interrupted at middle (Figure 23E), distal abscissa of CU distinct.

Legs. Fore tarsal claw with four pectinae (Figure 23A), mid tarsal claw with four pectinae (Figure 23B), hind tarsal claw with six pectinae (Figure 23C), pectinae of hind tarsal claw of different size (Figure 23C).

Metasoma (Figure 22E,F). Strongly rugose punctate. T1–T3 with a pair of lateromedian longitudinal carinae, T1 transversely wrinkled between cariane, T2 rugulose between carinae and with a wealy defined longitudinal wrinkle between the carinae, T3 centrally with a weak longitudinal carina between the lateromedian longitudinal carinae (Figure 22 E). T4 rugose punctate, medially with irregular longitudinal wrinkles (Figure 22F). Ovipositor sheath 0.58 times as long as hind basitarsomere.

Colour (Figure 21A). Face mostly yellow (centrally brownish yellow) with lower 0.25 black. Mandible basally and apically black, medially brown. Palpi yellowish brown. Antenna dorsally fuscous, ventrally yellowish brown, scape (Figure 22A) yellow with a lateral black stripe, pedicel and base of first flagellomere black. Tegula and subtegular ridge yellowish brown. Scutellum yellow with anterior 0.3 yellowish brown. Metasoma mainly black, T1–T3 posterior laterally with brownish yellow triangular spots, centrally interrupted by the lateromedian longitudinal carinae (Figure 22E,F), T4–T5 posteriorly with a transverse brownish yellow band. Fore leg with coxa black, trochanter with outer side black and inner side brownish yellow, femur fuscous, outer side with basal 1/2 black, tibia and tarsus brownish yellow. Middle leg with coxa, trochanter and femur black, tibia fuscous with basal 0.4 brownish yellow, tarsus dark brown. Hind leg (Figure 21) with coxa, trochanter and femur black, tibia blackish brown with proximal 0.3 brown, tarsus brown. All tibial spurs whitish yellow. Wings hyaline, veins blackish brown. Ovipositor sheath brownish yellow.

Male. Unknown.

Distribution. Oriental region (China).

Etymology. The species is named after Prof. Li Ma, a hymenoptera taxonomist at Yunnan Agricultural University.

Differential diagnosis. The colour of the new species is very similar to *A. clavata* and *A. fusiformis*. It can be separated from *A. clavata* by fore and mid tarsal claws with four pectinae (fore and middle claws with more than six pectinae in *A. clavata*) and scape yellow dorsally (scape dorsally black in *A. clavata*). It differs from *A. fusiformis* by hind tarsal claw with seven pectinae (hind claw with three or four pectinae in *A. fusiformis*). The status of the new species is also supported by *COI* gene sequence; the genetic distances between this species and *A. clavata* and *A. fusiformis* range from 12.9% to 14.4%. (Table S2)

Figure 21. *Acerataspis maliae* Liu & Chen, sp. nov. holotype. (**A**) habitus, lateral view; (**B**) head frontal view. (Note: Facial colouration was altered after the DNA extraction experiment; the original colour of all species is yellow).

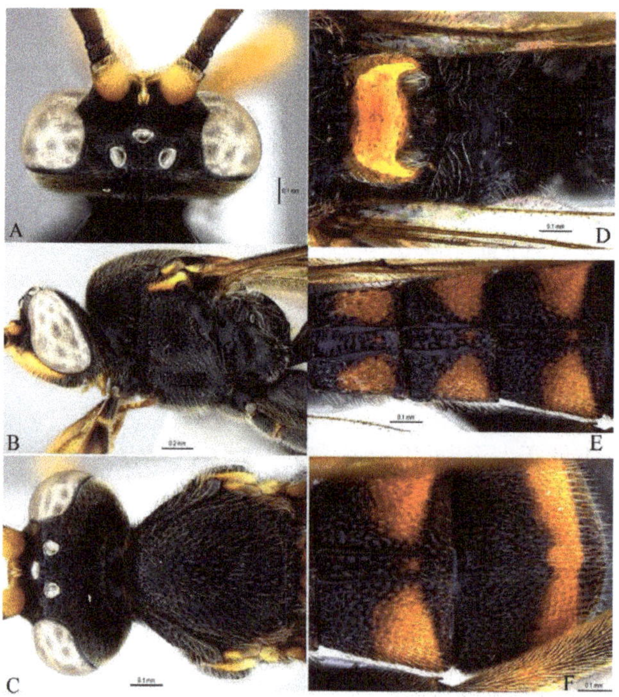

Figure 22. *Acerataspis maliae* Liu & Chen, sp. nov. holotype, female. (**A**) Head, dorsal view; (**B**) head and mesosoma, lateral view; (**C**) head and anterior part of mesosoma, dorsal view; (**D**) scutellum, metanotum and propodeum, dorsal view; (**E**) T1–T4, dorsal view; (**F**) T3–T4 dorsal view.

Figure 23. *Acerataspis maliae* Liu & Chen, sp. nov. holotype, female. (**A**) Fore tarsus; (**B**) middle tarsus; (**C**) hind tarsus; (**D**) fore wing; (**E**) hind wing; (**F**) posterior part of mesopleuron and metapleuron, lateral view.

3.4.7. *Acerataspis separata* Liu, Reshchokv & Chen, sp. nov. (Figures 24–29)

Materials examined. Holotype, 1♀, CHINA: Yunnan Province, Xishuangbanna, Menghai County, Bu Lang Shan Town, grass land, site A. 21°44.981′ N, 100°26.907′ E, alt. 1710 m, 15.VIII.2018, Li Ma leg., Malaise trap, SCAU-3011982 (SCAU). Paratype. 1♂, China: Yunnan Province, Xishuangbanna, Menghai County, Bu Lang Shan Town, grassland, site D, 21°44.745′ N, 100°26.070′ E, alt. 1621 m, Li Ma leg., Malaise trap, 16.VII–14.IX.2018, SCAU-3011978 (SCAU); 4♂, Yunnan Province, Xishuangbanna Botanical Garden, rubber forest, alt. 543 m, 21°912.439′ N 101°26.84′ E, 16.III–11.IV.2021, Li Ma leg., Malaise trap, SCBG-E0000044, SCBG-E0000051, SCBG-E0000052, SCBG-E0000054 (SCAU); 1♂, Yunnan, Xishuangbanna, Menghai, Bulangshan Village, alt. 1,621 m, Area E1, forest, malaise trap, 21°44.944′ N 100°26.182′ E, 15.IX–15.X.2021, Li Ma leg., Malaise trap, SCBG-E0000050 (SCAU).

Diagnosis. Fore wing (Figure 26D) with 1cu-a postfurcal to M&RS, areolet of fore wing sessile above. Propodeum with anterior transverse carina joining area superomedia at middle (Figure 25D). All tarsal claws densely pectinate. T1–T4 with yellow marks posteriorly, T4 with apical transverse band interrupted dorso-medially (Figure 25F).

Description. Female holotype. Body length 5.7 mm, fore wing length 4.3 mm.

Head. Face (Figure 24B) rugose punctate, punctures closed and forming some transverse wrinkles, combined face and clypeus 1.3 times as high as wide. Mandible slightly twisted, with upper tooth longer than lower tooth, mandible with a flange-like projection subasally. Malar space 0.5 times the width of mandibular base. Head 1.6 times as wide as long. Gena strongly narrowed behind eyes. POL:OD:OOL = 6:6:5. Antenna with 47 flagellomeres, first flagellomere 1.8 times as long as its apical width, 1.4 times as long as the second. Occiput strongly punctate and setiferous. Occipital carina complete.

Mesosoma. Pronotum (Figure 25A) with dorsal 0.3 strongly punctate and setiferous; lower 0.6 polished and glabrous, crenulate, with several transverse wrinkles along the posterior margin of pronotum. Epomia weak. Mesoscutum (Figure 25B) densely and closely punctate, and setose; posterior lateral corner of scutum with a lamella flange just behind the tegula that is 1.2 times the length of the tegula. Scuto-scutellar groove deeply crenulate, with four short longitudinal carinae. Scutellum transverse, punctate, with lateral posterior corner projected. Propedeum (Figure 25D) with area superomedia weakly anteriorly curved outward, polished and glabrous; anterior transverse carina joining area superomedia at middle; area externa sparsely punctate and with inner posterior corner polished; area dentipara with outer 0.6 rugose punctate and setose, inner 0.3 polished and glabrous; area lateralis weakly rugose, setose. Spiracle oval, connected to pleural carina. Mesopleuron (Figure 25C) strongly convex, densely and strongly punctate; sternaulus broadly impressed. Epicnemial carina reaching to upper 0.7 of mesopleuron; mesopleuron with a narrow triangular polished area along the position of mesopleural furrow (which is not indicated) (Figure 25C). Metapleuron weakly convex, with minute setiferous punctures, polished in between, anterior margin strongly crenulate, juxtacoxal carina complete, upper side with 1–2 wrinkles. Submetapleural carina complete, anteriorly with a projection.

Wings. Fore wing (Figure 26D) with 1cu-a postfurcal to M&RS, areolet sessile, ratio of 2rs-m and 3rs-m = 10:8, areolet receiving 2m-cu before middle. Hind wing (Figure 26E) with 12 hamuli, nervellus interrupted at lower 0.3, distal abscissa of CU weakly pigmented.

Legs. Fore femur swollen, 2.5 times as long as its maximum width; fore tarsal claw (Figure 26A) densely pectinate, with seven pectinae, basitarsus 0.8 times the combined length of tarsomeres 2–5. Middle femur 3.2 times as long as its maximum width, middle claw densely pectinate (Figure 26B). Hind femur (Figure 25E) 4.0 times as long as its maximum width, the longer tibial spur 0.7 times the length of basitarsus. Hind tarsal claw densely pectinate, teeth of equal size (Figure 26C).

Metasoma (Figure 25F). Strongly punctate. T1–T3 with a pair of latero-median carinae; area between latero-median carinae of T1 weakly transversely rugose, that of T2 weakly rugose, and with a longitudinal wrinkle reaching to posterior 0.8 of T2, that of T3 weakly rugose and with a longitudinal wrinkle reaching to posterior 0.8 of T3. Dorso-lateral carina of T1 absent. T4 densely punctate, medially longitudinally convex, without carina. T1 1.3 times as long as its apical width, T2 0.7 times as long as its apical width, T3 0.5 times as long as its apical width. T5 and T6 strongly punctate reticulate. Ovipositor sheath as long as hind basitarsus.

Colour. Body black with white maculations (Figure 24A). Face with upper 0.8 white (color altered after DNA extraction), lower 0.2 black. Mandible black. Palpi whitish. Antenna black, inner side of scape whitish. Tegula and subtegular ridge white. Scutellum white. Metanotum black. Fore coxa black with anterior side white; trochanter white, dorsally with a brown mark on basal 0.6; fore femur anteriorly white, posteriorly dark brown with apex white; fore tibia white, inner side with basal 0.7 dark brown; tarsus dark brown except for white basitarsus. Middle coxa black with white spot on anterior side; mid trochanter white, dorsally with basal 0.8 dark brown; mid femur black except for proximal base and extreme apex white; mid tibia white, inner side with basal 0.6 dark brown, mid tarsus with basitarsus mostly white (extreme apex dark brown), 2nd–5th tarsomeres dark brown. Hind coxa black, trochanter black with anterior side dark brown, hind femur black with proximal base white; tibia with proximal base and apical half black, basal half white; tarsus black. All tibial spurs white. T1 with a pair of large triangular white marks laterally; T2 with a pair of white marks on lateral posterior half; T3 with a medially interrupted subapical transverse white band; T4 with a medially interrupted subapical transverse white band; T5 with a complete transverse white band subapically. Wings hyaline, veins and pterostigma blackish brown.

Male (Figures 27–29). Similar to female, except: body length 7.1 mm, fore wing length 5.4 mm. Epicnemial carina strong, reaching to upper end of mesopleuron (Figure 28B). Propodeum (Figure 28E) with area dentipara strongly rugose, area superomedia rugulose. Tarsal claws strongly pectinate (Figure 29A–C). White marks on T1–T3 smaller than female (Figure 28F). Scutellum white.

Distribution. Oriental region (Yunnan province of China).

Etymology. The specific name is derived from Latin *separatus*, referring to the separated maculation of T4.

Comments. This species can be separated from other congeneric species mainly by combinations of: T4 with separated maculations, densely pectinate claws, and fore wing with 1cu-a postfurcal to M&RS. This new species and *A.clavata* from Thailand (SCAU-3013718) share the separated maculation of T4, with differences between them as follows: *A. separata* sp. nov. with scape dorsally black (scape dorsally yellow in Thailand *A. clavata*) and metapleuron with three foveae on anterior margin (metapleuron with only two foveae in Thailand *A. clavata*). This new species was also supported by the interspecific distance of the *CO1* gene.

Figure 24. *Acerataspis separata* Liu, Reshchokv & Chen, sp. nov., holotype, female. (**A**) habitus, dorsal view; (**B**) head, frontal view. (Note: Facial colouration was altered after the DNA extraction experiment; the original colour of all species is yellow).

Figure 25. *Acerataspis separata* Liu, Reshchokv & Chen, sp. nov., holotype, female. (**A**) Head and mesosoma, lateral view; (**B**) head and mesosoma, dorsal view; (**C**) mesosoma, lateral view; (**D**) scutellum, metanotum and propodeum, dorsal view; (**E**) hind leg, lateral view; (**F**) metasoma, dorsal view.

Figure 26. *Acerataspis separata* Liu, Reshchokv & Chen, sp. nov., holotype, female. (**A**) Fore tarsus; (**B**) middle tarsus; (**C**) hind tarsus; (**D**) wings; (**E**) hind wing.

Figure 27. *Acerataspis separata* Liu, Reshchokv & Chen, sp. nov., paratype, male, habitus, dorsal view.

Figure 28. *Acerataspis separata* Liu, Reshchokv & Chen, sp. nov., paratype, male. (**A**) Head and anterior part of mesosoma, lateral view; (**B**) mesosoma, lateral view; (**C**) head and mesosoma, dorsal view; (**D**) wings; (**E**) posterior part of scutellum, metanotum and propodeum, dorsal view; (**F**) metasoma, dorsal view.

Figure 29. *Acerataspis separata* Liu, Reshchokv & Chen, sp. nov., paratype, male. (**A**) Fore tarsus; (**B**) middle tarsus; (**C**) hind tarsus; (**D**) T6 and paramere, lateral view.

3.4.8. *Acerataspis similis* Liu, Reshchokv & Chen, sp. nov. (Figures 30–32)

Materials examined. Holotype, 1♀, CHINA: Yunnan Province, Xishuangbanna, Menghai County, Bulangshan Town, grass, 21°44.521′ N, 100°26.647′ E, alt. 1646 m. 16.VIII–14.IX. 2018, Malaise trap, Li Ma leg., SCAU-3011981 (SCAU).

Diagnosis. This species can be distinguished from other species by the following combined characters: tarsal claws (Figure 32A–C) weakly and sparsely pectinate, fore tarsal claw with three pectinae, mid tarsal claw with four pectinae, hind tarsal claw with four pectinae (basal two slightly bigger than the apical two); fore wing with 1cu-a opposite M&RS (Figure 31D), areolet sessile and receiving 2m-cu before middle; metanotum black; hind leg black, hind tibia with a narrow whitish band (Figure 31E); T4 without carina (Figure 31F).

Description. Female, holotype (Figure 30A), body length 8.3 mm, fore wing length 6.3 mm.

Head. Face rugose punctate (Figure 30B), punctures close, some forming transverse wrinkles; moderately densely setose; combined face and clypeus 1.5 times as high as wide. Mandible slightly twisted, with upper margin longitudinally convex, upper tooth longer than the lower one. Malar space 0.5 times as long as the mandibular basal width. Head in dorsal view 2.2 times as wide as long. Temple very short, strongly narrowed behind eyes, 0.2 times as long as eye in dorsal view. POL:OD:OOL = 10:10:6. Vertex sloping vertically from posterior ocellus. Antenna with 58 flagellomeres, first flagellomere 1.1 times as long as its apical wide, 1.25 times as long as the second flagellomere.

Mesosoma. Pronotum (Figure 31A) with dorsal 0.3 strongly coarsely punctate and setose; lower 0.6 polished and glabrous, with several transverse carinae along the posterior margin (crenulate). Mesoscutum (Figure 31B) coarsely and closely punctate, lateral side of scutum with punctures closed and more or less forming wrinkles. Scuto-scutellar groove deeply crenulate, medially with three short longitudinal carinae. Scutellum transverse, punctate, lateral posterior corner with a sharp projection. Propodeum with area superomedia rectangular, 1.1 times as long as wide, anterior half weakly rugose, posterior half polished; anterior transverse carina joining area superomedia near middle; lateromedian longitudinal carinae subparallel; area externa with setiferous punctures, area dentipara laterally weakly rugose and setose; lateral longitudinal carina complete and strong; area spiracularis convered with long setae, pleural carina complete. Spiracle oval, connected to pleural carina. Mesopleuron (Figure 31C) strongly convex, densely and coarsely punctate, sternaulus broadly impressed. Epicnemial carina reaching to upper 0.7 of mesopleuron, with anterior end separated from the anterior margin of mesopleuron; mesopleuron posteriorly with a narrow longitudinal polished area along mesopleural furrow; mesoplerual furrow smooth, without fovea. Metapleuron (Figure 31C) convex, shiny, with fine setiferous punctures, anterior margin coarsely crenulate; juxtacoxal carina weak, interrupted on anterior 0.3. Submetapleural carina with an anterior flange.

Wings. Fore wing (Figure 31D) with 1cu-a, weakly inclivous, opposite to M&RS; 2r &RS slightly arched just behind the junction with pterostigma; 2rs-m:3rs-m = 13:11; areolet sessile and receiving 2m-cu before middle; 1/Cu:2cu-a = 16:7. Hind wing with nervellus interrupted at lower 0.3, distal abscissa of CU weakly pigmented.

Legs. Fore femur swollen, 1.88 times as long as its maximum width; fore tarsal claw sparsely pectinate, with three or four pectinae (Figure 32A), second to fifth tarsomeres ventrally flattened, shortly and densely pubescent. Middle femur 3.3 times as long as its maxium width, middle tarsal claw with four pectinae (Figure 32B). Hind femur (Figure 31E) 5.0 times as long as its maximum width, longest spur of hind tibia 0.63 times the length of basitarsus. Hind claw with four pectinae (Figure 32C), the basal two slightly longer and stronger than the apical two pectinae.

Metasoma. Strongly punctate. T1–T3 (Figure 31F) with a pair of latero-median carinae, area between latero-median carinae of T1 rugose punctate, that of T2 irregularly longitudinally wrinkled, that of T3 longitudinally rugose. Dorso-lateral carina of T1 absent behind spiracle. T1 as long as apical width, T2 0.88 times as long as apical width. T3 0.62 times as long as apical width, basal width:apical width = 8:13. T4 strongly punctate, transverse. Ovipositor sheath 0.73 times the length of hind basitarsus.

Colour. Body black with yellow maculation (Figure 30A). Face with upper 0.6 yellow (colour altered following DNA extraction), with lower 0.3 black. Mandible blackish brown with teeth apically lighter. Palpi yellowish brown. Antenna ventrally fuscous and dorsally blackish brown, scape yellow, laterally blackish, pedicel and annellus blackish. Tegula brown. Mesoscutum with posterior lateral corner yellow. Scutellum reddish brown. Subtegular ridge reddish. Wings hyaline, veins and pterostigma blackish brown. T1–T3 each with a pair of yellow spots subapically, the yellow maculation of T1 and T2 weakly connected medially, that of T3 interrupted dorso-medially (Figure 31F). T4 and T5 subapically with a yellow band. All coxae black (extreme apex of fore coxa brown); fore leg with trochanter dark brown, femur dark brown with apex yellowish brown, tibia and tarsus yellowish brown; middle leg with trochanter blackish brown, femur black, tibia blackish brown with basal 0.2 and apex yellowish brown; tarsus fuscous. Hind leg black, (except tibia subasally with a yellowish-brown band). Ovipositor sheath yellowish brown.

Distribution. Oriental region (China, Yunnan Province).

Differential diagnosis. This new species differs from *A. clavata* by fore wing with 1cu-a opposite to M&RS (postfurcal to M&RS of *A. clavata*), fore tarsal claw with 3–4 pectinae (with 8–10 pectinae in *A. clavata*). The new species is also quite similar to *A. fusiformis*, different in hind tarsal claw with four pectinae (hind tarsal claw with three pectinae in *A. fusiformis*) and black metanotum (metanotum yellow in *A. fusiformis*).

Etymology. The species name is derived from the Latin *similis*, referring to its colour pattern that is similar to *A. clavata* (Uchia, 1934).

Figure 30. *Acerataspis similis* Liu, Reshchokv & Chen, sp. nov. holotype, female. (**A**) habitus lateral view; (**B**) head, frontal view. (Note: Facial colouration was altered after the DNA extraction experiment; the original colour of all species is yellow).

Figure 31. *Acerataspis similis* Liu Reshchokv & Chen, sp. nov., holotype, female. (**A**) Head and anterior part of mesosoma, lateral view; (**B**) head and mesosoma, dorsal view; (**C**) mesosoma, lateral view; (**D**) fore wing; (**E**) hind leg; (**F**) metasoma, dorsal view.

Figure 32. *Acerataspis similis* Liu Reshchokv & Chen, sp. nov., holotype, female. (**A**) Fore tarsal claw; (**B**) mid tarsal claw; (**C**) hind tarsal claw.

3.4.9. *Acerataspis sinensis* Michener, 1940 (Figures 33 and 34)

Acerataspis sinensis Michener, 1940, 123 [9]. Holotype from China (CAS).

Materials examined. Holotype, 1♀, labelled "Yim Na San E. Kwangtung, S. China VI-14–36; HOLOTYPE, *Acerataspis sinensis,* C. D. Michener" (CAS). **Other materials:** 1♀, CHINA: Zhejiang Province, West Mt. Tianmushan, 1982, Rui-liang Wang, No. 825,916 (ZJUH). labeled "*Acerataspis sinensis* Michener/He, 1983.XI.8".

Diagnosis. Scuto-scutellar groove crenulate, with three median longitudinal carina, the central one strong. Scutellum strongly punctate (Figure 34A). Propodeum with the anterior transverse carina joining the area superomedia at apex. Fore wing with 1cu-a postfurcal to M&RS (Figure 34D); areolet slightly sessile (Figure 34B). Tarsal claws strongly pectinate, with 6–7 pectinae (Figure 34F). T1–T3 (Figure 34C) with a pair of median longitudinal carinae, area between carinae rugose punctate. T4 without longitudinal carina.

Distribution. Oriental and Palearctic regions [6].

3.4.10. *Acerataspis szechuanensis* Chao, 1962 (Figures 35 and 36)

Acerataspis szechuanensis Chao, 1962: 165 [3]. Holotype from China (IOZ).

Materials examined. Type materials, label (in Chinese) "Sichuan, E'mei Shan Qingyin Ge, 800–1000 m, 19857.IV.22; HOLOTYPE; *Acerataspis szechuanensis*, Chao ♀, Holotype" (IOZ).

Diagnosis. Propodeum with anterior transverse carina joining area superomedia at apex. Area between median longitudinal carina of T2 and T3 with an irregular weak carina (Figure 36A), T4 basally with a very short carina or wrinkle (Figure 36B). Fore tarsal claw with 6–7 pectinae (Figure 36C). Hind leg entirely blackish brown (Figure 35), hind tibia without white band basally.

Distribution. Oriental region (China, Sichuan Province).

Differential diagnosis. This species is very similar to *A. sinensis* in colour pattern; it is only different from *A. sinensis* in having scutellum black (yellow in *A. sinensis*) and recticulate punctation (normal punctuation in *A. sinensis*). It is probably a variant of *A.*

sinensis; however, as there is no molecular data for either species, the status of *Acerataspis szechuanensis* is reserved.

Figure 33. *Acerataspis sinensis* Michener, holotype, female, habitus, lateral view (Image: Robert Zuparko).

Figure 34. *Acerataspis sinensis* Michener, holotype, female. (**A**) Head and mesosoma, dorsal view; (**B**) apical part of fore wing; (**C**) metasoma, dorsal view; (**D**) wings; (**E**) hind tarsal claw; (**F**) labels (Images: Robert Zuparko).

Figure 35. *Acerataspis szechuanensis* Chao, holotype, female, lateral view (Image: Huang Zhengzhong).

Figure 36. *Acerataspis szechuanensis* Chao, holotype, female. (**A**) Habitus, dorsal view; (**B**) T4 and T5, dorsal view; (**C**) fore tarsal claw, ventral view; (**D**) labels (Image: Huang Zhengzhong).

4. Discussion

Some studies have suggested that ichneumonid wasps have some characteristics— such as frequently occurring genetic introgression caused by *Wolbachia* endosymbionts and the low evolutionary rate of the *CO1*-marker—that make the relation between morphospecies and molecular species delimitation more complicated than in other Hymenoptera [33–35]. Therefore, the combination of morphological, molecular and some other characters have been proposed as a reliable method for species delimitation. In this study, we revised the species of *Acerataspis* using an integrative approach combining morphology and DNA barcoding techniques. In general, the morphological and molecular data complemented each other quite well in delimiting species. Compared with the *CO1* sequences of *A. clavata* newly generated in this study, the two sequences labeled as *A. clavata* downloaded from GenBank have an 8% basepair difference, suggesting that these two sequences may represent a different species or a haplotype given that these two sequences—together with the newly generated sequences—form a monophyletic clade (Figure 1). Similar situations where sequence data in public databases (e.g., Genbank and BOLD) might be based on misidentified specimens have been found in the genus *Enicospilus* (Ichnumonidae) [35].

In previous studies, the following characters were mainly used to separated different species in the key by various authors: the lateromedian longitudinal carinae of T1–T3 with or without an additional longitudinal carina [3,4,8,9,13,28]; the position where the anterior transverse carina of propodeum joins the area superomedian of propodeum; whether there is a pair of lateromedian longitudinal carinae on T4; the colour patterns of T1–T3; and the hind tibia is entirely black or black with a yellow band. However, besides the characters mentioned above, we found some additional morphological characters overlooked by previous studies that are also useful for the identification of *Aceratаspis* species, especially in females: (1) the number of pectinae of claw pectination; (2) position between 1cu-a and M&RS of fore wing (opposite or postfurcal); (3) form of areolet in fore wing (sessile or petiolate); (4) apical yellow or white band of T4 complete or interrupted medially.

Supplementary Materials: The following supporting information can be downloaded at: https://www.mdpi.com/article/10.3390/insects14040389/s1, Table S1: Interspecific pairwise distance of *Aceratаspis* based on *COI* sequences; Table S2: Genetic distance of *CO1* within species under K2P model.

Author Contributions: Conceptualization, J.-X.L. and H.-Y.C.; methodology, all authors; software, J.-X.L. and H.-Y.C.; investigation, all authors; data curation, all authors; writing—original draft preparation, all authors; writing—review and editing, all authors; supervision, J.-X.L. and H.-Y.C.; funding acquisition, J.-X.L. and A.R. All authors have read and agreed to the published version of the manuscript.

Funding: This research was funded by the Natural Science Foundation of Guangdong Province (No. 2021A15150730) for Jing-Xian Liu and Yunnan Intelligence Union Program (202203AM140015) for Alexey Reshchikov.

Data Availability Statement: All data are available in this paper.

Acknowledgments: We are grateful to the following colleagues for providing photos of type material: Kuiyan Zhang, Hongbin Liang, Hong Liu and Zhengzhong Huang (IOZ), Andreas Taeger (DEI), Maria Tavano (MCSN), Christopher Grinter, Rachel Diaz-Bastin (CAS), Chi-Feng Lee (TARI), and Hege Vårdal (NHRS) for providing label data of *Aceratаspis*. Many thanks to Chen Xue-xin, Tang Pu and Han Yuanyuan (Zhejiang University, China), Qiang Li and Li Ma (Yunnan Agricultural University), Wichai Srisuka (QSBG) and Christophe Barthélémy (Hong Kong) for provided materials. We are deeply grateful to Kenneth Rimdahl (the Monsoon Tea Company) for assistance in sampling in Northern Thailand during the Tea Fauna Project (https://www.teafauna.com).

Conflicts of Interest: The authors declare no conflict of interest.

References

1. Morley, C. *The Fauna of British India including Ceylon and Burma, Hymenoptera, Vol. 3. Ichneumonidae*; British Museum: London, UK, 1913; p. 531.
2. Townes, H.K.; Townes, M. Ichneumon-flies of America north of Mexico: 1 Subfamily Metopiinae. *Bull. Am. Mus. Nat. Hist.* **1959**, *216*, 1–318. [CrossRef]
3. Chao, H.F. Description of a new species of the genus Aceratаspis Uchida from Szechuan Province, China (Hymenoptera: Ichneumonidae). *Acta Entomol. Sin.* **1962**, *11*, 165–168.
4. Chiu, S.C. The Taiwan Metopiinae (Hymenoptera: Ichneumonidae). *Bull. Taiwan Agric. Res. Inst.* **1962**, *20*, 1–37.
5. He, J.H.; Chen, X.X.; Ma, Y. *Hymenoptera: Ichneumonidae, Economic Insect Fauna of China*; Science Press: Beijing, China, 1996; p. 697, (In Chinese with English Summary).
6. Yu, D.S.; van Achterberg, C.; Horstmann, K. *Taxapad 2016. World Ichneumonoidea 2015. Taxonomy, Biology, Morphology and Distribution*; Taxapad Database on Flash-Drive: Vancouver, BC, Canada, 2016.
7. Spasojevic, T.; Broad, G.R.; Klopfstein, S. Revision of 18 ichneumonid fossil species (Hymenoptera, Ichneumonidae) highlights the need for open nomenclature in palaeontology. *Foss. Rec.* **2022**, *25*, 187–212. [CrossRef]
8. Cushman, R.A. H. Sauter's Formosa-collection: Ichneumonidae. *Arb. Über Morphol. Und Taxon. Entomol.* **1937**, *4*, 283–311.
9. Michener, C.D. A synopsis of the genus Aceratаspis (Hymenoptera, Ichneumonidae). *Psyche* **1940**, *47*, 121–124. [CrossRef]

10. Sheng, M.L.; Sun, S.P.; Ding, D.S.; Luo, J.G. *Ichneumonid Fauna of Jianxi (Hymenoptera: Ichneumonidae)*; Science Press: Beijing, China, 2013; pp. 1–569, (In Chinese with English Summary).
11. Huang, Q.T.; Sun, S.P.; Li, T. A new Chinese record of genus Acerataspis (Hymenoptera, Ichneumonidae) from Guangxi Zhuang Autonomus Region. *South China For. Sci.* **2019**, *47*, 67–70.
12. Uchida, T. Acerataspis nom. nov. (Hym. Ichneum. Metopiinae). *Insecta Matsumurana* **1934**, *9*, 23.
13. Momoi, S. On the species of Acerataspis Uchida occurring in Japan (Hymenoptera, Ichneumonidae). *Insecta Matsumurana* **1960**, *23*, 52–54.
14. Tolkanitz, V.I. Metopiinae. In *Key to the Insects of Russia Far East*; Neuropteroidea, Mecoptera, Hymenoptera; Lelej, A.S., Ed.; Dalnauka: Vladivostok, Russia, 2007; Volume IV, pp. 638–667. (In Russian)
15. Sheng, M.L.; Sun, S.P. *Ichneumonid Fauna of Liaoning*; Science Press: Beijing, China, 2014; p. 464, (In Chinese with English Summary).
16. Quicke, D.; Laurenne, N.M.; Fitton, M.G.; Broad, G.R. A thousand and one wasps: A 28S rDNA and morphological phylogeny of the Ichneumonidae (Insecta: Hymenoptera) with an investigation into alignment parameter space and elision. *J. Nat. Hist.* **2009**, *43*, 1305–1421. [CrossRef]
17. Broad, G.R.; Shaw, M.R.; Fitton, M.G. Ichneumonid wasps (Hymenoptera: Ichneumonidae): Their classification and biology. In *Handbooks for the Identification of British Insects*; Field Studies Council: Shrewsbury, UK, 2018; Volume 7, pp. 1–418.
18. Quicke, D.L.J. *The Braconid and Ichneumonid Parasitoid Wasps: Biology, Systematics, Evolution and Ecology*; Wiley Blackwell: Hoboken, NJ, USA, 2015; p. 553.
19. Zheng, X.F.; Reshchikov, A.; Liu, J.X. Two new species of Drepanoctonus Pfankuch, 1911 (Hymenoptera, Ichneumonidae) from the Oriental region. *J. Hymenopt. Res.* **2021**, *83*, 1–19. [CrossRef]
20. Reshchikov, A.; Santos, B.F.; Liu, J.X.; Barthélémy, C. Review of Palpostilpnus Aubert (Hymenoptera, Ichneumonidae, Phygadeuontinae), with the description of ten new species. *Eur. J. Taxon.* **2019**, *582*, 1–63. [CrossRef]
21. Folmer, O.; Black, M.; Hoeh, W.; Lutz, R.; Vrijenhoek, R. DNA primers for amplification of mitochondrial cytochrome C oxidase subunit I from diverse metazoan invertebrates. *Mol. Marine Biol. Biotechnol.* **1994**, *3*, 294–299.
22. Tamura, K.; Stecher, G.; Kumar, S. MEGA11: Molecular evolutionary genetics analysis version 11. *Mol. Biol. Evol.* **2021**, *38*, 3022–3027. [CrossRef] [PubMed]
23. Kalyaanamoorthy, S.; Minh, B.Q.; Wong, T.K.F.; von Haeseler, A.; Jermiin, L.S. ModelFinder: Fast model selection for accurate phylogenetic estimates. *Nat. Methods* **2017**, *14*, 587–589. [CrossRef] [PubMed]
24. Uchida, T. Eine neue Gattung und eine neue Art der Unterfamilie Metopiinae (Hym. Ichneum.). *Trans. Sapporo Nat. Hist. Soc.* **1934**, *13*, 275–277.
25. Kusigemati, K. Taxonomic studies on the subfamiliy Metopiinae of Japan (Hymenoptera: Ichneumonidae). *Mem. Fac. Agric. Kagoshima Univ.* **1971**, *8*, 205–298.
26. Chao, H.F. Records of Ichneumon-flies from Fukien Province with description of a new species (Hymenoptera: Ichneumonidae). *Acta Entomol. Sin.* **1957**, *7*, 105–112, (In Chinese with English Summary).
27. Fea, L. Net Carin Indipendenti (con una Cartina nel testo e due disegni). *Strat. Dal Boll. Della Soc. Geogr. Ital.* **1888**, 1–15.
28. Gestro, A.; Fea, L. Annali del Museo Civico di Storia Naturale di Genova, Third Series 1. *Ann. Mus. Civ. Stor. Nat. Giacomo Doria* **1904**, *21–40*, 95–152.
29. Takahashi, M. Ichenumonoidea of Hachijo-jima Island, Izu Islands. *Gensei* **1995**, *67*, 7–8.
30. He, J.H.; Chen, X.X.; Fan, J.J.; Li, Q.; Liu, C.M.; Lou, X.M.; Ma, Y.; Wang, S.F.; Wu, Y.R.; Xu, Z.H.; et al. *Hymenopteran Insect Fauna of Zhejiang*; Science Press: Beijing, China, 2004; pp. 1–1373, (In Chinese with English Summary).
31. Watanabe, K.; Takeo, Y. Records of ichneumonid wasps (Hymenoptera) from Yakushima Island, the Ryukyu Islands, Japan. *Jpn. J. Syst. Entomol.* **2014**, *20*, 81–90.
32. Cubey, R.; Royal Botanic Garden Edinburgh Herbarium (E). Royal Botanic Garden Edinburgh. Occurrence Dataset. Available online: https://www.gbif.org/occurrence/1038281488 (accessed on 20 December 2022).
33. Klopfstein, S.; Kropf, C.; Baur, H. Wolbachia endosymbionts distort DNA barcoding in the parasitoid wasp genus Diplazon (Hymenoptera: Ichneumonidae). *Zool. J. Linn. Soc.* **2016**, *177*, 541–557. [CrossRef]
34. Johansson, N.; Cederberg, B. Review of the Swedish species of Ophion (Hymenoptera: Ichneumonidae: Ophioninae), with the description of 18 new species and an illustrated key to Swedish species. *Eur. J. Taxon.* **2019**, *550*, 1–136.
35. Shimizu, S.; Broad, G.R.; Maeto, K. Integrative taxonomy and analysis of species richness patterns of nocturnal Darwin wasps of the genus Enicospilus Stephens (Hymenoptera, Ichneumonidae, Ophioninae) in Japan. *ZooKeys* **2020**, *990*, 1–144. [CrossRef]

insects

MDPI

Article

Nematode and Strepsipteran Parasitism in Bait-Trapped and Hand-Collected Hornets (Hymenoptera, Vespidae, *Vespa*)

Natsumi Kanzaki [1,*], Shun'ichi Makino [2], Hajime Kosaka [3], Katsuhiko Sayama [4], Keiko Hamaguchi [1] and Shinji Narayama [1]

[1] Kansai Research Center, Forestry and Forest Products Research Institute (FFPRI), 68 Nagaikyutaroh, Momoyama, Fushimi, Kyoto 612-0855, Japan; hamaguti@ffpri.affrc.go.jp (K.H.); narayama@ffpri.affrc.go.jp (S.N.)

[2] Department of Forest Entomology, Forestry and Forest Products Research Institute, 1 Matsunosato, Tsukuba, Ibaraki 305-8687, Japan; makino@ffpri.affrc.go.jp

[3] Department of Mushroom Science and Forest Microbiology, FFPRI, 1 Matsunosato, Tsukuba, Ibaraki 305-8687, Japan; hkosaka@ffpri.affrc.go.jp

[4] Kyushu Research Center, Forestry and Forest Products Research Institute, 4-11-16 Kurokami, Chuo, Kumamoto 860-0862, Japan

* Correspondence: nkanzaki@ffpri.affrc.go.jp

Simple Summary: Parasites sometimes manipulate host behavior to effectively utilize their hosts. The parasitism of host-manipulating parasites of hornets (*Vespa* spp.) was examined for baited or hand-collected hosts. There, two groups of endoparasites that sterilize their hosts, a nematode species, *Sphaerularia vespae*, and two species of parasitic insects, *Xenos* spp., were recovered during the survey. Although the nematode parasitism was scarcely observed (4 out of 1641 host hornets), the parasitism of *Xenos* spp. varied among host species between 0 and 17%. Interestingly, the rate of hornets parasitized by *Xenos* spp. was higher in baited individuals than in hand-collected ones. Because the bait trap attracts feeding (not foraging for colony) individuals, the parasitism manipulates the host to visit a feeding site, and the result is in good accordance with previous observations; i.e., parasitized hosts frequently visit and stay at feeding sites instead of foraging and reproducing. The present study could be the first case to quantify the *Vespa* host manipulation of *Xenos* spp. Genetically, both the *Xenos* spp. had four haplotypes in the mitochondrial cytochrome oxidase subunit I gene, and these were close to Japanese and some other East Asian genotypes.

Abstract: The parasitism of two groups of host-manipulating parasites of hornets was examined in Kyoto, Japan. *Vespa mandarinia* (661 individuals), *V. simillima* (303), *V. analis* (457), *V. ducalis* (158), *V. crabro* (57), and *V. dybowskii* (4) were collected either by bait trap or hand collection with an insect net, and examined for their parasites. An endoparasitic nematode, *Sphaerularia vespae* was isolated from three overwintered gynes of *V. mandarinia* and a gyne of *V. ducalis*. While endoparasitic insects, *Xenos* spp., were recovered from 13 *V. mandarinia*, 77 *V. analis*, two *V. ducalis*, and three *V. crabro*, and those recovered from *V. analis* and others were molecularly identified as *X. oxyodontes* and *X. moutoni*, respectively. Comparing *Xenos* parasitism level and capturing methods, the parasitism level was significantly higher in trapped hosts than in hand-collected ones, suggesting that stylopized hosts are more strongly attracted to the food source (bait trap) compared with unparasitized hosts. The genotypes of *S. vespae* were identical to each other, and near identical to its type population. While each of the two *Xenos* spp. showed four mitochondrial DNA haplotypes. A phylogenetic comparison suggested that *Xenos* haplotypes found in the present study are close to those previously reported from Japan and other Asian countries.

Keywords: behavior manipulation; parasitism; *Sphaerularia*; *Vespa*; *Xenos*

1. Introduction

Hornets, or the eusocial wasps of the genus *Vespa*, have great importance to humans not only as natural enemies of agricultural and sanitary pest insects [1], but also as dangerous stinging agents sometimes causing casualties [2]. In addition, several *Vespa* species are important invasive species, such as *V. velutina* Lepeletier, 1836, and *V. mandarinia* Smith, 1852, of Asian origins, expanding their distributions in Europe [3] and North America [4], respectively. Therefore, considering their status as a group of social insects, and practical importance as beneficial and pest insects, the genus can be regarded as an interesting group of insects from many aspects [1].

As are other insect groups [5], *Vespa* wasps are associated with several different groups of parasites [1,6,7]. Some parasites manipulate their host's behavior for their reproduction, and host manipulation by parasites has been intensively studied from ecological and evolutionary viewpoints [8]. If the hosts are serious pests, the parasites become important potential biocontrol agents.

Two groups of parasites of *Vespa* spp., i.e., nematodes in the genus *Sphaerularia* (Nematoda: Sphaerularioidea) and parasitic insects in the genus *Xenos* (Strepsiptera, Stylopidae), have been reported to manipulate the wasps' behavior to complete their life cycle. *Sphaerularia vespae* Kanzaki, Kosaka, Sayama, Takahashi, and Makino, 2007, was described from *V. simillima* Smith, 1868, in Hokkaido, Japan [9]. The life cycle of the nematode has been examined by Sayama et al. [10]. The inseminated infective (free-living) females infest the body cavity (hemocoel) of hibernating gynes in their hibernating sites, e.g., inside the rotten wood, and develop into the parasitic female called 'uterium', which is the hypertrophied female reproductive system specialized to oviposition. Thereafter, the nematode eggs hatch in the host's body cavity, and develop into juveniles taking nutrients from the host's body. The parasitized gynes are sterilized, i.e., they cannot establish a colony, and fly around the feeding sites and potential hibernating sites for their own feeding and inoculate the nematode juveniles at the hibernating sites for the next generation gynes. The inoculated nematode juveniles develop into male and female adults, and mate at the hibernation site. Thereafter, the males die out there, and the inseminated females parasitize the overwintering gynes. Thus, the nematode manipulates the hosts to disperse their juveniles in potential hibernating sites [10] in a similar way as reported for its congener *S. bombi* Dufour, 1837, which parasitizes *Bombus* spp. [11,12]. In addition to these two nominal parasite species, an undescribed *Sphaerularia* species has been reported from paper wasps, *Parapolybia* spp. [13]

The strepsipteran parasites, *Xenos* spp., parasitize hornets and paper wasps [1,13–20]. The life cycle and the host manipulation of *Xenos* spp. have been illustrated by several authors [1,13–20]. The parasite shows remarkable sexual dimorphism, i.e., winged male and magot-like neotenic female: the female stays in the host's body, and the male flies for mating. The host hornets (workers or gynes) parasitized by an inseminated female parasite(s) overwinter like unparasitized gynes, and frequently visit and stay at feeding sites to release the first instar infective larvae (triangulin) of the parasites at their feeding sites, e.g., sap flow and flower honey, in spring. The triangulin larvae probably sneak into nests of the hosts by attaching to adult hornets, and invade host larvae in the nests. Infected ("stylopized") hosts, even though emerging as adults, do not conduct normal behavior as workers or reproductives, but release the triangulins of the parasites at their feeding sites. This infection cycle could be repeated once or more during the active season of the host wasps/hornets. Because the nematode infection occurs during hibernation, stylopized gynes could be infested by the nematode. However, so far, their co-parasitism has not been reported.

Therefore, both groups of parasites potentially affect *Vespa* populations by manipulating the wasps. However, the basic ecological information for these parasites is still limited. In this study, we investigated the distribution and host range of these parasites in the Kyoto area, in the middle of Japan, where records on parasitism are sparse, to better understand the prevalence of the parasites, and make detailed molecular analyses, e.g., genomic and transcriptomic analyses of the host–parasite interactions. During the collections, several new findings emerged on the capturing and parasitism patterns of wasps, which are reported herein.

2. Materials and Methods

2.1. Collection and Dissection of Wasps

The hornets were collected with two different methods, bait traps and hand collection. Seven traps with liquid bait (sweet potato liquor: orange juice = 1:1) [16] were placed at a height of ca. 1.2 m above ground in the experimental stands of the Kansai Research Center, Forestry and Forest Products Research Institute, Kyoto, Japan, at the end of March 2021, before the wasps' active season, to the middle December, after their active season. During this collection, bait was replaced approximately every 10 days. The wasps captured by the bait trap were collected almost daily, 5–7 times a week. Hand collection was performed during the same period with the bait traps in the collection area where the traps were installed. Hornets were collected with an insect net 5–7 times a week while they were feeding or flying where one or two collector(s) walked throughout the collection site, and collected the wasps using an insect net with a handle of maximum 4.0 m long. Feeding individuals were sitting on tree trunks (for tree sap) of mostly *Quercus* spp. or on flowers (nectar) of *Cayratia japonica* (Thunb., 1784), *Dendropanax trifidus* (Thunb., 1784) and unidentified herbs; flying individuals were often holding prey, or flying in open spaces, around flower or tree trunks without carrying prey. Thus, 'flying' may include some hunting and feeding individuals.

Collected wasps were morphologically identified for species, caste, and sex, stored in refrigerator, and dissected under a stereomicroscope (SMZ1500, Nikon, Tokyo, Japan) within a day after the collection. The queen and worker were distinguished during the dissection by the body size and the development of ovaries. Because the parasites are supposed to be in the host abdomen, the abdomen of the wasps was taken and cut open with a pair of tweezers or dissecting scissors to examine the presence of their parasites. The dissected wasps were individually vouchered in a glass vial filled with 99.5% ethanol in the laboratory. When found, nematodes and strepsipteran parasites were stored in a glass vial filled with 99.5% ethanol for further molecular identification, and the number and sex of strepsipteran parasites were recorded for each host. In addition, some nematode materials were killed by heat (55 °C for 30 s), and fixed in TAF (2% triethnolamine, 3.5% formalin) as morphological vouchers. The number, caste, and parasitism level of wasps were compared among species, and between the two collection methods. The statistical analyses were conducted using the Chi-square test.

2.2. Identification of Parasites

When parasites were found from the host wasps, they were molecularly identified. Approximately 4 kb of ribosomal DNA (rDNA) including near-full length of the small subunit (SSU), internal transcribed spacer region (ITS), and D1–D4 extension segments of the large subunit (D1–D4 LSU) was determined for nematodes, and ca. 0.6 kb of partial code of mitochondrial cytochrome oxidase (mtCOI) was determined for the strepsipteran parasites.

Several (four to six) parasitic juveniles of the nematode were digested using nematode digestion buffer [21,22] as the template DNA, and the rDNA sequence was determined with PCR-direct (Sanger) sequencing using several first PCR and internal primers [23] (Supplementary Table S1). The sequence analysis was conducted for all parasitized hosts.

For the strepsipteran parasites, only females were examined for their sequences because almost all male parasites were already emerged (dispatched from host body), and their cocoons retained in the host body were tangled with host tissue. A small piece (less than 10 µL in volume) of parasite body tissue was taken with a pair of tweezers, transferred to 100 µL of DNA extraction solution (PrepMan Ultra Sample Preparation Reagent, ThermoFisher, Waltham, MA, USA), and kept at 95 °C for 10 min. The supernatant was used as PCR template, and the mtCOI of each sample was amplified and sequenced using a universal primer set (Supplementary Table S1).

Determined sequences were then compared with those deposited in the GenBank database, https://blast.ncbi.nlm.nih.gov/Blast.cgi?PROGRAM=blastn&PAGE_TYPE=BlastSearch&LINK_LOC=blasthome (accessed on 4 February 2022) using BLAST program, https://blast.ncbi.nlm.nih.gov/Blast.cgi?PROGRAM=blastn&PAGE_TYPE=BlastSearch&LINK_LOC=blasthome (accessed on 4 February 2022) for their species identification. In addition, their intraspecific genotypic diversity was examined using the program BioEdit [24], where the conspecific sequences were aligned, and the variations were examined manually. Thereafter, to compare *Xenos* spp. sequenced obtained in the present study and previously reported ones, an unrooted phylogenetic tree was generated using MrBayes 3.2 software [25,26] with a substitution model (GTR + G) selected using MEGA X software [27], where four chains were run for 4×10^6 generations, and the Markov chains were sampled at intervals of 100 generations [28]. Two independent runs were performed, and after confirming convergence and discarding the first 2×10^6 generations as burn-in, the remaining topologies were used to generate a 50% majority-rule consensus tree.

3. Results

In addition to the following data, row data for collection are provided as a spreadsheet (Supplementary Data S1).

3.1. Hornet Captures

In total, 1641 wasps of six *Vespa* species were captured during April and November (Figure 1, Table 1, Supplementary Data S1). The most abundant species was *V. mandarinia* (661 individuals), followed by *V. analis* Fabricius,1775 (457), *V. simillima* (303), *V. ducalis* Smith, 1858 (158), and *V. crabro* Linnaeus 1758 (57). A social parasite, *V. dybowskii* André, 1884, was rare, and only five individuals were captured during the collection season (Figure 1, Table 1, Supplementary Data S1). Thus, *V. dybowskii* is excluded from the following analysis.

The capturing pattern varied among the species. Although all six species were captured by both collection methods, the majority of the dominant four species (*V. mandarinia, V. analis, V. simillima,* and *V. ducalis*) were captured with an insect net, and this tendency was conspicuous in *V. simillima*, where 290 individuals out of 303 total individuals were obtained by hand collection. Although detailed location data were not taken, the species was mostly collected from flowering trees. Contrastingly, *V. crabro* was collected more by bait trapping (41 out of 57 individuals).

Vespa mandarinia, V. analis, V. simillima, and *V. crabro* were collected throughout the collection season (April to November), but *V. ducalis,* which exclusively preys on larvae and pupae of paper wasps [1], was only collected from May to October; i.e., the active season of the species was shorter than the other four species because of their prey preference.

The rate of *Xenos* parasitism was different between collection methods; i.e., the rate was generally higher in bait-trapped hornets than for hand-collected ones. *Xenos oxyodontes* Nakase and Kato, 2013, exclusively obtained from *V. analis* parasitized 36.3% and 4.7% of bait-trapped and hand-collected hosts, respectively (Figure 1, Table 1, Supplementary Data S1) ($p < 0.01$). In addition, although the differences were not statistically significant, *X. moutoni* (du Buysson, 1903) parasitized 3.3% and 1.5% of bait-trapped and hand-collected *V. mandarinia*, respectively; the parasites were found only from trapped individuals in *V. ducalis* and *V. crabro* (Figure 1, Table 1, Supplementary Data S1).

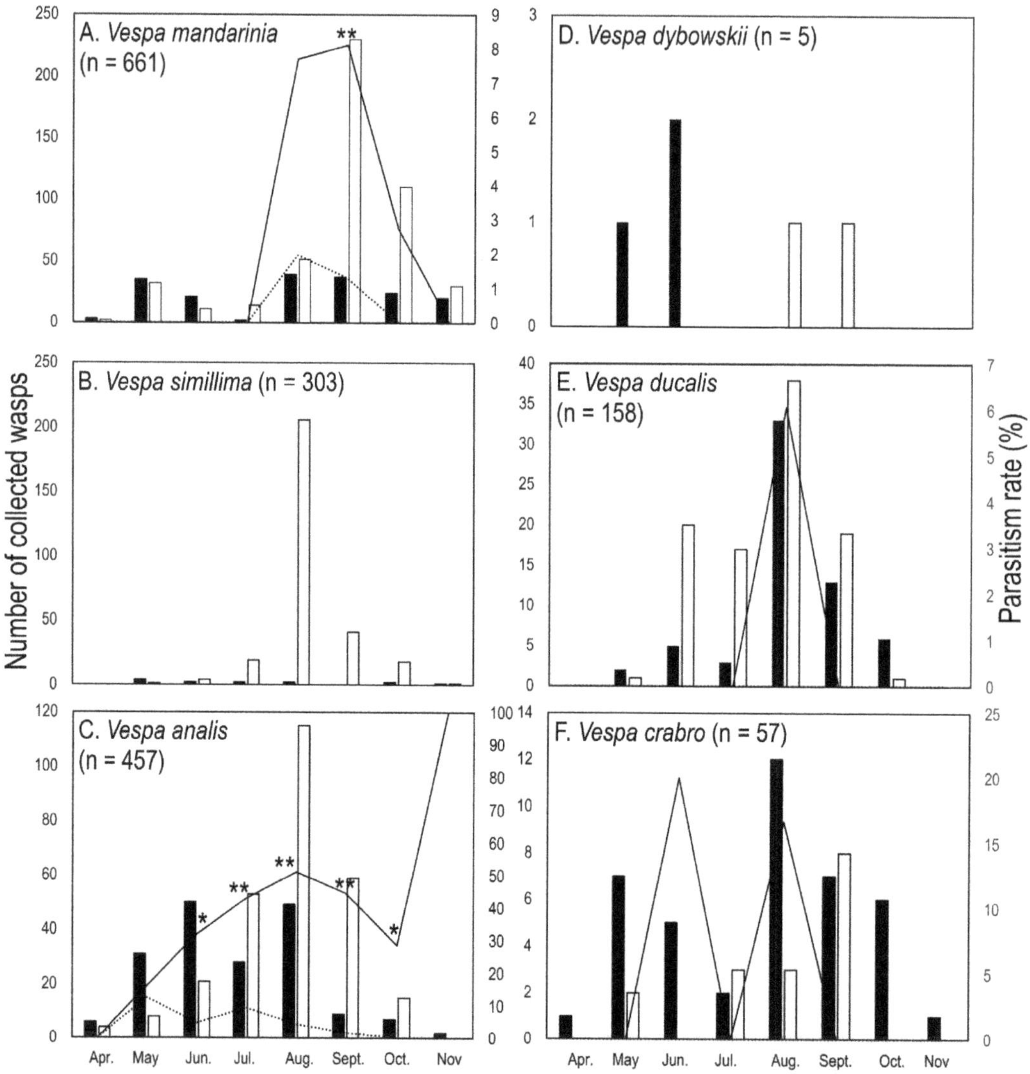

Figure 1. Number of collected wasps and rate of *Xenos* parasitism: (**A**): *Vespa mandarinia*; (**B**): *V. simillima*; (**C**): *V. analis*; (**D**): *V. dybowskii*; (**E**): *V. ducalis*; (**F**): *V. crabro*. Number of collected wasps (bar graph) and the parasitism rate (line graph) of each month are shown on left and right axis, respectively. Black and white bars are the numbers of trapped and hand-collected wasps, respectively, and solid and dotted line graphs indicate the parasitism ratio of trapped and hand-collected wasps, respectively. Asterisks on the bar graph indicate that the rate is significantly different between trapped and hand-collected wasps (*: $p < 0.05$; **: $p < 0.01$). Parasitism was not observed in *V. simillima* and *V. dybowskii*.

Table 1. Number of collected and parasitized wasps and their rate of parasitism.

Host Species (Total Number)	Collection Method	Number of Wasps		
		Collected	Parasitized by *Sphaerularia vespae*	Parasitized by *Xenos* spp.
Vespa mandarinia (661)	Trap	181	1 (0.6)	6 (3.3)
	Net	480	2 (0.4)	7 (1.5)
Vespa simillima (303)	Trap	13	0 (0.0)	0 (0.0)
	Net	290	0 (0.0)	0 (0.0)
Vespa analis (457)	Trap	182	0 (0.0)	66 (36.3)
	Net	275	0 (0.0)	13 (4.7)
Vespa dybowskii (4)	Trap	3	0 (0.0)	0 (0.0)
	Net	2	0 (0.0)	0 (0.0)
Vespa ducalis (158)	Trap	62	1 (1.6)	2 (3.2)
	Net	96	0 (0.0)	0 (0.0)
Vespa crabro (57)	Trap	41	0 (0.0)	3 (7.3)
	Net	16	0 (0.0)	0 (0.0)

3.2. Nematode Parasitism

Parasitic nematodes were found from four (three *V. mandarinia* and a *V. ducalis*) queens in May (Table 2). The nematodes were all identified as *Sphaerularia vespae* based on molecular sequences; i.e., the SSU, ITS, and D2-D3 LSU of the nematodes were identical to each other, and to the type material of the species (SSU + ITS: AB300595; D2–D3 LSU: AB300596).

Table 2. Collection information of *Sphaerularia vespae*.

Host Species	Host Indiv. ID	Collection Method	Date
Vespa mandarinia	2021-0021	Trap	4 May
	2021-0069	Net	18 May
	2021-0122	Net	27 May
Vespa ducalis	2021-0137	Trap	31 May

3.3. Strepsipteran Parasitism

Strepsipteran parasites (*Xenos* spp.) were recovered from four species (*V. mandarinia, V. analis, V. ducalis,* and *V. crabro*) with various parasitism rates (Figure 1, Table 1, Supplementary Data S1). The total rate was variable among four species (Chi-square test, $p < 0.01$), and the rate was significantly higher in *V. analis* (17.2%) than in *V. mandarinia* (2.0%) and *V. ducalis* (1.3%) ($p < 0.01$), but the difference is not significant between *V. analis* and *V. crabro* (5.3%) ($p = 0.062$) and other combinations ($p > 0.6$) (Holm's multiple comparison test).

Based on mtCOI sequence analysis, the parasites were identified as two different species, *X. moutoni* parasitizing *V. mandarinia, V. ducalis,* and *V. crabro*, and *X. oxyodontes*, which was exclusively recovered from *V. analis* (Figure 2; Table 3). Both species showed intraspecific variation in four genotypes (haplotypes) for each species (Figure 2; Table 3). The phylogenetic analysis suggested that these two species are clearly separate, and genotypes found in the present study are included in each species' clade, and close to other Japanese isolates (Figure 2).

The sex ratio (percentage of female) of *Xenos* spp. varied between parasite species and among host wasps. The ratio was more female-biased in *X. moutoni* (90%), where only two males were found from *V. crabro*, and only female parasites were found from *V. mandarinia* and *V. ducalis*. The number of parasites was mostly one for these three species (Tables 4 and 5, Supplementary Data S1). Contrastingly, the sex ratio in *X. oxyodontes* was approximately equal (54.3%), and the number of parasites per host varied from one to nine (Table 4). The sex ratio of the species was female-biased in the early season, and the ratio gradually decreased toward the late season (Table 5, Supplementary Data S1).

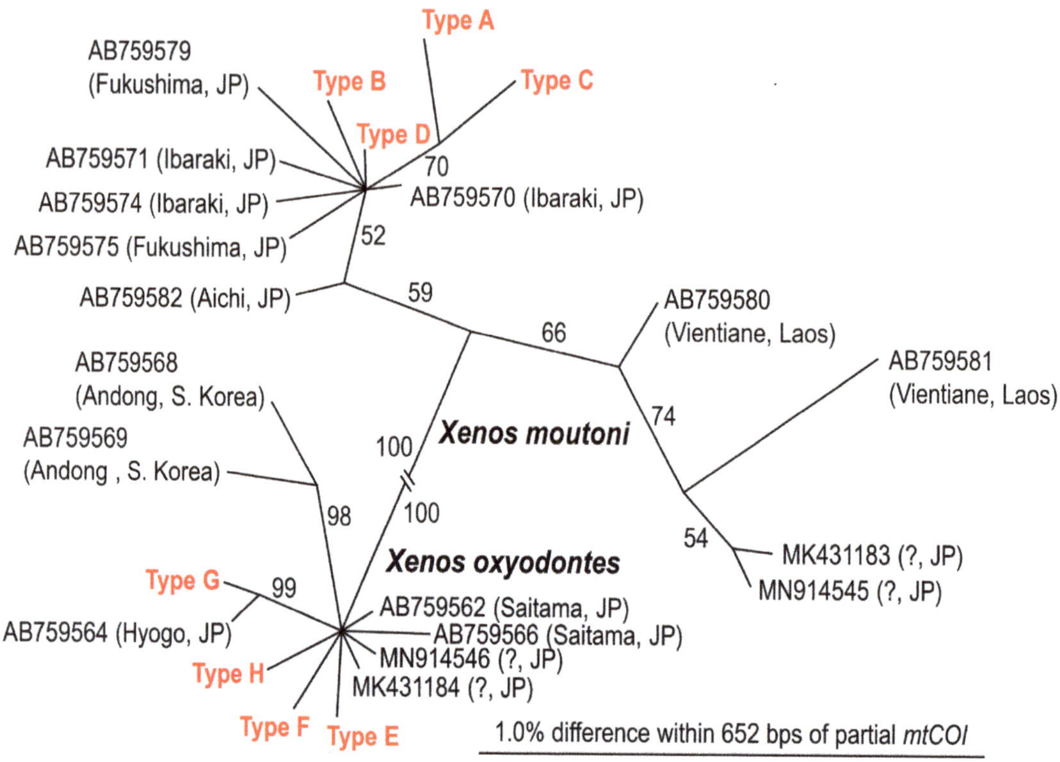

Figure 2. Phylogenetic relationship among haplotypes of two *Xenos* spp. Unrooted tree was generated with Bayesian analysis under GTR + G model (lnL = −1244.934; freqA = 0.33, freqC = 0.15, freqG = 0.10, freqT = 0.42; R(a) = 0.93, R(b) = 8.67, R(c) = 1.45, R(d) = 0.48, R(e) = 11.29, R(f) = 1.00; Pinva = n/a; Shape = 0.07). Posterior probability values exceeding 50% are given on appropriate clades.

Table 3. Variation and host of mitochondrial haplotypes of *Xenos* spp.

(A): *Xenos moutoni*						
Code (Number)	**GenBank Accession Number**	**Host Species**	**Base Count from 5′ End in 652 bps Segment**			
			109	112	436	544
A (1)	LC764843	*Vespa mandarinia*	C	A	A	A
B (1)	LC764844	*Vespa mandarinia*	T	T	A	G
C (1)	LC764845	*Vespa mandarinia*	T	A	G	A
D (13)	LC764846	*Vespa mandarinia, V. ducalis, V. crabro*	T	A	A	G
(B): *Xenos oxyodontes*						
Code (Number)	**GenBank Accession Number**	**Host Species**	**Base Count from 5′ End in 652 bps Segment**			
			286	340	391	463
E (1)	LC764847	*Vespa analis*	A	T	A	C
F (2)	LC764848	*Vespa analis*	G	C	A	C
G (19)	LC764849	*Vespa analis*	G	T	T	C
H (30)	LC764850	*Vespa analis*	G	T	A	A

Table 4. Average number and sex ratio of *Xenos* spp. parasitizing an infected host.

Host Species (Number of Host Individuals)	*Xenos* Species	Female	Male	Total	Sex Ratio
Vespa mandarinia (13)	*Xenos moutoni*	1	0	1	100
Vespa analis (77)	*Xenos oxyodontes*	0.82 ± 0.74 (0–5)	0.69 ± 0.98 (0–4)	1.51 ± 1.25 (1–9)	54.3
Vespa ducalis (2)	*Xenos moutoni*	1.5 ± 0.71 (1–2)	0	1.5 ± 0.71 (1–2)	100
Vespa crabro (3)	*Xenos moutoni*	0.33 ± 0.56 (0–1)	0.66 ± 0.56 (0–1)	1	33

Table 5. Seasonal variation in total number and sex ratio (% of female) of *Xenos* spp.

Species	Host	Sex	May	Jun.	Jul.	Aug.	Sept.	Oct.	Nov.	Total
Xenos moutoni	*Vespa mandarinia, V. ducalis, V. crabro*	Female	0	0	0	8	6	3	0	17
		Male	0	1	0	1	0	0	0	2
		Sex ratio	-	0	-	88	100	100	-	89
Xenos oxyodontes	*Vespa analis*	Female	6	17	11	24	4	1	0	63
		Male	0	0	11	27	11	2	2	53
		Sex ratio	100	100	50	47	27	33	0	54

Xenos parasitism was not detected in April, and thus, the table starts from May. *Xenos* parasitism was confirmed only in trapped hosts in *Vespa ducalis* and *V. crabro*. *Xenos moutoni* is significantly female-biased ($p < 0.01$).

4. Discussion

Several different types of data were presented in this study, and some are dependent on each other. Therefore, these data will be discussed separately as a general capturing pattern in relation to capturing method in Sections 4.1 and 4.2, nematode parasitism in Section 4.3, the biology of *Xenos* spp. in Sections 4.4 and 4.5, and genetic diversity of *Xenos* spp. in the research site in Section 4.6, followed by additional remarks in Section 4.7.

4.1. Capture Outcomes of Bait-Trapping Versus Hand-Collecting Hornets

Out of the six hornet species collected, more individuals were obtained by hand-collection than bait traps in four hornet species (*V. mandarinia*, *V. simillima*, *V. analis*, and *V. ducalis*) (Figure 1, Table 1, Supplementary Data S1). While *V. crabro* was collected more by bait traps (Figure 1, Table 1, Supplementary Data S1). This difference in the capturing pattern probably reflects the intensiveness of hand-collection and behavioral characteristics of hornet species. Hand collection could have covered a wider area than the traps, because the traps were settled on the trees, and the hornets that visited these trees were collected, but the hand-collection was conducted through all areas including the tree stands. While the hand collection was conducted only during the daytime, the *V. crabro* exhibits foraging and feeding activity in the early morning and late evening into the night as well, i.e., the periods when the hand-collection cannot cover [29].

Interestingly, 95.7% (290 out of 303) of *V. simillima* were hand-collected, and only 13 individuals (4.3%) were bait-trapped. Based on previous bait trap-based surveys [16,30–32] and their own survey, Kudô et al. [33] suggested that *V. simillima* is dominantly collected by bait traps in cooler regions, but not in warm areas, which our trapping study supports; i.e., *V. simillima* was not trapped in Kyoto, which is in a relatively warm temperate zone. However, a fair number of *V. simillima* were captured by hand mostly from flowering trees, suggesting that although the species was commonly distributed in the collection site, the traps failed to attract it. The bait traps used in the present and previous studies in Japan mainly contain liquor and/or fruit juice [16,17,31–33], which emit odors similar to fermented tree sap or fruits. Thus, *V. simillima* possibly prefers other food sources, e.g., flower honey, in warmer regions. Otherwise, considering the body size, i.e., *V.*

simillima is the smallest *Vespa* species in Japan, *V. simillima* might be outcompeted from tree sap by the larger *Vespa* spp., and utilize other available resources, i.e., flower honey. This difference in the capturing pattern suggests that multiple types of traps are necessary to monitor the *Vespa* fauna in warmer regions [34].

4.2. Comparison of Collection Methods

In all *Vespa* species where *Xenos* occurred, the parasitism level was consistently higher in the bait-trapped individuals than in hand-collected ones (Figure 1, Table 1, Supplementary Data S1). This was shown in the most marked way by *V. analis* in which the level was significantly higher in bait-trapped wasps than in hand-collected ones both in total number and monthly collections during June to October (Figure 1, Table 1, Supplementary Data S1).

In the present study, trapped hosts are considered to be attracted to odors similar to tree sap, which seems to account for the higher levels of *Xenos* parasitism in bait-collected as compared to hand-collected individuals. Further, almost all parasitized hornets in the hand-collected samples were obtained while they were feeding at or near tree sap holes. These results are consistent with the assumption that *Vespa*-associated *Xenos* spp. manipulate host adults by making them frequent feeding sites such as tree sap holes where the parasitic females may release the first-instar larvae [1,16].

A similar, or basically the same, behavioral modification has been reported in a paper wasp, *Polistes dominula* (Christ, 1791)—*Xenos vesparum* Rossius, 1793 system [19]. There, the parasitized host desists from the colony task and aggregates their own food source, trumpet creeper bushes, to feed and release infective larvae of the parasite [19]. This behavioral modification could be a common characteristic of these *Xenos* species.

4.3. Nematode Parasite

Sphaerularia vespae was collected from two host species, three queens of *V. mandarinia* and a queen of *V. ducalis* (Table 1, Supplementary Data S1). The nematode was originally recorded from *V. simillima* in Hokkaido, Japan [9,35], and subsequently found from other *Vespa* spp. from various areas in Japan [36]. Sayama et al. [35] reported that the parasitism rate of the nematode was very high in *V. simillima* from the Hokkaido region; i.e., more than 60% of overwintered gynes (new queens) collected with bait traps were infected by the nematode in some populations. However, in the present study, the nematode was found only from *V. mandarinia* and *V. ducalis*, and the parasitism rate was much lower, 1.6% (three out of 182) and 2.0% (one out of 50) of overwintered gynes, respectively, and none of nine overwintered gynes of *V. simillima* was infested by the nematode (Supplementary Data S1). We do not have a clear explanation for this difference in the infection rate between the Hokkaido and Kyoto populations, because the data for the infection rate have not been examined in different localities. The nematodes are disseminated to the hosts' hibernation site because of the parasitized hosts' aberrant behavior, and infest new hosts during hibernation [10]. Therefore, the difference in environmental conditions, i.e., subarctic (Hokkaido) and temperate (Kyoto) zones, could affect wasps' behavior, e.g., segregation of hibernation sites between *V. simillima* and others as observed in feeding preference.

4.4. Levels of Xenos Parasitism in Bait-Trapped Hosts

The parasitism rate varied among host species, especially in trapped individuals; i.e., 36.3% (66 out of 182) of *V. analis* were parasitized by *X. oxyodontes*; 3.3% (6/181) of *V. mandarinia*, 3.2% (2/62) of *V. ducalis*, and 7.3% (3/41) of *V. crabro* were parasitized by *X. moutoni*, and no parasitism was found in *V. simillima* (Table 1, Supplementary Data S1).

The *Xenos* parasitism rate of *Vespa* spp. has been examined in several previous studies using bait traps (Table 6). Makino and Yamashita [16] collected the wasps in Miyazaki, Japan, and reported that four species (3457 individuals) of the wasps were captured, and the parasitism rate was high (10.6%) in *V. analis*, followed by *V. mandarinia* (3.8%), *V. ducalis* (3.8%), and *V. simillima* (1.4%). Subsequently, Makino [30] reported a similar pattern based on three seasons of trapping in Ibaraki, Japan; i.e., the parasitism rate was higher in *V. analis* (15–20%) than that in *V. simillima*, *V. ducalis*, and *V. crabro* (less than 2%), and no parasitism was observed in *V. mandarinia*. Similarly, Sayama [17] examined the parasitism rate for 10 years in Hokkaido, Japan, and collected five species of *Vespa* spp. There, approximately 8% of *V. analis* and less than 0.5% of *V. simillima* and *V. crabro* were stylopized, but no parasitism was confirmed in *V. mandarinia* and *V. dybowskii* (only one case of *Xenos* parasitism in *V. dybowskii* was observed in a worker of a collected nest). On the other hand, Oyaizu and Kudô [32] reported that the rate was rather high in *V. mandarinia* (16.7–23.5%), and *V. analis* (8.8–16.7%) was not clearly different from those of *V. ducalis* (12.5%) and *V. crabro* (10.1%), but higher than that of *V. simillima* (0.4–1.2%), based on two years of collections in Niigata, Japan.

Table 6. Summary of parasitism of *Xenos* spp. on the *Vespa* spp. captured by bait traps provided in the previous studies.

Host	Expected *Xenos* Species	Locality (Climate) *	Parasitism (%)	Reference
Vespa mandarinia	*Xenos moutoni*	Hokkaido (SA-CT)	0	[17]
		Niigata (CT)	16.7–23.5	[32]
		Ibaraki (T)	0	[30]
		Kyoto (T-WT)	3.3 (2.0)	This study
		Miyazaki (WT)	3.8	[16]
Vespa ducalis	*Xenos moutoni*	Hokkaido (SA-CT)	Not collected	[17]
		Niigata (CT)	12.5	[32]
		Ibaraki (T)	<2.0	[30]
		Kyoto (T-WT)	3.2 (1.3)	This study
		Miyazaki (WT)	3.8	[16]
Vespa dybowskii	*Xenos moutoni*	Hokkaido (SA-CT)	0	[17]
		Niigata (CT)	Not collected	[32]
		Ibaraki (T)	Not collected	[30]
		Kyoto (T-WT)	0 (0)	This study
		Miyazaki (WT)	Not collected	[16]
Vespa crabro	*Xenos moutoni*	Hokkaido (SA-CT)	<0.5	[17]
		Niigata (CT)	10.1	[32]
		Ibaraki (T)	<2.0	[30]
		Kyoto (T-WT)	7.3 (5.3)	This study
		Miyazaki (WT)	Not collected	[16]
Vespa simillima	*Xenos moutoni* or *X. oxyodontes*	Hokkaido (SA-CT)	<0.5	[17]
		Niigata (CT)	0.4–1.2	[32]
		Ibaraki (T)	<2.0	[30]
		Kyoto (T-WT)	0 (0)	This study
		Miyazaki (WT)	1.4	[16]
Vespa analis	*Xenos oxyodontes*	Hokkaido (SA-CT)	ca 8.0	[17]
		Niigata (CT)	8.8–16.7	[32]
		Ibaraki (T)	15–20	[30]
		Kyoto (T-WT)	36.3 (16.8)	This study
		Miyazaki (WT)	10.6	[16]

* The localities are ordered from north to south, and abbreviations for climate are as follows: CT = cool temperate; SA: subarctic; T: temperate; WT: warm temperate. A case of *X. moutoni* parasitism of *V. dybowskii* found in a hand-collected individual was noted in Sayama [17]. Parasitism ratio for the present study is written in a form ratio for bait-trapped hosts (total collected hosts).

Although the levels of parasitism not only vary among the host species but among the studies, the level is consistently high (ca 8.0–36.3%) in *V. analis*; a low level of parasitism has been found in *V. ducalis* (3.2–12.5%) and *V. crabro* (<0.5 to 10.1%); the level of parasitism is consistently low in *V. simillima* (<2.0%); and the level is highly variable in *V. mandarinia* (0–23.5%) in bait-trapped hosts (Table 6) [16,17,30,32]. This suggests that the affinity between *V analis* and *X. oxyodontes* is higher than that between the other *Vespa* spp. and *X. moutoni*, and that *V. simillima*, which has two strepsipteran parasite species [37], has a low affinity with both parasite species. Our results are in good accordance with the previous studies for high and low parasitism rates in *V. analis* and *V. simillima*, respectively, and generally similar to the pattern in Miyazaki. Although the number of examined localities was not sufficient, both Kyoto and Miyazaki belong to the relatively warm temperate zone, and thus, environmental conditions could affect the pattern of parasitism via the wasps' behavioral pattern. For example, *V. simillima* was mostly collected by hand from the flower honey; i.e., the feeding (and possibly hunting) site of *V. simillima* is segregated from other hornet species in the experimental site, and the segregation allowed *V. simillima* to avoid the parasites derived from the other *Vespa* spp.

4.5. Number and Sex Ratio of *Xenos* spp.

The number of *Xenos* parasites per host varied between one to nine, and in most cases, one male or female was recovered from a parasitized host. A single female of *X. moutoni* mostly parasitized *V. mandarinia*, *V. ducalis*, and *V. crabro* (Table 4, Supplementary Data S1). While the number varied in *X. oxyodontes*, five females and four males were found in an individual host at maximum (Table 4, Supplementary Data S1).

Oyaizu and Kudô [32] reported the number of *Xenos* spp. per host varied from one to two, which is in good accordance with our data for *X. moutoni*, but multiple parasitisms of *X. oxyodontes* were not observed in the study. While, Makino and Yamashita [16] reported that the number varied from one to four *X. oxyodontes* in *V. analis* one to five *X. moutoni* in the other species, and 22 out of 102 (22%) of *V. analis* and 16 out of 84 (19%) of other *Vespa* spp. (*V. mandarinia* and *V. ducalis*) were parasitized by multiple individuals of *Xenos* spp.; i.e., the rate of multiple parasitisms is similar between two *Xenos* spp. In the present study, multiple parasitisms were observed in 17 out of 77 (22.1%) *V. analis* and none of the 17 others; i.e., the rate of multiple parasitisms is similar in *X. oxyodontes*, but the rate is lower in *X. moutoni*. The rate could be affected by the population density, behavior, and interspecific interaction of host wasps, but further investigation is necessary to explain these differences.

The overall sex ratio of *Xenos* individuals on the host throughout the season was greatly different between the two species, i.e., strongly female-biased in *X. moutoni* (female ratio = 90% in total), and mostly even (54.3%) in *X. oxyodontes* (Table 5, Supplementary Data S1). The female ratio of *X. oxyodontes* gradually decreased according to the season; i.e., only females that overwintered with host wasps were found in the early season (April and May), and the female ratio decreased to zero by the end of collections (Table 5, Supplementary Data S1). The pattern in *X. oxyodontes* is in accordance with the observation by Makino and Yamashita [16] and Makino [30]. There, although the *Xenos* species was all regarded to be *X. moutoni*, because *X. oxyodontes* was not separated at the time, males appear in the late season (mid-August) in Miyazaki [16], and *X. oxyodontes* males appear in July in Ibaraki [30]. Matsuura and Yamane [1] suggested that female parasites overwinter parasitizing the host, and reproduce three or more generations during the host's active season. While although the parasite species is not specified, Sayama [17] considered that the *Xenos* parasite is univoltine in Hokkaido. The data in the present study, i.e., the female ratio decreased to zero more or less linearly, suggest that the parasite is univoltine at the study site. There, female parasites are collected from overwintered hosts in the early season, and next-generation parasites showing an even sex ratio appear in the middle season. Thereafter, host hornets parasitized by female parasites hibernate earlier than a regular queen, and the hosts harboring the male parasite cocoon can be captured in the late season. Although the number of generations (univoltine or bivoltine; i.e., the parasite could have reproduced two generations in the middle season) is not clearly examined, the number could be affected by the environment, especially the temperature, because the observation by Matsuura and Yamane [1] (possibly three or more generations) and Sayama [17] (one generation) were yielded from cold and warm areas, respectively. Moreover, a similar bivoltine cycle has been confirmed in *Polistes dominula* paper wasp and its parasite, *Xenos vesparum* based on field observations [19]. To confirm the generation time of *Vespa*-associated *Xenos* spp. in relation to environmental conditions, further field observations in the different areas are necessary.

We do not have a clear explanation for the female-biased sex ratio in *X. moutoni*; although, this could be simply because of the small (insufficient) number of collected materials. Because *X. oxyodontes* was separated from *X. moutoni* recently [37], two *Xenos* spp. were not distinguished in the previous studies. Therefore, the difference in the sex ratio is not determined whether it is derived from genetic differences related to the species-specific reproductive strategy or phenotypic variation in the species; e.g., the sex ratio of parasites is often variable among conspecific populations, and sometimes among subpopulations depending on the population density [38,39]. To understand the seasonal pattern of sex ratio in each *Xenos* spp. re-examination of previous data and further collections with species identification is necessary. We also need *Xenos* sex ratios in hornet nests, from which stylopized adults disperse into the field.

4.6. Molecular Identification and Genotypes of Xenos spp.

Two *Xenos* spp. were obtained from four host species, i.e., *X. moutoni* from *V. mandarinia*, *V. ducalis*, and *V. crabro*, and *X. oxyodontes* from *V. analis*.

Xenos oxyodontes was recently separated from *X. moutoni* based on molecular and morphological characters by Nakase and Kato [37]. Nakase and Kato [37] examined host-parasite associations of the two *Xenos* spp. and concluded that the two species segregate their hosts: *X. moutoni* is parasitic on *V. mandarinia*, *V. ducalis*, *V. crabro*, and *V. simillima*, while *X. oxyodontes* mainly parasitizes *V. analis*, and much less frequently *V. simillima*. The isolation pattern of *Xenos* spp. in the present study corroborates the observations of Nakase and Kato [37]; i.e., *X. oxyodontes* was exclusively found from *V. analis*, and those obtained from the other host species were all *X. moutoni* (Tables 2 and 3; Supplementary Data S1).

Both *Xenos* spp. showed intraspecific variations in mtCOI sequences, and four haplotypes were recognized within each species (Figure 2; Table 3), suggesting that multiple (at least four) female lines are present in the collection site. The sequences of these haplotypes were phylogenetically compared with those deposited in the GenBank database, and visualized as an unrooted tree (Figure 2). Four haplotypes of *X. moutoni* (Types A–D) belonged to a phylogenetic clade including other conspecific Japanese materials; although, the posterior probability (PP) support was not high (59%) (Figure 2). *X. oxyodontes* haplotypes found in Kyoto were close to the other Japanese materials, and two Korean isolates were separated from them with relatively high (98%) PP support (Figure 2). However, in both species, a clear phylogeographical pattern was not recognized among Japanese materials, possibly because the sequence variation was relatively low, with only four variable sites within 652 bps in the specimens collected in the present study (Table 3). In *X. moutoni*, three haplotypes (A–C) were found only once, and all were isolated from *V. mandarinia*, and the dominant haplotype (D) from 13 individuals was found from three wasp species (Table 3A). The relationship between haplotype and host species is not clear because the number of sampled specimens was not sufficient.

4.7. Additional Remarks

In the present study, two groups of host-manipulating parasites of hornets were examined using two different collection methods, bait trap and hand-collection, mostly focusing on *Xenos* spp. and the capturing pattern was in good accordance with the manipulation of *Xenos* spp. In addition, the biological characteristics of *Xenos* spp. were mostly consistent with previous studies conducted in Japan [1,15–17,30–33]. However, considering the wide distribution area of both *Vespa* spp. and *Xenos* spp., further studies of their biological interaction are necessary to be conducted in different localities, i.e., different climate conditions. Biological and genetic studies have intensively proceeded regarding the *P. dominula*–*X. vesparum* system [18–20]. There, detailed field observations and genomic/transcriptomic analysis have revealed the life cycle and mechanisms of their host manipulation [18–20,40,41].

Considering the high parasitism ratio and accessibility, the *V. analis–X. oxyodontes* system could be a good comparison of the *P. dominula–X. vesparum* system.

Supplementary Materials: The following supporting information can be downloaded at: https://www.mdpi.com/article/10.3390/insects14040398/s1, Table S1. PCR and sequencing primers used in the present study.; Supplementary Data S1. Collection information of *Vespa* spp.

Author Contributions: Conceptualization, N.K. and K.H.; methodology, N.K.; validation, N.K. and S.M.; formal analysis, N.K.; investigation, N.K.; resources, N.K., K.H. and S.N.; data curation, N.K.; writing—original draft preparation, N.K., S.M., H.K., K.S. and K.H.; writing—review and editing, N.K., S.M., H.K., K.S., K.H. and S.N.; visualization, N.K. and K.H.; supervision, S.M.; project administration, N.K.; funding acquisition, N.K., S.M., H.K. and K.S. All authors have read and agreed to the published version of the manuscript.

Funding: This study was supported in part by Grants-in-Aid for Scientific Research (B) (Nos. 20H03026, 22H02690) and (C) (No. 21K05693) from The Japan Society for the Promotion of Science (JSPS).

Data Availability Statement: All original data are provided either in Supplementary Data S1 (collection information) or the GenBank database (molecular sequences determined in the present study).

Acknowledgments: The authors sincerely thank Robin Giblin-Davis, Emeritus at the University of Florida, for critical reading of an early version of the manuscript, and Yoshiko Shimada, Kansai Research Center, FFPRI for assistance in the collection and dissection of wasps and molecular analyses.

Conflicts of Interest: The authors declare no conflict of interest.

References

1. Matsuura, M.; Yamane, S.K. *Biology of the Vespine Wasps*; Springer: Berlin, Germany, 1990; p. 323.
2. Otaki, N. Allergic symptoms to stings of wasps and bees. In *Preventions and Treatments of Stings by Wasps and Bees*, 2nd ed.; Matsuura, M., Ed.; Forestry and Timber Manufacturing Safety & Health Association: Tokyo, Japan, 2005; pp. 211–257. (In Japanese)
3. Budge, G.E.; Hodgetts, J.; Jones, E.P.; Ostojá-Starzewski, J.C.; Hall, J.; Tomkies, V.; Semmence, N.; Brown, M.; Wakefield, M.; Stainton, K. The invasion, provenance and diversity of *Vespa velutina* Lepeletier (Hymenoptera: Vespidae) in Great Britain. *PLoS ONE* **2017**, *12*, e0185172. [CrossRef] [PubMed]
4. Zhu, G.; Illan, J.G.; Looney, C.; Crowder, D.W. Assessing the ecological niche and invasion potential of the Asian giant hornet. *Proc. Natl. Acad. Sci. USA* **2020**, *117*, 24646–24648. [CrossRef] [PubMed]
5. Poinar, G.O. *Entomogenous Nematodes: A Manual and Host List of Insect-Nematode Associations*; Brill: Leiden, The Netherlands, 1975; p. 303.
6. Edwards, R. *Social Wasps. Their Biology and Control*; Rentokil: East Grinstead, UK, 1980; p. 398.
7. Ono, M. *Science of Japanese Hornets (Vespa spp.)*; Kaiyusha: Tokyo, Japan, 1997; p. 174. (In Japanese)
8. Heil, M. Host manipulation by parasites: Cases, patterns, and remaining doubts. *Front. Ecol. Evol.* **2016**, *4*, 80. [CrossRef]
9. Kanzaki, N.; Kosaka, H.; Sayama, K.; Takahashi, J.-I.; Makino, S. *Sphaerularia vespae* sp. nov. (Nematoda, Tylenchomorpha, Sphaerularioidea), an endoparasite of the common hornet, *Vespa simillima* Smith (Insecta, Hymenoptera, Vespidae). *Zool. Sci.* **2007**, *24*, 1134–1142. [CrossRef]
10. Sayama, K.; Kosaka, H.; Makino, S. Release of juvenile nematodes at hibernation sites by overwintered queens of the hornet *Vespa Simillima*. *Insectes Soc.* **2013**, *60*, 383–388. [CrossRef]
11. Poinar, G.O.; Van der Laan, P.A. Morphology and life history of *Sphaerularia bombi*. *Nematologica* **1972**, *18*, 239–252. [CrossRef]
12. Bedding, R.A. Nematode parasites of Hymenoptera. In *Plant and Insect Nematodes*; Nickle, W.R., Ed.; Marcel Dekker: New York, NY, USA, 1984; pp. 755–795.
13. Morooka, F. The prevalence of the parasitic nematode *Sphaerularia* sp. in the overwintering gynes of *Parapolybia* spp. (Hymenoptera, Polistinae). *J. Hymenopt. Res.* **2014**, *38*, 37–43. [CrossRef]
14. Kathirithamby, J. Biodiversity of Strepsiptera. In *Insect Biodiversity: Science and Society*, Vol II. ed.; Foottit, R.G., Adler, P.H., Eds.; John Wiley & Sons: New York, NY, USA, 2018; pp. 673–703. [CrossRef]
15. Matsuura, M. *Social Wasps of Japan in Color*; Hokkaido University Press: Sapporo, Japan, 1995; p. 353, (In Japanese with English Summary).
16. Makino, S.; Yamashita, Y. Levels of parasitism by *Xenos moutoni* du Buysson (Strepsiptera, Stylopidae) and their seasonal changes in hornets (Hymenoptera: Vespidae, *Vespa*) caught with bait traps. *Entomol. Sci.* **1998**, *1*, 537–543.

17. Sayama, K. Host utilization and seasonal occurrence of the hornet parasite *Xenos moutoni* in Hokkaido, northern Japan. *Nat. Insects* **2012**, *47*, 21–24. (In Japanese)
18. Hughes, D.P.; Kathirithamby, J.; Turillazzi, S.; Beani, L. Social wasps desert the colony and aggregate outside if parasitized: Parasite manipulation? *Behav. Ecol.* **2004**, *15*, 1037–1043. [CrossRef]
19. Beani, L.; Cappa, F.; Manfredini, F.; Zaccaroni, M. Preference of *Polistes dominula* wasps for trumpet creepers when infected by *Xenos vesparum*: A novel example of co-evolved traits between host and parasite. *PLoS ONE* **2018**, *13*, e0205201. [CrossRef] [PubMed]
20. Beani, L.; Dallai, R.; Cappa, F.; Manfredini, F.; Zaccaroni, M.; Lorenzi, M.C.; Mercati, D. A Strepsipteran parasite extends the lifespan of workers in a social wasp. *Sci. Rep.* **2021**, *11*, 7235. [CrossRef] [PubMed]
21. Kikuchi, T.; Aikawa, T.; Oeda, Y.; Karim, N.; Kanzaki, N. A rapid and precise diagnostic method for detecting the pinewood nematode *Bursaphelenchus xylophilus* by loop-mediated isothermal amplification (LAMP). *Phytopathology* **2009**, *99*, 1365–1369. [CrossRef] [PubMed]
22. Tanaka, R.; Kikuchi, T.; Aikawa, T.; Kanzaki, N. Simple and quick methods for nematode DNA preparation. *Appl. Entomol. Zool.* **2012**, *47*, 291–294. [CrossRef]
23. Kanzaki, N.; Masuya, H.; Ichihara, Y.; Maehara, N.; Aikawa, T.; Ekino, T.; Ide, T. *Bursaphelenchus carpini* n. sp., *B. laciniatae* n. sp. and *B. cryphali okhotskensis* n. subsp. (Nematoda: Aphelenchoididae) isolated from Japan. *Nematology* **2019**, *21*, 361–388. [CrossRef]
24. Hall, T.A. BioEdit: A user-friendly biological sequence alignment editor and analysis program for windows 95/98/NT. *Nucleic Acids Symp. Ser.* **1999**, *41*, 95–98.
25. Huelsenbeck, J.P.; Ronquist, F. MRBAYES: Bayesian inference of phylogenetic trees. *Bioinformatics* **2001**, *17*, 1754–1755. [CrossRef]
26. Ronquist, F.; Teslenko, M.; van der Mark, P.; Ayres, D.L.; Darling, A.; Höhna, S.; Larget, B.; Liu, L.; Suchard, M.A.; Huelsenbeck, J.P. MrBayes 3.2: Efficient Bayesian phylogenetic inference and model choice across a large model space. *Syst. Biol.* **2012**, *61*, 539–542. [CrossRef]
27. Kumar, S.; Stecher, G.; Li, M.; Knyaz, C.; Tamura, K. MEGA X: Molecular evolutionary genetics analysis across computing platforms. *Mol. Biol. Evol.* **2018**, *35*, 1547–1549. [CrossRef]
28. Larget, B.; Simon, D.L. Markov chain Monte Carlo algorithms for the Bayesian analysis of phylogenetic trees. *Mol. Biol. Evol.* **1999**, *16*, 750–759. [CrossRef]
29. Kelber, A.; Jonsson, F.; Wallén, R.; Warrant, E.; Kornfeldt, T.; Baird, E. Hornets can fly at night without obvious adaptations of eyes and ocelli. *PLoS ONE* **2011**, *6*, e21892. [CrossRef] [PubMed]
30. Makino, S. Seasonal changes in levels of parasitism and sex ratio of *Xenos moutoni* du Buysson (Strepsiptera, Stylopidae) in the Japanese hornet, *Vespa analis insularis* Dalla Torre (Hymenoptera, Vespidae), collected with attractant traps. *Tijdschr. Voor Entomol.* **2001**, *144*, 217–222. [CrossRef]
31. Makino, S.; Sayama, K. Species compositions of vespine wasps collected with bait traps in recreation forests in northern and central Japan (Insecta, Hymenoptera, Vespidae). *Bull. Forestry For. Prod. Res. Inst.* **2005**, *4*, 283–289.
32. Oyaizu, W.; Kudô, K. Seasonal changes in the number of vespine wasps and levels of parasitism by *Xenos moutoni* (Strepsiptera, Stylopidae) collected with attractant traps in Matsunoyama forest, Tokamachi city, Japan. *Bull. Facul. Educat. Nat. Sci. Niigata Univ.* **2013**, *6*, 49–57.
33. Kudô, K.; Oyaizu, W.; Kusama, R.; Yamagishi, K.; Yamaguchi, Y.; Koji, S. Yearly and seasonal changes in species composition of hornets (Hymenoptera: Vespidae) caught with bait traps on the Sea of Japan coast. *Far East. Entomol.* **2021**, *426*, 10–18. [CrossRef]
34. Lioy, S.; Laurino, D.; Capello, M.; Romano, A.; Manino, A.; Porporato, M. Effectiveness and selectiveness of traps and baits for catching the invasive hornet *Vespa velutina*. *Insects* **2020**, *11*, 706. [CrossRef]
35. Sayama, K.; Kosaka, H.; Makino, S. The first record of infection and sterilization by the nematode *Sphaerularia* in hornets (Hymenoptera, Vespidae, *Vespa*). *Insectes Soc.* **2007**, *54*, 53–55. [CrossRef]
36. Kosaka, H.; Sayama, K.; Kanzaki, N.; Takahata, Y.; Makino, S. Host range and geographical distribution of *Sphaerularia vespae*, the nematode parasite of queen hornets. In Proceedings of the 7th International Congress of Nematology, Antibes Juan-les-Pins, Antibes, France, 1–6 May 2022.
37. Nakase, Y.; Kato, M. Cryptic diversity and host specificity in giant *Xenos* strepsipterans parasitic in large *Vespa* hornets. *Zool. Sci.* **2013**, *30*, 331–336. [CrossRef]
38. Werren, J.H. Sex ratio adaptations to local mate competition in a parasitic wasp. *Science* **1980**, *208*, 1157–1159. [CrossRef]
39. Poulin, R. Population abundance and sex ratio in dioecious helminth parasites. *Oecologia* **1997**, *111*, 375–380. [CrossRef]
40. Carapelli, A.; Vannini, L.; Nardi, F.; Boore, J.L.; Beani, L.; Dallai, R.; Frati, F. The mitochondrial genome of the entomophagous endoparasite *Xenos vesparum* (Insecta: Strepsiptera). *Gene* **2006**, *376*, 248–259. [CrossRef] [PubMed]
41. Geffre, A.C.; Liu, R.; Manfredini, F.; Beani, L.; Kathirithamby, J.; Grozinger, C.M.; Toth, A.L. Transcriptomics of an extended phenotype: Parasite manipulation of wasp social behaviour shifts expression of caste-related genes. *Proc. R. Soc. B* **2017**, *284*, 20170029. [CrossRef] [PubMed]

insects

MDPI

Article

A revision of the *Encarsia mexicana* species-group (=*Dirphys* Howard) (Hymenoptera: Aphelinidae), gregarious endoparasitoids of whiteflies (Hemiptera: Aleyrodidae) in the Neotropical Region †

Andrew Polaszek [1,*], Estrella Hernández-Suárez [2], Robert L. Kresslein [3], Paul Hanson [4], Yvonne M. Linton [5], Jacqueline MacKenzie-Dodds [6] and Stefan Schmidt [7]

1 Insects Division, Natural History Museum, London SW7 5BD, UK
2 Unidad Protección Vegetal, Instituto Canario de Investigaciones Agrarias, 38270 San Cristóbal de La Laguna, Spain
3 Department of Entomology, University of California, Riverside, CA 92521, USA
4 Escuela de Biología, Universidad de Costa Rica, San Jose 11501-2060, Costa Rica
5 Smithsonian Institution, Museum Support Center, Walter Reed Biosystematics Unit, Suitland, MD 20746, USA
6 Molecular Collections Facility, Natural History Museum, London SW7 5BD, UK
7 SNSB-Zoologische Staatssammlung München, 81247 Munich, Germany
* Correspondence: a.polaszek@nhm.ac.uk
† urn:lsid:zoobank.org:pub:2CE58923-A39A-412A-896E-DCFD4CC01FD7.

Simple Summary: *Encarsia* (Family Aphelinidae) is a genus of minute parasitoid wasps that target a diversity of agricultural pest insects including whiteflies and armoured scale insects. Since the genus was described in 1878, 450+ species of *Encarsia* have been described. Historically, it has been difficult to provide a subgeneric classification for the species of *Encarsia*. As a result, researchers have divided the group into numerous informal species-groups. Our work uses an alignment of ribosomal DNA sequences (sequences that code for ribosomal RNA) to construct a phylogenetic tree that shows that species of the genus *Dirphys* are correctly placed within *Encarsia*. With these results, we establish the *Encarsia mexicana* species-group (for the six species previously placed in *Dirphys*) and describe 14 new species. We also briefly discuss morphological characters that may correspond to the relationships recovered in the molecular phylogeny. With this information, we can better understand the patterns of evolution which brought about the present diversity within *Encarsia* and provide a more accurate classification of the genus.

Abstract: The genus *Dirphys* Howard 1914 **syn. n.** is synonymized with *Encarsia* Förster, and treated as a species-group of *Encarsia*, referred to henceforth as the *Encarsia mexicana* species-group. The monophyly of *Encarsia* is discussed in relation to *Dirphys*. The new synonymy is based on phylogenetic analyses of the nuclear ribosomal 28S-D2 gene region (43 taxa, 510 bp). The *Encarsia mexicana* species-group is recovered as strongly monophyletic within *Encarsia*. All species of the *Encarsia mexicana* species-group are revised. The group includes six previously described species, and fourteen newly described species. All species are described (or redescribed) and illustrated. Detailed distributional data, and, where available, plant associate and host records are provided for all species. *Encarsia myartsevae* Kresslein and Polaszek **nom. nov.** is here proposed as a replacement name for *Encarsia mexicana* Myartseva, now preoccupied by *Encarsia mexicana* (Howard). A dichotomous identification key, supplemented by an online multiple-entry key, is provided for all species.

Keywords: Aleyrodidae; Aleurodicinae; parasitoid; biological control; new world

1. Introduction

The genus *Dirphys* was initially described by Howard [1] for *Mesidia mexicana* Howard [2]. Where known, species in this genus are primary endoparasitoids of Aleyrodidae [3–5] and are gregarious, with up to 16 developing in a single host [3,4]. This behavior is unknown in any other chalcid parasitoids of whiteflies. *Dirphys* has been regarded as occupying a transitional zone between *Coccophagus* Westwood and *Encarsia* Förster [3] due to its having intermediate characters of those genera, especially with regard to setation of the mesoscutal mid lobe—more setose than most *Encarsia*, less setose than most *Coccophagus*. However, it displays a unique morphological synapomorphy of the sculpture of the dorsal mesosoma, which is always markedly rugose, irrespective of whether the pattern is aciculate (Figures 6E and 11E), transverse (Figure 10E) or longitudinal (Figures 4C and 20E), or contains combined elements of these patterns (Figures 12E and 13E). Importantly, reticulate mesosomal sculpture is unknown in *Dirphys*. A second apparent autapomorphy concerns the division of the mesoscutal side lobes (see e.g., Figures 4C, 6E, 18E and 23E). These apparent autapomorphies notwithstanding, analyses of the 28S-D2 ribosomal DNA (Kresslein et al. unpublished), and loci recovered with Anchored Hybrid Enrichment (Cruaud et al. unpublished, Kresslein et al. unpublished) show *Dirphys* nested within an otherwise monophyletic *Encarsia*. Further confusion about the relationship between *Dirphys* and *Encarsia* arose with the description of *Encarsiella* Hayat [6], which bears a strong superficial resemblance to *Dirphys* and was at one time synonymized with it [7,8]. *Encarsiella* was synonymized with *Encarsia* by Shafee and Rizvi [9] and is here regarded as the *Encarsia noyesi* species-group [10,11].

A preliminary study into phylogenetic relationships within the subfamily Coccophaginae was undertaken by Polaszek and Hayat [3] based on 24 morphological characters. In that work, the monophyly of *Dirphys* was supported by a single synapomorphy, the mesoscutal and scutellar sculpture. Another character supporting the monophyly of *Dirphys* was the proximity of the scutellar sensilla, although this character was known to have evolved independently many times within *Encarsia* [12]. In the same work, Polaszek and Hayat revised the species of *Dirphys* known at that time, describing three new species, *D. diablejo* Polaszek and Hayat, *D. encantadora* Polaszek and Hayat, and *D. mendesi* Polaszek and Hayat. Chavez [4] described a fifth species, *D. larensis* Chavez from Venezuela, and Polaszek added a sixth, *D. aphania* Polaszek [5].

In the present manuscript, we synonymize *Dirphys* (hereinafter referred to as the *mexicana* species-group) with *Encarsia*. Using maximum likelihood analysis of 28s D2 rDNA (43 taxa, 510 bp), we recover the *Encarsia mexicana* species-group as strongly monophyletic within *Encarsia*. We provide a comprehensive revision of the known species of the *Encarsia mexicana* species-group with species description (or redescription), illustrations, distributional data, and, where available, plant associate and host records. Fourteen species are described here as new: *Encarsia acusa* Polaszek and Hernández-Suárez **sp. n.**, *Encarsia aisha* Polaszek and Hernández-Suárez **sp. n.**, *Encarsia avida* Polaszek and Hernández-Suárez **sp. n.**, *Encarsia catula* Polaszek and Hernández-Suárez **sp. n.**, *Encarsia cylindrica* Polaszek and Hernández-Suárez **sp. n.**, *Encarsia dichaeta* Polaszek and Hernández-Suárez **sp. n.**, *Encarsia erwini* Polaszek and Hernández-Suárez **sp. n.**, *Encarsia fredbennetti* Polaszek and Hernández-Suárez **sp. n.**, *Encarsia inbioa* Polaszek and Hernández-Suárez **sp. n.**, *Encarsia napo* Polaszek and Hernández-Suárez **sp. n.**, *Encarsia marynoyesae* Polaszek and Hernández-Suárez **sp. n.**, *Encarsia noora* Polaszek and Hernández-Suárez **sp. n.**, *Encarsia svetlana* Polaszek and Hernández-Suárez **sp. n.**, and *Encarsia venia* Polaszek and Hernández-Suárez **sp. n.** The name, *Encarsia myartsevae* Kresslein and Polaszek **nom. nov.** is here proposed as a replacement name for *Encarsia mexicana* Myartseva [13], now preoccupied by *Encarsia mexicana* (Howard). We also provide a dichotomous identification key, supplemented by an online multiple-entry key for all species of the *Encarsia mexicana* species-group.

2. Materials and Methods

2.1. Specimen Depositories: Abbreviations

Material examined as a part of this investigation is deposited at the following institutions.

NHMUK: Natural History Museum, London, UK.

UCRC: University of California, Riverside, USA.

USNM: National Museum of Natural History, Smithsonian, Washington D.C., USA.

MZUCR: Museo de Zoología Universidad de Costa Rica.

2.2. Morphological Study

Populations of the *Encarsia mexicana* species-group were studied from different localities (Table 1). Host-reared material was collected in Costa Rica, Ecuador, Mexico, and Trinidad and Tobago, as part of intensive foreign exploration efforts to search for parasitoids of whitefly pests (Hemiptera: Aleyrodidae), mostly in the subfamily Aleurodicinae. For morphological analysis, female specimens were mounted on microscope slides following Noyes [14] with some modifications as follows: no maceration in 10% KOH was needed after DNA extraction. Specimens were washed in distilled water for one hour and then dehydrated for 5 min in graded ethanol of the following concentrations: 35%, 70%, 85%, and 100%. After clearing in clove oil and allowing alcohol evaporation, specimens were dissected in Canada balsam. The wings, antennae, head, and remaining body parts were mounted separately on a single slide.

In total, 110 females and 4 males of 20 species were examined, including the extensive recording of measurements and ratios. Males are rare or unknown for most species and were not therefore included, except for *Encarsia diablejo* which is known only from the male. Measurements were taken with a Leitz Dialux 20EB microscope from slide-mounted material following Heraty and Polaszek [12] with the following five measurements added: scape length, pedicel length, submarginal vein length, marginal vein length and length of the mid basitarsus. (Figure 1). All measurements of antennae, fore wings, and legs refer to the maximum length of the structure in lateral view. The terminology of morphological characters follows Kim and Heraty [15] and Hayat [16].

Specimens were imaged using a Leitz Dialux 20EB (Wetzlar, Germany) compound microscope using Nomarski Differential Interference Contrast illumination (DIC) and photographed with a MicroPublisher 5.0 RTV (QImaging, Surrey, Canada) camera. Additional images (claval sensorial area; mandibles) were imaged with an Olympus BX63 microscope (Olympus, Tokyo, Japan) also utilizing DIC. Scanned sections were stacked and combined using Synoptics AutoMontage Pro® ver. 5.03 software (Leitz Dialux images) and Helicon Focus software (Olympus BX63 images). The final images were edited with Adobe Photoshop CC®.

2.3. DNA Extraction, Amplification, and Sequencing

Genomic DNA was extracted from single, whole specimens using a non-destructive genomic DNA extraction protocol developed by Chao-Dong Zhu, John Noyes, and others at the Natural History Museum, London [17]. Occasionally 2–3 specimens were extracted together when known to be conspecific.

Specimens were softened in 70% ethanol (to reduce potential damage during subsequent steps) at room temperature for a minimum of 2 h. Seventy percent ethanol was removed carefully by pipette and specimens were allowed to air-dry briefly. DNA was extracted using the Qiagen DNeasy Blood and Tissue Kit (250) 69506 (Qiagen, Hilden, Germany). Specimens were immersed in 180 µL of Lysis Buffer ATL, premixed with 20 µL Proteinase K and incubated at 55 °C overnight (8 h minimum) with no mixing, taking care that the specimen was submerged/floating in the buffer and not adhered to the side of the tube.

After digestion, the lysis buffer was carefully removed by pipette into a clean 1.5 mL microfuge tube. The specimen was immediately washed by adding 500 µL distilled water for a minimum of 30 min, then replaced with 500 µL 70% ethanol for a minimum of 30 min, then finally stored in 100% ethanol until slide-mounted in Canada balsam.

DNA was extracted from the lysis buffer using the Qiagen QUIA quick PCR Purification Kit (250) 28106 following the protocol: 'Isolation of total DNA from Animal Tissues' (step 3 onwards). Standard PCR reactions were then carried out in a thermal cycler using 2.0 µL DNA extract, Taq buffer (1.5 mM MgCl2), 1.5 U Taq polymerase (Roche), 10 nmol dNTPs (Amersham Pharmacia Biotech; APB) and 20 mol of each primer at the Natural History Museum's DNA sequencing facility.

Table 1. Ingroup taxa (species identity, voucher IDs, accession numbers, locality, plant associate and host).

Species	Dna Code	Accesion Number	Locality	Plant Associate	Host(s)
Encarsia acusa	DNA0148 DNA0212	OQ683554	Costa Rica: Heredia, Est. Biol. La Selva Peru: Loreto, Iquitos, Barillal		
Encarsia aisha	DNA0146 DNA0164	OQ683562 OQ683562	Costa Rica: Heredia, Est. Biol. La Selva Costa Rica: Alajuela, Est. Caribe		
Encarsia aphania	DNA0218 DNA0213	OQ683546 OQ683545	Belize: Cayo, Las Cuevas Belize: Cayo, Las Cuevas Belize: Cayo, Chiquibul Costa Rica: Puntarenas	*Inga* sp.	*Aleurodicus pulvinatus* *Azuraleurodicus pentarthus* *Nealeurodicus altissimus*
Encarsia avida	DNA0143	OQ683547	Costa Rica: Heredia, Est. Biol. La Selva		
Encarsia catula	DNA0278	OQ683547	Costa Rica: Limon, Hitoi-Cerere		
Encarsia cylindrica	 DNA0209 DNA0211 DNA0149		Brazil: Minas Gerais, Vicosa, Costa Rica: Puntarenas RF, Piedras Blancas Costa Rica: San Juan, Ciudad Colon Costa Rica: Heredia, Est. Biol. La Selva Jamaica: Fair Prospect	*Citrus* sp.	*Aleurothrixus floccosus* (?) *Aleurodicus jamaicensis*
Encarsia diablejo			Peru: Loreto, Iquitos		
Encarsia dichaeta	 DNA0132-0135 DNA0133-0136 DNA0208 DNA0151	 OQ683550 OQ683552 OQ683551 OQ683549	Brazil: Bahia Costa Rica: Alajuela, P.N. Arenal, Pilon Costa Rica: Alajuela, P.N. Arenal, Pilon Costa Rica: Guanacaste, Pitilla Costa Rica: Heredia, Est. Biol. La Selva Ecuador: Napo, Anangucocha		*Aleurodicus flavus* *Aleurodicus* sp.
Encarsia encantadora			Ecuador: Napo River Mexico: Tabasco	*Lippia myriocephala*	*Nealeurodicus altissimus*
Encarsia erwini			Ecuador: Napo River		
	DNA0216		Trinidad and Tobago: Trinidad, Mount St Benedict		
Encarsia fredbennetti	DNA0215B	OQ683559	Trinidad and Tobago: Trinidad, St Augustine	*Theobroma cacao*	Aleurodicinae
Encarsia inbioa	DNA0128	OQ683553	Costa Rica: Alajuela, P.N. Arenal, Pilon		
Encarsia larensis			Venezuela: Cabudare, Lara	*Hura crepitans*	*Aleurodicus pulvinatus*
Encarsia marynoyesae	DNA0163 DNA0167	OQ683563	Costa Rica: Alajuela, Est. Caribe Costa Rica: Alajuela, Est. Caribe		
Encarsia mendesi			Brazil: São Paolo, Mogi-Guazu	*Bauhinia holophylla*	*Aleurodicus maritimus*
Encarsia mexicana	 DNA0144 DNA0166 DNA0129	 OQ683560	Mexico: Tabasco, San Francisco del Peal Costa Rica: Heredia, Est. Biol. La Selva Costa Rica: Limon Costa Rica: Alajuela, Est. Caribe R. Costa Rica: Alajuela, P.N. Arenal, Pilon	*Lippia myriocephala*	*Nealeurodicus altissimus*
Encarsia napo			Ecuador: Napo River, Camp. Res. Waorani		
Encarsia noora	DNA0126	OQ683561	Costa Rica: Limon		
Encarsia svetlana	DNA0305	OQ683558	Guyana: Dubulay Ranch		
Encarsia venia	DNA0298 DNA0267	OQ683557	Costa Rica: Limon, Parque Nacional Cahuita Costa Rica: Limon, Hitoy-Cerere Costa Rica: Heredia La Selva		
E. mexicana-group sp.	DNA0165 D2672	OQ683556 OQ683555	Costa Rica: CR Alajuela Est. Caribe R. Rincon Forestal Ecuador: Orellana, Tiputini Biodiversity Sta.		

The D2 region of 28S rDNA was amplified using the following primers:

28SFW 5′-AGTACCGTGAGGGAAAGTTG-3′
28SRev 5′-TTGGTCCGTGTTTCAAGACGG 3′

PCR conditions were as follows: an initial denaturation of 94 °C for 3 min, then 35 cycles of denaturation at 94 °C for 1 min, annealing at 50 °C for 1 min, and extension at 72 °C for 2 min, followed by a final extension at 75 °C for 10 min, then samples were held at 4 °C until they could be analyzed. PCR products were run on a 1% agarose gel to confirm PCR success (clean bands of the expected size), then the remaining products were cleaned and sequenced. Removal of dye terminators was done by ethanol precipitation prior to sequencing.

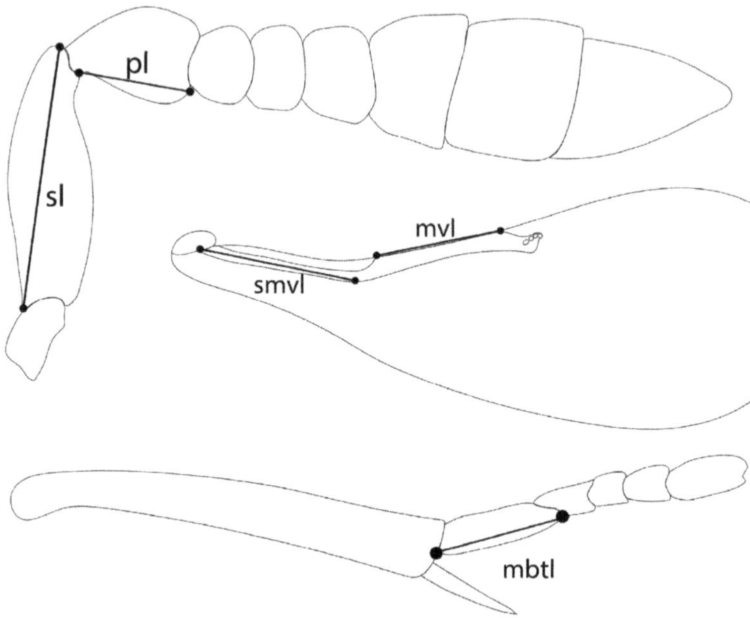

Figure 1. Additional measurements used in the present study: scape length (sl); pedicel length (pl); submarginal vein length (smvl); marginal vein length (mvl); mid basitarsus length (mbtl). Illustrated from type material of *Encarsia aisha*.

The DNA analyzer system ABI PRISM 3730 and 377 DNA sequencer were used, the samples were loaded onto the system's vertical polyacrylamide gel where they underwent electrophoresis, laser detection and computer analysis. Sequence editing and alignment were performed using Sequencher TM 4.8 (Gene Codes Corporation, Ann Arbor, USA) on a Macintosh computer. The resulting molecular dataset includes 18 sequences representing 13 species. Sequences have been deposited in the GenBank database under accession numbers OQ683545–OQ683576 (Tables 1 and 2).

2.4. Phylogenetic Analyses

Captured sequences were combined with previously published sequence data (Table 2) and aligned using the E-INS-I algorithm in MAFFT v7.490 [18]. Ten independent iterations of maximum likelihood were reconstructed using IQ-TREE version 2.0.7 [19], implementing a General Time Reversible model with invariant sites and gamma distributed rate variation (-m GTR + I + G). Bootstrap support was estimated from 1000 bootstrap trees constructed using ultrafast bootstrapping (-b 1000) [20]. Outgroups are comprised of a broad range of *Encarsia* species, a diversity of recognized species-groups, as well as *Coccophagus* Westwood (Aphelinidae: Coccophaginae) and *Aphytis* Howard (Aphelinidae: Aphelininae).

2.5. Nomenclatural Acts

The electronic edition of this article conforms to the requirements of the amended International Code of Zoological Nomenclature, and hence the new names contained herein are available under that Code from the electronic edition of this article. This published work and the nomenclatural acts it contains have been registered in ZooBank, the online registration system for ICZN. The ZooBank LSIDs (Life Science Identifiers) can be resolved, and the associated information can be viewed through any standard web browser by appending the LSID to the prefix "http://zoobank.org/ (accessed on 6 June 2023)". The LSID for this publication is urn:lsid:zoobank.org:pub:2CE58923-A39A-412A-896E-DCFD4CC01FD7.

The electronic edition of this work was published in a journal with an ISSN and has been archived and is available from the following digital repositories: PubMed Central, LOCKSS.

Table 2. Outgroup taxa (species identity, voucher IDs, accession numbers, and locality).

Species	DNA Code	Accesion Number	Locality
Encarsia luteola	D0243	AF223369	USA: California, Brawley
Encarsia formosa	D0231	AF223372	Egypt
Encarsia cubensis	DNA270	OQ683567	Costa Rica: Limon
Encarsia azimi	DNA259P17	AF254229	Australia: Queensland
Encarsia inaron	D0465	AY599399	New Zealand
Encarsia lounsburyi	DNA017	OQ683568	Costa Rica: Puntarenas
Encarsia citrina	DNA376	OQ683569	U.K.: London, Barnes Common
Encarsia boswelli	DNAAE534	OQ683570	India
Encarsia perplexa	D0296	AF254243	Guatemala: Coatepeque
Encarsia opulenta	DNA387	OQ683571	Mexico: Los Tuxtlas
Encarsia lutea	D0235	AF254238	Cyprus
E. noyesi group sp.	DNA0091	OQ683566	Costa Rica
Encarsia tamaulipeca	DNA0123	OQ683564	Ecuador
E. noyesi group sp.	DNA0089	OQ683565	Ecuador: Napo
Encarsia sophia	D0219	AF254198	Find in Heraty Lab
Encarsia protransvena	D0136	AF254208	USA: California, Orange Co.
Coccophagus sp.	DNA010	OQ683572	Costa Rica
Coccophagus sp.	DNA0185	OQ683573	Costa Rica: La Selva
Coccophagus lycimnia	DNAA1-006A	OQ683574	Costa Rica
Coccophagus semicircularis	DNAA3-023D	OQ683575	Costa Rica: Puntarenas
Coccophagus sp.	DNA034	OQ683576	Costa Rica: Puntarenas
Aphytis yanonensis	D0446	AY635336	UCR Culture: Originally from Japan, Fukuoka
Aphytis melinus	D0445	AY635342	UCR Culture: Originally from China, Fujian, Fuzhou

3. Results

3.1. Phylogenetic Analysis of Molecular Data

A maximum likelihood tree was constructed from partial sequences of 28S D2 ribosomal DNA of 13 species (from 18 specimens) and 23 outgroup taxa (Figure 2). The *Encarsia mexicana* species-group was recovered as a strongly supported clade within *Encarsia. Encarsia dichaeta* forms the sister clade to all remaining *E. mexicana*-group species. The *Encarsia mexicana* species-group was not placed sister to the *noyesi* species-group; however, backbone support in the recovered phylogeny is insufficient to confidently resolve inter- and intra-species-group relationships.

3.2. Taxonomy of the Encarsia mexicana Species-Group

Encarsia mexicana **species-group**

= ***Dirphys*** **Howard, 1914**

Etymology of *Dirphys*. *Dirphys* (Διρφυς) is a Greek feminine noun. Hence the modification by Hayat of Howard's (1914) combination *Dirphys mexicana* (Howard) to *Dirphys mexicanus* (Howard) was an unjustified emendation [21]. Hayat attributed the new combination to Howard [1], but this is not the case.

Diagnosis: head dorsally transverse. Frontovertex at narrowest wider than dorsal eye width. Facial lines evident, often broadly expanded; mediofrontal and transfacial lines developed. Eyes with evident setae. Mandibles usually with two teeth and a truncation (Figure 10A), the truncation sometimes reduced, and the teeth often strongly developed so mandibles appear to have only two teeth (Figure 22A). A bidentate upper tooth may be present in addition to the well-developed ventral tooth (Figure 4A).

Maxillary palps two-segmented. Antenna eight-segmented in both sexes, antennal formula variable (1, 1, 3, 3 or 1, 1, 0, 6), claval sensorial complex present (Figure 17B) or absent, suture between F5 perpendicular or oblique. Pronotum medially membranous. Mesoscutum with more than 20 setae. Side lobes divided (Figure 23E). Axillae large, strongly projecting forwards and separated medially by less than the maximum length of one axilla. Each axilla usually with a single seta, in *E. dichaeta* with two setae (Figure 11E). Thoracic sculpture aciculate, longitudinal, transverse or a combination of these types, never reticulate, always strongly rugose. Mesoscutellar sensilla close together, separated by about the width of one sensillum. Fore wings with two large setae on the submarginal vein, plus a variable number of smaller setae at the distal end of the submarginal vein. Linea calva present or absent. Mid basitarsus with a variable number of robust, spine-like setae, tarsi five-segmented.

Remarks: The *Encarsia mexicana* species-group (Figure 3) is restricted to the Neotropical zone, with species reaching as far south as the State of Bahia (Brazil), and as far north as southern Mexico.

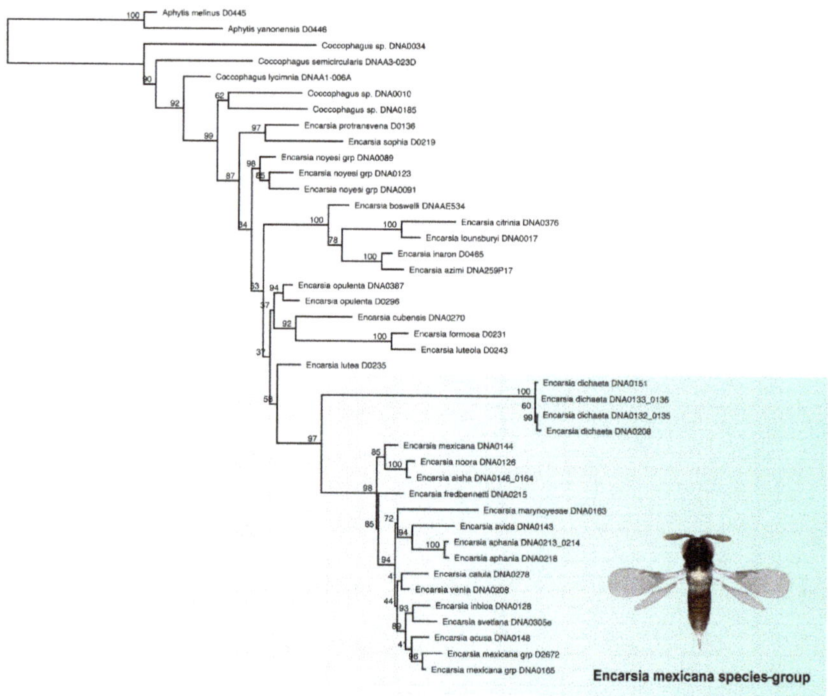

Figure 2. Maximum likelihood tree (IQ-TREE 2) based on 28S D2 ribosomal DNA (509 bp) from 42 taxa (19 ingroups, 23 outgroups); support values from 1000 ultrafast bootstrap replicates.

Figure 3. Lateral habitus of a female of the *Encarsia mexicana* species-group.

3.3. *Species Descriptions*

3.3.1. *Encarsia acusa* Polaszek and Hernández-Suárez **sp. n.**

(Figure 4A–F)

urn:lsid:zoobank.org:act:F179717C-0FF0-4189-90BE-4851A8625431

Female. Colour: Antennae light brown; radicle, scape, base of pedicel and F6 darker. Head dark brown, paler along the sutures and frons. Mesosoma and metasoma dark brown with posterior 80% of the mesoscutellum and sides of metanotum yellow. Legs yellow with most of the mid and hind femora brown, fore femora and tibiae brown, all tarsi pale. Fore wings hyaline, slightly infuscate below marginal vein, submarginal and marginal veins dark.

Morphology: Head (Figure 4A) with mediofrontal line complete; transfacial line obscure; facial lines narrow. Scrobes with longitudinal aciculate sculpture. Antennal formula (Figure 4D): 1, 1, 3, 3; scape 2.4x pedicel length; pedicel 1.9x F1; F1 0.8x F2; F2 equal to F3; funicle 0.56x clava; F6 slightly oblique, claval sensorial area present. Flagellum with the following number of longitudinal sensilla: F1: 0; F2: 1; F3: 1; F4: 3; F5: 4; F6: 3–4 (both counts present in holotype). Mandibles (Figure 4A) with two teeth and a broad truncation. Maxillary palps two-segmented. Mid-lobe of mesoscutum (Figure 4C) with 34–40 setae; each lateral lobe with two setae; each axilla with one seta; scutellum with four setae. Sculpture of mesoscutum longitudinal; sculpture of axillae and scutellum longitudinal. Fore wing (Figure 4E) with two large setae and 2–3 smaller setae on the submarginal vein, three setae in the basal cell, 6–7 setae on the anterior margin of the marginal vein, and one seta at the junction of the submarginal vein and parastigma. Linea calva present. Submarginal 0.75x marginal vein. Maximum length of fore wing 2.7x fore wing width; maximum width of wing 5.3x longest setae on marginal fringe. Ovipositor (Figure 4B) 1.85x mid tibial length; third valvulae 0.44x ovipositor length; second valvifer 1.3x third valvula. Mid tibial spur (Figure 4F) 0.86x corresponding basitarsus. Metasomal terga T1–T7 with 0, 2 + 2, 2 + 2, 1 + 1, 1 + 2 + 1, 1 + 2 + 1 and four setae, respectively. T7 (Figure 4B) extremely extended, almost covering the ovipositor.

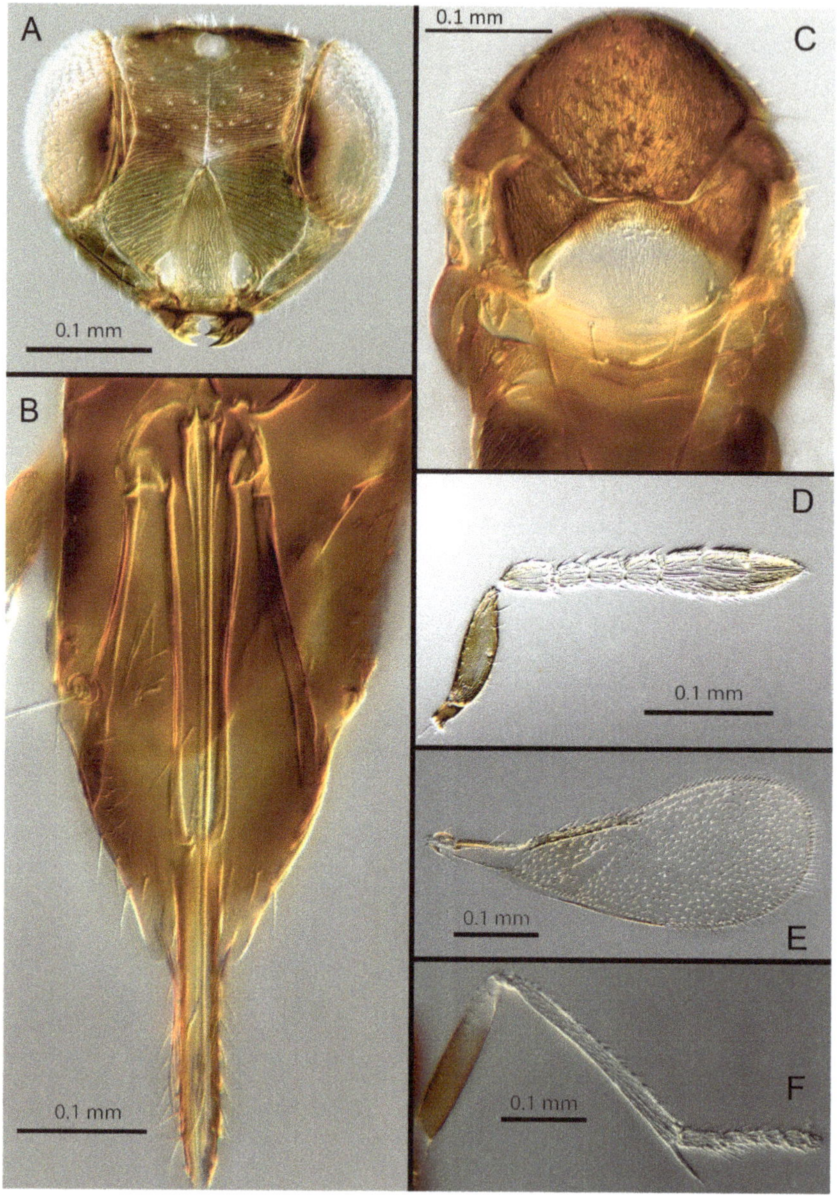

Figure 4. ***Encarsia acusa***: (**A**) head; (**B**) ovipositor; (**C**) dorsal mesosoma; (**D**) antenna; (**E**) fore wing; (**F**) mid leg.

Distribution. COSTA RICA: Heredia, Limon; PERU: Iquitos.

Material examined: Holotype ♀ COSTA RICA, Heredia Estación Biológica La Selva, 75 m, 10°26′ N 84°01′ W 27–28.ii.2003 (J.S. Noyes) [DNA148: OQ683554] (NHMUK). Paratypes: 1♀ COSTA RICA, Limon RB, Hitoy-Cerere 100 m, 14–19.i.1991 (J.S. Noyes) (MZUCR). 1♀ PERU, Iquitos, Barillal, 10.ii.1984 (L. Huggert BM 1984.337) [DNA 212] (NHMUK).

Remarks: T7 extremely extended, covering the ovipositor. *Encarsia acusa* appears to be most closely related to *E. inbioa* and *E. svetlana*, but is easily distinguished from those (and all other) species by the extremely long ovipositor and T7. DNA sequences from holotype deposited under GenBank accession number: OQ683554.

Etymology. From "acus" Latin for needle or pin, referring to the elongated T7.

3.3.2. *Encarsia aisha* Polaszek and Hernández-Suárez **sp. n.**

(Figure 5A–F)

Female. Colour: Antennae brown with F1 and F2 paler. Head dark brown, paler along the sutures. Mesosoma uniformly dark brown. Legs yellow with mid and hind femora, coxae and anterior third of mid tibiae brown, all tarsi pale. Wings hyaline, slightly infuscated below marginal vein; submarginal and marginal veins darker in contrast with the paler stigmal vein.

Morphology: Head (Figure 5A) with mediofrontal line complete; transfacial line evident; facial lines very broad along their entire lengths. Scrobes with longitudinally aciculate sculpture. Antennal formula (Figure 5B): 1, 1, 3, 3; scape 2.3x pedicel length; pedicel 2x F1; F1 1.2x F2; F2 0.75x F3; funicle 0.45x clava; F6 oblique, claval sensorial area present. Flagellum with the following number of longitudinal sensilla: F1: 0; F2: 0; F3: 1; F4: 2–3; F5: 4–5; F6: 3. Mandibles (Figure 5A) with one large ventral tooth and a bidentate upper tooth. Maxillary palps two-segmented. Mid-lobe of mesoscutum (Figure 5E) with about 34–40 setae; each lateral lobe with two setae; each axilla with one seta; scutellum with four setae and two apparent vestigial setal bases. Sculpture of mesoscutum aciculate; sculpture of axillae and scutellum longitudinal. Fore wing (Figure 5C) with two large setae and four smaller setae on submarginal vein, four setae in basal cell, 6–7 setae on anterior margin of marginal vein, and one seta at the junction of the submarginal vein and parastigma. Linea calva present. Submarginal 0.85x marginal vein. Maximum length of fore wing 2.47x fore wing width, maximum width of wing 5.72x longest seta on marginal fringe. Ovipositor (Figure 5F) 1.47x mid tibial length; third valvulae 0.42x ovipositor length; second valvifer 1.5x third valvula. Mid tibial spur (Figure 5D) 1.07x corresponding basitarsus. Metasomal terga T1–T7 with 0, 2 + 2, 2 + 2, 2 + 2, 1 + 2 + 1, 1 + 2 + 1 and four setae, respectively. T7 (Figure 5F) is extended, as long as ovipositor.

Distribution: COSTA RICA: Alajuela, Heredia.

Material examined: Holotype ♀ COSTA RICA, Alajuela, Est. Biol. Caribe, R. Rincon Forestal 10°53′ N 85°18′ W 400 m 19–20.ii.2003 (J.S. Noyes) [DNA 164: OQ683562] (NHMUK). 1♀ COSTA RICA, Heredia, Estacion Biologica La Selva, 10°26′ N 84°01′ W. 75 m 27–28.ii.2003 (J.S. Noyes) [DNA 146: OQ683562] (ZMUCR).

Remarks: *Encarsia aisha* is morphologically very similar to *E. marynoyesae* in many respects (though distant to it based on DNA). The species can be distinguished from *E. marynoyesae* by the second valvifers almost 2x (1.8) the third valvulae; while they are 1.5x as long in *E. aisha*. In *E. marynoyesaei* the clava is well over 2x the length of the funicle; in *E. aisha* it is less than 2x as long. DNA sequence from holotype and paratype (pooled extraction) deposited under GenBank accession number: OQ683562.

Etymology: Named for Aisha, daughter of the second author (EHS), and sister to Noora; see *E. noora*, below.

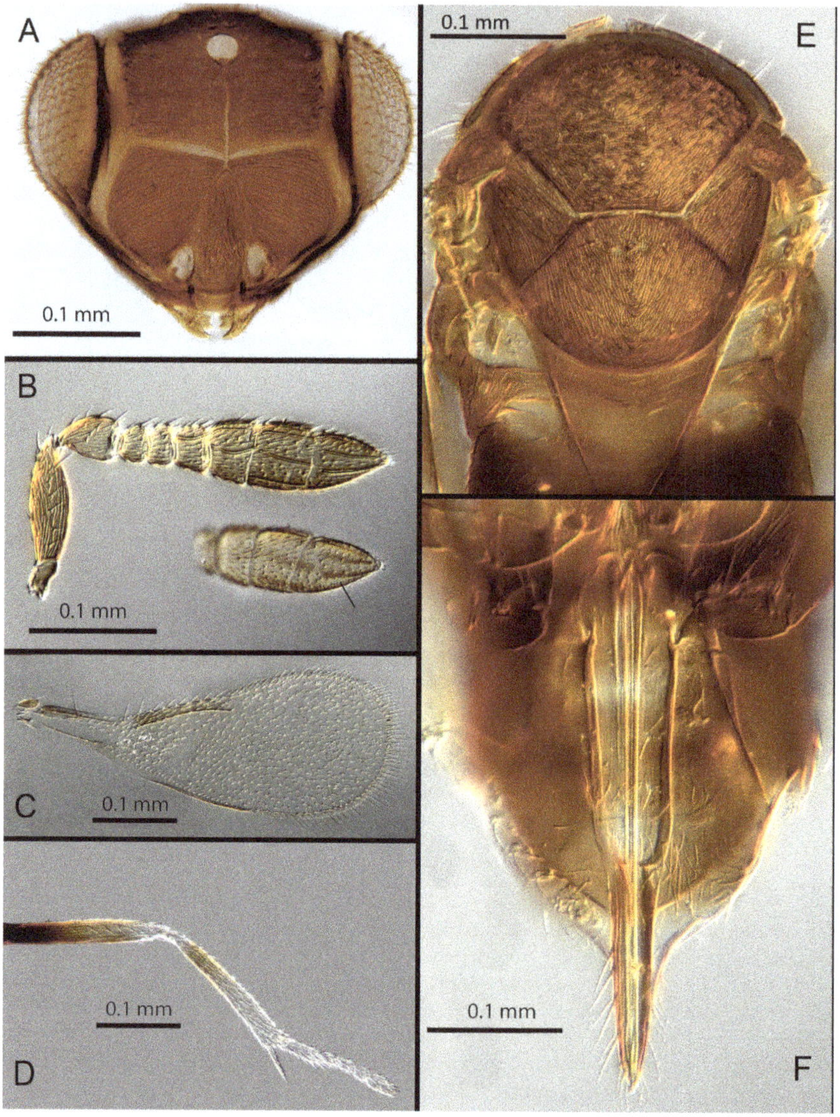

Figure 5. ***Encarsia aisha***: (**A**) head; (**B**) antenna, arrow: claval sensorial area; (**C**) fore wing; (**D**) mid leg; (**E**) dorsal mesosoma; (**F**) ovipositor.

3.3.3. *Encarsia aphania* (Polaszek) 1999 (in Martin and Polaszek, 1999: 1556). **comb. nov.**

(Figure 6A–F)

Female. Colour: Antennae pale brown with scape very dark. Head dark brown with pale lines bordering the eyes and extending along the genae towards clypeus, antennal scrobes, a line from the apex of the scrobes to the median ocellus, and a transverse line midway between the antennal scrobes. Mesosoma and metasoma uniformly dark brown. Legs yellow except all coxae and mid and hind femora which are brown. Fore wings faintly infuscate along the submarginal and the marginal veins; submarginal and marginal veins darker in contrast to the paler stigmal vein.

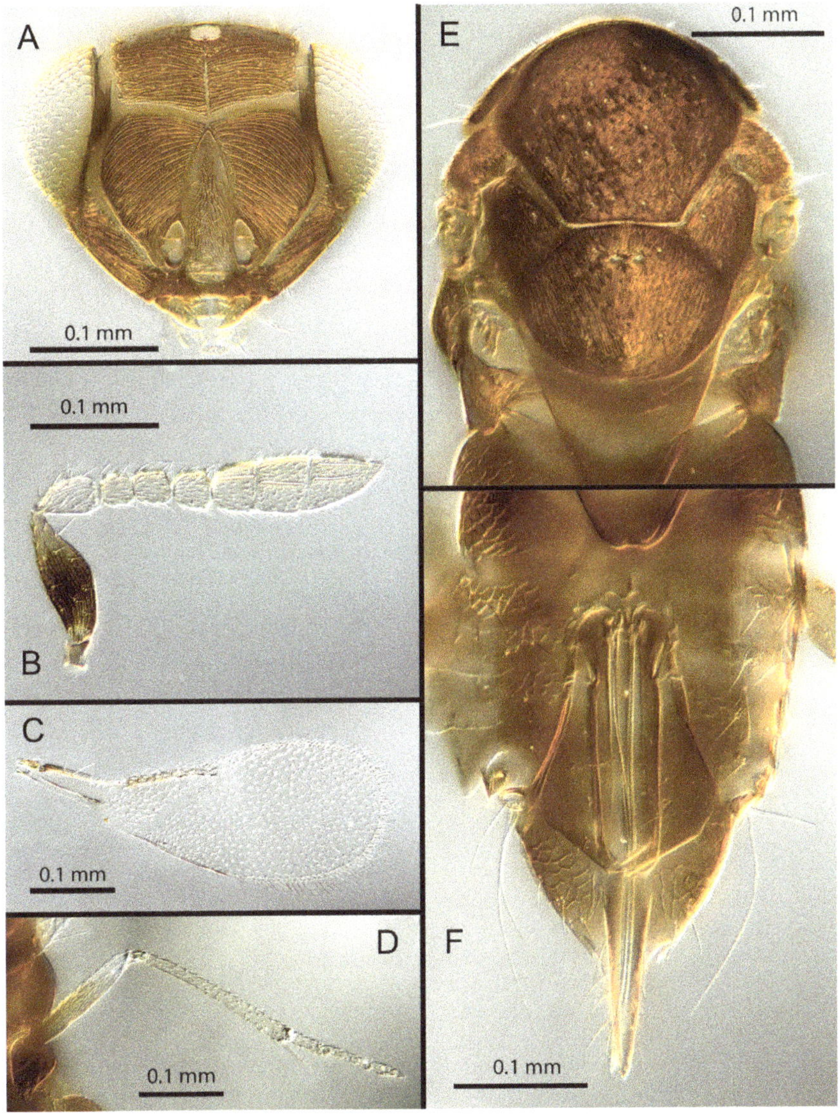

Figure 6. ***Encarsia aphania***: (**A**) head; (**B**) antenna; (**C**) fore wing; (**D**) mid leg; (**E**) dorsal mesosoma; (**F**) ovipositor.

Morphology: Head (Figure 6A) with mediofrontal line complete, though fading towards anterior ocellus; transfacial and facial lines very broad along their entire lengths. Scrobes with longitudinally aciculate sculpture centrally, smooth laterally. Antennal formula (Figure 6B): 1, 1, 3, 3; scape 2.39x pedicel; pedicel 2x F1; F1 0.8x F2; F2 equal to F3; funicle 0.67x clava; F4 and F5 partly fused; F6 broadly oblique; claval sensorial area present, indistinct. Flagellum with the following number of longitudinal sensilla: F1: 0; F2: 1; F3: 1; F4: 2–3; F5: 4; F6: 4. Mandibles (Figure 6A) with one large ventral tooth and a broad upper truncation; Maxillary palps two-segmented. Mid-lobe of mesoscutum (Figure 6E) with 30–40 setae; each lateral lobe with one seta; each axilla with one seta; scutellum with four setae. Sculpture of mesoscutum aciculate; sculpture of axillae and scutellum longitudinal. Fore wing (Figure 6C) with two large setae and 4–5 smaller setae on submarginal vein, 4–5 setae in basal cell, 5–7 setae on anterior margin of marginal vein, and 2–3 setae at the distal part of the base. Linea calva present. Submarginal 0.9x marginal vein. Maximum length of fore wing 2.6x fore wing width; maximum width of fore wing 5.2x longest setae on marginal fringe. Ovipositor (Figure 6F) 1.6x mid tibial length; third valvula 0.4x ovipositor length; second valvifer 1.3x third valvula. Mid tibial spur (Figure 6D) as long as corresponding basitarsus. Metasomal terga T1–T7 with 0, 2 + 2, 2 + 2, 1 + 1, 1 + 2 + 1, 1 + 2 + 1 and 5–6 setae, respectively. T7 (Figure 6F) with a pointed, extended apex covering ovipositor.

Distribution: BELIZE: Cayo District; COSTA RICA: Puntarenas.

Hosts: Aleurodicinae: *Aleurodicus pulvinatus* (Maskell), *Azuraleurodicus pentarthrus* Martin; *Nealeurodicus altissimus* (Hempel).

Material examined: Holotype ♀ BELIZE, Cayo District, Chiquibul Forest Reserve, Las Cuevas-Monkey Tail trail, 5.iii.1996 (J.H. Martin 6747) ex *Azuraleurodicus pentarthus* (NHMUK). 13♀ BELIZE, Cayo Las Cuevas, monkey tail trail, 5.iii.1996 (J.H. Martin 6747) ex *Azuraleurodicus pentarthus* [s27, s22, DNA218: OQ683546]. 1♀ BELIZE, Cayo Chiquibul Fr., Monkey tail trail, 21.iii.2003 (J.H. Martin 7768) ex *Nealeurodicus altissimus* on *Inga* sp. (all NHMUK). 1♀ COSTA RICA, Puntarenas, Est. Altamira send. Los Gigantes, 9.vii.2001 (D. Rubi 63984) [DNA213: OQ683545] 1.460 m, LS 331800 572100 (MZUCR)

Remarks: *Encarsia aphania* presents a unique combination of characters and appears to have no very close relatives. Morphologically it is closest to *E. larensis* but is easily distinguishable by the much longer third valvulae relative to the second valvifers (compare Figures 5F and 15F). DNA sequences were obtained from two specimens from Belize (type locality) and Costa Rica, Puntarenas; deposited under GenBank accession numbers: OQ683546, OQ683545.

3.3.4. *Encarsia avida* Polaszek and Hernández-Suárez **sp. n.**

(Figure 7A–F)

Female. Colour: Antennae pale brown, darker on F5–F6 and the scape, pedicel, and radicle. Head brown, paler along the sutures. Mesosoma dark brown with posterior three-quarters of scutellum pale. Metasoma uniformly dark brown. Legs yellow with all coxae, femora, and anterior half of fore leg tibiae brown; all tarsi pale. Wings infuscate below marginal vein; submarginal and marginal veins darker in contrast with the stigmal vein paler.

Morphology: Head (Figure 7A) with mediofrontal line complete; transfacial line obscure; facial lines present, narrow. Scrobes with longitudinally aciculate sculpture centrally, smooth laterally. Antenna formula (Figure 7B): 1, 1, 3, 3; scape 2.5x pedicel length; pedicel 1.95x F1; F1 0.66x F2; F2 equal to F3; funicle 0.6x clava; F6 perpendicular. Claval sensorial area present, distinct. Flagellum with the following number of longitudinal sensilla: F1: 0; F2: 1–2; F3: 1; F4: 2; F5: 3; F6: 3. Mandibles (Figure 7A) with two small ventral teeth and a truncation. Maxillary palps two-segmented. Mid-lobe of mesoscutum (Figure 7E) with 40–50 setae; each lateral lobe with two setae; each axilla with one seta; scutellum with four setae. Sculpture of mesoscutum aciculate; sculpture of axillae and scutellum longitudinal. Fore wing (Figure 7C) with two large setae and four smaller setae on submarginal vein, four setae in basal cell, eight setae on anterior margin of marginal vein, and one seta at the junction of the submarginal vein and parastigma. Linea calva present. Submarginal 0.67x marginal vein. Maximum length of fore wing 2.8x fore wing width; maximum width of wing 4.7x longest setae on marginal fringe. Ovipositor (Figure 7F) 1.6x mid tibial length; third valvulae 0.4x ovipositor length; second valvifer 1.5x third valvula. Mid tibial spur (Figure 7D) 0.8x corresponding basitarsus. Metasomal terga T1–T7 with 0, 2 + 2, 2 + 2, 2 + 2, 1 + 2 + 1, 1 + 2 + 1 and 4 setae, respectively. T7 (Figure 7F) extended and covering ovipositor (damaged in holotype).

Distribution: COSTA RICA: Heredia.

Material examined: Holotype ♀ COSTA RICA, Heredia, Est. Biol. La Selva, 10°26′ N 84°01′ W 75 m 27–28.ii.2003 (J.S. Noyes) [DNA143: OQ683547] (NHMUK).

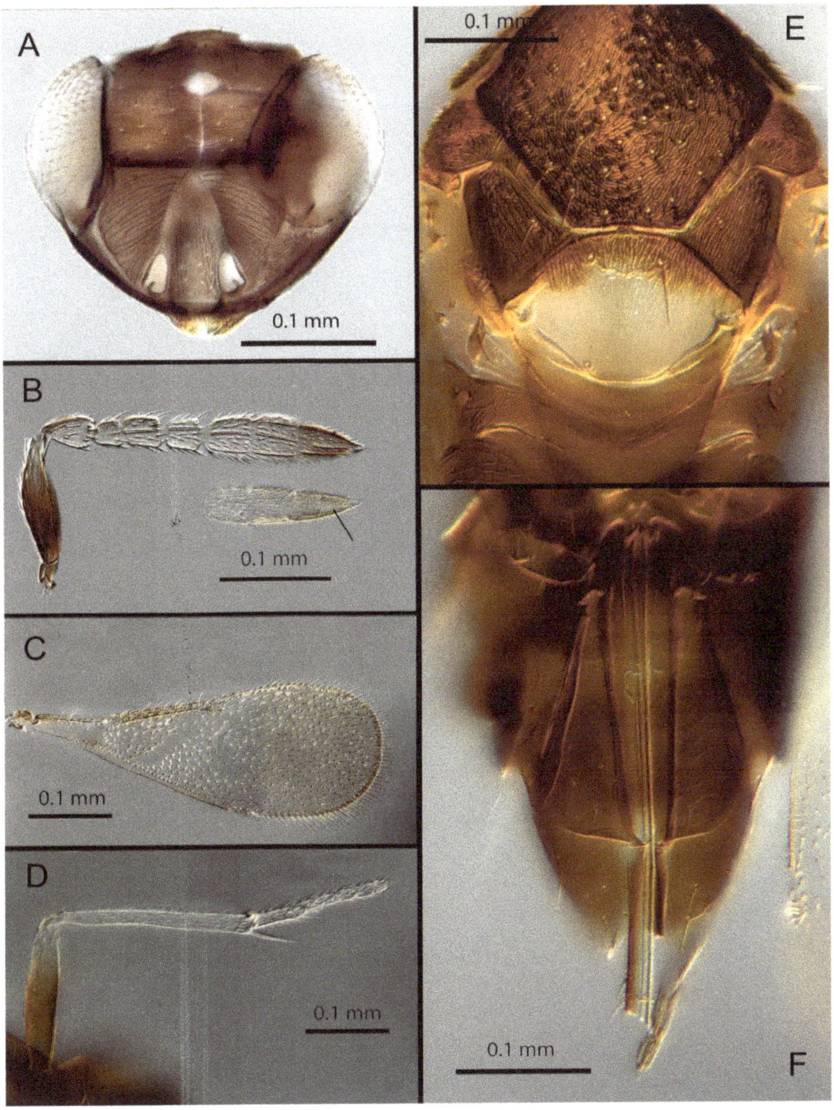

Figure 7. ***Encarsia avida***: (**A**) head; (**B**) antenna, arrow: claval sensory area; (**C**) fore wing; (**D**) midleg; (**E**) dorsal mesosoma; (**F**) ovipositor.

Remarks: *Encarsia avida* appears morphologically close to *E. acusa* with which it shares the color pattern (mesoscutellum anteriorly dark) and wing and antennal morphology.

The ovipositor in *E. acusa* is longer (1.8x mid tibia; 1.6x in *E. avida*). The most easily appreciated difference is in the sculpture of the frons: *E. avida* has scattered, shallow horizontal grooves (Figure 7A) while *E. acusa* has very dense horizontal grooves (Figure 4A). A similar difference in sculpture is evident on the lateral face. The two species are well-separated based on DNA (Figure 1) with *E. avida* coming out as sister to *E. aphania* with high support (95%). DNA sequence from holotype deposited under GenBank accession number: OQ683547.

Etymology: From "avida -us" meaning "greedy" (Latin).

3.3.5. *Encarsia catula* Polaszek and Hernández-Suárez **sp. n.**

(Figure 8A–F)

Female. Colour: Antennae brown. Head dark brown. Mesosoma uniformly dark brown. Legs yellow with mid and hind femora, coxae, and anterior half of tibiae brown, fore leg femora, coxae, and tibiae dark, all tarsi pale. Wings infuscate below marginal vein, submarginal and marginal veins dark, stigmal vein paler.

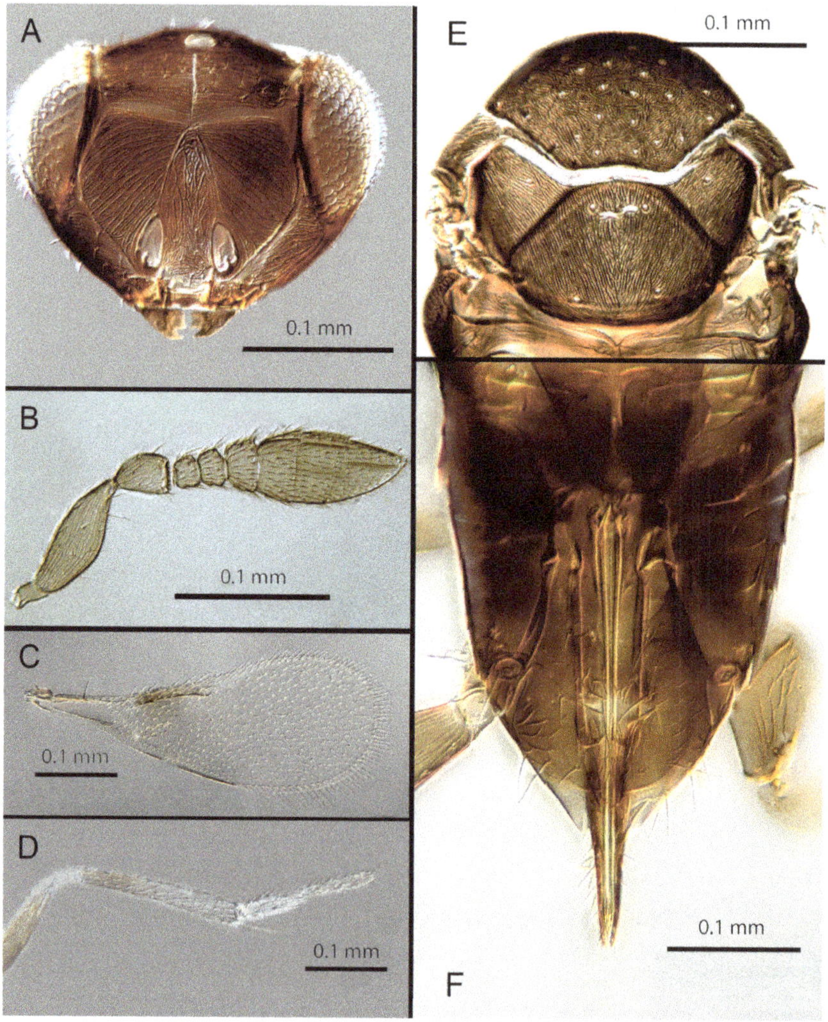

Figure 8. ***Encarsia catula***: (**A**) head; (**B**) antenna; (**C**) fore wing; (**D**) midleg; (**E**) dorsal mesosoma; (**F**) ovipositor.

Morphology: Head (Figure 8A) with mediofrontal line complete; transfacial line broad; facial lines present, narrow. Scrobes with longitudinally aciculate sculpture centrally, irregularly aciculate basally. Antenna formula (Figure 8B): 1, 1, 3, 3; scape expanded, 3.1x pedicel length; pedicel 2x F1; F1 0.9x F2; F2 0.8x F3; funicle 0.48x clava; F6 oblique, claval sensorial area present. Flagellum with the following number of longitudinal sensilla: F1: 0; F2: 0; F3: 1; F4: 3; F5: 3; F6: 3. Mandibles (Figure 8A) with one large ventral tooth and a broad truncation. Maxillary palps two-segmented. Mid-lobe of mesoscutum (Figure 8E) with fewer than 30 setae; each lateral lobe with two setae; each axilla with one seta; scutellum with four setae. Sculpture of mesoscutum aciculate; sculpture of axillae and scutellum longitudinal. Fore wing (Figure 8C) with two large setae and four smaller setae on submarginal vein, four setae in basal cell, six setae on anterior margin of marginal vein, and one seta at the junction of the submarginal vein and parastigma. Linea calva present. Submarginal equal to marginal vein. Maximum length of fore wing 2.38x fore wing width, maximum width of wing 4.64x longest setae on marginal fringe. Ovipositor (Figure 8F) 1.48x mid tibial length; third valvulae 0.46x ovipositor length; third valvulae 0.46x ovipositor length; second valvifer 1.2x third valvula. Mid tibial spur (Figure 8D) 1.1x corresponding basitarsus. Metasomal terga T1–T7 with 0, 2 + 2, 2 + 2, 2 + 2, 1 + 2 + 1, 1 + 2 + 1 and four setae, respectively. T7 (Figure 8F) extended although apparently not covering ovipositor.

Distribution: COSTA RICA: Limon.

Material examined: Holotype ♀ COSTA RICA, Limon, Hitoy-Cerere 90°40′ N 83°02′ W, 21–22.iii.2006 (J.S. Noyes) [DNA 278: OQ683547] (NHMUK).

Remarks: *Encarsia catula* shares aspects of its morphology with *E. marynoyesae*, but can be distinguished by having fewer than 30 setae on the mesoscutum, and third valvulae more than $\frac{1}{2}$ the length of second valvifers (less than $\frac{1}{2}$ as long in *E. marynoyesae*). The two species are relatively close based on DNA (Figure 1). DNA sequence from the holotype is deposited under GenBank accession number: OQ683547.

Etymology: From "catula" meaning dog/whelp (Latin).

3.3.6. *Encarsia cylindrica* Polaszek and Hernández-Suárez **sp. n.**

(Figure 9A–F)

Female. Colour: Antennae pale brown, slightly darker on F6, F1 and the base of the scape, pedicel, and radicle. Head dark brown. Mesosoma uniformly dark brown. Legs yellow with femora and coxae brown, fore legs with dark tibiae, all tarsi pale. Wings infuscate below submarginal vein, marginal and stigmal veins pale.

Morphology: Head (Figure 9A) with mediofrontal line incomplete, extending halfway to anterior ocellus; transfacial line obscure; facial lines very broad along their entire lengths. Scrobes with faint longitudinal sculpture apically, irregular/lateral sculpture basally. Antennal formula (Figure 9B): 1, 1, 3, 3; scape 2.75x pedicel length; pedicel equal to F1; F1 0.85x F2; F2 equal to F3; funicle 0.86x clava; F6 perpendicular. Flagellum with the following number of longitudinal sensilla: F1: 2; F2: 2; F3: 2; F4: 2; F5: 3; F6: 3. Mandibles (Figure 9A) with two small teeth and a broad truncation. Maxillary palps two-segmented. Mid-lobe of mesoscutum (Figure 9E) with about 40–50 setae; each lateral lobe with two setae; each axilla with one seta; scutellum with four setae. Sculpture of mesoscutum aciculate; sculpture of axillae and scutellum longitudinal. Fore wing (Figure 9C) with two large setae on submarginal vein and 11 smaller setae, six setae in basal cell, 11 setae on anterior margin of marginal vein, and one seta at the junction of the submarginal vein and parastigma. Linea calva absent. Submarginal 0.62x times marginal vein. Maximum length of fore wing 2.7x fore wing width, maximum width of wing 5.87x longest setae on marginal fringe. Ovipositor (Figure 9F) equal to mid tibial length; third valvula 0.45x ovipositor length; second valvifer 1.3x third valvula. Mid tibial spur (Figure 9D) 1.1x corresponding basitarsus. Metasomal terga T1–T7 with 0, 1 + 1, 1 + 1, 1 + 1, 1 + 2 + 1, 1 + 2 + 1 and six setae, respectively. T7 (Figure 9F) rounded, not extended but covering ovipositor.

Distribution: BRAZIL: Minas Gerais; COSTA RICA: Puntareñas, San Juan; JAMAICA.

Host: Aleurodicinae: *Aleurodicus jamaicensis* Cockerell.

Material examined: Holotype ♀ COSTA RICA, San Juan, Ciudad Colon, Heredia El Rodeo coll parataxonomist 16.ii.1991 [DNA 211] (NHMUK). Paratype 1♀ COSTA RICA, Puntareñas R.F. Golfo Dulce, 24 km W. Piedras Blancas [DNA 209] (MZUCR). 4♀ JAMAICA, Fair Prospect, xii.1968 (K. Heinze) ex *Aleurodicus jamaicensis* [s10] (on 1 slide, USNM). 8♀ BRAZIL, Vicosa, Minas Gerais, 6.xi.1935 (E.J. Hambleton) ex *Aleurothrixus floccosus* (?) (on one slide, one head missing, USNM).

Remarks: *Encarsia cylindrica* appears to be most closely related to *E. erwini*, with which it shares the elongate antenna and lack of a linea calva. It differs from *E. erwini* in having many more setae on the mesoscutum.

Etymology. "cylindrica" refers to the almost uniformly elongate antenna.

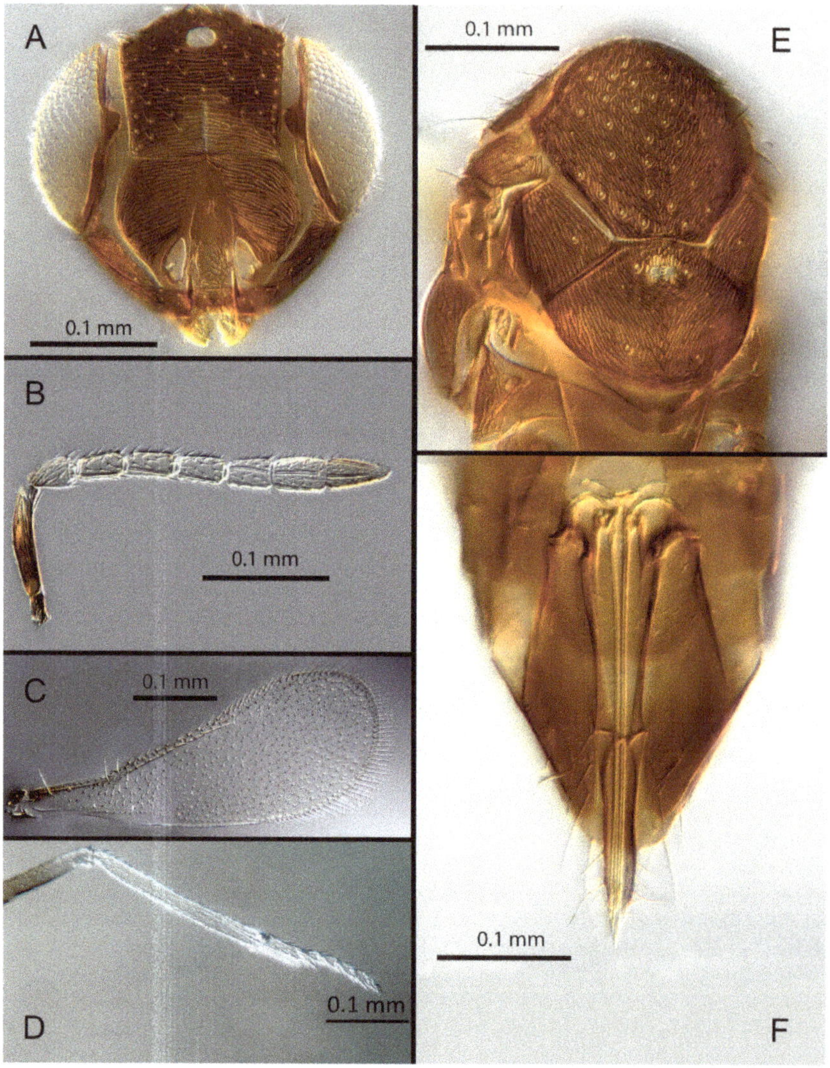

Figure 9. ***Encarsia cylindrica***: (**A**) head; (**B**) antenna; (**C**) fore wing; (**D**) midleg; (**E**) dorsal mesosoma; (**F**) ovipositor.

3.3.7. *Encarsia diablejo* (Polaszek and Hayat) **comb. n.**

(Figure 10A–F)

Dirphys diablejo Polaszek and Hayat, 1992: 189

Female. Unknown. This species is known only from the holotype.

Male. Colour: Antennae uniformly light brown, slightly darker on the base of the scape, pedicel, and radicle. Head brown with paler areas bordering the eyes and extending along the genae towards the clypeus. Mesosoma and metasoma uniformly brown. Legs light brown, the mid and hind tibia pale in contrast to the dark femora; all tarsi pale. Fore wings hyaline, stigmal vein pale in contrast with a darker marginal vein.

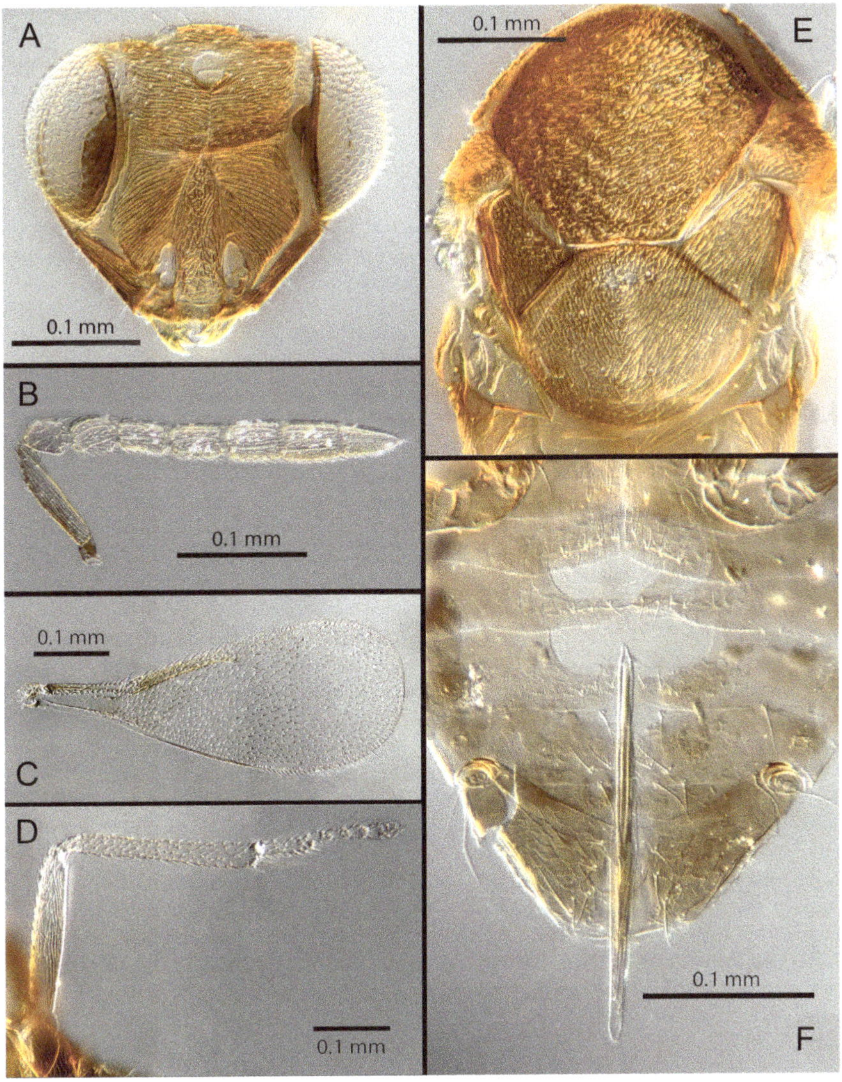

Figure 10. ***Encarsia diablejo***: (**A**) head; (**B**) antenna; (**C**) fore wing; (**D**) midleg; (**E**) dorsal mesosoma; (**F**) male genitalia.

Morphology:. Head (Figure 10A) with mediofrontal line complete, though fading towards anterior ocellus; transfacial line complete, narrow; facial lines very broad along their entire lengths. Scrobes entirely with irregular aciculate sculpture. Antennal formula (Figure 10B): 1, 1, 3, 3; scape 2.94x pedicel, F1 subequal to pedicel, F1 0.88X F2, F2 and F3 subequal; funicle 0.89x clava length. Flagellum with the following number of longitudinal sensilla: F1: 7; F2: 6; F3: 7; F4: 8; F5: 9; F6: 7. Mandibles (Figure 10A) with two large pointed teeth and a truncation; maxillary palps two-segmented. Mid-lobe of mesoscutum (Figure 10E) with more than 60 setae; each lateral lobe with one seta; each axilla with one seta; scutellum with four setae and two vestigial setal bases. Sculpture of mesoscutum transverse. Fore wing (Figure 10C) with two large setae and four smaller setae on submarginal vein, five setae in basal cell, 11 setae on anterior margin of marginal vein. Linea calva absent. Submarginal 0.79x marginal vein. Maximum length of fore wing 2.48x fore wing width, maximum width of fore wing 7x longest setae on marginal fringe. Mid tibial spur (Figure 10D) as long as corresponding basitarsus.

Distribution: PERU: Loreto.

Host: Unknown.

Material examined: Holotype ♂ PERU, Loreto, Iquitos, Granja Unap, 9.ii.1984 (L. Huggert BM 1984-337) [s26] (NHMUK).

Remarks: For the purposes of the identification key, we have assumed that the (unknown) female of *E. diablejo* shares the wing and mesosomal sculpture characters with the male; the combination of which is unique in the *Encarsia mexicana* species-group.

3.3.8. *Encarsia dichaeta* Polaszek and Hernández-Suárez **sp. n.**

(Figure 11A–F)

Female. Colour: Antennae light brown, slightly darker on the base of the scape, pedicel, and radicle. Head dark brown with pale lines bordering the eyes and extending along the genae towards clypeus. Mesosoma and metasoma uniformly dark brown. Legs yellow with femora and coxa brown, tibia, and all tarsi pale. Wings infuscate below submarginal and marginal veins, stigmal vein pale.

Morphology: Head (Figure 11A) with mediofrontal line complete; transfacial line narrow; facial lines relatively narrow along their entire lengths. Scrobes with irregular aciculate sculpture. Antennal formula (Figure 11B): antennal formula: 1, 1, 3, 3; scape 2.59x pedicel length; pedicel 1.85x F1; F1 0.85x F2; F2 0.8x F3; funicle 0.65x clava length; F6 perpendicular. Flagellum with the following number of longitudinal sensilla: F1:0; F2:1; F3:1; F4:1–2; F5:1–2; F6:1–2. Mandibles (Figure 11A) with two small teeth and a truncation. Maxillary palps two-segmented. Mid-lobe of mesoscutum (Figure 11E) with more than 50 setae; each lateral lobe with two setae; each axilla with two setae scutellum with four setae andtwo vestigial setal bases. Sculpture of mesoscutum aciculate; sculpture of axillae and scutellum longitudinal. Fore wing (Figure 11C) with two large setae and 7–8 smaller setae on submarginal vein, 14 setae in basal cell, eight setae on anterior margin of marginal vein, and one seta at the junction of the submarginal vein and parastigma. Linea calva absent. Submarginal 0.85x times marginal vein. Maximum length of fore wing 2.48x fore wing width, maximum width of wing 5.45x longest setae on marginal fringe. Ovipositor (Figure 11F) 0.93x mid tibial length; third valvulae 0.45x ovipositor; second valvifer 1.6x third valvula. Mid tibial spur (Figure 11D) equal to corresponding basitarsus. Metasomal terga T1–T7 with 0, 1 + 1, 1 + 1, 1 + 1, 1 + 2 + 1, 1 + 2 + 1 and 12 setae, respectively. T7 (Figure 11F) rounded, not extended but covering ovipositor.

Distribution: BRAZIL: Bahia; COSTA RICA: Guanacaste, Alajuela, Heredia; ECUADOR: Napo River.

Host: Aleurodicinae: *Aleurodicus flavus* Hempel, *Aleurodicus* sp.

Material examined: Holotype 1♀ COSTA RICA, Alajuela, P.N. Arenal Sendero Pilon, 10°27′ N 84°45′ W 600 m 25.ii.2003 (J.S. Noyes) [DNA 136: OQ683552] (NHMUK). Paratypes 3♀ COSTA RICA, Alajuela, P.N. Arenal, Sendero Pilon, 10°27′ N 84°45′ W 600 m 25.ii.2003 (J.S. Noyes), [DNA 132: OQ683550, 133: OQ683552, 135: OQ683550]

(2♀ NHMUK, 1♀ MZUCR). 1♀ COSTA RICA, Heredia, Est. Biol. La Selva, 10°26′ N 84°01′ W 75 m 27–28.ii.2003 (J.S. Noyes) [DNA151: OQ683549] (UCRC); 1♀ COSTA RICA, P.N. Guanacaste, Est. Pitilla (ACG), 11°00′ N. 85°26′ W. 700 m MT/YPT (J.S. Noyes) [DNA 208: OQ683551] (NHMUK). 27♀♀ BRAZIL, Bahia (Gregorio Bondar n°65b) ex *Aleurodicus flavus* (on 5 slides; USNM). 5♀ ECUADOR, Napo, Camino Añangucocha, 29.iii.04 (H. Evans) ex *Aleurodicus* sp. (NHMUK).

Remarks: There are some colour differences between the Costa Rican specimens and those from Brazil, the latter having the metasoma distally paler. Further studies on fresh material, in particular DNA sequencing, will be needed to confirm their status. DNA sequences were obtained from the holotype and five paratypes, deposited under GenBank accession numbers: OQ683550, OQ683552, OQ683551 and OQ683549 (three paratype specimens were pooled for extraction).

Etymology. "*dichaeta*" refers to the two setae on each axilla, unique for the genus.

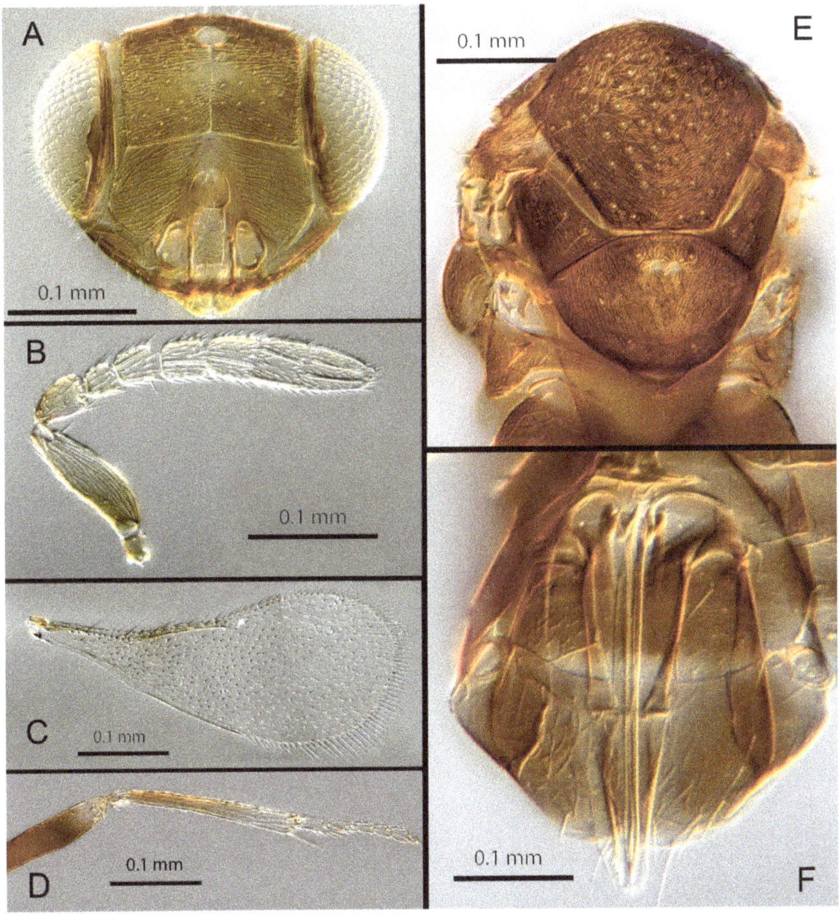

Figure 11. ***Encarsia dichaeta***: (**A**) head; (**B**) antenna; (**C**) fore wing; (**D**) midleg; (**E**) dorsal mesosoma; (**F**) ovipositor.

3.3.9. *Encarsia encantadora* (Polaszek and Hayat) **comb. n.**
(Figure 12A–F)

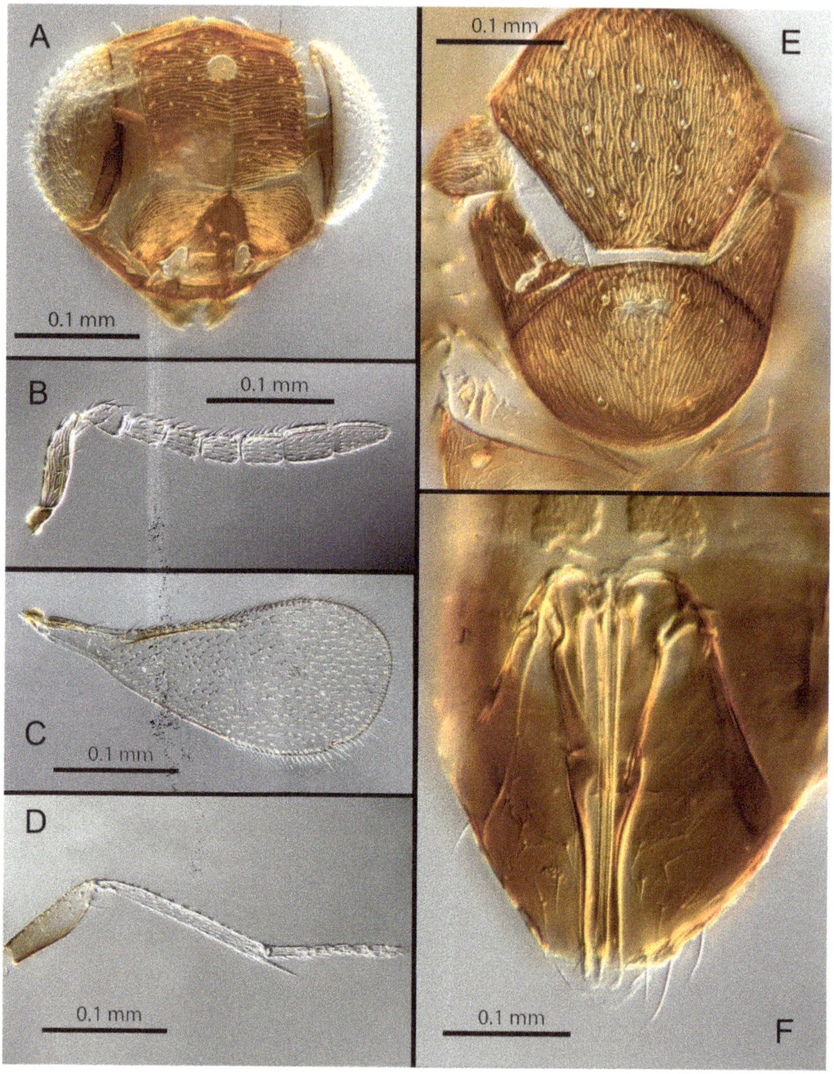

Figure 12. ***Encarsia encantadora***: (**A**) head; (**B**) antenna; (**C**) fore wing; (**D**) midleg; (**E**) dorsal mesosoma; (**F**) ovipositor.

Dirphys encantadora Polaszek and Hayat, 1992: 191.

Female. Colour: Antennae pale brown/yellow with dark scape and radicle. Head brown with paler areas bordering the eyes and extending along the genae towards the clypeus, antennal scrobes, a line from the apex of the scrobes to the median ocellus, and a transverse line midway between the antennal sockets and the median ocellus. Mesosoma brown in holotype but with posterior three-quarters of scutellum pale in Mexican specimens. Legs light brown, the mid and hind tibiae pale in contrast to the dark femora and coxa, all tarsi pale. Wings hyaline, faintly infuscate below the marginal vein, stigmal vein pale in contrast with a darker marginal vein.

Morphology: Head (Figure 12A) with mediofrontal line complete; transfacial line evident; facial lines very broad along their entire lengths. Scrobes with longitudinally aciculate sculpture. Antennal formula (Figure 12B): 1, 1, 3, 3; scape 2.3x pedicel length; pedicel equal to F1; F1 to F3 funicle segments all subequal in length; funicle 0.75x clava length; F6 perpendicular. Flagellum with the following number of longitudinal sensilla: F1: 1–2; F2: 1; F3: 1; F4: 2–3; F5: 3; F6: 4. Mandibles (Figure 12A) with two small ventral teeth and a broad truncation dorsally. Maxillary palps two-segmented. Mid-lobe of mesoscutum (Figure 12E) with fewer than 30 setae; each lateral lobe with three setae; each axilla with one seta; scutellum with four setae. Sculpture of mesoscutum aciculate; sculpture of axillae and scutellum longitudinal. Fore wing (Figure 12C) with two setae on submarginal vein, 7–9 setae in basal cell, 7–11 setae on anterior margin of marginal vein, and one seta at the junction of the submarginal vein and parastigma. Linea calva absent. Submarginal 0.73x times marginal vein. Maximum length of fore wing 2.52x fore wing width, maximum width of fore wing 6.9x longest setae on marginal fringe. Ovipositor (Figure 12F) 0.82x mid tibial length; third valvula 0.55x ovipositor length; second valvifer 0.79x third valvula. Mid tibial spur (Figure 12D) 0.87x corresponding basitarsus. Metasomal terga T1–T7 with 0, 1 + 1, 1 + 1, 1 + 1, 1 + 2 + 1, 1 + 2 + 1 and four setae, respectively. T7 (Figure 12F) rounded and covering ovipositor third valvula.

Distribution: ECUADOR: Napo; MEXICO: Tabasco.

Host: Aleurodicinae: *Nealeurodicus altissimus* (Quaintance)

Material examined: Holotype ♀ ECUADOR, Napo, Sacha, 5.iii.1983 (L. Huggert) (NHMUK). 1♀, fragments of a second ♀: MEXICO, Tabasco, San Francisco del Peal, 1.vii.1897 (C.H.T. Townsend) ex *Nealeurodicus altissimus* (Quaintance) on *Lippia myriophala* Schltdl. and Cham. (USNM; on slide with *E. mexicana* type material).

Remarks: *Encarsia encantadora* is morphologically closest to *E. erwini*, differing from that species mainly in having the third valvulae longer than the second valvifers. The fore wing is also broader in *E. encantadora*, especially measured relative to the longest wing fringe setae (compare Figures 11C and 12C).

3.3.10. *Encarsia erwini* Polaszek and Hernández-Suárez **sp. n.**

(Figure 13A–F)

Female. Colour: Antennae entirely pale, only the scape and radicle dark. Head dark brown. Mesosoma uniformly dark brown. Legs entirely pale except all coxae brown (female paratype with some infuscation on the hind femora).

Morphology: Head (Figure 13A) with mediofrontal line complete; transfacial and facial lines very broad along their entire lengths. Scrobes with irregularly aciculate sculpture centrally. Antennal formula (Figure 13B): 1, 1, 3, 3; scape 2.4x pedicel length; pedicel 1.2x F1 equal to F2; F2 equal to F3; funicle 0.77x clava; F6 perpendicular. Flagellum with the following number of longitudinal sensilla: F1: 1; F2: 1; F3: 1; F4: 1; F5: 2; F6: 2. Mandibles (Figure 13A) with 2 ventral teeth and a truncation dorsally. Maxillary palps two-segmented. Mid-lobe of mesoscutum (Figure 13E) with about 18 setae; each lateral lobe with three setae; each axilla with one seta; scutellum with four setae. Sculpture of mesoscutum and axillae longitudinally aciculate; sculpture of scutellum longitudinal, transverse apically. Fore wing (Figure 13C) with two large setae on submarginal vein and five smaller setae above, six setae in basal cell, seven setae on anterior margin of marginal vein, and one large seta at the junction of the submarginal vein and parastigma. Linea calva absent. Submarginal vein approximately equal in length to marginal vein. Maximum length of fore wing 2.9x fore wing width, maximum width of wing 3.75x longest seta on marginal fringe.

Ovipositor (Figure 13F) equal to mid tibial length; third valvula 0.44x ovipositor length; second valvifer 1.3x third valvula. Mid tibial spur (Figure 13D) 1.0x corresponding basitarsus. Metasomal terga T1–T7 with 0, 1 + 1, 1 + 1, 1 + 1, 1 + 2 + 1, 2 + 2 and seven setae, respectively. T7 (Figure 13F) conical, not extended but just covering ovipositor.

Distribution: ECUADOR: Napo.

Material examined: Holotype ♀ ECUADOR, Napo, transect ent. 1 km S. Onkone Gare Camp, Res. Etnica Waorani 220 m 0°39′10″ S 76°26′00″ W TL Erwin et al. fogging t.f. forest. Lot 1255 8.x.1995 (UCRC; 52715). Paratype ♀, same data as holotype except 4.x.1996 (NHMUK).

Remarks: *Encarsia erwini* appears to be most closely related to *E. cylindrica*, with which it shares the elongate antenna and lack of a linea calva. It differs from *E. cylindrica* in having far fewer setae on the mesoscutum. *E. erwini* is also morphologically close to *E. encantadora*, differing from that species mainly in having the third valvulae shorter than the second valvifers. The fore wing is also broader in *E. encantadora*, especially measured relative to the longest wing fringe setae.

Etymology: Named for the late Terry Erwin (1940–2020), prolific collector of insects, especially in the rain forest canopy of Ecuador.

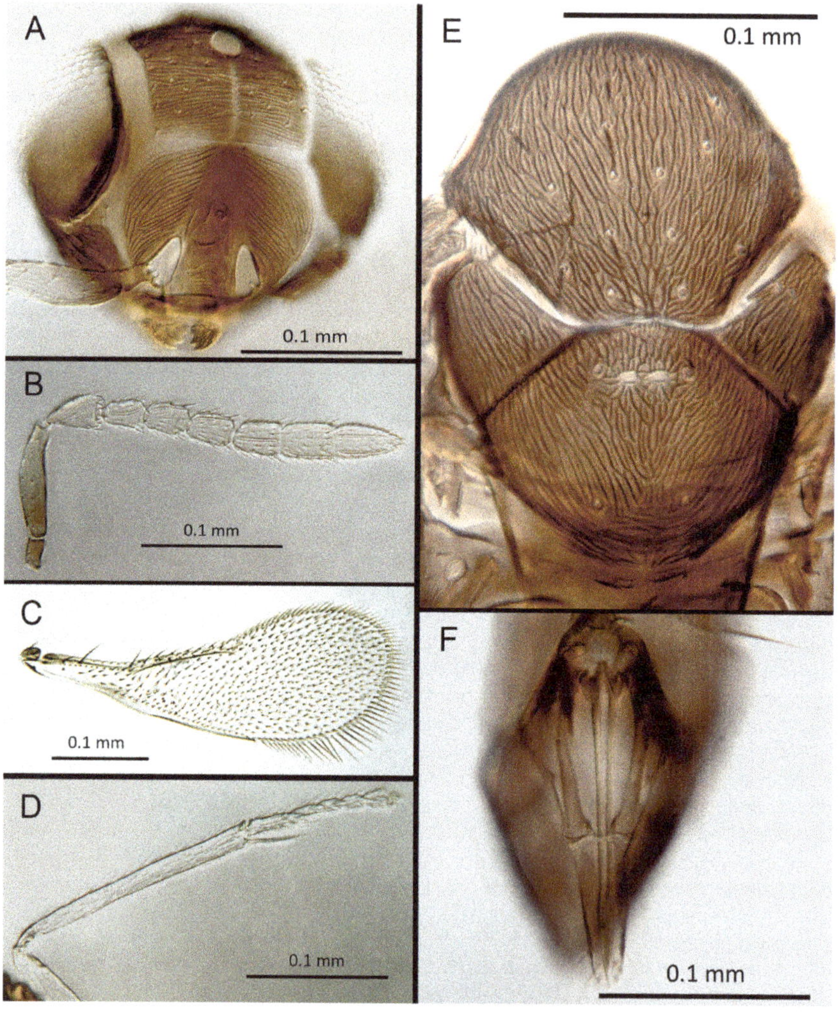

Figure 13. ***Encarsia erwini***: (**A**) head; (**B**) antenna; (**C**) fore wing; (**D**) midleg; (**E**) dorsal mesosoma; (**F**) ovipositor.

3.3.11. *Encarsia fredbennetti* Polaszek and Hernández-Suárez **sp. n.**
(Figure 14A–E)

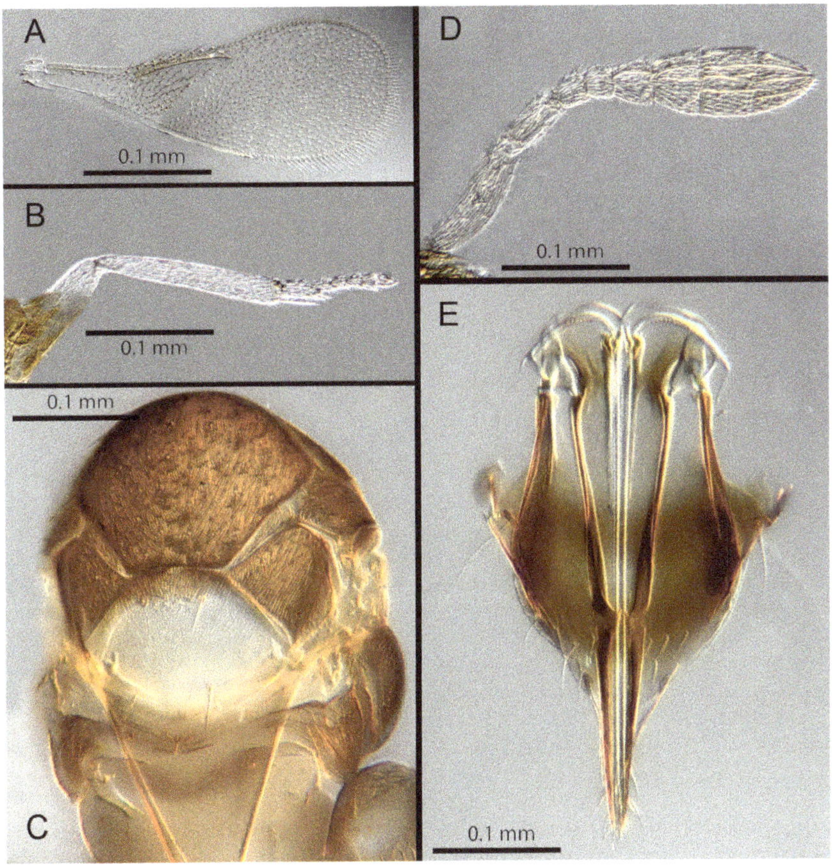

Figure 14. ***Encarsia fredbennetti***: (**A**) fore wing; (**B**) mid leg; (**C**) dorsal mesosoma; (**D**) antenna; (**E**) ovipositor.

Female. Colour: Antennae uniformly pale brown. Head brown with pale lines bordering the eyes and extending along the genae towards clypeus. Mesosoma dark brown with most of scutellum and post-scutellum pale; metasoma uniformly brown. Legs yellow with dark coxae, femora and anterior third of hind leg tibia. Wings infuscate below marginal vein, stigmal vein pale in contrast with darker marginal vein.

Morphology: Head (not shown) with all facial lines obscure in holotype (head absent in paratypes). Scrobes with longitudinally aciculate sculpture. Scrobes with longitudinally aciculate sculpture. Antennal formula (Figure 14D): 1, 1, 3, 3 (though could be interpreted as 1, 1, 2, 4); scape 2.39x pedicel length; pedicel 1.9x F1; F1 0.8x F2; F2 0.9x F3; funicle 0.6x clava; F6 perpendicular. Flagellum with the following number of longitudinal sensilla: F1: 0; F2: 1; F3: 1; F4: 2–3; F5: 3; F6: 3. Mandibles with one ventral tooth and a truncation. Maxillary palps two-segmented. Mid-lobe of mesoscutum (Figure 14C) with 30–40 setae; each lateral lobe with two setae; each axilla with one seta; scutellum with four setae. Sculpture of mesoscutum aciculate; sculpture of axillae and scutellum longitudinal. Fore wing (Figure 14A) with two large setae and four smaller setae on submarginal vein, 4–5 setae in basal cell, seven setae on anterior margin of marginal vein. Linea calva present. Submarginal 0.93x marginal vein. Maximum length of fore wing 3.8x fore wing width, maximum width of wing 3.7x longest setae on marginal fringe. Ovipositor (Figure 14E) 1.3x mid tibial length; third valvulae 0.5x ovipositor length; second valvifer equal to third valvula. Mid tibial spur (Figure 14B) equal to corresponding basitarsus. Metasomal terga T1–T7 with 0, 2 + 2, 2 + 2, 2 + 2, 1 + 2 + 1, 1 + 2 + 1 and four setae, respectively. T7 (Figure 14E) extended, covering ovipositor.

Distribution: TRINIDAD: St Augustine.

Host: Aleurodicinae.

Material examined: Holotype ♀ TRINIDAD, [St Augustine] ICTA [Imperial College of Tropical Agriculture] xii.1953 FD Bennett ex whitefly on cocoa (NHMUK); Paratype ♀ TRINIDAD, St Augustine, ex Aleurodicinae [DNA215: OQ683559] (NHMUK); Paratype ♀ TRINIDAD, Mt St Benedict, ex whitefly Coll. M. Jagroep [DNA216] (NHMUK).

Remarks: *Encarsia fredbennetti* is morphologically closest to *E. mexicana* and *E. inbioa* from which it differs by the enlarged clava. It is also molecularly closest to *E. mexicana*. Deposited under GenBank accession number: OQ683559.

Etymology: Named for the late Fred D. Bennett (1925–2021), former Director of the Commonwealth Institute of Biological Control and avid collector of parasitoids during much of his long life.

3.3.12. *Encarsia inbioa* Polaszek and Hernández-Suárez **sp. n.**

(Figure 15A–F)

Female. Colour: Antennae light brown slightly darker at F5 and F6, the base of the scape, pedicel, and radicle. Head brown with pale lines bordering the eyes and extending along the genae towards clypeus. Mesosoma dark brown but with posterior quarter of scutellum pale; metasoma uniformly dark brown. Legs yellow with dark coxae, femora and anterior third of hind leg tibia. Wings infuscate below marginal vein, stigmal vein pale in contrast with the darker marginal vein.

Morphology: Head (Figure 15A) with mediofrontal line incomplete, reaching to less than half the distance to the frontal ocellus; transfacial line evident, narrow; facial lines very broad along their entire lengths, particularly at the level of the lower eye. Scrobes with aciculate/reticulate sculpture basally and centrally, smooth apically and apico-laterally. Antennal formula (Figure 15C): 1, 1, 3, 3; scape 2.39x pedicel length; pedicel 1.9x F1; F1 0.8x F2; F2 0.9x F3; funicle 0.6x clava; F6 perpendicular. Flagellum with the following number of longitudinal sensilla: F1: 0; F2: 1; F3: 1; F4: 2–3; F5: 3; F6: 3. Mandibles (Figure 15A) with two large teeth and a truncation. Maxillary palps two-segmented. Mid-lobe of mesoscutum (Figure 15E) with 30–40 setae; each lateral lobe with two setae; each axilla with one seta; scutellum with four setae. Sculpture of mesoscutum aciculate; sculpture of axillae and scutellum longitudinal. Fore wing (Figure 15B) with two large setae and four smaller setae on submarginal vein, 4–5 setae in basal cell, seven setae on anterior margin of marginal vein. Linea calva is present. Submarginal 0.93x marginal vein. Maximum length of fore wing 3.8x fore wing width, maximum width of wing 3.7x longest setae on marginal fringe. Ovipositor (Figure 15F) 1.3x mid tibial length; third valvulae 0.5x ovipositor length; second valvifer equal to third valvula. Mid tibial spur (Figure 15D) equal to corresponding basitarsus. Metasomal terga T1–T7 with 0, 2 + 2, 2 + 2, 2 + 2, 1 + 2 + 1, 1 + 2 + 1 and four setae, respectively. T7 (Figure 15F) extended and covering ovipositor.

Distribution: COSTA RICA: Alajuela.

Material examined: Holotype ♀ COSTA RICA, Alajuela, P.N. Arenal, Sendero Pilon, 26.ii.2003 (J.S. Noyes) [DNA 128: OQ683553] 600 m 10°27′ N 84°43′ W (NHMUK).

Remarks: *Encarsia inbioa* is morphologically closest to *E. fredbennetti* from which it differs by the non-enlarged clava. It is, perhaps surprisingly, molecularly closest to *E. svetlana*. Deposited under GenBank accession number: OQ683553.

Etymology: Named for INBio (Instituto Nacional de Biodiversidad) the national institute for biodiversity and conservation in Costa Rica.

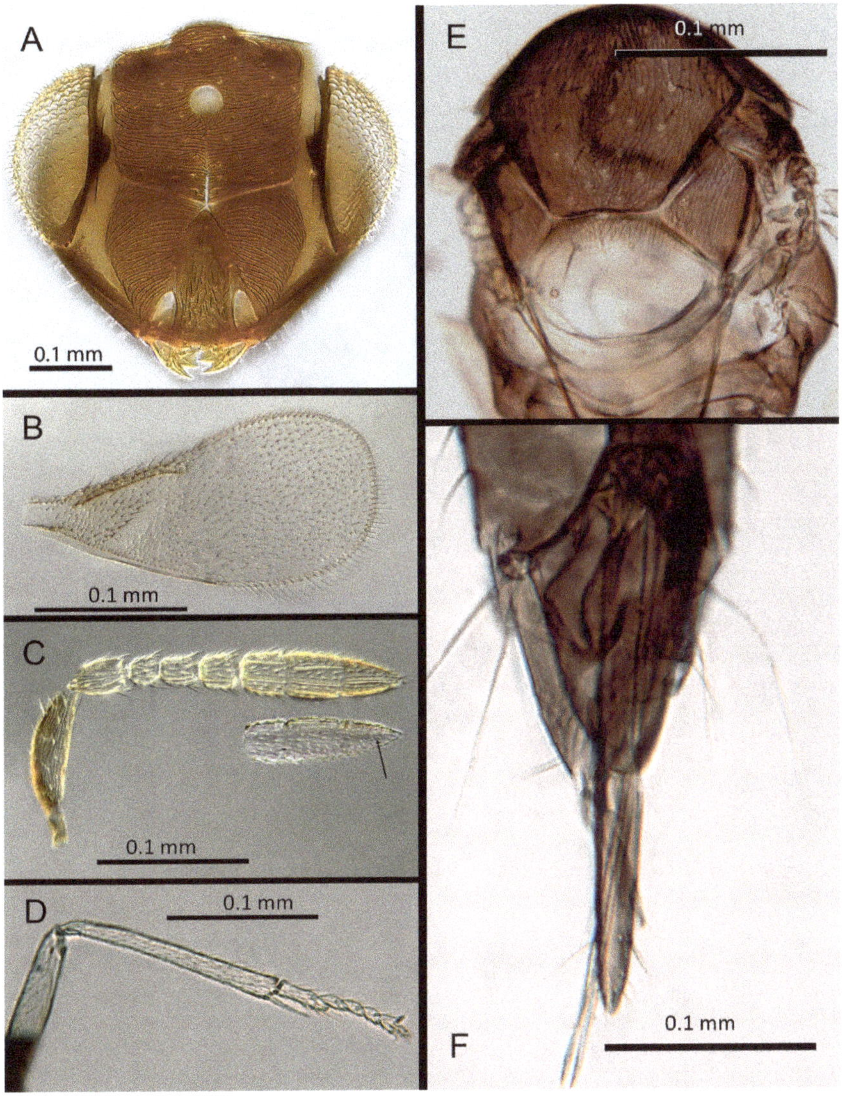

Figure 15. ***Encarsia inbioa***: (**A**) head; (**B**) fore wing; (**C**) antenna, arrow: claval sensory area; (**D**) mid leg; (**E**) dorsal mesosoma; (**F**) ovipositor.

3.3.13. *Encarsia larensis* (Chavez) **comb. n.**

Dirphys larensis Chavez, 1996: 11

(Figure 16A–F)

Female. Colour:. Antennae pale brown with slightly darker clava and pedicel. Head dark brown with paler areas bordering the eyes and extending along the genae towards the clypeus. Mesosoma and metasoma uniformly dark brown, third valvulae dark brown contrasting with the rest of ovipositor. Legs pale, mid and hind femur, and coxae brown, anterior third of mid leg tibia brown. Wings infuscate below the submarginal and marginal vein; marginal and stigmal veins dark.

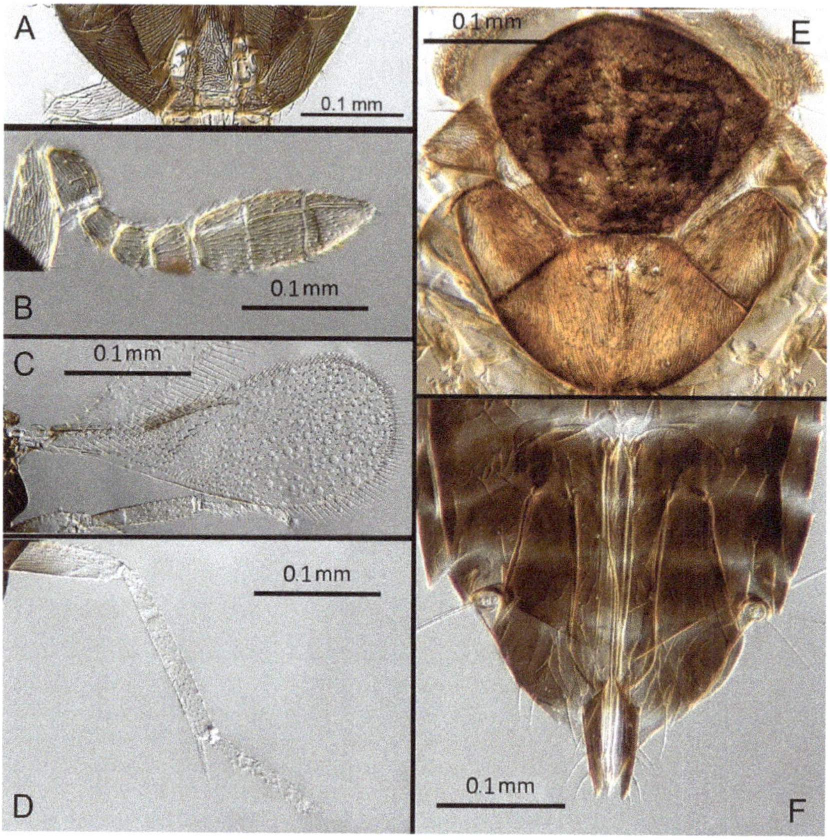

Figure 16. ***Encarsia larensis***: (**A**) head; (**B**) antenna; (**C**) fore wing; (**D**) mid leg; (**E**) dorsal mesosoma; (**F**) ovipositor.

Morphology: Head (Figure 16A) with mediofrontal line complete, reaching to the frontal ocellus; other facial lines obscure in paratypes examined due to mounting method. Scrobes with coarse longitudinal aciculate sculpture becoming irregular towards clypeus. Antennal formula (Figure 16B): 1, 1, 3, 3; scape expanded, 2.4–2.5x pedicel length; pedicel 1.9x F1; F1 equal to F2; F2 0.8x F3; funicle 0.58x clava; F6 oblique.

Flagellum with the following number of longitudinal sensilla: F1: 0; F2: 0; F3: 1; F4: 4; F5: 5–6; F6: 4. Mandibles (not shown) with one large ventral tooth and a bidentate upper tooth.

Maxillary palps two-segmented. Mid-lobe of mesoscutum (Figure 16E) with 46–60 setae; each lateral lobe with two setae; each axilla with one seta; scutellum with four setae and two vestigial setal bases. Sculpture of mesoscutum aciculate; sculpture of axillae and scutellum longitudinal. Fore wing (Figure 16C) with two large setae and 3–4 smaller setae on submarginal vein, 7–9 setae on anterior margin of marginal vein, and one seta at the junction of the submarginal vein and parastigma. Linea calva present. Submarginal equal to marginal vein; Maximum length of fore wing 2.54x fore wing width; maximum width of fore wing 6.7x longest setae on marginal fringe. Ovipositor (Figure 16F) 1.30x mid tibial length; third valvula 0.28x ovipositor length; second valvifer 2.4x third valvula. Mid tibial spur (Figure 16D) 1.15x corresponding basitarsus. Metasomal terga T1–T7 with 0, 2 + 2, 2 + 2, 2 + 2, 1 + 2 + 1, 1 + 2 + 1 and four setae, respectively. T7 (Figure 16F) rounded not covering ovipositor third valvula.

Male. Colour: Head light brown. Mesosoma and gaster uniformly brown but posterior third of mesoscutum, anterior third of axillae and scutellum yellow. Legs brown. Fore wings hyaline.

Morphology: Similar to that of female, except antennal formula, flagellum with longitudinal sensilla on all segments and funicle segments subequal in length.

Distribution: VENEZUELA: Cabudare, Lara.

Host: Aleurodicinae: *Aleurodicus pulvinatus* (Maskell).

Material examined: 1♀, 1♂: VENEZUELA, Cabudare, Lara, i.1994 (A. Chavez and F. Díaz) ex *Aleurodicus pulvinatus* on *Hura crepitans* L. (NHMUK).

Remarks: *Encarsia larensis* appears morphologically closest to *E. marynoyesae* from which it differs in having a much longer funicle, and shorter ovipositor. No molecular data were available for this species. Chavez (1996) recorded 16 individuals of *E. larensis* within a single whitefly host.

3.3.14. *Encarsia marynoyesae* Polaszek and Hernández-Suárez **sp. n.**

(Figure 17A–F)

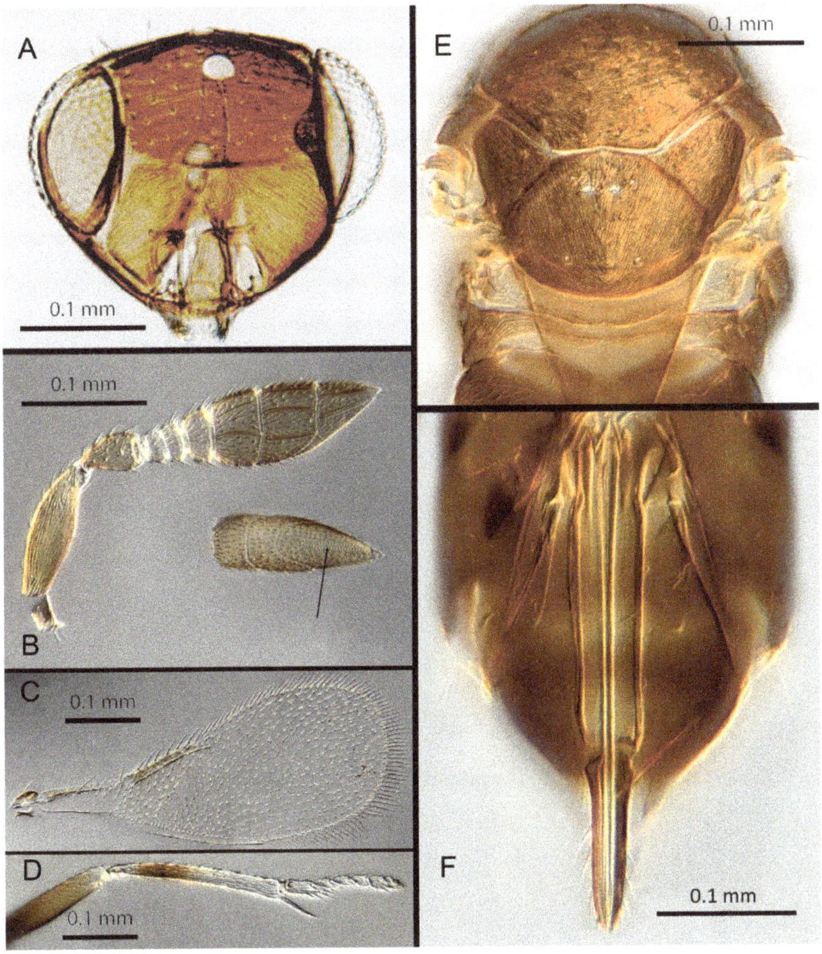

Figure 17. ***Encarsia marynoyesae***: (**A**) head; (**B**) antenna, arrow: claval sensory area; (**C**) fore wing; (**D**) mid leg; (**E**) dorsal mesosoma; (**F**) ovipositor.

Female. Colour: Antennae brown with F1 and F2 paler. Head dark brown with pale lines bordering the eyes and extending along the genae towards clypeus. Mesosoma uniformly dark brown. Legs yellow with mid and hind femora, coxa and anterior third of tibia brown, all tarsi pale. Wings slightly infuscated below anterior half of submarginal vein, stigmal vein pale in contrast with a darker marginal vein.

Morphology: Head (Figure 17A) with mediofrontal line complete; transfacial line narrow; facial lines very broad along their entire lengths. Scrobes with irregular aciculate sculpture. Antennal formula (Figure 17B): 1, 1, 3, 3; scape slightly expanded, 2.53x pedicel length; pedicel 2.85x F1; F1 equal to F2; F2 0.8x F3; funicle 0.36x clava; F5 and F6 strongly oblique, claval sensorial complex developed. Flagellum with the following number of longitudinal sensilla: F1: 0; F2: 0–1; F3: 1; F4: 3; F5: 4; F6: 4. Mandibles (Figure 17A) with one small ventral tooth and a bidentate upper tooth. Maxillary palps two-segmented. Mid-lobe of mesoscutum (Figure 17E) with approximately 40 setae; each lateral lobe with two setae; each axilla with one seta; scutellum with four setae and two vestigial bases. Sculpture of mesoscutum aciculate; sculpture of axillae and scutellum longitudinal. Fore wing (Figure 17C) with two large setae and four smaller setae on submarginal vein, four setae in basal cell, 6–7 setae on anterior margin of marginal vein, and one seta at the junction of the submarginal vein and parastigma. Linea calva present. Submarginal equal to marginal vein. Maximum length of fore wing 2.57x fore wing width, maximum width of wing 6.2x longest setae on marginal fringe. Ovipositor (Figure 17F) 1.4x mid tibial length; third valvulae 0.35x ovipositor length; second valvifer 1.9x third valvula. Mid tibial spur (Figure 17D) 1.07x corresponding basitarsus. Metasomal terga T1–T7 with 0, 1 + 1, 1 + 1, 1 + 1, 1 + 2 + 1, 1 + 2 + 1 and 4 setae, respectively. T7 (Figure 17F) extended although apparently not covering ovipositor.

Distribution: COSTA RICA (Alajuela).

Material examined: Holotype ♀ COSTA RICA, Alajuela, Est. Caribe Reserva Rincón Forestal, 19–20.ii.2003 (J.S. Noyes) [DNA 163: OQ683563] 400 m, 10°53′ N 85°18′ W (NHMUK). Paratype 1♀ COSTA RICA, Alajuela, Est. Caribe Reserva Rincón Forestal, 19–20.ii.2003 (J.S. Noyes) [DNA167] (NHMUK).

Remarks: *Encarsia marynoyesae* is morphologically very similar to *E. aisha* in many respects (though distant to it based on DNA). The species can be distinguished by *E. marynoyesae* having the second valvifers almost 2x (1.8) the third valvulae; while they are 1.5x as long in E. *aisha*. In E. *marynoyesae* the clava is well over 2x the length of the funicle; in E. *aisha* it is less than 2x as long. *E. marynoyesae* also shares aspects of morphology with *E. catula*, but can be distinguished by having more than 30 setae on the mesoscutum, and V3 less than $\frac{1}{2}$ the length of V2 (much more than $\frac{1}{2}$ as long in *E. catula*). Sequence deposited under GenBank accession number: OQ683563.

Etymology. Named for Mary Noyes MBE (Member of the Order of the British Empire).

3.3.15. *Encarsia mendesi* (Polaszek and Hayat) **comb. n.**

(Figure 18A–F)

Dirphys mendesi, Polaszek and Hayat 1992: 191

Female. Colour: Antennae brown, paler on their ventral halves. Head dark brown with pale lines bordering the eyes and extending along the genae towards clypeus, antennal scrobes, a line from the apex of the scrobes to the median ocellus, and a transverse line midway between the antennal sockets and the median ocellus centrally bordering the dorsal end of antennal scrobes. Mesosoma and metasoma uniformly dark brown. Legs pale yellow, with coxae and hind femora dark brown. Wings infuscated below the submarginal and marginal vein; stigmal vein pale in contrast with a darker marginal vein.

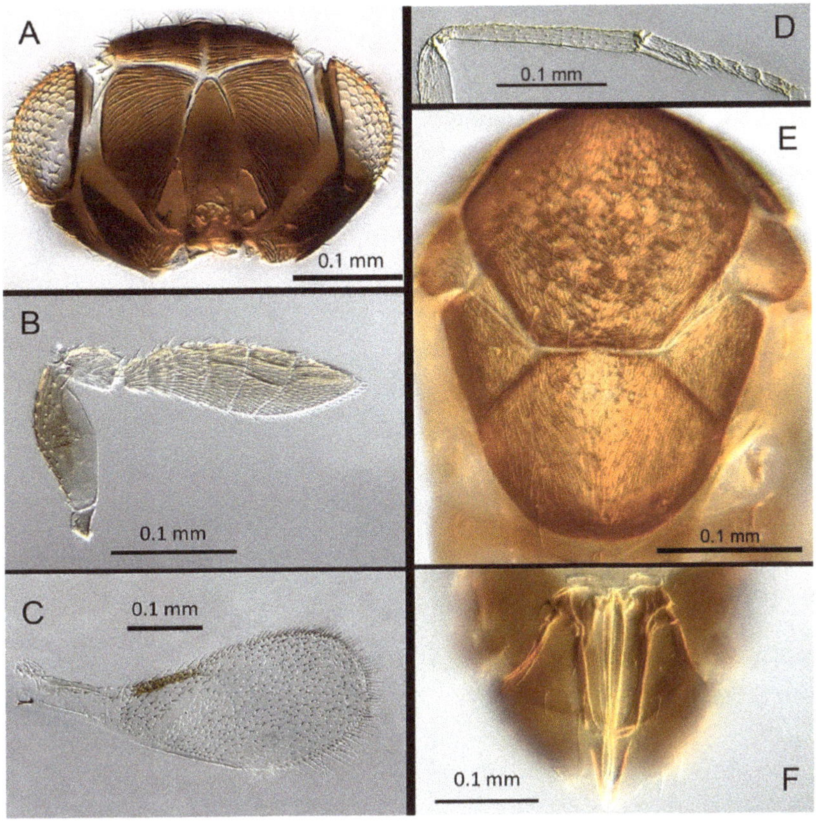

Figure 18. ***Encarsia mendesi***: (**A**) head; (**B**) antenna; (**C**) fore wing; (**D**) mid leg; (**E**) dorsal mesosoma; (**F**) ovipositor.

Morphology: Head (Figure 18A) with mediofrontal line complete check, ocellus; transfacial line evident; facial lines very broad along their entire lengths, especially at level of lower eye and adjacent to genae. Scrobes largely smooth, some irregular sculpture centrally. Antenna (Figure 18B): with funicle apparently absent, so the entire flagellum clavate (antennal formula therefore 1, 1, 6); scape expanded, 2.3–2.8x pedicel length; pedicel 2.4x F1; F1 0.9x F2; F2 0.8x F3; funicle 0.5x clava; F5 and F6 broadly oblique, claval sensorial complex developed. Flagellum with the following number of longitudinal sensilla: F1: 0; F2: 1; F3: 1; F4: 3; F5: 4; F6: 3. Mandibles missing from holotype; paratype (male) apparently with 2 teeth. Maxillary palps two-segmented. Mid-lobe of mesoscutum (Figure 18E) with fewer than 30 setae; each lateral lobe with two setae; each axilla with one seta; scutellum with four setae. Sculpture of mesoscutum transverse. Fore wing (Figure 18C) with two large setae and 2–3 smaller setae on submarginal vein, 3–5 setae in basal cell, 7–11 setae on anterior margin of marginal vein, and one seta at the junction of the submarginal vein and parastigma. Linea clava present. Submarginal equal to marginal vein; maximum width of fore wing 2.56x fore wing width, maximum width of wing 4.6x longest seta on marginal fringe. Ovipositor (Figure 18F) 0.8x mid tibial length; third valvulae 0.3x ovipositor length; second valvifer 2.13x third valvula. Mid tibial spur (Figure 18D) equal to corresponding basitarsus. Metasomal terga T1–T7 with 0, 1 + 1, 1 + 1, 1 + 1, 1 + 2 + 1, 1 + 2 + 1 and six setae, respectively. T7 (Figure 18F) rounded covering ovipositor third valvula.

Male. All aspects of coloration and morphology as for female, except the antennae and genitalic characters.

Distribution: BRAZIL: São Paulo.

Host: Aleurodicinae: *Aleurodicus maritimus* Hempel.

Material examined: Holotype ♀ BRAZIL, São Paulo, Mogi-Guazu, 12.v.1981 (M. Cytrynowicz) 84/8 ex *Aleurodicus maritimus* (NHMUK). Paratype 1♂ BRAZIL, São Paulo, Mogi-Guazu, 12.v.1981 (M. Cytrynowicz) 84/8 ex *Aleurodicus maritimus*. 1♂ 1♀ BRAZIL, São Paulo, E.E. Mogi-Guazu, 12.v.1981 (M. Cytrynowicz) 94 ex *Aleurodicus maritimus* on *Bauhinia holophylla* (Bong.) (Fabaceae) (all NHMUK).

Remarks. Morphologically *E. mendesi* appears closest to *E. marynoyesae* having the entire flagellum more or less clavate. It differs from that species by the very short ovipositor. No molecular data were available for *E. mendesi*.

3.3.16. *Encarsia mexicana* (Howard) **comb. n.**

(Figure 19A–F)

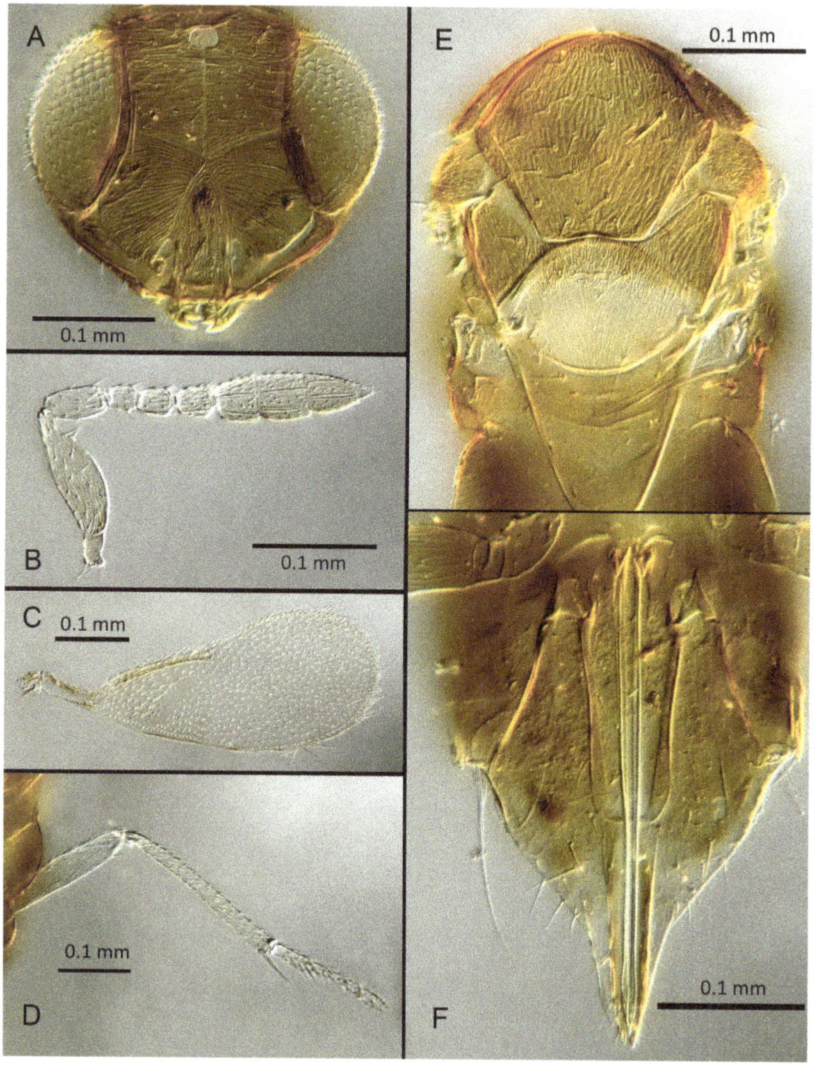

Figure 19. *Encarsia mexicana*: (**A**) head; (**B**) antenna; (**C**) fore wing; (**D**) mid leg; (**E**) dorsal mesosoma; (**F**) ovipositor.

Mesidia mexicana Howard, 1907: 74

Dirphys mexicana (Howard, 1914): 81

Female. Colour: Antennae pale brown, slightly darker on the base of the scape and radicle. Head dark brown with pale lines bordering the eyes and extending along the genae towards clypeus, antennal scrobes, a line from the apex of the scrobes to the median ocellus, and a transverse line midway between the antennal sockets and the median ocellus centrally bordering the dorsal end of antennal scrobes. Mesosoma and metasoma dark brown, with the posterior two-thirds of the scutellum and sides of the metanotum yellow. Legs yellow, with mid and hind coxae and femora partly brown. Wings slightly infuscated below the marginal vein; submarginal, marginal and stigmal veins dark.

Morphology: Head (Figure 19A) with mediofrontal line complete; transfacial line narrow; facial lines very broad adjacent to genae. Scrobes almost entirely with longitudinal sculpture. Antennal formula (Figure 19B): 1, 1, 3, 3; scape slightly expanded, 2.45x pedicel length; pedicel 1.9x F1; F1 0.8X F2; F2 0.9x F3; funicle 0.57X clava length; F6 slightly oblique, claval sensorial complex developed. Flagellum with the following numbers of longitudinal sensilla: F1: 0; F2: 1; F3: 1; F4: 3; F5: 3–4; F6: 4–5. Mandibles (Figure 19A) with two teeth and a truncation. Maxillary palps two-segmented. Mid-lobe of mesoscutum (Figure 19E) with 30–40 setae; each lateral lobe with two setae; each axilla with one seta; scutellum with four setae. Sculpture of mesoscutum aciculate; sculpture of axillae and scutellum longitudinal. Fore wing (Figure 19C) with two large setae and 3–4 smaller setae on submarginal vein, 3–7 setae in basal cell, 7–10 setae on anterior margin of marginal vein, and one seta at the junction of the submarginal vein and parastigma. Linea calva present. Submarginal 0.85x marginal vein. Maximum length of fore wing 2.7x fore wing width, maximum width of wing 4.65x longest setae on marginal fringe. Ovipositor (Figure 19F) length 1.3x mid tibial length; third valvula 0.5x ovipositor; second valvifer 1.3x third valvula; Mid tibial spur (Figure 19D) 0.9x corresponding basitarsus. Metasomal terga T1–T7 with 0, 2 + 2, 2 + 2, 2 + 2, 1 + 2 + 1, 1 + 2 + 1 and four setae, respectively. T7 (Figure 19F) elongate covering third valvula of ovipositor.

Distribution: COSTA RICA: Limon, Heredia, Arenal, Alajuela; MEXICO: Tabasco.

Host: Nealeurodicus altissimus (Quaintance) (=Ceraleurodicus altissimus).

Material examined: Holotype. ♀ MEXICO, Tabasco, San Francisco del Peal. 1.vii.1887 (C.H. Townsend) (USNM). Examined. 6♀ [compared with type series AP xi.90] MEXICO, Tabasco, San Francisco del Peal, 1.vii.1897 (C.H.T. Townsend) ex *Nealeurodicus altissimus* [reared from *Lippia myriocephala*] (NHMUK, USNM). 1♀ COSTA RICA, Heredia, Est. Biol. La Selva, 27–28.ii.2003 (J.S. Noyes) 75 m 10°26′ N 84°01′ W. [DNA 144: OQ683560] (NHMUK); 1♀ COSTA RICA, Arenal, Sen. Pilon, 26.ii.2003 (J.S. Noyes) [DNA 52] (NHMUK); 1♀ COSTA RICA, Alajuela, Est. Caribe R. Rincón Forestal, 19–20.ii.2003 (J.S. Noyes) 400 m, 10°53′ N 85°18′ W [DNA 166] (NHMUK); 1♀ COSTA RICA, Alajuela, P.N. Arenal, send. Pilon, 26.ii.2003 (J.S. Noyes), 600 m 10°27′ N 84°43′ W; [DNA 129] (MZUCR).

Remarks: *Encarsia mexicana* is morphologically closest to *E. fredbennetti* from which it differs by the non-enlarged clava. It is also molecularly closest to *E. fredbennetti*. Sequence data deposited at GenBank accession number: OQ683560.

3.3.17. *Encarsia napo* Polaszek and Hernández-Suárez **sp. n.**

(Figure 20A–F)

Female. Colour: Antennae pale brown, slightly darker on the base of the clava, scape, pedicel, and radicle. Head dark brown. Mesosoma and metasoma uniformly dark brown with the posterior two-thirds of the scutellum and sides of the metanotum yellow. Legs yellow, with hind femur, posterior half of mid femur and anterior half of fore femur light brown, all tarsi pale. Wings hyaline, submarginal vein pale in contrast with darker marginal and stigmal veins.

Morphology: Head (Figure 20A) with mediofrontal line complete; transfacial line evident, broad laterally and tapering towards the middle; facial lines extremely broad along their entire lengths. Scrobes largely smooth, some irregular sculpture centrally. Antennal formula (Figure 19B): 1, 1, 3, 3; scape 2.2x pedicel length; pedicel 1.3x F1; F1 0.88x F2; F2 equal to F3; funicle 0.7x clava; F6 perpendicular. Mandibles (Figure 20A) with two minute teeth and a truncation. Flagellum with the following number of longitudinal sensilla: F1: 1; F2:1; F3: 1 F4: 2 F5: 3; F6: 3. Maxillary palps two-segmented. Mid-lobe of mesoscutum (Figure 20E) with fewer than 20 setae; each lateral lobe with two setae; each axilla with one seta; scutellum with four setae and two vestigial setal bases. Sculpture of mesoscutum, axillae and scutellum longitudinal. Fore wing with two large setae and four smaller setae on submarginal vein, 4–5 setae in basal cell, 7–8 setae on anterior margin of marginal vein, and one seta at the junction of the submarginal vein and parastigma. Linea calva absent. Submarginal 0.80x times marginal vein. Maximum length of fore wing 2.68x fore wing width, maximum width of wing 4.79x longest setae on marginal fringe. Ovipositor (Figure 20F) 0.8x mid tibial length; third valvulae 0.45x ovipositor length; second valvifer 1.2x third valvula. Mid tibial spur (Figure 20D) 0.9x corresponding basitarsus. Metasomal terga T1–T7 with 0, 1 + 1, 1 + 1, 1 + 1, 1 + 2 + 1, 1 + 2 + 1 and six setae, respectively. T7 (Figure 20F) rounded, not extended but covering ovipositor.

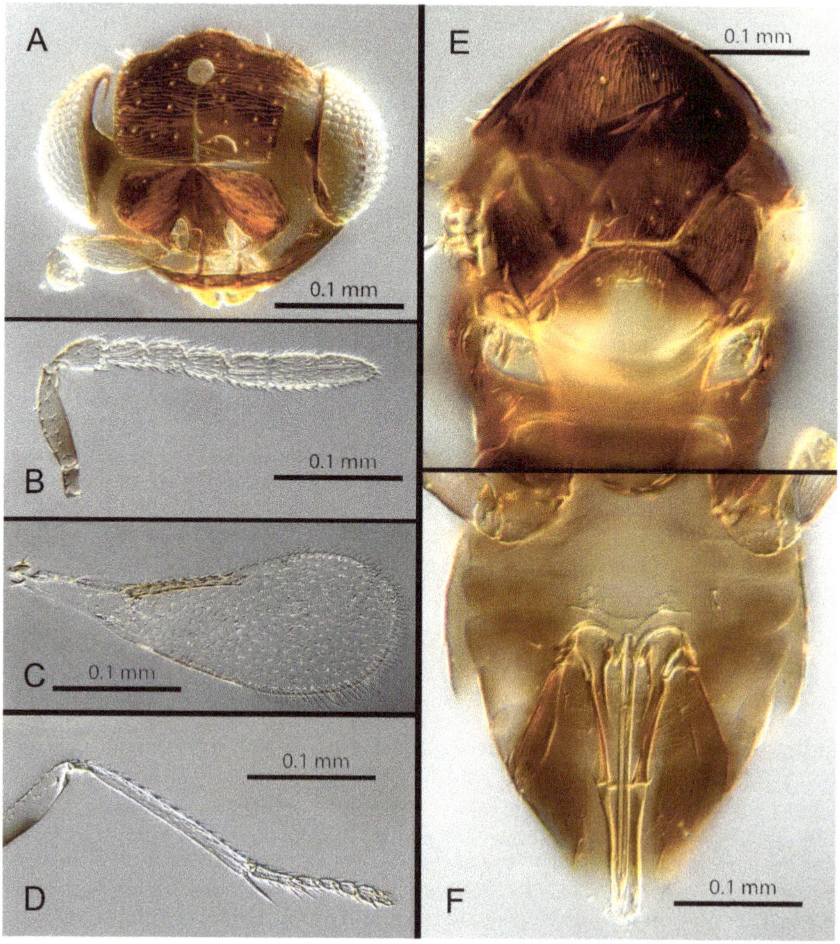

Figure 20. ***Encarsia napo***: (**A**) head; (**B**) antenna; (**C**) fore wing; (**D**) mid leg; (**E**) dorsal mesosoma; (**F**) ovipositor.

Distribution: ECUADOR: Napo River.

Material examined: Holotype ♀ ECUADOR, Napo, transect Ent. 1 km S Onkone Gare Camp, Res. Etnica Waorani, 220 m 0°39′10″ S 76°26′00″ W T.L. Erwin et al. fogging tf forest lot 1193 5.v.1995 [DNA314] (NHMUK). Paratypes 2 slides ♀ ECUADOR, Napo, transect Ent. 1 km S Onkone Gare Camp, Res. Etnica Waorani, 220 m 0°39′10″ S 76°26′00″ W T.L. Erwin et al. fogging tf forest lot 1193 5.v.1995 [DNA 312, 313] (NHMUK).

Remarks: *Encarsia napo* is morphologically closest to *E. erwini* from which it differs by the partly pale mesoscutellum. No molecular data were available for *E. napo*.

Etymology: Named for the Napo River (Rio Napo) on which the type locality is located.

3.3.18. *Encarsia noora* Polaszek and Hernández-Suárez **sp. n.**

(Figure 21A–E)

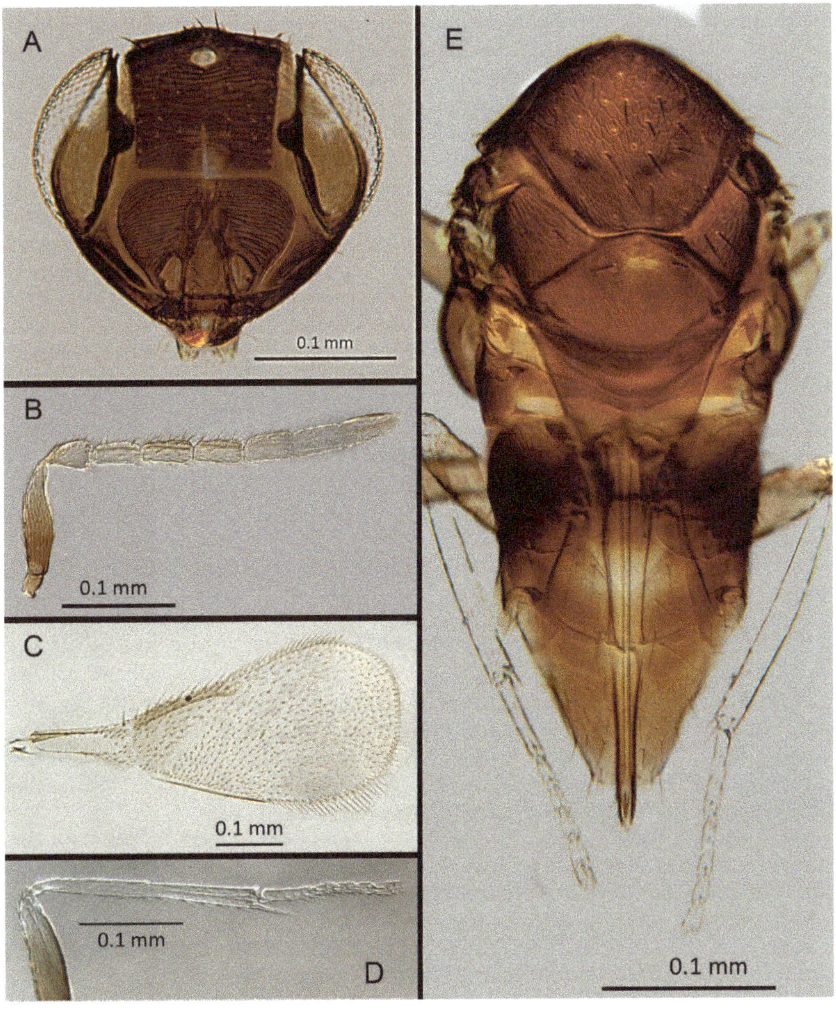

Figure 21. ***Encarsia noora***: (**A**) head; (**B**) antenna; (**C**) fore wing; (**D**) mid leg; (**E**) dorsal habitus.

Female. Colour: Antennae pale yellow. Head brown with pale lines bordering the eyes and extending along the genae towards clypeus. Mesosoma brown but with posterior three-quarters of scutellum pale. Metasoma uniformly light brown. Legs pale yellow; hind coxae and femora brown. Wings hyaline, slightly infuscated below marginal vein; stigmal vein pale in contrast with a darker marginal vein.

Morphology: Head (Figure 21A) with mediofrontal line incomplete, reaching less than halfway to frontal ocellus, transfacial line evident; facial lines very broad along their entire lengths. Scrobes almost entirely smooth, some irregular transverse sculpture basally. Antennal formula (Figure 21B): 1, 1, 3, 3; scape 2x pedicel length; pedicel 2.2x F1; F1 1.2x F2; F2 0.7x F3; funicle 0.6x clava; F6 slightly oblique, claval sensorial complex apparently present. Flagellum with the following number of longitudinal sensilla: F1: 0; F2: 0; F3: 1–2; F4: 3; F5: 4; F6: 4. Mandibles (Figure 21A) with two min teeth and a truncation. Maxillary palps two-segmented. Mid-lobe of mesoscutum (Figure 21E) with fewer than 30 setae; each lateral lobe with two setae; each axilla with one seta; scutellum with four setae. Sculpture of mesoscutum aciculate; sculpture of axillae and scutellum longitudinal. Fore wing (Figure 21C) with two large setae and four smaller setae on submarginal vein, eight setae in basal cell, 7–8 setae on anterior margin of marginal vein, and one seta at the junction of the submarginal vein and parastigma; a group of long setae below marginal vein. Linea calva present. Submarginal 0.8x marginal vein. Maximum length of fore wing 2.58x fore wing width, maximum width of wing 6.39x longest setae on marginal fringe. Ovipositor (Figure 21E) 1.34x mid tibial length; third valvulae 0.43x ovipositor length; second valvifer 1.2x third valvula. Mid tibial spur (Figure 21D) 0.9x corresponding basitarsus. Metasomal terga T1–T7 with 0, 2 + 2, 2 + 2, 2 + 2, 1 + 2 + 1, 1 + 2 + 1 and four setae, respectively. T7 (Figure 21E) extended although apparently not covering ovipositor.

Distribution: TRINIDAD: Mt St Benedict, St Augustine.

Material examined: Holotype ♀ TRINIDAD, I.C.T.A. [Imperial College of Tropical Agriculture, St. Augustine] xii.1953 (F.D. Bennett) [s9] ex whitefly on cocoa. ?*Coccophagus* sp. Det. FDB [identified as *Coccophagus*, F.D. Bennett] Imperial Parasite Service (NHMUK). Paratypes 2♀ TRINIDAD, Mt. St. Benedict, 18.x.1997 (M. Jagroep) [DNA 126: OQ683561] ex whitefly; [DNA 215b] ex Aleurodicinae (NHMUK).

Remarks: *Encarsia noora* is morphologically closest to *E. catula*, differing by the longer F1 in *E. noora*. Molecularly sister species to *E. aisha*. Sequence deposited at GenBank accession number: OQ683561.

Etymology: Named for Noora, daughter of the second author (EHS), and sister to Aisha; see *E. aisha*, above.

3.3.19. *Encarsia svetlana* Polaszek and Hernández-Suárez **sp. n.**

(Figure 22A–F)

Female. Colour: Antennae light brown slightly darker on F6 and radicle. Head brown with pale lines bordering the eyes and extending along the genae towards clypeus. Mesosoma dark brown but with scutellum and sides of the metanotum yellow. Metasoma light brown. Legs uniformly pale yellow. Wings infuscated below marginal vein, stigmal vein pale in contrast with a darker marginal vein.

Morphology: Head (Figure 22A) with mediofrontal line complete; transfacial line narrow; facial lines very broad at level of lower eyes and adjacent to genae. Scrobes with longitudinal aciculate sculpture in the upper half, transverse aciculate sculpture in the lower half. Antennal formula (Figure 22B): 1, 1, 3, 3; scape 2.25x pedicel length; pedicel 2.36x F1; F1 0.86x F2; F2 0.9x F3; funicle 0.49x clava length; F6 broadly oblique, claval sensorial complex developed. Flagellum with the following number of longitudinal sensilla: F1: 0; F2: 1; F3: 1; F4: 1; F5: 4; F6: 4. Mandibles (Figure 22A) with 2 very large ventral teeth and a small truncation, appearing bidentate. Maxillary palps 2-segmented. Mid-lobe of mesoscutum (Figure 22E) with fewer than 30 setae; each lateral lobe with two setae; each axilla with one seta; scutellum with four setae. Sculpture of mesoscutum, axillae and scutellum longitudinal. Fore wing (Figure 22C) with two large setae and four smaller setae on submarginal vein, four setae in basal cell, 6–7 setae on anterior margin of marginal vein, and one seta at the junction of the submarginal vein and parastigma. Linea calva present. Submarginal 0.9x marginal vein. Maximum length of fore wing 2.5x fore wing width, maximum width of wing 3.85x longest setae on marginal fringe. Ovipositor (Figure 22F) 0.6x mid tibial length; third valvulae 0.95x ovipositor length; second valvifer 1.1x third valvula. Mid tibial spur (Figure 22D) 0.9x corresponding basitarsus. Metasomal terga T1–T7 with 0, 2 + 2, 2 + 2, 1 + 1, 1 + 1 + 1, 1 + 1 and four setae, respectively. T7 (Figure 22F) apex rounded not covering ovipositor.

Distribution: GUYANA: Dubulay Ranch.

Material examined: Holotype ♀ GUYANA, Dubulay Ranch, 17–22.i.1999 (M. Sharkey and B. Brown) [DNA 305: OQ683558] Univ. Calif. Riverside, Ent. Res. Museum UCR ENT 182858 5°40.954′ N 57°51.524′ W (UCRC).

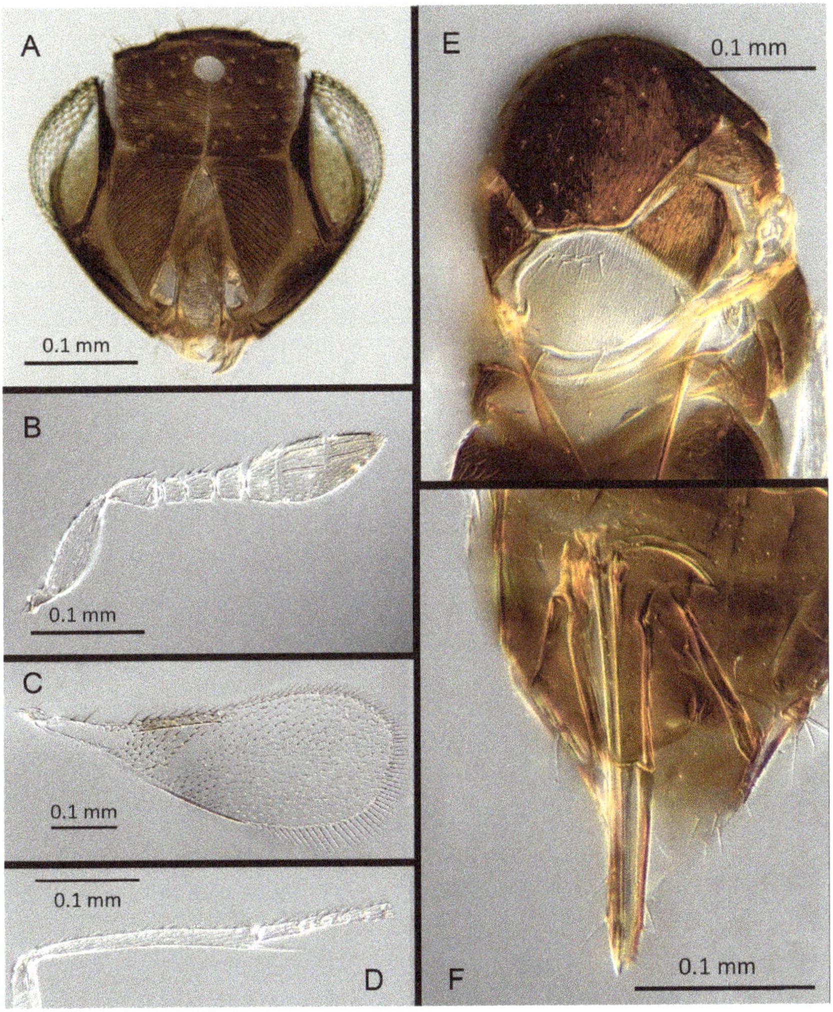

Figure 22. ***Encarsia svetlana***: (**A**) head; (**B**) antenna; (**C**) fore wing; (**D**) mid leg; (**E**) dorsal mesosoma; (**F**) ovipositor.

Remarks: *Encarsia svetlana* is morphologically closest to *E. venia* from which it differs in having an F3 transverse, and its scape is entirely pale. Sequence data deposited under GenBank accession number: OQ683558.

Etymology: Named for Svetlana Myartseva, prolific describer of Mexican Aphelinidae, including many *Encarsia* species.

3.3.20. *Encarsia venia* Polaszek and Hernández-Suárez **sp. n.**

(Figure 23A–F)

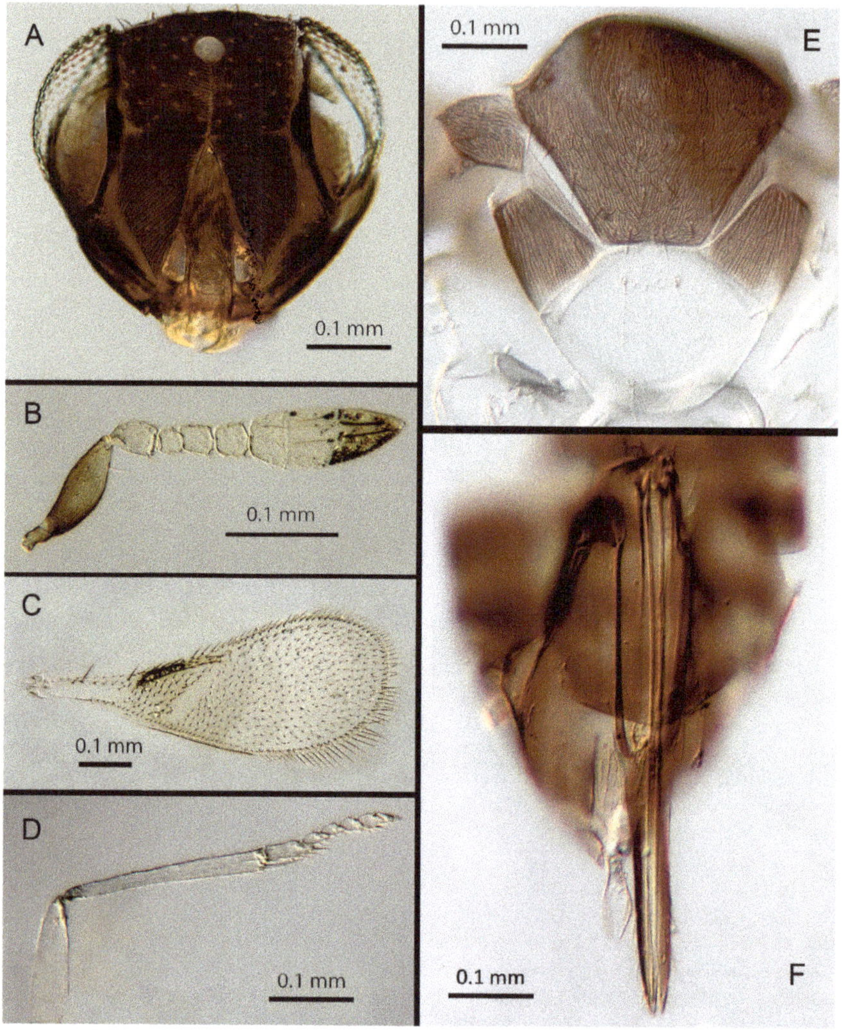

Figure 23. ***Encarsia venia***: (**A**) head; (**B**) antenna; (**C**) fore wing; (**D**) mid leg; (**E**) dorsal mesosoma; (**F**) ovipositor.

Female. Colour: Antennae light brown, slightly darker on scape and radicle. Head dark brown. Mesosoma and metasoma dark brown, with entire scutellum and sides of the metanotum yellow. Legs uniformly pale yellow. Wings hyaline, slightly infuscated below marginal vein.

Morphology: Head (Figure 23A) with mediofrontal line complete, very narrow towards frontal ocellus; transfacial line narrow; facial lines broad along their entire lengths. Scrobes with longitudinal aciculate sculpture in the upper half, transverse aciculate sculpture in the lower half. Antennal formula (Figure 23B): 1, 1, 3, 3; scape 2.4x pedicel length; pedicel 2x F1; F1 0.7x F2; F2 1.1x F3; funicle 0.6x clava; F6 slightly oblique, claval sensorial complex apparently present. Flagellum with the following number of longitudinal sensilla: F1: 0; F2: 1; F3: 1; F4: 2; F5: 3; F6: 3. Mandibles (Figure 23A) with one very large ventral tooth and an upper bidentate tooth. Maxillary palps two-segmented. Mid-lobe of mesoscutum (Figure 23E) with fewer than 30 setae; each lateral lobe with two setae; each axilla with 1 seta; scutellum with four setae. Sculpture of mesoscutum aciculate; sculpture of axillae and scutellum longitudinal. Fore wing with two large setae and four smaller setae on submarginal vein, 3–4 setae in basal cell, 7–8 setae on anterior margin of marginal vein, and one seta at the junction of the submarginal vein and parastigma. Linea calva present. Submarginal equal to marginal vein. Maximum length of fore wing 2.8x fore wing width, maximum width of wing 4.5x longest seta on marginal fringe. Ovipositor (Figure 23F) 1.4x mid tibial length; third valvulae 0.4x ovipositor length; second valvifer 1.3x third valvula. Mid tibial (Figure 23D) spur equal to corresponding basitarsus. Metasomal terga T1–T7 with 0, 1 + 1, 1 + 1, 2 + 2, 1 + 2 + 1, 1 + 2 + 1 and four setae, respectively. T7 (Figure 23F) apex extended but not covering ovipositor entirely. Distribution: COSTA RICA: Heredia.

Material examined: Holotype ♀ COSTA RICA, Limon, Parque Nacional Cahuita 2 m, 26.ii.2004 (J.S. Noyes) [DNA 298: OQ683557] 9°43′ N 82°49′ W (NHMUK). 1♀ COSTA RICA, Limon, Hitoy-Cerere, 21–22.iii.2006 (J.S. Noyes) [DNA 267] 90°40′ N 83°02′ W (NHMUK). 1♀ COSTA RICA, Heredia La Selva, 22.i–2.ii. 1991 (J.S. Noyes).

Remarks: Morphologically closest to *E. svetlana* from which it differs in having F3 quadrate (transverse in *E. svetlana* and a dark scape. Sequence deposited under GenBank accession number: OQ683557.

Etymology: From Latin *venia* meaning "kindness".

3.3.21. *Encarsia myartsevae* Kresslein and Polaszek **nom. nov.**

Remarks: Myartseva (2007) originally described this species as *Encarsia mexicana* Myartseva. In synonymizing *Dirphys* and *Encarsia*, this species name becomes preoccupied by the new combination *Encarsia mexicana* (Howard, 1907). As a result, we rename this species *Encarsia myartsevae* **nom. nov.** in honor of the species' author, Svetlana Myartseva, for her contributions to the systematics of *Encarsia* and in particular her dedication to providing accounts of their host associations. *Encarsia myartsevae* **nom. nov.** is not a member of the *Encarsia mexicana* species-group, instead belonging to the *opulenta* species-group, and is not discussed further in this revision.

3.4. Key to the species of the Encarsia mexicana species-group: Females

1.	Linea calva absent (Figure 9C)	2
-	Linea calva present (Figure 4E)	7
2.	Two setae on axilla (Figure 11E)	*E. dichaeta*
-	One seta on axilla (Figure 4C)	3
3.	Mesoscutum with transverse sculpture (Figure 10E)	*E. diablejo* *
-	Mesoscutum with aciculate (Figure 9E) or longitudinal (e.g., Figure 12E) sculpture	4
4.	Mid lobe of mesoscutum with more than 30 setae (Figure 9E)	*E. cylindrica*
-	Mid lobe of mesoscutum with fewer than 30 setae (e.g., Figure 12E)	5
5.	Third valvulae longer than second valvifers (Figure 12F)	*E. encantadora*
-	Thirds valvulae shorter than second valvifers (e.g., Figure 13F).	6
6.	Mesoscutellum entirely dark (Figure 13E); hind femora pale	*E. erwini*
-	Mesoscutellum pale in lower half (Figure 20E); hind femora dark	*E. napo*
7.	Mid lobe of mesoscutum with fewer than 30 setae (e.g., Figure 18E).	8
-	Mid lobe of mesoscutum with more than 30 setae (e.g., Figure 17E)	12
8.	Ovipositor shorter than mid tibia	*E. mendesi*
-	Ovipositor longer than mid tibia	9
9.	Mesoscutellum entirely pale (e.g., Figure 22E)	10
-	Mesoscutellum entirely dark (e.g., Figure 8E)	11
10.	F3 wider than long; scape pale (Figure 22B)	*E. svetlana*
-	F3 quadrate; scape dark (Figure 23B)	*E. venia*
11.	F1 shorter than, or equal to, F2 (Figure 8B)	*E. catula*
-	F1 longer than F2 (Figure 21B)	*E. noora*
12.	F1 equal in length to F2, or longer	13
-	F1 shorter than F2	15
13.	Suture between F5 and F6 oblique (e.g., Figure 17B)	14
-	Suture between F5 and F6 perpendicular (Figure 16B)	*E. larensis*
14.	Second valvifers almost 2x length third valvulae (Figure 17F)	*E. marynoyesae*
-	Second valvifers about 1.5x length third valvulae (Figure 5F)	*E. aisha*
15.	Mesoscutellum entirely dark	*E. aphania*
-	Mesoscutellum at least partly pale (e.g., Figure 14C)	16
16.	Ovipositor more than 1.8x mid tibia (Figure 4B)	*E. acusa*
-	Ovipositor less than 1.6x mid tibia	17
17.	Fore wing infuscate posterior to the marginal vein, and with robust setae proximad linea calva (e.g., Figure 15B)	18
-	Fore wing lacking the above characters states	19
18.	Antenna with clava enlarged; 3.5x longer than wide (Figure 14D)	*E. fredbennetti*
-	Antenna with clava not enlarged; 2.3x longer than wide (Figure 15C)	*E. inbioa*
19.	Antenna with scape and F6 dark (Figure 7B)	*E. avida*
-	Antenna with scape and F6 pale (Figure 19B)	*E. mexicana*

* Unknown female of *E. diablejo* assumed to share mesosomal sculpture with the male. An accompanying multiple entry key can be found at: https://rkres001.github.io/rkres001.encarsiamexicanakey/ (accessed on 6 June 2023).

4. Discussion

All taxa of *Encarsia mexicana* species-group recovered sister to *Encarsia dichaeta* have a linea calva present on the fore wing (absent in *E. dichaeta*; unknown D2672, and DNA.0165). The presence or absence of a linea clava may represent a local synapomorphy for these two clades within the *mexicana*-group. All taxa which lack a linea calva also lack the enlargement of the clava of the antenna common in this species-group, though *Encarsia noora* possesses a linea calva in the absence of an enlarged clava. Further phylogenetic analyses and sequence capture for the five other species without linea calva (*E. diablejo*, *E. cylindrica*, *E. encantadora*, *E. erwini* and *E. napo*) will be necessary to determine the informativeness of these characters. Other evident characters appear phylogenetically uninformative. Some surprises include *E. svetlana* appearing well-removed from *E. venia* while their morphology is extremely similar. *E. aisha* and *E. noora* despite being DNA sister species have markedly different flagellar shapes.

With the placement of *Dirphys* **syn. n.** in the middle of *Encarsia*, the genus expands to 473 described species. Establishment of the *Encarsia mexicana* species-group further increases the difficulty with which the genus can be recognized. In particular, the genus now contains species with a linea calva on the fore wing, and parasitoids with a gregarious life history, further expanding the already broad host range of the genus. The future classification of *Encarsia* at large will rely upon its clarification by robust phylogenetic hypotheses built upon large molecular datasets (Kresslein et al. unpublished). As the phylogeny of *Encarsia* is further resolved, it will be necessary to revisit the current classification of the genus and determine whether alternative generic classifications would allow for greater diagnosability of the taxa therein.

Author Contributions: Conceptualization, A.P. and E.H.-S.; methodology, A.P., E.H.-S., J.M.-D., Y.M.L. and S.S.; software, S.S., R.L.K. and Y.M.L.; validation, A.P., E.H.-S. and P.H.; formal analysis, S.S. and R.L.K.; investigation, A.P., E.H.-S. and S.S.; resources, A.P., J.M-.D., Y.M.L. and E.H.-S.; data curation, A.P. and J.M.-D.; writing—original draft preparation, A.P. and E.H.-S.; writing—review and editing, A.P., E.H.-S., J.M.-D., P.H., Y.M.L., S.S. and R.L.K.; visualization, A.P., E.H.-S. and R.L.K.; supervision, A.P.; project administration, A.P.; funding acquisition, E.H.-S. and A.P. All authors have read and agreed to the published version of the manuscript.

Funding: This work was supported in part by grants from FUNDESIMCA Fundación para el Desarrollo e Impulso de las Ciencias Agrarias en Canaria.

Data Availability Statement: All DNA sequence data that support the results and conclusions of this study can be found at GenBank.

Acknowledgments: We would like to thank Miguel Alonso-Zarazaga for advice on the gender of *Dirphys*; all lenders and donors of material, especially John Noyes; and Krissy Dominguez and Zach Lahey for many useful comments on the manuscript.

Conflicts of Interest: The authors declare no conflict of interest.

References

1. Howard, L.O. Concerning some Aphelininae. *Proc. Entomol. Soc. Wash.* **1914**, *16*, 79–85.
2. Howard, L.O. New genera and species of Aphelinidae with a revised table of genera. *Tech. Ser. Bur. Entomol. United States Dep. Agric.* **1907**, *12*, 69–88.
3. Polaszek, A.; Hayat, M. A revision of the genera *Dirphys* Howard and *Encarsiella* Hayat (Hymenoptera: Aphelinidae). *Syst. Entomol.* **1992**, *17*, 181–197. [CrossRef]
4. Chavez, T.A. A new species of *Dirphys* Howard (Hymenoptera: Aphelinidae). *Bol. Entomol. Venez.* **1996**, *11*, 12–16.
5. Martin, J.; Polaszek, A. A new genus of Neotropical whitefly, secreting blue-iridescent wax (Sternorrhyncha, Aleyrodidae, Aleurodicinae), and its parasitoids (Hymenoptera, Aphelinidae). *J. Nat. Hist.* **1999**, *33*, 1545–1559. [CrossRef]
6. Hayat, M. The genera of Aphelinidae (Hymenoptera) of the world. *Syst. Entomol.* **1983**, *8*, 63–102. [CrossRef]
7. Hayat, M. Notes on some species of *Coccophagoides*, *Dirphys* and *Encarsia* (Hymenoptera: Aphelinidae). *Orient. Insects* **1989**, *23*, 286–290. [CrossRef]
8. Hayat, M. A revision of the species of *Encarsia* Foerster (Hymenoptera, Aphelinidae) from India and the adjacent countries. *Orient. Insects* **1989**, *23*, 1–131. [CrossRef]

9. Shafee, S.A.; Rizvi, S. Taxonomic notes on some Indian Aphelinidae (Hymenoptera: Chalcidoidea). *Mitt. Der Schweiz. Entomol. Ges.* **1984**, *57*, 379–381.
10. Babcock, C.S.; Heraty, J.M.; De Barro, P.J.; Driver, F.; Schmidt, S. Preliminary phylogeny of *Encarsia* Forster (Hymenoptera: Aphelinidae) based on morphology and 28S rDNA. *Mol. Phylogenet. Evol.* **2001**, *18*, 306–323. [CrossRef] [PubMed]
11. Schmidt, S.; Polaszek, A. *Encarsia* or *Encarsiella*? Redefining generic limits based on morphological and molecular evidence (Hymenoptera, Aphelinidae). *Syst. Entomol.* **2007**, *32*, 81–94. [CrossRef]
12. Heraty, J.M.; Polaszek, A. Morphometric analysis and descriptions of selected species in the *Encarsia strenua* group (Hymenoptera: Aphelinidae). *J. Hymenopt. Res.* **2000**, *9*, 142–169.
13. Myartseva, S.N. Species of genus *Encarsia* Forster (Hymenoptera: Aphelinidae)—Parasitoids of whiteflies (Hemiptera: Aleyrodidae) associated with *Psidium guajava* L. in Mexico, with key and description of new species. *Biosystematica* **2007**, *1*, 7–19.
14. Noyes, J.S. Collecting and preserving chalcid wasps (Hymenoptera, Chalcidoidea). *J. Nat. Hist.* **1982**, *16*, 315–334. [CrossRef]
15. Kim, J.-W.; Heraty, J.M. A phylogenetic analysis of the genera of Aphelininae (Hymenoptera: Aphelinidae), with a generic key and descriptions of new taxa. *Syst. Entomol.* **2012**, *37*, 497–549. [CrossRef]
16. Hayat, M. *Aphelinidae of India (Hymenoptera: Chalcidoidea): A Taxonomic Revision*; Associated Publishers: Gainesville, FL, USA, 1998; Volume 13, p. viii+416-.
17. Polaszek, A.; Ayshford, T.; Yahya, B.E.; Fusu, L. *Wallaceaphytis*: An unusual new genus of parasitoid wasp (Hymenoptera: Aphelinidae) from Borneo. *J. Nat. Hist.* **2013**, *48*, 1111–1123. [CrossRef]
18. Katoh, K.; Standley, D.M. MAFFT multiple sequence alignment software version 7: Improvements in performance adn usability. *Mol. Biol. Evol.* **2013**, *30*, 772–780. [CrossRef]
19. Minh, B.Q.; Schmidt, H.A.; Chernomor, O.; Schrempf, D.; Woodhams, M.D.; von Haeseler, A.; Lanfear, R. IQ-TREE 2: New models and efficient methods for phylogenetic inference in the genomic era. *Mol. Biol. Evol.* **2020**, *37*, 1530–1534. [CrossRef] [PubMed]
20. Hoang, D.T.; Chernomor, O.; von Haeseler, A.; Minh, B.Q.; Vinh, L.S. UFBoot2: Improving the Ultrafast Bootstrap Approximation. *Mol. Biol. Evol.* **2017**, *35*, 518–522. [CrossRef]
21. Hayat, M. On the identity and systematic position of the genus *Dirphys* [Hym.: Aphelinidae]. *Entomophaga* **1981**, *26*, 441–444. [CrossRef]

insects

MDPI

Article

Phylogeny of *Rhynchium* and Its Related Genera (Hymenoptera: Eumeninae) Based on Universal Single-Copy Orthologs and Ultraconserved Elements

Min Dai, Shu-Lin He, Bin Chen and Ting-Jing Li *

Chongqing Key Laboratory of Vector Insects, Institute of Entomology and Molecular Biology, College of Life Science, Chongqing Normal University, Chongqing 401331, China; 2021110513042@stu.cqnu.edu.cn (M.D.); shulinhe@cqnu.edu.cn (S.-L.H.); bin.chen@cqnu.edu.cn (B.C.)
* Correspondence: ltjing1979@cqnu.edu.cn; Tel.: +86-159-2297-9264

Simple Summary: The eumenine wasps of the genus *Rhynchium* Spinola, 1806; *Allorhynchium* van der Vecht, 1963; *Anterhynchium* de Saussure, 1863; and *Pararrhynchium* de Saussure, 1855 are related and sometimes hard to tell apart from each other. In this study, we first reconstructed the phylogenetic relationships of these genera based on universal single-copy orthologs and ultraconserved elements extracted from 10 newly sequenced low-coverage whole genomes. The results showed that *Allorhynchium* and *Lissodynerus* are distinct from the other four taxa. The genus *Rhynchium* was recovered as monophyletic, whereas *Anterhynchium* was recovered as paraphyletic, with *Anterhynchium* (*Dirhynchium*) as a sister to *Rhynchium* and hence deserving a separate genus *Dirhynchium;* and within the genus *Pararrhynchium*, *P. septemfasciatus feanus* and *P. venkataramani* were separated, not clustered on a branch. It is suggested that the genus *Lissodynerus* should be restituted as a valid genus, not a synonym of the *Pararrhynchium.* The results are consistent with previous eumenine mitochondrial genome phylogenetic analyses. This paper confirms the feasibility of low-coverage whole genome eumenine wasp phylogenetics and provides a reference for subsequent research in Eumeninae.

Abstract: The subfamily Eumeninae is a large group of fierce predatory insects that prey mainly on the larvae of Lepidoptera pests. Because of the highly similar morphologies of the genus *Rhynchium* and its related genera in the subfamily, including *Rhynchium* Spinola, *Allorhynchium* van der Vecht, *Anterhynchium* de Saussure, *Pararrhynchium* de Saussure, it is essential to delineate their relationships. A previous phylogenetic analysis based on mitochondrial genomes suggested the inconsistent relationships of these genera under traditional classification based on morphological characters. In this study, we first used single-copy orthologs [USCO] and ultraconserved elements [UCE] extracted from 10 newly sequenced low-coverage whole genomes to resolve the phylogenetic relationships of the above genera. The newly sequenced genomes are 152.99 Mb to 211.49 Mb in size with high completeness (BUSCO complete: 91.5–95.6%) and G + C content (36.31–38.76%). Based on extracted 5811 USCOs and 2312 UCEs, the phylogenetic relationships of *Rhynchium* and its related genera were: ((*Allorhynchium* + *Lissodynerus*) + (*Pararrhynchium* + (*Anterhynchium* + (*Dirhynchium* + *Rhynchium*)))), which was consistent with the mitochondrial genome results. The results supported the genus *Rhynchium* as monophyletic, whereas *Anterhynchium* was recovered as paraphyletic, with *Anterhynchium* (*Dirhynchium*) as a sister to *Rhynchium* and hence deserving generic status; In addition, in the genus *Pararrhynchium*, *P. septemfasciatus feanus* and *P. venkataramani* were separated, not clustered on a branch, just as *P. septemfasciatus feanus* was not together with *P. striatum* based on mitochondrial genomes. Since *Lissodynerus septemfasciatus*, the type species of the genus *Lissodynerus*, was transferred to *Pararrhynchium*, it is considered that the genus *Lissodynerus* should be restituted as a valid genus, not a synonym of *Pararrhynchium.*

Keywords: Eumeninae; low-coverage whole genomes; universal single-copy orthologs [USCO]; ultraconserved elements [UCE]; phylogeny

1. Introduction

The subfamily Eumeninae (Hymenoptera: Vespidae) is a large group of fierce predatory insects that prey mainly on the larvae of Lepidoptera pests and can be used for biological control [1,2]. It contains seven genera with roots *rhynchium* worldwide, including *Allorhynchium* van der Vecht, 1963, *Anterhynchium* de Saussure, 1863, *Emeryrhynchium* Gusenleitner, 2007, *Gibberhynchium* Gusenleitner, 2002, *Pararrhynchium* de Saussure, 1855, *Rhynchium* Spinola, 1806, and *Xenorhynchium* van der Vecht, 1963. They are restricted to the Old World [3,4], with four, *Rhynchium*, *Anterhynchium*, *Pararrhynchium* and *Allorhynchium*, found in China. Hereafter, we refer to these four genera as the *Rhynchium* genera group in this study. The *Rhynchium* genera group has similar morphology, with a body size ranging from 8.9 mm to 15.1 mm, relatively stout, and slightly narrower first than the second gastral segment [5–14]. Because of the highly similar morphologies of these genera, it is sometimes challenging to tell them apart from each other, and therefore delineating their relationships is a significant exploration.

The phylogenetic relationships within the *Rhynchium* genera group have been discussed in a previous study based on morphological features and mitochondrial genomes [15], but uncertainties and conflicting classification schemes remain among genera and subgenera and on the taxonomic status of certain species. Firstly, the taxonomic status of *Anterhynchium* (*Dirhynchium*) is controversial. *Dirhynchium* van der Vecht, 1963 has long been regarded as a subgenus of *Anterhynchium* [8–11] based on morphology. This, however, conflicts with the mitochondrial phylogenetic analysis which suggested that the subgenus *Dirhynchium* should be elevated to genus [16]. Meanwhile, the taxonomy of wasps in *Anterhynchium* is still poor. As Selis and Carpenter systematically revised African *Anterhynchium* and did a phylogenetic analysis based on morphological characters to move several species from this genus to other genera [17], are there more revisions including the division of subgenera in this genus that needs to be improved in other regions? Secondly, the taxonomic status of the species *Pararrhynchium septemfasciatus feanus* (Giordani Soika 1941) remains uncertain. *Pararrhynchium septemfasciatus feanus* (Giordani Soika 1941) was transferred from the genus *Lissodynerus* with the synonymy of the genus *Lissodynerus* as *Pararrhynchium* by Carpenter and Brown based on morphological characters [18]. Originally, *P. septemfasciatus feanus* (Giordani Soika 1941) was described as *Ancistrocerus septemfasciatus var. feanus* Giordani Soika, 1941 in the genus *Ancistrocerus*, and then was combined within the genus *Lissodynerus* by Giordani Soika [19,20], which was supported by Tan et al. and Li et al. [5,21]. In a systematic taxonomic collection of the subfamily Eumeninae in China, Tan et al. placed *P. septemfasciatus feanus* in the genus *Lissodynerus* based on the prestigma of the forewing, which is longer than half the length of the pterostigma [5]. Meanwhile, in our previous study of the subfamily Eumeninae based on mitochondrial genomes [16], the species *P. septemfasciatus feanus* ran out of the genus *Pararrhynchium*. Resolving these above uncertainties requires more in-depth research.

With advancing sequencing technologies and increasing genomic data, low-coverage whole genome sequencing is becoming a popular technique in phylogenetic studies due to its simple sample preparation and low cost [22]. Universal single-copy orthologs (USCOs) are selected from OrthoDB orthologous groups that contain genes present as single-copy orthologs in at least 90% of the species and are of particular importance in providing the basis to infer species phylogenetics [23]. Due to their slow evolutionary rate, less saturation than other markers, and not crossing over with most types of paralogous genes, Ultra-conserved elements (UCEs) are another set of increasingly used markers for phylogenetic analysis [24,25]. Therefore, reconstructing phylogenetic relationships with USCOs and/or UCEs has been successfully applied to a range of species, such as springtails, bees, rodents, and monocots [26–29]. At present, although systematic phylogenetic analysis of all major lineages of Hymenoptera based on protein-coding genes has been conducted [30], phylogenetic studies of the subfamily Eumeninae based on nuclear data are still poorly. In this paper, we aim to investigate the phylogenetic relationships of *Rhynchium* and its related genera in the subfamily Eumeninae using low-coverage whole genome sequencing with

USCOs and UCEs, which could also serve as new molecular markers for the subfamily Eumeninae.

2. Materials and Methods

2.1. Sampling and Sequencing

Here, we sampled 10 species of the *Rhynchium* genera group with two species for each of the genera *Rhynchium*, *Allorhynchium*, and *Pararrhynchium*, and four species of the two subgenera *Anterhynchium* and *Dirhynchium* in the genus *Anterhynchium* (detail information is shown in Table 1, more information see Table S1 in the Supplementary Materials). Specimens of the 10 species were netted, smothered in collection tubes, immersed in 95% ethanol, and stored at −20 °C at the Institute of Entomology and Molecular Biology, Chongqing Normal University. The head and gaster were removed to reduce contamination of the intestinal flora; and the remaining tissue was sent to Novogene Bioinformatics Technology Co., Ltd. (Tianjin, China) for sequencing. Genomic DNA was extracted and purified using the DNeasy DNA Extraction Kit (Qiagen). The qualified libraries were pooled and sequenced on Illumina platforms with PE150.

Table 1. Summary for 10 newly-sequenced low-coverage whole genomic samples.

Species	Accession Number	Average Read Coverage (X)	BUSCO Completeness (%)	The Number of Scaffolds	Assembly Size (Mb)	Max Read Length (kb)	N50 Scaffold (kb)	GC (%)
Rhynchium brunneum	SAMN36845277	78.85	94.6	42,348	194.79	1170.58	134.32	37.24
Rhynchium quinquecinctum	SAMN36845333	78.26	94.0	48,619	196.27	1281.6	105.73	37.24
Allorhynchium argentatum	SAMN36845335	84.95	95.3	49,682	180.81	894.85	160.54	37.27
Allorhynchium chinense	SAMN36845336	72.63	91.5	187,058	211.49	318.76	23.39	36.31
Anterhynchium (*A.*) *abdominale*	SAMN36845337	83.57	94.0	60,405	183.79	586.96	63.68	38.61
Anterhynchium (*A.*) *mellyi*	SAMN36845339	90.69	94.7	23,487	169.37	842.5	112.22	38.66
Anterhynchium (*Dirhynchium*) *flavomarginatum*	SAMN36845338	86.10	94.4	23,366	178.40	786.82	74.28	37.83
Anterhynchium (*Dirhynchium*) *coracinum*	SAMN36845334	86.89	94.3	21,837	176.77	739.09	68.99	37.69
Pararrhynchium septemfasciatus feanus	SAMN36845341	100.40	95.6	15,712	152.99	4723.7	670.26	38.53
Pararrhynchium venkataramani	SAMN36845340	85.82	92.5	51,762	178.97	728.23	51.56	38.76

We chose two species in Vespinae as outgroups, including the genomic assemblies of *Dolichovespula media* (GCA_911387685.1) and *Vespa crabro* (GCA_910589235.2) from NCBI.

2.2. Genome Assembly and Matrix Generation

BBTools v38.96 [31] was used for quality control. The duplications were removed by clumpify.sh; low-quality reads were trimmed with bbduk.sh; reads shorter than 15 bp and containing more than 5 consecutive Ns after trimming were discarded; remaining reads were then normalized with bbnorm.sh to reduce the complexity and speed up the assembling. The reads were assembled de novo and scaffolded by using SPAdes v3.15.5 [32]; the gaps were filled by GapCloser v1.12 [33] in the SOAPdenovo2 to replace base N with base A\T\G\C. The completeness of the assemblies was evaluated by using BUSCO v5.4.3 with the Hymenoptera database (https://busco-data.ezlab.org/v5/data/lineages/hymenoptera_odb10.2020-08-05.tar.gz, accessed on 28 June 2021).

Two different marker sets (USCO and UCE) were extracted separately from the 10 genomes. Universal single-copy orthologs (USCOs) were extracted with BUSCO v5.4.3 [34] against Hymenoptera reference gene sets ($n = 5991$). USCO amino acid sequences were used for subsequent analyses. Each protein-coding gene sequence was aligned with the L-INS-I method in MAFFT v7.490 [35], and each alignment was trimmed with BMGE [36] to remove unreliable homologous regions and decrease compositional heterogeneity. Those alignments that violated the stationary, reversible, and homogeneous (SRH) assumptions [37] were further excluded using '--symtest' implemented in IQ-TREE v2.0 [38]. The trimmed alignments were subsequently concatenated using FASConCAT-g v1.05 [39]; three USCO amino acid matrices (USCO50, USCO90, and USCO100) with the completeness of 50%, 90%, and 100% were constructed, representing the lowest ratio of taxa for all partitions (see S1 in the Supplementary Materials). Ultraconserved element probes have been designed in several insect orders, including Hymenoptera. The Hymenoptera 2.5 kv2 bait set [40], containing 31,827 baits targeting 2590 loci, was used as the reference probe set. The UCEs of the 10 assembled genomes were extracted by PHYLUCE v1.6.3 [41]. Three UCE matrices, including UCE60, UCE90, and UCE100, were generated in a similar way to the construction of USCO matrices, including MAFFT alignment, BMGE trimming, FASConCAT-g concatenation, and IQ-TREE filtering for SRH model violations, for subsequent analyses (see S1 in the Supplementary Materials).

2.3. Phylogenetic Analyses

We used partitioned maximum likelihood (ML), heterologous model, and multi-species coalescent model (MSC) methods to reconstruct the phylogenetic relationships of the *Rhynchium* genera group. For the partitioned ML analyses, we used the MODELFINDER [42] to select the most appropriate substitution model for each partition by employing the relaxed hierarchical clustering algorithm '-rclusterf 10' in IQ-TREE [43]. The best protein substitution model for the USCO matrices was 'LG + I + G', and the best substitution model for the UCE matrices was 'GTR + I + G'. The full IQ-TREE command line used was 'iqtree2 -s example.phy -p example.nex -m MFP + MERGE --symtest-remove-bad -B 1000 -alrt 1000 --prefix -rcluster 10'. To avoid the influence of heterotachous evolutionary sequences on phylogenetic inference, we also used the General Heterogeneous Evolution on a Single Topology (GHOST) [44] model in IQ-TREE with '-m LG + FO*H4' for USCO amino acids and with '-m GTR + FO*H4' for UCE nucleotides. Simultaneously, 1000 replicates of SH-aLRT [45] and 1000 replicates of UFBoot2 [46] were performed for all ML analyses to obtain species trees. For the multi-species coalescent model analyses, individual gene trees were estimated from the previous steps of matrix generation, and ASTRAL-III v5.7.1 [47] was used with default parameters to infer species trees, with the advantage of highly accurate support values and fast computation speed [48]. Although support values such as bootstrap or posterior probability give statistical confidence in the species tree topology, there may still be significant underlying gene tree discordance [49] even for branches with 100% support. Therefore, genealogical consistency between species and gene trees was quantified with the gene concordance factor (gCF) and the site concordance (sCF) using IQ-TREE.

In addition, site heterogeneity has been highlighted as more important than protein-wise heterotactic (partitioning) in phylogenomic analyses [50]. To avoid long-branch attraction (LBA) artifacts, we also applied site-heterogeneous models in ML inference. The posterior mean site frequency (PMSF) [51] model was performed for each USCO matrix by specifying a profile mixture model LG + C20 + F + G in IQ-TREE.

2.4. Hypothesis Testing

Following the phylogenetic analyses, one node showed inconsistent results with analyses based on information from individual genes (ASTRAL and partitioned ML) and analyses of the different matrices using profile mixture models (GHOST and PMSF). Gene tree heterogeneity, such as incomplete lineage sorting, is a known source of erroneous inference in summary coalescent methods [52]. We suspected that the discordance between our results was due to uninformative genes in gene-wise partitioning and gene trees. First, using likelihood-mapping (LMAP) [53] as implemented in IQ-TREE, we evaluated whether individual genes contained sufficient information to distinguish between different topologies at the two competing nodes. For the recalcitrant node, four clusters of taxa were defined (Vespinae, *Rhynchium* s.s., *Pararhynchium venkataramani*, and *Allorhynchium* s.s.) (Figure 1). The topologies 1 and 2 were of interest; the topology 3, which was equally conceivable given the unrooted tree with four clusters but had not been in any of the earlier analyses, could have been recovered. Then, we tested the resultant two alternative topologies (T1, T2) with the matrix UCE90 using approximately the unbiased (AU), weighted Kishino–Hasegawa (WKH), and weighted Shimodaira–Hasegawa (WSH) tests under the partitioned model '-m GTR + I + G' in IQ-TREE: T1, (*Allorhynchium s.s.* + *Pararrhynchium venkataramani*) + remaining taxa; T2, *Allorhynchium s.s.* + (*Pararrhynchium venkataramani* + remaining taxa). The tree estimated from previous ASTRAL inference on matrix UCE90 was used as the guide tree and the starting tree.

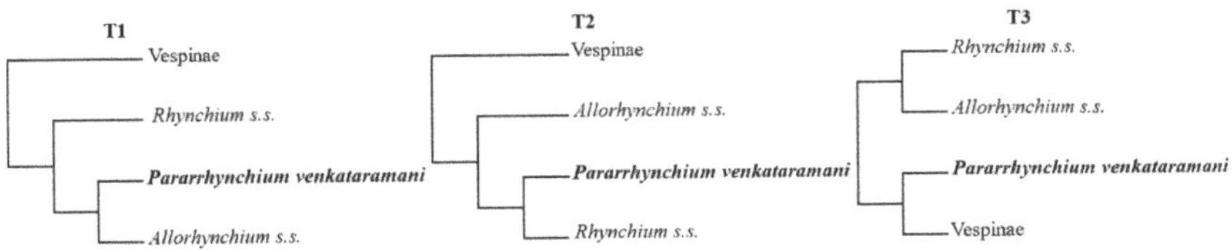

Figure 1. Likelihood mapping analyses (LMAP). Possible topologies for the position of *Pararrhynchium venkataramani* based on four clusters of taxa.

2.5. Divergence Time Analyses

Divergence times were estimated using MCMCTree in PAML v4.9j [54] based on the nucleotide sequences of UCEs. We performed divergence analyses for three matrices (UCE60, UCE90, and UCE100) representing three different data sizes (2,396,740, 2,159,911, and 1,933,308 sites, respectively, Table 2) to evaluate the potential influence of the data size on the branch lengths. For input trees, topology T2 estimated from previous ASTRAL analyses was preferred, and calibration fossils were searched on paleobiodb (https://www.paleobiodb.org/, accessed on 5 August 2023). The fossil record of *Rhynchium andrei* (61.6–59.2 Ma) [55] was used as a calibration point for the crown age of *Rhynchium*. To reduce the computational burden, approximate likelihood calculation and ML estimation of branch lengths was performed by using 'usedata = 2 and 3' [56]. We used the independent rates clock model (clock = 2) and the GTR substitution model (model = 7) to calculate the Hessian matrices. We set the parameters of the mean substitution rate and the rate drift as 'rgene_gamma = 2 20 1' and 'sigma2_gamma = 1 10 1'. The estimation was run for 1 billion MCMC generations sampled every 1,000,000 generations, with the first 0.2 million as burn-in. Convergence was checked using Tracer v1.7.2 [57] to ensure the Effective Sample Size was over 200.

Table 2. Summary of USCOs (amino acid) and UCEs (nucleotide) matrices.

Matrix	Average Missing Taxa per Locus (%)	Number of Loci	Number of Sites	Missing Sites (%)	Average Locus Length
USCO50	2.99	5714	3,467,928	3.19	606.92
USCO90	1.05	5115	2,945,110	1.18	575.78
USCO100	0	4398	2,423,609	0.01	551.07
UCE60	2.56	2138	2,396,740	2.56	1121.02
UCE90	0.77	1928	2,159,911	0.85	1120.29
UCE100	0	1726	1,933,308	0.24	1120.11

3. Results

3.1. Genome Assembly and Matrix Generation

We sequenced approximately 15 G raw data for each sample, with 72 to 100 sequence coverage. The sizes of the assemblies range from 152.99 Mb (*Pararrhynchium septemfasciatus feanus*) to 211.49 Mb (*Allorhynchium chinense*). The number of scaffolds ranges from 15,712 to 187,058, the N50 of the assemblies from 23.39 to 670.26 kb, the max read length from 842.5 kb to 4723.7 kb, and the GC content from 36.31% to 38.76%. Additional statistics, including average read coverage, the number of scaffolds, assembly size, maximum read length, and N50, are provided in Table 1.

The completeness of genome assemblies was 91.50–95.60%, of which 0.10–0.20% were duplications, and 1.40–3.60% were fragmented (Figure 2 and Table 1). The sequences of the same ortholog were merged into a FASTA file, resulting in 5811 USCOs with amino acid sequences ranging from 2,423,609 to 3,551,689. For UCE, a total of 2312 UCEs with 2,585,262 nucleotide sites were extracted. After sequence alignment, screening, and trimming, the six matrices used for phylogenetic analysis included 4398–5560 USCOs and 1726–2312 UCEs with 2,423,609–3,327,612 amino acid sites and 2,159,911–2,475,217 nucleotide sites, respectively (Table 2).

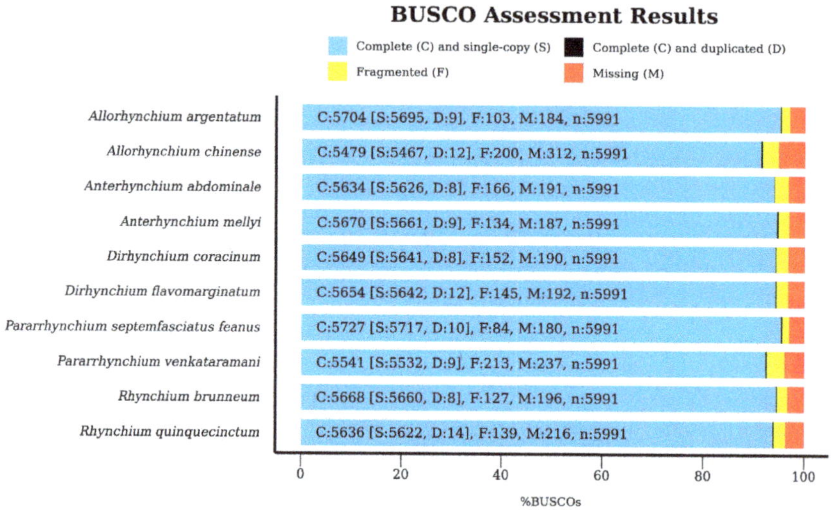

Figure 2. BUSCO completeness: complete (C), complete single-copy (S, light blue), complete duplicated (D, black), fragmented (F, yellow), and missing (M, red).

3.2. Phylogenetic Relationships among the Rhynchium Genera Group

Two topologies of the *Rhynchium* genera group were inferred based on the matrices and inference models, with the majority of nodes consistent except for the position of *Pararrhynchium venkataramani* in both topologies (see S2 in the Supplementary Materials). Under the first topology (T1), *Pararrhynchium venkataramani* was sister to a clade comprising *Pararrhynchium septemfasciatus feanus*, and *Allorhynchium*; under the second topology (T2), *Pararrhynchium venkataramani* was sister to a clade comprising (*Anterhynchium* + (*Dirhynchium* + *Rhynchium*)). The partitioning analyses of all matrices and the GHOST model and PMSF model of all USCO matrices supported topology 1 (T1). However, the GHOST model of UCE matrices supported topology 2 (T2), and ASTRAL analyses of UCEs and USCOs (USCO100 excluded) recovered topology 2 (T2) (Table 3). All tree files are available in the Supplementary Materials S3.

Table 3. Summary of topological hypotheses recovered under different analyses.

Matrix	Partitioning	GHOST	ASTRAL	PMSF
USCO50	T1	T1	T2	T1
USCO90	T1	T1	T2	T1
USCO100	T1	T1	T1	T1
UCE60	T1	T2	T2	-
UCE90	T1	T2	T2	-
UCE100	T1	T2	T2	-

Furthermore, we conducted an LMAP test for the relationships of *Pararrhynchium venkataramani* with others. The T2 was supported by USCOs matrices (more than 79.5%) and UCEs matrices (100%) (Figure 3). Meanwhile, T2 was strongly supported by all hypothesis tests (AU, WKH, and WSH) and T1 was strongly rejected (Table 4). Combining LMAP tests and hypothesis tests, the results supported the monophyletic *Anterhynchium* as a sister group to a clade comprising two genera *Dirhynchium* and *Rhynchium*. Meanwhile, *Pararrhynchium septemfasciatus feanus* and *Pararrhynchium venkataramani* were separated, not clustered on a branch (Figure 4).

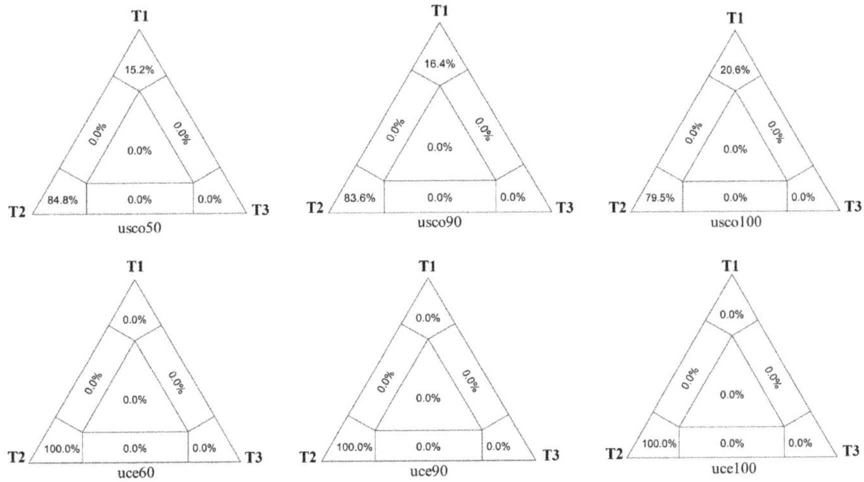

Figure 3. Likelihood maps for the position of *Pararrhynchium venkataramani* from two matrices with contrasting distributions of quartets.

Table 4. Results of the hypothesis tests.

Topologies	log*L*	delta*L*	bp-RELL	*P*-WKH	*P*-WSH	*P*-AU
T1	−8,408,349.19	964.1	0 *	0 *	0 *	2.88×10^{-6}
T2	**−8,407,385.093**	**0 ***	**1**	**1**	**1**	**1**

Best-fit hypotheses are given in boldface. 0 * means that the value < 0. AU, approximately unbiased; RELL, resampled estimated log-likelihood; WKH, weighted Kishino–Hasegawa; WSH, weighted Shimodaira–Hasegawa.

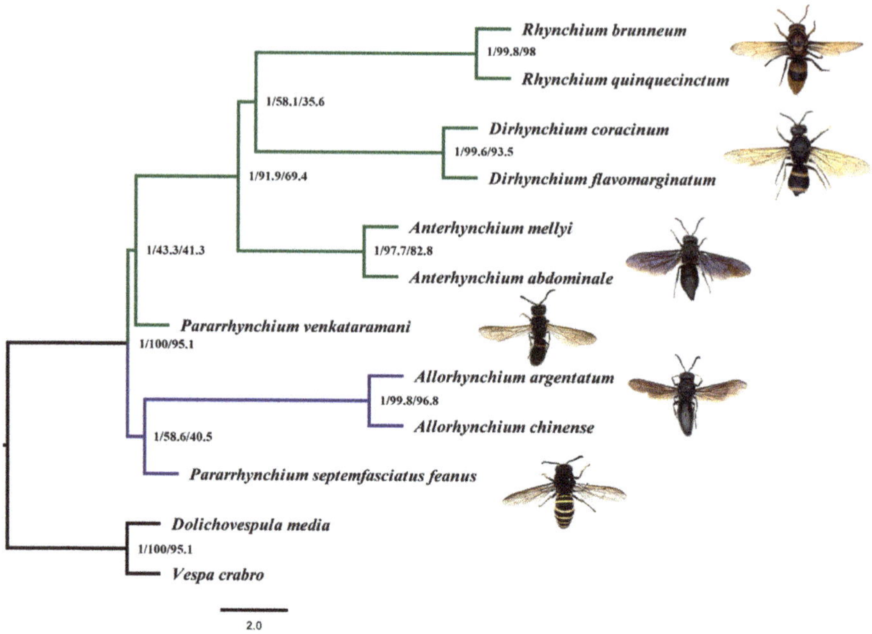

Figure 4. Phylogeny of the *Rhynchium* genera group inferred from matrix UCE90 using the MSC model implemented in ASTRAL. Node labels show BPP/gCF/sCF.

3.3. Molecular Dating of the Rhynchium Genera Group

All the runs of data matrices have ESS values greater than 200, and a strong correlation (R2 > 0.99) of the convergence plots indicates good convergence between parallel runs (Figure 5). The MCMCTree results (Figure 6) show that the most recent common ancestor (MRCA) of the *Rhynchium* genera group was inferred to be the Cretaceous (95.74 million years (Ma), 83.09–107.58 at 95% highest posterior density (HPD)). The crown groups of *Allorhynchium* and *Pararrhynchium septemfasciatus feanus* originated during the Cretaceous–Paleogene (71.93 Ma, 41.27–96.98); the crown age of *Allorhynchium* was estimated at the Paleogene–Quaternary (13.98 Ma, 1.42–33.35); the origins of *Pararrhynchium* and the clade containing *Dirhynchium, Rhynchium,* and *Anterrhynchium* occurred during the Cretaceous (88.35 Ma, 76.29–100.04); the origins of the clade comprising *Dirhynchium* and *Rhynchium* also occurred during the Cretaceous–Paleogene (71.47 Ma, 62.96–80.86); the crown age of *Rhynchium* falls at the Paleogene (60.24 Ma, 59.15–61.54), consistent with the timing of the fossil node; *Dirhynchium* and *Anterrhynchium* occurred during the Paleogene–Quaternary (14.57 Ma, 1.11–36.5; 13.98 Ma, 1.42–33.35, respectively).

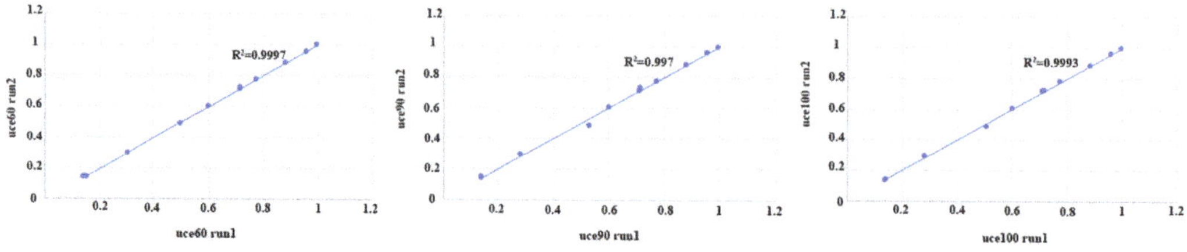

Figure 5. The convergence plot of MCMCtree result.

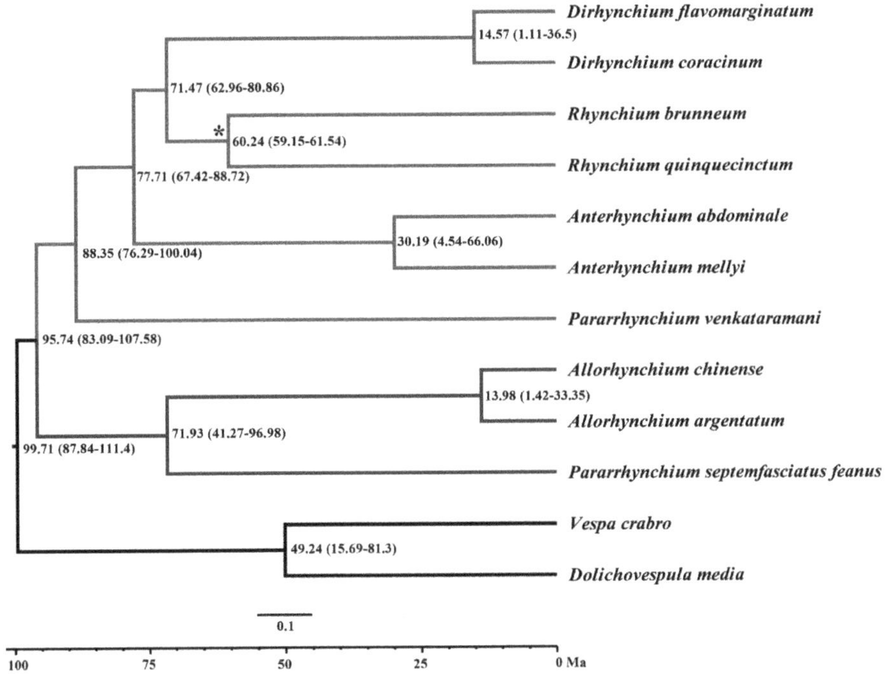

Figure 6. The *Rhynchium* genera group divergence time estimation performed by using MCMCTree on the ASTRAL topology. Node numbers and node bars represent 95% CIs of the estimated divergence times integrated from all MCMC runs. Asterisk mark node constrained with the fossil.

4. Discussion

4.1. Sources of Gene Tree Discordance

In our studies, phylogenetic results were incongruent depending on the inferred models. Partitioned ML, GHOST (USCO matrices), and PMSF analyses favored topology 1, while GHOST (UCE matrices) and a coalescent approach based on gene trees recovered topology 2. In the two topological trees, the support values such as bootstrap were high, but the gene concordance factor (gCF) and the novel site concordance factor (sCF) values ($gCF < 30$, $sCF < 35$, see S2 in the Supplementary Materials) for the inconsistent node was relatively low, demonstrating that systematic errors due to gene tree conflicts are the main reason for the inconsistency of the two phylogenetic trees.

Gene tree conflicts in closely related species are commonly due to genealogical discordances, such as ILS [58–60], which have been detected in hominids, pines, cichlids, finches, and grasshoppers [61–65]. MSC implemented by ASTRAL incorporates a higher degree of biological realism than the conventional concatenation-based models, which is expected to produce more plausible reconstructions [47]. In the present study, most of the ASTRAL analyses (except the USCO100 matrix) support T2, *Pararrhynchium venkataramani* was sister to a clade comprising (*Anterhynchium* + (*Dirhynchium* + *Rhynchium*)). In addition, accounting for the heterogeneity of substitution patterns between sites and genes in proteins [51], we also applied site-heterogeneous models to mitigate possible long-branch attraction (LBA) artifacts. Interestingly, the phylogenetic trees obtained from the site-heterogeneous model (partitioned and GHOST) and the site-homogeneous model (LG + PMSF(C20)) based on the USCO matrices were congruent, suggesting that there's little LBA bias in this study. The inconsistent phylogenetic trees obtained by partitioned ML, GHOST, MSC, and PMSF models, were tested by LMAP tests, which support T2 by both the USCO and UCE matrices, so ILS is the main source of conflict in the gene tree of this study.

UCEs have grown in appeal among hymenopteran systematists who employ hybrid enrichment or occasionally transcriptome sequencing techniques [66–68], while the application of USCOs as tools for phylogenetic inference in Hymenoptera is still in its infancy. The successful application of USCOs and UCEs in bees [27] suggests that low-coverage whole genomes could resolve uncertainty in phylogenetic analyses of Hymenoptera, providing a new set of molecular markers. We used both USCOs and UCE markers from low-coverage whole genomes for phylogenetic reconstruction and finally obtained consistent results with maximum support for all nodes. Once again, it has been proven that phylogenetic reconstructions in Hymenoptera using USCOs and UCEs from low-coverage whole genomes are feasible.

4.2. Phylogenetic Relationships

The best-supported topology shows the clade consisting of *Pararrhynchium, Anterhynchium, Dirhynchium,* and *Rhynchium* is monophyletic; *Allorhynchium* and *Lissodynerus* are more distantly related to the other four genera.

In one clade, a monophyletic *Anterhynchium* is the sister to a clade comprising *Dirhynchium* and *Rhynchium. Anterhynchium, Dirhynchium,* and *Rhynchium* form a monophyletic lineage that shares a common feature: propodeum and metanotum at different levels in lateral view [5,8–11,14,15]. This supports the elevation of *Dirhynchium* to a valid genus, as it is proposed based on a mitogenomic analysis [16] and corroborated by morphological differences. Current diagnostic characters of *Dirhynchium* are as follows: basal narrow part of the first gastral sternum with dense transverse striae through the whole width, the third and following gastral terga very coarsely punctate at the base, male mandible not deeply emarginated on the inner side. However, the diagnostic characters of the nominotypical subgenus *Anterhynchium* are distinctly different from *Dirhynchium*: narrow basal part of the first gastral sternum smooth or with only a narrow median band of fine, short, and shiny transverse striae, gastral terga only finely punctate at the base, male mandible deeply emarginated on the inner side near the middle [8–11]. Therefore, we propose a new genus, *Dirhynchium* van der Vecht, 1963.

In the phylogeny, the species *Pararrhynchium septemfasciatus feanus* is more closely related to *Allorhynchium* van der Vecht, 1963 rather than *Pararrhynchium venkataramani*, indicating the species does not belong to the genus *Pararrhynchium*. Considering *P. septemfasciatus* was transferred from *Lissodynerus* as a synonym *Pararrhynchium* [18], and *Lissodynerus septemfasciatus* was the type species of the genus *Lissodynerus*, the genus *Lissodynerus* should be restituted as a valid genus, not a synonym of the genus *Pararrhynchium*. The result is consistent with the analysis of mitochondrial genomes [16]. Further analyses with more specimens are needed to investigate whether the rest of the species previously included in the genus *Lissodynerus* should be all returned to *Lissodynerus* or some of them should be transferred to the genus *Pararrhynchium* as it already happened in the study of Carpenter & Brown [18].

4.3. Effects of Continental Drifts on the Distribution of Rhynchium Genera Group

The *Rhynchium* genera group is distributed in the Old World only, including the Australian, Ethiopian, Oriental, and Palearctic regions. The most recent common ancestor (MRCA) of the *Rhynchium* genera group was inferred in the Cretaceous (95.74 Ma), which is consistent with the results of the evolutionary history of the Hymenoptera [30]. The *Rhynchium* genera group might have spread in various plates according to the migration history of the Old World. During the Cretaceous, the Pangaea [69] split up and Africa collided with Europe, leading to the formation of the Alps [70]. In addition, following the separation of the Gondwanaland during the Late Jurassic (~135 Ma) to India and Oceania, the Eurasian subcontinent was formed after the collision between India and Asia in Southern Tibet in the Eocene (56–33.9 Ma) [71]. Consequently, the continents of the Old World except Australia had been connected from the Cretaceous–Paleogene (84–45 Ma) until the present day [71], which might be the reason that the *Rhynchium* genera group is distributed in the Old World. Moreover, the presence of many islands between Australia and the Eurasian subcontinent, such as the Philippine Islands and New Guinea, might raise the dispersal of the *Rhynchium* genera group to Australia through these islands just as the *Ropalidia* species groups in the vespid wasps [72,73].

Meanwhile, the ocean might be a limiting factor in the distributions of the *Rhynchium* genera group in the New World. During the Late Jurassic/Early Cretaceous (~110 Ma), the second breakup of Gondwana rifting of South America and Southern Africa [74] led to the formation of the South Atlantic basin [75,76]. In addition, during the Jurassic (~169.7 Ma) the Pacific Ocean basin was born and is by far the oldest preserved oceanic crust within the global oceans [77]. The estimated age of the MRCA of *Rhynchium* genera group in the Cretaceous is later than the formation of the ocean, which might consequently lead to the no distribution of this group in the New World.

5. Conclusions

Overall, our phylogenetic analyses based on different combinations of data matrices and models yielded consistent trees. However, the inconsistency of one node between analyses reaffirms the necessity of a large data set and model selection for phylogenetics [26]. Combining methodological suitability with morphological evidence, the topology recovered using the MSC model was preferred. *Rhynchium*, *Dirhynchium*, *Anterhynchium*, and *Pararrhynchium* comprise a monophyletic group; *Allorhynchium* and *Lissodynerus* are distinct from the other four taxa. Of these, it was corroborated that *Dirhynchium* van der Vecht, 1963 should be upgraded to a valid genus, and *Lissodynerus* Giordani Soika, 1993 should be restituted as a valid genus, not a synonymy of *Pararrhynchium* de Saussure, 1855. Considering the limitations of the sample, the attribution of the other species previously placed in *Lissodynerus* requires further investigation. This study is the first phylogenetic study using USCOs and UCEs extracted from low-coverage whole genomes in Eumeninae, providing a reference for further studies on the phylogeny of the subfamily Eumeninae.

Supplementary Materials: The following supporting information can be downloaded at: https://www.mdpi.com/article/10.3390/insects14090775/s1, Supplementary S1: data matrices; Supplementary S2: tree topologies; Supplementary S3: resultant tree files; Supplementary Table S1: collection information of the 10 newly sequenced samples.

Author Contributions: Conceptualization, T.-J.L. and S.-L.H.; methodology, S.-L.H. and M.D.; software, S.-L.H. and M.D.; validation, T.-J.L.; formal analysis, M.D.; investigation, T.-J.L.; resources, B.C.; data curation, M.D.; writing—original draft preparation, M.D.; writing—review and editing, T.-J.L., S.-L.H. and B.C.; visualization, M.D.; supervision, T.-J.L.; project administration, T.-J.L.; funding acquisition, T.-J.L. All authors have read and agreed to the published version of the manuscript.

Funding: This research was funded by Science & Technology Fundamental Resources Investigation Program (Grant No. 2022FY202100) and the National Natural Science Foundation of China (Nos: 31772490, 31372247, 31000976).

Data Availability Statement: The data presented in this study are openly available from the National Center for Biotechnology Information at https://www.ncbi.nlm.nih.gov (accessed on 5 August 2023), accession numbers: PRJNA1002480 (https://dataview.ncbi.nlm.nih.gov/objects?linked_to_id=PRJNA1002480, accessed on 5 August 2023).

Acknowledgments: We are extremely grateful to Yi-Fan Liu (Beijing Institute of Zoology, Chinese Academy of Sciences) for providing the alcohol specimens of *Pararrhynchium septemfasciatus feanus*.

Conflicts of Interest: The authors declare no conflict of interest.

References

1. Jennings, D.T.; Houseweart, M.W. Predation by Eumenid Wasps (Hymenoptera: Eumenidae) on Spruce Budworm (Lepidoptera: Tortricidae) and Other Lepidopterous Larvae in Spruce-Fir Forests of Maine. *Ann. Èntomol. Soc. Am.* **1984**, *49*, 39–45. [CrossRef]
2. Cowan, D.P. *The Social Biology of Wasps*; Cornell University Press: Ithaca, NY, USA, 1991; pp. 33–73.

3. Carpenter, J.M.; Gusenleitner, J.; Madl, M. A catalogue of the Eumeninae (Hymenoptera: Vespidae) of the Ethiopian Region excluding Malagasy Subregion. Part I: Introduction, Key to Genera, Genera *Aethiopicodynerus* Gusenleitner 1997 to *Cyrtolabulus* van der Vecht 1969. *Linz. Biol. Beitr.* **2009**, *41*, 513–638.
4. Carpenter, J.M.; Gusenleitner, J.; Madl, M. A Catalogue of the Eumeninae (Hymenoptera: Vespidae) of the Ethiopian Region excluding Malagasy Subregion. Part II: Genera *Delta* de Saussure 1885 to *Zethus* Fabricius 1804 and Species Incertae Sedis. *Linz. Biol. Beitr.* **2010**, *42*, 95–315.
5. Tan, J.L.; Carpenter, J.M.; Achterberg, C. An illustrated key to the genera of Eumeninae from China, with a checklist of species (Hymenoptera, Vespidae). *ZooKeys* **2018**, *740*, 109–149. [CrossRef] [PubMed]
6. Girish Kumar, P.; Sharma, G. A review of the genus *Allorhynchium* van der vecht, 1963 (Hymenoptera: Vespidae: Eumeninae) from the Indian subcontinent. *Prommalia* **2015**, *3*, 20–34.
7. Luo, L.; Zhang, Q.H.; Chen, B.; Li, T.J. Additions to the knowledge of the genus *Allorhynchium* van der Vecht, 1963 from China (Hymenoptera, Vespidae, Eumeninae). *J. Hymenopt. Res.* **2020**, *77*, 119–137. [CrossRef]
8. Kim, J.K. Taxonomic review of the genus *Anterhynchium* Saussure (Eumeninae, Vespidae, Hymenoptera) from East Asia. *Korean J. Entomol.* **2003**, *33*, 105–117. [CrossRef]
9. Kumar, P.G. A taxonomic study on the genus *Anterhynchium* de Saussure (Hymenoptera: Vespidae: Eumeninae) from Indian subcontinent. *Rec. Zool. Surv. India* **2013**, *113*, 139–158.
10. Selis, M. The genus *Anterhynchium* in the Philippine Islands (Hymenoptera: Vespidae, Eumeninae). *Fragm. Entomol.* **2017**, *49*, 175–178. [CrossRef]
11. van der Vecht, J. Studies on Indo–Australian and East Asiatic Eumenidae (Hymenoptera: Vespoidea). *Zool. Verh.* **1963**, *60*, 3–116.
12. Li, T.J.; Chen, B. Descriptions of four new species of *Pararrhynchium* de Saussure (Hymenoptera: Vespidae: Eumeninae) from China, with one newly recorded species and a key to Chinese species. *Orient. Insects* **2017**, *52*, 175–189. [CrossRef]
13. Kumar, P.G.; Carpenter, J.M.; Kishore, L. A review of the genus *Pararrhynchium* de Saussure (Hymenoptera: Vespidae: Eumeninae) from India with the description of a new species. *Halteres* **2017**, *8*, 85–91. [CrossRef]
14. Kumar, P.G.; Sharma, G. A taxonomic study on the genus *Rhynchium* Spinola (Hymenoptera: Vespidae: Eumeninae) from the India subcontinent. *Rec. Zool. Surv. India* **2013**, *113*, 105–122.
15. Pham, P.H.; Kumar, P.G. Taxonomic notes on the genus *Rhynchium* Spinola (Hymenoptera: Vespidae: Eumeninae) from Vietnam, with a first checklist of the species worldwide. *Ecol. Montenegrina* **2016**, *7*, 530–537. [CrossRef]
16. Luo, L.; Carpenter, J.M.; Chen, B.; Li, T.J. First Comprehensive Analysis of Both Mitochondrial Characteristics and Mitogenome-Based Phylogenetics in the Subfamily Eumeninae (Hymenoptera: Vespidae). *Insects* **2022**, *13*, 529. [CrossRef] [PubMed]
17. Selis, M.; Carpenter, J.M. Revision of the Afrotropical species of the genus *Anterhynchium* de Saussure (Hymenoptera: Vespidae: Eumeninae). *Zootaxa* **2023**, *5233*, 1–102. [CrossRef]
18. Carpenter, J.M. Catalogue of the Australian Eumeninae (Hymenoptera: Vespidae). *Zootaxa* **2021**, *4919*, 1–68. [CrossRef]
19. Giordani Soika, A. Contributo alla conoscenza degli Eumenidi dell'Australia e della Nuova Guinea (Hym. Vespoidea). *Bolletino Mus. Civ. Stor. Nat. Venezia* **1993**, *42*, 125–149. [CrossRef]
20. Giordani Soika, A. Ricerche sistematiche su alcuni generi di eumenidi della regione orientale e della Papuasia (Hymenoptera, Vespoidea). *Ann. Mus. Civ. Stor. Nat. Giacomo Doria* **1994**, *90*, 1–384. [CrossRef]
21. Li, T.J.; Chen, B. Two newly recorded genera *Stenodyneriellus* and *Lissodynerus* with three new species from China (Hymenoptera, Vespidae, Eumeninae). *J. Hymenopt. Res.* **2016**, *49*, 111–127. [CrossRef]
22. Lemmon, E.M.; Lemmon, A.R. High-Throughput Genomic Data in Systematics and Phylogenetics. *Annu. Rev. Ecol. Evol. Syst.* **2013**, *44*, 99–121. [CrossRef]
23. Scornavacca, C.; Delsuc, F.; Galtier, N. Phylogenetics in the Genomic Era 2020, Open Access Book. Available online: https://hal.inria.fr/PGE (accessed on 25 February 2021).
24. Derti, A.; Roth, F.P.; Roth, F.P.; Wu, C.T. Mammalian ultraconserved elements are strongly depleted among segmental duplications and copy number variants. *Nat. Genet.* **2006**, *38*, 1216–1220. [CrossRef] [PubMed]
25. McCormack, J.E.; Faircloth, B.C.; Crawford, N.G.; Gowaty, P.A.; Brumfield, R.T.; Glenn, T.C. Ultraconserved elements are novel phylogenomic markers that resolve placental mammal phylogeny when combined with species-tree analysis. *Genome Res.* **2012**, *22*, 746–754. [CrossRef] [PubMed]
26. Yu, D.Y.; Ding, Y.H.; Tihelka, E.; Chen, Y.C.; Hu, F.; Liu, M.Q.; Zhang, F. Phylogenomics of Elongate-Bodied Springtails Reveals Independent Transitions from Aboveground to Belowground Habitats in Deep Time. *Syst. Biol.* **2022**, *71*, 1023–1031. [CrossRef] [PubMed]
27. Zhang, D.; Niu, Z.Q.; Luo, A.R.; Orr, M.C.; Ferrari, R.R.; Jin, J.F.; Wu, Q.T.; Zhang, F.; Zhu, C.D. Testing the systematic status of Homalictus and Rostrohalictus with weakened cross-vein groups within Halictini (Hymenoptera: Halictidae) using low-coverage whole-genome sequencing. *Insect Sci.* **2022**, *29*, 1819–1833. [CrossRef] [PubMed]
28. Parada, A.; Hanson, J.; D'Eiía, G. Ultraconserved Elements Improve the Resolution of Difficult Nodes within the Rapid Radiation of Neotropical Sigmodontine Rodents (Cricetidae: Sigmodontinae). *Syst. Biol.* **2021**, *70*, 1090–1100. [CrossRef]
29. Timilsena, P.R.; Wafula, E.K.; Barrett, C.F.; Ayyampalayam, S.; McNeal, J.R.; Rentsch, J.D.; McKain, M.R.; Heyduk, K.; Harkess, A.; Villegente, M.; et al. Phylogenomic resolution of order- and family-level monocot relationships using 602 single-copy nuclear genes and 1375 BUSCO genes. *Front. Plant Sci.* **2022**, *13*, 876779. [CrossRef]

30. Peters, R.S.; Krogmann, L.; Mayer, C.; Donath, A.; Gunkel, S.; Meusemann, K.; Kozlov, A.; Podsiadlowski, L.; Petersen, M.; Lanfear, R.; et al. Evolutionary History of the Hymenoptera. *Curr. Biol.* **2017**, *27*, 1013–1018. [CrossRef]
31. Bushnell, B. BBtools. 2014. Available online: https://sourceforge.net/projects/bbmap/ (accessed on 25 February 2022).
32. Bankevich, A.; Nurk, S.; Antipov, D.; Gurevich, A.A.; Dvorkin, M.; Kulikov, A.S.; Lesin, V.M.; Nikolenko, S.I.; Pham, S.; Prjibelski, A.D.; et al. SPAdes: A new genome assembly algorithm and its applications to single-cell sequencing. *J. Comput. Biol.* **2012**, *19*, 455–477. [CrossRef]
33. Xu, M.; Guo, L.; Gu, S.; Wang, O.; Zhang, R.; Peters, B.A.; Fan, G.; Liu, X.; Xu, X.; Deng, L.; et al. TGS-GapCloser: A fast and accurate gap closer for large genomes with low coverage of error-prone long reads. *Gigascience* **2020**, *9*, giaa094. [CrossRef]
34. Waterhouse, R.M.; Seppey, M.; Simao, F.A.; Manni, M.; Ioannidis, P.; Klioutchnikov, G.; Kriventseva, E.V.; Zdobnov, E.M. BUSCO Applications from Quality Assessments to Gene Prediction and Phylogenomics. *Mol. Biol. Evol.* **2018**, *35*, 543–548. [CrossRef]
35. Katoh, K.; Standley, D.M. MAFFT multiple sequence alignment software version 7: Improvements in performance and usability. *Mol. Biol. Evol.* **2013**, *30*, 772–780. [CrossRef] [PubMed]
36. Criscuolo, A.; Gribaldo, S. BMGE (Block Mapping and Gathering with Entropy): Selection of phylogenetic informative regions from multiple sequence alignments. *BMC Evol. Biol.* **2010**, *10*, 210. [CrossRef]
37. Naser-Khdour, S.; Minh, B.Q.; Zhang, W.; Stone, E.A.; Lanfear, R. The prevalence and impact of model violations in phylogenetic analysis. *Genome Biol. Evol.* **2019**, *11*, 3341–3352. [CrossRef] [PubMed]
38. Minh, B.Q.; Schmidt, H.A.; Chernomor, O.; Schrempf, D.; Woodhams, M.D.; von Haeseler, A.; Lanfear, R. IQ-TREE 2: New models and efficient methods for phylogenetic inference in the genomic era. *Mol. Biol. Evol.* **2020**, *37*, 1530–1534. [CrossRef] [PubMed]
39. Kück, P.; Longo, G.C. FASconCAT-G: Extensive functions for multiple sequence alignment preparations concerning phylogenetic studies. *Front. Zool.* **2014**, *11*, 81. [CrossRef] [PubMed]
40. Branstetter, M.G.; Longino, J.T.; Ward, P.S.; Faircloth, B.C. Enriching the ant tree of life: Enhanced UCE bait set for genome-scale phylogenetics of ants and other Hymenoptera. *Methods Ecol. Evol.* **2017**, *8*, 768–776. [CrossRef]
41. Faircloth, B.C. PHYLUCE is a software package for the analysis of conserved genomic loci. *Bioinformatics* **2016**, *32*, 786–788. [CrossRef]
42. Kalyaanamoorthy, S.; Minh, B.Q.; Wong, T.K.F.; von Haeseler, A.; Jermiin, L.S. ModelFinder: Fast model selection for accurate phylogenetic estimates. *Nat. Methods.* **2017**, *14*, 587–589. [CrossRef]
43. Lanfear, R.; Calcott, B.; Kainer, D.; Mayer, C.; Stamatakis, A. Selecting optimal partitioning schemes for phylogenomic datasets. *BMC Evol. Biol.* **2014**, *14*, 82. [CrossRef]
44. Crotty, S.M.; Minh, B.Q.; Bean, N.G.; Holland, B.R.; Tuke, J.; Jermiin, L.S.; von Haeseler, A. GHOST: Recovering historical signal from heterotachously-evolved sequence alignments. *Syst. Biol.* **2020**, *69*, 249–264. [CrossRef]
45. Guindon, S.; Dufayard, J.F.; Lefort, V.; Anisimova, M.; Hordijk, W.; Gascuel, O. New algorithms and methods to estimate maximum–likelihood phylogenies: Assessing the performance of PhyML 3.0. *Syst. Biol.* **2010**, *59*, 307–321. [CrossRef] [PubMed]
46. Hoang, D.T.; Chernomor, O.; von Haeseler, A.; Minh, B.Q.; Vinh, L.S. UFBoot2: Improving the ultrafast bootstrap approximation. *Mol. Biol. Evol.* **2018**, *35*, 518–522. [CrossRef]
47. Zhang, C.; Rabiee, M.; Sayyari, E.; Mirarab, S. ASTRAL-III: Polynomial time species tree reconstruction from partially resolved gene trees. *BMC Bioinform.* **2018**, *19*, 153. [CrossRef] [PubMed]
48. Sayyari, E.; Mirarab, S. Fast coalescent-based computation of local branch support from quartet frequencies. *Mol. Biol. Evol.* **2016**, *33*, 1654–1668. [CrossRef] [PubMed]
49. Salichos, L.; Rokas, A. Inferring ancient divergences requires genes with strong phylogenetic signals. *Nature* **2013**, *497*, 327–331. [CrossRef] [PubMed]
50. Wang, H.C.; Susko, E.; Roger, A.J. The relative importance of modeling site pattern heterogeneity versus partition-wise heterotachy in phylogenomic inference. *Syst. Biol.* **2019**, *68*, 1003–1019. [CrossRef]
51. Wang, H.C.; Minh, B.Q.; Susko, E.; Roger, A.J. Modeling site heterogeneity with posterior mean site frequency profiles accelerates accurate phylogenomic estimation. *Syst. Biol.* **2018**, *67*, 216–235. [CrossRef]
52. Molloy, E.K.; Warnow, T. To include or not to include: The impact of gene filtering on species tree estimation methods. *Syst. Biol.* **2018**, *67*, 285–303. [CrossRef]
53. Strimmer, K.; von Haeseler, A. Likelihood-mapping: A simple method to visualize phylogenetic content of a sequence alignment. *Proc. Natl. Acad. Sci. USA* **1997**, *94*, 6815–6819. [CrossRef]
54. Yang, Z. PAML 4: A program package for phylogenetic analysis by maximum likelihood. *Mol. Biol. Evol.* **2007**, *24*, 1586–1591. [CrossRef]
55. Piton, L. *Paléontologie du Gisement Éocéne de Menat (Puy-de-Dôme) (Flore et Faune)*; Pascal and Francis: Clermont-Ferrand, France, 1940; pp. 1–303.
56. dos Reis, M.; Yang, Z. Approximate Likelihood Calculation on a Phylogeny for Bayesian Estimation of Divergence Times. *Mol. Biol. Evol.* **2011**, *28*, 2161–2172. [CrossRef] [PubMed]
57. Rambaut, A.; Drummond, A.J.; Xie, D.; Baele, G.; Suchard, M.A. Posterior summarisation in Bayesian phylogenetics using Tracer 1.7. *Syst. Biol.* **2018**, *67*, 901–904. [CrossRef] [PubMed]
58. Oliver, J.C. Microevolutionary processes generate phylogenomic discordance at ancient divergences. *Evolution* **2013**, *67*, 1823–1830. [CrossRef] [PubMed]

59. Pollard, D.; Iyer, V.N.; Moses, A.M.; Eisen, M.B. Widespread discordance of gene trees with species tree in Drosophila: Evidence for incomplete lineage sorting. *PLoS Genet.* **2006**, *2*, e173. [CrossRef]
60. Pag'es, M.; Bazin, E.; Galan, M.; Chaval, Y.; Claude, J.; Herbreteau, V.; Michaux, J.; Piry, S.; Morand, S.; Cosson, J.F. Cytonuclear discordance among Southeast Asian black rats (Rattus rattus complex). *Mol. Ecol.* **2013**, *22*, 1019–1034. [CrossRef]
61. Ebersberger, I.; Galogoczy, P.; Taudien, S.; Taenzer, S.; Platzer, M.; von Haeseler, A. Mapping Human Genetic Ancestry. *Mol. Biol. Evol.* **2007**, *24*, 2266–2276. [CrossRef]
62. Syring, J.; Farrell, K.; Businský, R.; Cronn, R.; Liston, A. Widespread Genealogical Nonmonophyly in Species of Pinus Subgenus Strobus. *Syst. Biol.* **2007**, *56*, 163–181. [CrossRef]
63. Takahashi, K.; Terai, Y.; Nishida, M.; Okada, N. Phylogenetic relationships and ancient incomplete lineage sorting among cichlid fishes in Lake Tanganyika as revealed by analysis of the insertion of retroposons. *Mol. Biol. Evol.* **2001**, *18*, 2057–2066. [CrossRef]
64. Jennings, W.B.; Edwards, S.V. Speciational history of Australian grassfinches (Poephila) inferred from thirty gene trees. *Evolution* **2005**, *59*, 2033–2047. [CrossRef]
65. Carstens, B.C.; Knowles, L.L. Estimating species phylogeny from gene-tree probabilities despite incomplete lineage sorting: An example from *Melanoplus* grasshoppers. *Syst. Biol.* **2007**, *56*, 400–411. [CrossRef]
66. Bossert, S.; Murray, E.A.; Almeida, E.A.B.; Brady, S.G.; Blaimer, B.B.; Danforth, B.N. Combining transcriptomes and ultraconserved elements to illuminate the phylogeny of Apidae. *Mol. Phylogenet. Evol.* **2019**, *130*, 121–131. [CrossRef] [PubMed]
67. Branstetter, M.G.; Müller, A.; Griswold, T.L.; Orr, M.C.; Zhu, C.D. Ultraconserved element phylogenomics and biogeography of the agriculturally important mason bee subgenus Osmia (Osmia). *Syst. Entomol.* **2021**, *46*, 453–472. [CrossRef]
68. Pisanty, G.; Richter, R.; Martin, T.; Dettman, J.; Cardinal, S. Molecular phylogeny, historical biogeography and revised classification of andrenine bees (Hymenoptera: Andrenidae). *Mol. Phylogenet. Evol.* **2021**, *170*, 107151. [CrossRef]
69. Cox, C.B.; Ladle, R.J.; Moore, P.D. *Biogeography, an Ecological and Evolutionary Approach*, 10th ed.; Science Ecology & Organismal Biology; John Wiley & Sons: Hoboken, NJ, USA, 2019.
70. Wright, N.M.; Seton, M.; Williams, S.E.; Müller, R.D. The Late Cretaceous to recent tectonic history of the Pacific Ocean basin. *Earth-Sci. Rev.* **2016**, *154*, 138–173. [CrossRef]
71. Willis, B. Principles of paleogeography. *Science* **1910**, *31*, 241–260. [CrossRef] [PubMed]
72. Saito, F.; Kojima, J. Taxonomy and biogeography of Australian species of the *Ropalidia stigma* group and *R. variegata* group (Hymenoptera: Vespidae). *Entomol. Sci.* **2005**, *8*, 179–188. [CrossRef]
73. Luo, L.; Huang, P.; Chen, B.; Li, T.J. Mitochondrial composition of and diffusion limiting factors of three social wasp genera Polistes, Ropalidia, and parapolybia (Hymenoptera: Vespidae). *BMC Ecol. Evol.* **2022**, *22*, 63. [CrossRef]
74. Heine, C.; Zoethout, J.; Müller, R.D. Kinematics of the South Atlantic rift. *Solid Earth* **2013**, *4*, 215–253. [CrossRef]
75. Perez-Diaz, L.; Eagles, G. Constraining South Atlantic growth with seafloor spreading data. *Tectonics* **2014**, *33*, 1848–1873. [CrossRef]
76. Granot, R.; Dyment, J. The Cretaceous opening of the South Atlantic Ocean. *Earth Planet. Sci. Lett.* **2015**, *414*, 156–163. [CrossRef]
77. Tominaga, M.; Sager, W.W.; Tivey, M.A.; Lee, S.M. Deep-tow magnetic anomaly study of the Pacific Jurassic Quiet Zone and implications for the geomagnetic polarity reversal timescale and geomagnetic field behavior. *Solid Earth* **2008**, *113*, B07710. [CrossRef]

insects

MDPI

Article

DNA Barcoding Unveils New Species of the Sexually Dimorphic Genus *Anteon* Jurine (Hymenoptera, Dryinidae) from China

Huayan Chen [1,*], Massimo Olmi [2], Frode Ødegaard [3], Leonardo Capradossi [4] and Jingxian Liu [5]

1 Key Laboratory of Plant Resources Conservation and Sustainable Utilization, South China Botanical Garden, Chinese Academy of Sciences, Guangzhou 510650, China
2 Tropical Entomology Research Center, Via De Gasperi 10, 01100 Viterbo, Italy; olmi@unitus.it
3 Department of Natural History, Norwegian University of Science and Technology (NTNU), NO-7491 Trondheim, Norway; frode.odegaard@ntnu.no
4 Independent Researcher, 01017 Tuscania, Italy; leonardocapradossi.lc@gmail.com
5 Department of Entomology, South China Agricultural University, Guangzhou 510642, China; liujingxian@scau.edu.cn
* Correspondence: huayanc@scbg.ac.cn

Simple Summary: *Anteon* Jurine, 1807 is a cosmopolitan genus of dryinid parasitoids that attack leafhoppers. Sexual dimorphism is extreme in *Anteon* species, rendering the morphological taxonomy of these parasitoids difficult. This paper employs an integrated taxonomic approach that combines morphology with DNA barcoding to investigate the species delimitation of *Anteon* from China. Of the twenty-nine morphospecies examined, twenty-four species were identified as previously known species and five were described here as new species. The sexual association of six species was confirmed by DNA barcoding, indicating that such tools are powerful to tackle the taxonomic challenges in parasitoids with sexual dimorphism.

Abstract: Species of *Anteon* Jurine, 1807 are a large group of parasitoids attacking leafhoppers, which are important insect pests. Despite their great potential in pest biological control, the taxonomy and biology of these parasitoids are far from clear. Sexual dimorphism is extreme in *Anteon* species and has hampered the taxonomy of these parasitoids, resulting in many species described based on a single sex. In this paper, we employed an integrated taxonomic approach for delimitating species, combining morphological examinations with DNA barcoding, to investigate *Anteon* species from China. In total, 53 COI sequences representing 29 species of *Anteon* were obtained and analyzed. On the basis of both morphology and DNA barcoding, five new species of *Anteon* were discovered and described: *A. clariclypeum* sp. nov., *A. maguanense* sp. nov., *A. parafidum* sp. nov., *A. shaanxianum* sp. nov., and *A. shandonganum* sp. nov. The neotype of *A. claricolle* Kieffer is designated. The sexual association of six species was confirmed by DNA barcoding, which led to the synonymy of *Anteon liui* Xu, Olmi & He 2010, new syn., under *Anteon meifenganum* Olmi, 1991. Keys to species of *Anteon* from the Oriental and Eastern Palaearctic are updated to contain the five new species. Our study demonstrates that DNA barcoding is a potent tool for tackling the taxonomic challenges in parasitoids with extreme sexual dimorphism.

Keywords: Chrysidoidea; morphological identification; key; molecular species delimitation

1. Introduction

The genus *Anteon* Jurine, 1807 (Hymenoptera, Dryinidae) is a large group of parasitoids attacking leafhoppers of the families Cicadellidae and Eurymelidae (Hemiptera, Cicadomorpha) [1]. Since many leafhoppers are important insect pests, these parasitoids are economically important in the biological control of leafhoppers [2]. The genus comprises 465 species distributed worldwide, except Antarctica [3]. In total, 125 species have been reported from China [2], with 116 species belonging to the Oriental region [4,5] and 9 species belonging to the Eastern Palaearctic region [6].

As has been pointed out by many studies, the taxonomy of Dryinidae is challenging due to the extreme sexual dimorphism in these parasitoids, resulting in many species described based on a single sex because the sexual association of the same species is difficult [7–10]. This situation is especially true in *Anteon*. However, this kind of taxonomic obstacle could be overcome by employing DNA barcoding methods such as the application of the mitochondrial cytochrome c oxidase 1 (*COI*) gene [9,10]. In fact, DNA sequences have been used in the taxonomy, phylogenetics and the confirmation of host associations in Dryinidae and have become an important tool [8,9,11–13].

As the second paper of an ongoing research campaign started by the first author (HC), with the objective to collect specimens of dryinids across China and identify the species using an integrated taxonomic approach that combines morphology and DNA barcoding, this study presents the results for the genus *Anteon*.

2. Materials and Methods

2.1. Insect Specimens and Identification

This study is based upon specimens collected by Malaise traps set up in many provinces of China (See Figure 1 in Olmi et al. [10]). Monthly collected specimens from Malaise traps were kept in 100% ethanol. Dryinidae specimens were sorted out and used for DNA sequencing and morphological examinations. The species of *Anteon* were identified by morphology with the keys provided by Xu et al. [2] and Olmi & Xu [6].

2.2. Repositories

The specimens examined in this study are deposited in the following collections, with the abbreviations used in the text: NTNU, Norwegian University of Science and Technology, Department of Natural History, NO-7491 Trondheim, Norway; SCAU, South China Agricultural University, Guangzhou, China; SCBG, South China Botanical Garden, Chinese Academy of Sciences, Guangzhou, China; ZJUC, Zhejiang University, Hangzhou, Zhejiang, China. Regarding the Scandinavian specimens, one mid leg of each individual was torn off and submitted to the University of Guelph in Canada for *COI* barcoding and the generated sequences were uploaded to the BOLD system [14]. All the types of Oriental and Palaearctic species of *Anteon* were examined.

2.3. Morphology

The terminology follows Olmi et al. [3]. Abbreviations and morphological terms used in the text are as follows: OL: the distance between inner edges of a lateral ocellus and the median ocellus; OOL: the distance from the outer edge of a lateral ocellus to the eye; OPL: the distance from the posterior edge of a lateral ocellus to the occipital carina; POL: the distance between inner edges of the lateral ocelli; and TL: the distance from the posterior edge of an eye to the occipital carina. Body length was measured from the head to abdominal tip and expressed in millimeters; other measurements are reported in relative values. The terms "metapectal–propodeal disc" and "dorsal surface of the propodeum" are here used in the sense of Kawada et al. [15] and respectively correspond to the "dorsal surface of propodeum" and "posterior surface of propodeum" used by Olmi and Xu [6] and Xu et al. [2].

2.4. Imaging

Multifocal images of the new species and representatives of the known species (Figures S1–S25) were made using a mirrorless camera Sony Alpha 6300 (Sony, Tokyo, Japan) with cross table Proxxon KT 70 (Proxxon, Wecker, Luxemburg) or Leica M205C multifocal equipment (Leica, Wetzlar, Germany) and a Nikon SMZ25 microscope mounted with a Nikon DS-Ri 2 digital camera system (Melville, NY, USA). Image plates were made with Adobe Photoshop CS6 Extended.

2.5. Sequence Analysis and Molecular Species Delimitation

Genomic DNA of the Chinese specimens was nondestructively extracted from the entire wasp using the TIANamp Micro DNA Kit (Tiangen Biotech, Beijing, China) following the methods used by Taekul et al. [16]. Voucher specimens (Table S1) are deposited in the insect collection of South China Botanical Garden, Chinese Academy of Sciences, Guangzhou, China (SCBG). The *COI* gene was amplified using the LCO1490/HCO2198 primer pair [17]. PCRs, sequencing and sequence analysis were conducted as in the Olmi et al. study [10]. All the newly generated sequences were deposited into GenBank (Table 1), and sequences of six *Anteon* species were downloaded from the BOLD system: *A. claricolle* Kieffer, 1906 (HYMNI1041), *Anteon collare* (Dalman, 1818) (HYMNI945), *Anteon ephippiger* (Dalman, 1818) (HYMNI1151), *Anteon exiguum* (Haupt, 1941) (HYMNI1251), *Anteon fulviventre* (Haliday, 1828) (HYMNI390), *Anteon gaullei* Kieffer, 1905 (HYMNI430). Sequences were aligned by codons using MUSCLE implemented in Geneious 11.0.3. The K2P distances within and between species were calculated in MEGA 11 [18]. A maximum likelihood (ML) tree was generated based on the aligned sequences using the RAxML plugin in Geneious 11.0.3. A sequence of *Aphelopus niger* Xu & He, 1999 (Hymenoptera, Dryinidae) (MZ151323) was downloaded from GenBank and used as an outgroup, according to the phylogenetic topologies present in Tribull [12].

Table 1. List of sequenced species and accession numbers.

Code	Species	Sex	GenBank Accession No.
SCAU 3040430	*Anteon abdulnouri* Olmi, 1987	female	OR818495
SCAU 3011714	*Anteon achterbergi* Olmi, 1991	female	OR818496
SCAU 3011677	*Anteon blanduscutum* Xu, He & Rui, 1996	female	OR818497
SCAU 3011678	*Anteon blanduscutum* Xu, He & Rui, 1996	male	OR818498
SCAU 3011669	*Anteon borneanum* Olmi, 1984	female	OR818499
SCAU 3011671	*Anteon clariclypeum* sp. n.	female	OR818500
SCAU 3011712	*Anteon claricolle* Kieffer, 1906	male	OR818501
SCAU 3040561	*Anteon claricolle* Kieffer, 1906	female	OR818502
SCAU 3040517	*Anteon confusum* Olmi, 1991	female	OR818503
SCAU 3044058	*Anteon confusum* Olmi, 1991	male	OR818504
SCAU 3011621	*Anteon exiguum* (Haupt, 1941)	female	OR818505
SCAU 3040512	*Anteon fidum* Olmi, 1991	male	OR818506
SCAU 3040522	*Anteon funiuense* Xu, He & Olmi, 2001	female	OR818507
SCAU 3040511	*Anteon henanense* Xu, He & Olmi, 2001	male	OR818508
SCAU 3044072	*Anteon henanense* Xu, He & Olmi, 2001	female	OR818509
SCAU 3011713	*Anteon hilare* Olmi, 1984	female	OR818510
SCAU 3011676	*Anteon hirashimai* Olmi, 1993	female	OR818511
SCAU 3011738	*Anteon maguanense* sp. n.	female	OR818512
SCAU 3040514	*Anteon meifenganum* Olmi, 1991	female	OR818513
SCAU 3040520	*Anteon meifenganum* Olmi, 1991	male	OR818514
SCAU 3040515	*Anteon mite* Olmi, 1996	male	OR818515
SCAU 3011661	*Anteon multicolor* Xu, He & Olmi, 1998	female	OR818516
SCAU 3040571	*Anteon multicolor* Xu, He & Olmi, 1998	male	OR818517
SCAU 3040573	*Anteon multicolor* Xu, He & Olmi, 1998	male	OR818518
SCAU 3040574	*Anteon multicolor* Xu, He & Olmi, 1998	female	OR818519
SCAU 3044011	*Anteon multicolor* Xu, He & Olmi, 1998	male	OR818520
SCAU 3044014	*Anteon multicolor* Xu, He & Olmi, 1998	female	OR818521
SCAU 3044065	*Anteon multicolor* Xu, He & Olmi, 1998	male	OR818522
SCAU 3044074	*Anteon multicolor* Xu, He & Olmi, 1998	female	OR818523
SCAU 3044114	*Anteon multicolor* Xu, He & Olmi, 1998	male	OR818524
SCAU 3048997	*Anteon multicolor* Xu, He & Olmi, 1998	male	OR818525
SCAU 3040524	*Anteon munitum* Olmi, 1984	male	OR818526
SCAU 3040521	*Anteon naduense* Olmi 1987	female	OR818527
SCAU 3040433	*Anteon nanlingense* Xu, Olmi & He, 2011	female	OR818528
SCAU 3048821	*Anteon nanlingense* Xu, Olmi & He, 2011	male	OR818529
SCAU 3011658	*Anteon parafidum* sp. n.	female	OR818530
SCAU 3011662	*Anteon parafidum* sp. n.	male	OR818531
SCAU 3040566	*Anteon parafidum* sp. n.	female	OR818532

Table 1. *Cont.*

Code	Species	Sex	GenBank Accession No.
SCAU 3040567	*Anteon parafidum* sp. n.	male	OR818533
SCAU 3044018	*Anteon parafidum* sp. n.	male	OR818534
SCAU 3044026	*Anteon parafidum* sp. n.	female	OR818535
SCAU 3044033	*Anteon parafidum* sp. n.	female	OR818536
SCAU 3044066	*Anteon parafidum* sp. n.	male	OR818537
SCAU 3044088	*Anteon parafidum* sp. n.	male	OR818538
SCAU 3044089	*Anteon parafidum* sp. n.	female	OR818539
SCAU 3040518	*Anteon pilosum* Xu, Olmi & He, 2010	male	OR818540
SCAU 3011631	*Anteon priscum* Olmi, 1991	female	OR818541
SCAU 3011723	*Anteon priscum* Olmi, 1991	male	OR818542
SCAU 3040519	*Anteon pteromaculatum* Xu, Olmi, Guglielmino & Chen, 2012	female	OR818543
SCAU 3040516	*Anteon shaanxianum* sp. n.	male	OR818544
SCAU 3011688	*Anteon shandonganum* sp. n.	female	OR818545
SCAU 3040523	*Anteon tengchongense* Xu, He & Olmi, 1998	female	OR818546
SCAU 3011682	*Anteon yuani* Xu, He & Olmi, 1998	male	OR818547

To explore the molecular species delimitation among the studied *Anteon* species, the distance-based barcode gap approach using the Automatic Barcode Gap Discovery (ABGD, Puillandre et al. [19]) and the updated Poisson tree processes model (bPTP, Zhang et al. [20]) were tested. The ABGD method sorts the sequences to hypothetical species by partitioning and comparing the difference between sequences to identify a "barcode gap (a given threshold distance)" (Puillandre et al. [19]). The ABGD analysis was performed on the web interface (https://bioinfo.mnhn.fr/abi/public/abgd/) (accessed on 15 November 2023) using the default priors, Pmin = 0.001, Pmax = 0.1, Steps 10, and with a barcode relative gap width = 1.00, Nb bins (for distance distribution = 20). The bPTP method tests species boundaries on non-ultrametric phylogenetic trees by detecting significant differences in the numbers of substitutions between species and within species (Zhang et al. [20]). For the bPTP analysis, the ML tree generated above was used to perform the analysis on the bPTP web server (https://species.h-its.org/ptp/) (15 November 2023) with default parameters.

3. Results

3.1. Morphological Identification

In this study, we identified 28 morphospecies based on 53 specimens collected from China and Thailand. Twenty-three species were identified as previously known species and five are described below as new species (Table 1). With the help of DNA barcoding, the sexual association of six species was confirmed, leading to the synonymy of *Anteon liui* Xu, Olmi & He 2010 to *Anteon meifenganum* Olmi, 1991.

3.2. Molecular Analysis

The 53 newly generated *COI* sequences in the present study ranged from 576 bp to 674 bp. When analyzed in the BOLD system and GenBank, the sequence of *A. exiguum* (Haupt, 1941) received a close match (around 99%) with multiple sequences labeled as *A. exiguum*, while the sequences of *Anteon claricolle* Kieffer, 1906 received close matches with *A. albidicolle* Kieffer, 1905, which was a synonym of *A. ephippiger* (Dalman, 1818). Other species received no similar sequences in BOLD or GenBank. The intraspecific pairwise distances ranged from 0 to 3% (Table S2). The interspecific pairwise distances ranged from 3.1% to 20.7% (Table S3). The ABGD method produced 31 groups at a priori genetic distance thresholds of 0.002–0.036. *Anteon achterbergi* Olmi, 1991 was grouped together with all three specimens of *A. claricolle*, but genetic distances between these two species ranged from 3.1% to 4.6%. Based on the ML tree, the bPTP method delimited 33 putative species. The two specimens of *Anteon nanlingense* Xu, Olmi & He, 2011, collected from south China and Thailand, respectively, were assigned as putative species. For all the other species, both ABGD and bPTP returned congruent results with the morphological identification, as shown in Figure 1.

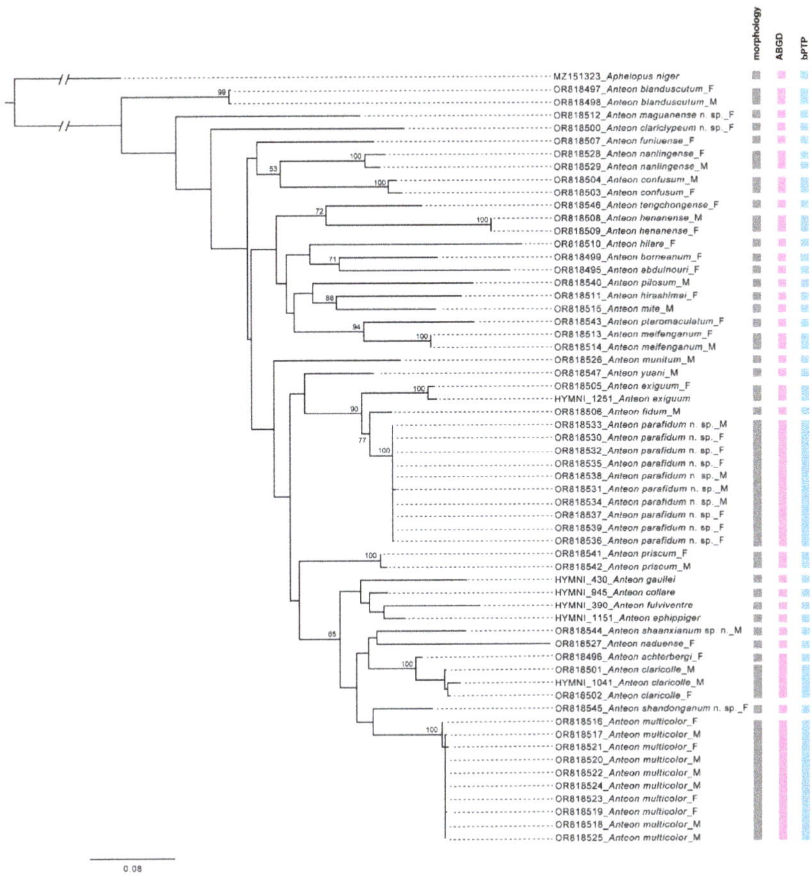

Figure 1. RAxML tree and result of the species delimitation of *Anteon* based on *COI* barcodes. Bootstraps values of 50 and above are indicated. The scale bar represents 0.08 substitutions per site.

3.3. Systematics

Anteon Jurine, 1807 [21].

See Xu et al. [2] and Olmi and Xu [6] for taxonomic details on the genus.

3.4. Species Descriptions

3.4.1. Anteon Blanduscutum Xu, He & Rui (Figures 2 and 3)

Anteon blanduscutum Xu, He & Rui 1996: 214 [22]; He & Xu 2002: 184 [23]; Xu et al. 2011 (Suppl.): 4 [24]; Xu et al. 2013: 78 [2].

Material examined. Type: holotype, female: CHINA: Zhejiang, Mt. Tianmushan, 10–12.ix.1983, Xingsheng Wan leg. (ZJUC). Other material: 1 female, CHINA: Fujian, Mt. Longqishan, 11.viii.1991, Changming Liu leg. (ZJUC); 1 female, Fujian, Mt. Longqishan, 8.VII.1991, Changming Liu leg. (ZJUC); 1 female, Hainan, Mt. Yinggeling, 17–20.vii.2010, Huayan Chen leg. (SCAU); 3 females, Hainan, Baisha County, Jiujialing, 17–20.VII.2010, Huayan Chen leg. (SCAU); 2 females, 1 male, Yunnan, Xishuangbanna, Menghai, Bulangshan Village, 1683 m, Area A2, grass, 21°44.981′ N 100°26.907′ E, 15.viii.2018, Li Ma leg., 3040566, 3011677, 3011678 (SCBG); 1 female, Zhejiang, Taishun County, Wuyanling Provincial Nature Reserve, 7.vii–5.viii.2005, (ZJUC); 1 female, Zhejiang, Lin'an, Mt. Qingliangfeng, 9.viii.2005, Min Shi leg. (SCAU).

Distribution. China (Zhejiang, Fujian, Hainan, Yunnan).

Remarks. *A. blanduscutum* was known only by females [2,22], the male of this species is here confirmed by DNA barcoding. Therefore, we present the following description and diagnosis of the male.

Description. Male. Body length 4.1 mm; fully winged (Figure 2A,B). Head black with mandible testaceous; antenna brown, but proximal half of scape testaceous; mesosoma black; metasoma brown; legs testaceous, except the basal extremity of metacoxa is darkened. Antenna filiform; antennomeres in the following proportions: 16:8:14:13:13:13:12:12:11:13. Head (Figure 2E) shiny, punctate, areas between punctures smooth; frontal line absent; frons with two short longitudinal keels near orbits directed towards antennal toruli; OL = 4; OOL = 9; OPL = 8; POL = 8; TL = 8; greatest breadth of lateral ocellus shorter than OL (3:4); occipital carina complete. Mesoscutum and mesoscutellum shiny, punctate, areas between punctures smooth. Notauli incomplete, present at the anterior half of mesoscutum. Metanotum shiny, largely smooth. Metapectal–propodeal complex with strong transverse keel between disc and propodeal declivity; disc reticulate rugose; propodeal declivity rugose, with two complete longitudinal keels. Forewing hyaline; distal part (Rs) of stigmal vein (2r-rs&Rs) more than 0.5, as long as the proximal (2r-rs) part (8:13). Paramere (Figure 2F) about as long as the aedeagus, without a distal inner pointed process. Tibial spurs 1/1/2.

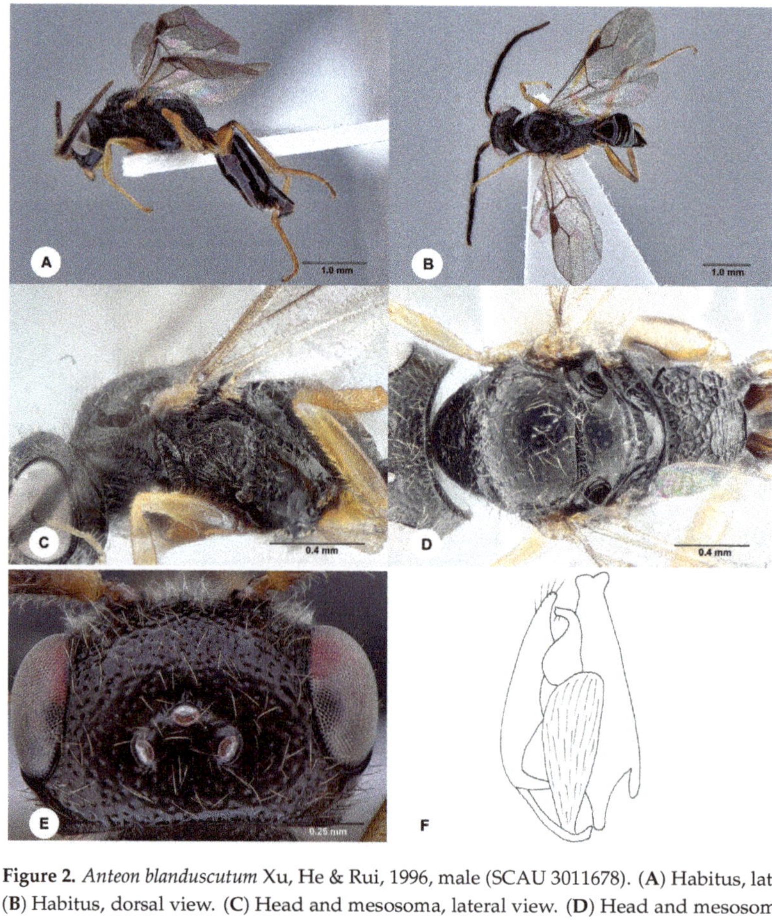

Figure 2. *Anteon blanduscutum* Xu, He & Rui, 1996, male (SCAU 3011678). (**A**) Habitus, lateral view. (**B**) Habitus, dorsal view. (**C**) Head and mesosoma, lateral view. (**D**) Head and mesosoma, dorsal view. (**E**) Head, dorsal view. (**F**) Genitalia (right half removed).

Diagnosis. The diagnosis of the male of *A. blanduscutum* is as follows: head black with mandible testaceous; head punctate, areas between punctures smooth; mesoscutum punctate, areas between punctures smooth; notauli present at the anterior half of the mesoscutum; propodeal declivity rugose, with two complete longitudinal keels; distal part of the stigmal vein (2r-rs&Rs) more than 0.5 as long as the proximal part; paramere (Figure 2F) about as long as the aedeagus, without a distal inner pointed process; inner side of the paramere not sculptured by papillae (Figure 2F). Because of the above diagnosis, the male of *A. blanduscutum* can be inserted in the key to the males of the Oriental *Anteon* published by Xu et al. [2] by modifying couplet 57 as follows:

57. Head black, except mandible, clypeus and anterior margin of face testaceous . . . *A. amabile* Xu, Olmi & He.

- Head black, except mandible testaceous. . .57′

57′. Distal part of stigmal vein (2r-rs&Rs) more than 0.5 as long as proximal part (Figure 2B) . . .*A. blanduscutum* Xu, He & Rui

- Distal part of stigmal vein (2r-rs&Rs) less than 0.5 as long as proximal part. . .*A. insertum* Olmi.

3.4.2. *Anteon clariclypeum* Chen, Olmi &, Ødegaard, sp. nov. (Figure 3)

urn:lsid:zoobank.org:act:88025100-DF2A-4058-8943-1A5F843CE18D

Material examined. Type: holotype, female: CHINA: Yunnan, Xishuangbanna, Menghai, Bulangshan Village, 21°44.761′ N 100°25.959′ E, 1595 m, Area D, forest, 16.VIII–14.IX.2018, Li Ma leg., MT, 3011671 (SCBG).

Distribution. China (Yunnan).

Etymology. The species is named *clariclypeum*, because its clypeus is testaceous (from the Latin adjective *clarus* (=clear) + the noun "clypeus").

Description. Female. Body length 3.9 mm; fully winged (Figure 3A,B). Head black, except clypeus and mouthparts testaceous; antenna yellow, except antennomeres 6–10 slightly darkened; mesosoma black; metasoma brown; legs yellow. Antenna clavate; antennomeres in the following proportions: 13:6:13:11:10:9:9:9:9:11. Head (Figure 3E) shiny, punctate, areas between punctures smooth; frontal line absent; frons without lateral keels around orbits; occipital carina complete; OL = 3; OOL = 8; OPL = 8; POL = 7; TL = 8; greatest breadth of lateral ocellus shorter than OPL (3:8). Pronotum shiny, punctate, areas between punctures smooth; pronotal tubercle reaching tegula; posterior surface of pronotum much shorter than mesoscutum (6:21). Mesoscutum and mesoscutellum shiny, punctate, areas between punctures smooth. Notauli incomplete, present at the anterior 0.6 of scutum (Figure 3D). Metanotum shiny, unsculptured. Mesopleuron shiny, punctate, areas between punctures smooth, except the proximal half is dull and rugose. Metapleuron transversely striate and rugose. Metapectal–propodeal complex with transverse keel between disc and propodeal declivity; disc reticulate rugose; propodeal declivity with two longitudinal keels, median area almost completely shiny and unsculptured and lateral areas reticulate rugose. Forewing hyaline, without dark transverse bands; distal part (Rs) of stigmal vein (2r-rs&Rs) shorter than proximal (2r-rs) part (8:13). Protarsomeres in following proportions: 11:3:3:7:19. Protarsomere 2 produced into hook. Enlarged claw (Figure 3F) with proximal prominence bearing one long bristle. Protarsomere 5 (Figure 3F) with the basal region about as long as the distal region, with two rows of about 38 lamellae without interruption to distal apex. Tibial spurs 1/1/2.

Figure 3. *Anteon clariclypeum* Chen, Olmi & Ødegaard, sp. nov., holotype, female (SCAU 3011671). (**A**) Habitus, lateral view. (**B**) Habitus, dorsal view. (**C**) Mesosoma, lateral view. (**D**) Mesosoma, dorsal view. (**E**) Head, dorsal view. (**F**) Chela.

Male. Unknown.

Diagnosis. Female with head (Figure 3E) punctate and unsculptured among the punctures; clypeus testaceous; frons without lateral keels along orbits; OL shorter than POL, OPL more than twice as long as the greatest breadth of lateral ocellus; mesosoma totally black; propodeal declivity with two longitudinal keels and median area almost completely unsculptured; protarsomere 4 shorter than 1; basal region of the protarsomere 5 (Figure 3F) about as long as the distal region. *A. clariclypeum* sp. nov. is similar to *A. kresli* Olmi, 2008. However, in *A. clariclypeum*, the clypeus is testaceous; the frons has no lateral keels along the orbits directed towards the antennal toruli; protarsomere 4 is shorter than 1; the basal region of the protarsomere 5 (Figure 3F) is about as long as the distal region (in *A. kresli*, the clypeus is black; the frons has two lateral keels along the orbits directed towards antennal toruli; the protarsomere 4 is longer than 1; the basal region of the protarsomere 5 is much shorter than the distal region (figure 32A in Xu et al. 2013) [2]). Following the description of *A, clariclypeum*, the key to the females of the Oriental species of *Anteon* published by Xu et al. (2013) [2] should be modified by replacing couplet 67 as follows:

67. Propodeal declivity with median region shiny, punctate and areas between punctures smooth. . .67′

- Propodeal declivity with median region dull and rugose. . .68

67′. Clypeus black; frons with two lateral keels along orbits directed towards antennal toruli; protarsomere 4 longer than 1; basal region of protarsomere 5 much shorter than the distal region (figure 32A in Xu et al. [2]) . . .*A. kresli* Olmi

- Clypeus testaceous; frons without two lateral keels along orbits directed towards antennal toruli (Figure 3E); protarsomere 4 shorter than 1; basal region of protarsomere 5 about as long as the distal region (Figure 3F) . . .*A. clariclypeum* sp. nov.

3.4.3. *Anteon claricolle* Kieffer (Figures 4 and 5)

Anteon claricollis Kieffer, in Kieffer & Marshall, 1906: 514 [25]; Richards, 1939: 268 [26].

Material examined. Type: female, neotype (here designated): NORWAY: Oslo, Gaustad "Jubileumsenga", 59.94841764° N, 10.711697° E, 110 m, 29.VI–1.VIII.2014, MT, K.M. Olsen leg. (NTNU). Other material: 1 male, NORWAY: EIS 12, Østfold, Hvaler, Søndre Sandøy, Ødegårdstranda, 59.00221° N, 11.07474° E, 8.V–15.VII.2014, Frode Ødegaard leg. (NTNU); 1 male, Østfold, Hvaler, Søndre Sandøy, 59.002° N, 11.075° E, 15.VII.2014, Frode Ødegaard leg. (NTNU); 1 male, CHINA: Inner Mongolia, Wulan Aodu Experimental Station, North, 38°46′15″ N, 108°46′41″ E, 20–30.VI.2013, MT, Yongming Luo leg., SCAU 3011712 (SCBG); 2 males, Ningxia Hui Nationality Autonomous Region, Mt. Liupanshan, 35°29′12″ N, 106°20′29″ E, 3–14.VII.2009, Huayan Chen leg., 3040792, 3040793 (SCBG); 1 male, Shandong, Linyi, Lanling County, MT, 34°51′ N, 118°4′ E, 25.VIII–8.IX.2014, Xuejun Yang leg., SCAU 3040677 (SCBG); 1 male, Sichuan, Chengdu, Longquanyi District, Baihe Town, Changsong Village, 30°31′17″ N, 104°17′0″ E, 8.X.2012, MT, SCAU 3019637 (SCBG).

Distribution. Norway, China (Inner Mongolia, Ningxia, Shandong, Sichuan).

Remarks. *Anteon claricolle* Kieffer, 1906 is a species ignored in all treatments of Palaearctic *Anteon*, including the main revisions led by Olmi [6,27]. The main reason for this omission is that the type was considered lost (see also Richards [26]) and the original description is unreliable. Kieffer examined six females of this species, all deposited in Marshall's collection [25]. In the labels, the localities of capture were indicated by the following abbreviations: Bugsby, Bfm., B.T. and N. (all localities apparently situated in England). However, these specimens are not deposited in the main European collections, including the Museum of Natural History of Paris (France) and the Natural History Museum of London (UK). Richards [26] wrote that the types were lost and considered *A. claricolle* as a junior synonym of *A. ephippiger* (Dalman, 1818) var. *collaris* (Dalman, 1818). In recent years, one of the authors (Frode Ødegaard) collected some female specimens with a pale yellow prothorax, whose morphology and color corresponded to the description of *A. claricolle*. COI sequences of these specimens supported that they were different from specimens of *A. ephippiger* var. *collaris* (i.e., females of *A. ephippiger* with head and mesosoma black, except that the prothorax is ferruginous or reddish). In addition, Frode Ødegaard also collected males of *Anteon* with COI sequences identical to those of the females of *A. claricolle* and different from the COI sequences of males of *A. ephippiger* var. *collaris*. Therefore, we decided to resurrect *A. claricolle* by designating a female neotype (see the description below).

Description. Female, neotype. Body length 2.4 mm; fully winged (Figure 4). Head black, except mandible and clypeus are testaceous; antenna yellow; mesosoma black, except prothorax is pale yellow (Figure 4A–D); metasoma brown; legs yellow. Antenna clavate; antennomeres in following proportions: 8:5:5:5:5:4:4:4:5:6. Head (Figure 4E) shiny, finely punctate, areas between punctures smooth; frontal line absent; frons with two lateral keels along orbits directed towards antennal toruli; occipital carina complete; OL = 3; OOL = 5; OPL = 4; POL = 4; TL = 4; greatest breadth of lateral ocellus shorter than

OPL (2:4). Pronotum shiny, finely punctate, areas between punctures smooth, with the posterior surface longer than mesoscutum (11:7); pronotal tubercle reaching tegula. Mesoscutum, mesoscutellum and metanotum shiny, punctate, areas between punctures smooth. Notauli incomplete, present at the anterior 0.6 of the mesoscutum (Figure 4D). Metapectal–propodeal complex with strong transverse keel between disc and propodeal declivity; disc reticulate rugose; propodeal declivity rugose, with two longitudinal keels. Forewing hyaline; distal part (Rs) of the stigmal vein (2r-rs&Rs) much shorter than the proximal (2r-rs) part (3:8). Protarsomeres in the following proportions: 5:2:2:7:14. Protarsomere 3 produced into a hook. Enlarged claw (Figure 4F) with a proximal prominence bearing one long bristle. Protarsomere 5 (Figure 4F) with one row of about 22 lamellae; distal apex with a group of about four lamellae. Tibial spurs 1/1/2.

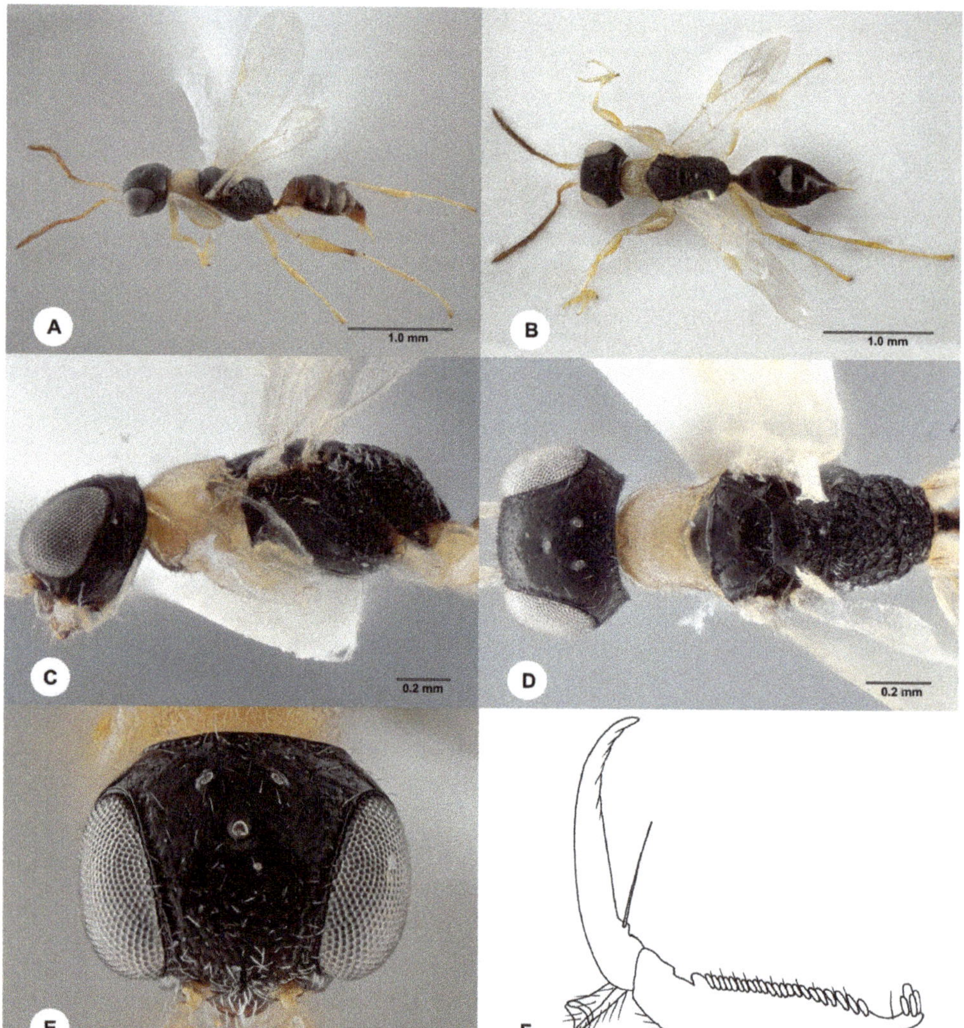

Figure 4. *Anteon claricolle* Kieffer, 1906, neotype, female. (**A**) Habitus, lateral view. (**B**) Habitus, dorsal view. (**C**) Head and mesosoma, lateral view. (**D**) Head and mesosoma, dorsal view. (**E**) Head and anterior mesosoma, dorsal view. (**F**) Chela.

Description. Male from China. Body length 3.1 mm; fully winged (Figure 5A,B). Head black, except mandible testaceous and clypeus ferruginous; antenna brown, except scape yellow; mesosoma black; metasoma brown; legs yellow. Antenna filiform; antennomeres in the following proportions: 8:5:6.5:6:6:6:6:6:6:9. Head (Figure 5E) shiny, punctate, areas between punctures smooth; frontal line absent; frons with median longitudinal furrow; OL = 3; OOL = 6; OPL = 4; POL = 5TL = 5; greatest breadth of lateral ocellus shorter than OL (2:3); occipital carina complete. Mesoscutum shiny, punctate, areas between punctures smooth. Notauli incomplete, reaching approximately 0.5 × the length of the mesoscutum (Figure 5D). Scutellum and metanotum shiny, smooth. Metapectal–propodeal complex with strong transverse keel between disc and propodeal declivity; disc reticulate rugose; propodeal declivity rugose, with two complete longitudinal keels (Figure 5D). Forewing hyaline; distal part (Rs) of the stigmal vein (2r-rs&Rs) shorter than the proximal (2r-rs) part (2.5:8). Paramere (Figure 5F) with a distal inner pointed process provided of inner distal margin not excavated; dorsal membranous band very large (Figure 5F). Tibial spurs 1/1/2.

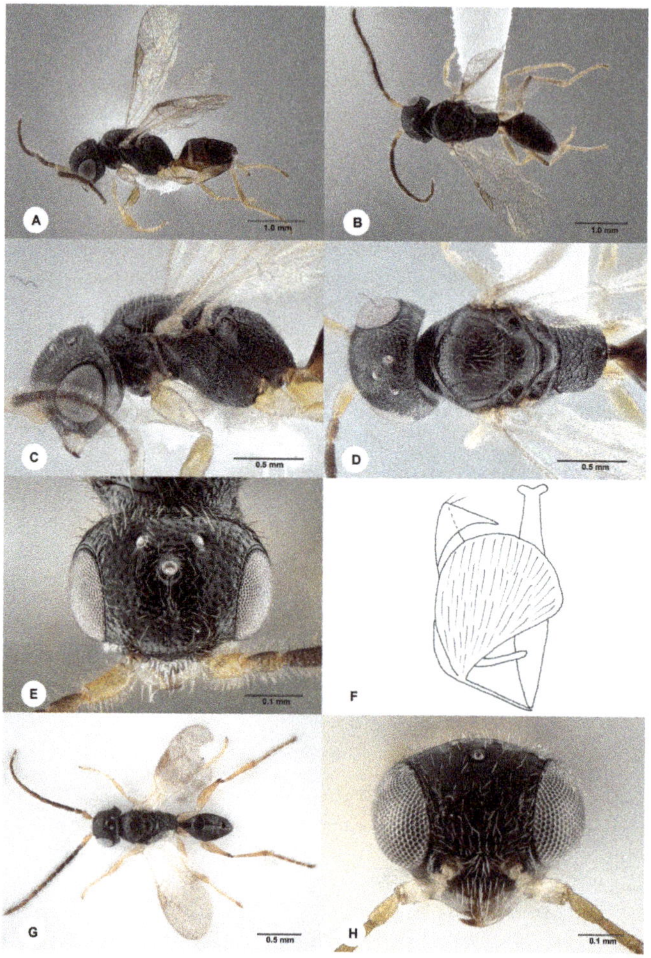

Figure 5. *Anteon claricolle* Kieffer, 1906, male, (**A–F**): SCAU 3011712) (**A**) Habitus, lateral view; (**B**) habitus, dorsal view; (**C**) head and mesosoma, lateral view; (**D**) head and mesosoma, dorsal view; (**E**) head and anterior mesosoma, dorsal view; (**F**) genitalia (right half removed); (**G**,**H**): from Norway) (**G**) habitus, dorsal view; (**H**) head, anterior view.

Description. Male from Norway. Body length 1.7 mm; fully winged (Figure 5G). Head black with mandible testaceous and clypeus ferruginous; antenna brown, except scape testaceous and pedicel brown-testaceous; mesosoma black; metasoma brown; legs yellow, except metacoxa proximally brown and distal extremity of metafemur brown. Antenna filiform; antennomeres in the following proportions: 8:5:6:5:7:7:7:7:6:9. Head (Figure 5H) shiny, finely punctate, areas between punctures smooth; frontal line incomplete, absent in anterior third of frons; frons without lateral keels around orbits directed towards antennal toruli; OL = 2; OOL = 5; OPL = 3; POL = 4; TL = 5; greatest breadth of lateral ocellus about as long as OL; occipital carina complete. Mesoscutum shiny, finely punctate, areas between punctures smooth, strongly punctate near anterior margin. Notauli incomplete, present at the anterior 0.6 of the mesoscutum (Figure 5G). Mesoscutellum unsculptured. Metanotum rugose. Metapectal–propodeal complex with strong transverse keel between disc and propodeal declivity; disc reticulate rugose; propodeal declivity rugose, with two longitudinal keels. Forewing hyaline, without dark transverse bands; distal part (Rs) of the stigmal vein (2r-rs&Rs) much shorter than the proximal (2r-rs) part. Genitalia with a distal inner pointed process and dorsal membranous process (shape like that shown in Figure 5F). Tibial spurs 1/1/2.

Diagnosis. Female (Figure 4A–D) with head black, finely punctate, areas between punctures smooth; mesosoma black, except prothorax pale yellow (Figure 4C,D); mesoscutum punctate, areas between punctures smooth (Figure 4D); notauli incomplete, present at the anterior 0.6 of the mesoscutum (Figure 4D); propodeal declivity rugose, with two longitudinal keels (Figure 4D); forewing hyaline (Figure 4A); protarsomere 5 (Figure 4F) with one row of about 22 lamellae. Male (Figure 5) with head black, except mandible testaceous and occasionally clypeus ferruginous; head (Figure 5E,H) punctate, areas between punctures smooth; propodeal declivity provided with two longitudinal keels and median area as rugose as lateral regions (Figure 5D); genitalia with a very large dorsal membranous band (Figure 5F); paramere (Figure 5F) with a very long distal inner pointed process extended apically, provided of inner distal margin not excavated; legs yellow (Figure 5A,G).

The female of *A. claricolle* can be included in the key to the *Anteon* females of Fennoscandia and Denmark [7] by replacing couplet 12 as follows:

12. Head dull, always clearly and strongly granulated; antennomere 1 approximately twice as long as 4. . . *A. fulviventre* (Haliday)

- Head shining, punctate, without sculpture among punctures (Figure 4E) or very weakly granulated; antennomere 1 approximately as long as or slightly longer than 4. . .13

13. Notauli incomplete, reaching about 0.3–0.5 length of the mesoscutum; mesosoma variably colored, occasionally black, except prothorax reddish or testaceous (never pale yellow); head variably colored, if black, mandible testaceous and clypeus black. . . *A. ephippiger* (Dalman)

- Notauli incomplete, reaching about 0.60 length of the mesoscutum; prothorax pale yellow (rest of mesosoma black); head black, except mandible and clypeus testaceous. . .*A. claricolle* Kieffer

The male of *A. claricolle* can be included in the key to the *Anteon* males of Fennoscandia and Denmark [7] by replacing couplet 11 as follows:

11. Paramere with dorsal membranous band very short (Plate 11H in Olmi and Xu [6]); head in part slightly granulated, in part rugose, in part punctate, occasionally alutaceous, with sculpture usually slightly distinct. . . *A. exiguum* (Haupt)

- Paramere with very long dorsal membranous band (figure 5F; plates 11C, 12E, 15H in Olmi and Xu [6]); head punctate, areas between punctures smooth; rarely slightly granulated among punctures; head surface never alutaceous. . . 11′

11′. Distal inner process of paramere very long, with inner margin not or slightly excavated (Figure 5F) . . . *A. claricolle* Kieffer

- Distal inner process of paramere shorter and with inner margin not excavated (figures 11C and 12E in Olmi and Xu [6]) . . . 12

12. Head more strongly punctate, areas between punctures smooth or very slightly granulated, with a short or long frontal line. . . *A. gaullei* Kieffer

- Head finely punctate, smooth, areas between punctures smooth or very slightly granulated, usually without frontal line. . . *A. ephippiger* (Dalman)

The female of *A. claricolle* can be included in the key to the *Anteon* females of the Eastern Palaearctic region [6] by replacing couplet 50 as follows:

50. Notauli incomplete, reaching about 0.3–0.5 length of the mesoscutum. . . *A. ephippiger* (Dalman)

- Notauli incomplete, reaching about 0.60–0.80 length of the mesoscutum. . . 51

51. Mesosoma completely testaceous, except disc of the metapectal–propodeal complex brown. . . *A. shandonganum* sp. Nov.

- Mesosoma totally black; occasionally black, except prothorax pale yellow (Figure 4D) . . . 52

52 Clypeus testaceous (Figure 4C); prothorax pale yellow (Figure 4D) . . . *A. claricolle* Kieffer

- Clypeus and prothorax black. . . *A. munitum* Olmi

The male of *A. claricolle* can be included in the key to the males of the *Anteon* males of the Eastern Palaearctic published by Olmi and Xu [6] by replacing couplet 38 as follows:

38. Propodeal declivity with median area almost completely smooth, not rugose. . . *A. sulawesianum* Olmi

- Propodeal declivity with median area rugose. . . 38′

38′. Distal inner process of paramere very long (figure 5F; Plate 26C in Xu et al. [2]) . . . 38″.

- Distal inner process of paramere shorter (Plates 11C, 12E in Olmi and Xu [6]) . . . 39

38″. Distal inner process of paramere with inner margin slightly excavated (Plate 26C in Xu et al. [2]) . . . *A. fidum* Olmi

- Distal inner process of paramere with inner margin not excavated (Figure 5F) . . . *A. claricolle* Kieffer

3.4.4. *Anteon maguanense* Chen, Olmi & Liu, sp. nov. (Figure 6)

urn:lsid:zoobank.org:act:1E2871CD-57E1-4343-A46B-A514F2EC62B7

Material examined. Type: holotype, female: CHINA: Yunnan, Muchang County, Maguan Town, coniferous forest, 22.91888° N, 104.162851° E, 1336 m, MT, VI.2017, Li Ma leg., 3011738 (SCBG).

Distribution. China (Yunnan).

Etymology. The species is named after Maguan Town, the main town near the capture locality.

Description. Female, holotype. Body length 2.8 mm; fully winged (Figure 6A,B). Head black with mandible testaceous; antenna testaceous; mesosoma black; metasoma brown; legs yellow with proximal extremity of metacoxa black. Antenna clavate; antennomeres in the following proportions: 11:5:7:7:6:6:6:6:5:8. Head (Figure 6E) shiny, punctate, areas between punctures smooth, except anterior third of frons rugose; vertex without two oblique keels connecting lateral ocelli to occipital carina; frontal line complete; frons with two lateral longitudinal keels along orbits directed towards antennal toruli; OL = 3; OOL = 5; OPL = 3; POL = 4; TL = 5; greatest breadth of lateral ocelli as long as OPL; occipital carina complete. Pronotum shiny, with the anterior surface about as long as the posterior surface; posterior surface shiny, punctate, areas between punctures smooth, much shorter than mesoscutum (6:16); pronotal tubercle reaching tegula. Mesoscutum and mesoscutellum shiny, punctate, areas between punctures smooth. Notauli incomplete, present at the anterior 0.7 of the mesoscutum (Figure 6D). Metanotum shiny, unsculptured. Metapectal–propodeal complex with strong transverse keel between disc and propodeal declivity; disc reticulate rugose; propodeal declivity rugose, with two complete longitudinal keels. Forewing hyaline; distal part (Rs) of the stigmal vein (2r-rs&Rs) shorter than the proximal (2r-rs) part (6:10). Protarsomeres in the following proportions: 6:3:3:9.5:19. Enlarged claw (Figure 6F) with proximal prominence bearing one long bristle. Protarsomere 5 (Figure 6F) straight before distal apex, with two rows of about 29 lamellae; distal apex with four lamellae. Tibial spurs 1/1/2.

Figure 6. *Anteon maguanense* Chen, Olmi & Liu, sp. nov., holotype, female (SCAU 3011738). (**A**) Habitus, lateral view. (**B**) Habitus, dorsal view. (**C**) Head and mesosoma, lateral view. (**D**) Head and mesosoma, dorsal view. (**E**) Head, dorsal view. (**F**) Chela.

Male. Unknown.

Diagnosis. Female with face partly reticulate rugose and partly punctate, areas between punctures smooth; head vertex without oblique keels connecting lateral ocelli to occipital carina; posterior surface of the pronotum about as long as the anterior surface, about 0.5 as long as the mesoscutum; forewing hyaline without dark transverse bands; propodeal declivity with two longitudinal keels and median area as rugose as lateral areas; protarsomere 4 longer than segment 1; protarsomere 5 with the basal part much shorter than the distal part; protarsomere 5 is straight before the distal apex (Figure 6F). *A. maguanense* sp. nov. is similar to *A. semirugosum* Xu, Olmi, Guglielmino & Chen, 2012. However, in the new species, the head vertex has no oblique keels connecting lateral ocelli to occipital carina and the protarsomere 5 is straight before the distal apex (Figure 6F) (in *A. semirugosum*, head vertex with two oblique keels connecting lateral ocelli to occipital carina and protarsomere 5 curved before the distal apex (Plate 44A in Xu et al. [2]). Following the description of *A. maguanense*, the key to the females of the Oriental species of *Anteon* published by Xu et al. [2] should be modified by replacing couplet 98 as follows:

98. Face completely punctate, areas between punctures smooth; posterior surface of pronotum about as long as mesoscutum. . . *A. wangi* Xu, He & Olmi

- Face partly rugose, with the rest of the surface punctate, areas between punctures smooth; posterior surface of pronotum about 0.5 as long as mesoscutum. . . 98′

98′. Vertex of head with two oblique keels connecting lateral ocelli to occipital carina; protarsomere 5 curved before distal apex (figure 44A in Xu et al. [2]) . . . *A. semirugosum* Xu, Olmi & Guglielmino

- Vertex of head without oblique keels connecting lateral ocelli to occipital carina; protarsomere 5 straight before distal apex (Figure 6F) . . . *A. maguanense* sp. nov.

3.4.5. *Anteon meifenganum* Olmi (Figures 7 and 8)

Anteon meifenganum Olmi 1991: 173 [28]; Xu et al. 2013: 135 [2].

Anteon liui Xu, Olmi & He 2010: 404 [29] (**new syn.**).

Material examined: 1 male, CHINA: Shaanxi, Yang County, Huayang Town, 33.675537° N, 107.349632° E, 922 m, 12.IV–12.V.2017, MT, Haoyu Liu leg., SCAU 3040520 (SCBG); 1 female, Shaanxi, Yang County, Huayang Town, 33.382156° N, 107.507905° E, 1154 m, 26.V–26.VI.2017, MT, Haoyu Liu leg., SCAU 3040514 (SCBG).

Distribution. China (Taiwan, Guizhou, Hunan, Shaanxi, Zhejiang), Myanmar, Thailand (Xu et al. 2013).

Remarks. *Anteon meifenganum* Olmi, 1991 and *A. liui* Xu, Olmi & He 2010 were described, respectively, by males [28] and females [6]. The *COI* sequences of the new material of these two species generated in this study are identical, indicating that they are just the opposite sex of the same species. Therefore, *A. liui* is here treated as a junior synonym of *A. meifenganum*.

Figure 7. *Anteon meifenganum* Olmi, 1991, female (SCAU 3040514). (**A**) Habitus, lateral view. (**B**) Habitus, dorsal view. (**C**) Head and mesosoma, lateral view. (**D**) Head and mesosoma, dorsal view. (**E**) Head, dorsal view. (**F**) Chela.

Figure 8. *Anteon meifenganum* Olmi, 1991, male (SCAU 3040520). (**A**) Habitus, lateral view. (**B**) Habitus, dorsal view. (**C**) Head and mesosoma, lateral view. (**D**) Head and mesosoma, dorsal view. (**E**) Head, anterior view. (**F**) Genitalia (left half removed).

3.4.6. *Anteon parafidum* Chen, Olmi & Ødegaard, sp. nov. (Figures 9 and 10)

urn:lsid:zoobank.org:act:06445F95-8DC6-4E6E-B027-695946992551

Material examined. Types: female, holotype: CHINA: Yunnan, Xishuangbanna, Menghai, Bulangshan Village, 1683 m, Area A, grass, 21°45.037′ N, 100°26.715′ E, 17.V–21.VI.2018, MT, Li Ma leg., SCAU 3011658 (SCBG). Paratypes: 1 male, same locality label as holotype, SCAU 3011662 (SCBG); 1 female, 1 male, Yunnan, Xishuangbanna, Menghai, Bulangshan Village, 1683 m, Area A2, grass, 21°44.981′ N, 100°26.907′ E, 15.VIII.2018, MT, Li Ma leg., SCAU 3040566, SCAU 3040567 (SCBG); 1 female, 1 male, same locality as holotype, but 21.VIII–20.IX.2019, SCAU 3044018, SCAU 3044026 (SCBG); 1 female, 1 male, same locality as holotype, but 20.VII–21.VIII.2019, SCAU 3044089, SCAU 3044088 (SCBG);

1 male, same locality as holotype, but 28.V–28.VI.2019, SCAU 3044066 (SCBG); 1 female, Yunnan, Xishuangbanna, Menghai, Bulangshan Village, 1683 m, Area A2, grass, 21°44.981′ N, 100°26.907′ E, 20.IV–28.V.2019, MT, Li Ma leg., 3044033 (SCBG).

Distribution. China (Yunnan).

Etymology. The species is named *parafidum*, because it is morphologically similar to *Anteon fidum* Olmi, 1991 (from the Greek prefix para- (meaning "close to") and *fidum*).

Description. Female, holotype. Body length 2.6 mm; fully winged (Figure 9A,B). Head black with mandible testaceous; antenna brown with scape and pedicel testaceous; mesosoma black; metasoma brown; legs testaceous. Antenna clavate; antennomeres in the following proportions: 11:5:7:6:5:5:5:5:5:7. Head (Figure 9E) shiny, punctate, areas between punctures smooth; frontal line complete; frons with two lateral longitudinal keels along orbits directed towards antennal toruli; occipital carina complete; OL = 2; OOL = 5; OPL = 4; POL = 5; TL = 4; greatest breadth of lateral ocellus as long as OL. Pronotum shiny, punctate, areas between punctures smooth, with the posterior surface shorter than the mesoscutum (9:12); pronotal tubercle reaching tegula. Mesoscutum shiny, punctate, areas between punctures smooth. Notauli incomplete, present at the anterior 0.5 of the mesoscutum (Figure 9D). Mesoscutellum and metanotum shiny, unsculptured. Metapectal–propodeal complex with strong transverse keel between disc and propodeal declivity; disc reticulate rugose; propodeal declivity rugose, with two complete longitudinal keels. Forewing hyaline; distal part (Rs) of the stigmal vein (2r-rs&Rs) much shorter than the proximal (2r-rs) part (2:10). Protarsomeres in the following proportions: 6:2:2.5:7:16. Enlarged claw (Figure 9F) with proximal prominence bearing one long bristle. Protarsomere 5 (Figure 9F) with the basal part slightly shorter than the distal part, with one row of 13 large lamellae; distal apex with three large lamellae. Tibial spurs 1/1/2.

Description of Male. Fully winged (Figure 10A,B); length 2.1 mm. Head black with mandible testaceous; antenna brown with ventral side of scape testaceous; mesosoma black; metasoma brown; legs brown with tarsi and protibia testaceous. Antenna filiform; antennomeres in the following proportions: 10:6:7:7:7:7:7:7:7:9. Head (Figure 10E) shiny, punctate, areas between punctures smooth; OL = 3; OOL = 5; OPL = 3; POL = 5; TL = 3; greatest breadth of lateral ocellus shorter than OPL (2:3); occipital carina complete; frontal line absent; frons without lateral longitudinal keels along orbits. Mesoscutum punctate, areas between punctures smooth. Notauli incomplete, present at the anterior 0.5 of the scutum (Figure 10D). Mesoscutellum and metanotum shiny, unsculptured. Metapectal–propodeal complex with strong transverse keel between disc and propodeal declivity; disc reticulate rugose; propodeal declivity rugose, with two longitudinal keels. Forewing hyaline; distal part (Rs) of the stigmal vein (2r-rs&Rs) much shorter than the proximal (2r-rs) part (2.5:8). Paramere (Figure 10F) with distal inner pointed process; inner distal margin of pointed process slightly excavated. Dorsal process of paramere shorter than volsella (Figure 10F). Tibial spurs 1/1/2.

Figure 9. *Anteon parafidum* Chen, Olmi & Ødegaard, sp. nov., holotype, female (SCAU 3011658). (**A**) Habitus, lateral view. (**B**) Habitus, dorsal view. (**C**) Head and mesosoma, lateral view. (**D**) Head and mesosoma, dorsal view. (**E**) Head, dorsal view. (**F**) Chela.

Figure 10. *Anteon parafidum* Chen, Olmi & Ødegaard, sp. nov., paratype, male (SCAU 3011662). (**A**) Habitus, lateral view. (**B**) Habitus, dorsal view. (**C**) Head and mesosoma, lateral view. (**D**) Head and mesosoma, dorsal view. (**E**) Head, anterior view. (**F**) Genitalia (left half removed).

Diagnosis. Female (Figure 10E) with head punctate and areas between punctures smooth; OL shorter than POL; OPL twice as long as the greatest breadth of lateral ocellus; mesosoma black; propodeal declivity with two complete longitudinal keels and median region as rugose as lateral regions; protarsomere 4 slightly longer than 1; protarsomere 5 provided with very large lamellae, with the basal part shorter than the distal part (Figure 10F). Male (Figure 10E) with head punctate and areas between punctures smooth; propodeal declivity with two complete longitudinal keels and median region as rugose as lateral regions; pointed distal inner process of paramere with inner distal margin slightly excavated; paramere with dorsal process shorter than volsella (Figure 10F).

The female of *A. parafidum* is very similar to *A. fidum* Olmi, 1991. However, in *A. parafidum*, protarsomere 5 has its basal part shorter than the distal part (Figure 9F) (in *A. fidum*, protarsomere 5 with the basal region about as long as the distal region (figure 26D in Xu et al. [2]). Following the description of *A. parafidum* sp. nov., the key to the females of the Oriental *Anteon* published by Xu et al. [2] should be modified by replacing couplet 68 as follows:

68. Protarsomere 5 with very large lamellae (Figure 9F) . . . 68′

- Protarsomere 5 with small lamellae (Plates 19F, 22A in Xu et al. [2]) . . . 69

68′. Protarsomere 5 with the basal region about as long as the distal region (Plate 26D in Xu et al. [2]) . . . *A. fidum* Olmi

- Protarsomere 5 with the basal part shorter than the distal part (Figure 9F) . . . *A. parafidum* sp. nov.

The male of *A. parafidum* sp. nov. is very similar to *A. fidum* Olmi, 1991, *A. priscum* Olmi, 1991, *A. shaanxianum* sp. nov. For the differences between *A. parafidum* and the other species, see updated keys for *A. shaanxianum* sp. nov.

3.4.7. *Anteon priscum* Olmi (Figure 11)

Anteon priscum Olmi 1991: 164 [28]; Xu et al. 2013: 155 [2].

Material examined: 1 male, CHINA: Jiangsu, Nanjing, Xianlin, Mt. Duoshan, 32°6′51″ N 118°54′43″ E, 9–15.IV.2012, MT, Jie Zhao leg., SCAU 3011723 (SCBG); 1 female, Jiangsu, Nanjing, Xianlin, Mt. Duoshan, 32°6′51″ N 118°54′43″ E, 30.IV–7.V.2012, MT, Jie Zhao leg., SCAU 3011631 (SCBG).

Distribution. China (Gansu, Henan, Shaanxi, Ningxia, Jiangsu, Zhejiang, Fujian, Guangdong, Taiwan, Guizhou, Xizang, Yunnan), India, Indonesia [2].

Remarks. *A. priscum* was known only by males [2,28]. The female of this species is here confirmed by *COI* sequences. Therefore, we present the following diagnosis and description of the female:

Description. Female. Body length 4.1 mm; fully winged (Figure 11A,B). Head black with mandible and clypeus testaceous; antenna brown with scape and pedicel testaceous; mesosoma black; metasoma brown; legs yellow. Antenna clavate; antennomeres in the following proportions: 12:6:9:8:8:8:7:7:7:9. Head (Figure 11E) shiny, punctate, areas between punctures smooth; frontal line incomplete, shortly present in front of anterior ocellus; frons with two lateral longitudinal keels along orbits directed towards antennal toruli; OL = 4; OOL = 7; OPL = 6; POL = 5; TL = 6; greatest breadth of lateral ocellus shorter than OPL (3:6); occipital carina complete. Pronotum smooth, shiny, punctate, areas between punctures smooth; posterior surface of pronotum much longer than the anterior surface, shorter than mesoscutum (15:20); pronotal tubercle reaching tegula. Mesoscutum shiny, punctate, areas between punctures smooth. Notauli incomplete, present at the anterior 0.9 of the mesoscutum (Figure 11D). Mesoscutellum and metanotum shiny, unsculptured. Metapectal–propodeal complex with strong transverse keel between disc and propodeal declivity; disc reticulate rugose; propodeal declivity rugose, with two complete longitudinal keels. Forewing (Figure 11A) hyaline; distal part (Rs) of the stigmal vein (2r-rs&Rs) much shorter than the proximal (2r-rs) part (5:12). Protarsomeres in the following proportions: 8:3:4:14:25. Enlarged claw (Figure 11F) with proximal prominence bearing one long bristle. Protarsomere 5 (Figure 11F) with two rows of about 37 small lamellae and basal part much shorter than the distal part; distal apex with about 8 lamellae. Tibial spurs 1/1/2.

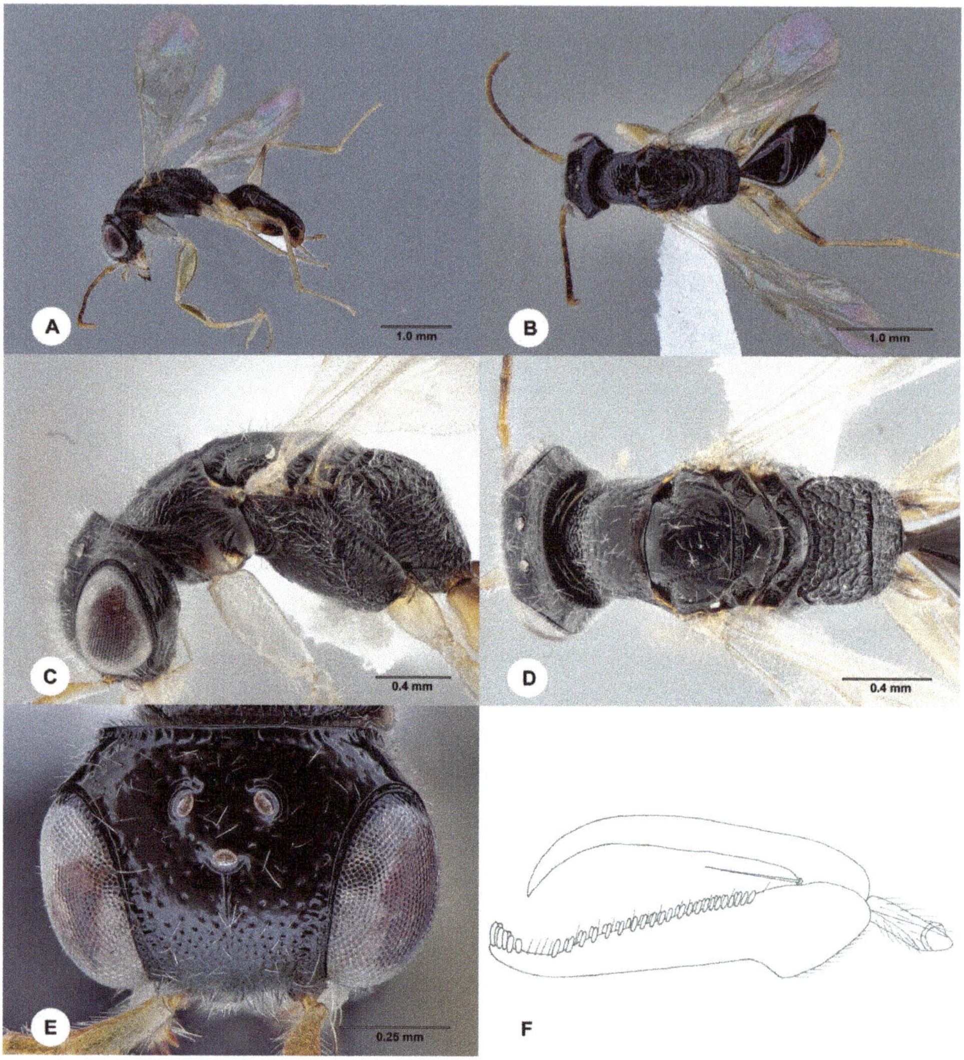

Figure 11. *Anteon priscum* Olmi, 1991, female (SCAU 3011631). (**A**) Habitus, lateral view. (**B**) Habitus, dorsal view. (**C**) Head and mesosoma, lateral view. (**D**) Head and mesosoma, dorsal view. (**E**) Head, dorsal view. (**F**) Chela.

Diagnosis of Female: Head black, except mandible and clypeus testaceous; head punctate, areas between punctures smooth (Figure 11E); frontal line incomplete, shortly visible in front of anterior ocellus (Figure 11E); mesoscutum punctate, areas between punctures smooth (Figure 11D); notauli reaching approximately 0.9 × length of the mesoscutum (Figure 11D); propodeal declivity with two complete longitudinal keels and median area as rugose as lateral areas; distal part of the stigmal vein (2r-rs&Rs) less than 0.5 as long as the proximal part (Figure 11A); protarsomere 5 very slender (Figure 11F). Because of the above diagnosis, the female of *A. priscum* can be inserted in the key to the females of the Oriental *Anteon* published by Xu et al. [2] by modifying couplet 69 as follows:

69. Protarsomere 5 very broad (Plate 22A in Xu et al. 2013) . . . *A. caii* Xu, He & Olmi
- Protarsomere 5 very slender (figure 11F; Plate 19F in Xu et al. [2]) . . . 69′

69′. Clypeus black; frontal line complete; notauli reaching about 0.5 × length of the mesoscutum. . . *A. blanduscutum* Xu, He & Rui

- Clypeus testaceous; frontal line incomplete, shortly visible in front of anterior ocellus (Figure 11E); notauli reaching about 0.9 × length of the mesoscutum (Figure 11D) . . . *A. priscum* Olmi

3.4.8. *Anteon shaanxianum* Chen, Olmi & Ødegaard, sp. nov. (Figure 12)

urn:lsid:zoobank.org:act:104C58C7-DA63-4DBA-BEE1-1F9C40F87488

Material examined. Type: male, holotype: CHINA: Shaanxi, Yang County, Huayang Town, 1154m, MT, 33.382156° N, 107.507905° E, 26.V–26.VI.2017, Haoyu Liu leg., SCAU 3040516 (SCBG).

Distribution. China (Shaanxi).

Etymology. The species is named after Shaanxi province, the locality of the holotype.

Description. Male. Body length 2.1 mm; fully winged (Figure 12A,B). Head black with mandible testaceous and clypeus ferruginous; antenna brown with scape yellow; mesosoma black; metasoma brown; legs yellow. Antenna filiform, with dense setae; antennomeres in the following proportions: 8:5:6:7:7:6:6:6:6:7. Head (Figure 12E) shiny, punctate, areas between punctures smooth; OL = 3; OOL = 5; OPL = 2; POL = 4; TL = 4; greatest breadth of the lateral ocellus as long as OPL; occipital carina complete; frontal line incomplete, present in front of the anterior ocellus and absent in the anterior half of frons; frons without lateral longitudinal keels along orbits directed towards antennal toruli. Mesoscutum punctate, areas between punctures smooth. Notauli incomplete, present at the anterior 0.7 of the scutum (Figure 12D). Mesoscutellum and metanotum shiny, unsculptured. Metapectal–propodeal complex with strong transverse keel between disc and propodeal declivity; disc reticulate rugose; propodeal declivity rugose, with two longitudinal keels. Forewing hyaline; distal part (Rs) of the stigmal vein (2r-rs&Rs) much shorter than the proximal (2r-rs) part (2.5:8). Paramere (Figure 12F) with distal inner pointed process; inner distal margin of pointed process slightly excavated. Dorsal process of paramere shorter than volsella (Figure 12F). Tibial spurs 1/1/2.

Female. Unknown.

Diagnosis. Male with head black and punctate, areas between punctures smooth; clypeus ferruginous; frontal line incomplete, present in front of the anterior ocellus and absent in the anterior half of frons; notauli present at the anterior 0.7 of mesoscutum; propodeal declivity with two complete longitudinal keels and median area as rugose as lateral areas; distal part of the stigmal vein (2r-rs&Rs) less than 0.5 as long as the proximal part; paramere with pointed distal inner process (Figure 12F), provided with inner distal margin slightly excavated; dorsal process of paramere shorter than volsella (Figure 12F); legs yellow.

Anteon shaanxianum sp. nov. is similar to *A. parafidum* sp. nov. and *A. fidum* Olmi, 1991, and the key to the males of the Oriental species of *Anteon* published by Xu et al. [2] should be modified by replacing couplet 59 as follows:

59. Paramere with distal inner rounded process (Plate 38I in Xu et al. [2]) . . . *A. parapriscum* Olmi

- Paramere with distal inner process more or less pointed (figures 10F and 12F; Plates 26C, 40B, C in Xu et al. [2]) . . . 60

60. Distal inner process of paramere with inner distal margin not excavated (Plate 40B, C in Xu et al. [2]) . . . *A. priscum* Olmi

- Distal inner process of paramere with inner distal margin slightly excavated (Figures 10F and 12F; Plate 26C in Xu et al. [2]) . . . 60′

60′. Frontal line complete. . . *A. fidum* Olmi

- Frontal line absent or incomplete. . . 60″

60″. Notauli present at about the anterior 0.5 of the mesoscutum (Figure 10D); frontal line absent; clypeus black. . . *A. parafidum* sp. nov.

- Notauli present at about the anterior 0.7 of the mesoscutum (Figure 12D); frontal line incomplete, present in front of the anterior ocellus and absent in the anterior half of frons; clypeus ferruginous. . . *A. shaanxianum* sp. nov.

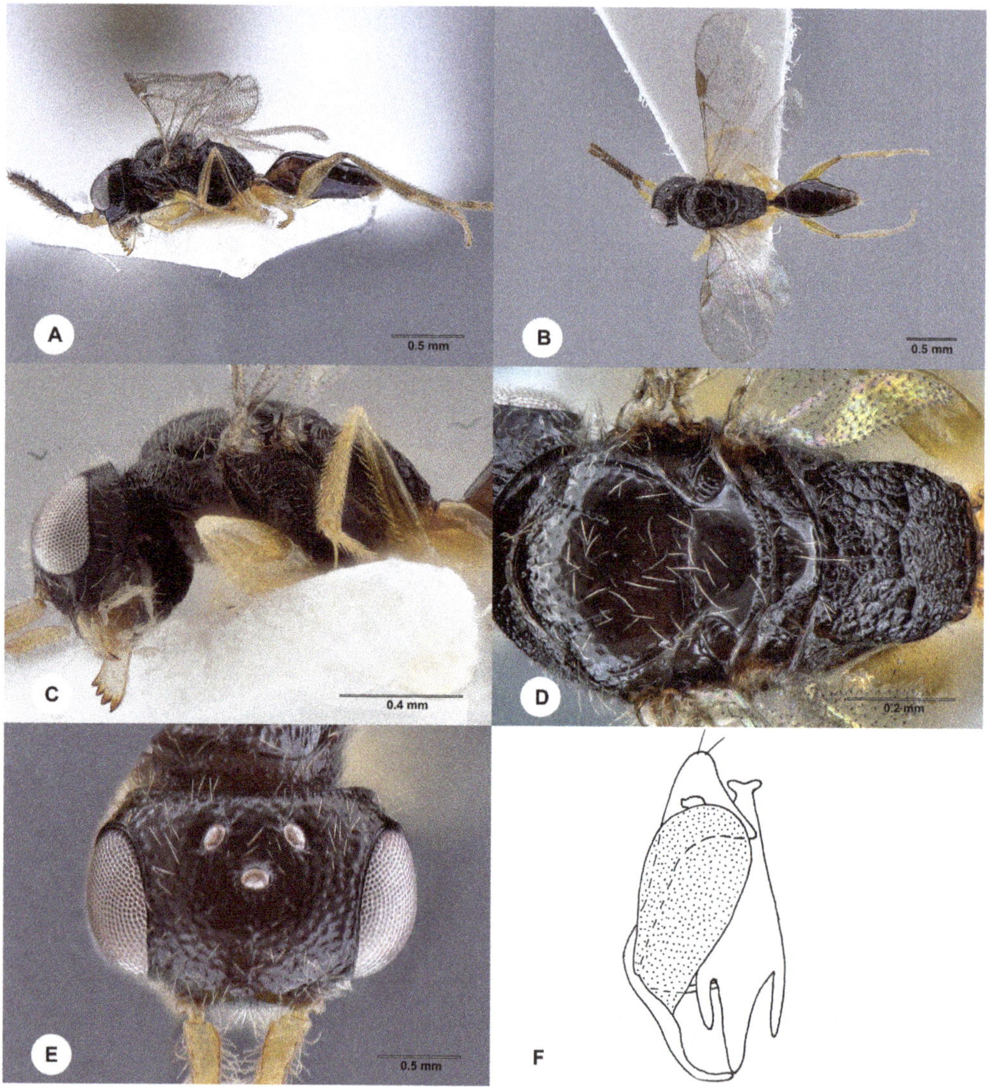

Figure 12. *Anteon shaanxianum* Chen, Olmi & Ødegaard, sp. nov., holotype, male (SCAU 3040516). (**A**) Habitus, lateral view. (**B**) Habitus, dorsal view. (**C**) Head and mesosoma, lateral view. (**D**) Mesosoma, dorsal view. (**E**) Head, dorsal view. (**F**) Genitalia (left half removed).

3.4.9. *Anteon shandonganum* Olmi, Chen & Liu, sp. nov. (Figure 13)

urn:lsid:zoobank.org:act:3BA2FB31-6A33-45F4-8BFE-DC90D3AF298A

Material examined. Type: holotype, female: CHINA: Shandong, Shanghe County, MT3, 37°16′4″ N 117°9′10″ E, 23–29.VI.2018, Jiahe Yan leg., SCAU 3011688 (SCBG).

Distribution. China (Shandong).

Etymology. The species is named after Shandong Province, the locality of the holotype.

Description. Female, holotype. Body length 2.2 mm; fully winged (Figure 13A,B). Body testaceous, except disc of metapectal–propodeal complex brown. Antenna clavate; antennomeres in the following proportions: 8:4:8:4:3:3:3:4:4:5. Head (Figure 13E) shiny, punctate, areas between punctures smooth; frontal line almost completely absent, only very shortly present in front of anterior ocellus; frons without lateral longitudinal keels along orbits directed towards antennal toruli; occipital carina complete; OL = 3; OOL = 4; OPL = 3; POL = 4; TL = 4; greatest breadth of the lateral ocellus shorter than OPL (2:3). Pronotum with the anterior surface much shorter than the posterior surface; posterior surface shiny, punctate, areas between punctures smooth, slightly shorter than mesoscutum (7:9); pronotal tubercle reaching tegula. Mesoscutum and mesoscutellum shiny, finely punctate, areas between punctures smooth. Notauli incomplete, present at the anterior 0.6 of the scutum (Figure 13D). Metapectal–propodeal complex with strong transverse keel between disc and propodeal declivity; disc reticulate rugose; propodeal declivity reticulate rugose, with two longitudinal keels. Forewing hyaline, without dark transverse bands; distal part (Rs) of the stigmal vein (2r-rs&Rs) much shorter than the proximal (2r-rs) part (3:8). Protarsomeres in the following proportions: 5:2:2:6:13. Enlarged claw (Figure 13F) with proximal prominence bearing one long bristle. Protarsomere 5 (Figure 13F) with the basal part much shorter than the distal part, with one row of about 21 lamellae; distal apex with four lamellae. Tibial spurs 1/1/2.

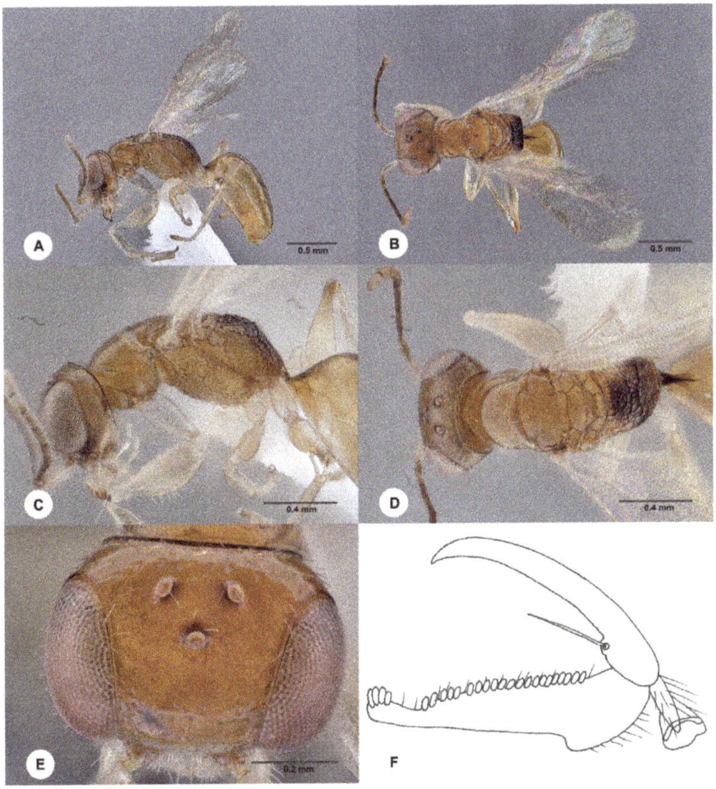

Figure 13. *Anteon shandonganum* Olmi, Chen & Liu, sp. nov., holotype, female (SCAU 3011688). (**A**) Habitus, lateral view. (**B**) Habitus, dorsal view. (**C**) Head and mesosoma, lateral view. (**D**) Head and mesosoma, dorsal view. (**E**) Head, dorsal view. (**F**) Chela.

Male. Unknown.

Diagnosis. Female with head testaceous, punctate and areas between punctures smooth; frontal line almost completely absent, only very shortly present in front of anterior ocellus; ocellar triangle not delimited by keels; mesosoma mostly testaceous; posterior surface of the pronotum about as broad as long; mesoscutum slightly longer than the posterior region of pronotum; notauli present at about the anterior 0.6 of the mesoscutum; propodeal declivity reticulate rugose, with two longitudinal keels; forewing hyaline; protarsomere 4 slightly longer than 1; protarsomere 5 with the basal part much shorter than the distal part (Figure 13F).

Anteon shandonganum sp. nov. is similar to *A. quatei* Olmi, 1991. However, in *A. shandonganum*, the frontal line is almost completely absent, the mesoscutum is slightly longer than the posterior region of the pronotum, the notauli (Figure 13D) reach about 0.6 × length of the mesoscutum (in *A. quatei*, the frontal line is complete; mesoscutum about twice as long as the posterior region of pronotum; notauli reaching about 0.9 × length of the mesoscutum). Following the description of *A. shandonganum*, the key to the females of the Oriental *Anteon* published by Xu et al. [2] should be modified by replacing couplet 63 as follows:

63. Head black with mandible, clypeus and anterior region of face between antennal toruli brown-testaceous; notauli reaching about 0.25 × length of mesoscutum. . . *A. zonarium* Xu, Olmi & He

- Head totally or almost totally testaceous, partly darkened; notauli reaching about 0.6–0.9 × length of the mesoscutum. . . 63′

63′. Frontal line complete; mesoscutum about twice as long as the posterior region of the pronotum; notauli reaching about 0.9 × length of the mesoscutum. . . *A. quatei* Olmi

- Frontal line almost completely absent; mesoscutum slightly longer than the posterior region of pronotum (Figure 13D); notauli reaching about 0.6 × length of the mesoscutum (Figure 13D) . . . *A. shandonganum* sp. nov.

The key to the females of the Eastern Palaearctic *Anteon* published by Olmi & Xu [6] should be modified by replacing couplet 50 as follows:

50. Notauli reaching about 0.65–0.80 × length of the mesoscutum. . . *A. munitum* Olmi

- Notauli reaching 0.3–0.6 × length of the mesoscutum. . . 50′

50′. Notauli reaching about 0.3–0.5 × length of the mesoscutum. . . *A. ephippiger* (Dalman)

- Notauli reaching about 0.6 × length of the mesoscutum (Figure 13D) . . . *A. shandonganum* sp. nov.

4. Discussion

The pronounced sexual dimorphism in Dryinidae makes it difficult to associate the opposite sexes of the same species. In the past taxonomic practice of the sexually dimorphic genus *Anteon*, many species were described based on a single sex. Such a practice might result in that one species would have been described twice, leading to the exaggeration of a higher diversity. On the other hand, while some species were described by both sexes, the female and male were often associated tentatively, i.e., by the comparison of some morphological characters, or because they were collected in the same locality. However, such tentative associations might result in two species that have been lumped into one species, underestimating the real diversity of these parasitoids. DNA barcode-based methods have become powerful in delimitating species boundaries and the confirmation of female–male associations for parasitoid groups with extreme sexual dimorphism [10,30,31]. Our present study on *Anteon* further demonstrates that DNA barcoding is useful to enhance species delimitation and the sexual association of the same species in Dryinidae. The interspecific distances are larger than intraspecific distance for the COI sequences, with the intraspecific distances generally lower than 3%, while the interspecific distances are higher than 3.1% (Tables S2 and S3). The results produced by the two molecular species delimitation methods, i.e., ABGD and bPTP, are mostly congruent with the morphological identification. In addition, the sexual association of six species was confirmed by DNA sequences and lead us to synonymize *A. liui* with *A. meifenganum*.

Given the great potential of *Anteon* in the biological control of leafhoppers, the confirmation of the host association is also important. As indicated by a few researchers, DNA barcoding can be useful for clarifying host–parasitoid relationships in Dryinidae [9,13], which requires a more complete DNA barcode database as a reference. This study represents another effort towards the achievement of this goal.

Supplementary Materials: The following supporting information can be downloaded at https://www.mdpi.com/article/10.3390/insects15010018/s1, Figure S1: *Anteon abdulnouri* Olmi, 1987, female (SCAU 3040430); Figure S2: *Anteon achterbergi* Olmi, 1991, female (SCAU 3011714); Figure S3: *Anteon blanduscutum* Xu, He & Rui, 1996, female (SCAU 3011677); Figure S4: *Anteon borneanum* Olmi, 1984, female (SCAU 3011669); Figure S5: *Anteon confusum* Olmi, 1991, female (SCAU 3040517); Figure S6: *Anteon confusum* Olmi, 1991, male (SCAU 3044058); Figure S7: *Anteon exiguum* (Haupt, 1941), female (SCAU 3011621); Figure S8: *Anteon fidum* Olmi, 1991, male (SCAU 3040512); Figure S9: *Anteon funiuense* Xu, He & Olmi, 2001, female (SCAU 3040522); Figure S10: *Anteon henanense* Xu, He & Olmi, 2001, female (SCAU 3044072); Figure S11: *Anteon henanense* Xu, He & Olmi, 2001, male (SCAU 3040511); Figure S12: *Anteon hilare* Olmi, 1984, female (SCAU 3011713); Figure S13: *Anteon hirashimai* Olmi, 1993, female (SCAU 3011676); Figure S14: *Anteon mite* Olmi, 1996, male (SCAU 3040515); Figure S15: *Anteon multicolor* Xu, He & Olmi, 1998, female (SCAU 3011661); Figure S16: *Anteon multicolor* Xu, He & Olmi, 1998, male (SCAU 3044011); Figure S17: *Anteon munitum* Olmi, 1984, male (SCAU 3040524); Figure S18: *Anteon naduense* Olmi 1987, female (SCAU 3040521); Figure S19: *Anteon nanlingense* Xu, Olmi & He, 2011, female (SCAU 3040433); Figure S20: *Anteon nanlingense* Xu, Olmi & He, 2011, male (SCAU 3048821); Figure S21: *Anteon pilosum* Xu, Olmi & He, 2010, male (SCAU 3040518); Figure S22: *Anteon priscum* Olmi, 1991, male (SCAU 3011723); Figure S23: *Anteon pteromaculatum* Xu, Olmi, Guglielmino & Chen, 2012, female (SCAU 3040519); Figure S24: *Anteon tengchongense* Xu, He & Olmi, 1998, female (SCAU 3040523); Figure S25: *Anteon yuani* Xu, He & Olmi 1998, male (SCAU 3011682); Table S1: Details of sequenced specimens; Table S2: Genetic distance of COI within Anteon species under K2P model; Table S3: Interspecific pairwise distance of Anteon based on COI sequences (%).

Author Contributions: Conceptualization, H.C. and M.O.; methodology, all authors; software, H.C.; investigation, H.C., M.O., F.Ø., L.C. and J.L.; data curation, H.C. and M.O.; writing—original draft preparation, all authors; writing—review and editing, all authors; supervision, H.C. and J.L.; funding acquisition, J.L. All authors have read and agreed to the published version of the manuscript.

Funding: This research was funded by the National Natural Science Foundation of China (No. 31472027) to Jing-Xian.

Data Availability Statement: All data are available in this paper.

Acknowledgments: Thanks to Li Ma (Yunnan Agricultural University) and Haoyu Liu (Hebei University) for providing some of the studied specimens, and to Lifang Xiao (Institute of Zoology, Guangdong Academy of Sciences) for making some of the image plates. We are grateful to Arnstein Staverløkk, Norwegian Institute for Nature Research, for producing photos of Norwegian specimens of *Anteon claricolle*. Many thanks to Toshiharu Mita (Kyushu University, Japan) for his precious comments and suggestions on the text of this paper. The material from Norway was obtained from the project INVENT-ART funded by the Norwegian Taxonomy Initiative. DNA barcode data from Norway were generated in collaboration with the Norwegian Barcode of Life Network (NorBOL), funded by the Research Council of Norway and the Norwegian Biodiversity Information Centre. This research was carried out in the frame of the MIUR (Italian Ministry for Education, University and Research) initiative "Department of Excellence" (Law 232/2016).

Conflicts of Interest: The authors declare no conflict of interest.

References

1. Guglielmino, A.; Olmi, M.; Bückle, C. An updated host-parasite catalogue of world Dryinidae (Hymenoptera: Chrysidoidea). *Zootaxa* **2013**, *3740*, 1–113. [CrossRef] [PubMed]
2. Xu, Z.; Olmi, M.; He, J. Dryinidae of the Oriental region (Hymenoptera: Chrysidoidea). *Zootaxa* **2013**, *3614*, 1–460. [CrossRef] [PubMed]
3. Olmi, M.; Copeland, R.S.; van Noort, S. Dryinidae of the Afrotropical region (Hymenoptera, Chrysidoidea). *Zootaxa* **2019**, *4630*, 1–619. [CrossRef] [PubMed]
4. Olmi, M.; Xu, Z.; Guglielmino, A. A new species of the genus *Anteon* Jurine (Hymenoptera, Dryinidae) from Thailand. *ZooKeys* **2015**, *504*, 141–147. [CrossRef] [PubMed]

5. Olmi, M.; Xu, Z.; Guglielmino, A.; Speranza, S. A new species of the genus *Anteon* Jurine (Hymenoptera, Dryinidae) from Laos. *ZooKeys* **2016**, *561*, 31–38. [CrossRef] [PubMed]
6. Olmi, M.; Xu, Z. Dryinidae of the Eastern Palaearctic region (Hymenoptera: Chrysidoidea). *Zootaxa* **2015**, *3996*, 1–253. [CrossRef] [PubMed]
7. Olmi, M. *The Dryinidae and Embolemidae (Hymenoptera: Chrysidoidea) of Fennoscandia and Denmark. Fauna Entomologica Scandinavica*; E.J. Brill: Leiden, The Netherlands, 1994; p. 100.
8. Mita, T.; Matsumoto, Y. First description of the male of *Gonatopus javanus* (R.C.L. Perkins) determined by mitochondrial COI sequence (Hymenoptera: Dryinidae). *Entomol. Sci.* **2012**, *15*, 214–218. [CrossRef]
9. Olmi, M.; Mita, T.; Guglielmino, A.; Vollaro, M.; Vári, G. Application of DNA barcoding confirms the female, male, larva and host of *Bocchus scobiolae* Nagy, 1967 (Hymenoptera: Dryinidae). *Zootaxa* **2021**, *4969*, 563–572. [CrossRef]
10. Olmi, M.; Chen, H.-Y.; Guglielmino, A.; Ødegaard, F.; Vollaro, M.; Capradossi, L.; Liu, J.-X. COI sequences of *Aphelopus* Dalman from China, with descriptions of four new species (Hymenoptera, Dryinidae). *Eur. J. Taxon.* **2022**, *794*, 40–71.
11. Mita, T.; Sanada–Morimura, S.; Matsumura, M.; Matsumoto, Y. Genetic variation of two apterous wasps *Haplogonatopus apicalis* and *H. oratorius* (Hymenoptera: Dryinidae) in East Asia. *Appl. Entomol. Zool.* **2013**, *48*, 119–124. [CrossRef]
12. Tribull, C.M. 2015. Phylogenetic relationships among the subfamilies of Dryinidae (Hymenoptera, Chrysidoidea) as reconstructed by molecular sequencing. *J. Hymenopt. Res.* **2015**, *45*, 15–29. [CrossRef]
13. Chen, H.-Y.; Olmi, M.; Pang, H.; Liu, J.-X. Application of DNA barcoding confirms the host of *Gonatopus viet* Olmi, 1986 (Hymenoptera, Dryinidae). *ZooKeys* **2020**, *944*, 115–120. [CrossRef] [PubMed]
14. Ratnasingham, S.; Hebert, P.D.N. BOLD: The Barcode of Life Data System (http://www.barcodinglife.org). *Mol. Ecol. Notes* **2007**, *7*, 355–364. [CrossRef] [PubMed]
15. Kawada, R.; Lanes, G.O.; Azevedo, C.O. Evolution of metapostnotum in flat wasps (Hymenoptera, Bethylidae): Implications for homology assessments in Chrysidoidea. *PLoS ONE* **2015**, *10*, e0140051. [CrossRef]
16. Taekul, C.; Valerio, A.A.; Austin, A.D.; Klompen, H.; Johnson, N.F. Molecular phylogeny of telenomine egg parasitoids (Hymenoptera: Platygastridae s.l.: Telenominae): Evolution of host shifts and implications for classification. *Syst. Entomol.* **2014**, *39*, 24–35. [CrossRef]
17. Folmer, O.; Black, M.; Hoch, W.; Lutz, R.; Vrijenoek, R. DNA primers for amplification of mitochondrial cytochrome c oxidase subunit I from diverse metazoan invertebrates. *Mol. Mar. Biol. Biotechnol.* **1994**, *3*, 294–299.
18. Tamura, K.; Stecher, G.; Kumar, S. MEGA11: Molecular evolutionary genetics analysis version 11. *Mol. Biol. Evol.* **2021**, *38*, 3022–3027. [CrossRef]
19. Puillandre, N.; Lambert, A.; Brouillet, S.; Achaz, G. ABGD, automatic barcode gap discovery for primary species delimitation. *Mol. Ecol.* **2012**, *21*, 1864–1877. [CrossRef]
20. Zhang, J.; Kapli, P.; Pavlidis, P.; Stamatakis, A. A general species delimitation method with applications to phylogenetic placements. *Bioinformatics* **2013**, *29*, 2869–2876. [CrossRef]
21. Jurine, L. *Nouvelle Méthode de Classer les Hyménoptères et les Diptères, 1. Hyménoptères*; Paschoud: Genève, Switzerland, 1807.
22. Xu, Z.; He, J.; Rui, K. Two new species of the genus *Anteon* Jurine (Hymenoptera: Dryinidae: Anteoninae). *Entomotaxonomia* **1996**, *18*, 213–215.
23. He, J.; Xu, Z. *Hymenoptera Dryinidae. Fauna Sinica, 29*; Science Press: Beijing, China, 2002; p. 464.
24. Xu, Z.; Olmi, M.; Guglielmino, A.; Chen, H. A new species of Dryinidae (Hymenoptera: Chrysidoidea) from China. *Fla. Entomol.* **2011**, *94*, 848–852. [CrossRef]
25. Kieffer, J.J.; Marshall, T.A. Proctotrypidae. In *Species des Hyménoptères d'Europe et d'Algerie, 9*; André, E., Ed.; Hermann: Paris, France, 1906; pp. 289–552.
26. Richards, O.W. The British Bethylidae (s.l.) (Hymenoptera). *Trans. R. Entomol. Soc. Lond.* **1939**, *89*, 185–344. [CrossRef]
27. Olmi, M. *A revision of the Dryinidae (Hymenoptera)*; Memoirs of the American Entomological Institute: Logan, UT, USA, 1984; p. 1913.
28. Olmi, M. Supplement to the revision of the world Dryinidae (Hymenoptera Chrysidoidea). *Frustula Entomol.* **1991**, *12*, 109–395.
29. Xu, Z.; Olmi, M.; He, J. Two new species of Anteon (Hymenoptera: Dryinidae) from China. *Fla. Entomol.* **2010**, *93*, 403–406. [CrossRef]
30. Liu, Z.; Yang, S.J.; Wang, Y.Y.; Peng, Y.Q.; Chen, H.Y.; Luo, S.X. Tackling the taxonomic chal-lenges in the family Scoliidae (Insecta, Hymenoptera) using an integrative approach: A case study from southern China. *Insects* **2021**, *12*, 892. [CrossRef]
31. Zhang, B.-L.; Yan, C.-J.; van Achterberg, C.; Peng, Y.-Q.; Chen, H.-Y. Integrated taxonomy unveils new species of Trigonalyidae (Insecta, Hymenoptera) from Yunnan, China. *J. Hymenopt. Res.* **2022**, *90*, 101–128. [CrossRef]

insects

MDPI

Article

Ecological Speciation without Morphological Differentiation? A New Cryptic Species of *Diodontus* Curtis (Hymenoptera, Pemphredonidae) from the Centre of Europe †

Eduardas Budrys *, Svetlana Orlovskytė and Anna Budrienė

Institute of Ecology, Nature Research Centre, Akademijos 2, 08412 Vilnius, Lithuania; svetlana.orlovskyte@gamtc.lt (S.O.); ana.budriene@gamtc.lt (A.B.)

* Correspondence: eduardas.budrys@gamtc.lt

† Zoobank link: urn:lsid:zoobank.org:pub: E09F09F9-93EA-4058-9780-167E767CD585.

Simple Summary: The application of molecular methods in the studies of biological diversity revealed that there are many more species than we can recognise by their appearance. Species pairs or groups, which are difficult to recognise and are known as cryptic species, may differ in their ecological traits, place in trophic networks, and their functional role in ecosystems. It is important to separate cryptic species, particularly if we apply species composition or other diversity metrics for an assessment or monitoring of the ecosystem state or human pressures and impacts. In our study, we revealed a new cryptic species of aphid-hunting wasp that is virtually indistinguishable from a well-known one by morphology. These two species can be easily segregated using molecular characters, and presumably they differ also by nesting habitat.

Abstract: Upon exploring the mitotype diversity of the aphid-hunting wasp, *Diodontus tristis*, we revealed specimens with highly divergent mitotypes from two localities in Lithuania and nesting in clayey substrate, while the specimens with typical mitotypes were found nesting in sandy sites. The comparison of inter- and intra-specific distances and application of delimitation algorithms supported the species status of the clay-nesting populations. Using a set of DNA markers that included complete or partial sequences of six mitochondrial genes, three markers of ribosomal operon, two homeobox genes, and four other nuclear genes, we clarified the phylogenetic relationships of the new cryptic species. The endosymbiotic bacteria infestation was checked, considering the option that the divergent populations may represent clades isolated by *Wolbachia* infection; however, it did not demonstrate any specificity. We found only subtle morphological differences in the new clay-nesting species, *D. argillicola* sp. nov.; the discriminant analysis of morphometric measurements did not reliably segregate it as well. Thus, we provide the molecular characters of the cryptic species, which allow confident identification, its phylogenetic position within the genus, and an updated identification key for the *D. tristis* species group.

Keywords: ecological speciation; endosymbionts; differentiation by habitat; multi-gene phylogeny; species delimitation

1. Introduction

Cryptic species, which are closely similar or identical in morphology, but different in molecular, behavioural, ecological, and other characteristics, represent high levels of undiscovered biodiversity [1–3]. Hence, the rigorous assessment of cryptic biodiversity is crucial for both a deeper understanding of the evolutionary processes of speciation and recognising species limits in order to inform their research or conservation actions [4].

Accelerating investigation into cryptic species' diversity in the past decades has revealed that modern molecular-based delimitation methods provide valuable complements to morphological taxonomy [2,5–7]. Especially useful is an integrative analytical approach, incorporating a range of species' traits including the behavioural, ecological, and life history aspects [3,8–10]. Recently, the integration of morphometrics and molecular methods has been used for the differentiation of cryptic species among the multiple taxa of wasps also [11–17].

Diodontus Curtis, 1834 (Pemphredonidae; family status after [18]), is a genus of predatory solitary apoid wasps with 78 described species worldwide, two thirds of which inhabit the Palearctic region [19]. Females of this genus dig unicellular or multicellular tunnel nests primarily in clay or sand, provisioning them with Aphididae (Hemiptera) as larval food [20,21]. Apoid wasps, including Pemphredonidae, fulfil ecosystem services of biological control (as important natural enemies of herbivorous insects) and to a lesser extent, of pollination [20,22–25]. Most of them depend on specific natural or early- or late-successional habitats, and thus they may serve as ecological indicators of anthropogenic impact [26].

At present, a total of 12 accepted species of the genus *Diodontus* are known in continental Europe [19,27,28]. However, a preliminary analysis of barcoding sequences available from the Barcode of Life Data System (BOLD) has revealed that the Mediterranean complex of species related to *D. insidiosus* Spooner, 1938, is represented by multiple divergent and geographically restricted mitotypes forming separate Barcode Index Numbers (BINs), which may correspond to undescribed cryptic species [28]. Upon exploring the mitotype diversity of *D. tristis* (Vander Linden, 1829) in Lithuania, we also revealed specimens with highly divergent mitotypes, which formed a separate cluster in phylogenetic analyses.

Among various potential drivers that have been associated with mitotype divergence in arthropods, ecological factors and the effects of maternally inherited endosymbiotic bacteria have been suggested [29]. Due to their general biological similarity, cryptic species may compete for limited ecosystem resources and exclude each other [14,30–32], particularly in cases of recent phylogenetic divergence [4]. Alternatively, they may coexist as a result of either sharing of unlimited common resources [12] or due to niche partitioning and ecological divergence along with behavioural or physiological reproductive isolation [33,34], particularly if their similarity is the result of morphological convergence or stasis [4].

Unlike the cryptic divergence generated by ecological factors, the contribution of endosymbiotic bacteria to the evolutionary diversification of the host species still raises intriguing questions and debates [35,36]. Reproductive manipulators, such as *Wolbachia* Hertig, 1936 (Alphaproteobacteria), *Arsenophonus* Gherna et al., 1991 (Gammaproteobacteria), and *Spiroplasma* Saglio et al., 1973 (Mollicutes), in order to enhance their own vertical transmission, may reduce the reproductive output of uninfected females relative to the infected ones due to induced cytoplasmic incompatibility, or cause female-biased sex ratios by killing male offspring [37]. Such reproductive manipulations may reduce the gene flow between populations, leading to increasing reproductive isolation and even speciation triggered by bacterial endosymbionts [38–41]. The prevalence and effects of endosymbionts on Hymenoptera, including solitary wasps, remains relatively little explored [42].

Therefore, the aim of this study is to test the hypotheses that the cluster of divergent *Diodontus tristis* mitotypes represents a reproductively isolated population infected by a specific strain of endosymbionts or that it is a new undescribed cryptic species.

2. Materials and Methods

2.1. Specimens

The specimens were collected using an entomological net at their nesting site on a clay wall of a daubed building or while feeding on the honeydew of *Panaphis juglandis* (Goeze, 1778) on *Juglans regia* leaves. They were fixed in 96% ethanol, and a part of them was pinned after DNA extraction. The specimens and the samples of DNA are deposited in the collection of the Nature Research Centre, Vilnius (NRCV).

2.2. Morphometric Measurements

The measurements were performed using an ocular micrometer in a Nikon SMZ1000 binocular microscope (www.nikon.com, accessed on 10 September 2020). Images were obtained by means of a digital camera (Nikon DS-Fi2) connected to the microscope; they were stacked using the software CombineZP, version 1.0 (Alan Hadley; https://alan-hadley.software.informer.com, accessed on 15 January 2023).

We used the morphometric measurements described in earlier studies [28,43]: width of head (WH) was defined as the maximum visible width in the dorsal or frontal view; length of face (LF)—the distance between the fore margin of fore ocellus and the midpoint of the lower margin of clypeus when viewed at a right angle; lower interocular distance (LID)—the shortest distance between the inner margins of the eyes approximately at the level of the antennal sockets in the frontal view; upper interocular distance (UID)—the shortest distance between the inner margins of the eyes approximately at the level of the fore ocellus in the frontal view; length of vertex (LV)—the shortest distance between the hind margin of the fore ocellus and the midpoint of the occipital carina when viewed at a right angle; post-ocellar distance (POD)—the shortest visible distance between the inner margins of the hind ocelli when measured at a right angle; oculo-ocellar distance (OOD)—the shortest visible distance between the outer margin of the hind ocellus and the margin of eye when measured at a right angle; intermandibular distance (IMD)—the distance between the outer margins of the genal swellings just above the fore mandibular condyles in the anterodorsal view; length of clypeus (LCL)—the distance between the midpoint of the frontoclypeal suture and the midpoint of the lower margin of the clypeus when viewed at a right angle; width of clypeal apex (WCA)—the distance between the tips of the lateral teeth on the lower margin of the clypeus; length of scape (LSC)—the maximum visible length of the scape; length of flagellomeres 1–3 (3FL)—maximum visible combined length of the first three flagellomeres without the pedicel; length of flagellomere 6 (L6F)—the maximum visible length; width of flagellomere 6 (W6F)—the maximum visible width without the layer of trichoid sensilla; width of pronotum (PRN)—the distance between the tips of the pronotal lobes in the dorsal view; width of pronotal collar (COL)—the distance between the lateral edges of the transverse carina of the pronotal collar in the anterodorsal view.

The Kolmogorov–Smirnov test for normality of measurement distribution, analysis of variance, and canonical discriminant analysis of the measurements were performed using Statistica, version 8 (Statsoft, Tulsa, OK, USA).

2.3. Selection of Molecular Characters

We used the following DNA marker set, a major part of which was applied in our earlier study [28]: two mitochondrial DNA markers—namely a continuous 2650–2664 base pair (bp) long sequence that included complete cytochrome c oxidase subunit 1 (*CO1*) and subunit 2 (*CO2*) and ATP synthase subunit 8 (*ATP8*) with transfer RNA genes (partial *tRNA-Trp*, complete *tRNA-Leu*, *tRNA-Asp*, and *tRNA-Lys*), and non-coding interspaces between them and a continuous 2022–2049 bp long sequence that included complete NADH dehydrogenase subunit 6 (*ND6*), cytochrome b (*CytB*) genes, and partial (3′ end) reverse NADH dehydrogenase subunit 1 (*ND1*) gene with the transfer RNA gene *tRNA-Ser* and non-coding interspaces between them. We also analysed the nuclear DNA markers, including the ribosomal RNA operon, namely partial small subunit (*18S*) and large subunit (*28S*) sequences and internal transcribed spacer 1 (*ITS1*) with a combined length 1973–2020 bp as well as selected protein-coding genes (Table 1). Among the latter, we studied the homeobox genes that control the early development of morphological character-rich anterior and posterior body segments, namely *proboscipedia* (*PB*) and *abdominal B* (*AbdB*). The other markers were chosen from a recently published set of sequences of 3256 protein-coding genes, obtained from the whole-body transcriptomes of 167 species of Hymenoptera [44]. After a preliminary assessment of the overall mean distance among the sequences as a proxy for the evolutionary rate, using MEGA version 7 [45], we selected exon-primed intron-crossing (EPIC) markers in two highly conserved genes, namely ubiquitin-conjugating enzyme E2 G1 (*Ube2g1*) and arginine kinase (*ArgK*), and two highly variable genes, namely mitochondrial ribonuclease P protein 3 (*MRPP3*) and pentatricopeptide repeat domain 2 (*PTCD2*). The exon-intron structure of these genes was reconstructed comparing them with the structure of orthologous genes of *Apis mellifera* Linnaeus, 1758 (Hymenoptera: Apidae), available at GenBank (www.ncbi.nlm.nih.gov/genbank, accessed on 25 September 2023) [46].

We attempted to study the internal transcribed spacer 2 (*ITS2*), as well, however, it could not be sequenced successfully even in homozygous male samples of *Diodontus argillicola* sp. nov. and the related *D. tristis* due to the presence of several tandem repeat regions, the length of which presumably varied among the multiple copies of the spacer sequence inside the same organism.

In the phylogenetic analysis based on the full set of used DNA markers, we included other *Diodontus* species with available ethanol-fixed specimens, namely *D. tristis*, *D. medius* Dahlbom, 1844, *D. minutus* (Fabricius, 1793), *D. polytylus* Budrys, 2019, and *D. guichardi* Budrys, 2019. The phylogenetic relationships of *D. argillicola* sp. nov. with the remaining *Diodontus* were clarified using the *CO1–5′* barcode sequences publicly available from the Barcode of Life Data System [47,48].

Table 1. DNA markers used as molecular characters for *Diodontus* species identification, phylogeny reconstruction, and assessment of endosymbiotic bacteria infestation.

Abbreviation	Description	Length (bp)	Source
	Mitochondrial protein-coding genes:		
CO1–5′	Cytochrome c oxidase, subunit 1, from 5′ end to 715 bp	715	[28]
CO1–3′	Cytochrome c oxidase, subunit 1, from 710 bp to the 3′ end	825–828	[28]
CO2	Cytochrome c oxidase, subunit 2, complete sequence	678	[28]
ATP8	ATP synthase, subunit 8, complete sequence	149	new
ND6	NADH dehydrogenase, subunit 6, complete sequence	502–535	[28]
CytB–5′	Cytochrome b, from 105 to 859 bp	754	[28]
ND1	Cytochrome b, 3′ end from 885 bp, and NADH dehydrogenase, subunit 1, 335 bp long 3′ end (considering reverse translation comparing to CytB)	665	[28]
	Nuclear ribosomal DNA operon:		
18S	18S rDNA partial sequence, including V2–V4 variable regions	797–798	[28]
28S	28S rDNA partial sequence, including D2–D3 variable regions	683–685	[28]
ITS1	Internal transcribed spacer 1, partial sequence	491–538	[28]
	Nuclear protein-coding genes:		
PB	*Proboscipedia* gene, partial sequence	500–513	[28]
AbdB	*Abdominal B* gene, partial sequence	648	[28]
ArgK	Arginine kinase gene, partial sequence	599–605	[28]
MRPP3	Mitochondrial ribonuclease P protein 3 gene, partial sequence	678–691	[28]
Ube2g1	Ubiquitin-conjugating enzyme E2 G1 gene, partial sequence	597–614	[28]
PTCD2	Pentatricopeptide repeat domain 2 gene, partial sequence	664–668	[28]
16S (Bact.)	16S rDNA of endosymbiotic bacteria, partial sequence	745–785	[28]
wsp	*Wolbachia* outer surface protein, partial sequence	577–589	[49]

2.4. DNA Extraction, Polymerase Chain Reaction, and Sequencing

The total genomic DNA was extracted from a piece of thoracic muscle tissue of air-dried or preserved in 96% ethanol specimens, according to the protocol of the GeneJet Genomic DNA Purification Kit (Thermo Fisher Scientific Baltics, Vilnius, Lithuania) or using the method described in [50]. The Polymerase Chain Reaction (PCR) mixture included 12.5 μL of 2× DreamTaq PCR Master Mix (Thermo Fisher Scientific Baltics, Vilnius, Lithuania), 2.5 μL of 10 pmol/μL each primer, 1 μL of genomic DNA, and deionised water up to a total volume of 25 μL. We used T3– and T7–tailed forward and reverse PGR primers, which were constructed using an online BiSearch Primer Design and Search Tool [51] and synthesised at Macrogen (Seoul, South Korea). The primers and PCR conditions are presented in [28,49], except for the new ATP8 marker. The latter was amplified using new primers, T3DioCO2-f2 (5′-attaaccctcactaaagttattggaaaycaatgataytgaag-3′) and T7PemATP6-r1 (5′-aatacgactcactatagatggatcaaaaatwgaaaataaattag-3′), under the following PCR conditions: an initial denaturation at 95 °C for 5 min; 50 cycles of denaturation at 94 °C for 1 min; annealing at 43 °C for 1 min; extension at 68 °C for 1.5 min with final extension at 68 °C for 5 min.

The PCR products were electrophoresised on 1.5% agarose gel (Thermo Fisher Scientific Baltics, Vilnius, Lithuania) with 10,000× GelRed (Biotium, Fremont, CA, USA) and purified following the protocol of Exonuclease I and FastAP Thermosensitive Alkaline Phosphatase (Thermo Fisher Scientific Baltics, Vilnius, Lithuania). The sequencing was performed at the Macrogen company using the BigDye Terminator version 3.1 Cycle Sequencing Kit (Applied Biosystems, Foster City, CA, USA). For sequencing, 17–mer universal primers T3 (5′-attaaccctcactaaag-3′) and T7 (5′-aatacgactcactatag-3′) were applied. The obtained new DNA sequences were deposited in GenBank (Table 2).

Table 2. Accession numbers of DNA sequences of *Diodontus* specimens (vouchers in NRCV) deposited in GenBank (www.ncbi.nlm.nih.gov/genbank, accessed on 10 November 2023).

Species	*D. argillicola* sp. nov.	*D. tristis*	*D. medius*	*D. minutus*
Collection place	54°13′09″ N 23°24′13″ E	55°52′39″ N 23°28′16″ E	55°52′41″ N 23°28′08″ E	55°35′32″ N 24°02′58″ E
Collection date	13.vi.2015	12.vii.2015	12.vii.2015	03.vi.2017
Voucher number	EBPEM-7338	EBPEM-7340	EBPEM-7339	EBPEM-7546
DNA markers:				
CO1–CO2–ATP8	PP025829–35	MK625008	PP025835	MK625005
ND6–CytB–ND1	PP025836–40	MK757260	PP025841	MK757257
18S	PP059061	MK640428	PP059062	MK640425
28S	PP059063	MK640440	PP059064	MK640437
ITS1	PP059065	MK640432	PP059066	MK640429
PB	PP025842–43	MK628914	PP025844	MK628911
AbdB	PP025845	MK628918	PP025846	MK628915
ArgK	PP025847	MK628922	PP025848	MK628919
Ube2g1	PP025849	MK628926	PP025850	MK628923
MRPP3	PP025851	MK634482	PP025852	MK634479
PTCD2	PP025853	MK634478	PP025854	MK634475

2.5. Phylogeny Reconstruction and Species Delimitation

The nucleotide sequences were aligned, using the BioEdit 7.2.5 software [52], either with the application of the internal two-sequence alignment or the ClustalW multiple alignment algorithm, with additional manual control. When needed, the sequences were converted from Fasta to Nexus format using Mesquite 3.51 [53]. The inter- and intra-specific Kimura 2-parameter (K2P) distances were estimated, and the most likely nucleotide substitution model was selected from the list of 24 available models, using MEGA version 11 [54] and applying the maximum likelihood method. In all cases, the substitution patterns were best described by the general time reversible model with gamma distribution evolutionary rates, applying four discrete categories, and a fraction of invariable sites (GTR+G+I). We applied this model for the reconstruction of phylogenetic relationships using the maximum likelihood algorithm, implemented in MEGA 11, and the method of Bayesian inference, implemented in MrBayes, version 3.2.7 [55]. We treated each gene and where present, each exon and intron of the gene as a separate partition. The nucmodel parameter was set to "codon" for the partitions of exons of mitochondrial and nuclear protein-coding genes; the nucmodel parameter was set to "4by4" for the partitions of tRNA, rRNA, internal transcribed spacer, and introns. For the reconstruction of phylogenetic relationships, Markov Chain Monte Carlo (MCMC) analyses were run for 5,000,000 generations, with tree sampling every 10,000 generations. The convergence between runs was considered sufficient after the standard deviation of the split frequencies reached a value of 0.01. In reconstructing the phylogeny with the application of the maximum likelihood algorithm, we used 10,000 bootstrap replicates.

For species delimitation, we applied three algorithms: the single threshold General Mixed Yule Coalescent model (GMYC) [56], implemented as *splits* package in R environment, the Bayesian Poisson Tree Processes method (bPTP) web server (species.h-its.org/ptp, accessed on 2 June 2023) [17], and the Assemble Species by Automatic Partitioning (ASAP) web server (bioinfo.mnhn.fr/abi/public/asap/asapweb.html, accessed on 2 June 2023) [57]. The ultrametric dichotomous trees for the GMYC analysis were obtained using the RelTime-ML algorithm, implemented in MEGA 11. The results of species delimitation were compared with the Barcode Index Number (BIN) database clustering available from the Barcode of Life Data System [47,48].

3. Results

3.1. Identification

The specimens with divergent mitotypes were morphologically hardly separable from the typical *Diodontus tristis*. The most useful differences are included in the identification key (Sections 3.4 and 3.5). The statistically significant morphometric differences are presented in the diagnosis (Section 3.4). The discriminant analysis of morphometric measurements did not allow a reliable segregation of either sex of the genetically divergent specimens from *D. tristis* (Figure 1).

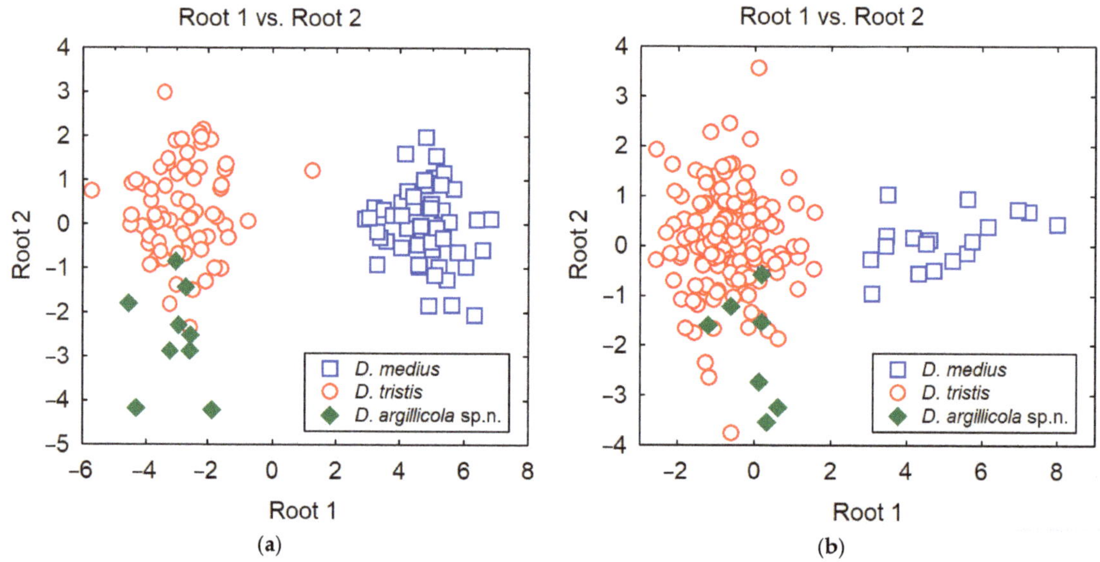

Figure 1. Results of discriminant (canonical) analysis of morphometric measurements (see Section 2.2) of *Diodontus argillicola* sp. nov., *D. tristis* and *D. medius*. Scatterplot of scores of canonical functions: (**a**) females; (**b**) males. Function coefficients are in Table 3.

However, these specimens could be reliably separated by multiple characters of mitochondrial and nuclear DNA. Among them, the *CO1–5′* barcode and its part—the 132 bp long "minibarcode" (Figure 2) that is easier to obtain from dry collection specimens than the full barcode—as well as the sequences of other studied mitochondrial genes (*CO2*, *ATP8*, *ND6*, *CytB*) and the nuclear markers, including the intron of a conserved *proboscipedia* (*PB*) homeobox gene that differs from that of *D. tristis*, in addition to several single nucleotide differences via a typical of the new species, 7 bp long deletion at position 130 (Figure 3).

Table 3. Coefficients of discriminant (canonical) functions best segregating the morphometric measurements (see Section 2.2) of *Diodontus argillicola* sp. nov., *D. tristis* and *D. medius*. Scatterplots of canonical scores are in Figure 1.

Female			Male		
Measurement	**Root 1**	**Root 2**	**Measurement**	**Root 1**	**Root 2**
WH	−10.85	22.37	WH	−15.30	0.41
LID	39.82	11.30	LID	21.09	39.24
UID	6.06	19.43	UID	−24.19	−3.51
IMD	−36.18	−7.64	WCA	−20.32	11.00
WCA	25.94	4.19	POD	−27.62	8.22
WLA	−1.73	17.74	OOD	84.73	−45.76
LM	−3.02	−4.06	WHO	7.62	−62.84
POD	−55.25	−23.05	LF	3.19	16.15
OOD	10.52	−9.02	LSC	−16.19	−52.17
WHO	32.81	−68.44	3FL	25.11	2.19
LV	−14.31	−2.81	L6F	−20.06	38.94
LF	17.08	−35.31	W6F	1.93	−55.94
LCL	−33.64	31.40	PRN	7.61	−7.89
LSC	0.18	−8.46	COL	−5.92	5.09
3FL	−4.39	23.45	Constant	2.83	−1.22
PRN	14.13	−19.10			
COL	−5.61	6.87			
Constant	2.73	6.22			

3.2. Phylogenetic Relationships and Species Delimitation

The phylogenetic analysis of all barcoding *CO1–5′* mitotypes of *Diodontus*, obtained in this study or available from the BOLD BIN database, using the Bayesian inference and the maximum likelihood algorithms, demonstrated that the divergent mitotype cluster is a sister group of the cluster of the remaining mitotypes of *D. tristis*. The application of three species delimitation algorithms, GMYC, bPTP and ASAP, to the available barcoding sequence set of genus *Diodontus* confirmed the distinctiveness of the cluster and its status as a separate cryptic species, *D. argillicola* sp. nov. (Figure 4).

The mean inter-specific distance between the two species, estimated using the 658 bp long barcoding sequence and K2P substitution model, was 0.116, while the intra-specific distances among the mitotypes were 0.006 for *D. argillicola* sp. nov. and 0.002 for *D. tristis*, thus implying the presence of a barcoding gap. The mitotype FBACA563-10 from the BOLD BIN database (Figure 4), most probably representing one more cryptic species of the *D. tristis* complex, was not included in this analysis.

The distinctiveness of the new species was validated via an application of the delimitation algorithms to the mitotypes of *D. argillicola* sp. nov. and *D. tristis* based on a 4719 bp long mtDNA sequence of complete genes *CO1*, *CO2*, *ATP8*, *ND6*, *CytB*, and partial *ND1* (Figure 5a). The mean inter-specific distance between the two species was 0.101, while the mean intra-specific distance among the studied mitotypes of *D. argillicola* sp. nov. was 0.006, and the distance among the available mitotypes of *D. tristis* was 0.001. The distinctiveness was confirmed by differences in all the studied nuclear markers, including a unique deletion in the *proboscipedia* (*PB*) gene's lesser intron at the position 130 (Figure 3).

The summarizing reconstruction of phylogenetic relationships using the Bayesian inference and the maximum likelihood algorithms and the 10,521 bp long combined set of mitochondrial and nuclear DNA markers confirmed the relatedness of *D. argillicola* sp. nov. and *D. tristis*; *D. medius* turned out to be a more phylogenetically distant member of the *D. tristis* species group (Figure 5b).

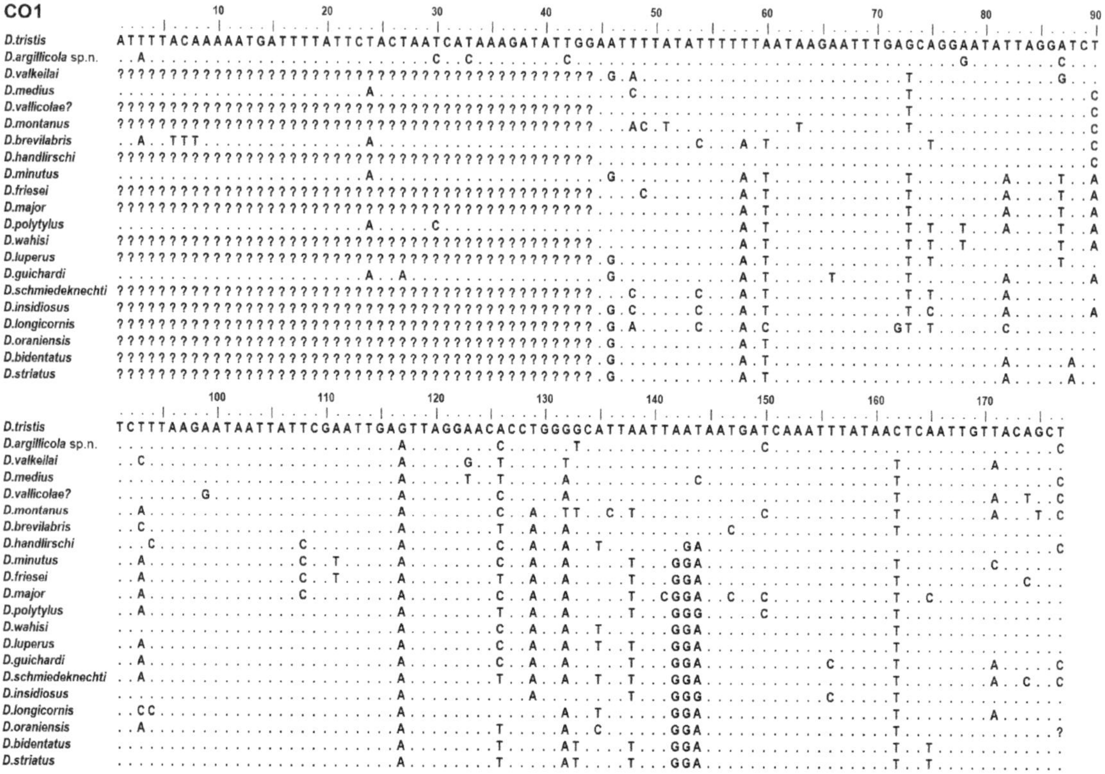

Figure 2. Comparison of the 5′ end of mitochondrial cytochrome c oxidase, subunit 1, and the "minibarcode" sequence of *Diodontus argillicola* sp. nov. with those of *D. tristis* and other species of the genus, where this molecular character is available. Dots—identical nucleotides; question marks—unknown sequences. *D. vallicolae?* represents the sequence of an unidentified specimen with BOLD processID AMCAO361-20 (see also Figure 4) that likely belongs to *D. vallicolae* Rohwer, 1909.

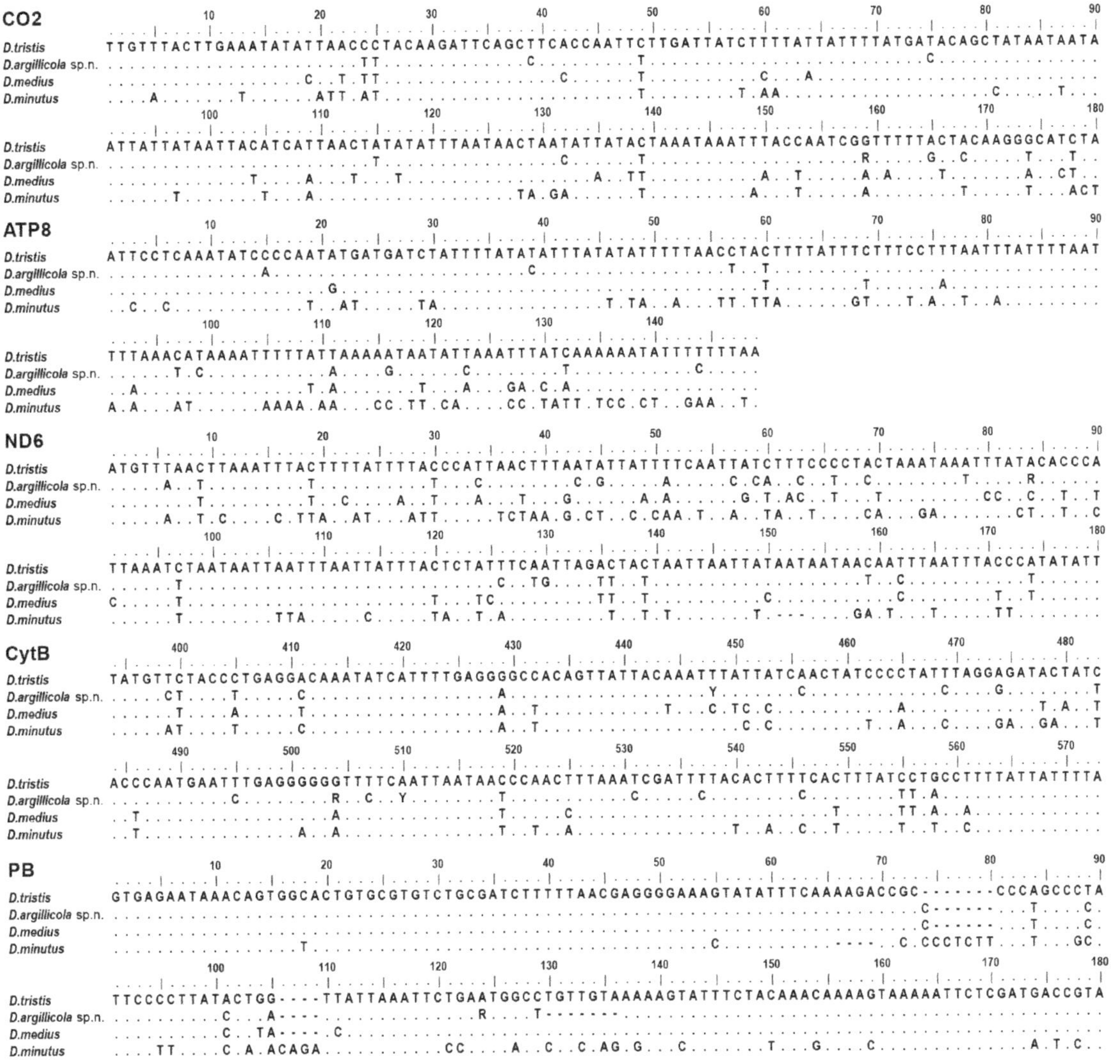

Figure 3. Comparison of selected molecular characters (180 bp long consensus sequences) of *Diodontus argillicola* sp. nov. with those of *D. tristis*, *D. medius*, and *D. minutus*: CO2, 5′ end of mitochondrial cytochrome c oxidase, subunit 2; ATP8, complete sequence of mitochondrial ATP synthase, subunit 8; ND6, 5′ end of mitochondrial NADH dehydrogenase, subunit 6; CytB, the fragment of mitochondrial cytochrome b, commonly used as the CytB barcode; PB, almost complete sequence of the lesser intron of *proboscipedia* homeobox gene. Dots—identical nucleotides; dashes—deletions.

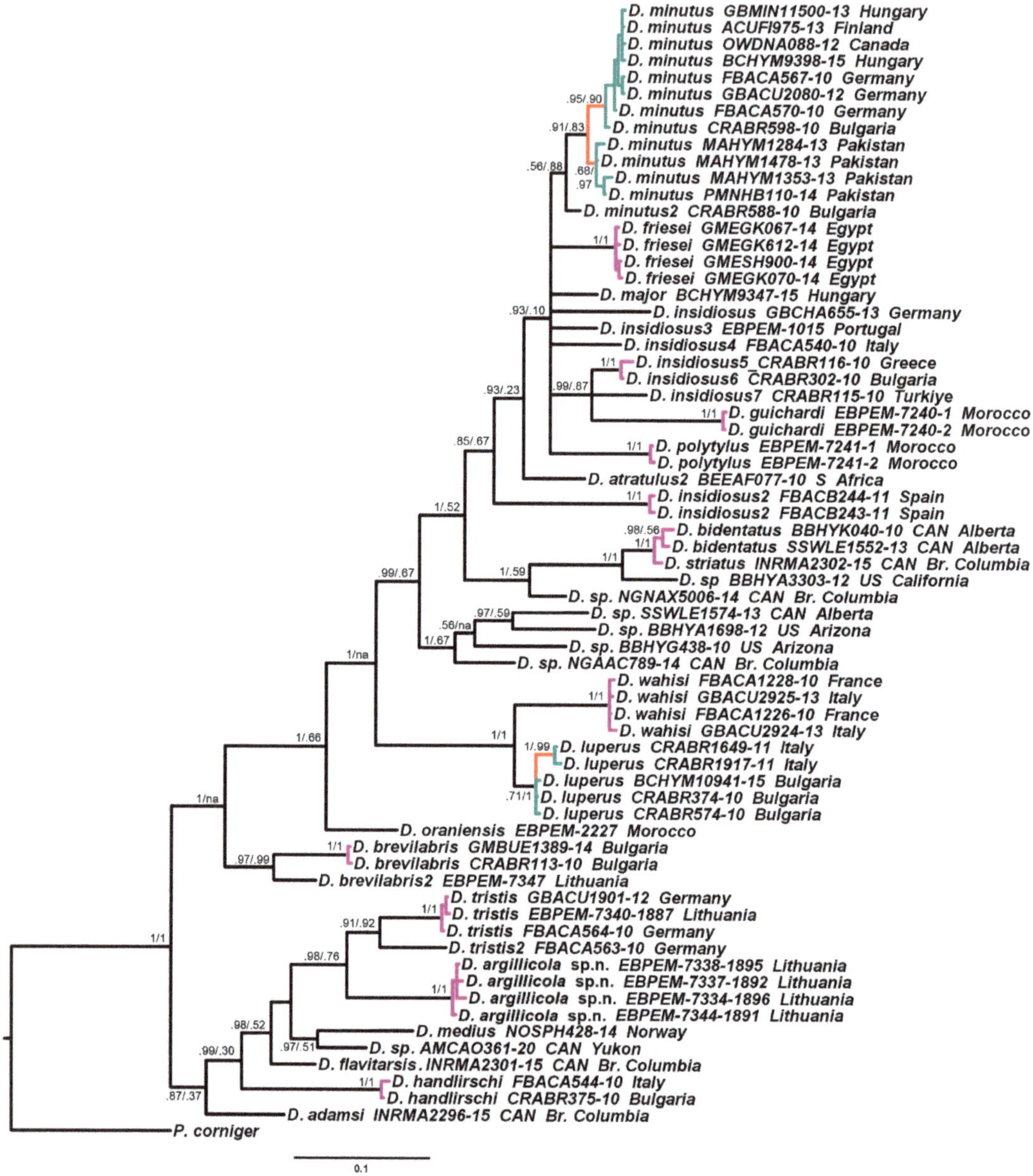

Figure 4. Phylogenetic relationships of *Diodontus*, reconstructed using GTR+G+I evolution model, applied to the barcoding *CO1–5′* sequence of mtDNA. Tree topology reconstructed using Bayesian inference; statistics of branches: Bayesian posterior probability (5,000,000 generations)/maximum likelihood probability (10,000 bootstrap replicates). Mitotype labels include the specimen process ID codes of the Barcode of Life Data System or (the labels including EBPEM) the unpublished barcode sequences from the database of the authors and the collection country. The mitotype clusters supported as species by delimitation algorithms: magenta—GMYC, bPTP, and ASAP; red—GMYC; teal—bPTP and ASAP. *Passaloecus corniger* Shuckard, 1837, was used as an outgroup. The scale: probability of expected nucleotide changes per site.

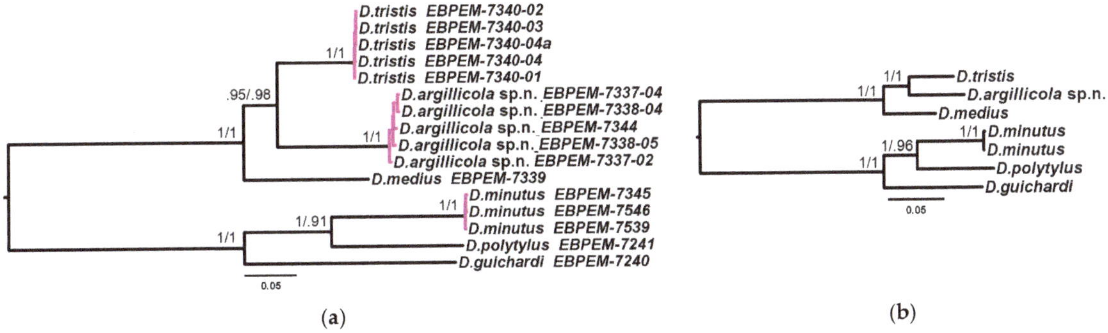

Figure 5. Phylogenetic relationships among *Diodontus argillicola* sp. nov. and closely related species *D. tristis* and *D. medius*. Tree topology reconstructed using Bayesian inference; branch statistics: Bayesian posterior probability (5,000,000 generations)/maximum likelihood probability (10,000 bootstrap replicates). *D. minutus* species group (*D. minutus*, *D. polytylus*, and *D. guichardi*) was used as an outgroup. (**a**) reconstruction using the markers of mitochondrial DNA, 4719 bp long sequence including complete *CO1*, *CO2*, *ATP8*, *ND6*, *CytB*, and partial *ND1* genes; the magenta haplotype clusters were supported as species by delimitation algorithms GMYC, bPTP, and ASAP. (**b**) reconstruction using full 10,521 bp long set of studied mitochondrial and nuclear markers (see Table 1). The scale: probability of expected nucleotide changes per site.

3.3. Endosymbiotic Bacteria Infestation

Using the *16S* marker, we explored the infestation of muscle tissues of *Diodontus* wasps by intra-cellular symbiotic bacteria. Nearly all studied specimens were infected by *Wolbachia*. One *D. tristis* specimen was infected by *Spiroplasma*, the *16S* partial sequence of which was very similar (99.7% similarity) to the sequence with GenBank ID LC388759.1 that was isolated from ticks. Three *D. minutus* specimens were infected by *Arsenophonus*; the *16S* partial sequence of which was most similar (99.9%) to the sequence with GenBank ID DQ115536.1, isolated from a hippoboscid fly.

Most *D. tristis* and *D. minutus* individuals had an infection of typical *Wolbachia pipientis* Hertig, 1936, with a partial sequence of outer surface protein (*wsp*) identical to that of, for instance, *Apanteles chilonis* Munakata, 1912 isolate (GenBank sequence ID KC161914.1). The 2 studied *D. tristis* and all 10 studied *D. argillicola* sp. nov. had a mixed *Wolbachia* infection, represented by two *wsp* gene strains, which were presumably identical to those with GenBank sequence IDs KC161914.1 and JN639221.1; the latter sequence was isolated, among others, from *Andrena wilkella* (Kirby, 1802).

We did not find any endosymbiotic bacteria strains specific only to *D. argillicola* sp. nov. or *D. tristis*. Therefore, we rejected the hypothesis that the new species represents a clade of *D. tristis*, which is reproductively isolated due to specific endosymbiont infestation.

3.4. *Diodontus argillicola* Budrys, Orlovskytė & Budrienė, New Species

Zoobank link: urn:lsid:zoobank.org:act:856E17E8-C310-4BF2-B459-D6DF115BD1ED

Figures 6, 7 and 8a,b,e.

Type material. Holotype: ♀, Lithuania, Ūdininkai, clay wall of daubed building (54°13′ N 23°24′ E), 13.vi.2015, E. Budrys leg. (deposited in NRCV, voucher No EBPEM-7338-6). Paratypes: Lithuania: same data as holotype, 6♀15♂, E. Budrys, S. Orlovskytė leg. (NRCV, voucher Nos EBPEM-7338-01–21); same locality and habitat as holotype, 5♀3♂, 04.vii.2013 (NRCV, voucher Nos EBPEM-7337-01–08); 1♂, 11.vii.2019 (NRCV, voucher No EBPEM-7631); Bilšiai (55°08′ N 25°16′ E), honeydew on *Juglans regia* leaves, 1♂, 27.viii.2019, all E. Budrys leg. (NRCV, voucher No EBPEM-7344).

Consensus barcode: aattttatatttttaataagaatttgagcagggatattaggctcttctttaagaataattattcga attgaattaggaacmcctgggtcattaattaataatgaccaaatttataactcaattgttacagcccatgcttttattataatttttttata gtaataccttttataattggagggtttggaaattgattaatccccataataattggrgcccctgatatggctttcccccgaataaataata taagattttgrttaatccccccctctcttttattttaattttaagaaacattttaaataatggggtagggacaggttgaacagtttaccccc ccttatccgctaatattagycataatggatcttcagtagatctcgctattttttctttacatattgcaggggtrtcatcaattataggrgct attaattttattgttacaattataaacataaaaaataaattcttaaattttgatcaaatacccttatttgtytgatcaatcttratcacaacc atccttttactcctatccttrccagtccttgcaggagcaattacaatattattaactgaccgaaatttaaatacaactttttttgacccatct ggaggggggggayccaattttatatcaacatttattt

Distribution. Currently known from two localities in Lithuania.

Etymology. Species name is derived from the substrate where the species was observed nesting, clay (Ancient Greek αργιλλος, "clay") and the Latin word *cola*, which means "inhabitant" or "dweller". A noun is used in apposition to the generic name.

Diagnosis. Both sexes of *D. argillicola* sp. nov. are very similar to *D. tristis* in body sculpture, punctation, and colouration. The morphometric differences are very weak (Figure 1; Table 3). A reliable tool for segregation of the two species is molecular barcoding (Figures 2 and 3). The two species seem to also differ behaviourally regarding their choice of nesting substrate: most of the specimens of the new species were collected from a colony nesting in a clay wall, while all barcoded specimens from the colonies nesting in sand or gravel substrate happened to be *D. tristis*.

(a) (b) (c) (d)

Figure 6. *Cont.*

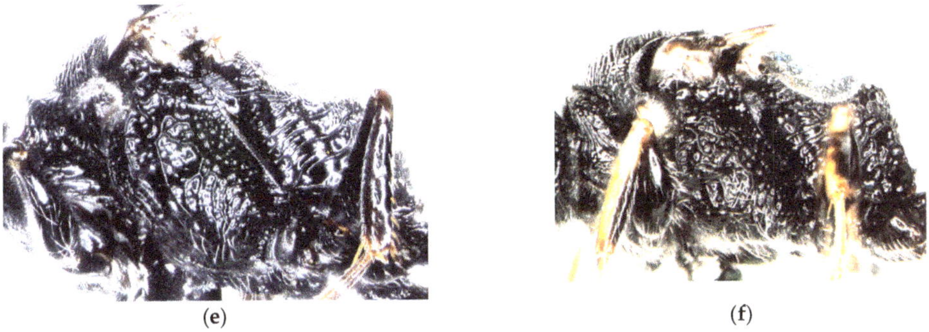

Figure 6. *Diodontus argillicola* sp. nov.: general view, (**a**) holotype female; (**b**) male; head, frontal view, (**c**) female; (**d**) male; thorax, lateral view, (**e**) female; (**f**) male.

Figure 7. *Diodontus argillicola* sp. nov., female: (**a**) foretarsus; (**b**) pygidial plate; male: (**c**) flagellum; (**d**) penial valve.

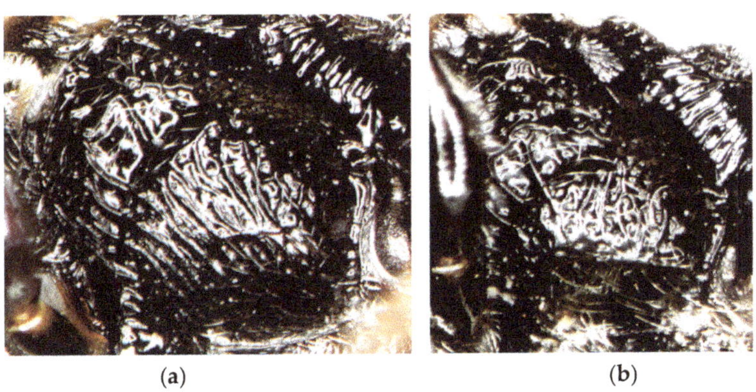

Figure 8. *Cont.*

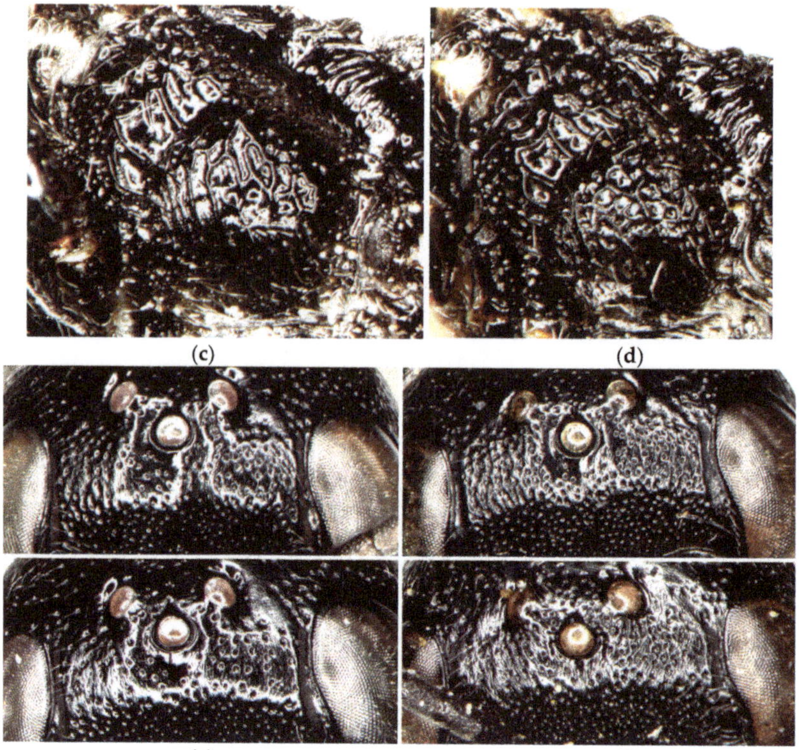

(c) (d) (e) (f)

Figure 8. Mesopleuron: *Diodontus argillicola* sp. nov., (**a**) female; (**b**) male; *D. tristis*, (**c**) female; (**d**) male. Female, upper frons: (**e**) *D. argillicola* sp. nov. (two specimens); (**f**) *D. tristis* (two specimens).

Diodontus argillicola sp. nov. differs from *D. tristis* by a generally slightly smoother sculpture and sparser punctation, although the variability of these characters overlaps in the two species. In *D. argillicola* sp. nov., the upper frons between the fore ocellus and the frontal gland is slightly smoother, shinier, and more sparsely punctate (Figure 8e) than in *D. tristis* (Figure 8f): the interspaces among punctures are mostly wider than or equal to but rarely narrower than punctures, while in *D. tristis*, the interspaces here are mostly narrower than the width of punctures, sometimes equal to them, and rarely wider. In *D. argillicola* sp. nov., the head of the female is slightly less transverse and more rounded than in *D. tristis*; the dorsal surface of the frons between the fore ocellus and anterior vertical surface is wider than the fore ocellus (Figure 8e), while in *D. tristis*, it is approximately equal to the fore ocellus (Figure 8f). The subscrobal area of the mesopleuron is more regularly obliquely to nearly transversely rugose; this rugosity is less regular below the posterior end of the scrobus (Figure 8a,b), however it is not so irregularly areolate as in *D. tristis* (Figure 8c,d).

Morphometric differences of *Diodontus argillicola* sp. nov. from *D. tristis*. In the female, head of *D. argillicola* sp. nov. is slightly less transverse with WH : LID = 1.564 $\pm$ 0.010 ($N = 9$), while it is 1.583 $\pm$ 0.003 ($N = 73$) in *D. tristis* (ANOVA: $F = 4.3$, $p = 0.04$). The oculo-ocellar distance in *D. argillicola* sp. nov. female is relatively smaller with POD : OOD = 0.900 $\pm$ 0.019 ($N = 9$), while it is 0.850 $\pm$ 0.005 ($N = 73$) in *D. tristis* (ANOVA: $F = 9.8$, $p = 0.002$). In the male of *D. argillicola* sp. nov., in contrast, the post-ocellar distance is comparatively smaller, POD : OOD = 0.912 $\pm$ 0.032 ($N = 7$), while it is 0.954 $\pm$ 0.004 ($N = 137$) in *D. tristis* (ANOVA: $F = 4.6$, $p = 0.03$). The flagellum of *D. argillicola* sp. nov. is slightly shorter: in the female, 3FL : LSC = 0.89 $\pm$ 0.01 ($N = 9$), while it is 0.95 $\pm$ 0.01

(N = 70) in *D. tristis* (ANOVA: F = 10.8, p = 0.002); in male, L6F : W6F = 1.30 ± 0.04 (N = 7), while it is 1.39 ± 0.01 (N = 127) in *D. tristis* (ANOVA: F = 5.0, p = 0.027).

Description.

Female (Figure 6a,c,e, Figure 7a,b and Figure 8a,e). The body length is 6.3–6.9 mm (holotype 6.6 mm). The body is black. In the holotype and a part of paratypes, the mandible is dark yellowish-brown with a dark red tip. The protibia medially, metatibia basally and apically, and tarsi are brown; in other paratypes, these body parts are dark brown. The pterostigma is nearly black with a dark brown strip along the edge of the wing.

The head is weakly transverse (frontal view; Figure 6c), WH : LF = 1.56. The face is moderately wide, LID : LF = 0.89. The inner orbits are not converging ventrad, with LID : UID = 1.02. The ocelli form an obtuse triangle, with POD : OOD = 0.90. The frontal line in front of the fore ocellus is weakly impressed. The area of the interantennal process is nearly flat. Orbital gland is concave, its upper part does not have distinct swelling medially; the mid part is approximately equal or slightly wider than the space between it and the inner orbit. The mandibular condyles are relatively widely separated, IMD : WH = 0.78. The clypeus is relatively long, LCL : LF = 0.26. The clypeal apex is tridentate, comparatively broad, WCA : IMD = 0.33, and the mid tooth is shorter than the lateral teeth. The labrum is subtriangular with a subtriangular apical notch. The scape is moderately long, LSC : LF = 0.46. The flagellum is moderately long, 3FL : LSC = 0.89. The frons is partly rugulose, weakly shiny, and densely punctate to scrobiculate; the interspaces are alutaceous and narrower than the punctures. The oculo-ocellar area of the upper frons is nearly smooth and with scattered punctation; a part of the interspaces is 1–2 times as wide as the punctures (Figure 8e). The subappressed pilosity of the upper frons is sparse and nearly as long as the width of the ocellus. The pilosity of the lower frons between the eye and antennal socket is pale grey, it does not conceal the underlying microsculpture. The upper part of the clypeus has relatively sparse short subappressed pilosity and multiple long setae; the shiny impunctate area above the lower edge covers more than half of the height of the clypeus.

The pronotal collar is moderately wide, COL : PRN = 0.58, it has narrowly rounded sides and weakly convex dorsal carina (frontal view). The lateral surface of the pronotal collar is separated from the dorsal surface by weak carina and rugulose. The anterior part of the scutum is densely punctate; it has interspaces narrower than the punctures. The disc of the scutum is smooth and shiny, with remote, irregularly scattered, and distinct punctation; the interspaces are 2–8 times as wide as the punctures. The scutellum is glabrous and shiny with a few shallow punctures. The episcrobal area of the mesopleuron is coarsely areolate. The subscrobal and ventral parts of the mesopleuron are coarsely, obliquely, and regularly rugose; this rugosity is less regular postero-dorsally below the scrobus (Figures 6e and 8a). The metapostnotum is irregularly rugose. The propodeum is coarsely areolate, with obtuse lateral angles and indistinct, uneven carina separating the lateral and posterior surfaces. The probasitarsal rake is weak, with preapical spines slightly longer than the width of the probasitarsus (Figure 7a).

The gastral terga are smooth and finely punctate: the tergum 1 is strongly shiny, without microsculpture between the punctures; the remaining terga are shiny, with fine alutaceous microsculptures. The pygidial plate is subtriangular, with strong scattered punctures, its surface is weakly shiny with distinct coriaceous to granulose microsculpture (Figure 7b).

Male (Figure 6b,d,f, Figure 7c,d and Figure 8b). The body length is 4.9–5.2 mm. Body is black with ivory spots on the tegula and spiracular lobe. The protibia anteriorly, mesotibia basally and antero-apically, and metatibia basally are pale yellow; in a part of the paratypes, the apical yellow marking of the mesotibia is strip-like elongated and nearly reaching the basal marking. The apical segment of the abdomen is reddish-brown.

The body structure is generally similar to that of the female. The head is moderately transverse, WH : LF = 1.62. The face is moderately narrow, LID : LF = 0.79; the inner orbits are distinctly converging ventrad, LID : UID = 0.85. The frontal line is indistinct,.

POD : OOD = 0.91. The clypeal apex is bidentate. The labrum is subtrapezoidal with a weakly emarginate apical margin. The scape is moderately long, LSC : LF = 0.39. The flagellum is comparatively long, 3FL : LSC = 1.39; the flagellomeres are more long than wide, L6F : W6F = 1.30. Flagellomeres 5–10 have angularly protruding apicoventral edges and dark, smooth, flat, ovoid tyloidea; very small tyloid-like tubercles are also discernible on flagellomeres 4 and 11 (Figure 7c). COL : PRN = 0.59. The probasitarsi and mesobasitarsi are nearly straight. The mesobasitarsus is not dilated. Tergum 7 has distinct lateral carinae, forming a subtrapezoidal pygidial plate. The penial valve has a wide denticulate apex (Figure 7d).

The sculpture and pilosity are similar to those in the female, with the following exceptions: punctation is generally denser; interspaces in the oculo-ocellar area are smaller than the punctures; the pilosity of the lower frons and clypeus is very dense, silver, it conceals the underlying microsculpture (Figure 6d). The disc of the scutum has close punctation; the interspaces are shining and 1–2 times as wide as the punctures. The scutellum is closely punctate; the interspaces are as wide as the punctures. The subscrobal part of mesopleuron is less regularly rugose (Figure 8b); its ventral part is nearly smooth and shining. Tergum 6 is without distinct small spines. The pilosity of sterna 3–5 is sparse without specific features.

3.5. *Key to the Palearctic Species of Diodontus tristis Species Group*

The *Diodontus tristis* species group is represented by relatively large and dark species with Palearctic or Nearctic distribution ranges, characterised by the even pilosity of the male's tergum 6, the lack of two groups of spines on it; such spines, which are clearly longer and thicker than the surrounding setae, are typical of males of the remaining *Diodontus* species.

1. In the male, tergum 6 has uniform pilosity and several longer preapical setae, without two distinct groups of thick spines. In the female, antenna, mandible, tegula, mid and hind tibiae are black or brown, without distinct yellow spots. In doubtful cases (in *D. montanus*, the tibiae are entirely pale brown), the face is very wide: the smallest distance between the eyes at the level of the antennal sockets is equal to the distance between the fore ocellus and the tip of the mid tooth of the clypeus; the distance between the lateral teeth of the clypeal apex is 1.4–2 times the length of the clypeus, and clearly larger than the distance between the lateral tooth and eye; the upper end of the frontal glands is flat or concave (*D. tristis* species group)....................................2

- In the male, tergum 6 has two preapical groups of small spines, which are clearly thicker than the surrounding pilosity. In the female, the combination of the listed characters is otherwise. If the mandibles and tibiae are dark, then at least one of the following is true: (a) the face is more rounded: the smallest distance between the eyes at the level of the antennal sockets is less than 0.85 times as wide as the distance between the fore ocellus and the tip of the mid tooth of the clypeus; and/or (b) distance between the lateral teeth of the clypeal apex is not larger than the distance between the lateral tooth and eye; and/or (c) the frontal gland is convex at the upper end ....(other species groups)

2. The frons and posterolateral part of the mesopleuron are smooth, they have sparse weak punctation and coriaceous interspaces. In the female, the pygidial plate is typically widely triangular, covering all the dorsal surface of tergum 6, without a distinct marginal rim; the frontal glands are narrow, with their upper part deeply impressed in relation to the oculo-ocellar area; the tibiae are brown to pale brown. In the male, the pilosity of the lower frons and clypeus is sparse and weakly shining, and it does not hide the surface of cuticle. (Mountains of Central Asia from SE Kazakhstan to Tajikistan) ....*D. montanus* Kazenas, 1992

- The central part of the frons is strongly and densely punctate. The posterolateral part of the mesopleuron is obliquely or irregularly rugose or areolate. In the female, the pygidial plate is narrower, it is subelliptic with a distinct marginal rim; the upper part of the frontal gland is flat or weakly concave; the mid and hind tibiae are black or dark brown. In the male, the pilosity of the lower frons and clypeus is dense, silvery, and it hides the surface of the cuticle...3

3. The scutum is finely and very densely punctate; the interspaces are mostly narrower than the punctures. In the female, the frontal gland is twice as wide as the space between the gland and the eye; all tibiae are laterally black and medially brown, without a yellow pattern. In the male, the tegula and spiracular lobe have ivory spots; the interspaces between the punctures on the oculo-ocellar area are coarsely granulose, matt. (Pan-palearctic boreal)...*D. medius* Dahlbom, 1844

- The scutum has more or less scattered punctation, particularly in the female; at least medially, the interspaces are wider than the punctures. In the female, the frontal gland is as wide as the space between the gland and eye. In the male, the tegula and spiracular lobe are black (*D. valkeilai* and *D. asiaticus*); if they are with ivory spots (most of *D. tristis* and *D. argillicola*), then the interspaces between the punctures on the oculo-ocellar area are alutaceous and shiny...4

4. The upper frons has dense erect setosity; the setae are longer than the width of the ocellus. In the female, the fore tibia anteriorly has a yellow strip. The face is very wide: the smallest distance between the eyes at the level of the antennal sockets is equal to the distance between the fore ocellus and the tip of the mid tooth of the clypeus. The body length is 8 mm in females and 6.5 mm in males. (Siberia: Chita oblast)...*D. valkeilai* Budrys, 1992

- The upper frons has sparser subappressed setosity; the setae are shorter than the width of the ocellus. In the female, the fore tibia is without a yellow strip; the face is narrower: the smallest distance between the eyes at the level of the antennal sockets is 0.8–0.95 as long as distance between the fore ocellus and the tip of the mid tooth of the clypeus. The body length is typically up to 7 mm in females and up to 6 mm in males...5

5. The oculo-ocellar area has scattered punctation: the interspaces are twice as wide as the punctures in females and 1.5 times as wide as the punctures in males. The posterolateral part of the mesopleuron is more or less regularly obliquely striate. In the female, the upper frons has moderately dense punctation, the interspaces are approximately as wide as the punctures and smooth. In the male, the tegula and spiracular lobe are black, while the hind tibia is black with a yellow base. (Mongolia)...*D. asiaticus* Tsuneki, 1972

- The oculo-ocellar area has dense punctation: the interspaces are as wide as the punctures or smaller. The posterolateral part of the mesopleuron is more irregularly rugose-areolate. In the female, the upper frons is very densely rugulose-punctate: the interspaces are smaller than the punctures. In the male, the tegula and spiracular lobe are usually (but not always) with ivory spots, and the hind tibia is either black with a yellow base or with more or less extended pale brown colouration antero-medially. (The following two species may be securely separated only using molecular characters)..6

6. The posterolateral part of the mesopleuron is more regularly rugose (Figure 8a,b). In the female, the dorsal declivous part of the upper frons between the fore ocellus and anterior vertical surface is slightly longer than the width of the fore ocellus; the area between the fore ocellus and frontal gland is smoother, shinier, and more sparsely punctate; its interspaces are mostly wider than or equal to the punctures (Figure 8e). ....*D. argillicola*, sp. nov.

- The posterolateral part of the mesopleuron is more irregularly areolate (Figure 8c,d). In the female, the dorsal declivous part of the upper frons between the fore ocellus and

anterior vertical surface is not longer than the width of the fore ocellus; the area between the fore ocellus and frontal gland is more distinctly undose, less shiny, and more densely punctate; its interspaces are mostly narrower than the punctures (Figure 8f). (Western Palearctic, southern Siberia, Mongolia)..
....*D. tristis* (Vander Linden, 1829)

4. Discussion

We have not found any morphological character that would allow a reliable separation of *Diodontus argillicola* sp. nov. from *D. tristis,* thus we may consider them as a cryptic species complex that needs further investigation. However, the two species can be easily segregated by virtually all studied DNA characters, including the *CO1* barcoding sequence (Figures 2 and 3). Thus, our study supports the possibility of a minimalist taxonomic approach, which relies on the application of DNA barcodes to quickly and efficiently identify and describe large numbers of newly discovered species from little studied and species-rich taxa [58]. Such an approach may be particularly useful in the clades where speciation and ecological adaptation and/or sexual selection processes are not related to morphological differentiation—for example, the trophic specialisation of emerging species solely includes the physiological adaptations to food plant chemical defences, or the mating partner is recognised by pheromone or epicuticle chemistry.

On the other hand, we agree that a different barcoding sequence of the mitochondrial *CO1* gene is an insufficient argument for considering the mitotype as a separate species. In addition to the assessment of genetic variability, in combination with species delimitation algorithms, the description should be ideally complemented by information on ecological, behavioural, or life history traits as well as the screening of endosymbionts' composition [10]. In arthropods, the prevalence of endosymbiotic bacteria has been documented in a range of cryptic species complexes, predominantly in agriculturally significant taxa, challenging the hypothesis that cryptic speciation may have been facilitated by endosymbiont infestations [59–62].

Our results did not demonstrate a peculiar infection pattern in the studied *D. tristis* and *D. argillicola* sp. nov., thus we currently cannot affirm a possible role of endosymbionts in the *Diodontus* species diversification. However, the discovered variety of infestation (*Arsenophonus*, *Spiroplasma*, and several strains of *Wolbachia*) allow us to imply the phylogenetic incongruence between the *Diodontus* hosts and their endosymbionts. This incongruence may indicate that, in addition to the vertical transmission (from mother to offspring), horizontal transmission events may have taken place, such as a transfer from prey to the predator or a transfer through the non-lethal probing of prey [61]. It has been reported that *Diodontus* females immobilize their aphid prey by malaxation with mandibles instead of stinging [20,63]. Since aphids are well-known hosts of bacterial endosymbionts, including *Wolbachia, Arsenophonus,* and *Spiroplasma* [64], the horizontal transfer of endosymbionts in *Diodontus* is notably possible, either during the malaxation of the provisioning female or through the consumption of aphids by the offspring.

Niche differentiation is typically associated with parapatric speciation, where the reduction of gene flow between populations is connected with habitat or mating ground segregation. The preliminary observations that *D. argillicola* sp. nov. nests in clay, while *D. tristis* prefers a sandy or gravel nesting substrate, could indicate the potential contribution of the ecological niche partitioning process in species differentiation. The variability of preferred soil characteristics has been considered as an important factor determining the parallel presence of viable populations of several closely related ground-nesting wasp species [65]. Differences in habitat use have also been demonstrated for sympatric cryptic species of cavity-nesting wasps [12,49], bumble bees [66,67], and mayflies [68].

Our results do not demonstrate any role of endosymbiont infection in the speciation of *D. tristis* and *D. argillicola* sp. nov. However, we cannot fully exclude that their differentiation has occurred through simultaneous niche separation and reproductive isolation due to the effects of endosymbionts, as our study of the endosymbiotic microorganisms with the exploration of just two markers is a preliminary one. There are examples in the ground-nesting bee, *Andrena proxima* (Kirby, 1802) species complex, of parapatric and sympatric distributions, where both *Wolbachia*–induced incompatibilities and ecological differences are thought to maintain species boundaries [69]. To better understand the role of bacterial endosymbionts in the diversification of *Diodontus* species, further studies focusing on the strains of each symbiont and the phylogenies of symbiont infections in relation to host phylogenies are required.

The currently known distribution range of *D. argillicola* sp. nov., restricted to Lithuania, is obviously underestimated, since the specimens of this cryptic species have not been separated from *D. tristis*. The lack of specimens of the new species in the BOLD barcode database implies that the new cryptic species may be missing from western Europe, where the available barcodes come from. This lack also implies that the name *Pemphredon tristis*, the nominal species described by Vander Linden from Belgium, most probably does not belong to the new species described here. The collection of Vander Linden, unfortunately, is lost, and therefore a designation of the neotype of *D. tristis* from the type territory, along with a study of its molecular characters, is needed. The presence of a specimen with deviating mitotype FBACA563-10 in the BOLD database, forming a separate BIN AAN4253, which has been collected in Brandenburg (Germany), along with the typical *D. tristis* (BIN AAN4254), suggests the possibility of a wider cryptic speciation in the *D. tristis* species group that needs further exploration.

In the assessments of anthropogenic pressures on early-successional habitats using the structure of digger wasp communities as an indicator, *D. tristis* has been recorded among common or even dominant species in central Europe [70,71]. Our study demonstrates that the data on the presence and abundance of *D. tristis* may need a reconsideration because the cryptic species of this complex may differ in their reaction to habitat change and human impacts.

Author Contributions: Conceptualization, E.B. and A.B.; methodology, E.B. and S.O.; validation, E.B., S.O. and A.B.; formal analysis, E.B. and S.O; investigation, E.B., S.O. and A.B.; resources, A.B.; data curation, S.O. and E.B.; writing—original draft preparation, A.B. and E.B.; writing—review and editing, E.B., S.O. and A.B.; visualization, E.B.; supervision, E.B.; project administration, E.B., S.O. and A.B.; funding acquisition, E.B. All authors have read and agreed to the published version of the manuscript.

Funding: This study was funded by the LIETUVOS MOKSLO TARYBA, contract No S-MIP-20-23.

Data Availability Statement: The data presented in this study are available in this article.

Acknowledgments: The authors are grateful to Artūras Matusevičius for an invitation to study the bees and wasps inhabiting the walls of wattle and daub buildings in his farm. They also appreciate the anonymous reviewers for valuable comments and suggestions on the manuscript.

Conflicts of Interest: The authors declare no conflicts of interest. The funders had no role in the design of the study, collection, analyses, or interpretation of data; in the writing of the manuscript, or in the decision to publish the results.

References

1. Loxdale, H.D.; Davis, B.J.; Davis, R.A. Known knowns and unknowns in biology. *Biol. J. Linn. Soc.* **2016**, *117*, 386–398. [CrossRef]
2. Gokhman, V.E. Integrative taxonomy and its implications for species-level systematics of parasitoid Hymenoptera. *Entomol. Rev.* **2018**, *98*, 834–864. [CrossRef]
3. Fišer, C.; Robinson, C.T.; Malard, F. Cryptic species as a window into the paradigm shift of the species concept. *Mol. Ecol.* **2018**, *7*, 613–635. [CrossRef]
4. Struck, T.H.; Feder, J.L.; Bendiksby, M.; Birkeland, S.; Cerca, J.; Gusarov, V.I.; Kistenich, S.; Larsson, K.-H.; Liow, L.H.; Nowak, M.D.; et al. Finding evolutionary processes hidden in cryptic species. *Trends Ecol. Evol.* **2018**, *33*, 153–163. [CrossRef]
5. Kress, W.J.; Garcia-Robledo, C.; Uriarte, M.; Erickson, D.L. DNA barcodes for ecology, evolution, and conservation. *Trends Ecol. Evol.* **2015**, *30*, 25–35. [CrossRef]
6. DeSalle, R.; Goldstein, P. Review and interpretation of trends in DNA barcoding. *Front. Ecol. Evol.* **2019**, *7*, 302. [CrossRef]

7. Lukhtanov, V.A. Species delimitation and analysis of cryptic species diversity in the XXI century. *Entomol. Rev.* **2019**, *99*, 463–472. [CrossRef]
8. Schlick-Steiner, B.C.; Steiner, F.M.; Seifert, B.; Stauffer, C.; Christian, E.; Crozier, R.H. Integrative taxonomy: A multisource approach to exploring biodiversity. *Annu. Rev. Entomol.* **2010**, *55*, 421–438. [CrossRef]
9. Lecocq, T.; Dellicour, S.; Michez, D.; Dehon, M.; Dewulf, A.; De Meulemeester, T.; Brasero, N.; Valterová, I.; Rasplus, J.-Y.; Rasmont, P. Methods for species delimitation in bumblebees (Hymenoptera, Apidae, *Bombus*): Towards an integrative approach. *Zool. Scr.* **2015**, *44*, 281–297. [CrossRef]
10. Zamani, A.; Dal Pos, D.; Fric, Z.F.; Orfinger, A.B.; Scherz, M.D.; Bartoňová, A.S.; Gante, H.F. The future of zoological taxonomy is integrative, not minimalist. *Syst. Biodivers.* **2022**, *20*, 2063964. [CrossRef]
11. Wilson, J.S.; Clark, S.L.; Williams, K.A.; Pitts, J.P. Historical biogeography of the arid-adapted velvet ant *Sphaeropthalma arota* (Hymenoptera: Mutillidae) reveals cryptic species. *J. Biogeogr.* **2012**, *39*, 336–352. [CrossRef]
12. Kurushima, H.; Yoshimura, J.; Kim, J.-K.; Kim, J.-K.; Nishimoto, Y.N.; Sayama, K.; Kato, M.; Watanabe, K.; Hasegawa, E.; Roff, D.A.; et al. Co-occurrence of ecologically equivalent cryptic species of spider wasps. *R. Soc. Open Sci.* **2016**, *3*, 160119. [CrossRef]
13. Orlovskytė, S.; Budrys, E.; Budrienė, A.; Radzevičiūtė, R.; Soon, V. Sibling species in the *Chrysis ignita* complex: Molecular, morphological and trophic differentiation of Baltic species, with a description of two new cryptic species (Hymenoptera: Chrysididae). *Syst. Entomol.* **2016**, *41*, 771–793. [CrossRef]
14. Darwell, C.T.; Cook, J.M. Cryptic diversity in a fig wasp community—Morphologically differentiated species are sympatric but cryptic species are parapatric. *Mol. Ecol.* **2017**, *26*, 937–950. [CrossRef] [PubMed]
15. Somavilla, A.; Barroso, P.C.S.; Aragão, M.; Mateus, S.; Menezes, R.S.T. An integrative taxonomic and phylogenetic approach reveals a new Neotropical swarm-founding social wasp, *Pseudopolybia cryptica* sp. n. (Vespidae: Polistinae: Epiponini). *Arthropod Syst. Phylogeny* **2021**, *79*, 25–35. [CrossRef]
16. Rudoy, A.; Zhu, C.-D.; Ferrari, R.R.; Zhang, Y.-Z. Integrative taxonomy based on morphometric and molecular data supports recognition of the three cryptic species within the *Encyrtus sasakii* complex (Hymenoptera, Encyrtidae). *J. Hymenopt. Res.* **2022**, *90*, 129–152. [CrossRef]
17. Zhang, Y.M.; Sheikh, S.I.; Ward, A.K.G.; Forbes, A.A.; Prior, K.M.; Stone, G.N.; Gates, M.W.; Egan, S.P.; Zhang, L.; Davis, C.; et al. Delimiting the cryptic diversity and host preferences of *Sycophila* parasitoid wasps associated with oak galls using phylogenomic data. *Mol. Ecol.* **2022**, *31*, 4417–4433. [CrossRef]
18. Sann, M.; Niehuis, O.; Peters, R.S.; Mayer, C.; Kozlov, A.; Podsiadlowski, L.; Bank, S.; Meusemann, K.; Misof, B.; Bleidorn, C.; et al. Phylogenomic analysis of Apoidea sheds new light on the sister group of bees. *BMC Evol. Biol.* **2018**, *18*, 71. [CrossRef]
19. Pulawski, W.J. Catalog of Sphecidae sensu lato (= Apoidea Excluding Apidae). 2018. Available online: https://www.calacademy.org/scientists/projects/catalog-of-sphecidae (accessed on 10 November 2023).
20. Kazenas, V.L. *Fauna and Biology of Sphecid Wasps (Hymenoptera, Sphecidae) of Kazakhstan and Central Asia*; Kazgos INTI: Almaty, Kazakhstan, 2001. Available online: https://zool.kz/wp-content/uploads/2020/10/kazenas_2001-1.pdf (accessed on 10 November 2023). (In Russian)
21. O'Neill, K. *Solitary Wasps: Behavior and Natural History*; Cornell University Press: Ithaca, NY, USA, 2001; ISBN 0-8014-3721-0.
22. Biddinger, D.J.; Surcică, A.; Joshi, N.K. A native predator utilizing the invasive brown marmorated stink bug, *Halyomorpha halys* (Hemiptera: Pentatomidae) as a food source. *Biocontrol Sci. Technol.* **2017**, *27*, 903–907. [CrossRef]
23. Bashir, N.H.; Yue, D.; Jiang, H.; Ma, L.; Li, Q. Taxonomic study of the subtribe Pemphredonina Dahlbom, 1835 (Hymenoptera: Crabronidae) with a new species and six new records from China. *J. Asia Pac. Entomol.* **2021**, *24*, 1055–1065. [CrossRef]
24. Noël, G.; Van Keymeulen, V.; Barbier, Y.; Smets, S.; Van Damme, O.; Colinet, G.; Ruelle, J.; Francis, F. Nest aggregations of wild bees and apoid wasps in urban pavements: A "street life" to be promoted in urban planning. *bioRxiv* **2021**. [CrossRef]
25. Vanoye-Eligio, M.; Horta-Vega, J.V.; Vanoye-Eligio, V.; Estrada-Ramírez, L.J.; Martí-Canché, B.R. Crabronidae (Hymenoptera: Apoidea) in Mexico: Occurrence and potential uses in biological control. *Rev. Colomb. Entomol.* **2022**, *48*, e11654. [CrossRef]
26. Poláček, R. The occurrence of Hymenoptera: Aculeata on mine spoil heaps after shale extraction (Jakartovice, Nízký Jeseník Mts). *Acta Mus. Sil. Sci. Nat.* **2020**, *69*, 55–63. [CrossRef]
27. Olszewski, P.; Ljubomirov, T.; Wiśniowski, B.; Kowalczyk, J.K.; Krzyżyński, M. New records of the genus *Diodontus* Curtis, 1834 (Hymenoptera: Crabronidae) from Bulgaria, Montenegro and Poland, with a key to central and eastern European species. *Zootaxa* **2016**, *4061*, 164–172. [CrossRef] [PubMed]
28. Budrys, E.; Budrienė, A.; Orlovskytė, S.; Soon, V. Two new species of *Diodontus* (Hymenoptera: Pemphredonidae) from the western Mediterranean and their phylogenetic relationships. *Can. Entomol.* **2019**, *151*, 558–583. [CrossRef]
29. Park, E.; Poulin, R. Extremely divergent COI sequences within an amphipod species complex: A possible role for endosymbionts? *Ecol. Evol.* **2022**, *12*, e9448. [CrossRef] [PubMed]
30. Hardin, G. The competitive exclusion principle. *Science* **1960**, *131*, 1292–1297. [CrossRef]
31. Pigot, A.L.; Tobias, J.A. Species interactions constrain geographic range expansion over evolutionary time. *Ecol. Lett.* **2013**, *16*, 330–338. [CrossRef] [PubMed]
32. Vodă, R.; Dapporto, L.; Dincă, V.; Vila, R. Cryptic matters: Overlooked species generate most butterfly beta-diversity. *Ecography* **2015**, *38*, 405–409. [CrossRef]
33. Leibold, M.A. Similarity and local co-existence of species in regional biotas. *Evol. Ecol.* **1998**, *12*, 95–110. [CrossRef]

34. Wellborn, G.A.; Cothran, R.D. Niche diversity in crustacean cryptic species: Complementarity in spatial distribution and predation risk. *Oecologia* **2007**, *154*, 175–183. [CrossRef]
35. Werren, J.H.; Baldo, L.; Clark, M.E. *Wolbachia*: Master manipulators of invertebrate biology. *Nat. Rev. Microbiol.* **2008**, *6*, 741–751. [CrossRef] [PubMed]
36. Weinert, L.A.; Araujo-Jnr, E.V.; Ahmed, M.Z.; Welch, J.J. The incidence of bacterial endosymbionts in terrestrial arthropods. *Proc. R. Soc. B Biol. Sci.* **2015**, *282*, 20150249. [CrossRef]
37. Feldhaar, H. Bacterial symbionts as mediators of ecologically important traits of insect hosts. *Ecol. Entomol.* **2011**, *36*, 533–543. [CrossRef]
38. Engelstädter, J.; Hurst, G.D.D. The ecology and evolution of microbes that manipulate host reproduction. *Annu. Rev. Ecol. Evol. Syst.* **2009**, *40*, 127–149. [CrossRef]
39. Brucker, R.M.; Bordenstein, S.R. Speciation by symbiosis. *Trends Ecol. Evol.* **2012**, *27*, 443–451. [CrossRef]
40. Cariou, M.; Duret, L.; Charlat, S. The global impact of *Wolbachia* on mitochondrial diversity and evolution. *J. Evol. Biol.* **2017**, *30*, 2204–2210. [CrossRef]
41. Kaur, R.; Shropshire, J.D.; Cross, K.L.; Leigh, B.; Mansueto, A.J.; Stewart, V.; Bordenstein, S.R.; Bordenstein, S.R. Living in the endosymbiotic world of *Wolbachia*: A centennial review. *Cell Host Microbe* **2020**, *20*, 879–893. [CrossRef]
42. Ramalho, M.D.O.; Kim, Z.; Wang, S.; Moreau, C.S. *Wolbachia* across social insects: Patterns and implications. *Ann. Entomol. Soc. Am.* **2021**, *114*, 206–218. [CrossRef]
43. Budrys, E. Morphometric similarity and summary of measurements of Palearctic species of the genus *Diodontus* Curtis (Hymenoptera, Sphecidae). In *Lietuvos Entomologų Darbai (Lietuvos Entomologų Draugijos 30-Mečiui)*; Jonaitis, V., Ed.; Lithuanian Entomological Society; Institute of Ecology: Vilnius, Lithuania, 1996; pp. 35–47.
44. Peters, R.S.; Krogmann, L.; Mayer, C.; Donath, A.; Gunkel, S.; Meusemann, K.; Niehuis, O. Evolutionary history of the Hymenoptera. *Curr. Biol.* **2017**, *27*, 1013–1018. [CrossRef]
45. Kumar, S.; Stecher, G.; Tamura, K. MEGA7: Molecular evolutionary genetics analysis version 7.0 for bigger datasets. *Mol. Biol. Evol.* **2016**, *33*, 1870–1874. [CrossRef]
46. Benson, D.A.; Karsch-Mizrachi, I.; Lipman, D.J.; Ostell, J.; Wheeler, D. GenBank. *Nucleic Acids Res.* **2005**, 33, D34–D38. [CrossRef]
47. Ratnasingham, S.; Hebert, P.D.N. BOLD: The barcode of life data system (www.barcodinglife.org). *Mol. Ecol. Notes* **2007**, *7*, 355–364. [CrossRef] [PubMed]
48. Ratnasingham, S.; Hebert, P.D.N. A DNA-based registry for all animal species: The barcode index number (BIN) system. *PLoS ONE* **2013**, *8*, e66213. [CrossRef] [PubMed]
49. Budrys, E.; Orlovskytė, S.; Lazauskaitė, M.; Budrienė, A. *Ancistrocerus* wasps (Hymenoptera, Vespidae) from the Centre of Europe: Phylogeny, cryptic species, neutral and non-neutral markers. *Zool. Scr.* **2023**, *52*, 454–474. [CrossRef]
50. Stunžėnas, V.; Petkevičiūtė, R.; Stanevičiūtė, G. Phylogeny of *Sphaerium solidum* (Bivalvia) based on karyotype and sequences of 16S and ITS1 rDNA. *Cent. Eur. J. Biol.* **2011**, *6*, 105–117. [CrossRef]
51. Tusnády, G.E.; Simon, I.; Váradi, A.; Arányi, T. BiSearch: Primer-design and search tool for PCR on bisulfite-treated genomes. *Nucleic Acids Res.* **2005**, *33*, e9. [CrossRef] [PubMed]
52. Hall, T.A. BioEdit: A user-friendly biological sequence alignment editor and analysis program for Windows 95/98/NT. *Nucleic Acids Symp. Ser.* **1999**, *41*, 5–98.
53. Maddison, W.P.; Maddison, D.R. Mesquite: A Modular System for Evolutionary Analysis. Version 3.04. 2015. Available online: https://mesquiteproject.org (accessed on 10 November 2022).
54. Tamura, K.; Stecher, G.; Kumar, S. MEGA11: Molecular evolutionary genetics analysis version 11. *Mol. Biol. Evol.* **2021**, *38*, 3022–3027. [CrossRef]
55. Ronquist, F.; Teslenko, M.; van der Mark, P.; Ayres, D.L.; Darling, A.; Höhna, S.; Larget, B.; Liu, L.; Suchard, M.A.; Huelsenbeck, J.P. MRBAYES 3.2: Efficient Bayesian phylogenetic inference and model selection across a large model space. *Syst. Biol.* **2012**, *61*, 539–542. [CrossRef]
56. Pons, J.; Barraclough, T.G.; Gomez-Zurita, J.; Cardoso, A.; Duran, D.P.; Hazell, S.; Kamoun, S.; Sumlin, W.D.; Vogler, A.P. Sequence-based species delimitation for the DNA taxonomy of undescribed insects. *Syst. Biol.* **2006**, *55*, 595–609. [CrossRef] [PubMed]
57. Puillandre, N.; Brouillet, S.; Achaz, G. ASAP: Assemble species by automatic partitioning. *Mol. Ecol. Resour.* **2020**, *21*, 609–620. [CrossRef] [PubMed]
58. Sharkey, M.J.; Janzen, D.; Hallwachs, W.; Chapman, E.G.; Smith, M.A.; Dapkey, T.; Brown, A.; Ratnasingham, S.; Naik, S.; Manjunath, R.; et al. Minimalist revision and description of 403 new species in 11 subfamilies of Costa Rican braconid parasitoid wasps, including host records for 219 species. *ZooKeys* **2021**, *1013*, 1–665. [CrossRef] [PubMed]
59. Amit, L.; Ben-Shlomo, R.; Chiel, E. Are microbial symbionts involved in the speciation of the gall-inducing aphid, *Slavum wertheimae*? *Arthropod Plant Interact.* **2017**, *11*, 475–484. [CrossRef]
60. Devescovi, F.; Conte, C.A.; Augustinos, A.; Martinez, E.I.C.; Segura, D.F.; Caceres, C.; Lanzavecchia, S.B.; Bourtzis, K. Symbionts do not affect the mating incompatibility between the Brazilian-1 and Peruvian morphotypes of the *Anastrepha fraterculus* cryptic species complex. *Sci. Rep.* **2019**, *9*, 18319. [CrossRef] [PubMed]
61. Andreason, S.A.; Shelby, E.A.; Moss, J.B.; Moore, P.J.; Moore, A.J.; Simmons, A.M. Whitefly endosymbionts: Biology, evolution, and plant virus interactions. *Insects* **2020**, *11*, 775. [CrossRef] [PubMed]

62. Andrianto, E.; Kasai, A. *Wolbachia* in black spiny whiteflies and their new parasitoid wasp in Japan: Evidence of the distinct infection status on *Aleurocanthus camelliae* cryptic species complex. *Insects* **2022**, *13*, 788. [CrossRef]
63. Lin, N. Sequential hypermalaxation in the digger wasp *Diodontus franclemonti*, Krombein (Hymenoptera: Sphecidae). *J. Kans. Entomol. Soc.* **1978**, *51*, 235–238.
64. Guo, J.; Hatt, S.; He, K.; Chen, J.; Francis, F.; Wang, Z. Nine facultative endosymbionts in aphids. A review. *J. Asia-Pac. Entomol.* **2017**, *20*, 794–801. [CrossRef]
65. Srba, M.; Heneberg, P. Nesting habitat segregation between closely related terricolous sphecid species (Hymenoptera: Spheciformes): Key role of soil physical characteristics. *J. Insect Conserv.* **2012**, *16*, 557–570. [CrossRef]
66. Waters, J.; Darvill, B.; Lye, G.; Goulson, D. Niche differentiation of a cryptic bumblebee complex in the Western Isles of Scotland. *Insect Conserv. Divers.* **2011**, *4*, 46–52. [CrossRef]
67. Scriven, J.J.; Whitehorn, P.R.; Goulson, D.; Tinsley, M.C. Niche partitioning in a sympatric cryptic species complex. *Ecol. Evol.* **2016**, *6*, 1328–1339. [CrossRef] [PubMed]
68. Leys, M.; Keller, I.; Robinson, C.T.; Räsänen, K. Cryptic lineages of a common alpine mayfly show strong life-history divergence. *Mol. Ecol.* **2017**, *26*, 1670–1686. [CrossRef] [PubMed]
69. McLaughlin, G.; Gueuning, M.; Genoud, D.; Frey, J.E.; Praz, C. Why are there so many species of mining bees (Hymenoptera, Andrenidae)? The possible roles of phenology and *Wolbachia* incompatibility in maintaining species boundaries in the *Andrena proxima*-complex. *Syst. Entomol.* **2023**, *48*, 127–141. [CrossRef]
70. Heneberg, P. Flagship bird species habitat management supports the presence of ground-nesting aculeate hymenopterans. *J. Insect Conserv.* **2012**, *16*, 899–908. [CrossRef]
71. Twerd, L.; Szefer, P.; Sobieraj-Betlińska, A.; Olszewski, P. The conservation value of Aculeata communities in sand quarries changes during ecological succession. *Glob. Ecol. Conserv.* **2021**, *28*, e01693. [CrossRef]

insects

MDPI

Article

A New Genus of Praeaulacidae (Hymenoptera: Evanioidea) from Mid-Cretaceous Kachin Amber: Insights into a Putative New Praeaulacinae Subclade

Jingtao Yang [1], Corentin Jouault [2,3,4], Longfeng Li [1,*], Chungkun Shih [5,6] and Dong Ren [5]

1 Institute of Vertebrate Paleontology, College of Life Science and Technology, Gansu Agricultural University, Lanzhou 730070, China; yangjt@gsau.edu.cn
2 Institut de Systématique, Évolution, Biodiversité (ISYEB), Muséum National d'Histoire Naturelle, CNRS, SU, EPHE-PSL, U A, CP50, 45 Rue Buffon, F-75005 Paris, France; jouaultc0@gmail.com
3 Institut des Sciences de l'Évolution de Montpellier (UMR 5554), Université de Montpellier, CNRS, F-34095 Montpellier, France
4 Univ. Rennes, CNRS, Géosciences Rennes, UMR 6118, F-35000 Rennes, France
5 College of Life Sciences, Capital Normal University, 105 Xisanhuanbeilu, Haidian District, Beijing 100048, China; chungkun.shih@gmail.com (C.S.); 4989@cnu.edu.cn (D.R.)
6 Department of Paleobiology, National Museum of Natural History, Smithsonian Institution, Washington, DC 20013, USA
* Correspondence: fenger4499@163.com

Simple Summary: A new fossil praeaulacid wasp, *Azygdellitha nova* gen. et sp. nov., is described and figured from mid-Cretaceous Kachin amber. This new taxon not only enriches the diversity of praeaulacid wasps from the Kachin amber biota but also provides additional evidence supporting the discussion of a distinct tribe within the subfamily Praeaulacinae.

Abstract: A new praeaulacid genus and species, *Azygdellitha nova* gen. et sp. nov., is described and illustrated based on a male specimen from mid-Cretaceous Kachin amber from Hukawng Valley, Myanmar. This newly discovered taxon increased the diversity of praeaulacid wasps during the Cretaceous period. While this new taxon shares similarities of wing venation with most species of the subfamily Praeaulacinae, it strongly differs from that of three genera previously described from mid-Cretaceous Kachin amber: *Mesevania*, *Paleosyncrasis*, and *Praegastrinus*. We explore the possibility that these genera constitute a distinct tribe within the Praeaulacinae, distinguished by their wing venation. We provide illustrations and emphasize the potentially diagnostic traits supporting this classification.

Keywords: Apocrita; new taxon; amber; wing venation

1. Introduction

Members of the superfamily Evanioidea are easily recognized by their metasoma, which is attached in a high position to the propodeum and far from hind coxae. This character is considered an autapomorphy for Evanioidea [1]. Evanioidea comprise three extant families (Aulacidae, Evaniidae, and Gasteruptiidae) and five extinct families (Andreneliidae, Anomopterellidae, Baissidae, Othniodellithidae, and Praeaulacidae) recorded in abundant fossil records [2,3].

The family Praeaulacidae encompasses three subfamilies: Cretocleistogastrinae, Praeaulacinae, and Nevaniinae. The Praeaulacidae is the earliest diverged family of Evanioidea and is represented by more than 80 fossil species. Fossil specimens of Praeaulacidae are often documented in compression or imprint fossil deposits but rarely in amber deposits [3,4]. In 2015, the first praeaulacid species found in Kachin amber was described by Li et al. [5], based on one specimen. Subsequently, several amber species have been reported from the same deposit (i.e., Kachin amber), including representatives of the Nevaniinae and Praeaulacinae subfamilies [4,6–8]. The Cretocleistogastrinae are documented from Lower Cretaceous deposits of East Asia and Australia [9,10]. The Praeaulacinae are found in the Middle Jurassic deposits of northeastern China and Kazakhstan and in mid-Cretaceous deposits from Kachin [5,10–13]. Finally, the Nevaniinae are documented from the Middle Jurassic deposits of northeastern China and Kazakhstan and from the mid-Cretaceous deposits of Kachin. The Nevaniinae wasps are unique among evanioid wasps because they possess the first and second petiole-like metasomal segments and are distinctly thinner than the remaining segments [1,6,13]. The decline of Praeaulacidae is a topic of interest but remains poorly understood. There are no documented praeaulacid species postdating the Cretaceous period, suggesting their extinction during the Late Cretaceous. One proposed hypothesis for this decline involves competition with Ichneumonoidea [3]; however, the limited number of Praeaulacidae and Ichneumonoidea species described from mid- and Late Cretaceous deposits impedes the testing of this theory. To address this gap and enhance our understanding of mid-Cretaceous praeaulacid wasp diversity, we describe a new praeaulacine wasp, *Azygdellitha nova* gen. et sp. nov., discovered in mid-Cretaceous Kachin amber.

2. Materials and Methods

The amber piece containing the specimen was derived from the deposits of the Hukawng Valley (Kachin, Northern Myanmar). The amber deposits are approximately 100 km southwest of the Village of Tanai (Figure 1). For Kachin amber, the geological age has been estimated at about an early Cenomanian age (98.79 ± 0.62 Mya) based on zircon U-Pb dating [14–16]. The mines have yielded abundant insect inclusions, which were later used to illustrate many key evolutionary characters and behaviors (e.g., the early evolution of wing scales in moths and butterflies, and ovipositor sensilla in Hymenoptera) [17–23]. In recent years, many well-preserved Evanioidea and insect fossils have been reported form this deposit [2–6,24–28].

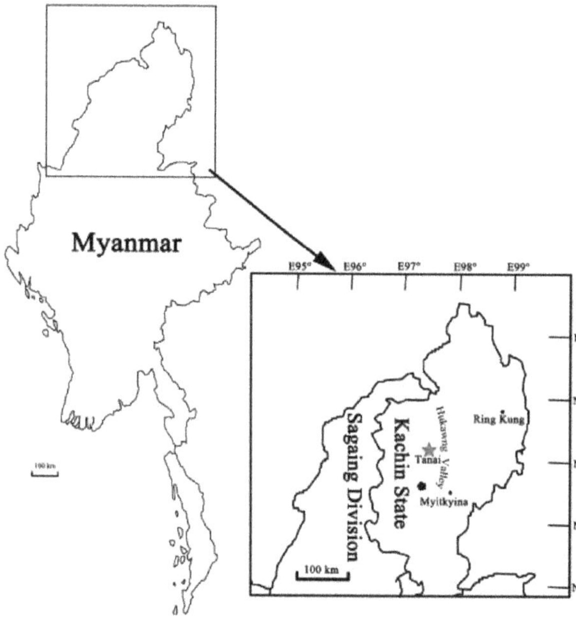

Figure 1. Map of the amber locality near Tanai Village in Hukawng Valley, red star refers to the amber deposit, hexagon refer to Kachin (after [5]).

The type specimen reported herein is housed in the College of Life Science and Technology, Gansu Agricultural University, Lanzhou City, Gansu Province, China. The male amber was examined and photographed using a Nikon SMZ 25 dissecting microscope (Nikon Precision Shanghai Co., Ltd., Shanghai, China) and a Nikon DS-Ri2 digital camera system (Nikon Precision Shanghai Co., Ltd., Shanghai, China). Morphological terminology used here is based on Jouault and Nel [4].

3. Results

Systematic palaeontology

Order Hymenoptera Linnaeus, 1758

Suborder Apocrita Gerstaecker, 1867

Superfamily Evanioidea Latreille, 1802

Remarks: The new specimen belongs to Evanioidea based on its metasoma articulated high on the propodeum and well above the metacoxae [29].

Clade Protoevanioides Jouault et al., 2022

Family Praeaulacidae Rasnitsyn, 1972

Remarks: The new specimen is placed within the family Praeaulacidae because of its complete fore wing venation (with two rs-m and two m-cu crossveins) and its hind wing venation with Cu, rs-m, and cu-a present. Additionally, it possesses a fused metasomal segment 1 (vs. metasomal segment 1 separated in Baissidae); a metasomal segment 1 with the anterior margin distinctly separated from the metanotum (vs. metasomal segment 1 with the anterior margin almost touching the metanotum in Gasteruptiidae); a long, cylindrical metasoma (vs. a short, ovate metasoma with distinct petiole in Evaniidae and a broad, oval metasoma in Aulacidae); a head without a cephalic horn and hind wings with Cu and cu-a present (vs. a head with a cephalic horn and hind wing venation lacking distinct Cu and cu-a in Othniodellithidae; [26]). The fore wing of the new specimen has a complete 2rs-m crossvein (vs. absent in anomopterellid wasps), which cannot be attributed to the family Anomopterellidae [30]. The new amber differs from Andreneliidae in that it has a comparatively thin and elongated pterostigma [31].

The new specimen cannot be placed within the Nevaniinae because of its one-segmented petiole [1,6,32]. It further differs from representatives of the latter subfamily because of its fore wing with a vein M + Cu longer than the combined lengths of the veins 1-Rs and 1-M [1]. We also ruled out affinities of the new specimen with the Cretocleistogastrinae because of its fore wing venation with a thin pterostigma (vs. wide in Cretocleistogastrinae), a 3rs-m crossvein that is completely preserved (vs. completely lost), and a short abscissa between Rs + M and 2-M (vs. absent) [9,33]. Therefore, we place the new fossil within the subfamily Praeaulacinae.

Subfamily Praeaulacinae Rasnitsyn, 1972

Azygdellitha Yang, Jouault, Li, Shih, and Ren, gen. nov.

urn:lsid:zoobank.org:act:02296CDB-2055-4319-BB90-8981E8F5E1BF

Etymology: The new generic name is a combination of the Greek words, '*azygos*' (sole, unpaired) and '*dellithos*' (meaning, a kind of "wasp" or "wasp-like"). The gender of the name is feminine.

Diagnosis: Large body size (longer than 9 mm). Head large elongated in length with large eyes, occupying most of head lateral surface; cephalic horn or projection above the antennal bases absent. Antenna with scape more than 3 times as long as pedicel. Fore wing with pterostigma long and thin; 1-M longer than 1-Rs; 1-Rs about as long as its distance to pterostigma, inclined, and reaching Sc+R close to pterostigma; 1cu-a interstitial with M+Cu fork; 2r-rs long and straight, reaching Rs at 2rs-m level, originating from pterostigma apical third; 2rs-m, 3rs-m, 2m-cu present; cell 2rm triangular, longer than wide; cell 3rm rectangular, longer and wider than cell 2rm, with subparallel sides; 2m-cu reaching M distad 2rs-m (i.e., in cell 3rm), closer to 2rs-m than 3rs-m along M. Hind wing with 1-Rs long and nearly straight; distal Rs abscissa fully developed; rs-m long and arched, longer than cu-a; 1-M longer than 1-Rs; cu-a straight, about 1/3 length of 1-M; distal M and

Cu abscissae fully developed. Mesosoma narrower than head, with propodeum about 3 times as long as metanotum; tibial spurs formula 2-2-1. Petiole widening after mid-length (slightly pear-shaped), longer than the second gaster segment; gaster long cylindric.

Composition: Type species only.

Remarks: Currently, six praeaulacine genera, *Archeogastrinus* Jouault and Rosse-Guillevic, 2023 [8]; *Habraulacus* Li, Rasnitsyn, Shih, and Ren, 2015 [5]; *Hadraulacus* Li, Shih, and Ren, 2023 [7]; *Mesevania* Basibuyuk and Rasnitsyn, 2000 [33]; *Paleosyncrasis* Poinar, 2019 [27]; and *Praegastrinus* Jouault and Nel, 2024 [4], are documented from mid-Cretaceous Kachin amber [4,5,7,8,27,34].

Azygdellitha nova Yang, Jouault, Li, Shih, and Ren, gen. et sp. nov. (Figures 2–4)

Figure 2. *Azygdellitha nova* gen. et sp. nov. Holotype GAU-HYM-MA-2016008. Habitus, lateral views. Scale bars = 1 mm.

urn:lsid:zoobank.org:act:B77A02D2-B3D3-4D11-B8F4-E3AFAE99CFD1

Type material: Holotype, Male, GAU-HYM-MA-2016008, except both of fore and hind wings folded and antenna partly preserved, body nearly well-preserved.

Etymology: The specific name means new and refers to the new species in Kachin amber, which is derived from the Latin adjective *novus*.

Type locality and horizon: Hukawng Village, Kachin State, Northern Myanmar; Mid-Cretaceous Cenomanian period.

Description: Male. Total body length about 9.3 mm long (from mandible to metasomal apex); fore wing length 5.6 mm; integument largely dark brown, wings clear and hyaline, veins brown in preserved color.

Head massive, elongated, with broadly rounded posterior corners, 1.81 mm long and 1.44 mm wide; compound eye ovoid, occupying more than half length of head, 0.89 mm long and 0.50 mm wide; antenna filiform (1.73 mm as preserved), with scape distinctly longer than pedicel; mandibles almost triangle-shaped, left mandible with four teeth, apical tooth the longest, first and second marginal teeth of similar length and shape, separated by broad v-shaped edges, third marginal tooth the shortest, blunt, right mandible with four teeth, apical tooth the longest, second marginal tooth slightly shorter than first, third minute and blunt; maxillary palpus elongated, with five palpomeres, about 1.64 mm, basal palpomere wider but shorter than remaining palpomeres, combined lengths of third to fifth palpomeres longer than eye length.

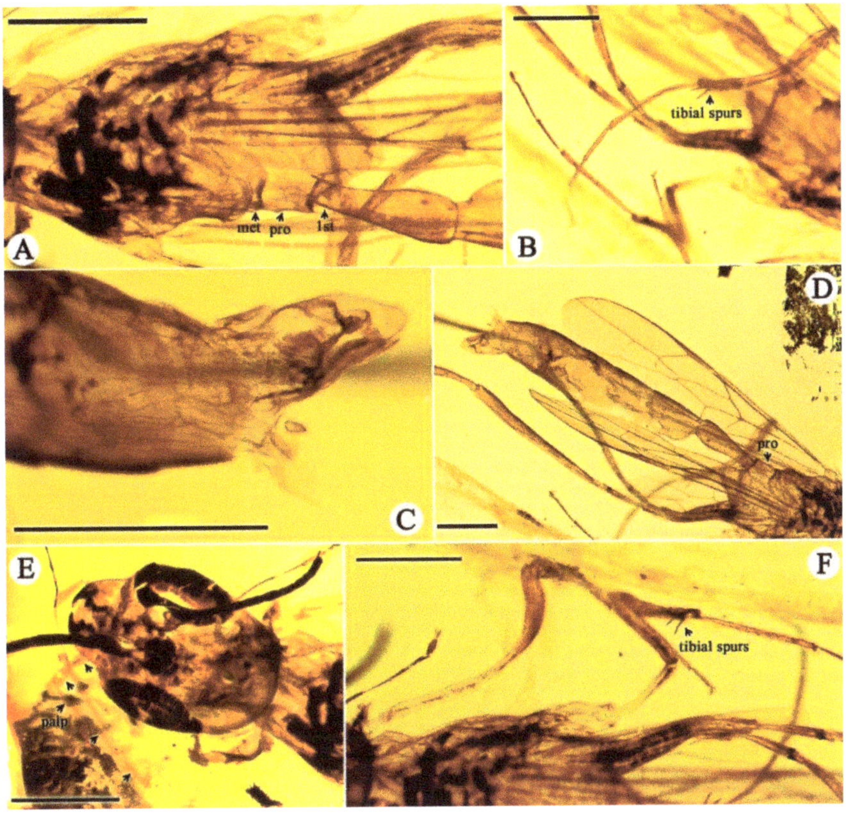

Figure 3. *Azygdellitha nova* gen. et sp. nov. Holotype GAU-HYM-MA-2016008. (**A**) Mesosoma with petiole. (**B**) Mid-leg. (**C**) Male genitalia. (**D**) Metasoma and wings. (**E**) Head. (**F**) Fore and mid legs. met, metanotum; pro, propodeum; 1st, first metasomal segment; palp, palpomeres. Scale bars = 1 mm.

Mesosoma laterally compressed, longer than high, 2.52 mm long and 1.58 mm high; mesosoma with propodeum weakly areolate, about 3 times as long as metanotum. Other parts of mesosoma poorly preserved. Legs slender, nearly completely preserved except for the tarsus of right hind leg. Hind legs distinctly longer than fore and mid legs, tarsi pentamerous; femora lengths (fore femur 0.98 mm, mid femur 1.46 mm, hind femur 2.85 mm); tibiae lengths (fore tibia 1.09 mm, mid tibia 1.66 mm, hind tibia 2.11 mm); tarsi lengths (fore tarsus 1.93 mm, mid tarsus 2.23 mm, hind tarsus 2.92 mm as preserved). Apex of fore and mid tarsi with a arolium, claws bidentate.

Fore wing long, about 6 mm long, 0.10 mm wide, distinctly longer than wide; 1-M (0.37 mm long) slightly curved, aligned with 1-Rs (0.27 mm long); M+Cu fork interstitial with 1cu-a; 2r-rs (0.48 mm long) straight and longer than 1-M; Rs abscissae from 2-Rs to 4-Rs nearly straight and aligned, reaching wing margin before apex; marginal cell elongated, more than three times as long as wide; two rs-m crossveins present, 2rs-m slightly anteriad 2r-rs along Rs, distinctly anteriad 2m-cu along M; 3rs-m (0.35 mm long) bent, distinctly longer than 2rs-m (0.18 mm long); third submarginal cell 1.4 times as long as second submarginal cell; Rs+M about twice as long as 2-Rs; 1m-cu shorter than Rs+M; first discal cell nearly slanted quadrangle with four sides not equal in lengths; 2m-cu present; A slightly longer than M+Cu, meeting 1cu-a. Hind wing about 3.09 mm long, with venation nearly complete, except lacking C; 1-Rs longer than rs-m, but shorter than 1-M; Cu and cu-a present, forming a wide 'Y' shaped angle; angle between 1-M and 1-Cu smaller than angle between first and second Cu abscissae; jugal lobe lacking.

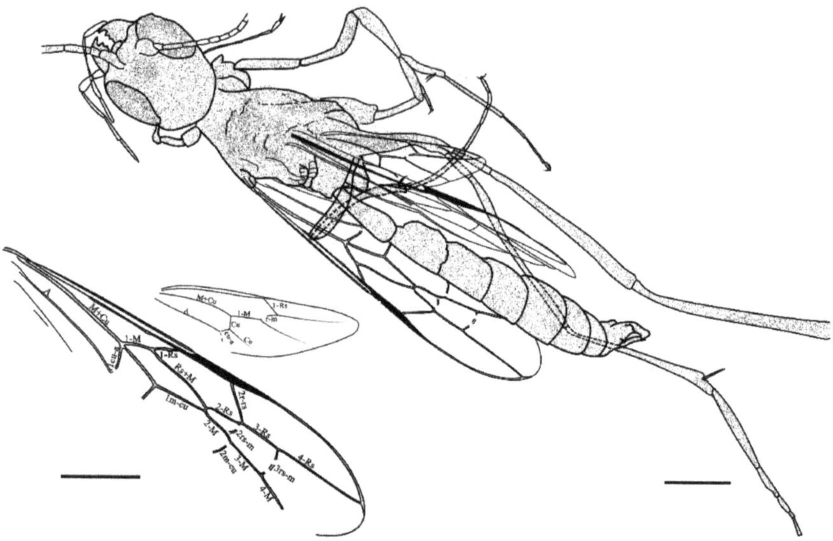

Figure 4. *Azygdellitha nova* gen. et sp. nov. Holotype GAU-HYM-MA-2016008. Line drawing of body and wings with names of veins labelled. Scale bars = 1 mm.

Metasoma 1.13 mm long (including petiole); gaster elongated, 4.16 mm long, second segment to fourth segment nearly equal in lengths and widths, fifth segment slightly wider, sixth and seventh gradually shorter in lengths. Male genitalia outside of the gaster, clasper broad and covered by setae.

4. Discussion

Comparisons with other Praeaulacidae from the Kachin amber biota

As shown in Figure 5, the new species can be easily differentiated from *Archeogastrinus* at least because its head and fore wing are more elongated (vs. rounded or quadrate in *Archeogastrinus*), its fore wing has an interstitial 1cu-a with 1-M (vs. 1cu-a postfurcal), and its hind wing has a long curved r-m (vs. short and straight), and its distal abscissae of Rs, M, and Cu are long (i.e., nearly reaching the wing margin vs. short) [8]. Recently, the genus *Hadraulacus* was described from mid-Cretaceous Kachin amber, and it stands out among the Praeaulacidae of this deposit because it lacks the vein 2rs-m in the fore wing [7]. The wing venation of *Habraulacus* can serve as a good example to differentiate it from the new specimen because it has a long 1-Rs vein (longer than 1-M), while it is shorter in the new specimen; 2r-rs meets Rs well before 2rs-m (vs. being close to 2rs-m in the new species); and 2-M is present and fully developed (vs. much shorter) [5]. Similarly, the hind wing venation of the *Habraulacus* strongly differs from that of the new species: 1-Rs is short (vs. long), and 1-Cu is shorter than cu-a (vs. cu-a longer shorter Cu) [5]. The rest of the fore wing venation being strongly deformed, we refrain from proposing a complete comparison of its venation with that of the new specimen.

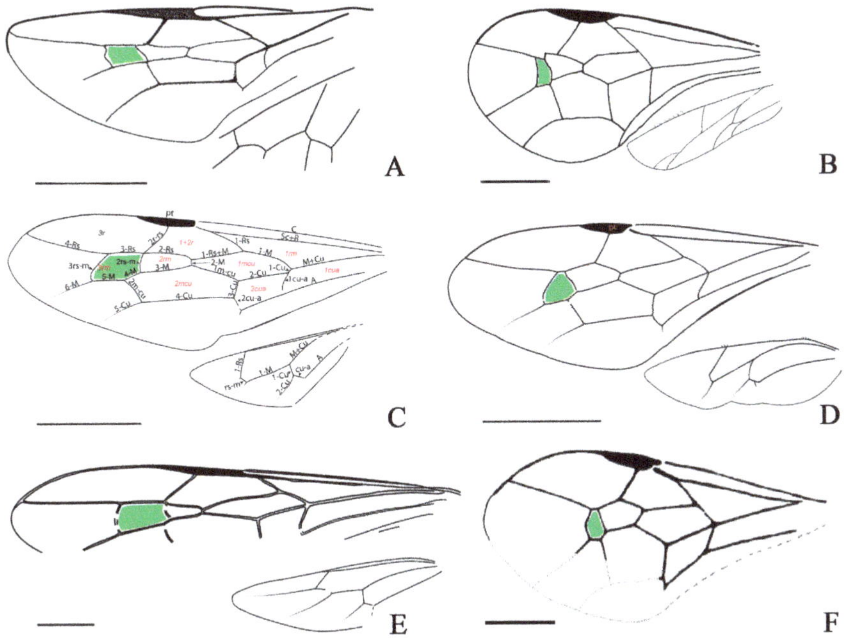

Figure 5. Wings of six species in Praeaulacidae: (**A**) *Habraulacus zhaoi* Li, Rasnitsyn, Shih, and Ren, 2015; (**B**) *Paleosyncrasis hongi* Poinar, 2019; (**C**) *Archeogastrinus kachinensis* Jouault and Rosse-Guillevic, 2023; (**D**) *Praegastrinus edithae* Jouault and Nel, 2019; (**E**) *Azygdellitha nova* Yang, Jouault, Li, Shih, and Ren, gen. et sp. nov.; (**F**) *Mesevania swinhoei* Basibuyuk and Rasnitsyn, 2000. Scale bars = 1 mm. Green color cells refer to 3rm.

The new specimen cannot be attributed to the genus *Mesevania* because of its elongated body with a tubular metasoma (vs. a short body and a metasoma virtually rounded in *Mesevania*), its fore wing with 1-Rs shorter than 1-M (vs. longer than 1-M), its crossvein 2r-rs meeting Rs close to 2rs-m (vs. well before), and its crossvein 3rs-m located just after the 2rs-m (vs. well after) [26]. Similarly, it cannot be attributed to the genus *Paleosyncrasis* because of its tubular and elongated metasoma (vs. short and rounded in *Paleosyncrasis*); its long and relatively thin fore wing (vs. short and broad), with 2r-rs meeting Rs close to 2rs-m (vs. well before) and its crossvein 3rs-m located well after 2rs-m (vs. just after 2rs-m); and its hind wing without Rs + M (vs. with a long Rs + M vein and a distal reemergence of both Rs an M) [20]. Finally, the new specimen differs from *Praegastrinus* at least because it lacks a frons with a bi-lobed ledge covering each torulus dorsally, and it has a fore wing with 1-Rs shorter than 1-M (vs. longer in *Praegastrinus*), a lack of a distinct abscissa of M between Rs+M and 1m-cu (vs. fully developed), the crossvein 2r-rs nearly interstitial with 2rs-m (vs. meeting Rs well before 2rs-m), the cell 3rm rectangular (vs. pear-shaped); and its hind wing with a well-developed 2-Rs (vs. absent) and a long and arched r-m crossvein (vs. short and vertical) [4].

A hidden phylogenetic signal in the fore wing venation of Praeaulacidae?

Praeaulacidae have a fore wing venation that is more complete than that of any other evanioid lineages, sometimes retaining a 1r-rs crossvein that delineates cells 1r and 2r ([1]: fig 9). Wing venation serves as a valuable character for delineating Hymenoptera clades due to its relatively stable nature (compared to Polyneoptera lineages) and relatively clear polarity. Across hymenopteran lineages, there is a general trend of reduction and simplification from the ancestral condition of complex wing venation, along with an enlargement of the pterostigma and reduction in the number of antennal flagellomeres [5,35–37]. Evanioidea are no exception to this trend, with early-diverged lineages (*Protoevanioides sensu* Jouault et al., 2022) [3] retaining the most plesiomorphic wing venation, the highest number of flagellomeres, and often larger body sizes (up to 2 cm), while the more derived *Neoevanioides* typically exhibit less complete wing venation (frequently lacking some rs-m and m-cu crossveins), fewer flagellomeres, and often a smaller size. These characters support the backbone topology of the Evanioidea phylogeny [2,3] and can provide insights into additional relationships within evanioid families.

Within Praeaulacinae, we observed that genera can be grouped into two distinct categories based on their fore wing venation. Specifically, the configuration of the cell enclosed by the crossvein 3rs-m (referred to as 3rm) and the relative proximity between 2rs-m and 3rs-m facilitate a straightforward differentiation of the genera *Mesevania*, *Paleosyncrasis*, and *Praegastrinus* from other members of Praeaulacinae (Figure 5). The only other genus bearing a slight resemblance to these aforementioned genera is *Evanigaster* Rasnitsyn, 1972, although the expression of this characteristic shape is not as pronounced as observed in the previously mentioned genera. Notably, *Mesevania*, *Paleosyncrasis*, and *Praegastrinus* are all documented from mid-Cretaceous Kachin amber. Consequently, this distinct wing venation pattern could either be interpreted as an apomorphy, justifying their classification into a new tribe within Praeaulacinae, or as an adaptation to a specific flight type or behavior. Given our current state of knowledge and the limited data for praeaulacid wasp diversity from Kachin amber, we refrain from proposing formal taxonomic revisions. However, with the continuous discovery of new species within the family Praeaulacidae in this deposit, we anticipate being able to test this hypothesis within a phylogenetic framework in the near future.

5. Conclusions

Based on a single well-preserved specimen of Praeaulacidae found in mid-Cretaceous Kachin amber, we described a new genus and species, *Azygdellitha nova* gen. et sp. nov. This new taxon not only enriches the diversity of praeaulacid wasps from the Kachin amber biota but also provides additional evidence supporting a putative distinct tribe within the subfamily Praeaulacinae. This new clade would be putatively supported by wing venation characters, notably the shape of the cell 3rm and the placement of the crossvein 3rs-m. Three genera placed in the Praeaulacinae share these wing venation patterns, which may suggest that they are closely related, potentially forming a unique clade exclusive to the Kachin amber biota.

Author Contributions: Conceptualization, L.L. and C.S.; resources, J.Y. and L.L.; draft preparation, L.L., C.J. and J.Y.; review and editing, C.J., D.R. and C.S.; validation, C.S., L.L. and D.R.; visualization, L.L. and C.S.; project administration, J.Y.; funding acquisition, L.L. and D.R. All authors have read and agreed to the published version of the manuscript.

Funding: This study is supported by the Scientific Research Start-up Funds for Openly Recruited Doctors (2017RCZX-28; 2017RCZX-27); the Fuxi Youth Talent Project of Gansu Agricultural University (Gaufx-03Y03); and the National Natural Science Foundation of China (No. 31900347, 32020103006).

Institutional Review Board Statement: Not applicable.

Informed Consent Statement: Not applicable.

Data Availability Statement: All data from this study are available in this paper and the associated papers.

Acknowledgments: We are grateful to the editor and reviewers for their constructive comments on an early version for our manuscript. Many thanks are extended to Jiajia Wang (Institute of zoology, Chinese Academy of sciences) for her kind help in taking photos of the amber.

Conflicts of Interest: The authors declare no conflicts of interest.

References

1. Zhang, H.C.; Rasnitsyn, A.P. Nevaniinae subfam. n., a new fossil taxon (Insecta: Hymenoptera: Evanioidea: Praeaulacidae) from the Middle Jurassic of Daohugou in Inner Mongolia, China. *Insect Syst. Evol.* **2007**, *38*, 149–166. [CrossRef]
2. Li, L.; Rasnitsyn, A.P.; Shih, C.; Labandeira, C.C.; Buffington, M.; Li, D.; Ren, D. Phylogeny of Evanioidea (Hymenoptera, Apocrita), with descriptions of new Mesozoic species from China and Myanmar. *Syst. Entomol.* **2018**, *43*, 810–842. [CrossRef]
3. Jouault, C.; Marechal, A.; Condamine, F.L.; Wang, B.; Nel, A.; Legendre, F.; Perrichot, V. Including fossils in phylogeny: A glimpse into the evolution of the superfamily Evanioidea (Hymenoptera: Apocrita) under tip-dating and the fossilized birth-death process. *Zool. J. Linn. Soc.* **2022**, *194*, 1396–1423. [CrossRef]
4. Jouault, C.; Nel, A. New evanioid wasps (Hymenoptera: Evaniidae; Praeaulacidae) from mid-Cretaceous Kachin amber. *Ann. Paléontol.* **2024**, *110*, 102662.
5. Li, L.; Rasnitsyn, A.P.; Shih, C.; Ren, D. A new genus and species of Praeaulacidae (Hymenoptera: Evanioidea) from Upper Cretaceous Myanmar amber. *Cretac. Res.* **2015**, *55*, 19–24. [CrossRef]
6. Jouault, C.; Nel, A.; Perrichot, V. New evanioid wasps (Hymenoptera: Praeaulacidae, Aulacidae) from Cenomanian Burmese amber. *Cretac. Res.* **2020**, *110*, 104407. [CrossRef]
7. Li, L.; Shih, C.; Yang, J.; Ren, D. A new praeaulacid genus with a new species in Praeaulacidae (Hymenoptera: Evanioidea) from mid-Cretaceous Myanmar amber. *Cretac. Res.* **2023**, *146*, 105483. [CrossRef]
8. Jouault, C.; Rosse-Guillevic, S. A new genus of praeaulacid wasp (Hymenoptera: Evanioidea) from the mid-Cretaceous Kachin amber (Myanmar). *Ann. Paléontol.* **2023**, *109*, 102599. [CrossRef]
9. Rasnitsyn, A.P. Hymenopterans, Vespida, in Pozdne-Mezozoyskie Nasekomye Vostochnogo Zabaykal'ya. *Tr. Paleontol. Instituta* **1990**, *239*, 177–205.
10. Rasnitsyn, A.P. Praeaulacidae (Hymenoptera) from the Upper Jurassic of Karatau. *Paleontol. Zh.* **1972**, *1*, 72–87. (In Russian)
11. Rasnitsyn, A.P. Hymenoptera Apocrita of Mesozoic. *Tr. Paleontol. Instituta* **1975**, *147*, 1–134. (In Russian)
12. Zhang, H.C.; Rasnitsyn, A.P. Middle Jurassic Praeaulacidae (Insecta: Hymenoptera: Evanioidea) of Inner Mongolia and Kazakhstan. *J. Syst. Palaeontol.* **2008**, *6*, 463–487. [CrossRef]
13. Rasnitsyn, A.P.; Zhang, H.C. Early evolution of Apocrita (Insect, Hymenoptera) as indicated by new findings in the Middle Jurassic of Daohugou, Northeast China. *Acta Geol. Sin.* **2010**, *84*, 834–873. [CrossRef]
14. Shi, G.H.; Grimaldi, D.A.; Harlow, G.E.; Wang, J.; Wang, J.; Yang, M.C.; Lei, W.Y.; Li, Q.L.; Li, X.H. Age constraint on Kachin amber based on U-Pb dating of zircons. *Cretac. Res.* **2012**, *37*, 155–163. [CrossRef]
15. Yu, T.T.; Thomson, U.; Mu, L.; Ross, A.; Kennedy, J.; Broly, P.; Xia, F.Y.; Zhang, H.C.; Wang, B.; Dilcher, D. An ammonite trapped in Burmese amber. *Proc. Natl. Acad. Sci. USA* **2019**, *116*, 11345–11350. [CrossRef] [PubMed]
16. Cruickshank, R.; Ko, K. Geology of an amber locality in the Hukawng Valley, Northern Myanmar. *J. Asian Earth Sci.* **2003**, *21*, 441–455. [CrossRef]
17. Wang, J.; Zhang, W.; Engel, M.S.; Sheng, X.; Shih, C.; Ren, D. Early evolution of wing scales prior to the rise of moths and butterflies. *Curr. Biol.* **2022**, *32*, 3808–3814.e2. [CrossRef] [PubMed]
18. Cui, Y.; Bardin, J.; Wipfler, B.; Demers-Potvin, A.; Bai, M.; Tong, Y.; Chen, G.N.; Chen, H.; Zhao, Z.; Ren, D.; et al. A winged relative of ice-crawlers in amber bridges the cryptic extant Xenonomia and a rich fossil record. *Insect Sci.* 2024. [CrossRef]
19. Lai, D.; Chen, P.; Li, S.; Xiang, X.; Ou, H.; Kang, N.; Yang, J.; Pang, H.; Shih, C.; Labandeira, C.C.; et al. The associated evolution of raptorial foreleg and mantispid diversification during 200 million years. *Natl. Sci. Rev.* **2023**, *10*, nwad278. [CrossRef]
20. Zhang, Y.; Rasnitsyn, A.P.; Zhang, W.; Song, F.; Shih, C.; Ren, D.; Wang, Y.; Li, H.; Gao, T. Stem chewing lice on Cretaceous feathers preserved in amber. *Curr. Biol.* **2024**, *34*, 916–922.e1. [CrossRef]
21. Wang, Z.; Vilhelmsen, L.; Rasnitsyn, A.P.; Viertler, A.; Shih, C.K.; Wen, S.S.; Yang, H.R.; Wu, Q.; Zhang, Y.J.; Ren, D.; et al. Specialized ovipositor sensilla of Cretaceous wasp (Insecta: Hymenoptera) possibly reveal a unique way of host detection. *Cladistics* **2024**. [CrossRef] [PubMed]
22. Han, N.; Zhang, W.; Fang, H.; Wang, J.; Shih, C.; Ren, D. Geometric morphometric analyses of Micropterigidae lineages (Lepidoptera)with two new species from mid-Cretaceous amber of northern Myanmar. *Cretac. Res.* **2024**, *160*, 105897. [CrossRef]
23. Ma, Y.-N.; DU, S.-L.; Ren, D.; Yao, Y.-Z. A new genus and species of Nabinae (Heteroptera: Cimicomorpha: Nabidae) from mid-Cretaceous Burmese amber. *Palaeoentomology* **2024**, *7*, 306–313. [CrossRef]
24. Engel, M.S. Two ensign wasps in Cretaceous amber from New Jersey and Myanmar (Hymenoptera: Evaniidae). *Pol. J. Entomol.* **2006**, *75*, 443–454.
25. Engel, M.S. New evanioid wasps from the Cenomanian of Myanmar (Hymenoptera: Othbiodellithidae, Aulacidae), with s summary of family-group names among Evanioidea. *Am. Mus. Novit.* **2017**, *3871*, 1–28. [CrossRef]
26. Engel, M.S.; Huang, D.; Alqarni, A.S.; Cai, C. A remarkable evanioid wasp in mid-Cretaceous amber from northern Myanmar (Hymenoptera: Evanioidea). *Cretac. Res.* **2016**, *60*, 121–127. [CrossRef]
27. Poinar, G., Jr. A new genus and species of wasp (Hymenoptera: Evanioidea: Praeaulacidae) associated with an angiosperm flower in Burmese amber. *Palaeoentomology* **2019**, *2*, 474–481. [CrossRef]
28. Shih, P.J.; Li, L.; Li, D.; Ren, D. Application of geometric morphometric analyses to confirm three new wasps of Evaniidae (Hymenoptera: Evanioidea) from mid-Cretaceous Myanmar amber. *Cretac. Res.* **2019**, *109*, 104249. [CrossRef]
29. Goulet, H.; Huber, J.T. *Hymenoptera of the World: An Identification Guide to Families*; Agriculture Canada Publication: Ottawa, ON, Canada, 1993; p. 680.

30. Li, L.; Rasnitsyn, A.P.; Shih, C.; Ren, D. Anomopterellidae Restored, with Two New Genera and Its Phylogeny in Evanioidea (Hymenoptera). *PLoS ONE* **2013**, *8*, e82587. [CrossRef]
31. Rasnitsyn, A.P.; Martínez-Delclòs, X. Wasps (Insecta: Vespida = Hymenoptera) from the Early Cretaceous of Spain. *Acta Geol. Hisp.* **2000**, *35*, 65–95.
32. Li, L.-F.; Shih, C.-K.; Ren, D. Two new species of *Nevania* (Hymenoptera: Evanioidea: Praeaulacidae: Nevaniinae) from the Middle Jurassic of China. *Alcheringa Australas. J. Palaeontol.* **2014**, *39*, 140–147. [CrossRef]
33. Zhang, H.C.; Zhang, J.F. Discovery of Praeaulacidae (Insecta Hymenoptera) in China and its significance. *Acta Micro-Palaeontol. Sin.* **2000**, *17*, 416–421.
34. Basibuyuk, H.H.; Rasnitsyn, A.P.; Fitton, M.G.; Quicke, D.L.J. An archaic new genus of Evaniidae (Insecta: Hymenoptera) and implications for the biology of ancestral evanioids. *Bull. Nat. Hist. Mus. Lond. (Geol.)* **2000**, *56*, 54–56.
35. Rasnitsyn, A.P. The origin and evolution of Lower Hymenoptera. *Trans. Paleontol. Inst.* **1969**, *123*, 1–196. (In Russian)
36. Rasnitsyn, A.P. Origin and evolution of Hymenoptera. *Trans. Paleontol. Inst.* **1980**, *174*, 1–192. (In Russian)
37. Jouault, C.; Ngô-Muller, V.; Pouillon, J.-M.; Nel, A. New Burmese amber fossils clarify the evolution of bethylid wasps (Hymenoptera: Chrysidoidea). *Zool. J. Linn. Soc.* **2021**, *191*, 1044–1058. [CrossRef]

insects MDPI

Article

Molecular Data Confirm Interspecific Limits of Four *Alloxysta* and One *Phaenoglyphis* Species of Parasitic Wasps within the Subfamily Charipinae (Cynipoidea: Figitidae)

Mar Ferrer-Suay [1,*], Mariana Bulgarella [2], George E. Heimpel [3], Ehsan Rakhshani [4] and Jesús Selfa [1]

[1] Departament de Zoologia, Facultat de Ciències Biològiques, Universitat de València, Campus de Burjassot-Paterna, Dr. Moliner 50, E-46100 Burjassot (València), Spain; jesus.selfa@uv.es
[2] School of Biological Sciences, Victoria University of Wellington, Wellington 6012, New Zealand; mariana.bulgarella@vuw.ac.nz
[3] Department of Entomology, University of Minnesota, St. Paul, MN 55108, USA; heimp001@umn.edu
[4] Department of Plant Protection, College of Agriculture, University of Zabol, Zabol P.O. Box 98615-538, Iran; rakhshani@uoz.ac.ir
* Correspondence: mar.ferrer@uv.es

Simple Summary: The interspecific limits within the Charipinae subfamily are still not clear, although many improvements have been made to the taxonomy of this group. Three genes (mitochondrial COI and 16S rDNA, and nuclear ITS2 rDNA) have been sequenced from five different species: *Alloxysta brevis* (Thomson, 1862), *A. castanea* (Hartig, 1841), *A. ramulifera* (Thomson, 1862), *A. victrix* (Westwood, 1833), and *Phaenoglyphis villosa* (Hartig, 1841). The phylogeny resulting from concatenating the three genes confirms that these molecular markers can be used to separate Charipinae species and also shows the close relationship between the genera *Alloxysta* and *Phaenoglyphis*. The genetic distances of these species are presented.

Abstract: The hymenopteran subfamily Charipinae (Cynipoidea: Figitidae) consist of a group of parasitic wasps that are exclusive hyperparasitoids of Hemipteran. The species boundaries in Charipinae have historically been unclear. While diagnostic morphological features have been established for the stepwise separation of species, it is recommended to confirm those limits using molecular data. Here, we focus on the genera *Alloxysta* Förster, 1869 and *Phaenoglyphis* Förster, 1869, both of which contain species that are hyperparasitoids of aphids. We sequenced three genes (mitochondrial COI and 16S rDNA, and nuclear ITS2 rDNA) from specimens that were identified as belonging to five species: *Alloxysta brevis* (Thomson, 1862), *A. castanea* (Hartig, 1841), *A. ramulifera* (Thomson, 1862), *A. victrix* (Westwood, 1833), and *Phaenoglyphis villosa* (Hartig, 1841). The phylogeny resulting from concatenating these genes supported the species status of the five morphologically identified taxa, with *P. villosa* nested within *Alloxysta*. Our study thus indicates that these molecular markers can successfully distinguish charipine species, and also indicates that the genera *Alloxysta* and *Phaenoglyphis* may be more closely related than previously hypothesized. We also present the first estimates of genetic distances for these species. Future studies that include more species, loci, and/or genomic data will complement our research and help determine species relationships within the Charipinae subfamily.

Keywords: *Alloxysta*; Charipinae; hyperparasitism; molecular markers; *Phaenoglyphis*; species boundaries

1. Introduction

Hymenopteran parasitoids are important in the biological control of pests [1–6]. However, biological control scientists often experience difficulty identifying specimens to the species level, and some biological control projects are further complicated by the presence of cryptic species [7–9]. There is therefore a great opportunity to improve the understanding of biological control interactions with the increased ability to incorporate molecular tools for identification of parasitoid species [10]. Members of the hymenopteran subfamily Charipinae (Cynipoidea: Figitidae) are secondary parasitoids (hyperparasitoids) of aphids via Aphidiinae (Braconidae) and Aphelinidae (Chalcidoidea), and of psyllids via Encyrtidae (Chalcidoidea) [11]. Charipine hyperparasitoids frequently attack biological control agents [12–16], and the suppression exerted on the populations of primary parasitoids has been demonstrated in greenhouses [17,18] and likely occurs in the field as well [11]. Thus, taxonomic studies within the Charipinae are important from the standpoint of biological control programs, in which accurate species identification can be problematic.

The effectiveness of molecular methods of species identification has led to a proposal of standardization and large-scale application whose purpose is to inventory as much life on the planet as possible: "The Barcode of Life" [19], this is a technique for characterizing species of organisms using a short DNA sequence from a standard and agreed-upon position in the genome. The cytochrome c oxidase subunit 1 mitochondrial region (COI) has emerged as the standard barcode region for higher animals [19]. This sequence is practically the same size in all animals, and a fragment of 650 nucleotides typically encompasses enough variation to allow discernment between two closely related species. It has been proposed that taxonomists continue with the morphological and ecological work they have been doing, but that they also include a DNA barcode sequence to allow non-specialists to quickly recognize species [19–24]. A major advantage of the barcode approach is that sequences of the COI region obtained from unidentified specimens can be compared with existing DNA sequence databases for identification. Also, COI sequences could be used to identify new or cryptic species. In addition, some authors argue that the evolution of COI is sufficiently rapid to investigate intraspecific diversity in some taxa [21]. Another mitochondrial gene, 16S ribosomal RNA encoding gene sequences (16S rDNA), is extensively used for phylogenetic analysis, and it complements COI for species-level identification. The nuclear-encoded internal transcribed spacer 2 (ITS2 rDNA) has been used for species-level identification alone [25] or in combination with other markers [26–28].

There is a very limited number of studies on the molecular identification of *Alloxysta* and *Phaenoglyphis* species. To the best of our knowledge, the only molecular study outside our research group aimed to differentiate two species of hyperparasitoids, *Alloxysta xanthopsis* (Ashmead 1896) and *Dendrocerus carpenteri* (Curtis, 1829) [29]. The majority of previous studies have focused on establishing species limits of charipines using DNA barcoding [30,31]. *Alloxysta* (Figitidae, Charipinae) is the most abundant and widely distributed genus within the subfamily and found on every continent, except for Antarctica [32]. There are few diagnostic features that allow species delimitation within *Alloxysta*, and even those available can be very difficult to recognize for the non-expert. The identification to species level has always been very difficult for this genus [33]. *Alloxysta* species have been found attacking numerous host species [34], and host specificity can vary greatly among species [35]. *Phaenoglyphis* is the second most diverse genus within Charipinae. Specimens of *Phaenoglyphis* are rarely encountered in the field, so information about this genus is more limited. The first molecular characterization of seven *Phaenoglyphis* species using COI sequences was conducted recently. The delimitation of these seven species was largely congruent with their morphological identifications [36].

The aim of this study is to corroborate the morphological identification of five charipine species from previous studies [30,31] with molecular data. We amplified three genes (COI, ITS2, and 16S) for individuals that we identified morphologically as belonging to five species: *Alloxysta brevis* (Thomson 1862), *A. castanea* (Hartig 1841), *A. ramulifera* (Thomson 1862), *A. victrix* (Westwood 1833), and *Phaenoglyphis villosa* (Hartig 1841).

2. Materials and Methods

2.1. Morphological Identification

The specimens of *Alloxysta* and *Phaenoglyphis* were collected from Yprès, Belgium (n = 35) during the 2022 summer season as part of another project. A total of 1025 specimens were identified using morphological characters and a published dichotomous key [37]. Thirty-five of these specimens preserved in absolute ethanol were selected from five species (seven of each): *Alloxysta brevis*, *A. castanea*, *A. ramulifera*, *A. victrix*, and *Phaenoglyphis villosa*. These 35 specimens were examined using a stereo microscope (OPTIKA ZSM-2, Ponteranica, Italy,) and environmental scanning electron microscope (FEI Quanta 200 ESEM) at the scientific technical services of the University of Barcelona and Hitachi S4800 from SCIE (Servicio Central de Soporte a la Investigación) at the Universitat de València.

2.2. DNA Extraction, PCR Amplification, and Sequencing

DNA extraction, PCR, and Sanger sequencing were performed by AllGenetics & Biology SL (www.allgenetics.eu, accessed on 27 February 2024). DNA extraction was performed non-destructively using whole specimens with the Quick-DNA Microprep Plus Kit (Zymo Research, Irvine, CA, USA), following the manufacturer's instructions. Briefly, the entire specimen was submerged in an extraction buffer without smashing or breaking it up. The kit employs a combination of enzymatic and chemical extraction mechanisms. This DNA extraction method leaves the insect integument intact so that the specimen can be morphologically studied. DNA was eluted in 12 µL of elution buffer. DNA was quantified using a Qubit High-Sensitivity dsDNA Assay (Thermo Fisher Scientific, Waltham, MA, USA). Sample yield was very low (see results). Three different genes were PCR-amplified: the mitochondrial cytochrome oxidase I gene (COI) and 16S ribosomal RNA encoding gene sequences (16S rDNA), and the nuclear internal transcribed spacer 2 (ITS2 rDNA).

PCR protocols varied for each gene as follows. For COI, we used the forward primer LCO1490 paired with the reverse primer HCO2198 [38]. PCRs were carried out in 12.5 µL volumes, containing 2 µL of template DNA solution, 0.5 µM of each primer, 6.25 µL of Supreme NZYTaq 2× Green Master Mix (NZYtech, Lisboa, Portugal), and ultrapure water. The reaction mixture was incubated as follows: an initial denaturation step at 95 °C for 5 min, followed by 35 cycles of 95 °C for 30 s, 49 °C for 45 s, 72 °C for 45 s, and a final extension step at 72 °C for 7 min. Samples that showed weak products were reamplified successfully. These PCRs were carried out in a final volume of 25 µL, containing 10 µL of the first PCR product, 0.5 µM of the primers, 6.25 µL of Supreme NZYTaq 2× Green Master Mix, and ultrapure water up to 25 µL. The reaction mixture was incubated as follows: an initial denaturation step at 95 °C for 5 min, followed by 5 cycles of 95 °C for 30 s, 49 °C for 45 s, 72 °C for 45 s, and a final extension step at 72 °C for 7 min.

For 16S, the forward primer 16SWbF [24] was paired with the reverse primer 16SWbR [39] to amplify a fragment around 500 bp long. The PCR reactions were carried out in a final volume of 12.5 µL, containing two slightly different reaction mixtures: (a) 6.25 µL of Supreme NZYTaq 2× Green Master Mix, 0.5 µM of the primers, 2 µL of template DNA solution, and ultrapure water up to 12.5 µL and (b) 3.13 µL of Supreme NZYTaq 2× Green Master Mix, 0.5 µM of the primers, 1.25 µL of template DNA, and ultrapure water up to 12.5 µL. Thermal cycling conditions included an initial denaturation step at 95 °C for 5 min, followed by 35 cycles of 95 °C for 30 s, 49 °C/51 °C (for different reaction mixtures: a and b, respectively) for 45 s, 72 °C for 45 s, and a final extension step at 72 °C for 7 min.

For ITS2, the primer pair used was ITS2F and ITS2R [25] which amplifies a region 400 bp long. PCRs were carried out in a final volume of 12.5 µL, containing 2 µL of template DNA solution, 0.5 µM of the primers, 6.25 µL of Supreme NZYTaq 2× Green Master Mix, and ultrapure water up to 12.5 µL. Cycling conditions included an initial denaturation step at 95 °C for 5 min, followed by 35 cycles of 95 °C for 30 s, 57 °C for 45 s, 72 °C for 45 s, and a final extension step at 72 °C for 7 min.

A negative control that contained no DNA was included in every PCR round to check for cross-contamination. PCR products were run on 2% agarose gels stained with

GreenSafe (NZYtech, Lisboa, Portugal) and imaged under UV light to verify the amplicon size. The PCR products were bi-directionally sequenced on an ABI 3730xl DNA Analyzer (Applied Biosystems, Waltham, MA, USA), with the same primers as those used in the PCR amplification. The amplification products that showed an intense background smear, primer dimers, or non-specific bands in the electrophoresis gel were purified using magnetic beads (Mag-Bind, Omega-Biotek, Norcross, GA, USA) prior to sequencing.

Initial electropherogram analysis was conducted by AllGenetics & Biology SL in Geneious 8.1.9 (https://www.geneious.com, accessed on 7 November 2023). The primer annealing regions and the low-quality regions at both ends of each electropherogram were trimmed (error probability limit of 0.03). Sequence reads were manually checked for sequencing errors or ambiguous base calls. Ambiguities and polymorphic positions were coded using the IUPAC ambiguity code.

2.3. Phylogenetic Analyses

The sequences were aligned for each gene region using the Geneious Prime v.2023.2.1 alignment algorithm with the default parameter settings (Biomatters, Auckland, New Zealand). The final length of COI retained for analyses was 612 bp, and all 35 individuals were amplified successfully. A region of 352 bp of 16S was amplified successfully for 24 specimens. For ITS2, a 415 bp region was retained for 28 individuals. We concatenated the three gene regions and included a COI sequence from a cynipid gall wasp, *Barbotinia oraniensis* (Barbotin) (GenBank accession number AF395179.1), as the outgroup to obtain a final alignment consisting of 1374 bp for 36 individuals (Supplemental Data File S1).

As previously found, *P. villosa* presented a 6 bp deletion in COI (Ferrer-Suay et al. 2018) at position 467–472 in our alignment, not present in any of the four *Alloxysta* species. For 16S and ITS2, the five species presented multiple gaps of variable lengths. We implemented the gap-coding program FastGap 1.2 (University of Aarhus, Aarhus, Denmark) [40] for the data on the three concatenated loci. This software codes gaps using the method of Simmons and Ochoterena [41], and gaps are added to the data file as separate partitions. In this case, the resulting alignment consisted of 1471 bp for the 36 individuals included (Supplemental Data File S2).

We built phylogenetic trees using the MrBayes v.3.2.6 plug-in [42] for Geneious Prime. Clade probabilities were obtained from the posterior distribution. Bayesian analyses were replicated twice, each with four Markov chains of 1.5 million generations, using a general time-reversible model and gamma-distributed rate variation, with a proportion of invariant sites (GTR + G + I). Trees were sampled every 1000 generations, and the first 500,000 generations were discarded as burn-in. We built phylogenetic trees for the original alignment with gaps uncoded (1374 bp) and for the alignment with gaps coded as separate partitions at the end (1471 bp). Sequences generated in this study have been archived in GenBank (accession numbers for COI: PP097396-PP097430, for 16S: PP109087-PP109110, and for ITS2: PP102679-PP102706).

Next, we downloaded all COI sequences available for *Alloxysta brevis*, *A. castanea*, *A. ramulifera*, *A. victrix*, and *Phaenoglyphis villosa* from GenBank. These sequences were aligned to the COI alignment we generated in this study and included *Barbotinia oraniensis* as the outgroup. The resulting alignment consisted of 615 bp long sequences for 239 individuals (Supplemental Data File S3). We used AIC [43] as implemented in Modeltest 3.7 [44] for PAUP* 4.0 [45] to determine the model of sequence evolution that best fits the mitochondrial COI data. The selected model was HKY85 + I + G. We obtained clade probabilities from the posterior distribution using MrBayes with four Markov chains of 1 million generations each. Trees were sampled every 1000 generations, and the first 500,000 generations were discarded as burn-in.

We calculated the intra- and interspecific pairwise distances between sequences for COI. Analyses were conducted using the Kimura 2-parameter model [46]. This analysis included 35 nucleotide sequences. All ambiguous positions were removed for each sequence pair (pairwise deletion option). The final dataset included 532 positions. Evolutionary analyses were conducted in MEGA11 [47]. We present the mean ± standard deviation of intra- and interspecific distances per species.

3. Results

Four *Alloxysta* and one *Phaenoglyphis* species were compared in this study. These five species can be distinguished by an expert based on five morphological characters. The genus *Phaenoglyphis* is currently being revised, and it is well differentiated from *Alloxysta* by the mesopleural sulcus and the scutellar foveae present (Figure 1). *Phaenoglyphis villosa* can be differentiated from other species of *Phaenoglyphis* by its partially open radial cell and mesoscutum without notauli. *Alloxysta castanea* has radial cells partially open (Figure 2), *A. victrix* does not have propodeal carinae (Figure 3), and *A. brevis* does not have pronotal carinae (Figure 4). *Alloxysta ramulifera* has a combination of features not present in the other *Alloxysta* species (Figure 5).

The following is a key to differentiate between these five species:

1. Lower part of mesopleuron with horizontal sulcus (Figure 1F). . . . . *P. villosa*
– Mesopleuron lacks horizontal sulcus. 2
2. Radial cell partially open (Figure 2A). . . . . *A. castanea*
– Radial cell closed. 3
3. Propodeal carinae absent (Figure 3C). . . . . *A. victrix*
– Propodeal carinae present, forming a plate (Figures 4C and 5C). 4
4. Pronotal carinae absent; F1 shorter than pedicel (Figure 4B). . . . . *A. brevis*
– Pronotal carinae present; F1 longer than pedicel (Figure 5B). . . . . *A. ramulifera*

In this study, we sequenced three partial molecular markers. For COI, all 35 specimens were amplified satisfactorily, although some samples needed to be reamplified to obtain the sequence of interest. For ITS2, 28 out of the 35 specimens produced good sequences. For 16S, we only obtained 24 sequences and no individual of *A. ramulifera* amplified via PCR, suggesting that species-specific primers need to be re-designed to amplify this locus in this species. The COI sequences presented 89.6% pairwise identity, and 74.5% of the 612 bp sites were identical. ITS2 presented 71.2% pairwise identity, and 37.4% of the sites were identical out of 415 bp. The 16S sequences presented 87.4% pairwise identity, with 70.2% of the sites being identical out of 352 bp. PCR amplification of these loci was not straightforward for these specimens. For example, a nested PCR approach was required to obtain 612 bp of COI of good quality. The primers for 16S amplify a fragment around 500 bp long, but we had to trim sequences to 352 bp to retain only good-quality base calls. We believe these difficulties are likely due to the extremely low concentrations of the DNA extractions (no specimen rendered > 3 ng/µL of DNA). These wasps are extremely small (1–2 mm), providing very limited starting material. Using the whole specimen material mechanically destroyed by a homogenizer is necessary to extract enough quantity of DNA for the subsequent studies.

The phylogenetic tree resulting from the three concatenated genes (when gaps were not coded as separate characters, 1374 bp alignment) shows that the 35 specimens studied separate into five monophyletic groups corresponding to the five species to which they were assigned based on morphology, and that the molecular markers combined can successfully distinguish charapine species. Considering these five species, *Alloxysta victrix* is the sister taxon to *P. villosa*, and this clade is more closely related to *A. brevis*, with *A. castanea* being more closely related to these three species than *A. ramulifera* (Figure 6A). It is worth highlighting the fact that *P. villosa* being nested within *Alloxysta* is an unexpected result. *Phaenoglyphis villosa* belongs in a different genus, based on morphological differences [32]. The Bayesian tree resulting from the concatenated sequences 1471 bp long with gaps coded as separate partitions uncovered the same relationships between the five species as the tree built with gaps not coded, albeit with lower posterior probability support (Figure 6B).

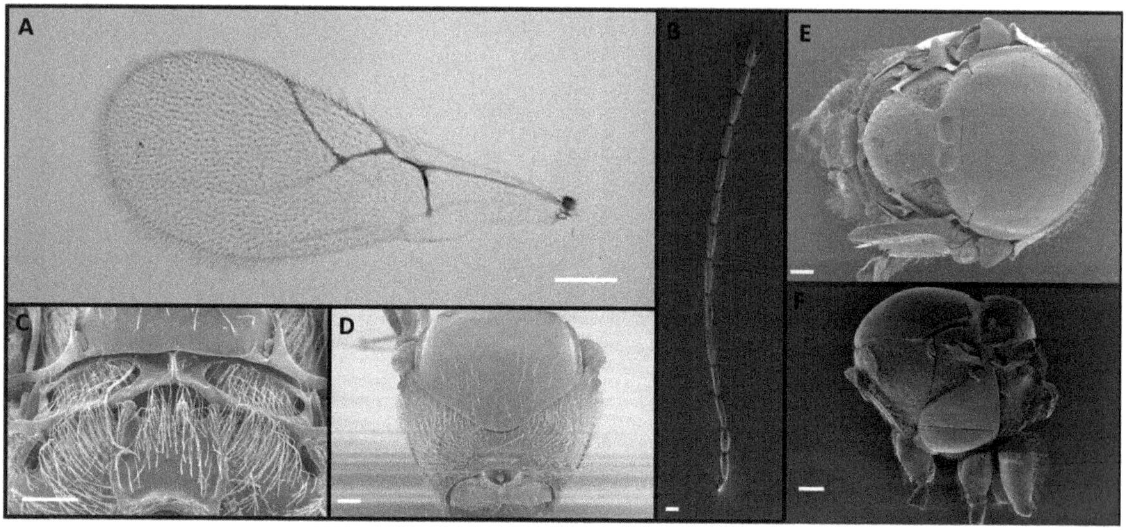

Figure 1. *Phaenoglyphis villosa* (Hartig, 1841), male: (**A**) forewing, with radial cell partially open; (**B**) antenna; (**C**) propodeum; (**D**) pronotum; (**E**) mesoscutum, with scutellar foveae marked with an arrow; (**F**) mesosoma, lateral view, with mesopleural sulcus marked with an arrow. Photographic credit: Jordi Paretas-Martínez. Scale: 50 µm.

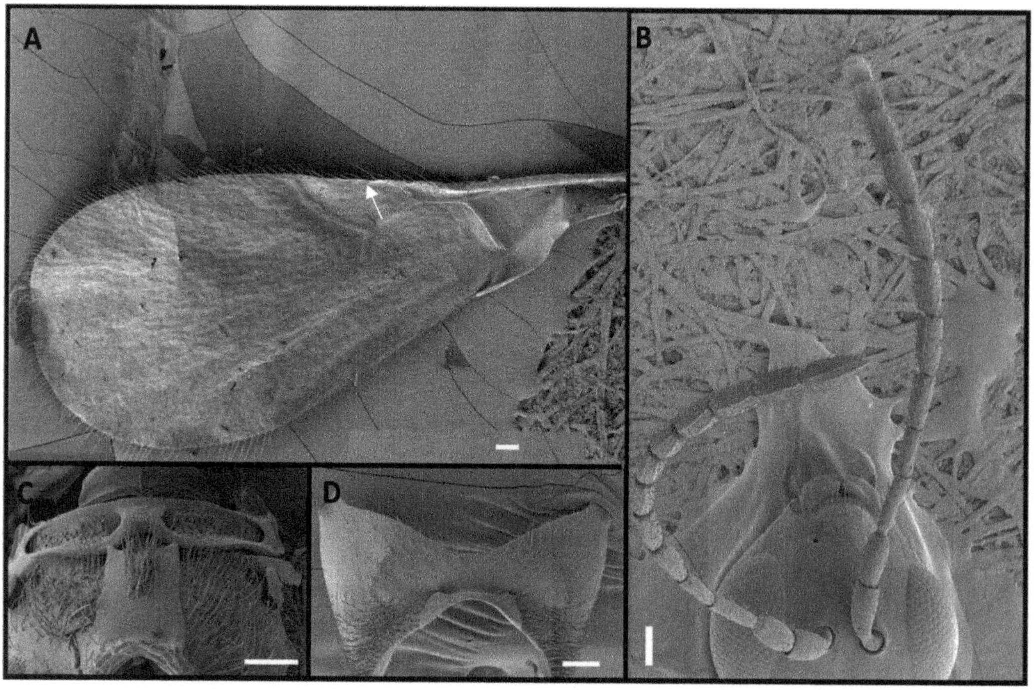

Figure 2. *Alloxysta castanea* (Hartig, 1841), female: (**A**) forewing, with radial cell partially open, marked with an arrow; (**B**) antenna; (**C**) propodeum; (**D**) pronotum. Photographic credit: Mar Ferrer-Suay. Scale: 50 µm.

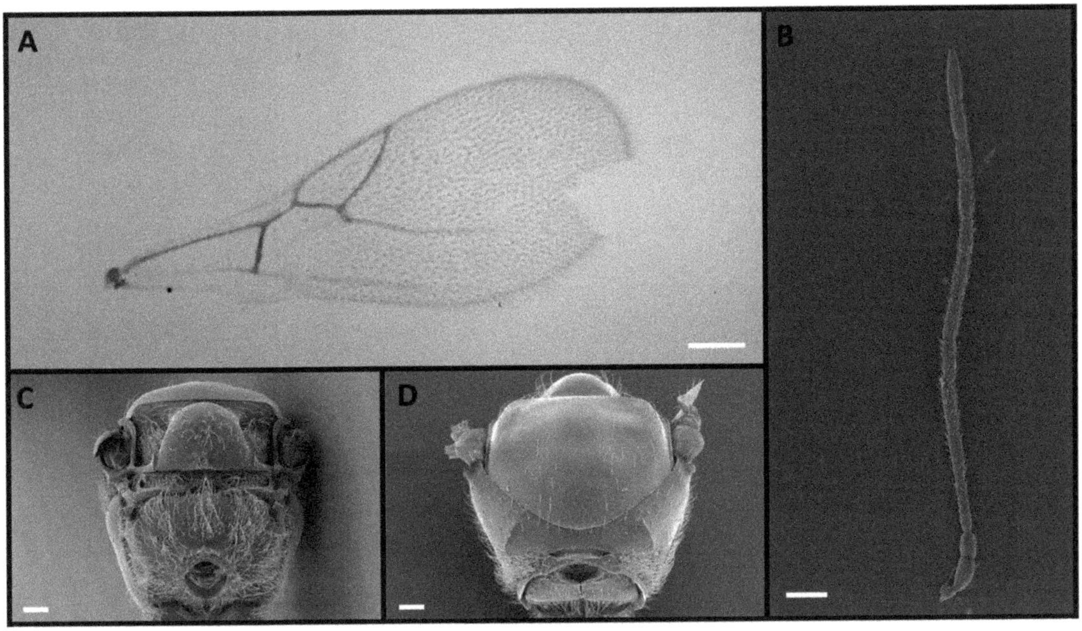

Figure 3. *Alloxysta victrix* (Westwood, 1833), female: (**A**) forewing, with closed radial cell; (**B**) antenna; (**C**) propodeum; (**D**) pronotum. Photographic credit: Jordi Paretas-Martínez. Scale: 50 µm.

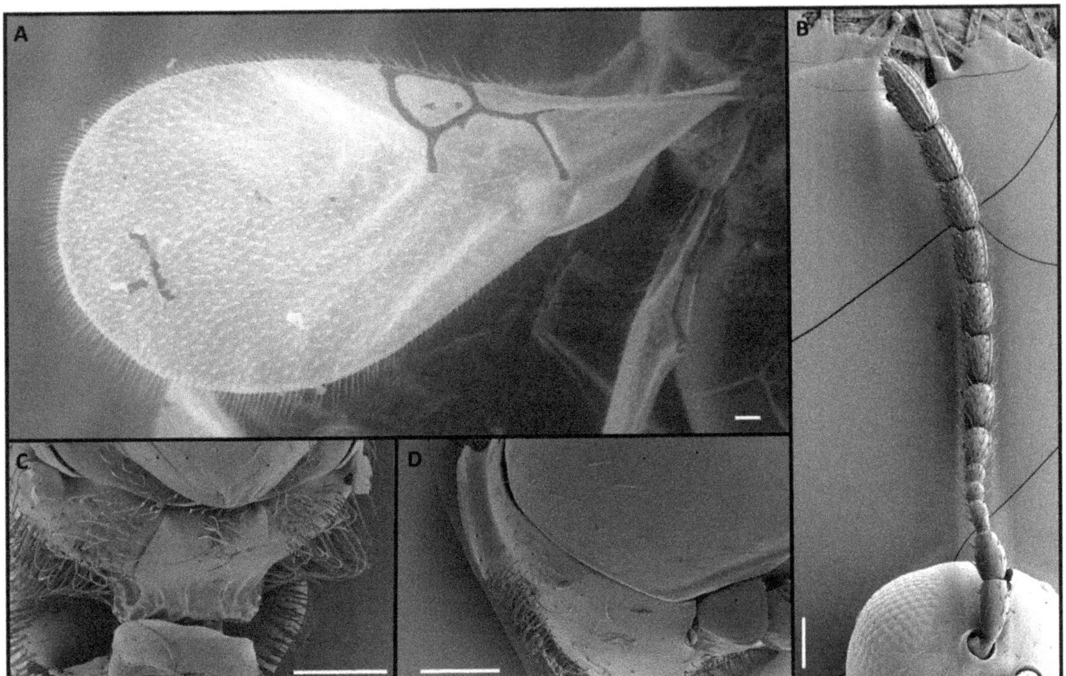

Figure 4. *Alloxysta brevis* (Thomson, 1862), female: (**A**) forewing; (**B**) antenna; (**C**) propodeum; (**D**) pronotum. Photographic credit: Mar Ferrer-Suay. Scale: 50 µm.

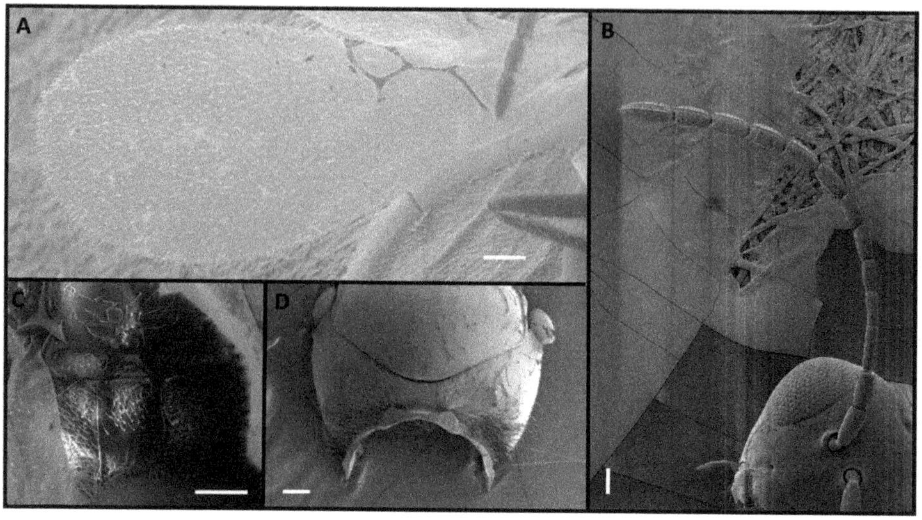

Figure 5. *Alloxysta ramulifera* (Thomson, 1862), female: (**A**) radial cell; (**B**) antenna; (**C**) propodeum; (**D**) pronotum. Photographic credit: Mar Ferrer-Suay. Scale: 50 µm.

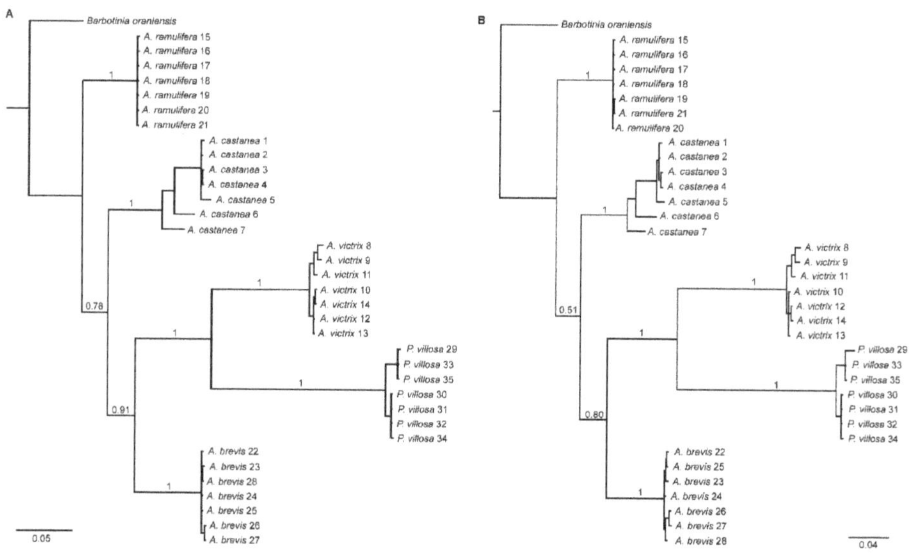

Figure 6. Bayesian phylogenies for four *Alloxysta* species and *Phaenoglyphis villosa*, based on concatenated analysis of three gene regions (**A**) with gaps not coded as separate characters. (**B**) with gaps coded as separate characters. Trees are rooted with *Barbotinia oraniensis* as the outgroup. Support for clades is indicated by the posterior probability values above branches.

Finally, the Bayesian tree built based on COI data only for 239 individuals from a number of countries shows approximately the same branch topographies, with *A. victrix* sister to *P. villosa* with 100% support, and both species sister to a clade formed by *A. brevis* and *A. castanea*. When more wasps from countries other than Belgium are included in the phylogeny, *A. brevis* is no longer monophyletic. Once again, *A. ramulifera* is the most distantly related species in this group (Figure 7).

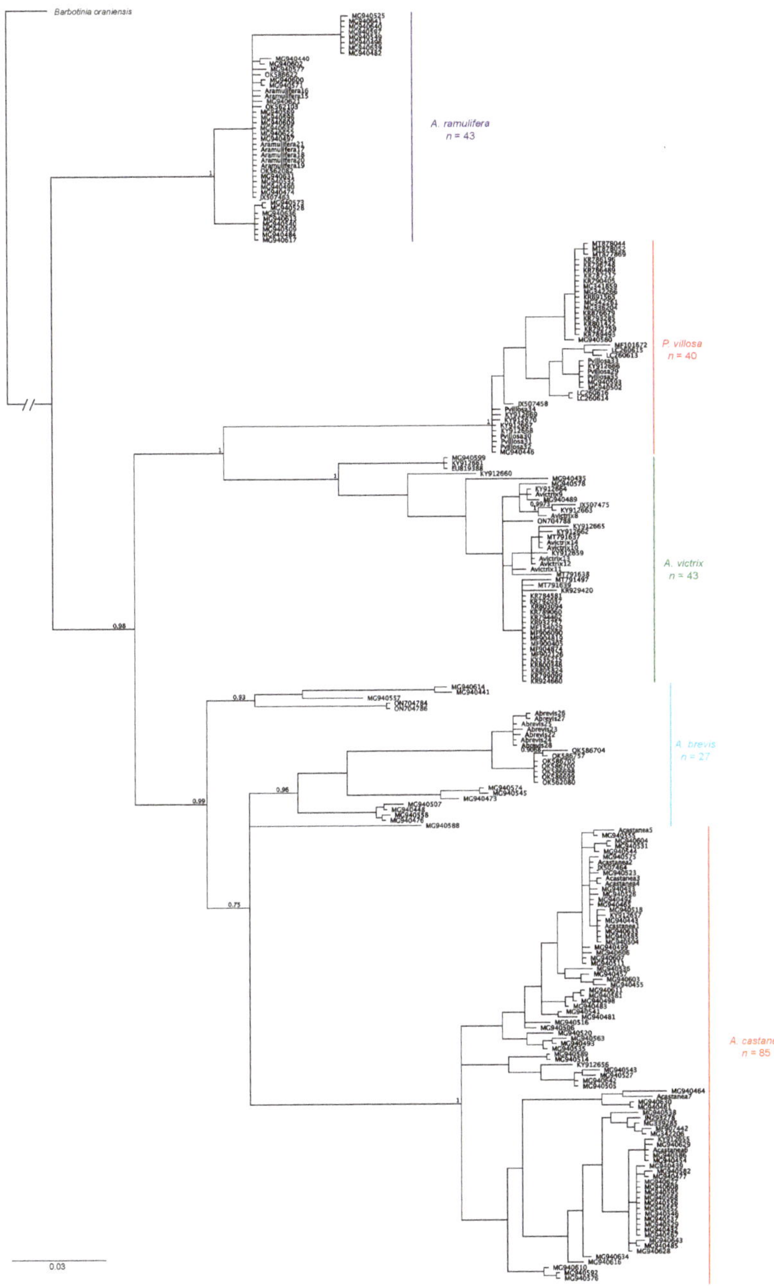

Figure 7. Bayesian phylogeny constructed combining COI sequences from the 35 individual wasps from this study (individuals labeled *Abrevis*, *Acastanea*, *Aramulifera*, *Avictrix*, and *Pvillosa*) plus sequences downloaded from the GenBank database for the five species studied (each individual labeled with their GenBank accession number). Alignment length was 615 bp for 239 specimens. *Barbotinia oraniensis* was selected as the outgroup. Support for clades is indicated by the posterior probability values above branches.

The estimated evolutionary sequence divergence based on Kimura 2-parameter distances shows that *A. castanea* has the highest intraspecific variability in COI sequences whereas *A. ramulifera* shows the least intraspecific variation. When comparing between species, the most divergent species pair is *A. castanea* and *P. villosa*, with *A. castanea* and *A. brevis* constituting the most similar species pair (Table 1).

Table 1. Mean pairwise distance matrix (Kimura 2-parameter distances ± standard deviation) for COI sequences within (on the diagonal) and between (below the diagonal) *Alloxysta* and *Phaenoglyphis* species.

	A. castanea	***A. victrix***	***A. ramulifera***	***A. brevis***	***P. villosa***
A. castanea	0.0247 ± 0.0329				
A. victrix	0.1452 ± 0.0017	0.0176 ± 0.0045			
A. ramulifera	0.1248 ± 0.0008	0.1409 ± 0.0009	0.0022 ± 0.0008		
A. brevis	0.1237 ± 0.0012	0.1367 ± 0.0011	0.1324 ± 0.0025	0.0035 ± 0.0035	
P. villosa	0.1640 ± 0.0012	0.1299 ± 0.0068	0.1431 ± 0.0093	0.1293 ± 0.0012	0.0145 ± 0.0113

4. Discussion

Species of *Alloxysta* are very small and exhibit a smooth, shiny body surface. They are characterized by relatively few diagnostic characters, as is the case for many microhymenopterans. This lack of morphological features makes their identification very difficult, a fact that is compounded by the high number of species within the subfamily, which currently stands at 108 described species [33]. Many papers have been published with the aim of revising the type material of *Alloxysta* and other members of Charipinae [48–54].

A previous study showed that most species limits established within the Charipinae based on morphological features seemed correct [30]. This conclusion was supported by a combined analysis of molecular and morphological information. However, there are still complexes of species in this subfamily that deserve attention, mainly within *Alloxysta*, which is the most speciose and diverse charipine genus. Another study compared brachypterous with fully winged species of *Alloxysta* to determine whether sexual dimorphism of wing length and shape could be used as a character to distinguish species [31]. This analysis was based on morphological characteristics, but the need to test the usefulness of wing dimorphism to establish species limits using breeding and genetic analyses was also discussed. More recently, we employed genetic analyses to support the hypothesis that some brachypterous forms that were previously considered to be separate species are part of fully winged species [55].

One interesting result from this study is the fact that *Phaenoglyphis villosa* was found to be nested within *Alloxysta*, suggesting that it should be placed within the genus *Alloxysta*. This species was originally placed in the genus *Phaenoglyphis* based on morphological characters [32]. A previous molecular phylogeny of *Alloxysta* from Norway included *P. villosa*, but it was considered the outgroup [30]. Our study shows that, *P. villosa* is genetically more closely related to *A. victrix*. Therefore, the molecular data are in discordance with the morphological data. We recommend revising the morphological features used to characterize the genus *Phaenoglyphis* to make sure these features are robust enough and obtaining further molecular data for more *Phaenoglyphis* species, in particular the novel metazoan-level universal single-copy orthologs (metazoan USCOs) which hold much promise for species delimitation and taxonomy [56]. Specimens of *Phaenoglyphis* are usually captured in a much lower proportion than *Alloxysta* specimens in the field, which is the main reason why we know little about this genus. A possible reason for *P. villosa* being nested within *Alloxysta* is that some species within the genus *Phaenoglyphis* have fewer diagnostic features, such as notauli or scutellar foveae, including *P. villosa*, which does not have notauli present, and this is similar to *Alloxysta*. Additionally, *Phaenoglyphis villosa* has radial cells partially open, as well as two other *Phaenoglyphis* species (*P. china*: Ferrer-Suay and Pujade-Villar 2013, and *P. asiatica*: Ferrer-Suay and Pujade-Villar 2013). Finally, cryptic species and complexes of species are very common for Hymenoptera that are parasitic [57], making their study more complicated. The novel relationship uncovered here warrants further exploration in future studies. According to a previous study [58], there is a group of *Phaenoglyphis* species that lack these features, and they are close to other Charipinae species, suggesting an evolutionary tendency towards feature loss and a simpler structure.

The genetic distances between species of *Alloxysta* and *Phaenoglyphis* had not been determined previously. Some studies found that a genetic distance of 8.9% was enough to separate species of the genus *Diplolepis* (Cynipidae) [59] and *Ephedrus* (Braconidae) [60]. For the four *Alloxysta* and the one *Phaenoglyphis* species included in this study, interspecific genetic distances ranged from 12.4% to 16.4%, based on COI data. It is not surprising that *A. castanea* has the highest intraspecific variability given that this species is very abundant in multiple biogeographic regions [34]. Having high haplotype variability might have contributed to the successful colonization of new habitats.

The three molecular markers included in our study, when combined, were useful in differentiating five Charipinae species. Each marker on its own was also able to separate the five species (not shown), although we are cautious not to draw conclusions from phylogenetic trees built based on as few base pairs as we retained after trimming the sequences for ITS2 and 16S.

In conclusion, the phylogenetic relationships of *Alloxysta* and *Phaenoglyphis* are more complex than previously thought, and these genera likely need a deep revision. It is worth noting that when more *Alloxysta* and *Phaenoglyphis* species are included to build the subfamily phylogenetic tree, these other species might fit in between the five species considered in this study. Ideally, future work should increase the number of samples and the number of species, include genomic data (such as USCOs), and reconcile morphology and genetics to help establish the evolution of Charipinae species limits and relationships unambiguously.

Supplementary Materials: The following supporting information can be downloaded at: https://www.mdpi.com/article/10.3390/insects15050354/s1, File S1: Fasta formatted alignment of 35 individual wasps in the Charipinae plus a cynipid gall wasp as outgroup, concatenating three genes, mitochondrial cytochrome oxidase I gene (COI), 16S ribosomal RNA encoding gene sequences (16S rDNA), and nuclear internal transcribed spacer 2 (ITS2 rDNA). Gaps in this alignment are not coded as separate partitions, the length of the alignment is 1374 base pairs; File S2: Fasta formatted alignment of 35 individual wasps in the Charipinae plus a cynipid gall wasp as outgroup, concatenating three genes, mitochondrial cytochrome oxidase I gene (COI), 16S ribosomal RNA encoding gene sequences (16S rDNA), and nuclear internal transcribed spacer 2 (ITS2 rDNA). Gaps are added to the data file as separate partitions, resulting in an alignment of 1471 base pairs; File S3: Fasta formatted alignment of 239 individual wasps in the Charipinae and a cynipid gall wasp as outgroup for mitochondrial cytochrome oxidase I gene (COI). Alignment length is 615 base pairs.

Author Contributions: Conceptualization, M.F.-S.; methodology, and formal analysis, M.B.; writing—original draft preparation, M.F.-S., G.E.H. and M.B.; writing—review and editing, M.F.-S., M.B., G.E.H., E.R. and J.S. All authors have read and agreed to the published version of the manuscript.

Funding: This research was funded by Council of Innovation, Universities, Science and Digital Society (Valencia, Spain): CIGE/2022/158. Contribution by Ehsan Rakhshani was supported by grant IR-UOZ-GR-6230, University of Zabol, Iran.

Data Availability Statement: The original contributions presented in the study are included in the article and Supplementary Materials, further inquiries can be directed to the corresponding author.

Acknowledgments: We would like to thank Wounter Dekonic and Fons Verheyde for providing the material studied here. We are also very grateful to Jordi Paretas-Martínez for taking some of the wasp photographs and for the scientific technical services of the University of Barcelona, and Robin Thomson of University of Minnesota for useful discussion.

Conflicts of Interest: The authors declare no conflict of interest.

References

1. Hassell, M.P. *The Spatial and Temporal Dynamics of Host-Parasitoid Interactions*; Oxford University Press: Oxford, UK, 2000.
2. Murdoch, W.W.; Briggs, C.J. *Nisbet Consumer-Resource Dynamics*; Princeton University Press: Princeton, NJ, USA, 2003.
3. Heimpel, G.E.; Mills, N.J. *Biological Control: Ecology and Applications*; Cambridge University Press: Cambridge, UK, 2017.
4. Ovruski, S.M.; Fidalgo, P. Use of parasitoids (Hym.) in the control of fruit flies (Dip.: Tephritidae) in Argentina: Bibliographic review (1937–1991). *IOBC-WPRS Bull.* **1994**, *17*, 84–92.
5. Rathman, R.J.; Johnson, M.W.; Tabashnik, B.E. Production of *Ganaspidium utilis* (Hymenoptera: Eucoilidae) for biological control of *Liriomyza* spp. (Diptera: Agromyzidae). *Biol. Control* **1991**, *1*, 256–260. [CrossRef]
6. Villeneuve, F.; Trottin Caudal, Y. La protection integree des cultures legumieres en France: Acquis et perspectives. *Infos* **1997**, *135*, 40–44.
7. Heraty, J. Molecular systematics, Chalcidoidea and biological control. In *Genetics, Evolution and Biological Control*; Ehler, L.E., Sforza, R., Mateille, T., Eds.; CABI Publishing: Oxon, UK, 2003; pp. 39–72.
8. Heraty, J.M.; Woolley, J.B.; Hopper, K.R.; Hawks, D.L.; Kim, J.W.; Buffington, M. Molecular phylogenetics and reproductive incompatibility in a complex of cryptic species of aphid parasitoids. *Mol. Phylogenet. Evol.* **2007**, *45*, 480–493. [CrossRef] [PubMed]
9. Desneux, N.; Stary, P.; Delebeque, C.J.; Gariepy, T.D.; Barta, R.J.; Hoelmer, K.A.; Heimpel, G.E. Cryptic species of parasitoids attacking the soybean aphid (Hemiptera: Aphididae) in Asia: *Binodoxys communis* (Gahan) and *Binodoxys koreanus* (Hymenoptera: Braconidae: Aphidiinae). *Ann. Entomol. Soc. Am.* **2009**, *102*, 925–936. [CrossRef]
10. Lue, C.H.; Abram, P.K.; Hrcek, J.; Buffington, M.L.; Staniszenko, P.P.A. Metabarcoding and applied ecology with hyper-diverse organisms: Recommendations for biological control research. *Mol. Ecol.* **2024**, *32*, 6461–6473. [CrossRef] [PubMed]
11. Menke, A.S.; Evenhuis, H.H. North American Charipidae: Key to genera, nomenclature, species checklists, and a new species of *Dilyta* Förster (Hymenoptera: Cynipoidea). *Proc. Entomol. Soc. Wash.* **1991**, *93*, 136–158.
12. Gomez-Marco, F.; Urbaneja, A.; Jaques, J.A.; Rugman-Jones, P.F.; Stouthamer, R.; Tena, A. Untangling the aphid-parasitoid food web in citrus: Can hyperparasitoids disrupt biological control? *Biol. Control* **2015**, *81*, 111–121. [CrossRef]
13. Villa-Ayala, P.; Sánchez-Rivera, G.; Rodríguez-Vélez, B.; Castrejón-Ayala, F. Parasitoids and Hyperparasitoids Associated with the Sugarcane Aphid, *Melanaphis sacchari* (Zehntner), in Morelos, Mexico. *Soc. Southwest. Entomol.* **2020**, *45*, 563–566. [CrossRef]
14. Zhang, Y.; Zhao, M.C.; Cheng, J.; Liu, S.; Yuan, H.B. Population dynamics and species composition of maize field parasitoids attacking aphids in northeastern China. *PLoS ONE* **2020**, *15*, e0241530. [CrossRef]
15. Bandyan, S.K.; Peters, R.S.; Kadir, N.B.; Ferrer-Suay, M.; Kirchner, W.H. A survey of aphid parasitoids and hyperparasitoids (Hymenoptera) on six crops in the Kurdistan Region of Iraq. *J. Hymenopt. Res.* **2021**, *81*, 9–21. [CrossRef]
16. Satar, S.; Kavallieratos, N.G.; Tüfekli, M.; Satar, G.; Athanassiou, C.G.; Papanikolaou, N.E.; Karacaoglu, M.; Özdemir, I.; Stary, P. *Capsella bursa-pastoris* Is a Key Overwintering Plant for Aphids in the Mediterranean Region. *Insects* **2021**, *12*, 744. [CrossRef]
17. Burton, R.L.; Starks, K.J. Control of a Primary Parasite of the Greenbug with a Secondary Parasite in Greenhouse Screening for Plant Resistance. *J. Econ. Entomol.* **1977**, *70*, 219–220. [CrossRef]
18. Nagasaka, K.; Takahasi, N.; Okabayashi, T. Impact of secondary parasitism on *Aphidius colemani* in the banker plant system on aphid control in commercial greenhouses in Kochi, Japan. *Appl. Entomol. Zool.* **2010**, *45*, 541–550. [CrossRef]
19. Ratnasingham, S.; Hebert, P.D.N. BOLD: The Barcode of Life Data System (http://www.barcodinglife.org). *Mol. Ecol. Notes* **2007**, *7*, 355–364. [CrossRef] [PubMed]
20. Rodrigues, M.A.; Morelli, K.A.; Jansen, A.M. Cytochrome c oxidase subunit 1 gene as a DNA barcode for discriminating *Trypanosoma cruzi* DTUs and closely related species. *Parasites Vectors* **2017**, *10*, 488. [CrossRef]
21. Hebert, P.D.N.; Cywinska, A.; Ball, S.L.; de Waard, J.R. Biological identifications through DNA barcodes. *Proc. Biol. Sci.* **2003**, *270*, 313–321. [CrossRef] [PubMed]
22. Moritz, C.; Cicero, C. DNA barcoding: Promise and pitfalls. *PLos Biol.* **2004**, *2*, e354. [CrossRef] [PubMed]
23. Dasmahapatra, K.K.; Mallet, J. DNA barcodes: Recent successes and future prospects. *Heredity* **2006**, *97*, 254–255. [CrossRef]
24. Dowton, M.; Austin, A.D. Molecular phylogeny of the insect order Hymenoptera: Apocritan relationships. *Proc. Natl. Acad. Sci. USA* **1994**, *91*, 9911–9915. [CrossRef]
25. Van Veen, F.F.; Belshaw, R.O.; Godfray, H.C.J. The value of the ITS2 region for the identification of species boundaries between *Alloxysta* hyperparasitoids (Hymenoptera: Charipidae) of aphids. *Eur. J. Entomol.* **2003**, *100*, 449–454. [CrossRef]
26. Wagener, B.; Reineke, A.; Löhr, B.; Zebitz, C.P. Phylogenetic study of Diadegma species (Hymenoptera: Ichneumonidae) inferred from analysis of mitochondrial and nuclear DNA sequences. *Biol. Control* **2006**, *37*, 131–140. [CrossRef]
27. Rahimi, S.; Hajizadeh, J.; Hosseini, R.; Sohani, M.M. Evaluation of COI and ITS2 sequences ability to distinguish Lysiphlebus fabarum and Lysiphlebus confusus (Hymenoptera: Aphidiidae). *J. Entomol. Soc. Iran* **2012**, *31*, 53–66.
28. Ptaszyńska, A.A.; Latoch, P.; Hurd, P.J.; Polaszek, A.; Michalska-Madej, J.; Grochowalski, Ł.; Strapagiel, D.; Gnat, S.; Załuski, D.; Gancarz, M.; et al. Amplicon Sequencing of Variable 16S rRNA from Bacteria and ITS2 Regions from Fungi and Plants, Reveals Honeybee Susceptibility to Diseases Results from Their Forage Availability under Anthropogenic Landscapes. *Pathogens* **2021**, *10*, 381. [CrossRef] [PubMed]
29. Chen, Y.; Pike, K.S.; Greenstone, M.H.; Shufran, K.A. Molecular Markers for Identification of the Hyperparasitoids *Dendrocerus carpenteri* and Alloxysta xanthopsis in *Lysiphlebus testaceipes* Parasitizing Cereal Aphids. *BioControl* **2006**, *51*, 183–194. [CrossRef] [PubMed]

30. Ferrer-Suay, M.; Staverløkk, A.; Selfa, J.; Pujade-Villar, J.; Naik, S.; Ekrem, T. Nuclear and mitochondrial markers suggest new species boundaries in *Alloxysta* (Hymenoptera: Cynipoidea: Figitidae). *Arthropod Syst. Phylogeny* **2018**, *76*, 463–473. [CrossRef]
31. Ferrer-Suay, M.; Selfa, J.; Cuesta-Porta, V.; Pujade-Villar, J. Intraspecific variation in the morphology of *Alloxysta fracticornis* (Thomson, 1862) (Hymenoptera: Figitidae: Charipinae). *J. Nat. Hist.* **2022**, *56*, 707–717. [CrossRef]
32. Ferrer-Suay, M.; Selfa, J.; Pujade-Villar, J. A review of the subfamily Charipinae (Hymenoptera: Cynipoidea: Figitidae): Hyperparasitoids potentially affecting the biological control of aphids. *Ann. Soc. Entomol. Fr.* **2021**, *57*, 481–498. [CrossRef]
33. Ferrer-Suay, M.; Selfa, J.; Pujade-Villar, J. Overview of the current status of Charipinae (Hymenoptera, Cynipoidea, Figitidae): Taxonomy, distribution, and diversity. *Zootaxa* **2023**, *5339*, 001–039. [CrossRef]
34. Ferrer-Suay, M.; Paretas-Martinez, J.; Selfa, J.; Pujade-Villar, J. Taxonomic and synonymic world catalogue of the Charipinae and notes about this subfamily (Hymenoptera: Cynipoidea: Figitidae). *Zootaxa* **2012**, *3376*, 1–92. [CrossRef]
35. Ferrer-Suay, M.; Janković, M.; Selfa, J.; van Veen, F.; Tomanović, Z.; Kos, K.; Rakhshani, E.; Pujade-Villar, J. Qualitative analysis of aphid and primary parasitoid trophic relations of genus *Alloxysta* (Hymenoptera: Cynipoidea: Figitidae: Charipinae). *Environ. Entomol.* **2014**, *43*, 1485–1495. [CrossRef]
36. Ferrer-Suay, M.; Vogel, J.; Peters, R.S.; Selfa, J. Characterizing the northwestern European species of *Phaenoglyphis* Förster, 1869 (Hymenoptera: Figitidae) with novel insights from CO1 barcoding data. *Biodivers Data J.* **2024**, *in press*.
37. Ferrer-Suay, M.; Selfa, J.; Pujade-Villar, J. A proposal of Charipinae's keys for the world fauna (Hymenoptera: Cynipoidea: Figitidae). *Zookeys* **2019**, *822*, 79–139. [CrossRef] [PubMed]
38. Folmer, O.; Black, M.; Hoeh, W.; Lutz, R.; Vrijenhoek, R. DNA primers for amplification of mitochondrial cytochrome c oxidase subunit I from diverse metazoan invertebrates. *Mol. Mar. Biol. Biotechnol.* **1994**, *3*, 294–299. [PubMed]
39. Whitfield, J.B. Molecular and morphological data suggest a single origin of the polydnaviruses among braconid wasps. *Naturwissenschaften* **1997**, *84*, 502–507. [CrossRef]
40. Borschsenius, F. FastGap 1.2. Department of Biosciences, Aarhus University, Denmark. 2009. Available online: http://www.aubot.dk/FastGap_home.htm (accessed on 1 March 2024).
41. Simmons, M.P.; Ochoterena, H. Gaps as characters in sequence-based phylogenetic analyses. *Syst. Biol.* **2000**, *49*, 369–381. [CrossRef] [PubMed]
42. Huelsenbeck, J.P.; Ronquist, F. MrBayes: Bayesian inference of phylogenetic trees. *Bioinformatics* **2001**, *17*, 754–755. [CrossRef]
43. Akaike, H. Information theory and an extension of the maximum likelihood principle. In *Proceedings of the 2nd International Symposium*; Petrov, B.N., Csaki, F., Eds.; Akademia Kiado: Budapest, Hungary, 1973; pp. 267–281.
44. Posada, D.; Crandall, K.A. Modeltest: Testing the model of DNA substitution. *Bioinformatics* **1998**, *14*, 817–818. [CrossRef]
45. Swofford, D.L. *PAUP*: Phylogenetic Analysis Using Parsimony (*and Other Methods)*; Version 4.0b10; Sinauer Associates: Sunderland, MA, USA, 2002.
46. Kimura, M. A simple method for estimating evolutionary rate of base substitutions through comparative studies of nucleotide sequences. *J. Mol. Evol.* **1980**, *16*, 111–120. [CrossRef]
47. Tamura, K.; Stecher, G.; Kumar, S. MEGA11: Molecular Evolutionary Genetics Analysis version 11. *Mol. Biol. Evol.* **2021**, *38*, 3022–3027. [CrossRef]
48. Ferrer-Suay, M.; Selfa, J.; Seco-Fernández, M.V.; Melika, G.; Alipour, A.; Rakhshani, E.; Talebi, A.A.; Pujade-Villar, J. A contribution to the knowledge of Charipinae (Hymenoptera: Cynipoidea: Figitidae) associated with aphids from Iran, including new records. *North-West. J. Zool.* **2013**, *9*, 30–44.
49. Ferrer-Suay, M.; Paretas-Martínez, J.; Pujade-Villar, J. Revision of *Apocharips* Fergusson (Hymenopterza: Figitidae: Charipinae) with description of three new species from Colombia. *Zootaxa* **2013**, *3646*, 487–500. [CrossRef] [PubMed]
50. Ferrer-Suay, M.; Selfa, J.; Pujade-Villar, J. Revision of Thomson and Zetterstedt collections of *Alloxysta* genus deposited in Lund Museum of Zoology (Sweden). *Entomol. Tidskr.* **2013**, *134*, 77–102.
51. Ferrer-Suay, M.; Selfa, J.; Notton, D.; Pujade-Villar, J. Revision of the types of species of *Alloxysta* Förster, 1869 described by Cameron and Fergusson (Hymenoptera: Figitidae: Charipinae) deposited in the Natural History Museum (London) including a key to the fauna of Great Britain. *Eur. J. Taxon.* **2013**, *53*, 1–27.
52. Ferrer-Suay, M.; Selfa, J.; Seco, M.V.; Pujade-Villar, J. Revision of Hellén types of *Alloxysta* Förster (Hymenoptera: Figitidae, Charipinae). *Entomol. Fenn.* **2014**, *25*, 86–101. [CrossRef]
53. Ferrer-Suay, M.; Selfa, J.; Pujade-Villar, J. Review of the Hartig type collection of *Alloxysta* (Hymenoptera: Figitidae: Charipinae) and other *Alloxysta* material deposited in the Zoologische Staatssammlung Museum (Munich). *Fragm. Faun.* **2014**, *57*, 75–116. [CrossRef]
54. Ferrer-Suay, M.; Selfa, J.; Pujade-Villar, J. New contribution to the knowledge of the genus *Alloxysta* (Hymenoptera: Cynipoidea: Figitidae): Revision of some type material. *Ann. Naturhist. Mus. Wien Ser. B* **2015**, *117*, 23–36.
55. Casiraghi, A.; Dregni, J.S.; Pérez Hidalgo, N.; Heimpel, G.E.; Selfa, J.; Ferrer-Suay, M. Brachyptery analysis in *Alloxysta* (Hymenoptera: Figitidae). *Proc. Entomol. Soc. Wash.* **2022**, *124*, 1–12. [CrossRef]
56. Dietz, L.; Eberle, J.; Mayer, C.; Kukowka, S.; Bohacz, C.; Baur, H.; Espeland, M.; Huber, B.A.; Hutter, C.; Mengual, X.; et al. Standardized nuclear markers improve and homogenize species delimitation in Metazoa. *Methods Ecol. Evol.* **2023**, *14*, 543–555. [CrossRef]
57. Gokhman, V.E. Chromosome Study of the Hymenoptera: History, Current State, Perspectives. *Biol. Bull. Rev.* **2023**, *13*, 247–257. [CrossRef]

58. Ferrer-Suay, M.; Selfa, J.; Mata-Casanova, N.; Hidalgo, N.P.; Pujade-Villar, J. Worldwide revision of the genus *Phaenoglyphis* Förster, 1869 (Hymenoptera, Cynipoidea, Figitidae, Charipinae). *Insect Syst. Evol.* **2019**, *50*, 235–296. [CrossRef]
59. Zhang, Y.; Gates, M.; Shorthouse, J. Testing species limits of Eurytomidae (Hymenoptera) associated with galls induced by *Diplolepis* (Hymenoptera: Cynipidae) in Canada using an integrative approach. *Can. Entomol.* **2014**, *146*, 321–334. [CrossRef]
60. Kocic, K.; Petrovic, A.; Ckrkic, J.; Mitrovic, M.; Tomanovic, Z. Phylogenetic relationships and subgeneric classification of European *Ephedrus* species (Hymenoptera, Braconidae, Aphidiinae). *Zookeys* **2019**, *878*, 1–22. [CrossRef] [PubMed]

insects

MDPI

Article

One Step beyond Species Description: Unveiling a Fine-Scale Diversity within the Genus *Dzhanokmenia* Kostjukov (Hymenoptera: Eulophidae) †

Bolormaa Ganbaatar [1,2], Qin Li [3,4], Ouyan Xi [3,4], Huanxi Cao [5,*] and Chaodong Zhu [1,2]

1 Key Laboratory of Zoological Systematics and Evolution, Institute of Zoology, Chinese Academy of Sciences, Beijing 100101, China; bolor88maa@gmail.com (B.G.); zhucd@ioz.ac.cn (C.Z.)
2 College of Life Sciences, University of Chinese Academy of Sciences, Beijing 101408, China
3 College of Life Science and Technology, Xinjiang University, Urumqi 830046, China; liqptero@163.com (Q.L.); xoy@stu.xju.edu.cn (O.X.)
4 Xinjiang Key Laboratory of Biological Resources and Genetic Engineering, Urumqi 830046, China
5 National Animal Collection Resource Center, Institute of Zoology, Chinese Academy of Sciences, Beijing 100101, China
* Correspondence: caohuanxi@ioz.ac.cn
† urn:lsid:zoobank.org:pub:CBEBCA76-C300-41E0-87F2-B535D91783C9.

Simple Summary: Chalcidoidea (chalcid wasps) are one of the most numerically, structurally, and biologically diverse groups in Hymenoptera. However, the conserved morphology coupled with a high intraspecific variability in certain key characters hinders research on chalcids. In this study, DNA barcoding with the COI gene was used to calibrate intra- and interspecific variation in diagnostic characters within the genus *Dzhanokmenia*. In addition, the nuclear gene, 28S D2, was used to infer a phylogeny to better understand the definition of *Dzhanokmenia* and its relationships to potentially close genera associated with *Tetrastichus*. A DNA barcode library that includes eight species was established for *Dzhanokmenia*. A new species, *D. brevifunis* Ganbaatar & Cao sp. nov., was described and illustrated. The results show that *Dzhanokmenia* is closely related to *Baryscapus* and suggest that the species diversity of *Dzhanokmenia* is underestimated. This study lays a foundation for further research on the diversity and classification of *Dzhanokmenia*.

Abstract: Although Chalcidoidea is one of the megadiverse superfamilies in Hymenoptera, numerous species are still being discovered and described. However, the difficulties in delimiting intra- and interspecific variation hinder this process. In this study, DNA barcoding methods using the COI gene were employed to investigate the morphological variation within *Dzhanokmenia* Kostjukov, 1977. The nuclear locus, 28S D2, was used to infer a phylogeny to gain an understanding of the relationship of *Dzhanokmenia* with other potentially close genera. Through a preliminary DNA barcode library established here, including eight species, we calibrated the intraspecific variation in certain diagnostic characters for the new species described here, *D. brevifunis* Ganbaatar & Cao sp. nov. Maximum likelihood results show that *Dzhanokmenia* is clustered with the genera associated with *Tetrastichus*, such as *Chaenotetrastichus* Graham, 1987, *Baryscapus* Förster, 1856, *Tetrastichus* Haliday, 1844, and *Oomyzus* Rondani, 1870 involved in this study. Our results indicate that the species diversity of *Dzhanokmenia* is understudied and tentatively confirm that *Dzhanokmenia* has a potential close relationship with *Baryscapus*. Along with the DNA barcode library, the referenced phylogeny datasets improve the understanding of the systematic position of *Dzhanokmenia* within the subfamily Tetrastichinae and the definition of this genus in terms of morphology, thereby facilitating species delimitation, discovery, and description within *Dzhanokmenia*.

Keywords: chalcids; taxonomy; DNA barcoding; species delimitation; intra- and interspecific variation

1. Introduction

Chalcidoidea, commonly known as chalcid wasps, represent one of the most diverse superfamilies within Hymenoptera, encompassing high species abundance and biological diversity. More than 22,000 species have been documented in Chalcidoidea, yet the total species count is estimated to surpass 500,000 [1]. Currently, Eulophidae are the largest family in terms of species diversity, comprising over 6000 valid species. Eulophidae are typically divided into four subfamilies, as confirmed by molecular data in Gauthier et al. [2]: Entedoninae, Entiinae (formerly known as Euderinae), Eulophinae, and Tetrastichinae. In 2011, the Opheliminae, which were previously recognized as a tribe within the Entedoninae, were elevated to subfamily status in a phylogenetic analysis that integrated molecular and morphological data [3]. This classification has been further supported by subsequent molecular studies [4,5]. Of the five subfamilies, the Tetrastichinae is the largest subfamily in terms of both the number of genera and species, with over 100 genera and more than 2000 species [1,6]. Two prominent genera in particular, *Aprostocetus* Westwood, 1833 and *Tetrastichus*, account for more than half of the total number of species within this subfamily [1].

Dzhanokmenia is one of nineteen genera within the subfamily Tetrastichinae that contains more than ten species. Until this study, a total of fifteen described species were known for this genus, all of which originated from Asia. Based on morphology and trophic associations, Kostjukov [7] established and treated *Dzhanokmenia* as one of seventeen subgenera within the genus *Tetrastichus* and designated *Tetrastichus bibikovae* Dzhanokmen, 1971 as the type species for *Dzhanokmenia*. Later, Kostjukov [8,9] described an additional six species and provided a key to the seven species of this subgenus. In a taxonomic revision of the European genera of Tetrastichinae, Graham [10] elevated *Dzhanokmenia* to generic status and considered this genus to be closely related to *Baryscapus*, especially the *evonymellae* group of *Baryscapus*, to which most species of this genus belong.

Eight further species were subsequently described within the genus. One was documented from Turkmenistan [11], two were found in the Asiatic region of Russia [12,13], and the remaining five were described from China [14,15]. To date, all fifteen known species of *Dzhanokmenia* originate from the arid and semi-arid regions of Asia, including China, Kazakhstan, southern Russia, and Turkmenistan. In addition to the five species described from China, two species, *D. demakovi* (Kostjukov, 1978) and *D. nikolskajae* (Kostjukov, 1984), have been reported from Xinjiang, China [16]. The host information for *Dzhanokmenia* has long remained unknown until recently, when *D. stefaniolae* Li, Wang & Hu, 2021, was documented to emerge from stem galls induced by *Stefaniola* sp. (Diptera: Cecidomyiidae) on *Haloxylon ammodendron* (C. A. Mey, 1829) (Chenopodiaceae) [15]. This discovery suggests that *Dzhanokmenia* species are associated with gall-forming insects inhabiting *Haloxylon* spp. (Tamaricaceae) and potentially with insects inhabiting *Tamarix* spp. (Tamaricaceae) as well [14,15].

Although considerable progress has been made in the systematics of the Chalcidoidea [3–5,17–21], fundamental taxonomic tasks such as species inventories, identifications, and descriptions remain essential for a comprehensive understanding of the biodiversity of Chalcidoidea [22–25]. Despite these taxonomic advances in the study of *Dzhanokmenia* research, the species diversity within this genus may be underestimated, given the wide distribution of its insect and plant hosts, particularly in arid and semi-arid regions. Limited knowledge of the relationships between *Dzhanokmenia* species and of intra- and interspecific variation hinder species discoveries in this genus. Furthermore, despite its elevation from a subgenus of *Tetrastichus* to a valid genus, much remains unknown about the relationships between *Dzhanokmenia* and genera that may be closely related to *Tetrastichus*.

One of the objectives of this study is to describe a new species of *Dzhanokmenia* collected in Qinghai and Xinjiang, China, and another objective is to investigate the phylogeny of *Dzhanokmenia* and closely related genera. In this study, the COI barcodes were generated to investigate intra- and interspecific variation within *Dzhanokmenia*. Additionally, the 28S

D2 gene of eight *Dzhanokmenia* species recorded from China was used to preliminarily discuss the relationship of *Dzhanokmenia* with other genera associated with *Tetrastichus*, by combining a filtered DNA matrix provided by Cao et al. [25]. Our DNA barcoding analysis revealed an additional species closely allied to *D. brevifunis*, suggesting that the species diversity, or at least the genetic diversity, of *Dzhanokmenia* may not yet be adequately explored. Phylogenetic analysis based on the 28S D2 gene confirmed the hypothesis that *Dzhanokmenia* may be closely related to *Baryscapus*.

2. Materials and Methods

2.1. Samples Collection

The specimens of the new species were collected in 2009, 2013, and 2017 in Xinjiang and Qinghai, China, using sweep nets and Malaise traps. Upon arrival at the laboratory in the Institute of Zoology, Chinese Academy of Sciences (IOZCAS), the Eulophidae specimens were separated from the bulk insect samples and preserved in 95–100% ethanol for subsequent analyses. The specimens of *Dzhanokmenia karamayica* Li, Wang & Zhu, 2016, and *D. gobica* Li, Wang & Zhu, 2016, were collected by sweeping herbaceous plants or shrubs in the semi-arid regions of Xinjiang, China, as documented in Li et al. [13]. Subsequently, the *Dzhanokmenia* specimens were sorted out in the laboratory at Xinjiang University and preserved in 95–100% ethanol before being transferred to the laboratory at IOZCAS for subsequent sequencing.

2.2. Specimen Preparations

The specimens used for morphological studies were dried with a Leica EM CPD300 automated critical point dryer (Leica Microsystems, Wetzlar, Germany). Subsequently, some specimens were mounted on cards, while others were dissected into head, mesosoma, and metasoma to obtain micrographs using a scanning electronic microscope (SEM). Specimens were examined using a Nikon SMZ 1500 stereomicroscope equipped with a 10 mm ocular grid with 100 divisions. The habitus images were taken with a Nikon D7000 digital camera connected to the stereomicroscope. The prepared parts used for the SEM were coated with gold using a Leica EM SCD050 super cool sputter coater. Micrographs were taken with an FEI Quanta 450 environmental scanning electron microscope. All color images were stacked using Helicon Focus software 8.2.2.

The terminology follows Gibson [26], except for substituting "scutellum" with "mesoscutellum". Abbreviations are as follows: F1–F3, funiculars 1–3; Gtn, gastral tergite number; MLM, midlobe of mesoscutum; MV, marginal vein; OOL, the shortest distance between an eye and a posterior ocellus; POL, the shortest distance between the posterior ocelli; SMV, submarginal vein. The type specimens of the new species described in this study are deposited in the National Animal Collection Resource Center of IOZCAS.

2.3. DNA Extraction, Amplification and Sequence Editing

Fifty-two specimens of *Dzhanokmenia* from Qinghai and Xinjiang, China, were processed for whole-genomic DNA extraction using the DNeasy Blood & Tissue Kit (Qiagen GmbH, Hilden, Germany), following the manufacturer's instructions (see successful sequencing specimens in Table 1). The specimens of *Dzhanokmenia* selected for the DNA extraction of the new species were chosen to capture all observed variation. To elucidate the phylogenetic relationship between *Dzhanokmenia* and *Baryscapus*, a representative species of *Baryscapus* was also chosen for DNA extraction. Subsequently, the barcoding region of the COI gene was generated for all specimens. Additionally, a specimen of *Tetrastichus howardi* (Olliff, 1893) was included in the amplification of the COI gene. Thirty-six specimens representing all analyzed species of *Dzhanokmenia* were selected to generate the 28S D2 region for phylogenetic analyses. All primers used in this study are listed in Table 2. The PCR reactions for COI were performed by following the protocol described by Huangfu et al. [27], while the reactions for 28S D2 were based on the protocol established by Cao et al. [25].

The raw sequences were assembled, edited, and aligned using BioEdit version 7.0.9.0 [28]. The COI matrix was translated into amino acids in MEGA7.0 [29] to identify any stop codons. In addition to these newly generated sequences, we downloaded publicly available sequences from GenBank (Table S1), as generated or utilized in Cao et al. [25]. Three DNA matrices were generated for subsequent maximum likelihood (ML) analyses. First, the generated 28S sequences were combined and aligned with a filtered matrix of 28S sequences provided by Cao et al. [25]. Identical sequences were then merged into one to form MatrixI. Second, based on the ML tree on MatrixI (Figure S1), MatrixII was refined from MatrixI to eliminate distant sequences from *Dzhanokmenia* according to the ML result of MatrixI, while retaining *Styotrichia pisoxylona* Cao & Zhu, 2024 as an outgroup. Lastly, the aligned COI and 28S sequences for specimens that had been successfully sequenced for both genes were concatenated, resulting in MatrixIII, which enables subsequent phylogenetic analyses based on two genes.

The voucher specimens used to generate the molecular data in this study have been deposited at IOZCAS. The DNA sequences obtained in this study have been submitted to GenBase, and the accession numbers can be found in Table 1. In addition, the information of other publicly available sequences used in the above three matrices has been provided in Table S1.

Table 1. List of specimens used in molecular analyses, with GenBase accession numbers.

Voucher Number	Species	Sex	Locality	COI	28S
CHX_DZH_117	*D. brevifunis*	female	Xinjiang, Aletai	C_AA071359.1	C_AA071403.1
CHX_DZH_118	*D.* cf. *brevifunis*	female	Xinjiang, Aletai	C_AA071360.1	C_AA071404.1
CHX_DZH_119	*D. brevifunis*	female	Xinjiang, Aletai	C_AA071361.1	C_AA071405.1
CHX_BAR_177	*Baryscapus* sp.	female	Beijing	C_AA071358.1	C_AA071402.1
CHX_DZH_737	*D. brevifunis*	female	Qinghai, Ge-Ermu	C_AA071362.1	C_AA071406.1
CHX_DZH_738	*D. brevifunis*	female	Qinghai, Ge-Ermu	C_AA071363.1	C_AA071407.1
CHX_DZH_739	*D. brevifunis*	female	Qinghai, Ge-Ermu	C_AA071364.1	C_AA071408.1
CHX_DZH_740	*D. brevifunis*	female	Qinghai, Ge-Ermu	C_AA071365.1	C_AA071409.1
CHX_DZH_741	*D. brevifunis*	female	Qinghai, Ge-Ermu	C_AA071366.1	C_AA071410.1
CHX_DZH_742	*D. brevifunis*	female	Qinghai, Ge-Ermu	C_AA071367.1	C_AA071411.1
CHX_DZH_743	*D. brevifunis*	female	Xinjiang, Tulufan	C_AA071368.1	C_AA071412.1
CHX_DZH_744	*D. brevifunis*	female	Xinjiang, Tulufan	C_AA071369.1	C_AA071413.1
CHX_DZH_747	*D. brevifunis*	female	Xinjiang, Tulufan	C_AA071370.1	C_AA071414.1
CHX_DZH_748	*D. brevifunis*	female	Xinjiang, Tulufan	C_AA071371.1	C_AA071415.1
CHX_DZH_888	*D.* cf. *brevifunis*	female	Xinjiang, Aletai	C_AA071372.1	C_AA071416.1
CHX_DZH_890	*D. brevifunis*	female	Xinjiang, Aletai	C_AA071373.1	-
CHX_DZH_891	*D.* cf. *brevifunis*	female	Xinjiang, Aletai	C_AA071374.1	C_AA071417.1
CHX_DZH_892	*D. brevifunis*	female	Xinjiang, Tulufan	C_AA071375.1	-
CHX_DZH_893	*D.* sp.1	female	Xinjiang, Tulufan	C_AA071376.1	-
CHX_DZH_894	*D.* cf. *brevifunis*	female	Xinjiang, Tulufan	C_AA071377.1	-
CHX_DZH_895	*D. brevifunis*	female	Xinjiang, Tulufan	C_AA071378.1	-
CHX_DZH_896	*D.* cf. *antonovae*	female	Xinjiang, Aletai	C_AA071379.1	C_AA071418.1
CHX_DZH_897	*D.* cf. *antonovae*	male	Xinjiang, Aletai	C_AA071380.1	C_AA071419.1
CHX_DZH_898	*D.* sp.2	female	Xinjiang Aletai	C_AA071381.1	C_AA071420.1
CHX_DZH_900	*D. brevifunis*	female	Xinjiang, Tulufan	C_AA071382.1	-
CHX_DZH_901	*D. brevifunis*	female	Qinghai, Germu	C_AA071383.1	-
CHX_DZH_902	*D.* cf. *brevifunis*	female	Xinjiang, Aletai	C_AA071384.1	C_AA071421.1
CHX_DZH_903	*D. brevifunis*	male	Xinjiang, Tulufan	C_AA071385.1	C_AA071422.1
CHX_DZH_904	*D. brevifunis*	male	Xinjiang, Tulufan	C_AA071386.1	C_AA071423.1
CHX_DZH_906	*D. karamayica*	female	Xinjiang, Karamay	C_AA071387.1	C_AA071424.1
CHX_DZH_907	*D. gobica*	female	Xinjiang, Shihezi	C_AA071388.1	C_AA071425.1
CHX_DZH_908	*D.* cf. *sugonjaevi*	female	Xinjiang, Aletai	C_AA071389.1	C_AA071426.1
CHX_DZH_909	*D. muleica*	female	Xinjiang, Karamay	-	C_AA071427.1
CHX_DZH_910	*D. gobica*	female	Xinjiang, Fukang	C_AA071390.1	C_AA071428.1
CHX_TET_054	*Tetrastichus howardi*	female	Brazil, Sete Lagoas	C_AA071391.1	OP538682.1
NT-04	*D. karamayica*	female	Xinjiang, Karamay	C_AA071392.1	C_AA071429.1

Table 1. *Cont.*

Voucher Number	Species	Sex	Locality	COI	28S
NT-05	*D. karamayica*	female	Xinjiang, Karamay	C_AA071393.1	C_AA071430.1
NT-06	*D. karamayica*	male	Xinjiang, Karamay	C_AA071394.1	C_AA071431.1
NT-07	*D. karamayica*	male	Xinjiang, Karamay	C_AA071395.1	C_AA071432.1
NT-08	*D. gobica*	male	Xinjiang, Shihezi	C_AA071396.1	C_AA071433.1
NT-09	*D. gobica*	female	Xinjiang, Shihezi	C_AA071397.1	C_AA071434.1
NT-10	*D. gobica*	male	Xinjiang, Shihezi	C_AA071398.1	C_AA071435.1
NT-11	*D. gobica*	male	Xinjiang, Shihezi	C_AA071399.1	C_AA071436.1
NT-13	*D. gobica*	female	Xinjiang, Shihezi	C_AA071400.1	C_AA071437.1
NT-16	*D. gobica*	male	Xinjiang, Shihezi	C_AA071401.1	C_AA071438.1

The symbol "-" indicates a failed case in PCR or sequencing. The 28S sequence of "CHX_TET_054" was deposited in GenBank by Cao et al. [25].

Table 2. List of primers used in this study.

Gene	Primer	Sequence (5′-3′)	References
COI	LCO1490	GGTCA ACAAA TCATA AAGAT ATTGG	[30]
COI	HCOout	CCAGG TAAAA TTAAA ATATA AACTTC	[31]
COI	FWPTF1	CCTGG TTCTT TRATT GGTAA TGATC	[32]
COI	Lep-R1	TAAAC TTCTG GATGT CCAAA AAATCA	[33]
28S	D2-3549F	AGTCG TGTTG CTTGA TAGTG CAG	[34]
28S	D2-3665F	AAGAGAGAGTTCAAGAGTACGTG	[35]
28S	D2-4068R	TTGGT CCGTG TTTCA AGACG GG	[34]
28S	D3-4083R	TAGTT CACCA TCTTT CGGGT CCC	[35]

2.4. Species Delimitation

Initially, we compiled a comprehensive list of all the examined specimens and meticulously documented their morphological variability. We assigned all examined specimens to morphospecies based on our understanding of intra- and interspecific variation, referring to the related taxonomy work of Kostjukov [7–9,11], Kostjukov and Kosheleva [12,13], Graham [10], and Li et al. [14,15]. Two molecular methods for species delimitation, Automatic Barcode Gap Discovery (ABGD) [36] and the Generalized Mixed Yule-Coalescent (GMYC) model [37], representing a distance-based and a tree-based coalescent approach, respectively, were employed to delimit intra- and interspecific variation and evaluate the species boundaries of these morphospecies.

ABGD automatically partitions DNA barcode sequences based on the barcode gap using iterative model-based confidence limits for intraspecific divergence [36]. Pairwise distances of *Dzhanokmenia* COI sequences were calculated based on the Kimura 2-parameter (K2P) model [38] in MEGA7.0 [29]. In addition, a Neighbor-Joining (NJ) tree [39] was constructed based on the K2P model with 1000 bootstrap replicates to calculate support values for the branches. The resulting K2P distance matrix was then analyzed using the ABGD online tool (https://bioinfo.mnhn.fr/abi/public/abgd/abgdweb.html, accessed on 20 April 2024) for partitioning. The relative gap width was set to 0.5, while other parameters were left at their default values.

The GMYC model determines intra- and interspecific branching events in a phylogenetic tree estimated from DNA sequence data [37]. For the COI gene, we performed a GMYC analysis using the single method implemented in the 'splits' package [37] with R 4.3.3 [40] to infer the putative species. After removing identical sequences to avoid terminal branches with zero length, ultrametric trees were estimated in BEAST v1.10.4 [41] under an uncorrelated log-normal relaxed clock model [42], HKY [43], and Gamma substitution with all other default priors. Markov chains were run for 5×10^7 generations, sampling every 5×10^4 generations. Tracer v1.7.2 [44] was used to visualize and evaluate the convergence of the analyses.

2.5. Phylogenetic Analysis

The ML analyses were conducted using IQ-TREE version 1.6.12 [45]. ModelFinder Plus (MFP) [46] was employed in all cases to automatically select the optimal substitution model, and 1000 ultrafast bootstrap replicates were performed to assess the robustness of the results. In the ML analysis of MatrixI, which contains almost all available 28S D2 sequences of Tetrastichinae that could be aligned, species from the tribe Cirrospilini (Eulophinae) were chosen as outgroups, as suggested by Gauthier et al. [2]. *Styotrichia pisoxylona* served as an outgroup in the ML analyses of MatrixII and MatrixIII. The resulting ML trees were visualized and annotated using iTOL [47].

3. Results

3.1. Species Identification and Delimitation

In the initial morphological identification, all specimens resembling *Dzhanokmenia brevifunis* were assigned to one morphospecies, with potential intraspecific variability documented. As a result, six morphospecies of *Dzhanokmenia* were identified: *D. brevifunis* sp. nov., *D. gobica*, *D. karamayica*, *D. muleica* Li, Wang & Hu, 2016, *D.* cf. *antonovae* and *D.* cf. *sugonjaevi*.

However, of the 52 individuals used for sequencing, 7 failed to yield COI sequences and were consequently excluded from the DNA barcoding analysis. The specimens used for the subsequent molecular analyses are summarized in Table 1. A total of 4 of these failed individuals belong to the species *Dzhanokmenia muleica*, resulting in its absence from the DNA barcode library. The resulting COI matrix for species delimitation comprised 42 individuals, with a sequence length of 434 bp, including aligned gaps. The results of the two molecular methods for species delimitation were consolidated in an NJ tree. Although ABGD failed to identify a barcoding gap within all available COI sequences of *Dzhanokmenia*, it suggested the presence of eight molecular units stably supported by COI sequences using this distance method. Furthermore, these eight putative species were corroborated by GMYC results. These molecular species delimitation results led to the separation of one morphospecies of *D. brevifunis* into three distinct species (see Figure 1).

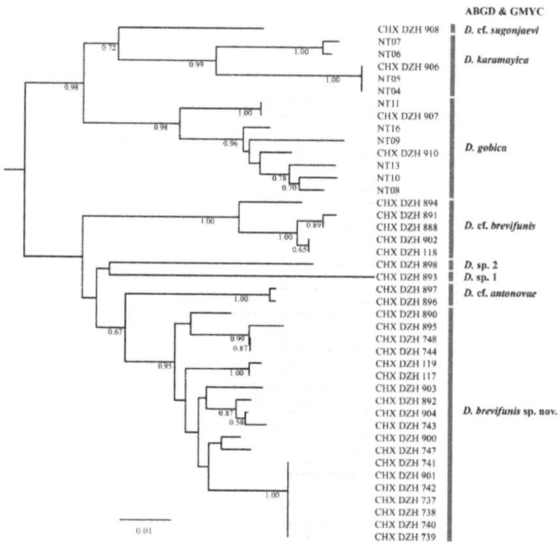

Figure 1. An NJ tree illustrating species delimitation results obtained using ABGD and GMYC methods based on K2P distances derived from the COI region. Bootstrap values are shown below the branches with values greater than 0.50. Each vertical bar on the right represents a distinct species delimitated by both ABGD and GMYC.

3.2. Phylogenetic Analyses

DNA barcodes were successfully generated for eight *Dzhanokmenia* species recorded from China (Table 1). We expanded the 28S matrix provided by Cao et al. [24] to include 97 sequences, incorporating 35 sequences from *Dzhanokmenia* species and 1 sequence from *Baryscapus* species generated in this study. After merging identical sequences, the following three matrices were generated for the phylogenetic analyses: (1) MatrixI: comprised 72 unique sequences with a length of 429 bp (28S, including aligned gaps); (2) MatrixII: comprised 30 unique sequences with a length of 419 bp (28S, including aligned gaps); (3) MatrixIII: comprised 35 unique concatenated sequences with a length of 852 bp (COI and 28S regions, including aligned gaps). The phylogenetic trees based on these three matrices are shown in Figures S1, S2 and 2, respectively.

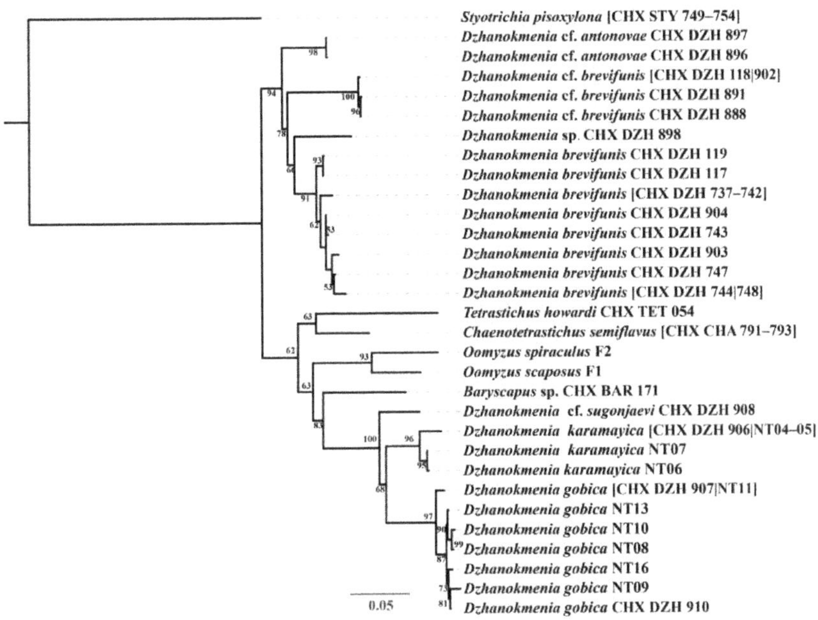

Figure 2. An ML tree based on MatrixIII, concatenating COI and 28S D2 regions, with *Styotrichia pixoxylona* as outgroup. Bootstrap values with values greater than 50 are shown to the left of the nodes.

3.3. Morphological Diagnosis and Species Treatments of Dzhanokmenia

3.3.1. Genus Dzhanokmenia Kostjukov, 1977

Dzhanokmenia Kostjukov, 1977 [7]: 189 (as subgenus of *Tetrastichus*). Type species: *Tetrastichus bibikovae* Dzhanokmen, 1971, by original designation.

Dzhanokmenia Kostjukov (as valid genus), Graham, 1991 [10]: 162–163.

Diagnosis. Mesosoma metallic green to blue, and metasoma with or without pale markings. Malar sulcus strongly curved. MLM with a single row of adnotaular setae on each side, exhibiting symmetrical or asymmetrical arrangement of setae. Antenna pale, although male pedicel sometimes darker dorsally at base; female funicle 3-segmented, male funicle 4-segmented. On hind leg, tarsomere 1 shortest, tarsomeres 2 and 3 subequal in length, and tarsomere 4 the longest. Fore wing except MV with extremely short marginal setae; MV and STV thickened, and SMV with one dorsal seta; hind wing knife-shaped, without marginal setae along upper margin.

Remarks. In terms of morphology, the most distinctive key character of *Dzhanokmenia* is the absence of marginal setae on the fore wing, except for MV, and the absence of marginal setae along the upper margin of the hind wing. While the MV and SMV are sometimes thickened in both the female and male, they are not as thick as those in *Dzhanokmenia*. In contrast, a key character of *Baryscapus* is the swollen scape in the male [10,48,49]. However, not all males of *Baryscapus* exhibit a swollen scape, and males are unknown in some species [48,49]. Furthermore, the genus *Oomyzus* exhibits a similar variation in male scape morphology, with some species possessing a swollen scape (*O. incertus* (Ratzeburg, 1844)) while others do not (e.g., *O. spiraculus* Song, Fei & Cao, 2020) [50]. Consequently, further studies are needed to determine whether the swollen male scape serves as a key character with phylogenetic significance. Additionally, in some *Baryscapus* species, the fore wing may also have short marginal setae on apical margin (e.g., *B. berhidanus* Erdős). We have also observed that certain species of Eulophidae (such as species of *Baryscapus* and *Neochrysocharis* Kurdjumov, 1912), Encyrtidae, Eurytomidae, and Pteromalidae, collected in semi-arid regions, possess relatively transparent fore wing discs with a reduced number of setae. This suggests that this character may be convergent due to adaptations to the environment, similar to the case of the enlarged mandibles in *Chaenotetrastichus* and *Styotrichia* LaSalle, which possibly evolved due to host adaptation [46]. While *Dzhanokmenia* species consistently have one dorsal seta on the SMV, *Baryscapus* species typically have two or more, very rarely with one seta. Therefore, the combination of characters outlined in the diagnosis of *Dzhanokmenia* is suggested to distinguish *Dzhanokmenia* species from closely related genera.

3.3.2. *Dzhanokmenia brevifunis* Ganbaatar & Cao, sp. nov.

Diagnosis. Body metallic green to metallic blue green, with weak to strong bronze tinge; antenna pale yellow, basal 3/4 of pedicel slightly infuscate dorsally; legs predominantly pale yellow, fore coxa pale yellow to brown at base, mid coxa infuscate basally to entirely metallic green, hind coxa metallic green with bronze tinge. Female antenna stout, and funicle 3-segmented; F1 distinctly transverse, 1.25–1.55× as broad as long; F2 subquadrate to slightly longer than broad, 1.05–1.45× as long as broad; F3 slightly transverse to subquadrate, 0.90–1.25× as broad as long; male antenna slender and funicle 4-segmented, with F1 distinctly transverse, about 2× as broad as long, and each remaining flagellomere longer than broad.

Female. Body length 0.9–1.5 mm. Body usually metallic green with a weak to strong bronze tinge, occasionally metallic blue, transscutal articulation along axilla pale yellow. Antenna pale yellow to brownish yellow, basal 3/4 of pedicel slightly infuscate dorsally. Legs predominantly pale yellow, except following parts: fore coxa pale yellow to brown at base, mid coxa infuscate basally to entirely metallic green, hind coxa metallic green with a bronze tinge, tarsomere 4 slightly infuscate, especially at apex, and claws brown. Wings hyaline, with pale yellow tegula and veins (Figure 3a,b).

Antenna with 1 transverse anellus, 3 funiculars, and 3 clavomeres (Figure 4); scape without plaque; pedicel distinctly longer than each funicular, usually shorter than and occasionally subequal in length with combined length of F1 and F2; F1 transverse, 1.25–1.55× as broad as long; F2 subquadrate to slightly longer than broad, 1.05–1.45× as long as broad, always longer than F1 and F3; F3 slightly transverse to subquadrate, 0.90–1.25× as broad as long; clava broader than funicle, 0.98–1.15× (without terminal spine) as long as funicle, clavomeres decreasing in length, clavomere 3 truncate apically, with an indistinct terminal spine (Figure 4). F1 with few scattered mushroom-shaped capitate peg sensilla apically, without longitudinal sensilla, remaining flagellomeres with longitudinal sensilla and apically with a circle of scattered, mushroom-shaped capitate peg sensilla (Figure 4).

Figure 3. *Dzhanokmenia brevifunis* sp. nov. (**a**) Habitus of female holotype in dorsolateral view; (**b**) Habitus of female paratype in dorsolateral view; (**c**) Habitus of male paratype in lateral view; (**d**) Habitus of male paratype in dorsolateral view. Scale bar: 0.5 mm.

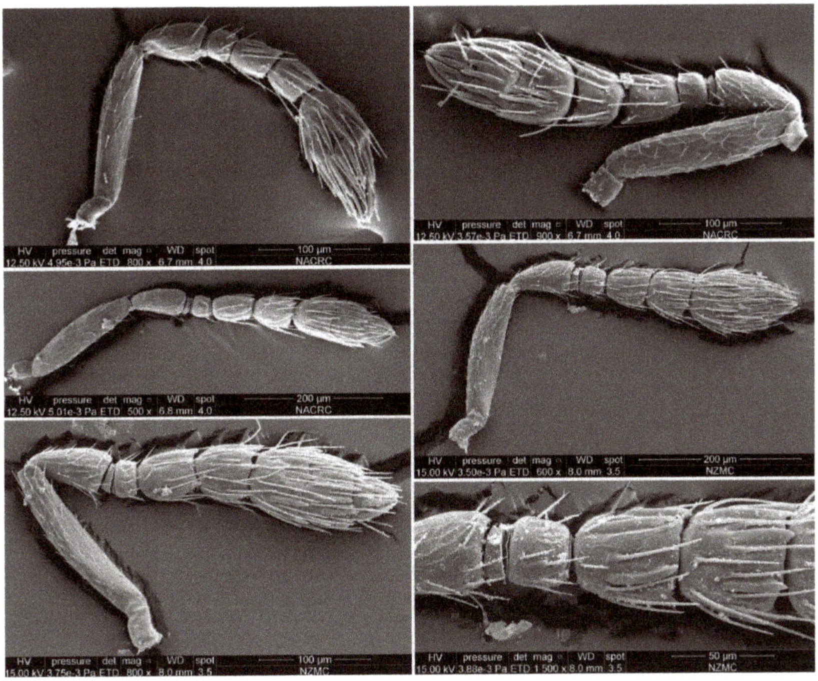

Figure 4. *Dzhanokmenia brevifunis* sp. nov., non-type females, variation in the antenna.

Head slightly broader than mesoscutum and very easily collapsing after death (even if preserved in ethanol). POL about 3× as long as OOL (Figure 5a). Ocelli arranged in an obtuse-angled triangle (Figure 5a). Frons with a broad V-shaped frontofacial suture connecting to ocellar area; upper face with a weak carina between scrobes; head easily collapsing along frontofacial suture, outer margin of scrobes, and outer margin of ocelli (Figure 6a–c). Face with weak reticulation, and even more weak on scrobes. Toruli inserted almost at the same level as lower margin of eyes (Figure 6a).

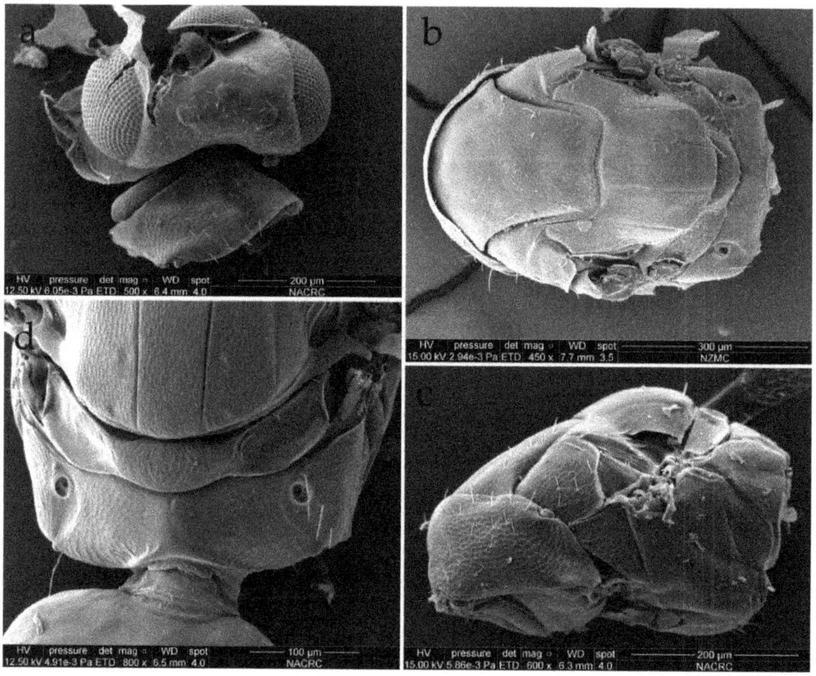

Figure 5. *Dzhanokmenia brevifunis* sp. nov., non-type females. (**a**) Head in dorsal view; (**b**) mesosoma in dorsal view; (**c**) metanotum and propodeum in dorsal view; (**d**) mesosoma in lateral view.

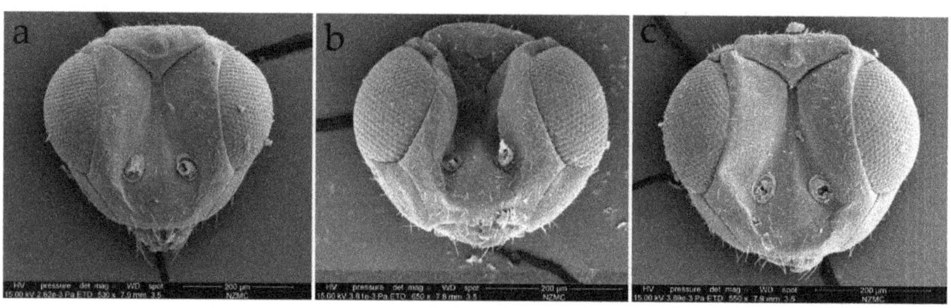

Figure 6. *Dzhanokmenia brevifunis* sp. nov., non-type females. (**a**) Head in front view of slightly collapsed specimen; (**b**) head in front view of specimen with collapsed face along scrobes; (**c**) head in front view of moderately collapsed specimen.

Vertex, occiput, and upper face with scattered short white setae (scrobal area bare), lower face with sparse longer white setae (Figures 5a and 6a–c). Eyes with short and sparse white setae, height of eye longer than malar space, about 1.8× as long as malar space. Malar space 0.8–0.9× as long as width of mouth opening, with malar sulcus strongly curved (Figure 6b). Anterior margin of clypeus bilobed; mandible bearing 3 teeth, with the length increasing (weak to strong) from the inner to the outer tooth (Figure 6b).

Pronotum strongly sloping and short in dorsal view, neck and collar not delimited, with scattered setae and raised reticulation (Figure 5a,c). Mesoscutum with raised reticulation that is slightly elongate; MLM nearly as long as maximum width, with a well-defined and complete median line, with a row of 2–4 adnotaular setae on each side, symmetric or asymmetric (i.e., with different number or position of setae on each side); axillae shifted strongly forward, with raised reticulation (Figure 5b). Mesoscutellum broader than long, about 1.3× as broad as long, convex in profile, with raised reticulation that is slightly elongate; mesoscutellum with a pair of strong submedian lines and a pair of weak but distinct sublateral lines (Figure 5b), and two pairs of setae, one pair situated slightly below the middle of mesoscutellum and the other pair situated near posterior margin; mesoscutellum with a narrow frenum, almost invisible in dorsal view because of convex mesoscutellum. Dorsellum convex, with reticulation as strong as that on mesoscutellum; lateral panels of metanotum depressed and smooth (Figure 5d).

Propodeum incised medially along anterior and especially posterior margins and thus short medially, 1.1–1.3× as long as median length of dorsellum; propodeum with raised reticulation, slightly weaker than that on dorsellum; median carina on propodeum hard to see but appears to be present due to elevated median area visible when viewed under a microscope; propodeum without plicae, but with curved paraspiracular carinae along inner margin of spiracles; spiracles medially large with entire rim exposed and separated from anterior margin of propodeum by half or more of the diameter of spiracle; callus reticulate, often with 3 short setae, occasionally with 2 setae (Figure 5b,d). Lateral panels of pronotum and prepectus with reticulation as strong as on pronotum, mesepimeron and mesepisternum with weak reticulation, almost smooth; acropleuron smooth; metapleuron reticulate (Figure 5c).

Petiole short and hidden in dorsal view (Figures 3b and 5d). Gaster slender, pointed apically, 1.8–2.0× as long as broad; gastral tergites each with weak raised reticulation; Gt_7 with 4 cercal setae, subequal in length; hypopygium nearly reaching to 1/2 the length of gaster (Figure 7c,d).

Figure 7. *Dzhanokmenia brevifunis* sp. nov., non-type females. (**a**) Fore wing; (**b**) hind wings; (**c**) gaster in dorsal view; (**d**) gaster in lateral view.

Legs with tarsomere 1 distinctly shorter than the other tarsomeres, tarsomere 2 and tarsomere 3 subequal in length and slightly shorter than tarsomere 4. Fore wing with only MV having long marginal setae, setae on margin of membrane are extremely short, which is challenging to discern even under a microscope, so it looks as if they are missing; fore wing disk without setae below SMV and MV; fore wing disk with short white setae becoming shorter towards base, and basal triangular area before the end of parastigma bare, and speculum is thus large and open below; PMV absent; SMV with 1 seta on dorsal surface (Figures 3a,b and 7a). Hind wing knife-shaped, with the upper margin lacking marginal setae, and the lower margin with soft long white setae that are challenging to discern under a microscope and even in a SEM (Figure 7b).

Male. The males have similar body length to the females and differ from females as follows (Figures 3b,c and 8a–e). Antenna more slender; the median part of scape slightly extending beyond the level of the scape; antenna with funicle 4-segmented; pedicel longer than broad, about 4.5× as long as F1; F1 distinctly transverse, about 2× as broad as long; F2 slightly longer than broad, about 2.5× as long as F1; F3 about 1.4× as long as broad, about 3.6× as long as F1; F4 about 1.2× as long as broad, and nearly as long as F3 (Figure 8c). Middle coxa predominantly metallic green, slightly pale brown apically (Figure 3b,c).

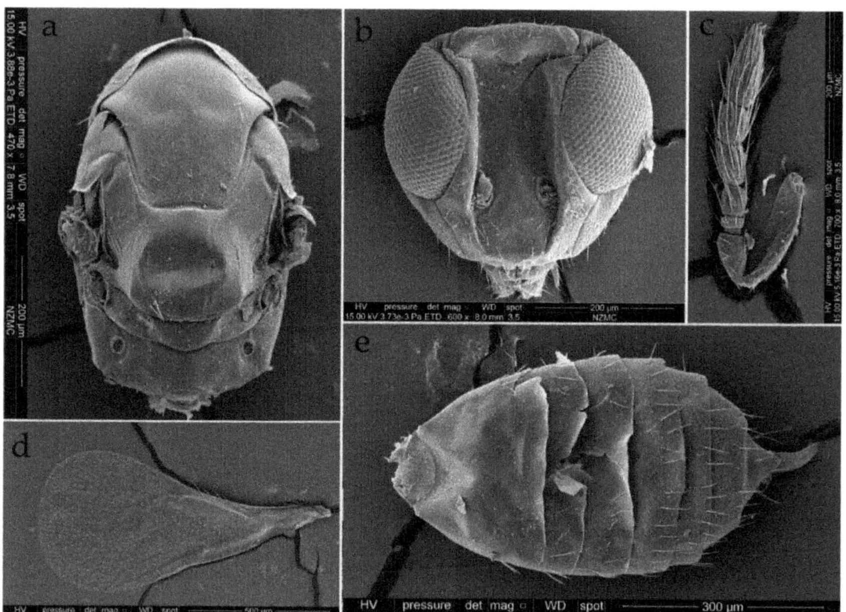

Figure 8. *Dzhanokmenia brevifunis* sp. nov., non-type male. (**a**) Mesosoma in dorsal view; (**b**) Head in front view; (**c**) Antenna; (**d**) Fore wing; (**e**) Gaster in dorsal view.

Etymology. From the Latin words 'brevis' (meaning 'short') and 'funis' (from 'funicle') (noun in apposition), referring to the transverse F1.

Type material. Holotype ♀, China, Xinjiang, the Desert Plant Garden of Tulufan (42.8542° N, 89.1928° E, −86 m), collected with Malaise Traps, 05-V-2013–15-V-2013, coll. Rui-Xia Liu (IOZCAS, IOZ(E)224671). Paratypes: 2♀2♂, same data as the holotype (IOZCAS, IOZ(E)224672–224675); 6♀, same data as the holotype, except for the collecting dates 25-IV-2013–05-V-2013 (IOZCAS, IOZ(E)224676–224681); 6♀, same data as the holotype, except for the collecting dates 05-V-2023–15-V-2013 (IOZCAS, IOZ(E)224682–224687); 3♀, China, Qinghai, Ge-Ermu, Baiyun Village on the Qinghai Province Road 303 (36.6103° N, 94.8506° E, 2805 m), swept on *Tamarix* sp., 17-VII-2017, coll. Qing-Tao Wu (IOZCAS,

IOZ(E)224688–224690); 5♀1♂, China, Xinjiang, A-Letai, Fuhai (47.14604° N, 87.5549° E, 500 m), swept on *Tamarix* sp., coll. Zhi-Liang Wang (IOZCAS, IOZ(E)224691–224696).

Additional material. 6♀voucher specimens for DNA barcodes, same data as the holotype; 4♀, same data as the holotype, except for the collecting dates 25-IV-2013–05-V-2013; 4♀, same data as the holotype, except for the collecting dates 05-V-2013–15-V-2013; 1♀1♂on slides, China, Qinghai, Ge-Ermu, Baiyun Village on the Qinghai Province Road 303 (36.6103° N, 94.8506° E, 2805 m), swept on *Tamarix* sp., 17-VII-2017, coll. Qing-Tao Wu; 6♀voucher specimens for DNA barcoding, China, Qinghai, Ge-Ermu, Baiyun Village on the Qinghai Province Road 303 (36.6103° N, 94.8506° E, 2805 m), swept on *Tamarix* sp., 17-VII-2017, coll. Qing-Tao Wu; 5♀voucher specimens for DNA barcoding, China, Xinjiang, A-Letai, Fuhai (47.14604° N, 87.5549° E, 500 m), swept on *Tamarix* sp., coll. Zhi-Liang Wang; 16♀preserved in 95%–100% ethanol, same data as holotype; 6♀, preserved in 95%–100% ethanol, same data as the holotype, except for the collecting dates 25-IV-2013–05-V-2013; 16♀, preserved in 95–100% ethanol, same data as the holotype, except for the collecting dates 05-V-2013–15-V-2013.

Host information. The holotype and some of the paratypes were collected by sweeping shrubs of *Tamarix* sp. in Xinjiang and Qinghai, China.

Distribution. Palearctic region (China: Qinghai, Xinjiang). The records of this species in semi-arid areas from both low and high altitudes suggest a potential distribution range spanning from about −100 to 2800 m above sea level.

Remarks. The specimens collected on the Tibetan Plateau at high altitudes have a larger body size than those collected in the Turpan Basin at low altitudes. *Dzhanokmenia brevifunis* shows variation in the coloration of the coxae on the fore and middle legs, as well as in the relative length of each funicular. In females, the fore coxa typically appears pale yellow, with a dark brown coloration that spreads to varying degrees, but never entirely covers the coxa or has a metallic green tint. However, the dark brown coloration of the middle coxa in females may extend from the base to the entire surface and sometimes have a metallic green tinge, similar to the color of the body. In addition, differences in the relative length of each funicular can be quite confusing due to their wide range of variation. However, F1 is consistently transverse and never subquadrate or longer than broad. In contrast, the fore coxa in males is typically pale yellow but may also have a dark brown color or even be entirely dark brown or metallic green. The relative length of each funicular appears to be relatively consistent compared to females. The undetectable variation in the male antenna may be attributed to the limited number of male individuals within a brood sample. The confirmation of these morphological variations was corroborated by the results of molecular species delimitation. For instance, the species represented by CHX_DZH_893 exhibits a fore coxa that is entirely metallic green and a subquadrate F1. Additionally, the species represented by CHX_DZH_898, which was also recognized from specimens swept on *Tamarix* sp., has antennae similar to *D. brevifunis* and entirely yellow fore and middle coxae but possesses a distinct gaster with yellow $Gt_{1–2}$.

The species referred to as "*D.* cf. *brevifunis*" in Table 1 and Figure 1 (as well as in Figures S1 and S2) is a smaller species with a body length of about 0.8 mm compared to *D. brevifunis*. It can be distinguished from *D. brevifunis* by the combination of the following characters: all coxae entirely metallic green; callus with 2 setae. Sometimes the antenna of *D.* cf. *brevifunis* can be distinguished from that of *D. brevifunis*, with transverse F1 being longer than 1/2 length of F2 and F2 being subquadrate and subequal to F3 in length. However, this species also shows variation in the relative length of each funicular, which complicates the comparison of interspecific variation between these two species, especially when dealing with a limited number of individuals of *D.* cf. *brevifunis*. Despite few distinctions, we treated *D.* cf. *brevifunis* as a species distinct from *D. brevifunis*. In *D. brevifunis*, the fore coxa is never entirely metallic green, and the callus typically has 3 setae; if the callus has 2 setae, then F2 and F3 are both slightly longer than broad.

Morphologically, *Dzhanokmenia brevifunis* appears to be closely related to species of this genus that have a completely metallic gaster. Among these species, *D. brevifunis* appears to be most closely related to *D. antonovae*, as confirmed by the ML trees (Figures S1 and S2). The clearly transverse F1 distinguishes *D. brevifunis* from these species within *Dzhanokmenia*. However, *D. karamayica*, another Chinese species with metallic gaster, was clustered with the other three species with partially yellow gaster (*D. gobia*, *D. muleica*, and *D.* cf. *sugonjaevi*) rather than with *D. brevifunis* in the ML trees (Figures S1 and S2). Compared to these two clusters, the cluster containing *D. antonovae* and *D. brevifunis* has F1 transverse or at most subquadrate, while the other cluster containing *D. karamayica* has F1 longer than broad.

4. Discussion

In this study, we describe a new species of *Dzhanokmenia* distributed in Xinjiang and Qinghai, China, and it represents the first record of the genus from the semi-arid region of the Tibetan Plateau in Qinghai Province, China. A preliminary DNA barcode library for *Dzhanokmenia* was also constructed, revealing the potential species diversity around *D. brevifunis* by delimiting variation. Although the topology of the ML tree varies depending on the taxa or DNA regions included, the phylogenetic results nevertheless indicate a potential close relationship between *Dzhanokmenia* and *Baryscapus*, as well as to other genera associated with *Tetrastichus*, such as *Chaenotetrastichus* and *Oomyzus*.

One of the major challenges in species delimitations based on morphology arises from the complexities involved in discerning intra- and interspecific variation in morphological characters. This study unveiled that *Dzhanokmenia* species can manifest two extreme conditions. In one scenario, they display minor morphological variation but significant interspecific genetic differences (e.g., *D. brevifunis* and *D.* cf. *brevifunis*). Under this condition, specimens associated with a cluster of plant hosts may potentially represent distinct species. In the other scenario, they exhibit substantial morphological diversity yet demonstrate limited intraspecific genetic variation (e.g., the relative length of each funicular in *D. brevifunis*). In this case, there is a risk of mistakenly dividing one species into multiple entities. Therefore, a large number of specimens are essential for discussing intra- and interspecific morphological variation, which helps prevent the erroneous inclusion of closely related species under one classification or the excessive division of species.

The DNA barcoding analysis conducted in this study revealed a species complex containing at least two species. For specimens swept from adjacent clusters of *Tamarix* sp. (Tamaricaceae) and collected at a specific site with a Malaise trap, each genetically contained *D. brevifunis* and another distinct but occasional species. This occasional species, represented by *D.* cf. *brevifunis* (see Table 1 and Figure 2), remains undescribed due to an insufficient amount of material and was, therefore, not described here. Further extensive surveys of *Dzhanokmenia* in semi-arid regions and the acquisition of adequate material are needed to determine whether the currently observed variation in *D.* cf. *brevifunis* in a limited number of specimens can be classified as intra- or interspecific.

In addition, the other *Dzhanokmenia* species examined in this study are valuable for verifying the significance of variation in the relative length of each funicular (especially F1), the presence of a median line on the MLM, and the color of legs and gaster, as key characters for species delimitation within *Dzhanokmenia*. It is noteworthy that the range of variation in delimiting *Dzhanokmenia* species can vary between species, potentially increasing the challenges of delimitation in the absence of molecular data. The two similar species, *D. gobica* and *D. muleica*, which are primarily distinguished by differences in vertex and gaster [14], were not differentiated in the ML trees based on the 28S gene (Figures S1 and S2). In addition, *D. karamayica*, with an entirely metallic green gaster, is clustered with *D. gobica* and *D. muleica*, which have a predominantly yellow gaster. Future work is needed to determine whether the color variation in the gaster or vertex can be considered intraspecific in these species. In contrast, the yellow Gt_{1-2} directs the species represented by CHX_DZH_898 away from *D. brevifunis* (Figures 1 and 2), while the variation in the above-mentioned characters fall within the range of *D. brevifunis*. All species have a slender F1 (longer than broad) in one cluster and a transverse F1 (at most subquadrate) in the other cluster in all three ML trees, suggesting that the antenna may take precedence over the other three characters mentioned above in grouping *Dzhanokmenia* species.

The topology of the branch containing *Dzhanokmenia* in the ML tree based on MatrixII (Figure S2) shows slight differences compared to that based on MatrixI (Figure S1). In the former scenario, the genus *Dzhanokmenia* is grouped in one cluster, whereas in the latter, it is divided into two clusters interspersed with four other genera: *Baryscapus*, *Chaenotetrastichus*, *Oomyzus*, and *Tetrastichus*. The ML tree based on MatrixIII, concatenating COI and 28S, shows a similar topology to the tree based on MatrixI. The deviations in the topologies of the ML trees based on different datasets show that the trees are not robust. Nevertheless, all scenarios suggest possible close relationships of these genera associated with *Tetrastichus* included in the ML analyses. Although confirming the monophyly of *Dzhanokmenia* or *Baryscapus* requires a well-resolved phylogeny of these genera, these trees provide clues for understanding the relationships among these genera. Further research on *Dzhanokmenia* and *Baryscapus* will be impeded without a comprehension of their relationships.

This study provides a preliminary DNA barcode library for the genus *Dzhanokmenia* that will serve as a valuable resource for the identification of both intra- and interspecific variation. It is anticipated that this library will aid in future efforts in species discovery and description within the genus. However, the construction of a more comprehensive DNA barcode library for *Dzhanokmenia* remains essential and requires urgent attention. Analyses of the limited DNA data suggest a potentially close relationship between *Dzhanokmenia* and other genera related to *Tetrastichus*, beyond *Baryscapus*. Nonetheless, the current dataset lacks the robustness necessary to definitively establish the monophyly of these genera. As Cao et al. [25] emphasized, a more comprehensive dataset that includes a broader range of taxa and incorporates multiple lines of evidence is necessary to elucidate the phylogenetic relationships of these genera.

Supplementary Materials: The following supporting information can be downloaded at: https://www.mdpi.com/article/10.3390/insects15060406/s1, Figure S1: An ML tree based on MatrixI of 28S D2 sequences, with species from the tribe Cirrospilini as outgroups; Figure S2: An ML tree based on MatrixII of 28S D2 sequences, with *Styotrichia pixoxylona* as the outgroup; Table S1: List of public sequences used in the ML analyses with accession numbers from GenBank.

Author Contributions: Conceptualization, H.C. and C.Z.; methodology, H.C.; software, H.C. and B.G.; validation, all authors; formal analysis, H.C.; investigation, all authors; resources, H.C., C.Z., Q.L. and O.X.; data curation, H.C. and B.G.; writing—original draft preparation, H.C. and B.G.; writing—review and editing, all authors; visualization, H.C. and B.G.; supervision, H.C. and C.Z.; project administration, H.C.; funding acquisition, C.Z. All authors have read and agreed to the published version of the manuscript.

Funding: This research was funded by the Key Program of the National Natural Science Foundation of China, grant number 32330013, and Chaodong Zhu's lab was also supported by the Key Laboratory of Zoological Systematics and Evolution, grant number 2008DP173354.

Data Availability Statement: DNA data generated in this study are available on GenBase under accession numbers (see Table 1) and other public DNA data downloaded from GenBank are available under corresponding accession numbers mentioned in Table S1. Other data used in this study are deposited at the Institute of Zoology, Chinese Academy of Sciences, Beijing, China.

Acknowledgments: We thank Kuiyan Zhang (IOZCAS) for helping with SEM. Xuemei Yang (IOZCAS), Qingtao Wu (IOZCAS), Boyu Zou (IOZCAS), Chao Wang (Xinjiang University), and Ruixia Liu collected the specimens examined in this study. We thank the four anonymous reviewers for their useful comments on this manuscript. Huanxi Cao extends special thanks to Christer Hansson for his personal correspondence on the discussions of the *Tetrastichus* group, which significantly enhanced H.C.'s understanding of this large group within Tetrastichinae. Additionally, Christer Hansson generously assisted in refining the English language in the final version of this manuscript.

Conflicts of Interest: The authors declare no conflicts of interest. The funders had no role in the design of the study; in the collection, analyses, or interpretation of data; in the writing of the manuscript; or in the decision to publish the results.

References

1. Noyes, J. Universal Chalcidoidea Database. 2019. Available online: http://www.nhm.ac.uk/chalcidoids (accessed on 18 April 2024).
2. Gauthier, N.; LaSalle, J.; Quicke, D.L.; Godfray, H. Phylogeny of Eulophidae (Hymenoptera: Chalcidoidea), with a reclassification of Eulophinae and the recognition that Elasmidae are derived eulophids. *Syst. Entomol.* **2000**, *25*, 521–539. [CrossRef]
3. Burks, R.A.; Heraty, J.M.; Gebiola, M.; Hansson, C. Combined molecular and morphological phylogeny of Eulophidae (Hymenoptera: Chalcidoidea), with focus on the subfamily Entedoninae. *Cladistics* **2011**, *27*, 581–605. [CrossRef] [PubMed]
4. Cruaud, A.; Rasplus, J.-Y.; Zhang, J.; Burks, R.; Delvare, G.; Fusu, L.; Gumovsky, A.; Huber, J.T.; Janšta, P.; Mitroiu, M.-D.; et al. The Chalcidoidea bush of life: Evolutionary history of a massive radiation of minute wasps. *Cladistics* **2023**, *40*, 34–63. [CrossRef] [PubMed]
5. Rasplus, J.Y.; Blaimer, B.B.; Brady, S.G.; Burks, R.A.; Delvare, G.; Fisher, N.; Gates, M.; Gauthier, N.; Gumovsky, A.V.; Hansson, C.; et al. A first phylogenomic hypothesis for Eulophidae (Hymenoptera, Chalcidoidea). *J. Nat. Hist.* **2020**, *54*, 597–609. [CrossRef]
6. UCD Community. Universal Chalcidoidea Database Website. 2023. Available online: https://ucd.chalcid.org (accessed on 18 April 2024).
7. Kostjukov, V.V. Comparative morphology of chalcids of the subfamily Tetrastichinae and the system of the genus *Tetrastichus* Haliday, 1844 (Hymenoptera, Eulophidae). *Entomol. Obozr.* **1977**, *56*, 177–194, (In Russian with English Summary).
8. Kostjukov, V.V. Subfam. 5. Tetrastichinae. In *Keys to the Insects of the European Part of the USSR—Volume III, Hymenoptera, Part 2*; Medvedev, G.S., Trjapitzin, V.A., Eds.; Nauka Leningrad Division: Leningrad, Russia, 1978; pp. 430–467. (In Russian)
9. Kostjukov, V.V. New species of the genus *Tetrastichus* Haliday, 1844 (Hymenoptera, Eulophidae) from Turkmenistan. In *Entomophages of the Orchards Pests*; Shtiintsa: Kishinev, Moldova, 1984; pp. 30–35. (In Russian)
10. Graham, M.W.R.d.V. *A Reclassification of the European Tetrastichinae (Hymenoptera: Eulophidae): Revision of the Remaining Genera*; Associated Publishers c/o American Entomological Institute: Gainesville, FL, USA, 1991; pp. 1–322.
11. Kostjukov, V.V. New species of *Dzhanokmenia* Kostjukov, 1977 and *Kolopterna* Graham, 1987 (Hymenoptera, Eulophidae, Tetrastichinae) from Turkmenistan. *Biol. Plant Prot. Basis Ecosyst. Stab. Agroecosyst.* **2014**, *8*, 84–91.
12. Kostjukov, V.V.; Kosheleva, O.V. New species of *Dzhanokmenia* Kostjukov and *Kolopterna* Graham (Hymenoptera: Eulophidae: Tetrastichinae) from Russia. *Proc. Russ. Entomol. Soc.* **2014**, *85*, 160–164.
13. Kostjukov, V.V.; Kosheleva, O.V. A new species of the eulophid-wasp genus *Dzhanokmenia* Kostjukov, 1977 (Hymenoptera, Eulophidae: Tetrastichinae) from Stavropol Territory. *Entomol. Obozr.* **2015**, *94*, 451–454. [CrossRef]
14. Li, Q.; Wang, C.; Hu, H.Y.; Kostjukov, V.V.; LaSalle, J.; Zhu, C.D. Descriptions of three new species of *Dzhanokmenia* (Hymenoptera: Eulophidae) from China. *Zootaxa* **2016**, *4121*, 447–457. [CrossRef]
15. Li, Q.; Wang, C.; Hu, H.Y. Two new species of *Dzhanokmenia* (Hymenoptera, Eulophidae) from China, with first report on a host association for the genus. *ZooKeys* **2021**, *1009*, 67–79. [CrossRef]
16. Guo, S.H.; Kang, N.; Li, Q.; Hu, H.Y. Species and identification of *Dzhanokmenia* in China (Hymenoptera:Eulophidae). *Sichuan Anim.* **2022**, *41*, 63–73.
17. Burks, R.; Mitroiu, M.-D.; Fusu, L.; Heraty, J.M.; Janšta, P.; Heydon, S.; Papilloud, N.D.-S.; Peters, R.S.; Tselikh, E.V.; Woolley, J.B.; et al. From Hell's Heart I Stab at Thee! A Determined Approach towards a Monophyletic Pteromalidae and Reclassification of Chalcidoidea (Hymenoptera). *J. Hymenopt. Res.* **2022**, *94*, 13–88. [CrossRef]
18. Munro, J.B.; Heraty, J.M.; Burks, R.A.; Hawks, D.; Mottern, J.; Cruaud, A.; Rasplus, J.Y.; Jansta, P. A Molecular Phylogeny of the Chalcidoidea (Hymenoptera). *PLoS ONE* **2011**, *6*, e27023. [CrossRef]
19. van Noort, S.; Mitroiu, M.-D.; Burks, R.; Gibson, G.; Hanson, P.; Heraty, J.; Jansta, P.; Cruaud, A.; Rasplus, J.-Y. Redefining Ormyridae (Hymenoptera, Chalcidoidea) with establishment of subfamilies and description of new genera. *Syst. Entomol.* **2024**, 1–48. [CrossRef]
20. Zhang, J.X.; Lindsey, A.R.I.; Peters, R.S.; Heraty, J.M.; Hopper, K.R.; Werren, J.H.; Martinson, E.O.; Woolley, J.B.; Yoder, M.J.; Krogmann, L. Conflicting signal in transcriptomic markers leads to a poorly resolved backbone phylogeny of chalcidoid wasps. *Syst. Entomol.* **2020**, *45*, 783–802. [CrossRef]
21. Cruaud, A.; Delvare, G.; Nidelet, S.; Sauné, L.; Ratnasingham, S.; Chartois, M.; Blaimer, B.B.; Gates, M.; Brady, S.G.; Faure, S.; et al. Ultra-conserved elements and morphology reciprocally illuminate conflicting phylogenetic hypotheses in Chalcididae (Hymenoptera, Chalcidoidea). *Cladistics* **2021**, *37*, 1–35. [CrossRef] [PubMed]
22. Hong, D.; Zhuang, W.; Zhu, M.; Ma, K.; Wang, X.; Huang, D.; Zhang, Y.; Ren, G.; Bu, W.; Cai, W.; et al. Positioning taxonomic research for the future. *Zool. Syst.* **2022**, *47*, 185–187.
23. Mitroiu, M.D.; Rasplus, J.Y.; van Noort, S. New genera of Afrotropical Chalcidoidea (Hymenoptera: Cerocephalidae, Epichrysomallidae, Pirenidae and Pteromalidae). *PeerJ* **2024**, *12*, e16798. [CrossRef]
24. Zhu, C.; Luo, A.; Bai, M.; Orr, M.C.; Hou, Z.; Ge, S.; Chen, J.; Hu, Y.; Zhou, X.; Qiao, G.; et al. A joint call for actions to advance taxonomy in China. *Zool. Syst.* **2022**, *47*, 188–197.
25. Cao, H.X.; Dale-Skey, N.; Guo, P.F.; Zhu, C.D. Parasitoid wasps with enlarged mandibles associated with aculeate Hymenoptera: An integrative taxonomy of *Chaenotetrastichus* Graham and *Styotrichia* LaSalle (Hymenoptera: Eulophidae). *Insect Syst. Evol.* **2024**, *53*, 1–26. [CrossRef]
26. Gibson, G.A.P. Morphology and terminology. In *Annotated Keys to the Genera of Nearctic Chalcidoidea (Hymenoptera)*; Gibson, G.A.P., Huber, J.T., Woolley, J.B., Eds.; NRC Research Press: Ottawa, ON, Canada, 1997; pp. 16–44.

27. Huangfu, N.; Cao, H.-X.; Zhu, C.-D. Notes on the Genus *Aceratoneuromyia* Girault (Hymenoptera: Eulophidae). *Insects* **2022**, *13*, 450. [CrossRef]
28. Hall, T. BioEdit: A user-friendly biological sequence alignment editor and analysis program for Windows 95/98/NT. *Nucleic Acids Symp. Ser.* **1999**, *41*, 95–98.
29. Kumar, S.; Stecher, G.; Tamura, K. MEGA7: Molecular evolutionary genetics analysis version 7.0 for bigger datasets. *Mol. Biol. Evol.* **2016**, *33*, 1870–1874. [CrossRef]
30. Folmer, O.; Black, M.; Hoeh, W.; Lutz, R.; Vrijenhoek, R. DNA primers for amplification of mitochondrial cytochrome c oxidase subunit I from diverse metazoan invertebrates. *Mol. Mar. Biol. Biotechnol.* **1994**, *3*, 294–299. [PubMed]
31. Carpenter, J.M. Towards simultaneous analysis of morphological and molecular data in Hymenoptera. *Zool. Scr.* **1999**, *28*, 251–260. [CrossRef]
32. Li, Y.W.; Zhou, X.; Feng, G.; Hu, H.Y.; Niu, L.M.; Hebert, P.D.N.; Huang, D.W. COI and ITS2 sequences delimit species, reveal cryptic taxa and host specificity of fig-associated *Sycophila* (Hymenoptera, Eurytomidae). *Mol. Ecol. Resour.* **2010**, *10*, 31–40. [CrossRef] [PubMed]
33. Hebert, P.D.N.; Penton, E.H.; Burns, J.M.; Janzen, D.H.; Hallwachs, W. Ten species in one: DNA barcoding reveals cryptic species in the neotropical skipper butterfly *Astraptes fulgerator*. *Proc. Natl. Acad. Sci. USA* **2004**, *101*, 14812–14817. [CrossRef]
34. Campbell, B.C.; Steffen-Campbell, J.D.; Werren, J.H. Phylogeny of the *Nasonia* species complex (Hymenoptera: Pteromalidae) inferred from an internal transcribed spacer (ITS2) and 28S rDNA sequences. *Insect Mol. Biol.* **1993**, *2*, 225–237. [CrossRef]
35. Gillespie, J.J.; Munro, J.B.; Heraty, J.M.; Yoder, M.J.; Owen, A.K.; Carmichael, A.E. A secondary structural model of the 28S rRNA expansion segments D2 and D3 for chalcidoid wasps (Hymenoptera: Chalcidoidea). *Mol. Biol. Evol.* **2005**, *22*, 1593–1608. [CrossRef]
36. Puillandre, N.; Lambert, A.; Brouillet, S.; Achaz, G. ABGD, Automatic Barcode Gap Discovery for primary species delimitation. *Mol. Ecol.* **2012**, *21*, 1864–1877. [CrossRef]
37. Pons, J.; Barraclough, T.G.; Gomez-Zurita, J.; Cardoso, A.; Duran, D.P.; Hazell, S.; Kamoun, S.; Sumlin, W.D.; Vogler, A.P. Sequence-Based Species Delimitation for the DNA Taxonomy of Undescribed Insects. *Syst. Biol.* **2006**, *55*, 595–609. [CrossRef] [PubMed]
38. Kimura, M. A simple method for estimating evolutionary rates of base substitutions through comparative studies of nucleotide sequences. *J. Mol. Evol.* **1980**, *16*, 111–120. [CrossRef] [PubMed]
39. Saitou, N.; Nei, M. The neighbor-joining method: A new method for reconstructing phylogenetic trees. *Mol. Biol. Evol.* **1987**, *4*, 406–425.
40. R Core Team. *R: A Language and Environment for Statistical Computing*; R Foundation for Statistical Computing: Vienna, Austria, 2024; Available online: https://www.R-project.org/ (accessed on 26 April 2024).
41. Suchard, M.A.; Lemey, P.; Baele, G.; Ayres, D.L.; Drummond, A.J.; Rambaut, A. Bayesian phylogenetic and phylodynamic data integration using BEAST 1.10. *Virus Evol.* **2018**, *4*, vey016. [CrossRef] [PubMed]
42. Drummond, A.J.; Ho, S.Y.W.; Phillips, M.J.; Rambaut, A. Relaxed phylogenetics and dating with confidence. *PLoS Biol.* **2006**, *4*, 699–710. [CrossRef] [PubMed]
43. Hasegawa, M.; Kishino, H.; Yano, T. Dating of human-ape splitting by a molecular clock of mitochondrial DNA. *J. Mol. Evol.* **1985**, *22*, 160–174. [CrossRef] [PubMed]
44. Rambaut, A.; Drummond, A.J.; Xie, D.; Baele, G.; Suchard, M.A. Posterior summarization in Bayesian phylogenetics using Tracer 1.7. *Syst. Biol.* **2018**, *67*, 901–904. [CrossRef] [PubMed]
45. Nguyen, L.T.; Schmidt, H.A.; von Haeseler, A.; Minh, B.Q. IQ-TREE: A fast and effective stochastic algorithm for estimating maximum-likelihood phylogenies. *Mol. Biol. Evol.* **2015**, *32*, 268–274. [CrossRef] [PubMed]
46. Kalyaanamoorthy, S.; Minh, B.Q.; Wong, T.K.; von Haeseler, A.; Jermiin, L.S. ModelFinder: Fast model selection for accurate phylogenetic estimates. *Nat. Methods* **2017**, *14*, 587–589. [CrossRef]
47. Letunic, I.; Bork, P. Interactive Tree of Life (iTOL) v6: Recent updates to the phylogenetic tree display and annotation tool. *Nucleic Acids Res.* **2024**, gkae268. [CrossRef]
48. LaSalle, J.; Graham, M.W.R.d.V. On the identity of *Baryscapus* Förster (Hymenoptera: Eulophidae: Tetrastichinae). *Entomol. Gaz.* **1990**, *41*, 121–128.
49. LaSalle, J. North American genera of Tetrastichinae (Hymenoptera, Eulophidae). *J. Nat. Hist.* **1994**, *28*, 109–236. [CrossRef]
50. Song, H.-T.; Fei, M.-H.; Li, B.-P.; Zhu, C.-D.; Cao, H.-X. A new species of *Oomyzus* Rondani (Hymenoptera, Eulophidae) reared from the pupae of *Coccinella septempunctata* (Coleoptera, Coccinellidae) in China. *ZooKeys* **2020**, *953*, 49–60. [CrossRef]

insects MDPI

Article

Chilean Darwin Wasps (Ichneumonidae): Biogeographic Relationships and Distribution Patterns

Diego G. Pádua [1], Andrés Moreira-Muñoz [2], Vanezza Morales-Fierro [3,4] and Rodrigo O. Araujo [1,*]

[1] Laboratorio de Entomología General y Aplicada, Centro de Investigación de Estudios Avanzados del Maule, Universidad Católica del Maule, Avenida San Miguel, 3605, Talca 3460000, Chile; paduadg@gmail.com
[2] Instituto de Geografía, Pontificia Universidad Católica de Valparaíso, Avenida Brasil, 2241, Valparaíso 2340025, Chile; andres.moreira@pucv.cl
[3] Herbario EIF & Laboratorio de Evolución y Sistemática, Facultad de Ciencias Forestales y de la Conservación de la Naturaleza, Universidad de Chile, Av. Santa Rosa, 11315, La Pintana, Santiago 8820808, Chile; vdmorales@gmail.com
[4] Museo Nacional de Historia Natural, Interior Quinta Normal, s/n, Santiago 8350410, Chile
* Correspondence: araujorodrigodeoliveira@gmail.com

Simple Summary: The global biodiversity crisis poses many challenges for humanity, and continuing to classify and gain knowledge of the "hidden biodiversity" of less studied groups considered hyper-diverse insect groups, such as the parasitoid wasp (Ichneumonidae), is one of them. There is a particular need for more taxonomic and distributional knowledge of the Neotropics and its neighboring regions. We assessed the current knowledge of the Ichneumonidae, otherwise known as Darwin wasps, in Chile, a country with a diverse ecogeographic gradient, from the southern cold and humid tip of South America approaching the Antarctic Circle to the arid Atacama beyond the Tropic of Capricorn. Following the most up-to-date taxonomic knowledge, we assessed biogeographic relationships at the genus level and the spatial patterns of biodiversity at the species and genera levels along the latitudinal gradient. The results show that biogeographic relationships are based on six faunistic elements (cosmopolitan; endemic; Neotropical; Holarctic–Oriental; south-temperate; and Australasian), reinforcing the early view of two different areas for Chilean ichneumonids: a northern province and a Neantarctic realm, with a high percentage of endemic genera and species. Spatial biodiversity patterns assessed at different scales show a bimodal distribution of richness: around 34° and 38° S in the Metropolitan and Araucanía Regions. From an ecoregional perspective, richness is concentrated in the Valdivian temperate forests, but when assessed at a 0.5×0.5 cell scale, several outstanding cells are in the contact zone between the temperate forests and the Chilean Matorral in the Central Chilean biodiversity hotspot. Interpreting richness involves two phenomena difficult to disentangle: genuine species diversity and collection bias closer to areas with larger human populations. In contrast, the Atacama Desert shows little or no presence of Darwin wasps, which is to be expected due to the lack of potential hosts. These results reinforce the need to continue sampling and studying available collections to help close the knowledge gaps already recognized as Wallacean and Linnean shortfalls in order to gain additional information on potential threats to endemic genera and species.

Abstract: Ichneumonidae, or Chilean Darwin wasps, are an important component of South American hymenopteran diversity, but the taxonomic and distributional knowledge on this insect is still deficient. Taking advantage of recently updated taxonomic knowledge, we assessed biogeographic relationships at the genus level and biodiversity spatial patterns along the latitudinal gradient. The results show the presence of 264 species in Chile, arranged in 102 genera and 22 subfamilies. Biogeographic relationships are based on six elements (cosmopolitan (n = 50; 36%), endemic (n = 29; 21%), Neotropical (n = 22; 16%), Holarctic–Oriental (n = 19; 14%), south-temperate (n = 16; 11%) and Australasian) and composed of just three genera: *Anacis*, *Labena*, and *Meringops*. Species and genera show a bimodal distribution along the latitudinal gradient: around 34° and 38° S. From an ecoregional perspective, richness is concentrated in the Valdivian temperate forests, but when assessed at a 0.5×0.5 cell scale, several outstanding cells are in the contact zone between the temperate forests and the Chilean Matorral. On the other hand, the Atacama Desert shows little or no presence of Darwin wasps. The results agree with Charles Porter, who identified a northern province composed of Neotropical and cosmopolitan genera with their own representatives in the far north (11 genera), a distributional gap in the core of the Atacama Desert, and around 128 genera in Porter's Neantarctic realm, covering all of Chile from 25° S to Cape Horn, including the Juan Fernandez islands. These results reinforce knowledge gaps and the need for more sampling and studies of available collections. Due to sampling gaps at this stage, identifying a continued increase or decrease in richness towards higher latitudes is not possible. More taxonomic and distributional information is also needed to assess potential threats to endemic genera and species.

Keywords: biodiversity; Chile; cosmopolitan; endemism; faunistic elements; Holarctic; Neotropics; parasitoid wasps; richness

1. Introduction

If we take a close look at the bark of a tree in a subtropical forest, we will have a high likelihood of finding a Darwin wasp (Ichneumonidae) (Figure 1) in the struggle to secure a host for its eggs. Ichneumonidae are parasitoid wasps, a major component of global biodiversity, and at the same time a poorly studied group [1]. They comprise the most speciose group of Hymenoptera (one of the most species-rich orders on Earth), participate in a wide range of ecological processes, and provide humanity with essential ecosystem services [2]. Nevertheless, there are various impediments to improving our knowledge of the diversity and functional roles of parasitoid wasps, such as the Linnean shortfall (most species have not been described) and Wallacean shortfall (the distribution of many described species is unknown) [3].

These various impediments have dramatic consequences for insect conservation [4,5], especially actions to protect and manage so-called "hidden biodiversity" [6], and apply especially to (a) laboriously identifiable species; (b) those with unknown socio-economic potential; and (c) those which can only be collected in areas difficult to access [7]. The knowledge gaps related to parasitoid wasps and their systematic underestimation "biases our understanding of multi-trophic tropical interactions and determination of large-scale biodiversity patterns" [7] (p. 4697).

The lack of specialized entomologists and accurate, standardized, and cost-effective sampling protocols are added impediments to the plight to advance our knowledge of parasitoid wasps [2].

Though accurate updated numbers are lacking, estimates suggest that Hymenoptera diversity in the Neotropics is greater than in the Nearctic, Palearctic, and Australian regions [8].

Ichneumonidae are the largest Hymenoptera family, and are currently divided into 42 subfamilies [9] with over 25,000 valid species [10]. Besides having the highest endemism rates, this family is also highly relevant for the practice of biological control since the species that comprise it obligatorily deposit their eggs in arthropods [8].

The Chilean biota has attracted much attention due to its connection with the Neotropics on the one hand, and its connection with Australasia as a remnant of an ancient Gondwanan biota on the other [11]. Indeed, the long latitudinal gradient and the presence of different environments along this gradient make Chile a biogeographic laboratory. The rapid uplift of the Andes since the Late Miocene prompted the isolation of biota, leading to remarkable levels of endemism [11].

Figure 1. Darwin wasp representatives from Chile (in vivo): (**a**) *Trachysphyrus* sp. (Cryptinae); (**b**) Ichneumoninae sp.; (**c**) *Dotocryptus* sp. (Cryptinae); (**d**) *Macrogrotea* sp. (Labeninae); (**e**) *Trachysphyrus* sp. (Cryptinae); (**f**) *Hoplismenus* sp. (Ichneumoninae).

Regarding species richness, in the northern hemisphere, there are indications of an inverse richness pattern; that is, an increase in richness towards higher latitudes. In the southern hemisphere, the lack of complete inventories creates difficulty in testing general-to-regional distribution patterns, and the richness of several groups tends to be concentrated in mid-latitudes. This pattern still needs to be tested regarding several explanations related to life history traits and attack strategies, mainly divided into different ovipositor lifestyles: idiobionts and koinobionts [12,13].

Chile has a very unique Ichneumonidae fauna, with differences in comparison to the Neotropics (see [14–16]). Porter [15] mentions that Chilean Darwin wasps differ significantly from the rest of South America and cannot be included as a sub-element of the Neotropical, due to "its exceptionally high number of endemic genera and its surprisingly few Neotropic genera for an area in geographic proximity to the American tropics" [15] (p. 38).

Indeed, Ichneumonidae are the family with the most endemic genera and species in Chile (including one endemic subfamily: Claseinae). The last published catalog reports a total of 36 endemic genera and 170 endemic species, out of a total of 88 genera and

193 recorded species [17]. The Chilean fauna includes representatives of cosmopolitan, Holarctic or Holarctic–Oriental, Neotropical, Andino-Patagonian, and Transantarctic (sharing species with Australia and New Zealand) genera, in addition to some genera that are widely but disjunctively distributed in both the Northern and Southern Hemispheres [15].

Porter [15] hypothesizes that the aberrant and endemic Chilean Darwin wasp fauna probably represents survivors that moved north from Antarctica before the glaciation and that evolved in isolation for the last 40 million years due to mountains, desertification, and a cold climate affecting the region's eastern and northern boundaries by the mid-Cenozoic.

Thorough research into Chilean ichneumonid fauna ecology is crucial to understanding interspecies interactions and distribution patterns. The main goal is to pinpoint endemic areas for the conservation and investigation of native species for the effective biological control of agricultural pests.

We took advantage of the most up-to-date taxonomic revision of the Chilean Ichneumonidae, allowing us to make progress towards two specific goals: (a) disentangling the biogeographic relations of Chilean Darwin wasps at the genus level; (b) discovering the spatial patterns of biodiversity along the latitude and altitude gradients.

2. Materials and Methods

2.1. Study Area

Due to its current geographical conditions, Chile is considered a biogeographic island, bordering the Sechura desert in Peru beyond the Capricorn Tropic to the north, the Andean highlands to the east, the Pacific Ocean to the west, and Cape Horn approaching the Antarctic Circle to the south. The latitudinal gradient spans from the northern dry areas of the most arid desert in the world, Atacama, toward subtropical scrubs, Mediterranean sclerophyllous forests, deciduous forests, and the temperate evergreen Valdivian Forest in the south. Further south are the subantarctic moors and dwarf forests of Magallanes approaching Cape Horn. This high diversity in environments and their temporal evolution give insects like hymenopteran, coleopteran, and other diverse groups opportunities to diversify, resulting in a high proportion of endemic species and genera. A useful map of Chilean environments, suitable for continental comparisons, is the map of ecoregions by Dinerstein et al. [18]. The main ecoregions in the country from north to south are the Atacama Desert, Chilean Matorral, Valdivian temperate forests, and the Magellanic subpolar forests. On the borders with Argentina and Peru, we also have representation of the Sechura Desert, the Central Andean dry Puna, the Southern Andes Steppe, and the Patagonian Steppe. Taking these ecoregions as a base, we plotted the number of species and genera (Figure 2).

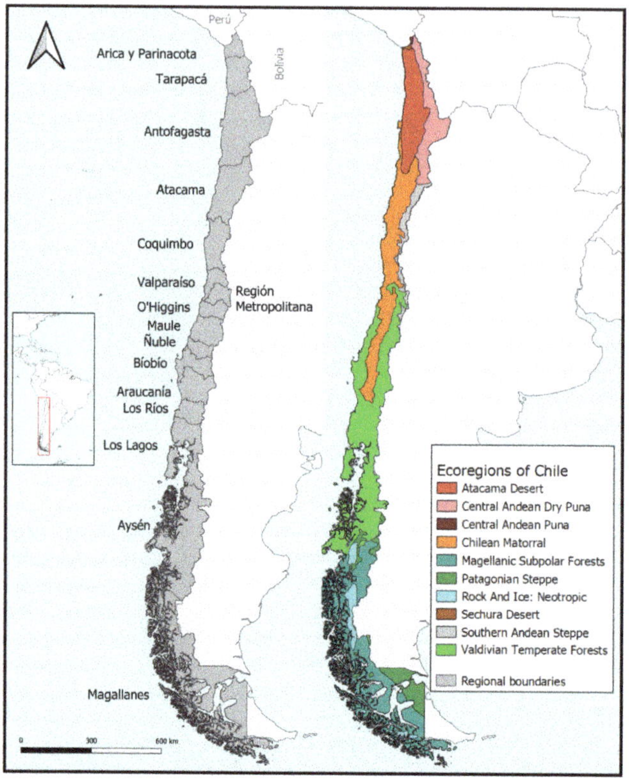

Figure 2. Study area: (**left**) administrative regions; (**right**) ecoregions according to Dinerstein et al. [18].

2.2. Biogeographic Relationships

For this analysis, 139 genera registered in Chile were considered according to Araujo et al. [19], except the introduced genera *Megarhyssa* and *Rhyssa*. We also included 35 genera with undetermined species (a total of 139 genera) (see Table S1). The genus *Stiboscopus* was not considered in the biogeographic classification because the genus is recorded by Porter [15] without species identification, and currently the genus has been divided into several genera, with *Stiboscopus* being synonymous with the genus *Lysibia*. It is also because we do not know which genus/genera the specimens belong to. The assessment of biogeographic relationships, including the classification of biotic elements, is a traditional task in biogeography [20–23]. We based our analysis on previous classifications of Chilean biota [11,15,24,25].

Porter [15], based on the knowledge available at the time, presented a fine description of biogeographic relationships among Chilean Ichneumonidae. He recognized 131 genera (including a couple undescribed) and arranged them in two main groups: (a) the genera restricted to the northern province, north of 25° S, including valleys in a desertic matrix and the high Andes adjacent to Peru and Bolivia; and (b) the Neantarctic realm south of 25° S encompassing all of central and southern Chile. Porter arranged 121 genera into five biogeographic elements in this Neantarctic realm: (a) endemic, (b) cosmopolitan, (c) Holarctic–Oriental, (d) Neotropical, (e) Australasian or Transantarctic, and (f) Holarctic–Neotropical–Australasian disjunct.

Following current taxonomic and distributional knowledge, we retrieved the following faunistic elements: (a) endemic, (b) cosmopolitan, (c) Holarctic–Oriental, and (d) Neotropical and Australasian. We added a specific element retrieved by Moreira-Muñoz [11] for Chilean flora: the south-temperate element, which encompasses Neotropical genera but occurs only south of 33° S in Chile and adjacent Argentina, mainly in temperate forests.

According to current knowledge, Porter's Holarctic–Neotropical–Australasian disjunct element is composed only of the genus *Isdromas*, but this genus can easily be considered subcosmopolitan.

2.3. Biodiversity Spatial Patterns

For this analysis, we only considered genera with determined species (a total of 102 genera, according to Araujo et al. [19]).

Data cleaning included reviewing the geo-referencing of 939 individuals. We used geographic name repositories (Geonames and Mapcarta) and our own localities database. We had to disregard 2.7% of the data due to misspellings or incomplete distributional information (e.g., a whole region, confused names, etc.). After database cleaning and the erasure of duplicates coordinated in individual collections, our database consisted of 914 records, encompassing 264 species in 101 genera (Table S2).

Maps were generated on two scales using ArcGis 10.3, considering ecoregions and cells of 0.5 × 0.5 degrees. This cell size has been shown to perform well at a national scale compared to 1 × 1 degree, which is better for visualization at the level of administrative regions (Figure 2). This has no biological meaning but is informative for richness and collection efforts.

3. Results

3.1. Biogeographic Relationships

For this analysis, all genera documented in Chile (totaling 139 genera) were taken into account following Araujo et al. [19].

The biogeographic relationships at the genus level showed a remarkable presence of 50 cosmopolitan (36%) and 29 endemic genera (21%) (Table 1). Porter [15] has already noted that the proportion of endemic genera is similar to or greater than Madagascar, New Zealand, and Australia [22]. Endemic genera (Neantarctic genera—sensu Porter) [14] are mostly represented by one unique species collected in a few localities.

Table 1. Darwin wasp genera elements from Chile (n = 139) (see Section 2).

	Elements	Genera (No.)	%
1	Cosmopolitan	50	36
2	Endemic	29	21
3	Neotropical	22	16
4	Holarctic–Oriental	19	14
5	South-temperate	16	11
6	Australasian	3	2

The other strong elements in the Chilean Darwin wasp genera are the Neotropical (n = 22; 16%), Holarctic–Oriental (n = 19), and south-temperate (n = 16) elements. The least represented element is the Australasian (Transantarctic in the sense of Porter [15]), with just three genera: *Anacis*, *Labena*, and *Meringops* (see Section 4).

3.1.1. Cosmopolitan Element

The cosmopolitan element of Chilean Darwin wasp genera is composed of 50 genera: *Habronyx*, *Parania*, and *Therion* (Anomaloninae); *Exetastes* and *Lissonota* (Banchinae); *Brachycyrtus* (Brachycyrtinae); *Campoletis*, *Campoplex*, *Casinaria*, *Diadegma*, *Dusona*, *Hyposoter*, *Meloboris*, and *Venturia* (Campopleginae); *Pristomerus* and *Trathala* (Cremastinae); *Cryptus* and *Mesostenus* (Cryptinae); *Diplazon*, *Syrphoctonus*, and *Woldstedtius* (Diplazontinae); *Diphyus*, *Eutanyacra*, *Hoplismenus*, *Ichneumon*, *Melanichneumon*, *Setanta*, *Dicaelotus*, *Tycherus*, and *Platylabus* (Ichneumoninae); *Cidaphus* and *Mesochorus* (Mesochorinae); *Colpotrochia* and *Hypsicera* (Metopiinae); *Enicospilus* and *Ophion* (Ophioninae); *Megastylus, Symplecis,* and *Stenomacrus* (Orthocentrinae); *Dichrogaster, Gelis, Xenolytus, Charitopes,* and *Atractodes* (Phygadeuontinae); *Clistopyga, Tromatobia, Itoplectis,* and *Pimpla* (Pimplinae); and *Netelia* (Tryphoninae). *Isdromas* (Phygadeuontinae) is distributed in Chile (Tarapacá region), Argentina, Brazil, Peru, and Ecuador in the Neotropics, Honduras in Central America, the United States in North America, and Australia. Hence, it can be considered subcosmopolitan (see Table S1). Indeed, most genera included in the element have a wide distribution in more than two continents or more than two main climatic zones (e.g., tropical and temperate). The element should be called *subcosmopolitan* (or semicosmopolitan, according to Porter [15]) because only half the genera occur in all continents, while the other half occur in the Nearctic, Oriental, Eastern and Western Palearctic, Neotropical, and European regions (Table S1).

3.1.2. Endemic Element

The endemic element is composed of genera distributed in continental Chile and the Juan Fernandez Islands. There is one endemic subfamily (Claseinae) and 29 endemic genera: *Archoprotus* and *Valdiviglypta* (Banchinae); *Clasis* and *Ecphysis* (Claseinae); *Caenopelte, Araucacis, Nothischnus,* and *Periplasma* (Cryptinae); *Pedinopa, Cacomisthus, Petilium,* and *Stipomoles* (Ctenopelmetinae); *Barronia* (Eucerotinae), *Chilelabus, Ithaechma,* and *Zophoplites* (Ichneumoninae); *Gauldianus* (Labeninae), *Chineater,* and *Latilumbus* (Mesochorinae); *Pedunculus* (Pedunculinae), *Acidnus, Rhabdosis, Surculus, Peumocryptus, Bilira,* and *Teluncus,* (Phygadeuontinae); and *Notophrudus* (Tersilochinae).

These genera are distributed in Central Chile (*Archoprotus, Periplasma,* and *Chilelabus*), Central–Southern Chile (*Araucacis* and *Ithaechma*), and most predominantly in Southern Chile (*Valdiviglypta, Ecphysis, Caenopelte, Nothischnus, Clasis, Pedinopa, Cacomisthus, Petilium, Stipomoles, Barronia, Zophoplites, Gauldianus, Chineater, Latilumbus, Pedunculus, Acidnus, Rhabdosis, Surculus, Peumocryptus, Bilira, Teluncus,* and *Notophrudus*).

3.1.3. Neotropical Element

The Neotropical element is composed of 22 genera, with Cryptinae the predominant one (*Dotocryptus Trachysphyrus Cyclaulus, Aeglocryptus, Cosmiocryptus, Hypsanacis, Itamuton, Neocryptopteryx, Phycitiplex,* and *Xylacis*), in addition to a few other genera of subfamilies such as *Cecidopimpla,* and *Diradops* (Banchinae); *Prochas* (Campoplegionae); *Coelorhachis* (Ctenopelmatinae); *Carinodes, Diacanthatius,* and *Thymebatis* (Ichneumoninae); *Alophophion* (Ophioninae), *Grotea* (Labeninae), *Calliephialtes,* and *Odontopimpla* (Pimplinae); and *Stethantyx* (Tersilochinae) (see Table S1).

3.1.4. Holarctic–Oriental Element

The Holarctic–Oriental element is composed of *Glypta* (Banchinae), *Cymodusa, Campoctonus, Microcharops, Nemeritis,* and *Phobocampe* (Campopleginae); *Sussaba* (Diplazongtinae), *Scolomus,* and *Seticornuta* (Metopiinae); *Apoclima, Helictes,* and *Gnathochorisis* (Orthocentrinae); *Aclastus, Ethelurgus, Stilpnus,* and *Distathma* (Phygadeuontinae); *Stenobarichneumon* (Ichneumoninae); and *Liotryphon* and *Polysphincta* (Pimplinae).

3.1.5. South-Temperate Element

The south-temperate element is composed of genera distributed in the Chilean Andes Mountains and adjacent Argentina. In Chile, 16 genera of Darwin wasps were found within this element: *Tatogaster* (Tatogastrinae); *Geraldus* (Banchinae); *Aglaodina, Chilecryptus, Myrmecacis, Oecetiplex, Picrocryptoides, Sciocryptus,* and *Xiphonychidion* (Cryptinae); *Cataptygma* and *Tetrambon* (Ctenopelmatinae); *Barythixis, Chilhoplites, Notophasma* (Ichneumoninae); *Torquinsha* (Labeninae); and *Lepidura* (Mesochorinae). All species of the genera classified under the south-temperate element have a Neantarctic distribution mainly in the Chilean part, and most south-temperate Darwin wasps with distribution in adjacent

Argentina are present in Neuquén Province. The only exception could be *Aglaodina* (Cryptinae), reaching Antofagasta at 23°17′ south latitude.

Petilium and *Notostilbops*, classified as endemic, may in the future be classified as south-temperate, mainly due to the species distributed in Natales and Punta Arenas that are probably also in Argentina.

3.1.6. Australasian Element

The Australasian element (Transantarctic in the sense of Porter [15]) is composed of genera distributed in Australia, South America, and the Pacific islands. In Chile, only three genera comprise this element: *Anacis* (Cryptinae), *Labena* (Labeninae), and *Meringops* (Phygadeuontinae).

3.2. *Spatial Pattern of Biodiversity along the Latitudinal Gradient*

For this section, we focused on genera with identified species as revised by Araujo et al. [19]. This database is composed of 922 records encompassing 264 species arranged in 102 genera. More than half of the genera (n = 65) are composed of only one species (Figure 3).

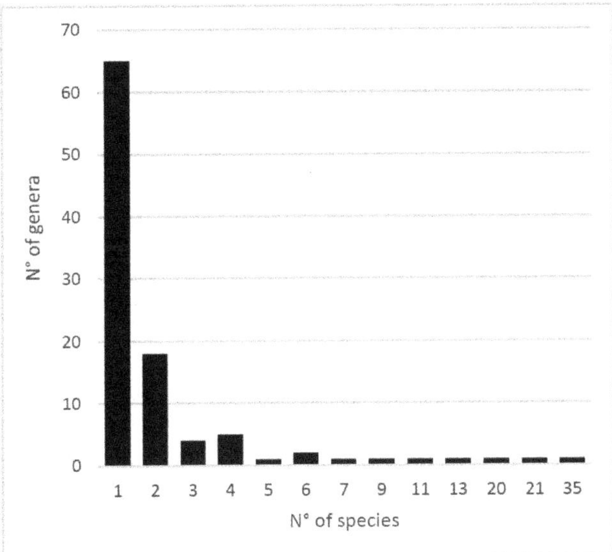

Figure 3. Number of species per genus among Chilean Ichneumonidae (only genera with determined species—n = 102).

Regarding the assessment of ecoregions, the largest number of species and genera (including 65 monospecific genera) are in the Valdivian temperate forests (167 species and 66 genera). The second ecoregion is the Chilean Matorral (with 89 species and 48 genera). Both ecoregions together make up the Chilean Winter Rainfall–Valdivian Forests biodiversity hotspot [26,27].

Regarding administrative regions following the latitudinal gradient, there is a bimodal distribution of species and genera richness: around 34° S in the Metropolitan Region and 38° S in the Araucanía Region (Figures 4 and 5). The regions least suited for ichneumonid wasps are the arid northern regions, especially the Antofagasta Region at the center of the Atacama Desert (Figures 3–5).

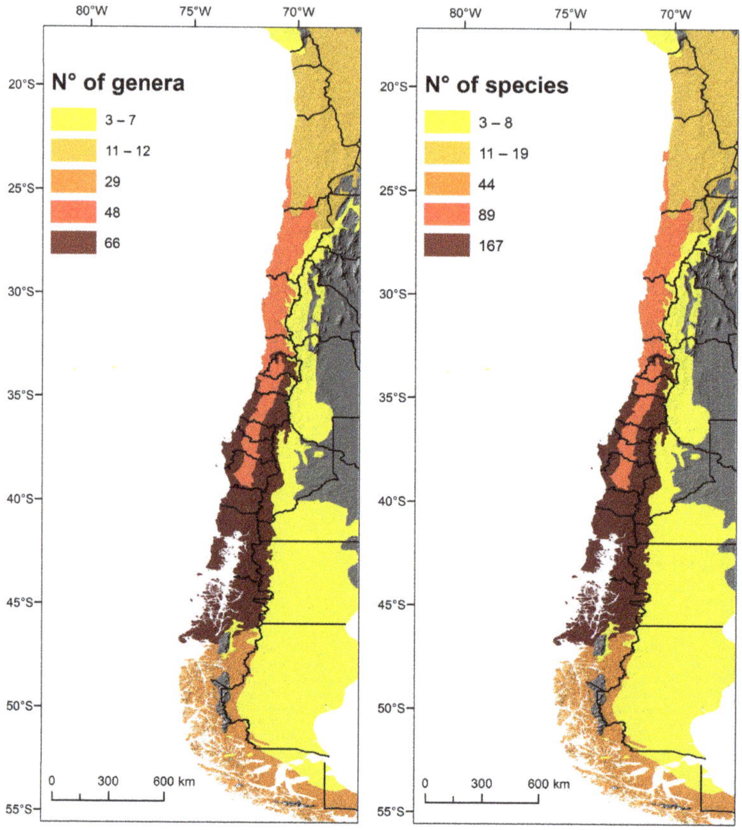

Figure 4. Ecoregions and species/genera richness. Ecoregions on the border with Argentina, Bolivia, and Peru only include data for Chile (only genera with determined species—n = 102).

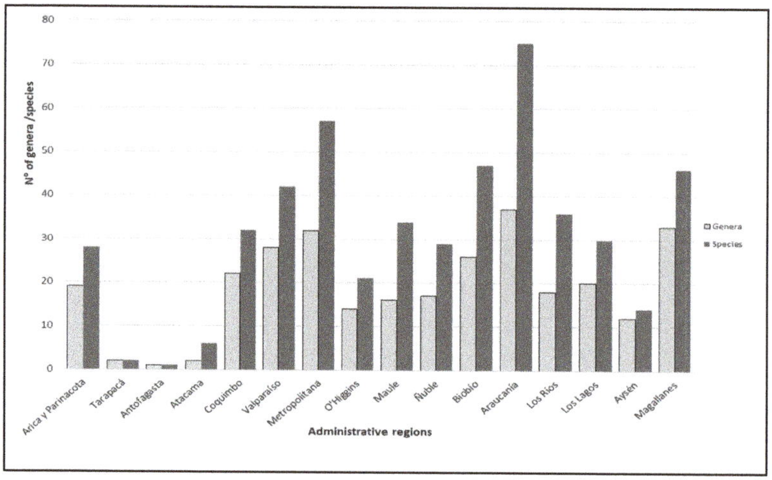

Figure 5. Species and genera richness for each administrative region (only genera with determined species—n = 102).

Plotted in cells of 0.5 × 0.5 degrees, the concentration of ichneumonid biodiversity is highest in the Santiago Andes (Figure 6). Several cells in the Andes of the Maule, Ñuble, and Araucanía Regions, as well as the coast of Biobío and Chiloé, also stand out (Figure 6).

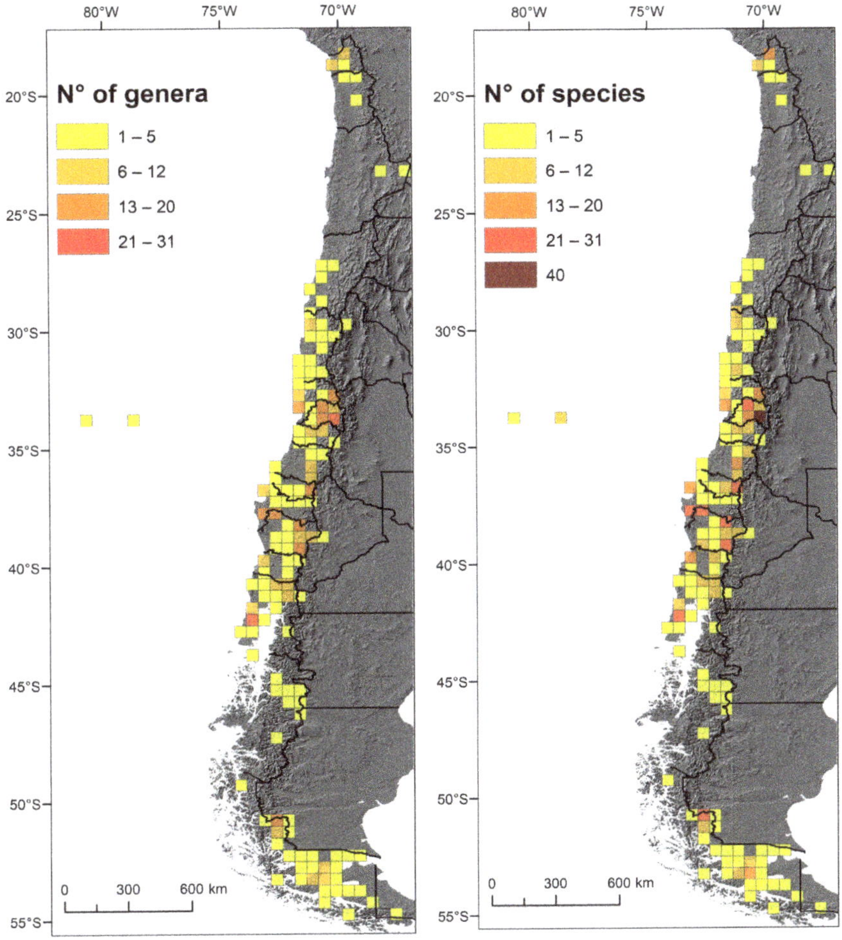

Figure 6. The richness of Chilean ichneumonid species plotted in 0.5 × 0.5 cells: (**left**) number of genera; (**right**) number of species.

4. Discussion

Current estimates suggest that the Neotropics host the highest Hymenoptera diversity globally [8]. By modeling host–parasitoid systems, Forbes et al. [28] proposed that there may be 2.5 to 3.2 times more Hymenoptera species than Coleoptera. This is not just a clue for entomologists, but also useful from ethological and ecological perspectives. As potential biological pest control agents, and in their still little-known ecological roles, the importance of these estimates to insect conservation is enormous.

Regarding Chilean ichneumonids, our understanding is still limited in terms of taxonomic and distributional aspects (Linnean and Wallacean shortfalls, respectively), even with the recent availability of an updated catalog [19]. There are currently a total of 139 genera classified in 23 subfamilies [19]. This represents 37% more than the last authoritative catalog [17] and is closer to the numbers produced by Porter [15] over 3 decades ago. Porter considered the Chilean ichneumonid fauna to be composed of 131 genera and over

170 species (see also [29]). At the time, Porter increased taxonomic knowledge by 100% over previous studies. He considered that increasing taxonomic knowledge would result in the discovery of 1000 to 1500 species. We are still far from these numbers, but they are continuously increasing. Still, 35 genera recognized by Porter and considered valid for Chile lack any specimens classified at the species level. The stability of numbers at the genus level and the gaps at the species level respond to two possible phenomena: the lack of specialists to increase these numbers (i.e., identification, collection with specific traps, etc.), and the presence of a depauperate fauna in Chile, rich in endemism but not as rich in species numbers as in the rest of the Neotropics. The country's prolonged biogeographic isolation accounts for the high proportion of endemism at the genus and species levels, but continuous environmental disturbances at a geological scale could also lead one to hypothesize innumerable extinctions [15].

The biogeographic relationships of Chilean ichneumonids have an intrinsic relationship to the evolution of Chilean biota as a whole, but one that is not well studied. Porter [15] arranged the Chilean genera in two main groups: (a) the genera restricted to the northern province, north of 25° S, including valleys in the Atacama Desert and the high Andes adjacent to Peru and Bolivia; and (b) the Neantarctic realm south of 25° S encompassing all of central and southern Chile.

According to this approach, the northern province is mainly composed of Neotropical and cosmopolitan genera, with their representatives in the far north shared with Peru and Bolivia, such as species from the genera *Brachycyrtus, Carinodes, Cosmiocryptus, Cyclaulus, Cymodusa, Hypsanasis, Isdromas, Itoplectis,* and *Mesostenus.* We can now add representatives of *Lissonota* and *Microcharops* to this list (one species each from the Azapa Valley).

All the other 120 genera belong to Porter's Neantarctic realm, encompassing all of Chile south of 25° S toward Cape Horn, including the Juan Fernandez islands. The debate between the existence of a Neotropical realm and a Neantarctic realm has been permanent in Austral biogeography (see [30,31]). At least for well-documented groups such as vascular plants, the existence of an "austral realm" has long been established [32], and more recently confirmed [11,33]. The terms Neantarctic, Holantarctic, Australasiatic, and Austral have similar meanings, emphasizing the biogeographical relationships of the disjunct distribution across the Pacific, mainly in southern South America and Australasia (New Zealand and Australia), with a minor relationship to the Cape Region in South Africa. Analyzing the distribution of vascular plants, Moreira-Muñoz [33] concluded that "there are 15 families and c. 60 genera that, under current taxonomic treatment, support the segregation of an Austral realm" [33] (p. 1657).

In Porter's concept of ichneumonid biogeography, 39 genera (32%) belong to the endemic element; 31 genera (25%) correspond to the cosmopolitan element; 20 genera (16%) belong to the Holarctic–Oriental element; 17 genera (14%) correspond to the Neotropical element; just 4 genera comprise the Transantarctic (Australasian) element; and 3 genera belong to the disjunct Holarctic–Neotropical–Australasian element.

The four main elements—endemic, cosmopolitan, Holarctic–Oriental, and Neotropical—are recognizable according to current knowledge as follows: cosmopolitan element (n = 50, 36%), endemic element (n = 29, 21%), Neotropical element (n = 22, 16%), Holarctic–Oriental (n = 19, 14%). We also now recognize a south-temperate element (n = 16, 11%) and, following taxonomic updating, the Australasian element is now only composed of three genera: *Anacis* (Cryptinae), *Labena* (Labeninae) and *Meringops* (Phygadeuontinae).

According to Porter [15], these endemic genera sharply define the Neantarctic realm. The Neantarctic realm has a variety of landscapes ranging from semi-desert with sclerophyllous woods to humid *Nothofagus* forests in the south. It also has a disjunct relationship with certain genera in Australia and the Holarctic region, revealing the absence of dominant Neotropical taxa. In addition, Porter [15] divided the Neantarctic realm into four provinces according to the biota and phytophysiognomy of its environments: Atacamense; Mediterranean or Central; Valdivian; and Magallanic. The Valdivian province is the Chilean region with the greatest diversity of Neantarctic Ichneumonidae. This is because Darwin wasps prefer to frequent humid environments or close forests. After all, most species are hygrophilous and sylvatic (but, especially under cool thermal regimes, they also tend to invade open forests and grasslands) [16]. Valdivian vegetation is characterized by temperate evergreen forest, typically with neatly developed herbaceous, shrubby, and arboreal strata. The dominant tree niche is occupied mainly by various *Nothofagus*, *Eucryphia*, *Gevuina*, *Embothrium*, *Lomatia*, and *Drimys* species. This province is one of the best-defined centers of endemism in South America [15,34].

Moreira-Muñoz [11] says that it would be better to refer to the Neotropical element as an "American" element that ranges from the Northwestern United States to Mexico, Central America, the Northern Andes, Amazonia, the Central Andes, and the Southern Andes, including Chile and Argentina. However, Porter [15] emphasizes the fact that Chilean ichneumonid fauna differs from the rest of the Neotropics due to having the highest presence of endemic taxa and the lack of a predominance of Neotropical taxa in the Neantarctic realm south of 27 degrees [15,16]. According to Araujo et al. [19], six neotropical genera have widely distributed species in northern and southern Chile (*Dotocryptus*, *Trachysphyrus*, *Itamuton*, *Thymebatis*, *Alophophion*, and *Calliephialtes*), five genera are exclusive to Northern Chile (*Cyclaulus*, *Cosmiocryptus*, *Hypsanacis*, *Microcharops*, and *Carinodes*), three genera are exclusive to Central Chile (*Phycitiplex*, *Prochas*, and *Diacanthatius*), five genera are distributed in Central–Southern Chile (*Aeglocryptus*, *Neocryptopterys*, *Cecidopimpla*, *Diradops*, *Coelorhachis*, and *Stethantyx*), one neotropical genus (*Xylacis*) is exclusive to Southern Chile, and the genus *Odontopimpla* has uncertain distribution in Chile.

The Neotropical element can be further split into several subgroups (described as generalized tracks in Moreira-Muñoz [11]): Wide Neotropical track—composed of genera occurring in NW United States, from Mexico to Chile. Among Darwin wasps, seven neotropical genera were identified: *Diacantharius*, *Diradops*, *Microcharops*, *Carinodes*, *Calliephialtes*, *Odontopimpla*, and *Stethantyx*. Two genera (*Diacantharius* and *Odontopimpla*) reach Mexico but not the United States. Wide Andean track—composed of genera occurring in Costa Rica and ranging from Colombia to Chile. Five genera were identified: *Cecidopimpla*, *Dotocryptus*, *Trachysphyrus*, *Hypsanacis*, and *Alophophion*. Only *Cecidopimpla* reached Costa Rica. Central Andean or Altiplano track—composed of genera occurring in the Andean Altiplano in Peru, Chile, Bolivia, and Argentina. Five genera were identified: *Cyclaulus*, *Cosmiocryptus*, *Itamuton*, *Phycitiplex*, and *Xylacis*. South Amazonian track—composed of genera occurring in the Andes and southern Amazonia. Five genera were identified: *Prochas*, *Aeglocryptus*, *Neocryptopteryx*, *Thymebatis*, and *Alophophion*. According to Fernandes et al. [35], in Brazil these five genera are distributed mainly in the South, Southwest, and Northeast regions. *Phycitiplex* is only distributed in Argentina, Chile, and Uruguay.

Regarding the Holarctic–Oriental element, Porter [15] mentions that the Holarctic genera arrived in the Eocene and Oligocene eras, mainly because the Oligocene and subsequent periods experienced cold pulses (the world has undergone repeated changes from cold to tropical and from humid to arid climatic regimes since the Oligocene) that allowed for the exchange of Holarctic biota across the Andes and Central American mountains to the south and from the Andean–Patagonian regions to the north.

Australasian (or Transantarctic) genera present in the New World and Australia are very similar to the oldest insect orders (e.g., Ephemeroptera, Odonata, Plecoptera, Mecoptera), as they frequently appear among vascular plants [11]. According to the evidence, Labeninae radiated throughout Gondwana after the separation of Africa, India, and Madagascar, but before the separation of Australia [36]. This indicates a plausible pathway for biotic exchange between South America and Australia, possibly via Antarctica. Biogeographic inference also reveals that North American groups underwent more recent range expansions before the formation of the Isthmus of Panama land bridge [37]. This implies a more intricate scenario for Labeninae biogeography than previously anticipated.

Porter [15] already proposed that Chilean Darwin wasps have a connection to Antarctica. According to the hypothesis, the major lineages of this family already existed during the Cretaceous era, which was characterized by a much warmer and more humid global climate than today, and South America was still connected to Antarctica. During this period, a diverse biota evolved in Antarctica, which was extensively shared with southern America and Australia. Antarctica and South America remained connected until about 25 mya [38]. The unusual and endemic biota of Neantarctic South America probably survived by moving north from Antarctica before its glaciation. The mid-Cenozoic era marked the isolation of the Neantarctic realm along its eastern and northern boundaries due to mountain building, desertification, and cold climatic changes. As a result, the region's insect biota has evolved in isolation for the last 25 million years (though more recent relations across the sea have also been proven) [39].

For the diversity analysis along the latitudinal gradient, we were able to plot occurrences for 264 species arranged in 102 genera, of which 65 were composed of just one species, which again indicates a depauperate fauna and/or knowledge gaps. Other regions such as Mexico also show a high proportion of genera (n = 123) composed of just one species [40].

In Chilean ichneumonid fauna, many species are represented by only one specimen, such as *Aeglocryptus nigricornis* (Brullé, 1846); *Acidnus ensifer* (Townes, 1970); *Aglaodina hyperbas* (Porter, 1967); *Barronia araucaria* (Gauld & Wahl, 2002), *Microcharops anticarsiae* (Gupta, 1987); and *Venturia porteri* (Brèthes, 1913). Others reach almost 20 species, such as *Nemeritis scaramozzinoi* Di Giovanni & Araujo, 2021 and *Trachysphyrus agenor* Porter, 1967, while others reach a dozen species, such as *Tycherus chileator* Diller, 2009 and *Trachysphyrus penai* Porter, 1967. Most genera are represented by few species, and these species are represented by few specimens. At this stage, it is not possible to discern between species that are indeed rare in the field from those that are under-collected due to limited collection efforts and a lack of specialists (a characteristic of the concept of "hidden biodiversity").

A total of 264 species and 102 genera is not a large number compared to megadiverse countries such as Brazil or Mexico (with 1066/239 and 1031/373 species/genera, respectively). Mexico registers 45% endemic species, while Brazil registers just 3.1% endemic species [35,40].

Comparing the richness of Chilean Ichneumonidae with the biogeographical regions of Tenebrionidae in Chile proposed by Peña [41], as well as the more recent ecoregions proposed by Morrone [42], Darwin wasps are mainly found in the central regions of coastal Cordillera and southern Andean Cordillera [41], currently classified by Morrone [42] as the districts of Santiago Province (33–37° S) in the Central Chilean subregion and the Northern Valdivian Forest and Valdivian Forest regions [41]. Both regions are classified by Morrone [42] as provinces (in part) of the Subantarctic Region, with the Northern Valdivian Forest in Maule Province (37–39° S) and the Valdivian Forest in the Valdivian Forest Province (39–47° S).

The distribution pattern of Chilean Darwin wasps at the regional level does not support an increasing richness towards the tropics nor an increasing richness towards the south, with the highest richness maintained at mid-latitudes, as has been shown for comparable groups such as Chilean bees [43]. Chilean butterflies also display the highest richness at mid-latitudes [44].

This pattern supports the traditional (but debated) latitudinal trend of increasing richness towards the tropics. But in the case of ichneumonids, the taxonomic and distributional knowledge is still too limited (Wallacean and Linnean shortfalls); however, changes in this tendency can confidently be expected in the coming years (if more specialists join the challenge). The pattern is similar in comparable megadiverse botanical groups such as the Asteraceae; Chile harbors much less diversity than Brazil or Peru [45]. This is a consistent pattern in this biotic group, where the taxonomic and distributional knowledge is much more reliable.

In the case of Ichneumonidae, even if the Linnean and Wallacean shortfalls are better filled in the coming years, the richness pattern will continue to have a bimodal distribution and be concentrated at mid-latitudes due to the presence of the Atacama Desert, which constitutes a physical barrier for the diversification of life. This fact does not imply a merely barren environment, but an evolutionary arena favoring several groups such as Cactaceae [11] (p. 202) and Nolaneae [46] in the case of plant groups adapted to aridity.

The inverse latitudinal pattern of ichneumonids and the concentration at mid-latitudes, if confirmed for the Southern Hemisphere, still need to be tested regarding several explanations related to life history traits and attack strategies, as revised by Santos and Quicke [12]. Attack strategies are divided into different ovipositor lifestyles: idiobionts and koinobionts. According to Santos and Quicke [12] (and references therein), the concentration of the richness towards the south could be based on certain non-mutually exclusive explanations, such as (a) the resource fragmentation hypothesis, proposing that the diversity of hosts rises towards the equator, but that the density of each host population is too low to support koinobiont species; (b) the predation on hosts hypothesis, suggesting that predation on herbivores in the tropics is greater than in temperate regions; (c) the interphyletic competition hypothesis that parasitoids have to compete for hosts with other parasitic organisms that are more diverse in the tropics; and (d) the "nasty hosts hypothesis," based on the tendency of tropical plants to have more chemical toxins than their temperate counterparts [12] (p. 374).

The concentration of Chilean ichneumonids at mid-latitudes is coincident with the highest presence of tree species at middle latitudes around 36° and 40° S, at the transition from the Mediterranean matorral to the temperate forest ecoregions [34]. This issue has been described as a "Gondwanan legacy" [47]. Nothofagus tree species, a potential habitat for parasitoid hosts, range from 33° S toward Cape Horn, but their richness is concentrated from 35° S to 42° S [11] (p. 256). Some studies note defoliator species such as the "sawfly from roble and raulí", parasited by *Clasis* sp. [48].

Though species and generic richness are concentrated in several cells at mid-altitudes in the contact zone between the Valdivian Forests ecoregion and the Chilean Matorral, the need for additional sampling and the explicit assessment of collection biases and gaps through richness estimators is evident (see [49]).

Comparing altitudinal patterns, Flinte et al. [50] discovered that low- and mid-altitude areas on a mountain in the Brazilian Atlantic Forest contained significant diversity in terms of Darwin wasps, unlike their high-altitude counterparts. Moreover, distinct species were observed at various elevations along the mountain. These results imply that tropical forests could potentially host concentrated populations of Darwin wasps and deforestation poses a substantial risk of losing this biodiversity. Prioritizing the conservation of forests at low-to-middle altitudes may prove most effective in safeguarding the diversity of these wasps, though ensuring protection across a broad altitude range is essential for the preservation of all species.

The knowledge of Chilean ichneumonid biodiversity richness is biased towards the most human-populated regions in the country. On the other hand, the highest levels of richness are in several cells in the contact zone between the Valdivian Forests ecoregion and the Chilean Matorral (Figure 7). It should be noted that several cells represent emblematic collection localities, such as the Santiago Range, the Nahuelbuta Coastal Range, the Araucanía Range, and Torres del Paine National Park in Magallanes. In contrast, the Atacama Desert shows little or no presence of Darwin wasps (Figure 5), which is to be expected due to the lack of a forest ecosystem rich in potential hosts and humidity. According to Porter [16], ichneumonids are generally restricted to forests or jungles (regions with frequent dew or rainfall), with semi-arid and arid zones being unfavorable for them and their hosts as, according to Townes [51], adult Darwin wasps need to drink water once a day (in the form of condensed dew on plant leaves). Meanwhile, Gauld [52] comments the following on the Costa Rican Darwin wasp: "Species-richness is generally greatest in forests or other humid areas, whilst there are relatively few species in more open, dry habitats [...]" [52] (p. 25). This may be one of the reasons for the scant richness of Darwin wasps in the Chilean North.

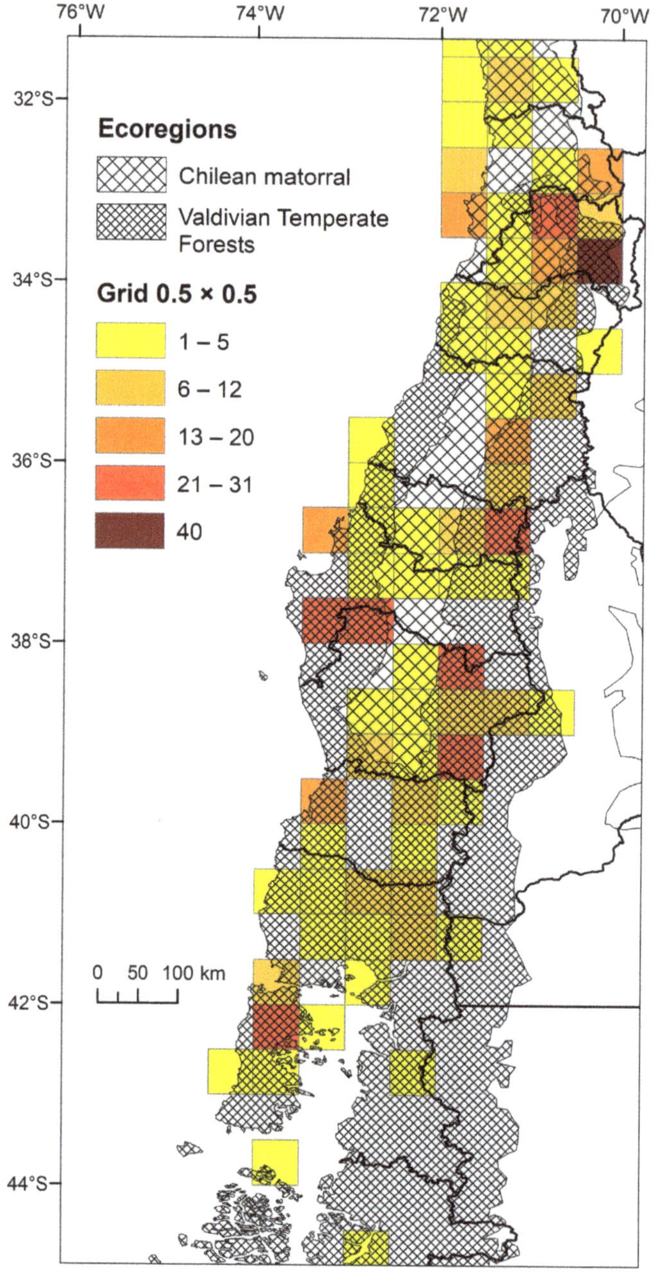

Figure 7. Species-level richness of Chilean ichneumonids is concentrated in mid-latitudes between 32° and 43° degrees south. Several cells (0.5 × 0.5 degrees) with the highest numbers occur in the contact zone between the Valdivian Forest and the Chilean Matorral ecoregions.

Several authors have already emphasized that the global distributional knowledge of Ichneumonidae is too limited to be able to reach conclusions about latitudinal and altitudinal trends in the family [7,53]. Our results reinforce the still limited taxonomic and distributional information on Chilean subtropical and temperate ichneumonid fauna. This also has implications for the conservation of "hidden biodiversity" [54]. We also have a long way to go regarding understanding the relationships between parasitoids and their hosts (e.g., [55]) so we can better understand functions within and across ecosystems and latitudinal gradients.

Supplementary Materials: The following supporting information can be downloaded at: https://www.mdpi.com/article/.../s1, Table S1: Chilean genera, ordered by Darwin wasp (Ichneumonidae) elements: (1) Cosmopolitan, (2) Endemic, (3) Neotropical, (4) Holarctic-Oriental, (5) South-temperate, (6) Australasian; Table S2: Chilean Ichneumonidae database [55–117].

Author Contributions: Conceptualization, D.G.P. and R.O.A.; methodology, D.G.P., R.O.A., V.M.-F. and A.M.-M.; software, A.M.-M. and V.M.-F.; validation, D.G.P., R.O.A., V.M.-F. and A.M.-M.; formal analysis, D.G.P., V.M.-F. and A.M.-M.; investigation, D.G.P., A.M.-M. and R.O.A.; resources, R.O.A. and A.M.-M.; data curation, D.G.P. and R.O.A.; writing—original draft preparation, D.G.P. and A.M.-M.; writing—review and editing, D.G.P., R.O.A. and A.M.-M.; visualization, D.G.P. and A.M.-M.; supervision, R.O.A. and A.M.-M.; project administration, R.O.A. and A.M.-M.; funding acquisition, A.M.-M. All authors have read and agreed to the published version of the manuscript.

Funding: This research was supported by the "Agencia Nacional de Investigación y Desarrollo (ANID)", Chile, through the project "FONDECYT Regular" n° 1221879.

Data Availability Statement: Details regarding the supporting data will be reported soon in the World Ichneumonidae Database. The data can also be requested directly from the corresponding author.

Acknowledgments: We thank Vilma A. Sarto Zeferino from USP/ESALQ/Divisão de Biblioteca (Comutação Bibliografica) for finding and sending us old hard-to-obtain papers. We also thank Diego Reyes, Guillermo Arenas, Jose Luis Inostroza, Pablo Nuñez, Paulo Valencia, and Pedro Vargas, who allowed us to reproduce their photographs of Chilean Darwin wasp species.

Conflicts of Interest: The authors declare no conflicts of interest.

References

1. Klopfstein, S.; Santos, B.F.; Shaw, M.R.; Alvarado, M.; Bennett, A.M.R.; Dal Pos, D.; Giannotta, M.; Herrera Florez, A.F.; Karlsson, D.; Khalaim, K.I.; et al. Darwin wasps: A new name heralds renewed efforts to unravel the evolutionary history of Ichneumonidae. *Entomol. Commun.* **2019**, *1*, ec01006. [CrossRef]
2. Saunders, T.E.; Ward, D.F. Variation in the diversity and richness of parasitoid wasps based on sampling effort. *Peerj* **2018**, *6*, e4642. [CrossRef] [PubMed]
3. Cardoso, P.; Erwin, T.L.; Borges, P.A.V.; New, T.R. The seven impediments in invertebrate conservation and how to overcome them. *Biol. Conserv.* **2011**, *144*, 2647–2655. [CrossRef]
4. New, T.R. *Invertebrate Surveys for Conservation*; Oxford University Press: Oxford, UK, 1998; pp. 1–240.
5. Sanchéz-Bayo, F.; Wyckhuys, K.A.G. Worldwide decline of the entomofauna: A review of its drivers. *Biol. Conserv.* **2019**, *232*, 8–27. [CrossRef]
6. Milić, D.; Radenković, S.; Ačanski, J.; Vujić, A. The importance of hidden diversity for insect conservantion: A case study in hoverflies (the *Merodon atratus* complex, Syrphidae, Diptera). *J. Insect Conserv.* **2019**, *23*, 29–44. [CrossRef]
7. Veijalainen, A.; Wahlberg, N.; Broad, G.R.; Erwin, T.L.; Longino, J.T.; Sääksjärvi, I.E. Unprecedented ichneumonid parasitoid wasp diversity in tropical forests. *Proc. R. Soc. B Biol. Sci.* **2012**, *279*, 4694–4698. [CrossRef] [PubMed]
8. Fernández, F.; Sharkey, M.J. *Introducción a los Hymenoptera de la Región Neotropical*; Sociedad Colombiana de Entomología y Universidad Nacional de Colombia: Bogotá, Colombia, 2006; pp. 1–894.
9. Bennett, A.M.R.; Cardinal, S.; Gauld, I.D.; Wahl, D.B. Phylogeny of the subfamilies of Ichneumonidae (Hymenoptera). *J. Hymenopt. Res.* **2019**, *71*, 1–156. [CrossRef]
10. Yu, D.S.K.; Achterberg, C.; Horstmann, K. (2016) Taxapad 2016, Ichneumonoidea 2015. Taxapad, Ottawa, Database on flashdrive. Available online: http://www.taxapad.com (accessed on 30 January 2023).
11. Moreira-Muñoz, A. *Plant and Vegetation 5: Plant Geography of Chile*; Werger, M.J.A., Ed.; Springer: New York, NY, USA, 2011; pp. 1–320.
12. Santos, A.M.C.; Quicke, D.L.J. Large-scale diversity patterns of parasitoid insects. *Entomol. Sci.* **2011**, *14*, 371–382. [CrossRef]
13. Kishineysky, M.; Keasar, T. Trait-based characterisation of parasitoid wasp communities in natural and agricultural areas. *Ecol. Entomol.* **2022**, *47*, 657–667. [CrossRef]
14. Porter, C.C. Zoogeografia de las Ichneumonidae Latino-Americanas (Hymenoptera). *Acta Zool. Lilloana* **1980**, *36*, 5–52.
15. Porter, C.C. Biogeographic of Chilean Ichneumonid Flies (Hymenoptera: Ichneumonidae). *Acta Entom. Chil.* **1991**, *16*, 37–68.

16. Porter, C.C. Guía de los Géneros de Ichneumonidae en la Región Neantática del Sur de Surdamérica. *Opera Lillo.* **1998**, *42*, 1–229.
17. Elgueta, M.; Rojas, F. Hymenoptera de Chile. In *Hacia un Proyecto CYTED para el Inventario y Estimación de la Diversidad Entomológica en Iberoamérica: PrIBES-2000*; Martín-Piera, F., Morrone, J.J., Melic, A., Eds.; m3m: Monografías Tercer Milenio: Zaragoza, Spain, 2000; Volume 1, pp. 245–251.
18. Dinerstein, E.; Olson, D.; Joshi, A.; Vynne, C.; Burgess, N.D.; Wikramanayake, E.; Hahn, N.; Palminteri, S.; Hedao, P.; Noss, R.; et al. An Ecoregion-Based Approach to Protecting Half the Terrestrial Realm. *BioScience* **2017**, *67*, 534–545. [CrossRef]
19. Araujo, R.O.; Dal Pos, D.; Fernandes, D.R.R.; Moreira-Muñoz, A.; Pádua, D.G. Unveiling the secrets of South American Darwin Wasps: A comprehensive Catalog of the Chilean Ichneumonidae (Hymenoptera: Ichneumonidae). *Cent. Investig. Estud. Av. Del Maule Univ. Católica Del Maule Talca Chile*, 2024; *manuscript in preparation*.
20. Darlington, P.J. *Biogeography of the Southern End of the World. Distribution and History of Far-Southern Life and Land, with an Assessment of Continental Drift*; Harvard University Press: Cambridge, MA, USA, 1965; pp. 1–246.
21. Holloway, J.D.; Jardine, N. Two approaches to zoogeography: A study based on the distributions of butterflies, birds and bats in the Indo-Australian area. *Proc. Linn. Soc.* **1968**, *179*, 153–188. [CrossRef]
22. Gauld, I.D. *An Introduction to the Ichneumonidae of Australia*; British Museum: London, UK, 1984; pp. 1–413.
23. Hausdorf, B.; Hennig, C. Biotic Element Analysis in Biogeography. *Syst. Biol.* **2003**, *52*, 717–723. [CrossRef] [PubMed]
24. Crisci, J.V.; Cigliano, M.M.; Morrone, J.J.; Roig-Junent, S. Historical Biogeography of Southern South America. *Syst. Biolo.* **1991**, *40*, 152–171. [CrossRef]
25. Crisci, J.; Katinas, L.; Posadas, P. *Historical Biogeography: An Introduction*; Harvard University Press: Cambridge, MA, USA, 2003; pp. 1–264.
26. Fuentes-Castillo, T.; Hernández, H.J.; Pliscoff, P. Hotspots and ecoregion vulnerability driven by climate change velocity in Southern South America. *Reg. Environ. Chang.* **2020**, *20*, 1–15. [CrossRef]
27. Moreira-Muñoz, A.; Scherson, R.; Luebert, F.; Román, M.J.; Monge, M.; Diazgranados, M.; Silva, H. Biogeography, phylogenetic relationships and morphological analyses of the South American genus *Mutisia* L.f. (Asteraceae) shows early connections of two disjunct biodiversity hotspots. *Org. Div. Evol.* **2020**, *20*, 639–656. [CrossRef]
28. Forbes, A.A.; Bagley, R.K.; Beer, M.A.; Hippee, A.C.; Widmayer, H.A. Quantifying the unquantifiable: Why Hymenoptera, not Coleoptera, is the most speciose animal order. *BMC Ecol.* **2018**, *18*, 21. [CrossRef]
29. Lanfranco, L.D. Contribución al conocimiento de los icneumónidos de Chile. *Rev. Chil. Entomol.* **1980**, *10*, 77–84.
30. Cox, C.B. The biogeographic regions reconsidered. *J. Biogeogr.* **2001**, *28*, 511–523.
31. Morrone, J.J. Biogeographical regions under track and cladistic scrutiny. A comment on C. B. Cox (2001). *J. Biogeogr.* **2002**, *29*, 149–152. [CrossRef]
32. Skottsberg, C. Remarks on the plant geography of the southern cold temperate zone. *Proc. R. Soc. B Biol. Sci.* **1960**, *152*, 447–457.
33. Moreira-Muñoz, A. The Austral floristic realm revisited. *J. Biogeogr.* **2007**, *34*, 1649–1660. [CrossRef]
34. Villagrán, C.; Hinojosa, L.F. Historia de los bosques del sur de Sudamérica, II: Análisis fitogeográfico. *Rev. Chil. Hist. Nat.* **1997**, *70*, 241–267.
35. Fernandes, D.R.R.; Santos, B.F.; Pádua, D.G.; Araujo, R.O. *Ichneumonidae in Catálogo Taxonômico da Fauna do Brasil*; PNUD, 2024. Available online: http://fauna.jbrj.gov.br/fauna/faunadobrasil/18131 (accessed on 7 May 2024).
36. Gauld, I.D.; Wahl, D.B. The Labeninae (Hymenoptera: Ichneumonidae): A study in phylogenetic reconstruction and evolutionary biology. *Zool. J. Linn. Soc.* **2000**, *129*, 271–347. [CrossRef]
37. Santos, B.F.; Sandoval, M.; Spasojevic, T.; Giannotta, M.M.; Brady, S.G. A Parasitoid Puzzle: Phylogenomics, Total-evidence Dating, and the Role of Gondwanan Vicariance in the Diversification of Labeninae (Hymenoptera, Ichneumonidae). *Insect Syst. Divers.* **2022**, *6*, 3. [CrossRef]
38. Grimaldi, D.; Engel, M.S. *Evolution of the Insects*; Cambridge University Press: New York, NY, USA, 2005; pp. 1–755.
39. Kuschel, G. Biogeography and Ecology of South American Coleoptera. In *Biogeography and Ecology in South America*; Fittkau, E.J., Illies, J., Klinge, H., Schwabe, G.H., Sioli, H., Eds.; Monographiae Bioloficae; Springer: Dordrecht, The Netherlands; Berlin/Heidelberg, Germany, 2012; Volume 2, pp. 709–722.
40. Ruíz-Cancino, E.; Kasparyan, D.R.; González-Moreno, A.; Khalaim, A.I.; Coronado-Blanco, J.M. Biodiversidad de Ichneumonidae (Hymenoptera) en México. *Rev. Mex. Biodivers.* **2014**, *85*, 385–391. [CrossRef]
41. Peña, L.E. A Preliminary Attempt to Divide Chile into Entomofaunal Regions, Based on the Tenebrionidae (Coleptera). *Postilla* **1966**, *97*, 1–18.
42. Morrone, J.J. Biogeographical regionalization of the Andean region. *Zootaxa* **2015**, *3936*, 207–236. [CrossRef] [PubMed]
43. Marshall, L.; Ascher, J.S.; Villagra, C.; Beaugendre, A.; Herrera, V.; Henríquez-Piskulich, P.; Vera, A.; Vereecken, N.J. Chilean bee diversity: Contrasting patterns of species and phylogenetic turnover along a large-scale ecological gradient. *Ecosphere* **2023**, *14*, e4535. [CrossRef]
44. Samaniego, H.; Marquet, P.A. Mammal and butterfly species richness in Chile: Taxonomic covariation and history. *Rev. Chil. Hist. Nat.* **2009**, *82*, 135–151. [CrossRef]
45. Moreira-Muñoz, A.; Monge, M.; Grossi, M.A.; Avila, F.A.; Morales-Fierro, V.; Heiden, G.; Britto, B.; Beck, S.; Nakajima, J.; Salgado, V.G.; et al. South America holds the greatest diversity of native daisies (Asteraceae) in the world: An updated catalogue supporting continental-scale conservation. *Front. Plant Sci.* **2024**, *15*, 1393241. [CrossRef]

46. Moreira-Muñoz, A.; Palchetti, M.V.; Morales-Fierro, V.; Duval, V.S.; Allesch-Villalobos, R.; Gonzalez-Orozco, C.E. Diversity and Conservation Gap Analysis of the Solanaceae of Southern South America. *Front. Plant Sci.* **2022**, *13*, 854372. [CrossRef] [PubMed]
47. Segovia, R.A.; Armesto, J.J. The Gondwanan legacy in South American biogeography. *J. Biogeogr.* **2015**, *42*, 209–217. [CrossRef]
48. Bauerle, P.; Rutherford, P.; Lanfranco, D. Defoliadores de roble (*Nothofagus obliqua*), raulí (*N. alpina*), coigüe (*N. dombeyi*) y lenga (*N. pumilio*). *Bosque* **1997**, *18*, 97–107. [CrossRef]
49. Pizarro-Araya, J.; Villalobos, E.V.; Alfaro, F.M.; Moreira-Muñoz, A. Conservation efforts in need of survey improvement in epigean beetles from the Atacama coast, Chile. *J. Arid Environ.* **2023**, *214*, 104995. [CrossRef]
50. Flinte, V.; Pádua, D.G.; Durand, E.M.; Hodgin, C.; Khattar, G.; da Silveira, L.F.L.; Fernandes, D.R.R.; Sääksjärvi, I.E.; Monteiro, R.F.; Macedo, M.V.; et al. Variation in a Darwin Wasp (Hymenoptera: Ichneumonidae) Community along an Elevation Gradient in a Tropical Biodiversity Hotspot: Implications for Ecology and Conservation. *Insects* **2023**, *14*, 861. [CrossRef]
51. Townes, H.K. The genera of Ichneumonidae, Part 4. *Mem. Amer. Ent. Inst.* **1971**, *17*, 1–372.
52. Gauld, I.D. The Ichneumonidae of Costa Rica, I. *Mem. Amer. Ent. Inst.* **1991**, *47*, 1–589.
53. Quicke, D.L.J. We Know Too Little about Parasitoid Wasp Distributions to Draw Any Conclusions about Latitudinal Trends in Species Richness, Body Size and Biology. *PLoS ONE* **2012**, *7*, e32101. [CrossRef] [PubMed]
54. Jerez, V.; Zúñiga-Reinoso, A.; Muñoz-Escobar, C.; Pizarro-Araya, J. Acciones y avances sobre la conservación de insectos en Chile. *Gayana* **2003**, *79*, 1–3. [CrossRef]
55. Jervis, M.A.; Ferns, P.N.; Heimpel, G.E. Body size and the timing of egg production in parasitoid wasps: A comparative analysis. *Funct. Ecol.* **2003**, *17*, 375–383. [CrossRef]
56. Townes, H.K. The genera of Ichneumonidae, Part 2. *Mem. Amer. Ent. Inst.* **1970**, *12*, 1–537.
57. Townes, H.K.; Townes, M. A catalogue and reclassification of the Neotropic Ichneumonidae. *Mem. Amer. Ent. Inst.* **1966**, *8*, 1–367.
58. Lanfranco, L.D. Contribución al conocimiento de la icneumonofauna de la región de Magallanes (Hymenoptera - Ichneumonidae). *An. Inst. Patagon.* **1974**, *5*, 199–208.
59. Porter, C.C. A revision of the south American species of *Trachysphyrus*. *Mem. Amer. Ent. Inst.* **1967**, *10*, 1–368.
60. Alvarado, M. Revision of the South American wasp genus *Alophophion* Cushman, 1947 (Hymenoptera: Ichneumonidae: Ophioninae). *Rev. Peru. Biol.* **2014**, *21*, 3–60. [CrossRef]
61. Jerez, V.; Lanfranco, L.D.; Andrade, B. Aspectos ecologicos de los ichneumonidos del Bosque de Quintero. *An. Mus. Hist. Nat. Valpso.* **1977**, *10*, 161–168.
62. Porter, C.C. A review of the Chilean genera of the tribe Mesostenini (Hym. Ichneumonidae). *Studia Ent.* **1967**, *10*, 369–418.
63. Porter, C.C. New species and records of *Anacis* (Hymenoptera: Ichneumonidae: Cryptini) from tropical and temperate Andean South America. *Insecta Mundi* **2004**, *17*, 119–127.
64. Porter, C.C. A revision of the Chilean Mesostenini (Hymenoptera: Ichneumonidae). *Contrib. Am. Entomol. Inst.* **1987**, *23*, 1–164.
65. Bordera, S.; Mazón, M.; Sääksjärvi, I.E. The Neotropical species of *Atractodes* (Hymenoptera, Ichneumonidae, Cryptinae), II: The *A. pleuripunctatus* species-group. *Zootaxa* **2016**, *4161*, 437–444. [CrossRef] [PubMed]
66. Gauld, I.D.; Wahl, D.B. The Eucerotinae: A Gondwanan origin for a cosmopolitan group of Ichneumonidae? *J. Nat. Hist.* **2002**, *36*, 2229–2248. [CrossRef]
67. Townes, H.K. The genera of Ichneumonidae, Part 3. *Mem. Amer. Ent. Inst.* **1970**, *13*, 1–307.
68. Spinola, M. *Icneumonitos. Zoologia. 6: "Historia física y politica de Chile."*; Gay, C., Ed.; Paris, France, 1851; pp. 471–550. Available online: https://www.biodiversitylibrary.org/item/130101#page/7/mode/1up (accessed on 30 January 2023).
69. Porter, C.C. Ichneumonidae de Tarapacá. 1. Subfamily Ephialtinae (Hymenoptera). *Idesia* **1979**, *5*, 157–187.
70. Broad, G.R.; Sääksjärvi, I.E.; Veijalainen, A.; Notton, D.G. Three new genera of Banchinae (Hymenoptera: Ichneumonidae) from Central and South America. *J. Nat. Hist.* **2011**, *45*, 1311–1329. [CrossRef]
71. Porter, C.C. Joppini (Hymenoptera: Ichneumonidae) of Tarapacá. *Fla. Entomol.* **1980**, *63*, 226–242. [CrossRef]
72. Townes, H.K. Revisions of twenty genera of Gelini (Ichneumonidae). *Mem. Amer. Ent. Inst.* **1983**, *35*, 1–281.
73. Wahl, D.B. Cladistics of the genera of Mesochorinae (Hymenoptera: Ichneumonidae). *Syst. Entomol.* **1993**, *18*, 371–387. [CrossRef]
74. Dasch, C.E. Neotropic Mesochorinae (Hymenoptera: Ichneumonidae). *Mem. Amer. Ent. Inst.* **1974**, *22*, 1–509.
75. Lee, J.W. A revision of the genus *Cidaphus* (Hymenoptera: Ichneumonidae: Mesochorinae). *Contrib. Am. Entomol. Inst.* **1991**, *5*, 1–48.
76. Roman, A. *Ichneumoniden von Juan Fernandez, 3: "The Natural History of Juan Fernandez and Easter Island"*; Skottsberg, C., Ed.; New York Botanical Garden: New York, NY, USA, 1920; pp. 289–295.
77. Bordera, S.; Palacio, E.; Martinez, J.J. The Neotropical species of *Clistopyga* (Hymenoptera, Ichneumonidae, Pimplinae). Part V: The *C. diazi* species group, with the description of three new species. *Zootaxa* **2019**, *4661*, 545–565. [CrossRef]
78. Porter, C.C. A revision of *Cosmiocryptus* in the coastal desert of Peru and north Chile (Hymenoptera: Ichneumonidae). *Psyche* **1985**, *92*, 463–492. [CrossRef]
79. Porter, C.C. *Cyclaulus* in the Peruvian coastal desert (Hymenoptera: Ichneumonidae). *Fla. Entomol.* **1976**, *59*, 353–360. [CrossRef]
80. Haliday, A.H. Descriptions of the Hymenoptera collected by Captain P.P. King, R.N., F.R.S., in the survey of the Straits of Magellan. *Trans. Linn. Soc. Lond.* **1836**, *17*, 316–331.
81. Guerrero, S.M.A.; Lamborot, C.L.; Arretz, V.P. Parasitic action of three Hymenopteran species on the larvae and pupae of *Plutella xylostella* in a cabbage field. *Rev. Chil. Entomol.* **1986**, *13*, 17–20.

82. Muriel, S.B.; Grez, A.A. Abundancia y parasitismo de *Plutella xylostella* L. (Lepidoptera: Plutellidae) en parches de *Brassica oleracea* con diferente forma y vegetacion circundante. *Actu. Biol.* **2003**, *25*, 99–103. [CrossRef]
83. Brèthes, J. Quelques Ichneumonidae Nouveaux. *Bol. Mus. Nac. Hist. Nat.* **1913**, *5*, 310–312. [CrossRef]
84. Diller, E.; Schoenitzer, K. Verbreitung neotropischer Phaeogenini der Gattung *Dicaelotus* Wesmael (1845), mit Beschreibungen neuer Taxa (Insecta, Hymenoptera, Ichneumonidae, Ichneumoninae, Phaeogenini). *Linz. Biol. Beitr.* **2009**, *41*, 1089–1102.
85. Dasch, C.E. The neotropic Diplazontinae. *Contrib. Am. Entomol. Inst.* **1964**, *1*, 1–77.
86. Ruiz, P.F. Himenopteros de la Provincia de Coquimbo. *Rev. Ch. Hist. Nat.* **1936**, *40*, 159–169.
87. Brèthes, J. Cueillette d'insectes au Rio Blanco. *Rev. Ch. Hist. Nat.* **1919**, *22*, 161–171.
88. Townes, H.K. The genera of Ichneumonidae, Part 1. *Mem. Am. Ent. Inst.* **1969**, *11*, 1–300. [CrossRef]
89. Broad, G.R. A review of the genus *Geraldus* Fitton (Hymenoptera: Ichneumonidae: Banchinae), with description of a new species. *J. Nat. Hist.* **2010**, *44*, 1419–1425. [CrossRef]
90. Quicke, D.L.J.; Laurenne, N.M.; Fitton, M.G.; Broad, G.R. A thousand and one wasps: A 28S rDNA and morphological phylogeny of the Ichneumonidae (Insecta: Hymenoptera) with an investigation into alignment parameter space and elision. *J. Nat. Hist.* **2009**, *43*, 1305–1421. [CrossRef]
91. Porter, C.C. A Taxonomic Revision of the Chilean Groteini. *Acta Entomol. Ch.* **1989**, *15*, 143–162.
92. Brèthes, J. Quelques Hyménopteres du Chili. *Rev. Ch. Hist. Nat.* **1916**, *20*, 83–89.
93. Porter, C.C. *Habronyx* Foerster (Hymenoptera: Ichneumonidae: Anomaloninae) in Andean and Neantarctic South America with description of new species from Bolivia and Chile. *Insecta Mundi* **2007**, *20*, 1–8.
94. Porter, C.C. Ichneumoninae of the genera *Hoplismenus* and *Platylabus* in Tarapacá (Chile) (Hymenoptera: Ichneumonidae). *Idesia* **1986**, *10*, 19–28.
95. Lanfranco, L.D. Ichneumonidos (Hymenoptera-Ichneumonidae) del Parque Nacional "Vicente Perez Rosales". *An. Mus. Hist. Nat.* **1974**, *7*, 261–267.
96. Porter, C.C. New Chilean *Itamuton* (Hymenoptera: Ichneumonidae: Mesostenini) reared from *Elicura litigator* (Neuroptera: Myrmeleontidae). *Fla. Entomol.* **1989**, *72*, 660–664. [CrossRef]
97. Porter, C.C. A revision of the South American species of *Itoplectis* (Hymenoptera, Ichneumonidae). *Acta Zool. Lill.* **1970**, *26*, 63–104.
98. Porter, C.C. The Transantarctic genus *Labena* (Hymenoptera: Ichneumonidae: Labenini) in Chile. *Insecta Mundi* **2005**, *19*, 177–185.
99. Araujo, R.O.; Vivallo, F. Taxonomic revision of *Lepidura* Townes, 1971 (Hymenoptera: Ichneumonidae: Mesochorinae) with the description of three new species, new distribution records and a key to the all known species. *Zootaxa* **2018**, *4514*, 215–229. [CrossRef] [PubMed]
100. Gupta, V.K. A revision of the genus *Microcharops* (Hymenoptera: Ichneumonidae). *Cont. Am. Ent. Inst.* **1987**, *23*, 1–42.
101. Araujo, R.O.; Di Giovanni, F. Description of the first species of *Nemeritis* Holmgren (Hymenoptera: Ichneumonidae: Campopleginae) from the Southern Hemisphere, with a key to the New World species. *Zootaxa* **2021**, *5023*, 263–272. [CrossRef] [PubMed]
102. Lamborot, L.; Arretz, P.; Guerrero, M.A.; Araya, J.E. Parasitism of eggs and larvae of *Copitarsia turbata* (Herrich & Schaffer) (Lepidoptera: Noctuidae) on horticultural crops in the Metropolitan Region of Chile. *Acta Ent. Ch.* **1995**, *19*, 129–133.
103. Porter, C.C. First record of Phrudinae (Hymenoptera: Ichneumonidae) from South America with notice of a new genus and species from Chile. *J. N. Y. Entomol. Soc.* **1993**, *101*, 130–134.
104. Porter, C.C. *Picrocryptoides*: A new genus of the tribe Mesostenini from southern South America (Hymenoptera, Ichneumonidae). *Psyche* **1965**, *72*, 167–174. [CrossRef]
105. Porter, C.C. A revision of the South American species of *Coccygomimus* (Hymenoptera, Ichneumonidae). *Studia Ent.* **1970**, *13*, 1–192.
106. Neira, C.M.; Ruff, F.J.; Mundaca, B.N. Natural enemies of *Pieris brassicae* L. (Lepidoptera: Pieridae) from cultivated crucifers in Valdivia. *Agro-Ciencia* **1989**, *5*, 5–10.
107. Lanfranco, L.D.; Cerda, L. *Coccygomimus fuscipes* (Hym.: Ichneumonidae): Un parasitoide nativo de la Polilla del Brote, *Rhyacionia buoliana* (Lep.: Tortricidae). *Bosque* **1986**, *7*, 36–37. [CrossRef]
108. Araujo, R.O.; Vivallo, F.; Santos, B.F. Discovery of two new Andean species of *Scolomus* (Townes & Townes), with a key to all known species (Hymenoptera: Ichneumonidae: Metopiinae). *Zootaxa* **2018**, *4429*, 189–194. [PubMed]
109. Walkley, L.M. A second species of Ichneumonidae belonging to *Scolomus* Townes (Hymenoptera). *Proc. Entomol. Soc. Wash.* **1962**, *64*, 231–233.
110. Cameron, P. On some Hymenoptera (chiefly undescribed) from Japan and the Pacific. *Proc. Trans. Nat. Hist. Soc. Glasg.* **1886**, *1*, 263–276.
111. Cameron, P. Descriptions of one new genus and some new species of parasitic Hymenoptera. *Mem. Proc. Manch. Lit. Phil. Soc.* **1887**, *26*, 117–136.
112. Porter, C.C. *Certonotus* Kriechbaumer (Hymenoptera: Ichneumonidae), an Australian genus newly recorded in South America. *Fla Entomol.* **1981**, *64*, 235–244. [CrossRef]
113. Porter, C.C. *Trachysphyrus* and the new genus *Aeliopotes* in the coastal desert of Peru and north Chile (Hymenoptera: Ichneumonidae). *Psyche* **1985**, *92*, 513–545. [CrossRef]

114. Diller, E.; Riedel, M.; Melzer, R.R.; Schoenitzer, K. Neotropische Phaeogenini mit Beschreibung neuer Arten der Gattung *Tycherus* Föerster, 1869 (Hymenoptera, Ichneumonidae, Ichneumoninae, Phaeogenini). *Mitt. Müench. Entomol. Ges.* **2009**, *99*, 3–95.
115. Diller, E. Neue Taxa der Gattung *Tycherus* Foerster, (1869), aus der Neotropis, ein Nachtrag (Insecta; Hymenoptera, Ichneumonidae, Ichneumoninae, Phaeogenini). *Entomofauna* **2010**, *31*, 413–428.
116. Porter, C.C. A new genus of the tribe Mesostenini from Chile (Hymenoptera, Ichneumonidae). *Psyche* **1963**, *70*, 117–119. [CrossRef]
117. Durán, M.L. Las cuncunas de los pinos, un problema de entomología forestal. *Agric. Tec.* **1944**, *4*, 17–25.

insects

MDPI

Article

Comparative SEM Study of Sensilla and Tyloid Structures in the Antennae of Vespinae (Hymenoptera: Vespidae)

Tong Zhou, Xiaojuan Huang, Hasin Ullah, Yan Tang, Danyang Zhu, Hongli Xu, Qian Wen, Xiaoxia Tian and Jiangli Tan *

Shaanxi Key Laboratory for Animal Conservation/Key Laboratory of Resource Biology and Biotechnology in Western China, Ministry of Education, College of Life Sciences, Northwest University, 229 North Taibai Road, Xi'an 710069, China; zhoutong@stumail.nwu.edu.cn (T.Z.); huangxiaojuan@stumail.nwu.edu.cn (X.H.); hasenullah888@yahoo.com (H.U.); tangyan1@stumail.nwu.edu.cn (Y.T.); zhudanyang@stumail.nwu.edu.cn (D.Z.); 202233084@stumail.nwu.edu.cn (H.X.); wenqian@stumail.nwu.edu.cn (Q.W.)

* Correspondence: tanjl@nwu.edu.cn

Simple Summary: Our study examined antennal structures in nine species of Vespinae wasps, providing insights into their morphology and sensilla diversity. Using SEM, we identified 19 sensilla types, including unique variations in pit organs and sensilla trichodea. The males of seven species exhibited tyloids. This research enhances our understanding of sensory systems in Hymenoptera, aiding taxonomy and evolutionary studies.

Abstract: This study investigates the distribution, morphology, and potential functions of antennal sensilla in various wasp species, including *Dolichovespula flora*, *D. intermedia*, *Vespula structor*, *Vl. vulgaris*, *Provespa barthelemyi*, *Vespa bicolor*, *V. ducalis*, *V. mocsaryana*, and *V. velutina* var. *nigothorax*. The study thoroughly analyzes the antennal structure of these species, representing all four genera of the yellow-jacket and hornet subfamily Vespinae. Using scanning electron microscopy (SEM), the study identifies a total of nineteen types of sensilla, including sensilla trichodea (ST-I, ST-II, ST-III), sensilla campaniform (SCF-I, SCF-II, SCF-III), pit organs (SCO-I, SCO-II, and SA), sensilla placodea (SP-I, SP-II), sensilla chaetica (SCH-I, SCH-II), sensilla basiconica (SB-I, SB-II), sensilla agmon (SAG-I, SAG-II), and sensilla coelocapitular (SCA). Additionally, tyloids were observed in the males of seven species, except for *Vl. structor* and *Vl. vulgaris*. The study provides insights into these sensilla types' morphology, abundance, and distribution. It discusses the variations in sensilla morphology among different species and the presence of gender-specific sensilla. This study provides new data about the morphology and distribution patterns of sensilla and tyloid.

Keywords: antennal sensilla; distribution; morphology; scanning electron microscope; Vespinae

1. Introduction

Insects heavily rely on their sensory organs to navigate their environment and interact with conspecifics, prey, and the surrounding ecosystem [1,2]. Among these sensory organs, antennae are crucial in detecting and processing external stimuli, including odors, vibrations, and mechanical cues [3,4]. The antennae of wasps, belonging to the family Vespidae, are of particular interest due to their diverse ecological roles and complex behaviors [5,6]. Understanding the structure and function of antennal sensilla, the specialized sensory structures found on the surface of the antenna, is essential for unraveling the sensory capabilities and adaptations of wasps [7,8]. Antennal sensilla are known to mediate various sensory modalities, including olfaction, gustation, hygroreception, thermoreception, and mechanoreception [9,10]. By analyzing the distribution and morphology of sensilla, as well as their potential functions, we can gain valuable insights into the sensory ecology of wasps and their interactions with the environment.

Several studies have investigated the antennal sensilla of wasps, contributing to our understanding of their sensory adaptations. For instance, Mohamed et al. (2013) [11] conducted a study on the sensory structure and behavioral ecology of the Oriental hornet (*Vespa orientalis* L.) within the Vespidae family. Tan et al. [12] focused on the potentially lethal social wasps and the fauna of Chinese Vespinae. Carpenter and Kojima [13] compiled a species checklist in the Vespinae subfamily, providing valuable taxonomic information. Kimsey and Carpenter [14] explored the Vespinae species in North America, offering insights into their diversity and distribution. These studies have shed light on the diversity and distribution of antennal sensilla in different wasp species.

Other studies have examined the antennal sensilla of insects beyond the Vespidae family, building on the pioneering work in this field. For example, several studies by Schneider [15], Boeckh [16], and Altner et al. [17] laid the foundational understanding of insect antennal sensilla structure, function, morphology, and classification. Renthal et al. [18] more recently investigated the structure and distribution of antennal sensilla in the red imported fire ant, expanding on this earlier work. Similarly, Lambin et al. [19] studied the antennal movements of worker honeybees as indicators of odor detection, building on our knowledge of insect olfactory systems. Said et al. [20] analyzed the structure and function of the antennal sensilla of the palm weevil *Rhynchophorus palmarum*, further contributing to our understanding of sensory adaptations in other insect species. These foundational and more recent studies provide valuable insights into insects' sensory capabilities and adaptations beyond the Vespidae family.

However, despite these valuable contributions, there is still much to learn about the antennal sensilla of wasps, particularly within the Vespidae family. This research manuscript aims to fill this knowledge gap by investigating antennal sensilla distribution, morphology, and potential functions in various wasp species. By employing scanning electron microscopy (SEM) and light microscopy, we examined the external structures of sensilla and their arrangements and densities along the antennae. Through careful observation and analysis, we identified and classified the different types of sensilla.

2. Material and Methods

The specimens were collected via hand sweeping. Most were from Shaanxi, besides *Dolichovespula intermedia* from Gansu and *Provespa barthelemyi* from Tibet, China. A total of 7 male and female individuals per species were observed on each species.

The antennae were carefully taken off from the antennal sockets using a pair of fine forceps under a stereomicroscope. For scanning electron microscopy, the antennae were fixed for 6 h in a 4% glutaraldehyde solution buffered with phosphate (pH 6.8), then dehydrated in a graded ethanol series (30%, 50%, 70%, 90%, 100%, 100%), replaced in isoamyl acetate twice for 30 min each time, dried in a CO_2 critical-point dryer, sputter-coated with gold, and observed with a Hitachi S3400N scanning electron microscope at 15 kV. At least seven sensilla of each type were calculated, and their average and standard deviation sizes were derived using the software package SPSS Version 16 (http://www.spss.com) accessed on 20 August 2020.

Table 1 examines different types of sensilla in various wasp species. The mean values with standard deviations reported in Tables 2 and 3 provide valuable insights into the size variations of sensilla among the different wasp species, enhancing our understanding of their sensory capabilities.

The morphological terminology follows Agmon et al. [21] and Mohamed et al. [11]. The voucher specimens were preserved in the College of Life Sciences, Northwest University, Xi'an, China (NWUX). A morphological cluster analysis was conducted on the antennal sensilla of the Vespinae species. The analysis utilized the cluster program Tree Analysis Technology (TNT). The dataset included sampling data for the studied taxa, and all morphological characteristics were treated as ordered. Each character state was assigned a specific numerical code to standardize the data. The data matrix (S 2) included multiple character states that represented the types and distributions of the sensilla based on their presence or absence in the specific taxonomic positions of the taxa. For the maximum parsimony (MP) analyses, new technology search algorithms were performed using TNT version 1.6.

Table 1. Different types of sensilla are examined in various species.

Species	Gender	ST-			SP-		SCF-		SCH-		SAG-		SB-		SA	SCO-		SCA
		I	II	III	I	II	I	II	I	II	I	II	I	II		I	II	
Dolichovespula intermedia	♀	+	+	+	+	+	−	−	−	+	+	−	+	+	+	+	−	+
	♂	+	+	+	+	+	−	+	+	−	−	+	+	−	+	−	−	+
Dolichovespula flora	♀	+	−	+	+	+	+	+	+	+	+	−	+	−	−	−	−	+
	♂	+	+	+	+	+	+	+	+	−	+	−	+	+	+	+	−	−
Provespa barthelemyi	♀	+	+	+	+	+	+	+	+	+	+	−	+	−	+	+	+	+
	♂	+	+	+	+	+	+	+	−	−	−	−	+	+	+	+	−	+
Vespula vulgaris	♀	+	+	+	+	+	+	−	+	−	+	−	−	−	+	−	−	−
	♂	+	+	+	+	+	+	−	+	−	−	−	+	−	+	−	−	−
Vespula structor	♀	+	+	+	+	+	+	+	−	+	+	+	+	+	+	−	−	+
	♂	+	+	+	+	+	+	+	−	−	+	+	+	+	+	−	−	+
Vespa mocsaryana	♀	+	+	+	+	+	−	−	+	+	−	−	−	+	+	+	−	+
	♂	+	+	+	+	+	−	+	−	−	+	−	+	+	+	−	−	−
Vespa bicolor	♀	+	+	+	+	+	−	+	+	−	+	−	+	−	+	−	−	+
	♂	+	+	+	+	+	−	+	−	−	−	−	+	+	+	+	−	+
Vespa ducalis	♀	+	+	+	+	+	+	+	−	−	+	+	+	+	−	−	−	−
	♂	+	+	+	+	+	−	+	+	−	+	+	+	+	+	−	−	−
Vespa velutina var. *nigrithorax*	♀	+	+	+	+	+	−	+	+	−	+	+	+	−	+	−	−	−
	♂	+	+	+	+	+	−	+	+	+	+	+	+	+	+	−	+	+

Notes: "+" indicates the presence of the sensillum; "−" indicates the absence of the sensillum.

Table 2. The size of ST and SP on antennae in the males and females of nine species (n = 7).

Species \ Sensilla		ST-I		ST-II		SP-I		SP-II	
		Basal Diameter	Length	Basal Diameter	Length	Width	Length	Width	Length
Vespula structor	M	2.186 ± 0.177	19.914 ± 1.489	1.25 ± 0.09	14.20 ± 0.83	3.118 ± 1.332	28.460 ± 5.295	4.456 ± 0.378	17.707 ± 0.722
	F	1.965 ± 0.160	13.682 ± 1.120	1.16 ± 0.15	12.57 ± 1.38	2.926 ± 0.485	24.202 ± 0.605	3.306 ± 0.387	17.667 ± 0.260
Dolichovespula intermedia	M	2.002 ± 0.129	14.855 ± 0.951	1.80	14.4	2.86 ± 0.18	28.725 ± 0.949	3.90 ± 0.13	15.26 ± 0.74
	F	1.719 ± 0.113	13.406 ± 1.49	1.71	8.19	3.192 ± 0.277	26.520 ± 1.233	4.041 ± 0.542	18.219 ± 0.972
Dolichovespula flora	M	2.145 ± 0.156	18.237 ± 1.439	1.5	13.70	3.208 ± 0.358	21.630 ± 0.166	4.022 ± 0.349	16.118 ± 1.509
	F	1.933 ± 0.224	14.610 ± 0.928	-	-	3.265 ± 0.192	25.223 ± 1.437	4.369 ± 0.343	19.819 ± 0.711
Vespula vulgaris	M	2.000 ± 0.202	19.250 ± 0.695	1.19	8.36	3.847 ± 0.299	35.898 ± 2.117	4.321 ± 0.214	19.611 ± 1.117
	F	2.000 ± 0.202	21.790 ± 2.595	-	-	3.680 ± 0.260	28.745 ± 0.242	4.038 ± 0.199	20.876 ± 0.864
Vespa ducalis	M	2.714 ± 0.200	20.143 ± 1.613	1.52 ± 0.07	15.56 ± 0.98	2.589 ± 0.247	26.375 ± 0.685	4.054 ± 0.142	19.661 ± 1.022
	F	2.754 ± 0.286	18.839 ± 2.789	1.54 ± 0.17	15.79 ± 0.63	3.461 ± 0.398	27.599 ± 1.821	4.463 ± 0.338	19.459 ± 0.577
Vespa bicolor	M	2.321 ± 0.159	22.018 ± 1.322	1.69	17.41	2.81 ± 0.86	32.21 ± 7.31	4.60 ± 0.26	21.23 ± 0.61
	F	2.529 ± 0.266	23.079 ± 1.814	-	-	3.552 ± 0.459	30.900 ± 1.348	5.230 ± 0.446	25.936 ± 1.634
Vespa velutina var. *nigrithorax*	M	2.373 ± 0.151	22.012 ± 0.841	1.45	17.42	2.589 ± 0.247	26.375 ± 0.685	4.054 ± 0.142	19.661 ± 1.022
	F	2.343 ± 0.199	16.986 ± 1.700	1.44 ± 0.09	15.08 ± 1.15	3.461 ± 0.398	27.599 ± 1.821	4.463 ± 0.338	19.459 ± 0.577
Vespa mocsaryana	M	2.577 ± 0.218	20.236 ± 1.35	1.32 ± 0.14	13.20 ± 1.67	3.999 ± 0.583	30.281 ± 3.312	4.856 ± 0.496	20.253 ± 0.964
	F	2.064 ± 0.192	21.978 ± 1.576	1.44 ± 0.14	15.66 ± 0.90	3.856 ± 0.266	30.481 ± 1.966	4.427 ± 1.006	22.761 ± 2.036
Provespabarthelemyi	M	2.322 ± 0.096	21.730 ± 1.804	1.5 ± 0.23	15 ± 1.03	2.85 ± 0.30	30.51 ± 2.46	4.11 ± 0.35	24.06 ± 1.39
	F	2.231 ± 0.163	25.607 ± 2.315	1.56 ± 0.19	16.87 ± 1.56	3.584 ± 0.206	32.694 ± 1.280	4.671 ± 0.301	25.043 ± 0.726

Notes: M, male. F, female. The size is presented as mean ± standard deviation. The "-" marks that we have not found this sensilla. The unit is "μm".

Table 3. The size of SAG, SCF, SB, PO, and SCA on the antenna (n = 8).

Species	Sensilla	SAG-I		SCF		SB		PO	SCA
		Wall Diameter	Length	Wall Diameter	Length	Basal Socket Diameter	Length	Inner Diameter	Bulbous Diameter
Vespula structor	M	3.28 ± 0.20	6.58 ± 0.70	10.357 ± 0.587	9.90 ± 1.64	2.52 ± 0.23	12.02 ± 2.85	1.856 ± 0.275	1.57
	F	3.276 ± 0.510	5.229 ± 0.720	8.250 ± 0.473	9.32 ± 1.17	2.22 ± 0.15	12.58 ± 2.20	1.796 ± 0.230	1.39
Dolichovespula intermedia	M	3.469 ± 0.078	5.400 ± 1.055	7.620 ± 1.043	9.31 ± 0.70	2.18 ± 0.17	9.12 ± 1.80	2.506 ± 0.163	1.366 ± 0.115
	F	4.40 ± 0.28	8.036 ± 0.367	13.406 ± 0.951	9.09 ± 0.37	2.43 ± 0.38	8.33 ± 1.38	2.337 ± 2.506	1.472
Vespula vulgaris	M	4.090 ± 0.404	8.0500 ± 0.792	12.409 ± 0.954	9.544 ± 1.813	2.94 ± 0.48	8.83 ± 1.39	2.025 ± 0.472	1.866 ± 0.223
	F	4.027 ± 0.404	9.386 ± 0.792	10.933 ± 0.987	10.933 ± 0.49	3.13 ± 0.38	9.13 ± 1.38	1.840 ± 0.203	1.199 ± 0.115
D. flora	M	4.295 ± 0.227	11.162 ± 0.6	7.935 ± 0.833	10.68 ± 0.48	3.05 ± 0.06	13.64 ± 1.04	1.902 ± 0.194	-
	F	4.016 ± 0.205	8.828 ± 0.782	9.200 ± 0.740	10.94 ± 0.85	3.24 ± 0.34	11.30 ± 1.48	2.41 ± 0.08	-
Vespa ducalis	M	4.49 ± 0.67	6.16 ± 1.34	9.401 ± 0.696	11.42 ± 0.95	2.93 ± 0.28	9.50 ± 1.05	-	-
	F	3.880 ± 0.246	9.463 ± 1.045	10.278 ± 0.743	11.25 ± 0.76	3.22 ± 0.24	10.60 ± 1.53	2.208 ± 0.177	-
Vespa bicolor	M			22.018 ± 1.322	5.71	3.49 ± 0.48	8.83 ± 0.14	2.421 ± 0.396	1.938 ± 0.088
	F	5.580 ± 0.308	8.323 ± 0.721	5.580 ± 0.308	9.78 ± 0.63	2.67 ± 0.18	8.80 ± 0.77	2.238 ± 0.338	1.970 ± 0.174
Vespa velutina var. *nigrithorax*	M	-	-	8.89 ± 0.16	10.22 ± 0.02	-	-	2.141 ± 0.374	1.35
	F	5.19 ± 0.17	5.55 ± 1.17	8.893 ± 1.118	11.37 ± 1.14	2.68 ± 0.21	10.57 ± 2.09	0.983	-
Vespa mocsaryana	M	4.316 ± 0.484	7.767 ± 2.504	10.162 ± 0.295	10.62 ± 0.58	2.73 ± 0.21	9.20 ± 1.76	3.06	-
	F	3.82 ± 0.95	4.37 ± 1.54	7.78 ± 0.74	10.36 ± 1.29	2.84 ± 0.24	10.90 ± 2.75	3.207 ± 0.188	1.83
Provespabarthelemyi	M	-	-	9.872 ± 0.412	10.48 ± 1.33	3.04 ± 0.07	10.88 ± 2.85	3.791 ± 0.253	1.956 ± 0.365
	F	4.92 ± 0.80	7.38 ± 0.83	10.914 ± 0.370	12.47 ± 1.65	3.26 ± 0.16	10.52 ± 2.41	3.763 ± 0.253	1.288 ± 0.148

Notes: M, male. F, female. The size is presented as mean ± standard deviation. The "-" marks that we have not found these sensilla. The unit is "µm".

3. Results

3.1. General Morphology of the Antenna

In general, the antennae of all the specimens are geniculate, consisting of three parts, namely the scape, pedicel, and flagellum (Figure 1). The scape is elongated with dense short setae and relatively sparse long setae (Figures 1 and 2A,B), having a basal radicle that shows a scaly cuticle and dense setae fitting into the antennal socket (Figures 1 and 2A). The pedicel is very short with scaly cuticles bearing numerous short setae and having an area with several pores apically (Figures 1 and 2B). At the basal radicle of the scape and basal pedicel, there is an area with more than ten Böhm bristles (BB) clustered (Figures 1 and 2C,D). The flagellum is long, consisting of 10 (female) or 11 (male) flagellomeres with various types of sensilla, as shown in the following part (Figures 1 and 2E,F). The tip of the apical flagellomere is generally pointed (Figure 1), except for the male of *Provespa barthelemyi*, which truncates apically (Figure 2F). In the male, a few tyloids are located on the ventral flagellum, with the *Vespula structor* and *Vl. vulgaris* as an exception.

3.2. Different Types and Morphology of Sensilla with Distribution

There are 19 types of sensilla being observed on the antenna of both sexes (Figure 1), including three types of sensilla trichodea (ST-I, II, III), Böhm bristles (BB), two types of sensilla placodea (SP-I, II), sensilla campaniformia (SCF-I, II), sensilla chaetica (SCH-I, II), sensilla basiconica (SB-I, II), sensilla agmon (SAG-I, II), four types of pit organs (PO) showing as two types of sensilla coeloconica (SCO-I, II), the sensilla ampullacea (SA), and the sensilla coelocapitular (SCA). Additionally, many pores are observed on the scape and apical pedicel, and few pores are scattered on the flagellum. The sensilla types of each species have been listed in Table 1 and Figure 1.

Sensilla trichodea (ST), also known as trichoid sensilla, are believed to serve various sensory functions, such as sensory perception, mechanoreception, and detecting physical stimuli, such as air currents or vibrations in the environment. ST are long and slender, taper towards the tip, hair-like, bending along the antennal axis toward the apex with spiral furrows, and found on the whole antennae (Figure 1). ST are divided into three subtypes according to the length and shape, as follows.

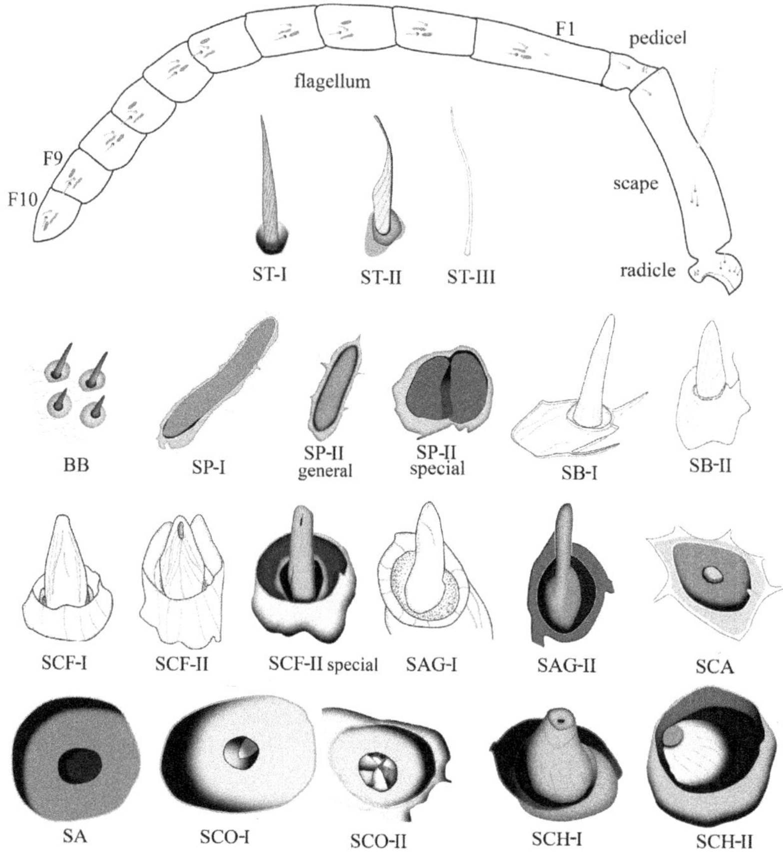

Figure 1. Antenna outline (F1 to F10 indicates the number of flagellomere) and different types of sensilla were observed in the Vespinae subfamily.

ST-I is mostly shorter and straighter than ST-II. The length is between 13.406 ± 0.951 μm and 25.607 ± 2.231 μm, the basal diameter is between 1.719 ± 0.113 μm and 2.754 ± 0.286 μm (Table 2). They are densely distributed, with the most common type being found in all studied species and being widespread on the scape, pedicel, and flagellomeres (Figures 1, 2 and 3A).

ST-II is similar to ST-I, but differs as its distal part being distinctly curved. The function is supposed to be the same as ST-I. Its size ranges from 12.57 ± 1.38 μm to 16.87 ± 1.56 μm long and 1.16 ± 0.15 μm to 1.56 ± 0.19 μm wide (Table 2). Their abundance became higher and higher towards the apical flagellum. ST-II was observed in all studied species. It is distinctly shorter and thinner than ST-I (Figures 1 and 3B). It mostly bears on the flagellomeres.

ST-III is longer when compared to ST-I and ST-II, often with a pronounced curve or bend. They are present mostly in the scape. ST-III is characterized by its more elongated structure. In general, ST-III only bears on the scape (Figure 1, Figure 2A,B and Figure 3C). In a few species, ST-III is also on the last flagellomere, such as the female *Vl. structor* and *P. barthelemyi*, and the male *D. intermedia*.

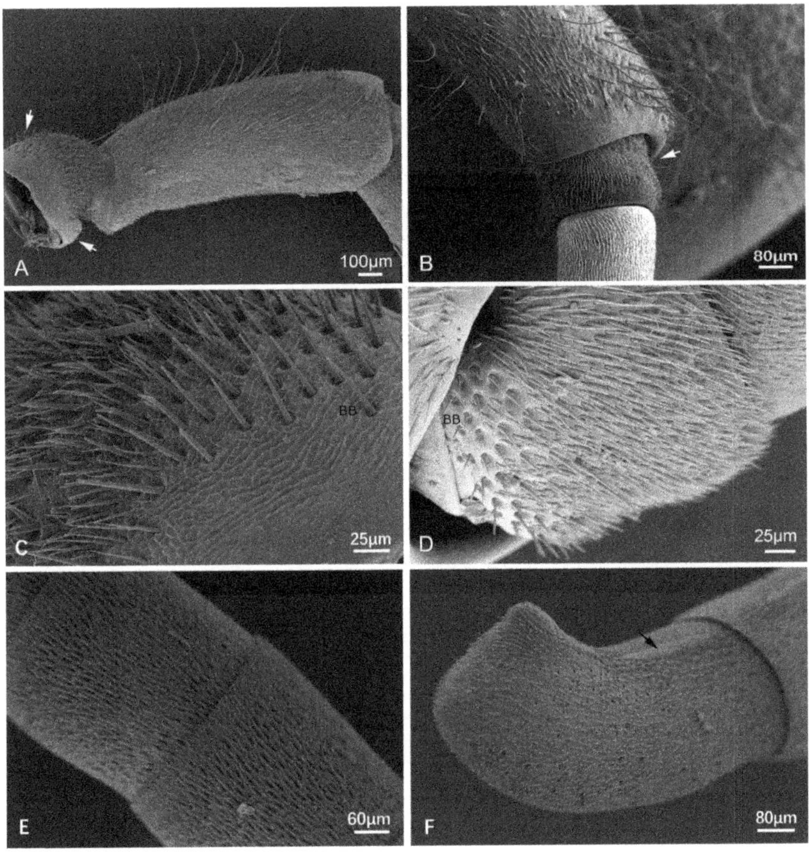

Figure 2. Scanning electron micrographs of antennal morphology and sensilla distribution. (**A**) the scape of male *Vespa velutina* var. *nigrithorax*, showing ST and BB on the basal radicle (white arrowed). (**B**) the pedicel of male *Dolichovespula flora*, showing many pores at the apical part and BB area (white arrowed). (**C**) the part of the basal radicle of male *V. velutina* var. *nigrithorax* amplified, showing BB and ST. (**D**) the pedicel of the female *Vespa bicolor*, showing BB and ST. (**E**) flagellomeres of female *Vespula structor*. (**F**) the terminal flagellomere of male *P. barthelemyi*, showing a truncated apex and a tyloid (black arrowed).

Böhm bristles (BB): These sensilla are similar to ST-I (Figure 2C) or SB-I (Figure 2D), with a socket depression present at the basal radicle of scape (12.23–63.19 µm long) and the basal pedicel (11.35–26.36 µm long). Normally, more than ten sensilla are clustered around one area and are without spiral furrows (Figure 2D). However, on the radicle of the scape, a few longer BB do have longitudinal furrows (Figure 2C). It is difficult to separate them from ST-I, except for the fact that the furrows are not spiral (Figures 1 and 2C,D). BB is a mechanical sensor as a proprioceptor, being able to sense changes in the position of the antennae themselves.

Sensilla placodea (SP) are only present on the flagellum, are aligned in parallel with the antennal axis, and can be found in all studied species, being very common and in great quantity, shaped as an elongated boat, and with a flat plate embedded in the same shaped socket. The plate structure has three different shapes, namely flat, sunken, and raised (Figure 3D–F). In our studied species, among the female specimens, the flat plates predominate, while SP existed in males and are mainly raised or sunken plates. However, *Vespula structor* and *Vespa ducalis* show sunken plates both in males and females. They are divided into two subtypes based on the distinctly different ratios of length and width. Except for the general structure, three special types of SP are present in male *D. flora,* and *Vl. structor* was found first. For instance, an "L"-shaped SP-I (Figure 3G), a hook-like SP-I (Figure 3H), and a circle plate with a crack in the middle SP-II (Figures 1 and 3I).

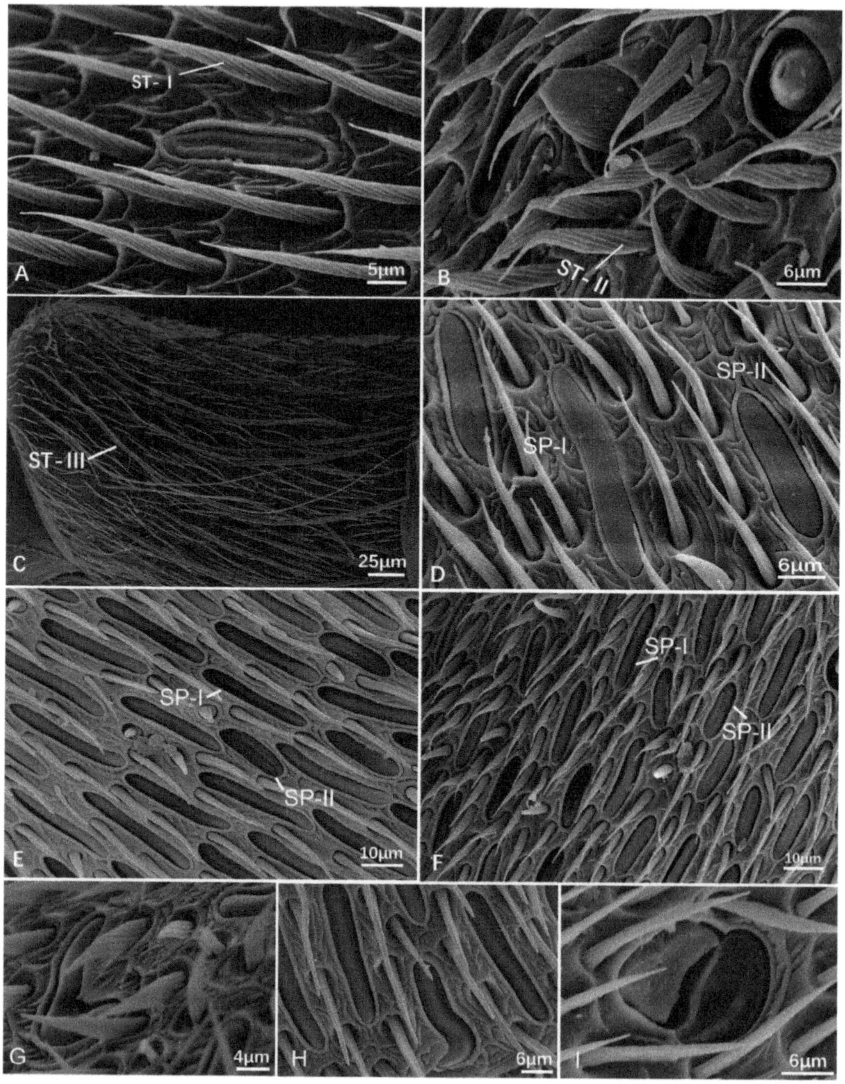

Figure 3. Scanning electron micrographs of the sensilla trichodea (ST) (**A–C**) and sensilla placodea (SP) (**D–I**). (**A**) Part of flagellum, female *Vespa velutina* var. *nigrithorax*. (**B**) Ibid, male *Vespa mocsaryana*. (**C**) Scape of female *Vespula structor*. (**D**) Female of *Provespa barthelemyi*. (**E**) Male of *Vespa mocsaryana*. (**F**) Female of *Vespa velutina* var. *nigrithorax*. (**G–I**) Showing special shaped SP: (**G**) a "L"-shaped SP-I, male *Dolichovespula flora*. (**H**) A hook L-shaped SP-I, male of *Vespa mocsaryana*. (**I**) A rounded bean-shaped SP-II, male of *Vespula structor*.

They are supposed to provide channels for volatile information substances and are generally believed to have an olfactory function. Some studies also suggest that it can sense infrared radiation, and is an important receptor for insects to perform long-distance chemical localization.

SP-I is relatively thin and long, being 21.630 ± 1.166 μm to 35.898 ± 2.117 μm long and 2.589 ± 0.247 μm to 4.600 ± 0.221 μm wide.

SP-II is relatively broad and short, being 15.26 ± 0.74 μm to 25.33 ± 0.38 μm long and 3.61 ± 0.27 μm to 4.87 ± 0.13 μm wide (Table 2).

Sensilla basiconica (SB) are distributed along the entire length of the antennal flagellum but are sparsely found among the dense ST-I/ST-II and SP. It is a cone with a blunt top and with inconspicuous longitudinal furrows on the surface, without a distinct basal wall-like structure. Instead, the basal cycle socket is raised from the cuticle (Figure 1). The size changes from 8.33 ± 1.38 μm to 13.64 ± 1.04 μm in length and from 2.18 ± 0.17 μm to 3.49 ± 0.48 μm in basal socket diameter (Table 3). Sensilla basiconica can be further categorized into two subtypes according to size.

SB-I is typically longer and thicker when compared to SB-II sensilla (Figure 4A,B).

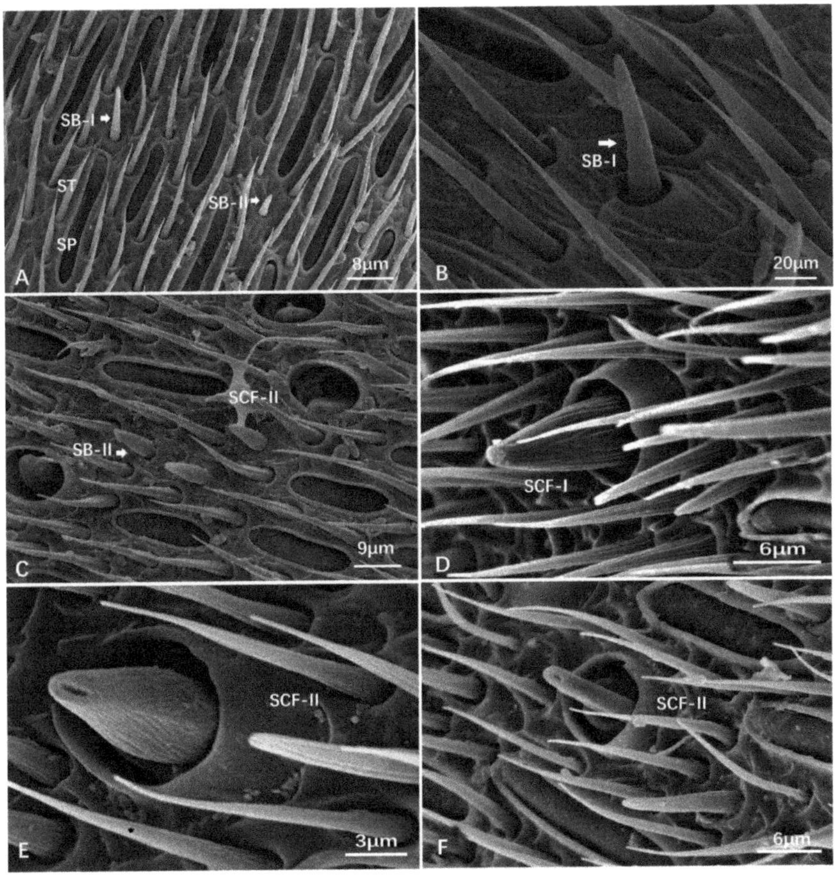

Figure 4. Scanning electron micrographs of the sensilla basiconica (SB) and sensilla campaniformia (SCF). (**A**) Female *Vespula structor*. (**B**) Female *Provespa barthelemyi*. (**C**) Female *Vespa ducalis*. (**D**) Male *D. flora*. (**E**) F8, male *Vespula structor*. (**F**) F9, ibid, female.

SB-II is shorter and thinner than SB-I sensilla (Figure 4A,C).

Sensilla Campaniformia (SCF) are dome-shaped structures with a central depression. They have a relatively big or high basal annulus-shaped wall structure and a stout volume cone with longitudinal furrows on the surface. The height of the cone ($n = 7$) varies from 9.09 ± 0.37 μm to 12.47 ± 1.65 μm, and the inner diameter of the wall structure ranges from 6.11 ± 0.47 μm to 9.19 ± 0.59 μm (Table 3). According to whether the aperture is on the top or not, the SCF can be further categorized into two subtypes as follows:

SCF-I, which looks like a solid cone without an aperture on the top (Figure 4D);

SCF-II, which looks like a hollow cone with an opening on the top (Figure 4E). Additionally, a special type of SCF-II was observed first, which is a column-like structure with a top opening (Figure 4F).

SCF is generally distributed on the flagellum. In *Provespa barthelemyi*, SCF-I is found on F6, and specialized SCF-II is found in females, while a small number of SCF are distributed on F11 in males. A unique SCF-II is observed on F9 in the female *Vespula structor*, and many SCF-II are present on F11 in male *Vespa mocsaryana*. In *Vespa bicolor*, SCF-II is found on F8 in females and F4 in males. A large number of SCF-I are distributed on the distal flagellomeres in female *Vespa ducalis*, with various sensilla, including a large number of SCF-II, on the tip of F11 in males. In female *Vespa velutina* var. *nigrithorax*, SCF-II is found on F4, and both SCF-I and SCF-II are on F9.

Agmon sensilla (SAG) exhibits a structure with a blunt top fitting into a thick annulus-shaped wall rising above the cuticle, smooth and lacking any obvious grooves. They are columnar in most of the species, except for male *Dolichovespula intermedia* and female *Vespa mocsaryana*. The female *Vespula structor* are special as we cannot see the obviously big bottom. Sometimes, SAG appear together and form a cluster. The plate ranges from 6.27 ± 0.61 μm to 10.46 ± 0.13 μm in length and the inner diameter of the wall is from 2.43 ± 0.14 μm to 3.98 ± 0.25 μm (Table 3). According to the shape, SAG can be divided into two subtypes as follows:

SAG-I: Small and conical sensilla with a slightly more rounded tip and a narrower base insertion into a shallow pit. Their surface texture is completely smooth, lacking visible microstructures. The embedded structure presents a big bottom and a thinner upper part; they are columnar (Figure 5A–C).

SAG-II: Small and conical sensilla with a slightly more pointed tip and a base insertion into a shallow pit with a broader base. Their surface texture is smooth, but may have faint microgrooves (Figure 5A,B,D).

SAG is generally present on the flagellum. In *Dolichovespula intermedia* and female *D. flora*, SAG-I is distributed on the proximal flagellum. In female *Provespa barthelemyi* and female *Vespula vulgaris*, SAG-I is observed on F4 and F6. Female *Vespula structor* has SAG-II on F5, while female *Vespa bicolor* has SAG-I on F4. Both SAG-I and SAG-II are distributed along the flagellum in female *Vespa ducalis*, with many SAG-II found on the F7 of males. In female *Vespa velutina* var. *nigrithorax*, SAG-I is present on F7. They were distributed on the segments of the antennae, and their density gradually increased toward the apex.

Pit organs (POs) have various shapes, including peg-like or hair-like structures inner side. They have a sensory pore or opening at the tip. The diameter of the bulbous structure was measured and listed in Table 3. Additionally, a special pit organ was found in male *V. velutina* var. *nigrithorax*, showing several raised structures around the hole (Figure 6D). There is no obvious size difference between these types. The diameter of the inner hole is from 2.23 ± 0.29 μm to 4.08 ± 0.25 μm. POs are further divided into sensilla coelocapitular (SCA), sensilla coeloconica (SCO-I, II), and sensilla ampullacea (SA). These pit organs serve as thermoreceptors, aiding the insects in regulating their body temperature, locating heat sources, or responding to environmental temperature changes.

Sensilla coelocapitular (SCA) are button-shaped sensilla (Figure 5E–H). SCA are generally rare compared to other types; however, they could be gathered on the tip of terminal flagellomere (F10/F11). For instance, quite a few SCA were observed on the F11 of male *Provespa barthelemyi* (Figure 5E). Additionally, in *Dolichovespula intermedia*, the concavity is heart shaped rather than circular (Figure 5H).

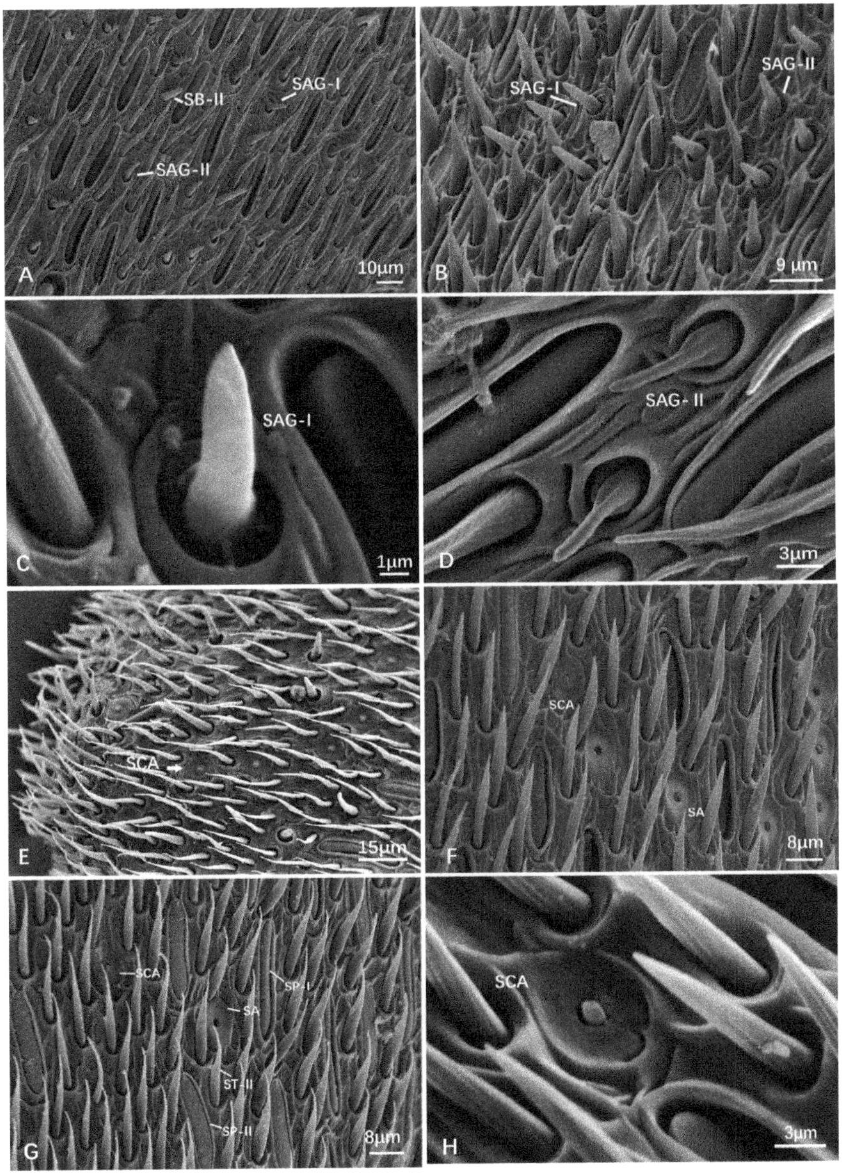

Figure 5. Scanning electron micrographs of the agmon (SAG) and sensilla coelocapitular (SCA). (**A**) F1, female *Vespa ducalis*. (**B**) Female *Vespula structor*. (**C**) Male *Dolichovespula intermedia*. (**D**) Male *Vespa mocsaryana*. (**E**) Apical F11, male *Provespa barthelemyi*, showing SCA clusters. (**F**) F4, male *Vespa velutina* var. *nigrithorax*. (**G**) F4, female *Vespa bicolor*. (**H**) F11, male *Dolichovespula intermedia*.

Sensilla ampullaceal (SA) has a completely hollow structure, a bulbous or flask-like shape, and a swollen base, known as the ampulla, which connects to the cuticle of the antenna. From the ampulla, a narrower sensory shaft extends outward.

Sensilla coeloconica (SCO) appears as cuticular openings. SCO-I appears with a peg-like protrude in the center (Figure 6B). SCO-II is special, with many peg-like protrudes in the center. Among the checking species of Vespinae, only one SCO-II was found in the F8 of male *Vespa velutinae* var. *nigrithorax* (Figure 6D). However, many SCO-II were found in the F10 of male *Eustenogaster micans* (Vespidae: Stenogasterinae) (unpublished).

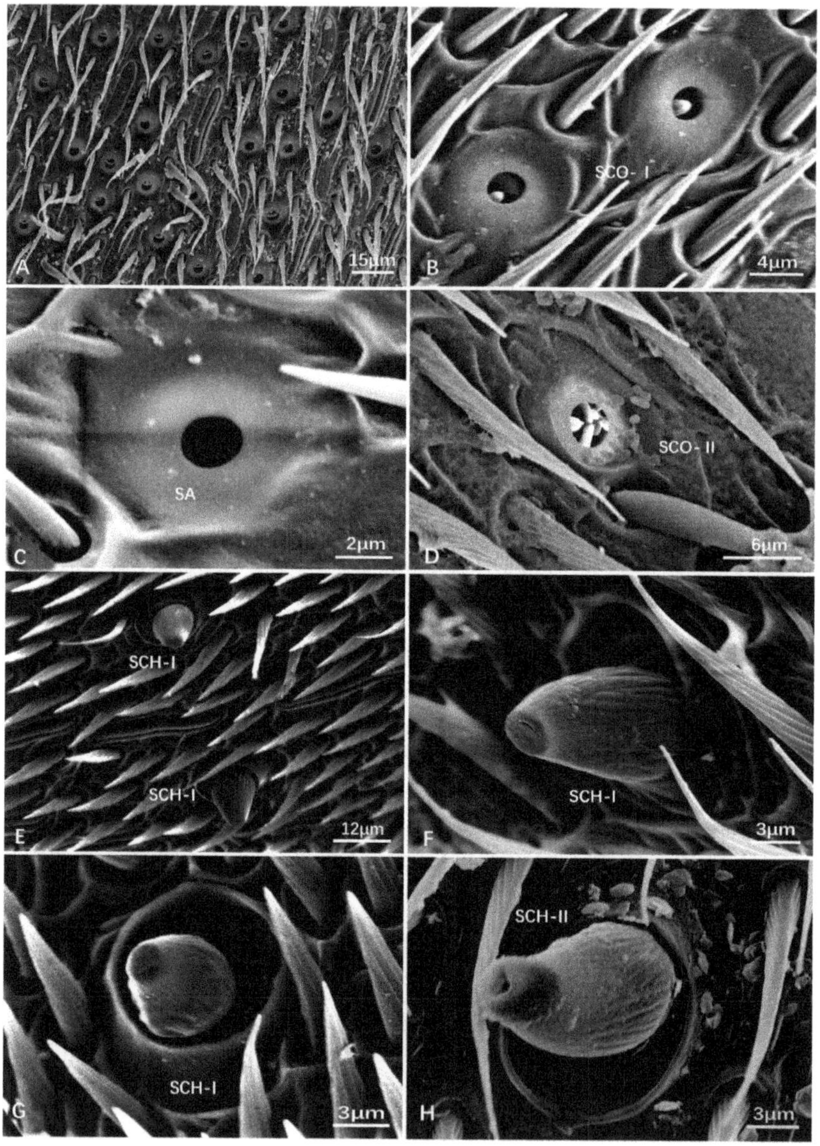

Figure 6. Scanning electron micrographs of the pit organs (PO) (**A–D**) and sensilla chaetica (SCH) (**E–H**). (**A**) a cluster of PO (SA, SCA, and SCO) on female *Provespa barthelemyi*. (**B**) female *Dolichovespula intermedia*. (**C**) male *D. flora*. (**D**) male *Vespa velutina* var. *nigrithorax*. (**E**) male *Dolichovespula flora*. (**F**) female *V. mocsaryana*. (**G**) female *D. intermedia*. (**H**) female *Provespa barthelemyi*.

SCO and SA are distributed on the flagellum. For instance, in *Vespa bicolor*, SCO-I is observed on F5 (Figure 5G), while, in male *V. velutina* var. *nigrithorax*, SCO-II is found on F8 (Figure 6D). SCO-I and SA are found on F4 (Figure 5F). Across all of the nine studied species, SCO and SA are prevalent types of sensilla, highlighting their commonality and potential significance in sensory functions.

Sensilla chaetica (SCH) consists of a short wall structure and stocky cone with irregular longitudinal striations on its surface, similar to SCF in terms of the cone shape protruding from the socket. The bottom of SCH can be seen from a dorsal view, while that of SCF cannot be observed. Most of the cones then shaped truncated cones (Figure 6E–H). SCH can be further categorized into two subtypes as follows:

SCH-I truncates apically without opening (Figure 6E–G). SCH-I is primarily mechanoreceptive and plays a role in detecting movement and tactile stimuli.

SCH-II with an opening on the tip (Figure 6H). SCH-II is believed to have a dual function, serving both mechanoreceptive and chemoreceptive roles. They may have structures such as pores that allow them to detect chemical stimuli.

SCH was observed on the flagellum. In female *Provespa barthelemyi*, SCH-I and SCH-II are observed on F10. In female *Vespula vulgaris*, SCH-I is found on F3 and F5. In the female *Vespula structor*, SCH-II is present on F6. In female *V. bicolor*, SCH-I is found on F4, F7, and F8. A special SCH-II was found in female *Provespa barthelemyi*, with a crateriform opening on the top of the cone (Figure 6D). These are a less distributed type of sensilla, occurring more in females than in males (Table 1).

3.3. Tyloids

Tyloids are only present on the ventral surface of the male flagellum. They are specialized, small, elongated, coarse, and irregular keratinized areas that protrude from the surface of the antenna with many small holes. They are believed to be a structure of cuticular glands and to be involved in antennal coils, in transferring pheromones secreted by glands, and in courtship behavior, not in sensory organ clusters. They were found on the inner side of flagella in males, except for *Vespula structor* and *Vespula vulgaris,* within the nine species. The amount is different in the different species. In most species, there is one tyloid on the first flagellomere (F1) and two in the others, with a few exceptions. There are two tyloids on the F1 of *Vespa mocsaryana,* while there is no tyloid on the FI of *V. ducalis.* In *Dolichovespula intermedia*, the first tyloid was found in the F8, with each then following flagellomere with two tyloids, except for only one tyloid being present on F11 (Table 4).

Table 4. The number of tyloids on each flagellomere(F) in the different species.

F \ Species	*Vespula structor*	*Vespula vulgaris*	*Dolichovespula intermedia*	*D. flora*	*Vespaducalis*	*Vespa bicolor*	*Vespa velutina v. nigrithorax*	*Vespa mocsaryana*	*Provespa barthelemyi*
F1	0	0	0	1	0	1	1	2	1
F2	0	0	0	2	1	2	2	2	2
F3	0	0	0	2	1	2	2	2	2
F4	0	0	0	2	1	2	2	2	2
F5	0	0	0	2	2	2	2	2	2
F6	0	0	0	2	2	2	2	2	2
F7	0	0	0	2	2	2	2	2	2
F8	0	0	2	2	2	2	2	2	2
F9	0	0	2	2	2	2	2	2	2
F10	0	0	2	2	2	2	2	2	2
F11	0	0	1	2	2	2	2	2	2
C/S	-	-	S	C	S	S	S	S	C

Notes: The "C/S" marks that two tyloids on each flagellomere (F) are separated or connected: C, connected; S, separated.

On the whole, there are two display modes of tyloids, viz. two tyloids being separate on one flagellomere (Figure 7B,D,F–H), or two tyloids being connected into one (Figure 7A,E). The ultrastructure of the separate tyloid types shows relatively smooth cuticula, with several pores and with few short and sparse setae (Figure 8A–C); the other type shows rough reticulate cuticula with pores (Figure 8D–H).

Figure 7. Light micrography of tyloids on the male antennae. The arrow represents tyloids. (**A**) *Provespa barthelemyi*. (**B**) *Dolichovespula intermedia*. (**C**) *Vespula structor*. (**D**) *Vespa ducalis*. (**E**) *Dolichovespula flora*. (**F**) *Vespa velutina* var. *nigrithorax*. (**G**) *Vespa mocsaryana*. (**H**) *Vespa bicolor*. (**I**) *Vespula vulgaris*. Scale bar = 500 μm.

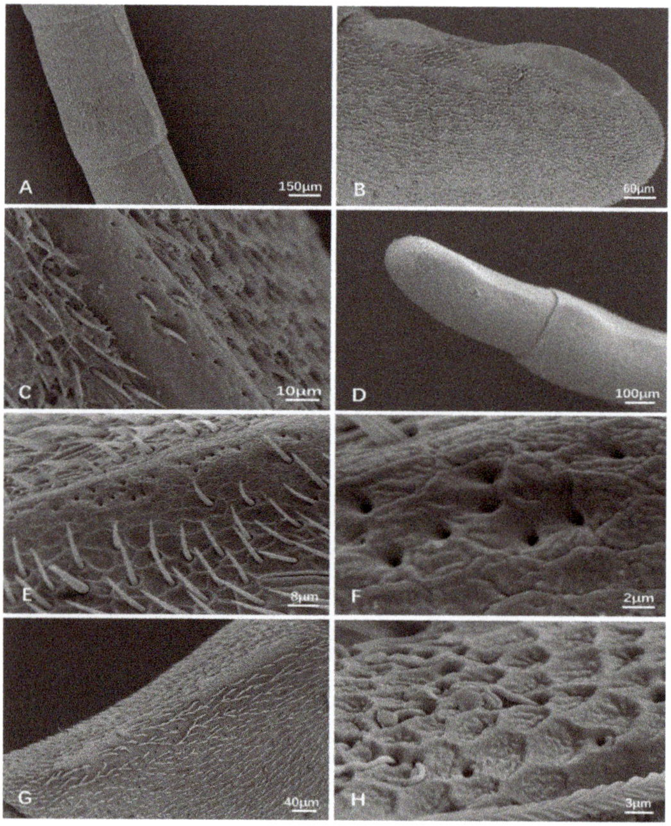

Figure 8. Tyloids on the male antennae, SEM. (**A–C**) *Vespa velutina* var. *nigrithorax*. (**D–F**) *Dolichovespula flora*. (**F**) Ibid, showing pores on the tyloid. (**G,H**) *Provespa barthelemyi*.

Vespula structor and *Vespula vulgaris* do not show the presence of tyloids (Figure 7C,I). *Dolichovespula intermedia* have very indistinct separated tyloids with two on F8, F9, and F10, and only one on F11 (Figure 7B). *Dolichovespula flora*, *Vespa bicolor*, *Vespa velutina* var. *nigrithorax*, and *Provespa barthelemyi* show the same tyloid numbers for each flagellomere (Figure 7A,E,F,H). *Vespa ducalis* displays little difference. F1 showed zero, F2, F3, and F4 showed one, and the remaining flagellomere showed two tyloids on each (Figure 7D). *Vespa mocsaryana* showed two tyloids on each flagellomere (7G).

3.4. Cluster Analysis Based on Antennal Sensilla and Tyloids

Based on the morphological data of the antennal sensilla of the eight vespids species (this study) and three species (unpublish data in our lab) representing all four subfamilies of Vespidae from China as outgroups, we used TNT for cluster analysis. The ordering of the sensilla characteristics is provided in Table S1, and the Matrix characteristics of the antennae used for the phylogenetic tree are provided in Table S2. The cluster results show that the antenna of *Polistes megei* is most different from all others. The studied species were separated into two groups (Figure 9) as follows: (1) *Vespa ducalis* sibling to *Vespa velutina* var. *nigrithorax*, their two combined with *Rhynchium quinquecinctum*, similar to *Vespula structor*; (2) *Vespa bicolor* is similar to *Vespa mocsaryana*, *Vespula vulgaris* is similar to *Eustenogaster micans*, and these four species combined with *Dolichovespula intermedia* are siblings to *Dolichovespula flora*. Surprisingly, the results poorly supported their taxonomic taxon (genus). More research is expected to understand the character's utility for phylogeny analysis.

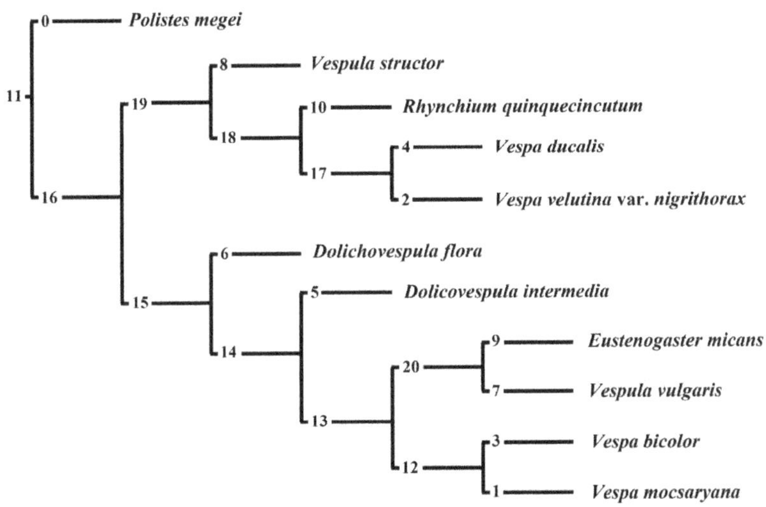

Figure 9. Cluster analysis based on antennal sensilla and tyloids. The digits on the branches indicate the total number of character changes (steps) along that branch.

4. Discussion

Insects are susceptible to a wide range of mechanical stimuli, including touch, sound, air currents, and body area deformations brought on by either self-generated movement or external forces [22,23]. Sensilla trichoidea (ST-I, ST-II, ST-III), sensilla placodea (SP-I, SP-II), sensilla chaetica (SCH-I, SCH-II), sensilla campaniformia (SCA), and sensilla basiconica (SB-I) are currently thought to be in charge of mechanoreception in Gerromorpha, according to the morphological study [24]. The current study describes several structures, with a specific focus on SP-I, SP-II, SCF-II, SCH-II, and SCA, which were notable findings in this study.

We observed that sensilla placodea (SP) are aligned in parallel with the antennal axis and can be found in all studied species. They consist of a plate structure and are divided into two types based on the distinctly different ratios of length and width. Additionally, our investigation revealed significant variation in the shape of SP among different insect families. For instance, in the Sphecidae family, SP appeared as a round chassis with an irregular convex structure in the middle, which may suggest adaptations for specific environmental interactions or sensory functions unique to this family [25,26]. In contrast, in the Braconidae family, the plates are extremely long and obtuse at both ends, potentially reflecting a different mode of sensory detection or structural requirements [27,28]. Similarly, in the Vespidae family, SP exhibited longitudinal shapes that were obtuse at both ends, indicating a possible adaptation for their ecological niche or behavioral traits. Conversely, in the Apidae family, SP took the form of a disc with a central depression, which might be associated with their distinct foraging behaviors and communication methods [29,30]. These variations underscore the diverse evolutionary adaptations of SP across different insect families, highlighting the importance of shape and structure in their respective sensory functions. Comparing these morphological differences with the results from various authors provides a comprehensive understanding of how SP have evolved to meet the specific needs of each family. These variations in shape were consistently observed across different taxonomic groups, highlighting the notable differences in SP morphology.

We reported a special SCH in female *Provespa barthelemyi*, namely a crateriform opening on the top of the cone (Figure 6H). They are a less distributed type of sensilla, occurring more in females than in males. Mohamed et al. [11] reported that *Vespa orientalis* (SCH) only exists in females, and the type SCF II (Figure 4E) is mere and only in males. However, in our study, SCH was also found in the males of *Vespula structor*, *Dolichovespula flora*, and *Vespa ducalis*; and several SCF II were observed in the females of *Vespa ducalis* and *Vespa velutina* var. *nigrithorax*.

We observed PO sensilla, characterized by irregular circular concavity with a bulbous or fastener-like structure, often present on the apical flagellomere. This sensilla type was abundant across most species, except in *Dolichovespula intermedia*, where heart-shaped concavities were occasionally seen (Figure 5H). The size of the bulbous structure was measured (Table 3). Interestingly, while Polidori et al. [31] reported the absence of a similar sensilla type (SCA) in the females of certain bee species, our study found SCA in both sexes of the studied wasp species. This suggests potential differences in the sensilla presence between bees and wasps. Additionally, Gerónimo et al. [32] found a higher abundance of sensilla-type ST-II in the females of Emphorini bees, but our study observed either a scarcity or absence of ST-II in certain female wasps, like *Dolichovespula flora* and *Vespa bicolor*. These variations in sensilla presence and abundance among species and sexes underscore the need for further comparative studies across a broader range of wasp species.

The phylogenetic relationships within the eusocial subfamily Vespinae remain controversial based on morphological behavioral and/or molecular data up to now [33,34]. More data are expected to help solve the problem. Therefore, we compared the size and morphological data of each type of sensilla and tried to analyze the kinship among the nine species. The results showed that sensilla placodea (SP) and tyloid may act as special characters for Vespidae. They are distinctly different from Andienidae (circular or oval plates), Apoidea (flat discs), Nyssoninae, and Philanthinae (the plates, situated in craters, are raised as low pegs) (Hymenoptera, Sphecidae) [25,31,35,36]. Tyloid has long been used in phylogenetic studies [37,38]. In our study, the conclusion of kinship based on the results of tyloid may be consistent with the conclusions of previous studies to some degree (Table 4). For instance, Perrard et al. [38] suggested that *Vespa velutina* var. *nigrithorax* was closer to *Vespa bicolor* than to its other congeners based on morphological and molecular data. Lopez-Osorio et al. [39] mentioned that the genus *Dolichovespula* was more closely related to *Vespa* than to *Vespula* via analyzing transcriptomic (RNA-seq) data. Our results on tyloids do not support their opinions. However, it disagreed with Perrard et al. [38] on the genus *Provespa* being closer to *Dolichovespula* than to *Vespa*. Furthermore, it does not align with the findings of

Fan et al. [40] and Wang et al. [34], who all support the clade (*Vespa* + *Vespula*) being a sister group to *Dolichovespula*. In our tyloids observation, *Dolichovespula* is closer to *Vespula* than to *Vespa* (Table 4). Obviously, as a morphological character, the tyloid could play a limited role in phylogeny. Future research should explore the functional significance of these sensilla types and their roles in species-specific behaviors and ecological adaptations within the Vespidae family. By conducting comprehensive studies, we can deepen our understanding of sensory adaptations and their implications in the biology of wasps.

Supplementary Materials: The following supporting information can be downloaded at https://www.mdpi.com/article/10.3390/insects15060448/s1, Table S1: Antennal characters selected and coded for matrix (Description of the character state); Table S2: Matrix characters of antennae used for phylogenetic tree.

Author Contributions: J.T. conceived and supervised this study; J.T., X.H., H.X. and T.Z. collected samples; T.Z., X.H. and X.T. performed experiments; T.Z., H.U., Y.T. and D.Z. analyzed the data; T.Z. wrote the manuscript draft; Q.W., H.U. and X.H. reviewed the data and revised the manuscript. All authors have read and agreed to the published version of the manuscript.

Funding: This research was supported jointly by the National Natural Science Foundations of China (NSFC, No. 32370476, 31872263, 31201732, 31572300).

Data Availability Statement: The data presented in this study are available in the article.

Acknowledgments: We gratefully acknowledge Cornelis van Achterberg for his wise comments and suggestions.

Conflicts of Interest: The authors declare no conflicts of interest.

References

1. Polidori, C.; Jorge-Garcia, A.; Nieves-Aldrey, J.L. Antennal sensillar equipment in closely related predatory wasp species (Hymenoptera: Philanthinae) hunting for different prey types. *Comptes Rendus Biol.* **2012**, *335*, 279–291. [CrossRef]
2. Perera, N.N.; Barrow, R.A.; Weston, P.A.; Rolland, V.; Hands, P.; Gurusinghe, S.; Weston, L.A.; Gurr, G.M. Characterization of antennal sensilla and electroantennography responses of the dung beetles Bubas bison, Onitis angulus, and Geotrupes spiniger (Coleoptera: Scarabaeoidea) to dung volatile organic compounds. *Insects* **2023**, *14*, 627. [CrossRef]
3. Joseph, R.M.; Carlson, J.R. Drosophila chemoreceptors: A molecular interface between the chemical world and the brain. *Trends Genet.* **2015**, *31*, 683–695. [CrossRef]
4. Steele, T.J.; Lanz, A.J.; Nagel, K.I. Olfactory navigation in arthropods. *J. Comp. Physiol. A* **2023**, *209*, 467–488. [CrossRef]
5. Lopez-Osorio, F.; Pickett, K.M.; Carpenter, J.M.; Ballif, B.A.; Agnarsson, I. Phylogenomic analysis of yellowjackets and hornets (Hymenoptera: Vespidae, Vespinae). *Mol. Phylogenet. Evol.* **2017**, *107*, 10–15. [CrossRef]
6. Post, D.C.; Jeanne, R.L.; Erickson, E.H. Variation in behavior among workers of the primitively social wasp Polistes fuscatus variatus. In *Interindividual Behavioral Variability in Social Insects*; CRC Press: Boca Raton, FL, USA, 2019; pp. 283–321.
7. Reynolds, A.M.; Reynolds, D.R.; Sane, S.P.; Hu, G.; Chapman, J.W. Orientation in high-flying migrant insects about flows: Mechanisms and strategies. *Philos. Trans. R. Soc. B Biol. Sci.* **2016**, *371*, 20150392. [CrossRef] [PubMed]
8. Taylor, G.K.; Krapp, H.G. Sensory systems and flight stability: What do insects measure and why? *Adv. Insect Physiol.* **2007**, *34*, 231–316.
9. Keil, T.A. Morphology and development of the peripheral olfactory organs. In *Insect Olfaction*; Springer: Berlin/Heidelberg, Germany, 1999; pp. 5–47.
10. Zacharuk, R.Y. Antennae and sensilla. In *Comprehensive Insect Physiology, Biochemistry, and Pharmacology*; Kerkut, G.A., Gilbert, L.I., Eds.; Pergamon Press: Oxford, UK, 1985; Volume 6, pp. 1–69.
11. Mohamed, M.; Khodairy, l.; Azza, A. A Study on the Sensory Structure, in Relation to Some Behavioral Ecology of the Oriental Hornet (*Vespa orientalis* L.) (Hymenoptera: Vespidae). *Life Sci. J.* **2013**, *10*, 48–52.
12. Tan, J.L.; van Achterberg, C.; Chen, X.X. *Potentially Lethal Social Wasps, Fauna of the Chinese Vespinae (Hymenoptera: Vespidae)*; Science Press: Beijing, China, 2015.
13. Carpenter, J.M.; Kojima, J. Checklist of the species in the subfamily Vespinae (Insecta: Hymenoptera: Vespidae). *Nat. Hist. Bull. Ibaraki Univ.* **1997**, *1*, 51–92.
14. Kimsey, L.; Carpenter, J. The Vespinae of North America (Vespidae, Hymenoptera). *J. Hymenopt. Res.* **2012**, *28*, 37–65. [CrossRef]
15. Schneider, D. Insect antennae. *Annu. Rev. Entomol.* **1964**, *9*, 103–122. [CrossRef]
16. Boeckh, J.; Kaissling, K.E.; Schneider, D. Insect olfactory receptors. In *Cold Spring Harbor Symposia on Quantitative Biology*; Cold Spring Harbor Laboratory Press: Cold Spring Harbor, NY, USA, 1965; Volume 30, pp. 263–280.

17. Altner, H.; Sass, H.; Altner, I. Relationship between structure and function of antennal chemo-, hygro-, and thermoreceptive sensilla in *Periplaneta americana*. *Cell Tissue Res.* **1977**, *176*, 389–405. [CrossRef]
18. Renthal, R.; Velasquez, D.; Olmos, D.; Hampton, J.; Wergin, W.P. Structure and distribution of antennal sensilla of the red imported fire ant. *Micron* **2003**, *34*, 405–413. [CrossRef] [PubMed]
19. Lambin, M.; DÉglise, P.; Gauthier, M. Antennal movements as indicators of odor detection by worker honeybees. *Apidologie* **2005**, *36*, 119–126. [CrossRef]
20. Said, I.; Tauban, D.; Renou, M.; Mori, K.; Rochat, D. Structure and function of the antennal sensillum of the palm weevil *Rhynchophorus palmarum* (Coleoptera, Curculionidae). *J. Insect Physiol.* **2003**, *49*, 857–872. [CrossRef]
21. Agmon, I.; Plotkin, M.; Ermakov, N.Y.; Barkay, Z.; Ishay, J.S. Antennal and cephalic organelles in the social wasp *Paravespula germanica* (Hymenoptera, Vespinae): Form and possible function. *Microsc. Res. Tech.* **2006**, *69*, 46–52. [CrossRef]
22. Strauß, J.; Stritih-Peljhan, N.; Nishino, H. Vibration receptor organs in the insect leg: Neuroanatomical diversity and functional principles. *Curr. Opin. Insect Sci.* **2024**, *61*, 101153. [CrossRef]
23. Keil, T.A. Functional morphology of insect mechanoreceptors. *Microsc. Res. Technol.* **1997**, *39*, 506–531. [CrossRef]
24. McIver, S.B. Structure of cuticular mechanoreceptors of arthropods. *Annu. Rev. Entomol.* **1975**, *20*, 381–397. [CrossRef]
25. Agren, L. Flagellar sensillum patterns in Nyssoninae and Philanthinae wasps (Hymenoptera, Sphecidae). *Zool. Scr.* **1989**, *18*, 527–535. [CrossRef]
26. Herzner, G.; Schmitt, T.; Linsenmair, K.E.; Strohm, E. Flagellar sensilla in male and female European beewolves, Philanthus triangulum F. (Hymenoptera: Sphecidae). *Entomologica Fennica.* **2003**, *14*, 237–247. [CrossRef]
27. Bai, J.C.; Chen, K.W.; Chen, L. Scanning electron microscopic studies of sensilla on the antennae of *Diachasmimorpha longicaudata*. *J. Environ. Entomol.* **2012**, *3*, 77–82.
28. Li, K.M.; Zhang, Y.J.; Wu, K.M.; Guo, Y.Y. Antenna Ultrastructure of Microplitis mediator. *Sci. Agric. Sin.* **2012**, *45*, 3522–3530.
29. Jung, J.W.; Park, K.W.; Oh, H.W.; Kwon, H.W. Structural and functional differences in the antennal olfactory system of worker honey bees of *Apis mellifera* and *Apis cerana*. *J. Asia-Pac. Entomol.* **2014**, *17*, 639–646. [CrossRef]
30. Shang, L.; Wang, Y.; Wang, P. Application of Rough Set Analysis in Species and Caste Discrimination of Bumblebees (Hymenoptera: Apidae: Bombus) Based on Antennal Sensilla. *Ann. Entomol. Soc. Am.* **2010**, *4*, 654–660. [CrossRef]
31. Jorge, A.; Polidori, C.; Nieves-Aldrey, J.L. Antennal sensilla in male gall-wasps (Hymenoptera: Cynipidae) and insights on the evolution of sexual dimorphism in cynipoid sensory equipment. *Zool. Anz.* **2019**, *283*, 213–230. [CrossRef]
32. Galvani, G.L.; González, A.; Roig-Alsina, A.H.; Settembrini, B.P. Distribution and morphometric studies of flagellar sensilla in Emphorini bees (Hymenoptera, Apoidea). *Micron* **2012**, *43*, 673–687. [CrossRef] [PubMed]
33. Carpenter, J.M. Phylogenetic relationship and classification of the Vespinae (Hymenoptera: Vespidae). *Syst. Entomol.* **1987**, *12*, 413–431. [CrossRef]
34. Wang, H.; Wang, H.; Wen, Q.; Wang, T.; Ran, F.; Wang, M.; Fan, X.; Wei, S.; Li, Z.; Tan, J. Next-Generation Sequencing of Four Mitochondrial Genomes of Dolichovespula (Hymenoptera: Vespidae) with a Phylogenetic Analysis and Divergence Time Estimation of Vespidae. *Animals* **2022**, *12*, 3004. [CrossRef]
35. Agren, L. Flagellar sensilla of two species of *Andrena*, (Hymenoptera: Andrenidae). *Int. J. Insect Morphol. Embryol.* **1978**, *7*, 73–79. [CrossRef]
36. Galvani, G.L.; Settembrini, B.P.; Roig-Alsina, A. Morphology of antennal sensilla in Doeringiella nobilis (Hymenoptera: Apidae). *J. Kans. Entomol. Soc.* **2008**, *81*(2), 142–152. [CrossRef]
37. Archer, M.E. *Vespine Wasp of the World, Behavior, Ecology, Taxonomy of the Vespinae*; Siri Scientific Press: Manchester, UK, 2012.
38. Perrard, A.; Pickett, K.; Villemant, C.; Kojima, J.; Carpenter, J. Phylogeny of hornets: A total evidence approach (Hymenoptera, Vespidae, Vespinae, Vespa). *J. Hymenopt. Res.* **2013**, *32*, 1–15. [CrossRef]
39. Lopez-Osorio, F.; Pickett, K.M.; Carpenter, J.M.; Ballif, B.A.; Agnarsson, I. Phylogenetic relationships of yellowjackets inferred from nine loci (Hymenoptera: Vespidae, Vespinae, *Vespula* and *Dolichovespula*). *Mol. Phylogenetics Evol.* **2014**, *73*, 190–201.
40. Fan, X.L.; Gong, Y.J.; Yan, P.; Chen, P.Y.; Tan, Q.Q.; Wei, S.J. Next-generation sequencing of the mitochondrial genome of dolichovespula panda (hymenoptera: Vespidae) with a phylogenetic analysis of vespidae. *J. Asia Pac. Entomol.* **2017**, *48*, 55–86. [CrossRef]

insects

MDPI

Article

Parasitoids of Insect Pests Feeding on *Scaevola taccada* (Goodeniaceae) from Yongxing Island in South China Sea

Huayan Chen [1,2,3], Cornelis van Achterberg [4], Yang Li [5], Zhen Liu [6], Jun Wang [1,2,3] and Shixiao Luo [1,2,3,*]

[1] Guangdong Provincial Key Laboratory of Applied Botany, South China Botanical Garden, Chinese Academy of Sciences, Guangzhou 510650, China; huayanc@scbg.ac.cn (H.C.); wxj@scbg.ac.cn (J.W.)
[2] State Key Laboratory of Plant Diversity and Specialty Crops, South China Botanical Garden, Chinese Academy of Sciences, Guangzhou 510650, China
[3] South China National Botanical Garden, Guangzhou 510650, China
[4] State Key Laboratory of Rice Biology and Ministry of Agriculture/Key Lab of Agricultural Entomology, Institute of Insect Sciences, Zhejiang University, Hangzhou 310058, China; kees@vanachterberg.org
[5] Sichuan Provincial Key Laboratory for Development and Utilization of Characteristic Horticultural Biological Resources, College of Chemistry and Life Sciences, Chengdu Normal University, Chengdu 611130, China; xly105@163.com
[6] Zoology Key Laboratory of Hunan Higher Education, College of Life and Environmental Sciences, Hunan University of Arts and Science, Changde 415000, China; qingniao8.27@163.com
* Correspondence: luoshixiao@scbg.ac.cn

Simple Summary: The beach naupaka, *Scaevola taccada,* is an important evergreen shrub on the islands in the South China Sea because it is one of the most common plant species inhabiting those small islands and plays an important role in vegetation restoration. However, the growth of this shrub on islands in the South China Sea has been seriously damaged by a few insect pests. In this study, we investigated the natural enemies of two main pests (*a* leaf borer and a leafminer) of the beach naupaka and assessed their potential for biological control of the pests. A braconid wasp, *Dolichogenidea stantoni,* killed almost half of the leaf borers, while a eulophid wasp, *Euderus albitarsis,* killed more than half of the leafminers. Therefore, these two parasitic wasp species have great potential to be used as biological control agents to reduce the damage caused by the two main pests of the beach naupaka.

Abstract: *Scaevola taccada* (Goodeniaceae) is an important evergreen coastal plant on islands in the South China Sea, which shows excellent tolerance for salty and drought conditions. Nevertheless, the growth of *S. taccada* populations on these islands in the South China Sea has been threatened by a few serious insect pests. However, we know little about the biology of these pests. In this study, we surveyed and identified the parasitoids of two main pests (*Herpetogramma submarginale* (Swinhoe, 1901) and *Ophiomyia scaevolana* Shiao and Wu, 1996) of *S. taccada* communities on Yongxing Island in the South China Sea, with the aim to assess their potential in biological control. *Dolichogenidea stantoni* (Ashmead, 1904) is a gregarious endoparasitoid of the larva of *H. submarginale* and contributes an average 48.9% parasitism rate on *H. submarginale. Opius biroi,* Fischer, 1960 and *Euderus albitarsis* (Zetterstedt, 1838) are both solitary endoparasitoids of the larva of *O. scaevolana,* with a respective 5.8% and 64.4% parasitism rate on *O. scaevolana.* We summarize the species diagnosis, biology, and distribution of the three parasitoid species. The potential of these parasitoids used in biological control is also discussed.

Keywords: biological control; natural enemies; DNA barcoding; parasitism

1. Introduction

The beach naupaka *Scaevola taccada* (Gaertn.) Roxb. (Goodeniaceae) is a perennial evergreen shrub or small tree commonly inhabiting almost the entire tropical and subtropical coastline of the Pacific and Indian Oceans [1,2]. Due to its strong adaptability to coral island environments, such as drought, salinity, and barrenness, *S. taccada* plays an important role in windbreak, sand fixation, and vegetation restoration of islands and coastal zones and has been widely used for preventing coastal erosion, for soil reclamation, and coastal landscaping [3,4].

Although unthoroughly investigated, a few insect pest species (Table S1) have been reported to feed on *S. taccada*, including leafminers [5–7], sawflies [7], beetles [7], scale insects [8], and thrips [9]. A pest survey and safety assessment study conducted on five islands of the Yongle Archipelago in the South China Sea found that *Ophiomyia scaevolana* Shiao and Wu, 1996 (Diptera, Agromyzidae) (misidentified as *Liriomyza sativae* Blanchard, 1938), *Arge geei*, Rohwer, 1912 (Hymenoptera, Argidae), *Phaedon brassicae*, Baly, 1874 (Coleoptera, Chrysomelidae), and *Colaphellus bowringi* (Baly, 1865) (Coleoptera, Chrysomelidae) caused very serious damage on *S. taccada* [7]. As the study suggests, these pests can easily cause serious problems, such as biosecurity and ecological environment security on the islands [7]. Currently, no specific pest management methods for *S. taccada* have been reported on the islands in the South China Sea, mainly due to the limited pest control resources available and the large population of *S. taccada* on these remote islands, although chemical pesticides are occasionally used to suppress the outbreak of insect pests (based on personal communication with local islanders). Given the importance of *S. taccada* in the vegetation restoration on islands and the vulnerable environment of those islands, sustainable pest management strategies, such as biological control using natural enemies, should be a favorable and cost-effective option. However, natural enemies of insect pests infesting *S. taccada* have not been investigated, nor have the application of natural enemies in the biological control against *S. taccada* pests.

During an expedition to Yongxing Island to survey natural enemies of insect pests, we found that *S. taccada* were heavily damaged by two insect pests: a leaf borer *Herpetogramma submarginale* (Swinhoe, 1901) (Lepidoptera, Crambidae) and a leafminer *Ophiomyia scaevolana* Shiao and Wu, 1996 (Diptera, Agromyzidae) (Figure 1). Leaves damaged by *H. submarginale* wilted, which can lead to the death of branches. Badly infested leaves by *O. scaevolana* became yellow and dropped off the plants. The growth of some *S. taccada* trees was apparently affected by these two pests (based on personal observation). The larvae of these two insect pests were also found to be attacked by parasitoid wasps. Thus, this study aims to identify and assess the potential of parasitoid species for biological control of key pests that feed on *S. taccada* on Yongxing Island in the South China Sea.

Figure 1. (**A**) Leaves of *Scaevola taccada* damaged by *Herpetogramma submarginale* larva. (**B**) *Herpetogramma submarginale*, female adult. (**C**) Leaves of *Scaevola taccada* damaged by *Ophiomyia scaevolana* larva. (**D**) *Ophiomyia scaevolana*, female adult.

2. Materials and Methods

2.1. Insect Collection and Rearing

The field survey was conducted in January 2024 on Yongxing Island (16.83° N, 112.34° E, ca. 5 m a.s.l.), South China Sea. The identity of *H. submarginale* was confirmed by a Lepidoptera expert (Dandan Zhang, Sun Yat-sen University) using the description provided by Wang et al. [10]. The identity of *O. scaevolana* was confirmed by a Diptera expert (Li Shi, Inner Mongolia Agricultural University) using the description provided by Shiao and Wu [5]. The larvae of *H. submarginale* were randomly collected from the infested leaves of *S. taccada* and were placed individually in a plastic petri dish (100 mm × 15 mm). The larvae were fed with fresh leaves of the *S. taccada* and checked daily for the emergence of parasitoids or the pupation of *H. submarginale*. The parasitoids that emerged from each caterpillar were preserved in a 1.5 mL tube with 100% ethanol until further morphological and molecular analyses. The number of parasitized larvae, the number of parasitoids that emerged from each caterpillar, and their sex ratio were recorded.

The larva (maggot) of *O. scaevolana* burrows into the leaves of *S. taccada* to feed at night, producing narrow winding mines in the leaf but hiding at daytime in the midvein of the leaf (based on personal observation). The leaves with new mines were randomly collected and brought back indoors for dissection. The midvein of the infested leaf was cut open with a cutter knife. The larvae or pupae of parasitoids beside an infested maggot were placed individually in a 10 cm glass tube and checked daily for the emergence of parasitoids. The maggots of *O. scaevolana* without symptoms of parasitism were also placed individually in a plastic petri dish (100 mm × 15 mm) and fed with fresh leaves of *S. taccada* and checked daily for the emergence of parasitoids or the pupation of *O. scaevolana*. The number of infested maggots, the number of parasitoids that emerged from each maggot, and their sex ratio were recorded.

The larvae of *H. submarginale* and leaves infested by *O. scaevolana* were collected from three spots at Yongxing Island, and 30 larvae and 30 leaves were collected from each spot.

2.2. Species Identification

Parasitoid species were first determined based on morphology. The identity of the Braconidae species was confirmed by authors CvA, YL, and ZL (Braconidae taxonomy expert) using the descriptions provided by Nixon [11] and Fischer [12]. The identity of the Eulophidae species was confirmed by HYC using the key and illustration provided by Askew [13]. To supplement morphological identification, we amplified the "barcode" region of the mitochondrial cytochrome oxidase subunit 1 (COI) using a non-destructive DNA extraction protocol, as described in Taekul et al. [14]. Genomic DNA was extracted from a female and a male of each parasitoid species using a TIANamp Micro DNA Kit (Tiangen Biotech, Beijing, China). The sequences of the *COI* gene were acquired as described by Yan et al. [15]. Briefly, the *COI* gene was amplified by polymerase chain reactions (PCRs) using the LCO1490/HCO2198 primer pair [16]. Amplicons were directly sequenced in both directions with forward and reverse primers using an Applied Biosystem (ABI) 3730XL from TsingKe Biological Technology (Beijing, China). Chromatograms were assembled into contigs using Geneious 11.0.3. The assembled sequences were translated to amino acids using the invertebrate mitochondrial code to check for stop codons and frameshifts and were blasted against the GenBank database to check for identification. All sequences generated in this study were deposited in the GenBank under accession numbers listed in Table S2. Voucher specimens are deposited in the insect collection of the South China Botanical Garden, Chinese Academy of Sciences, Guangzhou, China (SCBG).

2.3. Imaging

Photographs of live insects were taken using a Canon 5D Mark IV (Tokyo, Japan) camera with a 100 mm macro lens. Multifocal images of mounted specimens were made using a Nikon SMZ25 microscope with a Nikon DS-Ri 2 digital camera system (Nikon

Corporation, Tokyo, Japan). Images were post-processed with Adobe Photoshop CS6 Extended (Adobe Inc., San Jose, CA, USA).

2.4. Data Analysis

Parasitism rate (%) = (number of parasitized host larvae/total number of host larvae) × 100.

The mean and standard deviation of the parasitism rate and the female ratio were calculated using SPSS software (version 25.0, IBM Corp., Armonk, NY, USA) based on the averages derived from the three sampling points.

3. Results

Of the 90 larvae of *H. submarginale* collected from three spots on Yongxing Island, 44 larvae (Table S3) were parasitized by one gregarious endoparasitoid species, *Dolichogenidea stantoni* (Ashmead, 1904) (Hymenoptera, Braconidae) (Figures 2 and 3), resulting in an approximate 48.90 (± 8.4)% parasitism rate (Table 1). The number of emerged *D. stantoni* adults from each infested caterpillar ranges from 6~19, with 56.67 (±2.08)% being females (Tables 1 and S3).

Table 1. Parasitism and sex ratio of parasitoids of two serious pests feed on *Scaevola taccada*.

Host Species	Parasitoid Species	% Parasitism	% Female Parasitoids
Herpetogramma submarginale	*Dolichogenidea stantoni*	48.90 ± 8.40	56.67 ± 2.08
Ophiomyia scaevolana	*Opius biroi*	5.80 ± 1.15	67.23 ± 7.51
	Euderus albitarsis	64.40 ± 1.61	55.57 ± 6.81

Figure 2. The *Dolichogenidea stantoni* (Ashmead) larvae emerged from a larva of the *Herpetogramma submarginale* caterpillar. (**A**) The parasitoid larvae begin to pupate on the host caterpillar. (**B**) The parasitoid cocoons surrounding the dead host caterpillar.

Of the 90 leaves infested by *O. scaevolana*, each leaf contains 2–4 infested or healthy maggots. From within parasitized maggots emerged two solitary endoparasitoid species, *Opius biroi*, Fischer, 1960 (Hymenoptera, Braconidae) (Figures 4 and 5) and *Euderus albitarsis* (Zetterstedt, 1838) (Hymenoptera, Eulophidae) (Figures 6 and 7). *Opius biroi* contributes an approximate 5.80 (±1.15)% parasitism rate, with 67.23 (±7.51)% females, while *E. albitarsis* is the dominant parasitoid of *O. scaevolana*, contributing an approximate 64.40 (±1.61)% parasitism rate, with 55.57 (±6.81)% females (Tables 1 and S4).

The three parasitoid species are summarized below.

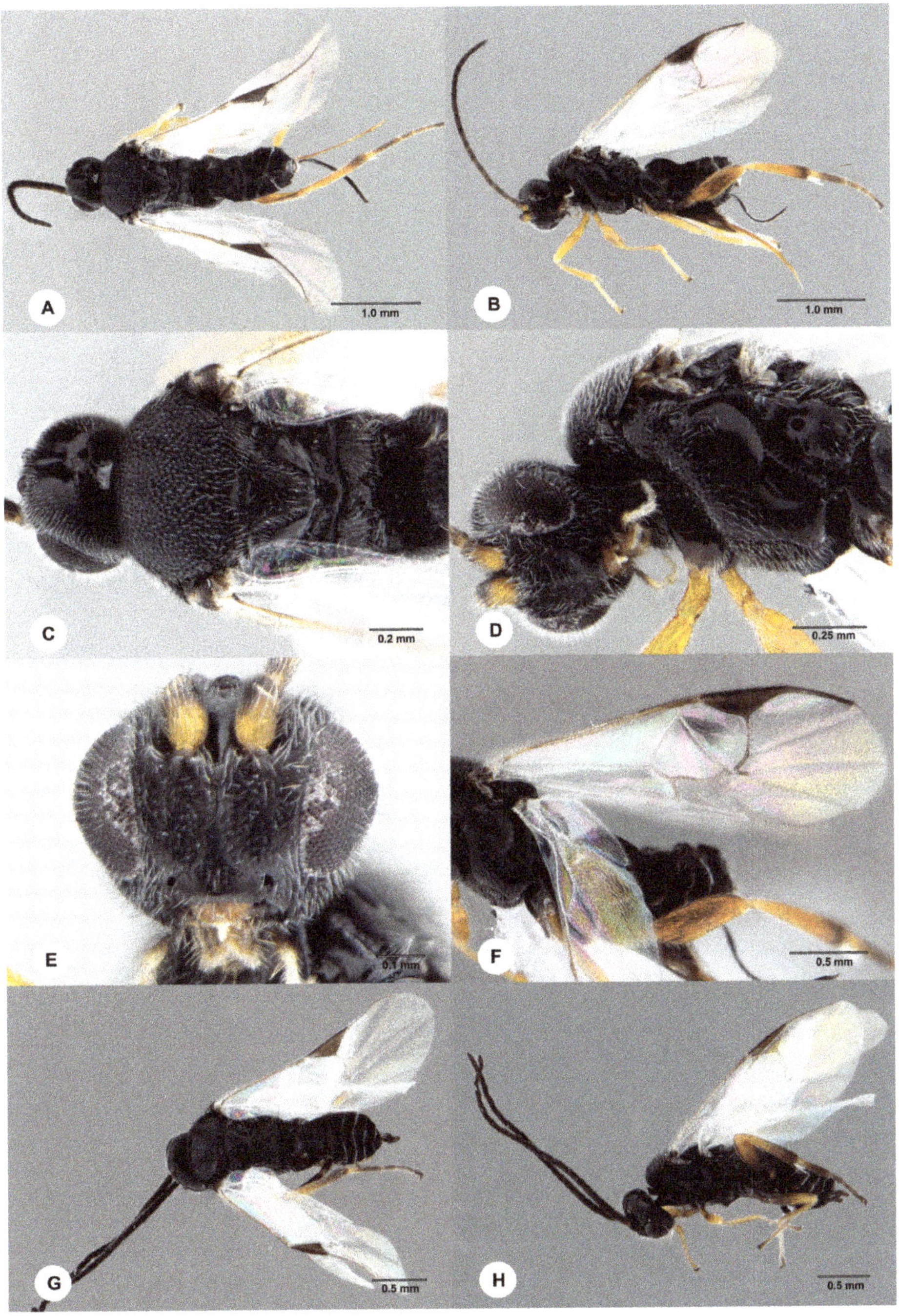

Figure 3. *Dolichogenidea stantoni* (Ashmead), (**A**–**F**) female. (**A**) Dorsal habitus. (**B**) Lateral habitus. (**C**) Head and mesosoma, dorsal view. (**D**) Head and mesosoma, lateral view. (**E**) Head, anterior view. (**F**) Wings (**G**,**H**) male. (**G**) Dorsal habitus. (**H**) Lateral habitus.

Dolichogenidea stantoni (Ashmead, 1904) (Braconidae)

Diagnosis. The female body length is about 2.5 mm. The body is black, the hind femur is yellow, the wings are hyaline, and the venation is proximal to the areolet, which is almost colorless. The propodeum has sharp and distinct areolation, and three posterior fields that are polished and smooth. The horizontal part of the first tergite is slightly transverse, being fully as long as wide, parallel-sided, and distinctly sculptured but often markedly smoother towards the apex. The second tergite has a weaker sculpture than the middle of the first tergite and is about half as long as the third tergite, and its medial area is as wide as the apex of the first tergite. The ovipositor sheath is about 1.3× longer than the hind tibia. The male is similar to the female, except for having a longer antenna (distinctly longer than the body, and the penultimate segment is 2.1× longer than wide). The first tergite has a weaker sculpture posteriorly than the female, and the second tergite is 2.8× wider than its midlength. **Note:** The identity of our specimens was mainly confirmed based on the descriptions provided by Nixon [11], but descriptions provided by Ashmead [17] and Wilkinson [18] were also compared. There are four COI sequences (Accession numbers: KJ564279, JQ849642, JQ848835, and JQ847315) labeled as *D. stantoni* present in the GenBank. The sequences of JQ849642, JQ848835, and JQ847315 match with the two sequences (Accession numbers: PQ530501 and PQ530502) we obtained in this study, with sequence similarity ranges between 98.2% and 100%. We compared our specimens with the images of the voucher specimens of JQ849642, JQ848835, and JQ847315 and found that they are highly similar. KJ564279 represents another species different from JQ849642, JQ848835, and JQ847315, and our sequences (PQ530501 and PQ530502) because of low sequence similarity (<92.3%), but the identity of this sequence cannot be confirmed because no voucher specimen is available.

Biology. The known hosts include: *Agonoxena pyrogramma* Meyrick (Agonoxenidae), *Cydalima laticostalis* (Guenée) (Crambidae), *Diaphania glauculalis* (Guenée) (Crambidae), *Diaphania indica* (Saunders) (Crambidae), *Diaphania pyloalis* Walker (Crambidae), *Glyphodes vertumnalis* Guenée (Crambidae), *Haritalodes derogata* (F.) (Crambidae), *Olene mendosa* (Hübner) (Erebidae), *Olethreutes codonectis* Meyrick (Tortricidae), *Palpita marginata* Hampson (Crambidae), *Parotis marginata* (Hampson) (Crambidae), *Pieris ida* Cramer, and *Tinea pachyspila* Meyrick (Tineidae) [19]. However, some of the above host records should be viewed with some caution as previous researchers had different opinions. For example, Wilkinson recorded *D. stantoni* as a solitary parasitoid of *D. glauculalis* (=*Margaronia glauculalis* (Guenée) [18]), and Nixon considered it an erroneous record [11]. *Herpetogramma submarginale* is recorded as its host for the first time. It is a gregarious endoparasitoid of the larva of *H. submarginale*, with multiple individuals emerging from a single host.

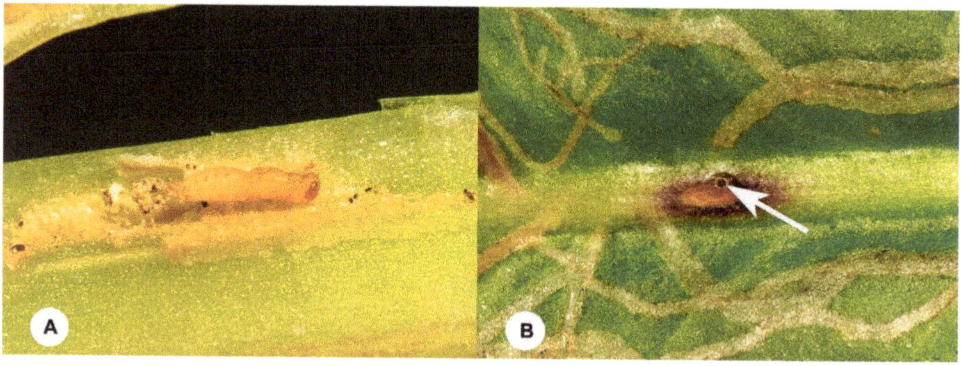

Figure 4. *Opius biroi* Fischer. (**A**) Pupa. (**B**) Emergence hole (white arrow).

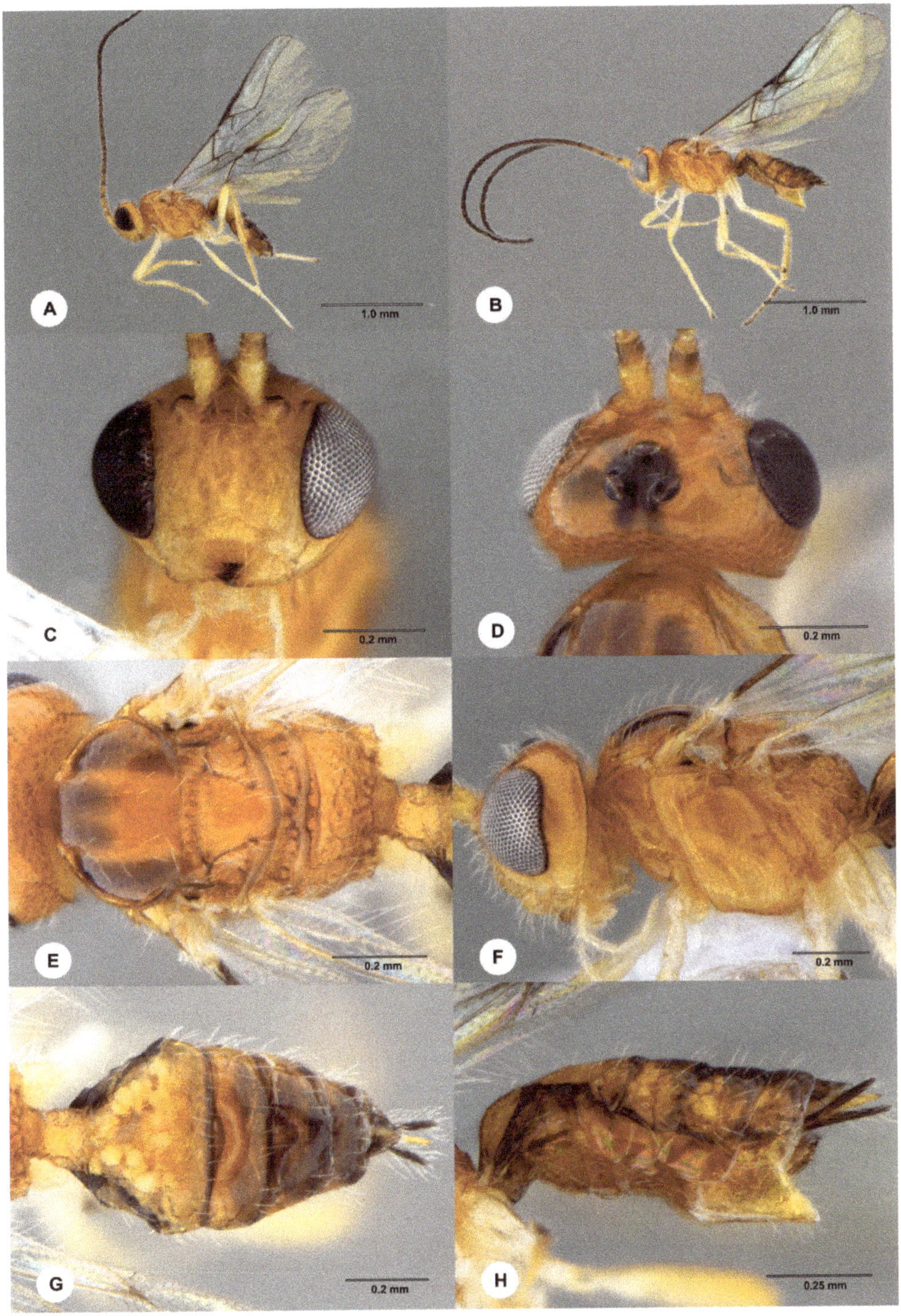

Figure 5. *Opius biroi* Fischer, (**A**) Male, lateral habitus. (**B–H**) Female, (**B**) Lateral habitus. (**C**) Head, anterior view. (**D**) Head, dorsal view. (**E**) Mesosoma, dorsal view. (**F**) Head and mesosoma, lateral view. (**G**) Metasoma, dorsal view. (**H**) Metasoma, lateral view.

Distribution. Their distribution includes China (Zhejiang, Fujian, Guangdong, Hainan, Guangxi, Guizhou, and Taiwan), Papua New Guinea, Fiji, India, Malaysia, the Philippines (type locality), and Vietnam [19].

Figure 6. *Euderus albitarsis* (Zetterstedt). (**A**) A larva during the last instar (red arrow) feeds as an ectoparasitoid on the larva (white arrow) of *Ophiomyia scaevolana,* Shiao and Wu. (**B**) A pupa. (**C**) An emergence hole (black arrow). (**D**) A female adult.

Opius biroi Fischer, 1960 (Braconidae)

Diagnosis. The female body length is 1.5–2.0 mm. The body is mostly yellowish brown, but the stemmaticum, antenna (except for a largely brownish-yellow basal four segments), mandible apically, mesoscutum medio-anteriorly and laterally, tarsal claws, fifth to seventh tergites, and ovipositor sheath are blackish-brown. The wings are hyaline, and the pterostigma and veins are dark brown. Their antenna has 23–27 segments. The occipital carina is moderately far from the hypostomal carina and mediodorsally absent. The frons and the face are smooth. The clypeus is flattened in the lateral view, depressed ventrally, and sickle-shaped. The malar suture is distinct and linear. The pronope is medium-sized and round. The precoxal sulcus is indistinct and is as smooth as the remainder of the mesopleuron. The notauli is almost completely absent on the disc and is only anteriorly shallowly impressed. The propodeum glabrous' surface is largely rugose, with a few crenulae posteriorly and a medio-longitudinal carina absent. The hind femur is about 4.0 times as long as it is wide, and the hind tibia is without an oblique carinula, basally. The vein SR1 of the fore wing is 3.0–3.4 times as long as vein 3-SR. The vein 1r-m of the hind wing is distinctly oblique and is 0.4 times the length of the vein 1-M. The basal cell of the hind wing is very narrow. The first tergite is as long as its apical width or slightly longer; its surface is medially smooth, slightly convex posteriorly, and rugose, and its dorsal carinae are developed, and the dorsope is absent. The second and third tergites are finely coriaceous and medio-anteriorly more or less rugulose, with the tergites being smooth. The length of the setose part of the ovipositor sheath is 0.3 times the length of the hind tibia. The male is similar to the female. **Note:** No COI sequences of this species are available in GenBank, and we here provide the first COI sequences (Accession numbers: PQ530503, PQ530504) for this species.

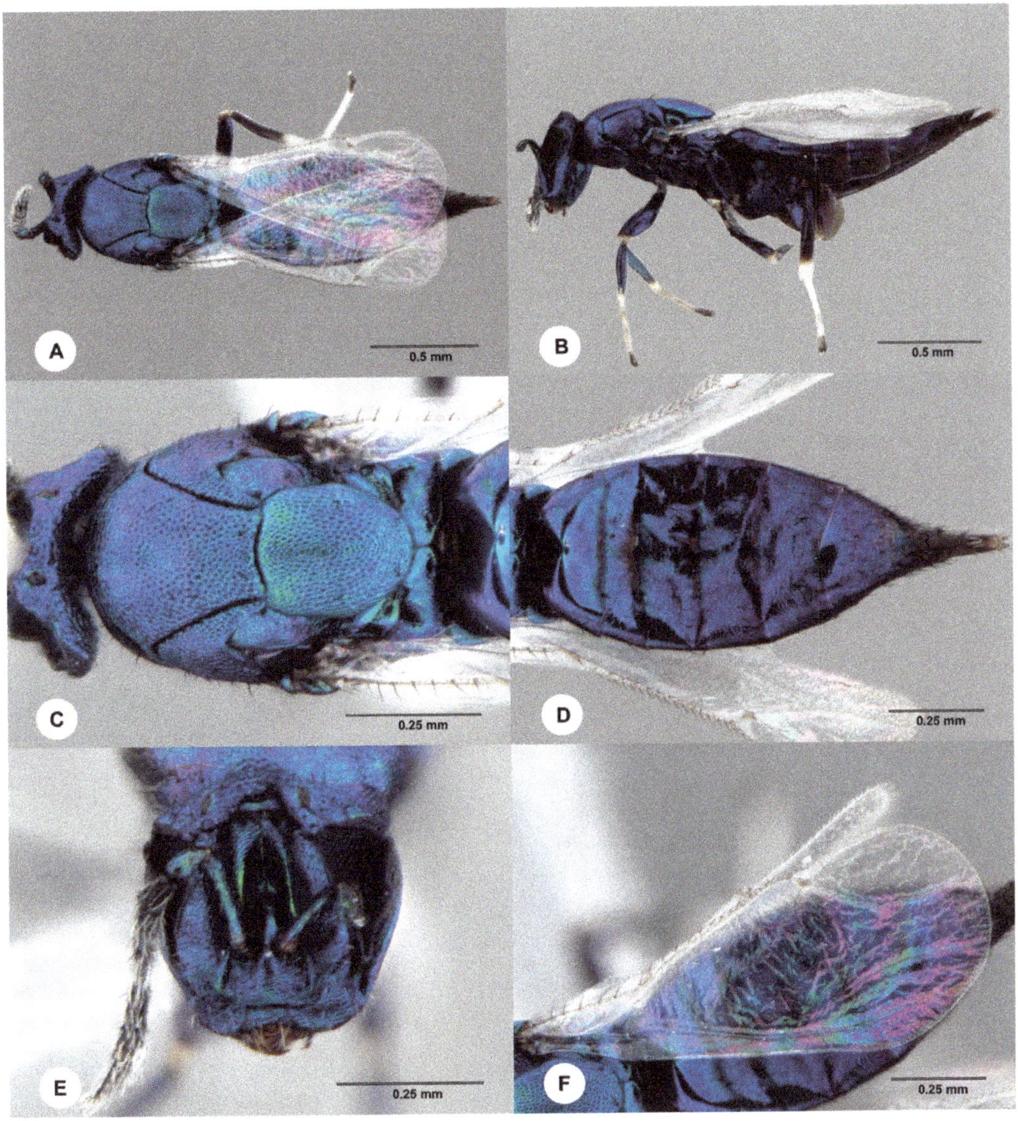

Figure 7. *Euderus albitarsis* (Zetterstedt), female. (**A**) Dorsal habitus. (**B**) Lateral habitus. (**C**) Head and mesosoma, dorsal view. (**D**) Metasoma, dorsal view. (**E**) Head, anterior view. (**F**) Forewing.

Biology. Of the solitary endoparasitoid of the larvae of several leafminer fly species with a final ectoparasitoid phase, a single individual emerged from a single host. The known hosts include *Chromatomyia horticola* (Goureau) (Agromyzidae), *Liriomyza bryoniae*

(Kaltenbach) (Agromyzidae), and *Liriomyza trifolii* (Burgess) (Agromyzidae) [20]. *Ophiomyia scaevolana* is recorded as its host for the first time.

Distribution. It is distributed in China (Beijing, Fujian, and Guizhou) [21], Hungary [22,23], Iran [24], Italy [25], Spain (including the Canary Islands) [26–28], and Turkey [29].

Euderus albitarsis (Zetterstedt, 1838) (Eulophidae)

Diagnosis. The female body length is 2.1–2.4 mm. The body is violet with metasoma, which is nearly black ventrally. The antenna is violet, becoming darker apically. The legs are violet to nearly black, except for the tip of the tibiae and the first three tarsi, which are white. The wings are hyaline, and the pterostigma and veins are light brown. The female antenna with clava is only slightly broader than the fourth funicle segment. The male antenna with setae is less outstanding, rather thicker and darker. The middle lobe of the mesoscutum has six adnotaular setae. The propodeum weakly reticulates with three setae on the callus, one-fourth as long as the scutellum. The forewings have four apical setal lines. The postmarginal vein is shorter than 1.5 times that of the stigmal vein. The female metasoma lanceolate exerted ovipositor sheaths longer than the last tergite propodeum, with three setae on the callus. **Note:** The COI sequences (Accession numbers: PQ530505 and PQ530506) we attained highly matched (around 99% similarity) with four sequences (Accession numbers: MG836466–MG836469) labeled as *E. albitarsis* in the GenBank.

Biology. Of the solitary endoparasitoid of larvae belonging to the four orders of insects, one single individual emerged from a single host. The known hosts include *Saperda populnea* (Linnaeus) (Coleoptera, Cerambycidae), *Ceutorhynchus constrictus* Schoenherr (Coleoptera, Curculionidae), *Ceutorhynchus obstrictus* (Marsham) (Coleoptera, Curculionidae), *Scolytus intricatus* (Ratzeburg) (Coleoptera, Curculionidae), *O. scaevolana* (Diptera, Agromyzidae), *Asphondylia pictipennis* Kieffer (Diptera, Cecidomyiidae), *Hartigia xanthostoma* (Eversmann) (Hymenoptera, Cephidae), *Phylloecus linearis* (Schrank) (Hymenoptera, Cephidae), *Zeuzera pyrina* L. (Lepidoptera, Cossidae), *Lymantria monacha* Moore (Lepidoptera, Erebidae), *Ectoedemia sericopeza* (Zeller) (Lepidoptera, Nepticulidae), *Trifurcula sericopeza* (Zeller) (Lepidoptera, Nepticulidae), *Thaumetopoea pityocampa* Denis and Schiffermüller (Lepidoptera, Notodontidae), *Adaina microdactyla* (Hübner) (Lepidoptera, Pterophoridae), *Rhyacionia buoliana* (Denis and Schiffermüller) (Lepidoptera, Tortricidae), and *Spilonota ocellana* (Denis and Schiffermüller) (Lepidoptera, Tortricidae) [30–32]. *Ophiomyia scaevolana* is recorded as its host for the first time.

Distribution. It is distributed in China (Jilin, Liaoning, Hebei, Ningxia, Gansu, Hunan, Fujian, Guangxi, Yunan, and Hainan) [32], South Korea, Japan, India, Israel, Kyrgyzstan, Tajikistan, Yemen, Russia, Hungary, Estonia, Poland, the Czech Republic, Slovakia, Croatia, Serbia, Montenegro, Moldova, Slovenia, Bulgaria, Romania, Austria, Norway, Finland, Sweden, the United Kingdom, the Netherlands, Germany, France, Italy, Spain, Canada, and the United States [33].

4. Discussion

Scaevola taccada is an important shrub in the vegetation restoration on islands in the South China Sea [34] but has been severely infested by several insect pests [7]. In this study, we identify three parasitoid species attacking two major pests of *S. taccada*, *H. submarginale*, and *O. scaevolana* on Yongxing Island. According to the host records, these parasitoids are all generalists (i.e., species with a broad host range). Both empirical and theoretical studies have found that island parasitoid communities are biased in favor of generalists when compared with mainland communities [35,36]. As an example of island syndrome [37], island hosts usually suffer higher attack rates by generalist parasitoids [35]. The three parasitoid species we found, especially *D. stantoni* and *E. albitarsis*, show high parasitism rates on their hosts, indicating that these parasitoids are potential biocontrol agents for pests.

A previous study found that *D. stantoni* could be used as a promising candidate for the biological control of the melon borer *Diaphania indica* (Saunders) (Lepidoptera: Crambidae) due to its aggregative response to host density [38]. *Dolichogenidea stantoni* is the only parasitoid species attacking the larvae of *H. submarginale* on Yongxing Island and contributes an average 48.90 (±8.4)% parasitism rate, which is aligned with its performance in infesting *D. indica*. The gregarious reproduction mode of *D. stantoni* should enable it to reproduce large numbers, even when limited host individuals are available. Therefore, *D. stantoni* should be considered a promising biocontrol agent against *H. submarginale*.

The maggots of *O. scaevolana* were infested by two solitary parasitoid species with distinctive different parasitism rates. *Opius biroi* is rare and only occurred in 5.80 (±1.15)% of the surveyed maggots, suggesting that this parasitoid species might not be a promising biocontrol agent against *O. scaevolana*. However, host records indicate that *O. biroi* is a generalist and could infest various leafminer fly species [20]. Since several leafminer species have been found on Yongxing Island [7] (based on personal observation), the effect of *O. biroi* on the populations of other leafminers requires further investigation. With an average parasitism rate of 64.40 (±1.61)%, *E. albitarsis* apparently plays a major role in regulating the population of *O. scaevolana* on Yongxing Island. The biology of *E. albitarsis* largely remains unknown despite its wide range of distribution and diverse host records [32]. Further studies on its biological characteristics are required to explore the possible application of this parasitoid in biocontrol programs.

5. Conclusions

Three parasitoid species attack two serious pests of *Scaevola taccada* on Yongxing Island in the South China Sea, viz., *D. stantoni* feeds on the larva of *H. submarginale* and shows a gregarious behavior. *O. biroi* and *E. albitarsis* are both solitary and feed on the larva of *O. scaevolana*. With high parasitism rates on their hosts, *D. stantoni* and *E. albitarsis* have great potential to be used in biocontrol.

Supplementary Materials: The following supporting information can be downloaded at: https://www.mdpi.com/article/10.3390/insects15120926/s1, Table S1. List of insect pests feed on *Scaevola taccada*; Table S2. List of sequenced species and accession numbers; Table S3. Total number of parasitized larvae of *Herpetogramma submarginale* out of 30 surveyed larvae, emerged parasitoids at three collecting spots; Table S4. Total number of parasitized maggots of *Ophiomyia scaevolana* of 30 leaves, emerged parasitoids at three collecting spots.

Author Contributions: Conceptualization: H.C.; Methodology: H.C.; Investigation: H.C., C.v.A., Y.L. and Z.L.; Data analyses: H.C., C.v.A., Y.L. and Z.L.; Writing—original draft preparation: H.C., C.v.A., Y.L. and Z.L.; Writing—review and editing: H.C., C.v.A., Y.L., Z.L., J.W. and S.L.; Supervision: J.W. and S.L.; Funding acquisition: J.W. and S.L. All authors have read and agreed to the published version of the manuscript.

Funding: This research was funded by the National Key R&D Program of China (Key Special Project for Marine Environmental Security and Sustainable Development of Coral Reefs 2021-403) and by the South China Botanical Garden, Chinese Academy of Sciences (Grant No: QNXM-01).

Data Availability Statement: All data are available in this paper.

Acknowledgments: We thank Li Shi (Inner Mongolia Agricultural University) and Dandan Zhang (Sun Yat-sen University) for their help in identifying *Ophiomyia scaevolana* and *Herpetogramma submarginale*, respectively. This study was supported by the National Key R&D Program of China and by the South China Botanical Garden, Chinese Academy of Sciences (Grant No: QNXM-01).

Conflicts of Interest: The authors declare no conflicts of interest.

References

1. Howarth, D.G.; Gustafsson, M.H.G.; Baum, D.A.; Motley, T.J. Phylogenetics of the Genus *Scaevola* (Goodeniaceae): Implication for Dispersal Patterns across the Pacific Basin and Colonization of the Hawaiian Islands. *Am. J. Bot.* **2003**, *90*, 915–923. [CrossRef]
2. Howarth, D.G.; Baum, D.A. Genealogical Evidence of Homoploid Hybrid Speciation in an Adaptive Radiation of *Scaevola* (Goodeniaceae) in the Hawaiian Islands. *Evolution* **2005**, *59*, 948–961.
3. Castillo-Campos, G.; García-Franco, J.; Martínez, L. First Record of Naturalization of *Scaevola taccada* (Gaertn.) Roxb. (Goodeniaceae) in Southeastern Mexico. *BioInvasions Rec.* **2021**, *10*, 425–435. [CrossRef]

4. Li, S.; Mao, X.; He, Z.; Xu, S.; Guo, Z.; Shi, S. Chromosomal-Scale Genome Assemblies of Two Coastal Plant Species, *Scaevola taccada* and *S. hainanensis*—Insight into Adaptation Outside of the Common Range. *Int. J. Mol. Sci.* **2023**, *24*, 7355. [CrossRef]
5. Shiao, S.F.; Wu, W.J. Four new agromyzid species from Taiwan (Diptera: Agromyzidae). *Trans. Am. Entomol. Soc.* **1996**, *122*, 213–226.
6. Wijesekara, A. Synopsis of the Agromyzidae (Diptera) of Sri Lanka. *Cey. J. Sci. (Bio. Sci.)* **2002**, *29*, 41–62.
7. Chen, Q.; Liang, X.; Wu, C.L.; Chen, Q. Pest survey and safety assessment on five Islands of Yongle Archipelago. *Chin. J. Trop. Crops* **2020**, *41*, 148–156. (In Chinese)
8. Hill, M.G. Susceptibility of *Scaevola taccada* (Gaertn.) Roxb. Bushes to attack by the coccid *Icerya seychellarum* Westwood: The effects of leaf loss. *Ecol. Entomol.* **1980**, *5*, 345–352. [CrossRef]
9. Masumoto, M.; Okajima, S. Three new species of the genus *Thrips* (Thysanoptera, Thripidae) in Japan. *Zootaxa* **2019**, *4614*, 575–584. [CrossRef]
10. Wang, X.; Li, M.; Huang, G. Notes on three *Herpetogramma* species feeding on *Alternanthera philoxeroides* (Mart.) Griseb. *Chin. Agric. Sci. Bull.* **2010**, *26*, 302–304. (In Chinese)
11. Nixon, G.E.J. The Indo-Australian species of the *ultor*-group of *Apanteles* Förster (Hymenoptera: Braconidae). *Bull. Br. Mus. (Nat. Hist.) Entomol. Ser.* **1967**, *21*, 1–34.
12. Fischer, M. Hymenoptera, Braconidae (Opiinae I). *Das Tierreich* **1972**, *91*, 1–620.
13. Askew, R.R. Handbooks for the identification of British Insects, Hymenoptera Chalcidoidea Section (b). *R. Entomol. Soc. Lond.* **1968**, *8*, 1–39.
14. Taekul, C.; Valerio, A.A.; Austin, A.D.; Klompen, H.; Johnson, N.F. Molecular phylogeny of telenomine egg parasitoids (Hymenoptera: Platygastridae s.l.: Telenominae): Evolution of host shifts and implications for classification. *Syst. Entomol.* **2014**, *39*, 24–35. [CrossRef]
15. Yan, C.J.; Talamas, E.; Lahey, Z.; Chen, H.Y. *Protelenomus* Kieffer is a derived lineage of *Trissolcus* Ashmead (Hymenoptera, Scelionidae), with comments on the evolution of phoresy in Scelionidae. *J. Hymenopt. Res.* **2022**, *94*, 121–137. [CrossRef]
16. Folmer, O.; Black, M.; Hoch, W.; Lutz, R.; Vrijenoek, R. DNA primers for amplification of mitochondrial cytochrome c oxidase subunit I from diverse metazoan invertebrates. *Mol. Mar. Biol. Biotechnol.* **1994**, *3*, 294–299.
17. Ashmead, W.H. A list of Hymenoptera of the Philippine Islands with descriptions of new species. *J. N. Y. Entomol. Soc.* **1904**, *12*, 1–22.
18. Wilkinson, D.S. A revision of the Indo-Australian species of the genus *Apanteles* (Hym. Bracon.). Part I. *Bull. Entomol. Res.* **1928**, *19*, 79–105, 109–146. [CrossRef]
19. Yu, D.S.; van Achterberg, C.; Horstmann, K. *Taxapad 2016: World Ichneumonoidea 2015*; Taxapad Database; Taxapad: Ottawa, ON, Canada, 2016.
20. Ku, D.S.; Belokobylskij, S.A.; Cha, J.Y. *Economic Insects of Korea 16: Braconidae (Hymenoptera)*; Insecta Koreana; National Institute of Agricultural Science and Technology: Suwon, Republic of Korea, 2001; 283p.
21. Wen, J.Z.; Lei, Z.R.; Wang, Y. Opiinae parasitoids of the leafminer *Liriomyza* spp. in China. *Entomol. Knowl.* **2002**, *39*, 14–16. (In Chinese)
22. Fischer, M. Die europäischen Arten der Gattung *Opius* Wesm. Teil 1Va. *Ann. Zool.* **1960**, *19*, 33–112.
23. Papp, J. A checklist of the Braconidae of Hungary (Hymenoptera). *Folia Entomol. Hung.* **2005**, *66*, 137–194.
24. Ghahari, H.; Fischer, M.; Sakenin, H.; Imani, S. A contribution to the Agathidinae, Alysinae, Aphidiinae, Braconinae, Microgastrinae and Opiinae (Hymenoptera: Braconidae) from cotton fields and surrounding grasslands of Iran. *Linz. Biol. Beitraege* **2011**, *43*, 1269–1276.
25. Fischer, M. Die Opiinae des Museo Civico di Storia Naturale in Genau. *Ann. Mus. Civ. Stor. Nat. Genova* **1962**, *73*, 71–97.
26. Báez, M.; Koponen, M.; Gracia, A.; Martin, E. *Lista de Especies Silvestres de Canarias (Hongos, Plantas y Animales Terrestres)*; Hymenoptera; Gobierno de Canarias: Canarias, Spain, 2001; pp. 267–279.
27. Jiménez Peydró, R. Opiinae de la provincia de Valencia (Hymenoptera, Braconidae). *Bol. Asoc. Española Entomol.* **1982**, *6*, 277–283.
28. Avinent, L.; Jiménez, R. Opiinae from the collection of the Zoology Department in the University of Valencia: I. Madrid, Palencia and Segovia (Spain). *Bol. Asoc. Española Entomol.* **1987**, *11*, 121–134.
29. Fischer, M.; Beyarslan, A. A survey of Opiinae (Hymenoptera: Braconidae) of Turkey. *Fragm. Faun.* **2005**, *48*, 27–62. [CrossRef]
30. Bouček, Z.; Askew, R.R. Hym. Chalcidoidea. Palearctic Eulophidae (excl. Tetrastichinae). In *Index of Entomophagous Insects*; Delucchi, V., Remaudière, G., Eds.; Le François: Paris, France, 1968; Volume 3, 260p.
31. Gibson, G.A.P.; Gates, M.W.; Buntin, G.D. Parasitoids (Hymenoptera: Chalcidoidea) of the cabbage seedpod weevil (Coleoptera: Curculionidae) in Georgia, USA. *J. Hymenopt. Res.* **2006**, *15*, 187–207.
32. Deng, S.; Li, C. Description of A New Species and A New Country Record Species of the Genus *Euderus* (Hymenoptera: Eulophidae) from China. *J. Northeast For. Univ.* **2022**, *50*, 121–124.
33. Yefremova, Z.A. An annotated checklist of the Eulophidae (excl. Tetrastichinae) (Hymenoptera: Chalcidoidea) of Israel. *Zootaxa* **2015**, *3957*, 001–036. [CrossRef]
34. Ren, H.; Jian, S.G.; Zhang, Q.M.; Wang, F.G.; Shen, T.; Wang, J. Plants and vegetation on South China Sea Islands. *Ecol. Environ. Sci.* **2017**, *26*, 1639–1648. (In Chinese)
35. Santos, A.M.C. Ecology and biogeography of island parasitoid faunas. *Front. Biogeogr.* **2012**, *4*, 19–24. [CrossRef]

36. Wyckhuys, K.A.G.; Sanchez Garcia, F.J.; Santos, A.M.C.; Canal, N.A.; Furlong, M.J.; Melo, M.C.; GC, D.Y.; Pozsgai, G. Island and Mountain Ecosystems as Testbeds for Biological Control in the Anthropocene. *Front. Ecol. Evol.* **2022**, *10*, 912628. [CrossRef]
37. Baeckens, S.; Van Damme, R. The island syndrome. *Curr. Biol.* **2020**, *30*, 338–339. [CrossRef] [PubMed]
38. Soumya, K.; Visalakshy, P.N.G.; Krishnamoorthy, A.; Pillai, K.G. *Dolichogenidea stantoni* (Hymenoptera: Braconidae) a potential biocontrol agent for melon borer, *Diaphania indica*. *Entomon* **2017**, *42*, 1–6.

insects

MDPI

Article

New Record of *Encarsia protransvena* and Confirmed Occurrence of *Encarsia hispida* (Hymenoptera: Aphelinidae) as Parasitoids of *Singhiella simplex* (Hemiptera: Aleyrodidae) in Italy

Giuliano Cerasa [1], Luigi Tomasello [1], Gianluca Melone [2], Elia Russo [3], Gaetano Siscaro [4], Carmelo Cavallaro [4], Annamaria Ienco [5], Francesca Laudani [5], Vincenzo Palmeri [5], Orlando Campolo [5], Francesca Garganese [6], Francesco Porcelli [6], Paolo A. Pedata [2], Vittorio Farina [1], Giovanni Gugliuzza [7], Roberto Rizzo [7], Stefania Laudonia [3,*] and Gabriella Lo Verde [1,*]

1 Department of Agricultural, Food and Forestry Science, University of Palermo (UNIPA), Viale delle Scienze, Ed. 5, 90128 Palermo, Italy; giuliano.cerasa@unipa.it (G.C.); ltomasello51@gmail.com (L.T.); vittorio.farina@unipa.it (V.F.)
2 National Research Council, Institute for Sustainable Plant Protection, P. le E. Fermi 1, 80055 Portici, Italy; gianluca.melone1998@gmail.com (G.M.); paolo.pedata@ipsp.cnr.it (P.A.P.)
3 Department of Agricultural Sciences, University of Naples Federico II (UNINA), Via Università 100, 80055 Portici, Italy; elia.russo@unina.it
4 Department of Agriculture, Food and Environment, University of Catania (UNICT), Via Santa Sofia, 100, 95123 Catania, Italy; gaetano.siscaro@unict.it (G.S.); carmelo.cavallaro@unict.it (C.C.)
5 Department of Agriculture, Mediterranean University of Reggio Calabria (UNIRC), Loc Feo di Vito, 89100 Reggio Calabria, Italy; annamaria.ienco@unirc.it (A.I.); francesca.laudani@unirc.it (F.L.); vpalmeri@unirc.it (V.P.); orlando.campolo@unirc.it (O.C.)
6 Department of Soil Sciences, Plants and Food, University of Bari Aldo Moro (UNIBA), Via Amendola 165/A, 70126 Bari, Italy; francesca.garganese@uniba.it (F.G.); francesco.porcelli@uniba.it (F.P.)
7 CREA-Research Centre for Plant Protection and Certification, c/o Department of Agricultural, Food and Forestry Science, University of Palermo (UNIPA), Viale delle Scienze, Ed. 5, 90128 Palermo, Italy; giovanni.gugliuzza@crea.gov.it (G.G.); roberto.rizzo@crea.gov.it (R.R.)
* Correspondence: stefania.laudonia@unina.it (S.L.); gabriella.loverde@unipa.it (G.L.V.)

Simple Summary: We report the detection of *Encarsia protransvena* and confirm *Encarsia hispida* as parasitoids of the exotic invasive whitefly *Singhiella simplex* (Hemiptera: Aleyrodidae) in Italy. Native to the Oriental region, *S. simplex* has now spread worldwide, causing significant damage to exotic fig plants in southern Italy, providing a unique opportunity to study its natural interactions. The whitefly and its parasitoids were collected in some south Italian regions, obtaining two species of Chalcidoidea Aphelinidae. Using a combination of morphological and molecular analyses, we identified *E. protransvena*, a species reported as the only known natural enemy in the Mediterranean region. This discovery increases our understanding of the biology of both the pest and its parasitoid, offering valuable insights for future biological control programs. As we write this note, we also have obtained specimens of *Encarsia hispida* from the host collected in Campania. We aim to provide bio-ethological information about the species in the future.

Abstract: *Encarsia protransvena* (Hymenoptera: Aphelinidae) is recorded here for the first time in Italy as a parasitoid of the whitefly *Singhiella simplex* (Hemiptera: Aleyrodidae), one of the most invasive alien pests of exotic *Ficus* species. *Singhiella simplex*, originating from the Oriental region, has established a global presence. Monitoring of the whitefly and its parasitoids was conducted in the southern areas of Italy, providing crucial insights into their distribution and interactions. The taxonomic identity of *E. protransvena*, was confirmed by scrutiny of morphological and molecular taxonomic characters. At the time of writing, we also obtained some specimens of *Encarsia hispida* from the host collected in Campania. We reserve the right to provide bio-ethological information on the species in

https://doi.org/10.3390/insects16010040

the future. Comprehensive illustrations and diagnostic features are provided for the host and the parasitoids. An identification key is included for all *Encarsia* species associated with *S. simplex*, which provides a valuable tool to distinguish these aphelinid wasps for future research and applications in biological control programs.

Keywords: fig whitefly; natural enemies; aphelinid wasps; morphological and molecular identification

1. Introduction

The fig whitefly *Singhiella simplex* (Singh, 1931) (Hemiptera: Aleyrodidae) is an invasive sap-feeding pest native to Southeast Asia that primarily infests ornamental plants of the genus *Ficus*, mainly *Ficus benjamina* L. and *Ficus microcarpa* L.f., but can also target lesser host plant species, such as *Hibiscus* spp. (Malvaceae) and *Mangifera indica* L. (Anacardiaceae) [1–6].

After its first record as an invasive species in the United States (Hodges, 2007), the pest spread to Central and South America and the Caribbean [5]. In 2014, the fig whitefly was found in Europe on several fig species in Cyprus [4] after its European recording, *S. simplex* spread across several Mediterranean countries [7–13], including Italy, where it has significantly impacted urban trees in roadsides, parks, and gardens [14–17]. Due to the significant phytosanitary risks associated with *S. simplex*, the pest was included in the EPPO Alert List from 2014 to 2018 [2].

Recent studies on the biology of *S. simplex* indicate that, under favourable climatic conditions such as those in Mediterranean countries, this whitefly can develop several overlapping generations per year [18].

The direct damage caused by *S. simplex* is mainly due to sap-sucking by the nymphs, which gradually weakens and yellows the infested leaves. Unlike most whitefly species, *S. simplex* feeds on both sides of the leaf [19]. It excretes considerable honeydew, promoting intense sooty mold that extensively coats the host plant. Infestations cause extensive defoliation and twig dieback, posing an additional threat to the overall plant health [6,7].

Several natural enemies of *S. simplex* have been reported in the literature, including predators and entomopathogens [6,20–24]. Different species of hymenopteran parasitoids belonging to the genus *Encarsia* Förster (Hymenoptera: Aphelinidae) are present worldwide and associated with the whiteflies, among them the worldwide species *Trialeurodes vaporariorum* (Westwood) [25] and *Aleurocanthus spiniferus* (Quaintance, 1903), recently introduced in Europe [26]. Three species of hymenopteran parasitoids in the genus *Encarsia* are reported as associated with the fig whitefly: *Encarsia singhiellae* Shih and Polaszek, 2015; *E. hispida* De Santis, 1948 and *E. protransvena* Viggiani, 1985 [5,27–29]. The recording of *E. tricolor* Förster, 1878 [30] has been considered a misidentification based on the aphelinid known species distribution [5]. The platigastrid *Amitus bennetti* Viggiani and Evans, 1992, and the eulophid *Baeoentedon balios* Wang, Huang and Polaszek, 2014, have also been reported [28,31]. Occasionally, males of *E. variegata* Howard, 1908, develop as hyperparasitoids on the eulophid previously mentioned [6,28]. *Encarsia protransvena* is the only confirmed parasitoid of *S. simplex* in the Mediterranean region [24]. At the same time, the recent report of *E. hispida* [17] deserves confirmation. In particular, the iconography reported in the article does not show morphological characteristics of *E. hispida* at all. A confirmation has come while we write this note, thanks to a recent collection of *S. simplex* in the Campania region with parasitization activity of *E. hispida*, currently in reduced numbers, contextual with that due to *E. protransvena*.

In 2024, surveys in southern and central Italy (Apulia, Basilicata, Calabria, Campania, Lazio, and Sicily regions) monitored the presence of the invasive *S. simplex*, and female parasitoids belonging to the genus *Encarsia* were detected in multi-site collections of the whitefly.

Here, we combined morphological analysis and molecular data from mitochondrial and nuclear markers to confirm the taxonomic identity of *E. protransvena* and *E. hispida*. We also documented the distribution of *E. protransvena* in Italy and provided a detailed morphological characterization to improve its identification. Finally, we propose an identification key for all *Encarsia* species associated with *S. simplex*, which will aid in distinguishing these aphelinid wasps in the following research and biocontrol programs management.

2. Materials and Methods

2.1. Hosts and Parasitoids Collection

Surveys have been conducted on *F. microcarpa* and *F. benjamina* trees planted in urban greeneries, parks, and roadsides in Sicily (Provinces of Palermo and Catania), Calabria (Province of Reggio Calabria), Apulia (Bari, metropolitan city), and on isolated plants of private gardens in Lazio (Province of Latina), in Campania (Provinces of Naples, Caserta and Salerno), and in Basilicata (Province of Matera). The collecting sites in which the whitefly was present are listed in Table 1. During each collection, infested leaves with *S. simplex* immature stages were gathered and then examined by a stereoscopic microscope in the laboratory. The collected material was isolated in small rearing cages and kept under laboratory conditions (26° C, 70% RH) until the whitefly or the parasitoid adult emergence. The whitefly nymphs and adults of both the host and the parasitoid were preserved in ethanol (EtOH 90%, *V*/*V* water). Adults of whitefly and parasitoids and leaves bearing the whitefly instars were photographed using a Wild-Heerbrugg M8 stereoscopic microscope equipped with a digital camera (Canon 7D, Canon Inc., Tokyo, Japan). Single montage images were obtained from image stacks with the freeware CombineZP [32] and then processed in Adobe Photoshop CS4.

Table 1. Sampling sites, coordinates, host plants, and dates of *Singhiella simplex*, *Encarsia hispida* and *Encarsia protransvena* monitoring activities in central and south Italy.

Sampling Sites (Province)	Coordinates	Host Plant	Date
Palermo (PA)	38°06′20″ N; 13°20′57″ E	*F. microcarpa*	9.II.2024 16.II.2024
Catania (CT)	37°31′09″ N; 15°06′19″ E	*F. microcarpa*	6.X.2024
Acicastello (CT)	37°33′28″ N; 15°09′01″ E	*F. microcarpa*	12.X.2024
Acireale (CT)	37°37′44″ N; 15°09′33″ E	*F. microcarpa*	7.X.2024
Reggio Calabria (RC)	38°07′21″ N; 15°39′32″ E 38°07′13″ N; 15°39′29″ E 38°07′32″ N; 15°39′17″ E	*F. microcarpa*	7.X.2024
Santa Maria Capua Vetere (CE)	41°05′01″ N; 14°15′56″ E 41°04′42″ N; 14°15′42″ E	*F. microcarpa* *F. benjamina*	30.IX.2024 30.IX.2024
Salerno (SA)	40°40′38″ N; 14°45′53″ E 40°40′12″ N; 14°47′15″ E 40°38′33″ N; 14°51′26″ E	*F. benjamina*	25.X.2024 18.X.2024 8.XI.2024
Naples (NA)	40°51′35″ N; 14°16′14″ E	*F. benjamina*	27.XI.2024
Formia (LT)	41°15′48″ N; 13°37′33″ E	*F. benjamina*	5.X.2024
Bari (BA)	41°05′46″ N; 16°53′20″ E 41°06′31″ N; 16°52′23″ E	*F. microcarpa*	6.III.2024; 5.VI.2024; 15.VIII.2024; 17.IX.2024; 14.X.2024; 21–22.XI.2024 (in both sites, only *S. simplex*)

2.2. Morphological Identification

Whitefly slide mounting followed standard methods reported by Martin [33] and Pizza and Porcelli [34], whereas slide mounting of parasitoids was carried out according to Noyes [35] and Polaszek et al. [36].

Mounted insects were examined in a bright field on a Zeiss Universal Photomicroscope III and photographed using a Leica DFC420C camera mounted on a Leica DM 2005 compound microscope (Leica Application Suite software, Leica Microsystems Ltd., Buccinasco (MI), Italy). CombineZP [32] produced image stacks post-processed in Adobe Photoshop CS4.

Identification of both adult and pupae of the whitefly was carried out according to Jensen [37], Hodges [30], Suh et al. [38], and Kondo and Evans [19]. The parasitoid identifications were carried out following the original description by Viggiani [39] and the identification keys reported by Heraty and Polaszek [40] and Polaszek et al. [41]. The specimens of *Encarsia* collected in different areas (see Examined material) and prepared on slides were compared with type material deposited in the Silvestri Museum of the University of Naples collection.

2.3. Molecular Identification

Sequencing of the parasitoid specimens emerged from *S. simplex* (see Examined material) targeted a portion of the mitochondrial cytochrome *c* oxidase subunit 1 (*COI*) gene and the nuclear D2 expansion segment of 28S (*28S-D2*) rDNA. These molecular markers were selected for their wide application in the taxonomy of Chalcidoidea [42–44]. Newly emerged parasitoids were individually killed in 96% ethanol, and genomic DNA was extracted using the non-destructive Chelex-proteinase K protocol by Gebiola et al. [45], preserving the morphological details of specimens.

The *COI* segment was amplified with the forward primers C1-J-2183 or C1-J-2195 paired with the reverse TL2-N-3014 [46], previously used for molecular characterization of *Encarsia* species [43,44]. The *28S-D2* region was amplified using the universal primers D2F and D2R developed by Campbell et al. [47]. Table 2 lists the oligonucleotide sequences and their characteristics. The polymerase chain reaction (PCR) runs were performed using DreamTaq Green PCR Master Mix (Thermo Fisher Scientific, Ferentino (FR), Italy) in a total volume of 25 µL, comprising 12.5 µL PCR Master Mix (2X), 2 µL DNA template, 1 µL of each primer (10 µM), and 8.5 µL nuclease-free water. Thermal cycling conditions included an initial denaturation at 95 °C for 3 min, followed by 40 cycles of 95 °C for 30″, 50–55 °C (depending on melting temperatures of the primer) for 30″, and 72 °C for 1′. The protocol concluded with a final extension at 72 °C for 7 min. PCR products were separated on a 1% agarose gel stained with SYBR™ Safe DNA Gel Stain (Thermo Fisher Scientific) and observed under a ChemiDoc (Bio-Rad, Segrate (MI), Italy) transilluminator. Positive amplicons were sent to Eurofins Genomics (Ebersberg, Germany) for purification and bi-directional sequencing.

Table 2. Primers used for PCR analysis targeting the DNA barcoding regions of *Encarsia protransvena* and *Encarsia hispida* for molecular identification.

Name	Nucleotide Sequence (5′-3′)	Melt. Temp. (°C)	Target Gene
C1-J-2183	CAACATTTATTTTGATTTTTTGG	55	
C1-J-2195	TTGATTTTTTGGTCATCCAGAAGT	55	*COI*
TL2-N-3014	TCCAATGCACTAATCTGCCATATTA	63	
D2F	AGTCGTGTTGCTTGATAGTGCAG	57	*28S-D2*
D2R	TTGGTCCGTGTTTCAAGACGGG	62	

Forward and reverse reads were manually edited using BioEdit version 7.2.5 [48] (Hall, 1999) to correct overlapping peaks in the electropherograms. Low-quality bases were trimmed from both ends, and consensus sequences were generated. The *COI* sequences were virtually translated into protein using the Transeq (EMBOSS) tool to check for frameshifts or premature stop codons. All sequences were submitted to NCBI's BLASTn protocol in GenBank to verify matches against standard databases using default parameters.

3. Results

3.1. Morphological Identification

The identity of *S. simplex* was confirmed based on the morphology of both adults and immature instars (Figures 1–4). The insect was found in all sampled sites only in Sicily and Calabria. In Apulia, the whitefly was recorded only in two out of six sites, where *Macrohomotoma gladiata* Kuwayama, 1908 (Hemiptera: Psylloidea: Homotomidae), another fig pest introduced in the EPPO area [48], was always present. Finally, no *S. simplex* was found in the only sampling site in the Basilicata region (Policoro, Matera province).

Parasitoids were obtained from all sampled sites where the whitefly was found, except the Apulia sites. Based on morphological features, the emerged parasitoid species were identified as *E. protransvena* and *E. hispida* (Hymenoptera: Aphelinidae). Their taxonomy, host association, and distribution are briefly discussed below.

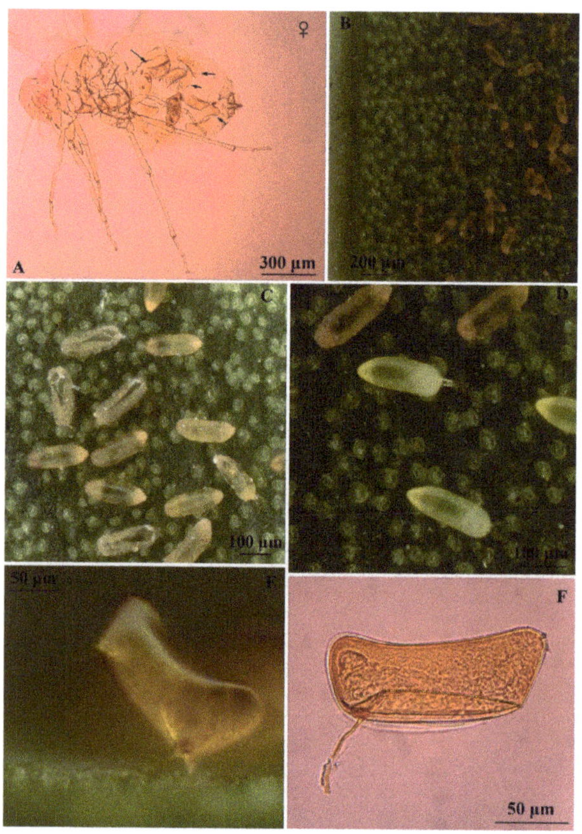

Figure 1. *Singhiella simplex*: (**A**) slide-mounted adult female, arrows show eggs retained in the abdomen; (**B–F**) whitefly eggs.

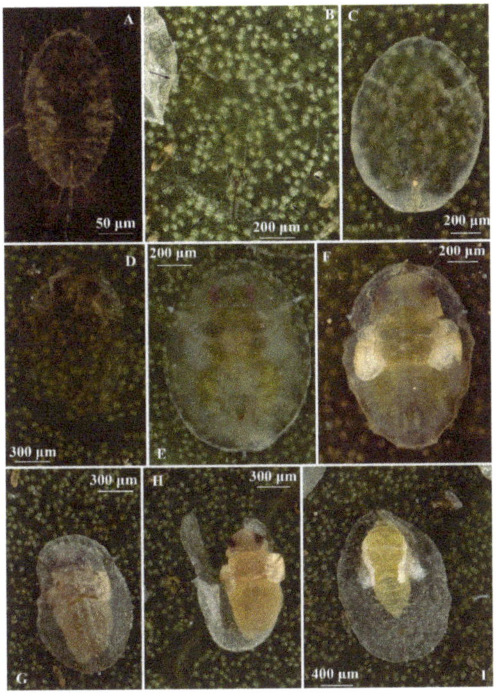

Figure 2. Immature stages of *Singhiella simplex*: (**A**) whitefly first instar; (**B–E**) second to fourth instar; (**F–I**) prepupa and pupa.

Figure 3. *Singhiella simplex*: (**A**,**B**) emerging adults; (**C**) exuvia (pupal case); (**D–G**) adults in dorsal, lateral and ventral view (scale bar: 300 μm, except in (**D**)).

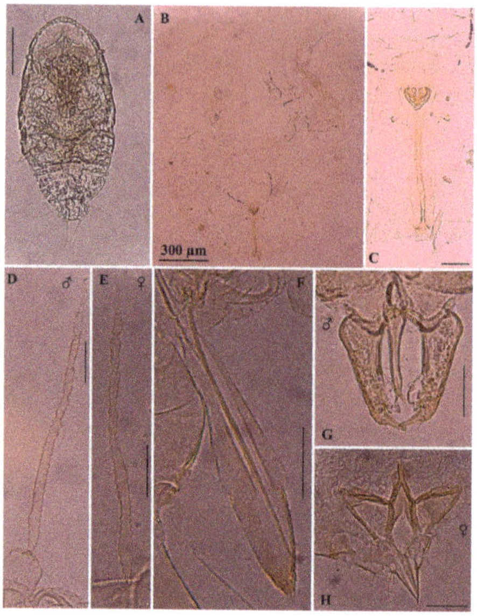

Figure 4. *Singhiella simplex*, microscope slide mounted: (**A**) first instar; (**B**,**C**) pupal case and enlargement shows vasiform orifice and caudal furrow; (**D**) antenna of male; (**E**) antenna of female; (**F**) needle-like mouthparts; (**G**) genital capsule, claspers and aedeagus of the male; (**H**) female genitalia (scale bar: 50 μm, except in (**B**)).

Encarsia protransvena Viggiani, 1985
(Figures 5–8)

Figure 5. *Encarsia protransvena*: (**A**–**C**) larvae; (**D**) pupa; (**E**), pupa ready to emerge; (**F**,**G**) dissection showing the emerging adult and pupal remains; (**H**,**I**) puparia cases of *S. simplex* with *E. protransvena* emerging hole; (**J**) meconium of *E. protransvena* (scale bar: 300 μm, except in (**J**)).

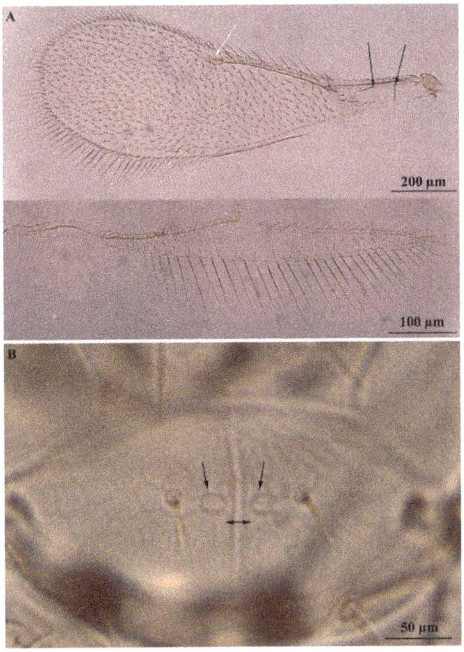

Figure 6. *Encarsia protransvena*: (**A**) forewing and hindwing, the black arrow indicating the presence of two large setae on the submarginal vein and the white arrow a bare area just above the stigmal vein; (**B**) scutellar sensillae (black arrows) closely placed or touching, ovoid and separated by less than their maximum diameter, rarely by a full diameter (left-right black arrow indicates the distance between the sensilla).

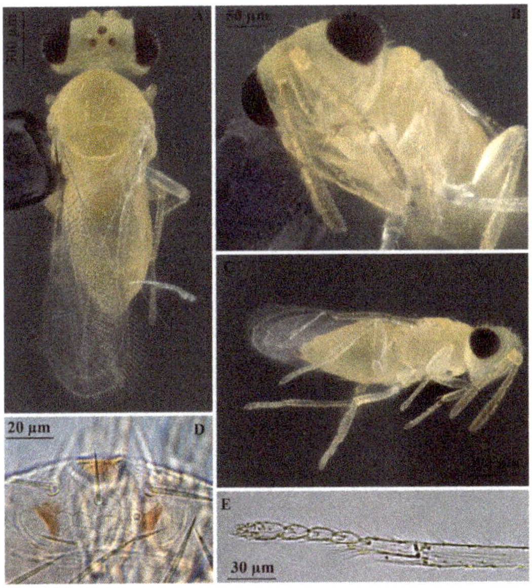

Figure 7. *Encarsia protransvena*: (**A**) body, dorsal view, mid-lobe of mesosoma with 4 pairs of setae, preapical pair not reaching the base of apical pair; (**B**) head, ventrolateral view; (**C**) body, lateral view; (**D**) stemmaticum, reticulate; (**E**) apex of middle tibia and tarsus.

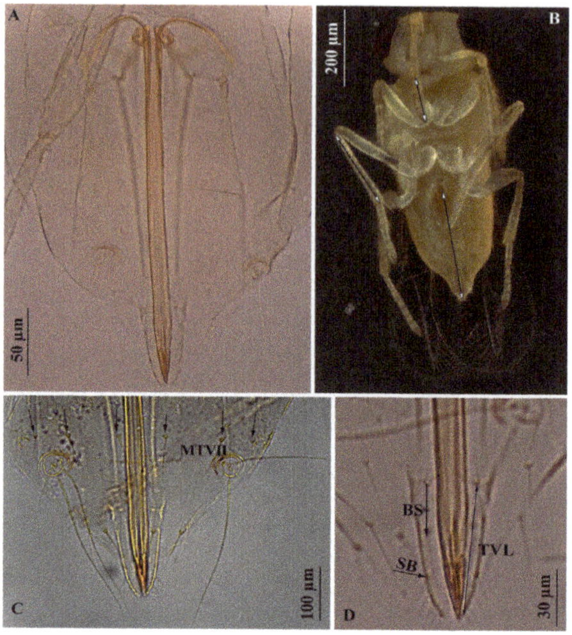

Figure 8. *Encarsia protransvena*: (**A**) ovipositor slender and with straight tip; (**B**) body, ventral view, the arrows indicate ovipositor, antennal clava and middle tibia length; (**C**) seventh metasomal tergum (Mt7) with 6 setae, 4 long setae medial to cerci; (**D**) basal seta (BS) of third valvula (TVL) not reaching base of subapical seta (SB); subapical seta located beyond halfway (0.65) between basal seta of third valvula and apex.

Examined material.

12 ♀♀ on slides, (ITALY: Sicily, Palermo, 9.II.2024, ex *S. simplex* (Singh, 1931) on *F. microcarpa* L. f., leg. G. Cerasa). 9 ♀♀ on slides, (ITALY: Sicily, Palermo, 16.II.2024, ex *S. simplex* (Singh, 1931) on *F. microcarpa* L. f., leg. L. Tomasello) (Department of Agricultural, Food and Forest Sciences, University of Palermo).

20 ♀♀ on slides, (ITALY: Calabria, Reggio Calabria, 7.X.2024, ex *S. simplex* (Singh, 1931) on *F. microcarpa* L. f., leg. V. Palmeri) (Department of Agriculture-Mediterranean University of Reggio Calabria).

2 ♀♀ in ethanol (ITALY: Sicily, Catania, 6.X.2024, ex *S. simplex* (Singh, 1931) on *F. microcarpa* L. f., leg. G. Siscaro). 10 ♀♀ on slides and 30 ♀♀ in ethanol (ITALY: Sicily, Acicastello (CT), 12.X.2024, ex *S. simplex* (Singh, 1931) on *F. microcarpa* L. f., leg. G. Siscaro). 10 ♀♀ in ethanol (ITALY: Sicily, Acireale (CT), 7.X.2024, ex *S. simplex* (Singh, 1931) on *F. microcarpa* L. f., leg. G. Siscaro) (Department of Agriculture, Food and Environment, University of Catania).

10 ♀♀ on slides and 11 ♀♀ in ethanol (ITALY: Sicily, Palermo, 9.II.2024, ex *S. simplex* (Singh, 1931) on *F. microcarpa* L. f., leg. G. Cerasa). 3 ♀♀ in ethanol (ITALY: Campania, Santa Maria Capua Vetere, 30.IX.2024, ex *S. simplex* (Singh, 1931) on *F. microcarpa* L. f., leg. P. A. Pedata). 5 ♀♀ in ethanol (ITALY: Lazio, Formia, 5.X.2024, ex *S. simplex* (Singh, 1931) on *F. benjamina* L., leg. P. A. Pedata). 2 ♀ in ethanol (ITALY, Campania, Naples, 27.XI.2024, ex *S. simplex* (Singh, 1931) on *F. benjamina* L. f., leg. P. A. Pedata) (Department of Agricultural Sciences, University of Naples Federico II).

Original description: Viggiani [39] described the species based on 1 ♀, holotype ex *Dialeurodes kirkaldyi* (Kotinsky, 1907) collected in U.S.A, Florida, Sept. '84, leg. C.R.R. Thomson; 6 ♀, paratypes, same data.

Distribution: Argentina, China, Colombia, Egypt, French Polynesia, Georgia, Honduras, Indonesia, Iran, Italy, Malaysia, Mexico, Puerto Rico, Spain, Taiwan, Turkey, USA, Western Australia [24,27,49]. We first report *E. protransvena* associated with the invasive whitefly *S. simplex* in Italy.

Hosts: The known hosts of *E. protransvena* are *Acaudaleyrodes citri* (Priesner et Hosny, 1934) [50], *Aleurotrachelus rubi* Takahashi, 1933, *Aleurothrixus floccosus* (Maskell, 1896), *Bemisia porteri* Corbett, 1935 [51], *B. tabaci* (Gennadius, 1889) [40,50,52,53], *Dialeurodes citri* (Ashmead, 1885) [40,53,54], *D. citrifolii* (Morgan, 1893) [40,50,52,53], *D. kirkaldyi* (Kotinsky, 1907) [39,40,50,52,53], *Parabemisia myricae* (Kuwana, 1927) [40,55,56], *Singhiella citrifolii* (Morgan, 1893) [56], *S. simplex* [24], *Tetraleurodes acaciae* (Quintance, 1900) [57], *Trialeurodes abutiloneus* (Haldeman, 1850) [40], *T. packardi* (Morrill, 1903) [40,50,53,58], *T. vaporariorum* Westwood, 1856 [52,54], *T. variabilis* (Quaintance, 1900) [40].

Diagnosis: The species has been included in the *strenua*-group [59], which includes about 40 species [40]. The authors previously mentioned redescribed *E. protransvena*, *E. citri* (Ishii, 1938), and *E. strenua* (Silvestri, 1927), adding two new species to the *strenua*-group: *E. neocala* Heraty and Polaszek, 2000 and *E. bimaculata* Heraty and Polaszek, 2000. The authors stressed that univariate and bivariate measures of morphological characters could not distinguish all five species because their ranges overlapped. However, by performing a multivariate morphometric analysis, all species could be discriminated. The authors then developed an identification key for females of the *strenua*-group species. Furthermore, they described the males of all these species except *E. neocala* but did not use them in the separation of the species, probably because their differences were minimal and sometimes the males are scarce, as for the thelytokous species *E. protransvena* (2 males out of 360 females examined) [55].

Members of the *strenua*-group are identified by the combination of two–three marginal setae along the dorsal margin of the costal cell at the apex, a bare area just above the stigmal vein, and scutellar sensillae that are closely placed or touching (Figure 6A,B). According to Heraty and Polaszek [40], within the *strenua*-group, *E. protransvena* is distinguished from closely related species of the same group by the following characteristics: mid-lobe of mesosoma usually with four pairs of setae, preapical pair not reaching the base of apical pair (Figure 7A); forewing (Figure 6A) usually more than 2.7 times as long as broad, if between 2.6 and 2.7 times, antennal club less than 0.14 mm in dorsal length; ovipositor slender with tip always straight (Figure 8A), clearly shorter than the length of gaster, less than 1.5 times as long as the middle tibia and more than 2.0 times as long as club (Figure 8B); apex of gaster (tergite 7 with spiracles) with six setae, four long setae medial to cerci (Figure 8C); basal seta of third valvula not reaching base of subapical seta, subapical seta located beyond halfway (0.65) between basal seta of third valvula and apex (Figure 8D).

Encarsia hispida De Santis, 1948
(Figure 9)
Examined material.

1 ♀ on slide and 1 ♀ in ethanol (ITALY, Campania, Naples, 27.XI.2024, ex *S. simplex* (Singh, 1931) on *F. benjamina* L. f., leg. P. A. Pedata) (Department of Agricultural Sciences, University of Naples Federico II).

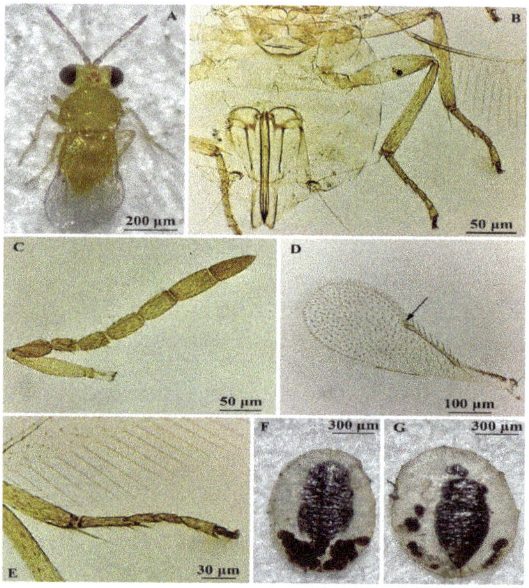

Figure 9. *Encarsia hispida* (**A–F**) and *Encarsia protransvena* (**G**): (**A**) adult female, dorsal view; (**B**) ovipositor, middle and hind tibia; (**C**) antenna; (**D**) forewing, the black arrow indicates the absence of a bare area around the stigmal vein; (**E**) apex of middle tibia, showing mid tarsus segments and the spur-to-basitarsus relative length; (**F**) pupa of *E. hispida* and (**G**) pupa of *E. protransvena*.

Original description: De Santis [60] described the species based on 1 ♀, holotype ex an unidentified aleyrodid on *Salvia splendens* (Schultes, 1820) [s/aleirodoideo en coral rojo] collected in Argentina, Santa Fe, Rosario, 1947 leg. Hack; 2 ♀, paratypes, same data.

Distribution: Argentina, Barbados, Brazil, Chile, Colombia, Dominican Republic, France (Guadeloupe, mainland), Guatemala, Honduras, Italy, Jamaica, Mexico, Netherlands (indoors), Peru, Portugal (Madeira), Puerto Rico, South Africa, Spain (Canary Islands, mainland), U.S.A., Venezuela [41].

Hosts: *Aleurodicus dugesii* Cockerell 1896, *Aleuroglandulus subtilis* Bondar 1923 [=*A. malangae* Russell 1944], *A. floccosus*, *Aleurothrixus porteri* Quaintance and Baker 1916, *A. rhamnicola*, *Aleurotrachelus trachoides* (Back 1912), *Aleyrodes spiraeoides* Quaintance 1900, *B. tabaci*, *Dialeurodes* sp., *Lipaleyrodes* sp., *Metaleurodicus minimus* (Quaintance 1900), *P. myricae* [41], *S. simplex* [17,27,29], *S. phillyreae*, *T. acaciae*, *T. abutiloneus*, *Trialeurodes floridensis* (Quaintance 1900), *T. vaporariorum*, *T. variabilis* (Quaintance 1900) [41].

Diagnosis: The species belongs to the *luteola*-group, which is characterized by four-segmented mid tarsus, fore wing without an asetose area around the stigmal vein, antennal formula 1:1:4:2 [61,62]. Polaszek et al. [41] included it in the *E. meritoria* species-complex of the *luteola* group, together with three more species (*E. dispersa* Polaszek, *E. haitiensis* Dozier and *E. meritoria* Gahan). They are all characterized by the typical female body color, generally mostly yellow, and by the tibial spur of the midleg, only slightly shorter than the corresponding basitarsus. Based on this latter character and molecular analysis of the 28S rDNA D2 region, a fifth yellow species, *E. californica* Polaszek, was not included in the species-complex and considered more closely related to *E. formosa* Gahan and *E. luteola* Howard. The authors could separate these very similar species based on the proportion of funicular segments, since these differences, even though very subtle, were consistently supported by multivariate and molecular analysis.

According to Polaszek et al. [41], the female of *E. hispida* can be separated from the other members of the *meritoria* complex by the following combination of characters: relative length of funicular segments, with F2 intermediate between F1 and F3; F6 1.2× as long as F5; scape less than 1.6× as long as F6; pedicel elongate, 1.2–1.3× as long as F1, mid tibial spur 0.75–0.8× as long as corresponding basitarsus (Figure 9A–E). The most closely related species in the complex is *E. meritoria*, whose males have antennae with F5 and F6 partially fused, while males of *E. hispida* have F5 and F6 separated.

Due to these similarities, these two species have passed through a complex taxonomic history since, after the synonymy proposed by Viggiani [63], Polaszek et al. [64] revalidated *E. hispida*. Considering that *E. meritoria* is known only from two series, both collected in Miami Beach, Florida, in 1918 and 1927, the taxonomic status of the two entities will not be fully understood as long as fresh material of *E. meritoria* from type locality will be available for molecular and biological analysis.

Moreover, *E. hispida* was successively considered conspecific with *E. brasiliensis* (Hempel) [41]. The authors proposed to maintain the name of the former species with respect to the senior synonym by reversal of precedence, considering that the type material of *E. brasiliensis* has been lost and that, due to a misidentification of Dozier [65], the name *E. brasiliensis* has been consistently applied to a distant taxonomic species belonging to the *E. opulenta* group.

Key to the females of *Encarsia* spp. reared from *S. simplex*:

1. Antennal formula 1.1.3.3 ... *E. protransvena* (Figures 6–8)

--Antennal formula 1.1.4.2 ... 2

2. Body entirely yellow; forewing hyaline; mid tarsi 4-segmented, tarsal formula 5-4-5 ... *E. hispida* (Figure 9)

--Body yellow with brown areas on mesosoma and metasoma; forewing slightly infuscate below marginal vein; mid tarsi 5-segmented, tarsal formula 5-5-5 ... *E. singhiellae*

3.2. Molecular Analysis

Two *E. protransvena* specimens collected in Palermo on 9.II.2024 and one *E. hispida* found in Naples on 27.XI.2024 (see Examined material) were further validated through sequence analysis of the mitochondrial *COI* locus and the nuclear *28S-D2* region. The expected sequences of parasitoids that emerged from *S. simplex* hosts were successfully amplified, sequenced, and uploaded to GenBank under accession numbers PQ451909 (*COI*) and PQ451935 (*28S-D2*) for *E. protransvena*, and PQ725685 (*COI*) and PQ725690 (*28S-D2*) for *E. hispida*.

The amplified *COI* portion and the *28S-D2* nuclear region were 774 and 598 bp long, respectively, in both parasitoids. The mitochondrial sequence of *E. protransvena* was truncated from its original size of 910 bp to exclude the non-transcribed region between the locus and the tRNA-leu gene, as previously described in molecular studies on *Encarsia* species [43].

For *E. protransvena*, the BLASTn analyses of the obtained *COI* sequences confirmed a complete match with the only sequence available in GenBank (AY264341.1) referring to specimens collected in Naples (Campania region, Italy) parasitizing *P. myricae* on *Citrus* spp. [43]. In addition, the *28S-D2* nuclear dataset showed 100% identity with the two sequences in the database (AF254209.1; AF254208.1) associated with Californian samples [61].

The identity of *E. hispida* was also confirmed, with its mitochondrial sequence perfectly aligning with a GenBank entry (EU488723.1) from specimens in Naples (Campania region, Italy) and San Diego (California) emerging from *Bemisia tabaci* Gennadius and *Aleurodicus dugesii* Cockerell hosts, respectively [66]. At the nuclear level, a single nucleotide substitution in the *28S-D2* region resulted in 99.83% similarity to Californian specimens reported in the NCBI database (AF223370.1). These mitochondrial and nuclear sequences provided strong genetic support for the molecular comparative identification of both *Encarsia* species associated with the fig whitefly.

4. Discussion

Since its first detection in Italy, *S. simplex* has caused significant damage to ornamental *Ficus* plants in several regions [14–17]. The insect was probably introduced through the trade of plants from infested areas, such as Southeast Asia [15]. Given the ornamental function of many *Ficus* plants and their location mainly in public places, where the application of pesticides is limited and problematic, it is essential to monitor the presence and effectiveness of natural enemies before implementing any management action, whether chemical or agronomic. In addition to the activities of predators, especially coccinellids, several A.A. report an average parasitization rate of whiteflies of 10% due to *Encarsia* spp. [22,29]. Although *B. balios* and *A. bennetti* appear to be the more effective parasitoids of *S. simplex* in the U.S.A. [6,28], *E. protransvena* provides a limited level of control in Turkey, where the highest natural parasitism rate was found to be 32.88% and 21.66% in October 2018 and 2019, respectively [24]. Preliminary data from sampling in the study site at Palermo (Sicily) from May 2024 confirm that the highest parasitism rate, about 20%, was recorded in October (unpublished data). The neotropical species *E. hispida* has been reported parasitizing *Aleurodicus dispersus* Russel, 1965 on *F. macrophylla* Pers. and *Ficus* spp. in the Canary Islands and Madeira [67]. This species, belonging to the *luteola*-group, was recently reported as a parasitoid of *S. simplex* in Northern Italy [17]. However, comparing the *E. hispida* description with the figures of the slide-mounted specimen reported, the identification appears to be doubtful. *Encarsia hispida*, as described by De Santis [60], is similar to *E. protransvena* in body color and easily distinguishable from the latter by the antennal and tarsal formulas (see the key to female species reported in the previous paragraph). The molecular characterization of *E. hispida* reported by Zugno et al. [17] and the slide image, clearly attributable to *E. protransvena,* suggest the possibility of a mixed population of the two parasitoids in the sampled area, as confirmed in Campania. Additionally, Zugno et al. [17] relied on *COI* sequences to support their identification of *E. hispida* but did not provide GenBank accession numbers for their data, instead directly using the BLASTn tool for sequence comparison, limiting reproducibility and broader comparative analyses. It is worth noting that these observations do not detract from the contributions of Zugno et al. [17] about the distribution of these parasitoids in Italy.

Integrating morphological and molecular analyses for the identification of parasitic Hymenoptera provides a robust and reliable method for enhancing taxonomic accuracy [68,69]. In this study, we applied a combined morphological and molecular approach to comprehensively identify *E. protransvena* and *E. hispida* using two informative molecular markers, *COI* and *28S*, corroborating our morphological findings. Molecular evidence derived from BLASTn analysis confirmed that the *E. protransvena* population we studied is genetically indistinguishable from the previously described in Italy [42] and California [61], as it shares identical mitochondrial and nuclear haplotypes.

This complete genetic identity among populations from geographically distant regions raises intriguing questions about dispersal mechanisms of this aphelinid wasp, highlighting the need for future investigation. Similarly, *E. hispida* showed high genetic homogeneity, with its mitochondrial sequence fully matching specimens from Italy and California [66]. A single nucleotide substitution in the *28S-D2* region indicated slight intraspecific nuclear variation (0.17% divergence), potentially reflecting regional adaptation or sequencing variability.

The high-resolution molecular dataset generated in this work represents a valuable resource for future research on the population genetics, phylogeography, and evolutionary dynamics of these aphelinid wasps, with important implications for ecological studies and potential biocontrol strategies against whiteflies.

Finally, the integrated taxonomy method allowed us to report for the first time in Italy the association of *E. protransvena* with the fig whitefly *S. simplex* and to confirm the association of *E. hispida*, already known in Italy, as a parasitoid of this invasive species.

Author Contributions: Investigation, experimental design, methodology, software, data curation, writing original draft preparation, writing and editing, supervision, G.C., S.L., G.S., V.P., O.C., F.P., P.A.P. and G.L.V.; investigation, experimental operation and revision of the article, G.M., E.R., C.C., A.I., F.L., F.G., R.R., G.G., V.F. and L.T. All authors have read and agreed to the published version of the manuscript.

Funding: This research was partially funded by Campania Region URCOFI IV "Unità di coordinamento e potenziamento delle attività di sorveglianza, ricerca, sperimentazione, monitoraggio e formazione in campo fitosanitario".

Data Availability Statement: The original contributions presented in this study are included in the article. Further inquiries can be directed to the corresponding author(s).

Conflicts of Interest: The authors declare no conflicts of interest.

References

1. Legaspi, J.C.; Mannion, C.; Amalin, D.; Legaspi, B.C., Jr. Biology, ecology and control of the Ficus whitefly, *Singhiella simplex* (Hemiptera: Aleyrodidae). *Potential Invasive Pests Agric. Crop.* **2013**, *3*, 363.
2. EPPO Global Database. Available online: https://gd.eppo.int/taxon/SINLSI/categorization (accessed on 21 September 2024).
3. EPPO—European and Mediterranean Plant Protection Organization. Mini Data Sheet on *Singhiella simplex* (Hemiptera: Aleyrodidae) Ficus Whitefly. 2018. Available online: https://gd.eppo.int/taxon/SINLSI/documents (accessed on 22 October 2022).
4. EPPO. First Report of *Singhiella simplex* in Cyprus: Addition to the EPPO Alert List. EPPO Reporting Service 11. 2014. Available online: https://gd.eppo.int/reporting/article-3306 (accessed on 21 September 2024).
5. Ko, C.C.; Shih, Y.T.; Schmidt, S.; Polaszek, A. A new species of *Encarsia* (Hymenoptera, Aphelinidae) developing on ficus whitefly *Singhiella simplex* (Hemiptera, Aleyrodidae) in China and Taiwan. *J. Hymenopt. Res.* **2015**, *46*, 85–90.
6. Ahmed, M.Z.; McKenzie, C.L.; Revynthi, A.M.; Evans, G.A.; Mottern, J.L.; Mannion, C.M.; Osborne, L.S. Pest status, suvey of natural enemies, and a management plan for the whitefly *Singhiella simplex* (Hemiptera: Aleyrodidae) in the United States. *J. Integr. Pest Manag.* **2022**, *13*, 1–14. [CrossRef]
7. Yükselbaba, U.; Topakcı, N.; Göçmen, H. A new record of Turkey Aleyrodidae fauna, ficus whitefly *Singhiella simplex* (Singh) (Hemiptera: Aleyrodidae). *Phytoparasitica* **2017**, *45*, 715–717. [CrossRef]
8. Germain, J.F.; Lemmet, S.; Balmès, V.; Streito, J.C. Incursion d'un nouvel aleurode nuisible aux ficus en France. *Phytoma* **2017**, *705*, 9–11.
9. Dader Alonso, B.; Viñuela Sandoval, E.; Medina Vélez, M.P.; Budia Marigil, M.F.; Adán del Río, A.; del Estal Padillo, P. Presencia en España de la mosca blanca del *Ficus* spp., *Singhiella simplex* (Singh, 1931) (Hemiptera: Aleyrodidae). Libro de Resúmenes XI Congreso de la Sociedad Española de Entomología Aplicada (Madrid, 2019-11-04/08), p. 108. Available online: https://www.upm.es/observatorio/vi/index.jsp?pageac=actividad.jsp&id_actividad=324992 (accessed on 21 September 2024).
10. Šimala, M.; Pintar, M.; Masten Milek, T. Interception of Ficus whitefly [*Singhiella simplex* (Singh, 1931)] in Croatia. *Glas. Biljn. Zaštite* **2020**, *20*, 540–547.
11. Malumphy, C.; Guillem, R. Seven species of whitefly new for Gibraltar, including fig whitefly *Singhiella simplex* (Singh) (Hemiptera: Aleyrodidae). *Entomol.'s Mon. Mag.* **2021**, *157*, 168–172. [CrossRef]
12. Kamel, M.K.B.H.; Zouari, S.; Adouani, R.; Ben Cheik, Z. Nouveau signalement de *Singhiella simplex* (Singh, 1931) (Aleyrodidae) sur Ficus en Tunisie. *EPPO Bull.* **2022**, *52*, 460462.
13. Laarif, A.; Bouslama, T. The Ficus Whitefly *Singhiella simplex* (Singh, 1931) (Hemiptera: Aleyrodidae): A new exotic whitefly found on urban Ficus species in Tunisia. *Orient. Insects* **2022**, *56*, 584–592. [CrossRef]
14. Longo, S.; Rapisarda, C.; Siscaro, G. La mosca bianca dei *Ficus*, *Singhiella simplex*, nuovamente alla ribalta. Georgofili. Info. 2019. Available online: https://www.georgofili.info/contenuti/la-moscabianca-dei-ficus-singhiella-simplex-nuovamente-alla-ribalta/13601 (accessed on 5 December 2024).

15. Laudani, F.; Giunti, G.; Zimbalatti, G.; Campolo, O.; Palmeri, V. *Singhiella simplex* (Singh) (Hemiptera: Aleyrodidae), a new aleyrodid species for Italy causing damage on *Ficus*. *EPPO Bull.* **2020**, *50*, 268–270. [CrossRef]
16. Fois, F.; Lallai, A.; Foddi, F.; Nannini, M. Prima segnalazione per la Sardegna di *Singhiella simplex* (Singh, 1931) (Hemiptera: Aleyrodidae) reperita su *Ficus benjamina* L. e *Ficus microcarpa* L.f. (Moraceae) [First report of *Singhiella simplex* (Singh, 1931) (Hemiptera: Aleyrodidae) from Sardinia, on *Ficus benjamina* L. and *Ficus microcarpa* L.f. (Moraceae)]. *Mediterr. Online/Nat.* **2020**, *35*, 35–39. (In Italian)
17. Zugno, M.; Tapparo, A.; Colombini, M.; Galimberti, G.; Sacchi, S.; Siena, F.; Cavagna, B.; Ciampitti, M.; Giordano, L. First report of the ficus whitefly *Singhiella simplex* (Singh, 1931) (Hemiptera: Aleyrodidae) in Northern Italy and first observation of its association with the parasitoid wasp *Encarsia hispida* De Santis, 1948 (Hymenoptera: Aphelinidae) in Europe. *EPPO Bull.* **2024**, *54*, 182–188. [CrossRef]
18. Šimala, M.; Pintar, M.; Milek, T.M. First records of new insect pests in Croatia between two Slovenian conferences on plant protection (2019–2022). In Proceedings of the Zbornik Predavanj in Referatov 15. Slovenskega Posvetovanja o Varstvu Rastlin z Mednarodno Udeležbo, Portorož, Slovenia, 1–2 March 2022.
19. Kondo, T.; Evans, G. *Singhiella simplex* (Singh) (Hemiptera: Aleyrodidae), a new aleyrodid invasive species for Colombia. *Bol. Mus. Entomol. Univ. Val.* **2012**, *13*, 31–34.
20. Torres-Barragán, A.; Anaya, A.L.; Alatorre, R.; Toriello, C. Entomopathogenic fungi from 'El Eden' Ecological Reserve, Quintana Roo, Mexico. *Mycopathologia* **2004**, *158*, 61–71. [CrossRef]
21. Mannion, C. Ficus Whitefly, Management in the Landscape. UF/IFAS Extension. Available online: https://cisr.ucr.edu/sites/default/files/2019-07/ficus_whitefly_feb2010_fact_sheet.pdf (accessed on 20 August 2024).
22. Avery, P.B.; Mannion, C.M.; Powell, C.A.; McKenzie, C.L.; Osborne, L.S. Natural enemies managing the invasion of the fig whitefly, *Singhiella simplex* (Hemiptera: Aleyrodidae), infesting a *Ficus benjamina* hedge. *Fla. Entomol.* **2011**, *94*, 696–698. [CrossRef]
23. Elliot, S.; Sabelis, M.; Janssen, A.; Der Geest, V.; Beerling, E.; Fransen, J. Can plants use entomopathogens as bodyguards? *Ecol. Lett.* **2000**, *3*, 228–235. [CrossRef]
24. Yükselbaba, U. Survey of the Ficus whitefly, *Singhiella simplex* (Hemiptera: Aleyrodidae), and its natural enemies in the Western Mediterranean Region of Turkey. *Fla. Entomol.* **2023**, *106*, 22–28. [CrossRef]
25. van Roermund, H.J.W.; van Lenteren, J.C. The parasite-host relationship between *Encarsia formosa* (Hymenoptera: Aphelinidae) and *Trialeurodes vaporariorum* (Homoptera: Aleyrodidae) XXXV. Life-history parameters of the greenhouse whitefly parasitoid *Encarsia formosa* as a function of host stage and temperature. *Wagening. Agric. Univ. Pap.* **1992**, *92–93*, 1–147.
26. Melone, G.; Ascolese, R.; Nugnes, F.; Porcelli, F.; Rapisarda, C.; Farina, A.; Picciotti, U.; Garganese, F.; Laudonia, S. An *Eretmocerus* species, parasitoid of *Aleurocanthus spiniferus*, was found in Europe: The secret savior of threatened plants. *Sustainability* **2024**, *16*, 2970. [CrossRef]
27. Myartseva, N.S.; González-Julián, P.; Ruíz-Cancino, E.; Coronado-Blanco, C.M.J.; Muñoz-Viveros, L.A. Parasitoides de la mosquita blanca *Singhiella simplex* (Singh, 1931) en México (Hymenoptera: Chalcidoidea: Aphelinidae). [Parasitoids of ficus whitefly Singhiella simplex (Singh, 1931) in Mexico]. *Entomol. Mex.* **2014**, *1*, 1113–1117. (In Spanish)
28. Ahmed, M.Z.; Hernandez, Y.; Francis, A.; Evans, G.; Rohrig, E.; Osborne, L.; Mannion, C. Balios eulophid Baeoentendon balios Wang, Huang & Polaszek (Insecta: Hymenoptera: Eulophidae). In *Featured Creatures, EDIS-IFAS EENY-676*; University of Florida: Gainesville, FL, USA, 2017.
29. Sánchez-Flores, O.A.; Coronado-Blanco, J.M. Parasitoide de *Singhiella simplex* (Singh, 1931) (Hemiptera: Aleyrodidae) en Tamaulipas, México. *Boletín Soc. Mex. Entomol.* **2020**, *6*, 9–14.
30. Hodges, G. The Fig Whitefly *Singhiella simplex* (Singh) (Hemiptera: Aleyrodidae): A New Exotic Whitefly Found on *Ficus* Species in South Florida. Pest Alert DPI-FDACS. 2007. Available online: http://www.freshfromflorida.com/pi/enpp/ento/Singhiella%20simplex.html (accessed on 21 September 2024).
31. Lahey, Z.J.; Polaszek, A. *Baeoentedon balios* (Hymenoptera: Eulophidae): A parasitoid of ficus whitefly, *Singhiella simplex* (Singh) (Hemiptera: Aleyrodidae), new to the United States. *Int. J. Pest Manag.* **2017**, *63*, 349–351. [CrossRef]
32. Hadley, A.; Combine, Z.P. 2011. Available online: https://combinezp.software.informer.com/ (accessed on 10 May 2019).
33. Martin, J.H. The whitefly fauna of Australia (Sternorrhyncha: Aleyrodidae), a taxonomic account and identification guide. *Tech. Pap. CSIRO Entomol.* **1999**, *38*, 1–197.
34. Pizza, M.; Porcelli, F. Sull'allestimento di preparati microscopici di pupari di Aleirodidi (Homoptera). *Boll. Soc. Entomol. Ital.* **1993**, *125*, 3–5.
35. Noyes, J.S. Collecting and preserving chalcid wasps (Hymenoptera: Chalcidoidea). *J. Nat. Hist.* **1982**, *16*, 315–334. [CrossRef]
36. Polaszek, A.; Ayshford, T.; Yahya, B.E.; Fusu, L. *Wallaceaphytis*: An unusual new genus of parasitoid wasp (Hymenoptera: Aphelinidae) from Borneo. *J. Nat. Hist.* **2014**, *48*, 1111–1123. [CrossRef]
37. Jensen, A.S. A cladistic analysis of *Dialeurodes*, *Massilieurodes*, and *Singhiella*, with notes and keys to the Nearctic species and descriptions of four new *Massilieurodes* species (Hemiptera: Aleyrodidae). *Syst. Entomol.* **2001**, *26*, 279–310. [CrossRef]

38. Suh, S.; Evans, G.A.; Oh, S. A checklist of intercepted whiteflies (Hemiptera: Aleyrodidae) at the Republic of Korea ports of entry. *J. Asia-Pac. Entomol.* **2008**, *11*, 37–43. [CrossRef]
39. Viggiani, G. Notes on a few Aphelinidae, with description of five new species of *Encarsia* Foerster (Hymenoptera: Chalcidoidea). *Boll. Lab. Entomol. Agrar. Filippo Silvestri* **1985**, *42*, 81–94.
40. Heraty, J.M.; Polaszek, A. Morphometric analysis and descriptions of selected species in the *Encarsia strenua* group (Hymenoptera: Aphelinidae). *J. Hymenopt. Res.* **2000**, *9*, 142–169.
41. Polaszek, A.; Manzari, S.; Quicke, D.L.J. Morphological and molecular taxonomic analysis of the *Encarsia* meritoria species-complex (Hymenoptera, Aphelinidae), parasitoids of whiteflies (Hemiptera, Aleyrodidae) of economic importance. *Zool. Scr.* **2004**, *33*, 403–421. [CrossRef]
42. Campbell, B.; Heraty, J.; Rasplus, J.Y.; Chan, K.; Steffen-Campbell, J.; Babcock, C. Molecular systematics of the Chalcidoidea using 28S-D2 rDNA. In *Hymenoptera Evolution, Biodiversity and Biological Control*; CSIRO Publishing: Collingwood, Australia, 2000; pp. 59–73.
43. Monti, M.M.; Nappo, A.G.; Giorgini, M. Molecular characterization of closely related species in the parasitic genus *Encarsia* (Hymenoptera: Aphelinidae) based on the mitochondrial cytochrome oxidase subunit I gene. *Bull. Entomol. Res.* **2005**, *95*, 401–408. [CrossRef]
44. Gebiola, M.; Monti, M.M.; Johnson, R.C.; Woolley, J.B.; Hunter, M.S.; Giorgini, M.; Pedata, P.A. A revision of the *Encarsia pergandiella* species complex (Hymenoptera: Aphelinidae) shows cryptic diversity in parasitoids of whitefly pests. *Syst. Entomol.* **2017**, *42*, 31–59. [CrossRef]
45. Gebiola, M.; Bernardo, U.; Monti, M.M.; Navone, P.; Viggiani, G. *Pnigalio agraules* (Walker) and *Pnigalio mediterraneus* Ferrière and Delucchi (Hymenoptera: Eulophidae): Two closely related valid species. *J. Nat. Hist.* **2009**, *43*, 2465–2480. [CrossRef]
46. Simon, C.; Frati, F.; Beckenbach, A.; Crespi, B.; Liu, H.; Flook, P. Evolution, weighting, and phylogenetic utility of mitochondrial gene sequences and a compilation of conserved polymerase chain reaction primers. *Ann. Entomol. Soc. Am.* **1994**, *87*, 651–701. [CrossRef]
47. Campbell, B.C.; Steffen-Campbell, J.D.; Werren, J.H. Phylogeny of the *Nasonia* species complex (Hymenoptera: Pteromalidae) inferred from an internal transcribed spacer (ITS2) and 28S rDNA sequences. *Insect Mol. Biol.* **1994**, *2*, 225–237. [CrossRef]
48. Hall, T.A. BioEdit: A user-friendly biological sequence alignment editor and analysis program for Windows 95/98/NT. *Nucleic Acids Symp. Ser.* **1999**, *41*, 95–98.
49. Universal Chalcidoidea Database (UCD) Curated in TaxonWorks. Available online: https://sfg.taxonworks.org (accessed on 20 September 2024).
50. Polaszek, A.; Abd-Rabou, S.; Huang, J. The Egyptian species of *Encarsia* (Hymenoptera: Aphelinidae): A preliminary review. *Zool. Meded.* **1999**, *73*, 131–163.
51. Shih, Y.T.; Ko, C.C.; Polaszek, A. *Encarsia* (Hymenoptera: Aphelinidae) parasitoids of *Bemisia* species in Taiwan (Hemiptera: Aleyrodidae). *J. Nat. Hist.* **2008**, *42*, 2923–2941. [CrossRef]
52. Nguyen, R.; Hamon, A.B. *Dialeurodes kirkaldyi* (Kotinsky), in Florida (Homoptera: Aleyrodidae: Aleyrodinae). *Entomol. Circ.* **1989**, *323*, 2.
53. Schmidt, S.; Naumann, I.D.; De Barro, P.J. *Encarsia* species (Hymenoptera: Aphelinidae) of Australia and the Pacific Islands attacking *Bemisia tabaci* and *Trialeurodes vaporariorum* (Hemiptera: Aleyrodidae)—A pictorial key and descriptions of four new species. *Bull. Entomol. Res.* **2001**, *91*, 369–387. [CrossRef]
54. Abd-Rabou, S. Revision of Aphelinidae (Hymenoptera) from Egypt. In *Proceedings of the Second International Conference of Plant Protection Research Institute*; Dokki: Giza, Egypt, 2002; Volume 1, p. 281.
55. Giorgini, M. Induction of males in thelytokous populations of *Encarsia meritoria* and *Encarsia protransvena*: A systematic tool. *BioControl* **2001**, *46*, 427–438. [CrossRef]
56. Viggiani, G. Host-range increase and phenology of Encarsia meritoria Gahan in Europe (Hymenoptera: Aphelinidae). In *Parasitic Wasps: Evolution, Systematics, Biodiversity and Biological Control, Proceedings of the International Symposium: "Parasitic Hymenoptera: Taxonomy and Biological Control", Köszeg, Hungary, 14–17 May 2001*; Melika, G., Thuróczy, C., Eds.; 2002; pp. 335–339. Available online: https://iris.cnr.it/handle/20.500.14243/103624 (accessed on 21 September 2024).
57. Myartseva, S.N.; Evans, G.A. Genus *Encarsia* Förster of Mexico (Hymenoptera: Chalcidoidea: Aphelinidae). A revision, key and description of new species. *Ser. Avis. Parasit. Plagas Otros Insectos* **2008**, *3*, 1–320.
58. Abd-Rabou, S.; Ghahari, H. A revision of *Encarsia* (Hymenoptera: Aphelinidae) species from Iran. *Egypt. J. Agric. Res.* **2004**, *82*, 647–684.
59. Hayat, M. A revision of the species of *Encarsia* Foerster (Hymenoptera: Aphelinidae) from India and the adjacent countries. *Orient. Insects* **1989**, *23*, 1–131. [CrossRef]
60. De Santis, L. Adiciones a la fauna Argentina de Afelínidos (Hymenoptera: Chalcidoidea). *Notas Mus. Plata Buenos Aires* **1948**, *13*, 45.

61. Babcock, C.S.; Heraty, J.M.; De Barro, P.J.; Driver, F.; Schmidt, S. Preliminary phylogeny of *Encarsia* Förster (Hymenoptera: Aphelinidae) based on morphology and 28S rDNA. *Mol. Phylogenet. Evol.* **2001**, *18*, 306–323. [CrossRef]
62. Heraty, J.M.; Polaszek, A.; Schauff, M.E. Systematics and Biology of *Encarsia*. In *Classical Biological Control of Bemisia tabaci in the United States—A Review of Interagency Research and Implementation*; Gould, J., Hoelmer, K., Goolsby, J., Eds.; Progress in Biological Control; Springer: Dordrecht, The Netherlands, 2008; Volume 4. [CrossRef]
63. Viggiani, G. Notes on some Nearctic and Neotropical *Encarsia* Förster (Hymenoptera: Aphelinidae). *Boll. Lab. Entomol. Agrar. Filippo Silvestri Portici* **1989**, *46*, 207–221.
64. Polaszek, A.; Evans, G.A.; Bennett, F.D. *Encarsia* parasitoids of *Bemisia tabaci* (Hymenoptera: Aphelinidae, Homoptera: Aleyrodidae): A preliminary guide to identification. *Bull. Entomol. Res.* **1992**, *82*, 375–392. [CrossRef]
65. Dozier, H.L. Two undescribed chalcid parasites of the woolly whitefly, *Aleurothrixus floccosus* (Maskell) from Haiti. *Proc. Entomol. Soc. Wash.* **1932**, *34*, 118–122.
66. Giorgini, M.; Monti, M.M.; Caprio, E.; Stouthamer, R.; Hunter, M.S. Feminization and the collapse of haplodiploidy in an asexual parasitoid wasp harboring the bacterial symbiont Cardinium. *Heredity* **2009**, *102*, 365–371. [CrossRef]
67. Hernández-Suárez, E.; Carnero, A.; Aguiar, A.; Prinsloo, G.; LaSalle, J.; Polaszek, A. Parasitoids of whiteflies (Hymenoptera: Aphelinidae, Eulophidae, Platygastridae; Hemiptera: Aleyrodidae) from the Macaronesian archipelagos of the Canary Islands, Madeira, and the Azores. *Syst. Biodivers.* **2003**, *1*, 55–108. [CrossRef]
68. Andreason, S.A.; Triapitsyn, S.V.; Perring, T.M. Untangling the *Anagyrus pseudococci* species complex (Hymenoptera: Encyrtidae), parasitoids of worldwide importance for biological control of mealybugs (Hemiptera: Pseudococcidae): Genetic data corroborates separation of two new, previously misidentified species. *Biol. Control* **2019**, *129*, 65–82.
69. Miller, K.E.; Polaszek, A.; Evans, D.M. A dearth of data: Fitting parasitoids into ecological networks. *Trends Parasitol.* **2021**, *37*, 863–874. [CrossRef] [PubMed]

insects

MDPI

Article

New Records of *Phenacoccus solenopsis* Natural Enemies in Europe and Taxonomic Additions on *Anagyrus matritensis*

Michele Ricupero [1,*], Emanuele Porcu [2], Agatino Russo [1], Lucia Zappalà [1] and Gaetano Siscaro [1]

[1] Department of Agriculture, Food and Environment, University of Catania, 95123 Catania, Italy; agatino.russo@unict.it (A.R.); lucia.zappala@unict.it (L.Z.); gaetano.siscaro@unict.it (G.S.)
[2] Department of Agriculture, University of Reggio Calabria, Feo di Vito, 89122 Reggio Calabria, Italy; emanuele.porcu@unirc.it
* Correspondence: michele.ricupero@unict.it

Simple Summary: *Phenacoccus solenopsis* (Hemiptera: Pseudococcidae) is threatening protected horticultural and ornamental crops in Mediterranean countries, and research into indigenous natural enemies is needed as a sustainable biological control tool. Field surveys were carried out to discover the natural enemies attacking the cotton mealybug in Sicily (Italy) where the pest is currently present. Hymenopteran parasitoids and generalist coccinellid predators were reported. *Aenasius arizonensis* (Hymenoptera: Encyrtidae) was reported for the first time in Europe, and *Anagyrus matritensis* (Hymenoptera: Encyrtidae) was recorded for the first time in association with *P. solenopsis*. The two parasitoid species were identified by morphological features and molecular tools using a partial sequence of the COI mitochondrial gene. Further morphological details were also added to the original description for *A. matritensis*. The generalist predators *Cryptolaemus montrouzieri*, *Hippodamia variegata* and *Parexochomus nigripennis* (Coleoptera: Coccinellidae) were also recorded preying on the cotton mealybug.

Abstract: The cotton mealybug *Phenacoccus solenopsis* (Hemiptera: Pseudococcidae) is a polyphagous invasive species native to America and considered one of the major cotton pests in Asia. It is currently threatening horticultural and ornamental protected crops in Mediterranean countries. Due to ecological and environmental concerns, the conventional chemical control of *P. solenopsis* in new areas of introduction is being replaced by exploring the potential of indigenous natural enemies as a sustainable biological control tool. After *P. solenopsis* introduction in Sicily (Italy), field surveys were conducted on native natural enemies attacking the mealybug to select promising biocontrol agents for field applications. For the first time, *Aenasius arizonensis* (Hymenoptera: Encyrtidae) was reported in Europe, and the native *Anagyrus matritensis* (Hymenoptera: Encyrtidae) was recorded in association with *P. solenopsis*. The two parasitoid species were identified by morphological features and molecularly using a portion of the mitochondrial cytochrome oxidase subunit I (mtCOI) gene. Because of missing information, additional morphological features were provided for the morphological identification of *A. matritensis*. In addition, the generalist predators *Cryptolaemus montrouzieri*, *Hippodamia variegata* and *Parexochomus nigripennis* (Coleoptera: Coccinellidae) were also recorded attacking the invasive mealybug.

Keywords: *Anagyrus*; *Aenasius arizonensis*; barcoding; biocontrol; morphological identification; parasitoid; cotton mealybug; tomato

1. Introduction

The unintentional introduction of invasive exotic insect species is an increasing phenomenon worldwide that is exacerbated by climate change and leads to an adverse impact on crop production [1]. After the dispersal and expansion of the geographical distribution of such organisms, new trophic associations can occur [2]. Among recent cases reported in the Mediterranean, noteworthy is the introduction of the cotton mealybug *Phenacoccus solenopsis* Tinsley (Hemiptera: Pseudococcidae). This invasive species is native to North America; it is currently present in all continents, and it is considered one of the key pests of cotton in Asia [3]. Within the Palaearctic region, the species has been reported on during the last ten years in several countries in the Mediterranean basin [4,5]. Although in these regions, *P. solenopsis* has been primarily recorded on wild hosts and ornamentals (e.g., *Hibiscus* sp. and *Lantana* sp.), in Israel and Egypt, the mealybug has become a serious pest in cotton fields and solanaceous protected crops, namely in bell pepper and tomato greenhouses [6].

P. solenopsis is a highly polyphagous sap-sucking pest that can feed on over 200 host plants belonging to about 60 botanical families. This mealybug feeds on phloem sap mainly on the aboveground plant parts and produces honeydew and waxy secretion. It is known to have high fecundity (from 150 to 600 eggs per female), and it can complete up to 15 overlapping generations under field conditions. Its feeding activity also causes chlorosis and can induce plant growth reduction and even plant death [3]. Although the honeydew emitted by the mealybug is not abundant, *P. solenopsis* establishes mutualistic relationships with several species of ants (Hymenoptera: Formicidae) including the invasive red imported fire ant *Solenopsis invicta* Buren [7]. This aspect could be detrimental for agricultural production and trade in areas where both species are present [4,8].

As the introduction of *P. solenopsis* raises ecological and environmental concerns also due to conventional chemical control [3,9], there is a growing interest in exploring the potential of indigenous natural enemies as sustainable control tools. Biological control using natural enemies, particularly parasitoids, has been thought to be the key factor in suppressing *P. solenopsis* populations in agricultural systems. More than twenty-five records of parasitoid species, mainly belonging to the hymenopteran family Encyrtidae, are reported as parasitizing on *P. solenopsis* [10]. A recent study recorded a total of 18 hymenopteran parasitoids belonging to 11 genera and 4 families in Southern China. Eight of these species are primary parasitoids, with *Aenasius arizonensis* (Girault) (Hymenoptera: Encyrtidae) being the dominant taxon, while the remaining ten species are most likely hyperparasitoids [11].

A further evaluation and use of parasitoid wasps in biological control programmes against *P. solenopsis* require accurate species identification in new invaded areas [3]. However, given the diversity of parasitoid wasps associated with *P. solenopsis*, species identification via morphology can be difficult. This aspect is particularly relevant for non-hymenopteran parasitoid specialists, as these organisms are typically small and morphologically similar, requiring a thorough examination of detailed morphological characters [12]. However, this limitation can be overcome by the DNA sequencing of mitochondrial cytochrome oxidase subunit I (mtCOI) since it has become an adequate tool for species identification in insects [13,14].

In this study, we used an integrated approach (morphological and molecular) to identify parasitoids associated with *P. solenopsis*, and we also recorded its predators for the first time in Italy. Moreover, since the morphological characterization of *Anagyrus matritensis* (Mercet) (Hymenoptera: Encyrtidae) was lacking information, we provided additional morphological details and molecular identifications of this species. The findings of this study may help to identify the natural enemies of *P. solenopsis*, as this is an important step towards the use of natural enemies as biological control agents against this serious insect pest in Mediterranean agroecosystems.

2. Materials and Methods

2.1. Insect Sampling

Between October 2023 and November 2024, we conducted periodic samplings of *P. solenopsis* from infested protected crops and ornamental gardens in Sicily, Italy (Table 1). Specifically, we inspected the aboveground parts of infested plants (both cultivated and wild) in four sites located in the southern coast of the island. We collected female mealybugs using a soft paintbrush and stored them in ventilated boxes for laboratory identification. Each sample was placed in a plastic bag and taken to the laboratory for examination. The colonies were maintained in an aerated plastic box in a climate cabinet (25 ± 1 °C and 60 ± 5% RU), and each parasitized mealybug specimen was singularly isolated in small vials until the emergence of adults. The newly emerged parasitoids were initially preserved in 80% alcohol. These were card-mounted, and at least one specimen of each species (or parts of one specimen) was mounted on slides.

Table 1. Location, sites and host plants where natural enemies were found in association with *Phenacoccus solenopsis* in Sicily between October 2023 and November 2024.

Location	Coordinates		Host Plant	Species
	Latitude (N)	Longitude (E)		
Marina di Ragusa	36°47′11.7″	14°34′21.1″	*Lycopersicon esculentum* Mill. (Solanaceae) *Hibiscus* sp. L. (Malvaceae)	*Cryptolaemus montrouzieri* *Parexochomus nigripennis*
Siracusa	36°58′34.8″	15°13′55.5″	*Hibiscus* sp. L. (Malvaceae)	*Aenasius arizonensis*
Catania	37°30′42.7″	15°04′55.2″	*Hibiscus* sp. L. (Malvaceae)	*Aenasius arizonensis*
	37°31′35″	15°04′15.9″	*Lantana camara* L. (Verbenaceae)	*Anagyrus matritensis* *Cryptolaemus montrouzieri* *Hippodamia variegata* *Parexochomus nigripennis*

The terminology used for the description refers to Noyes and Hayat [12]. All measurements and pictures reported were obtained using a digital image acquisition and processing system connected to a microscope; observations were made using a Leica Flexacam C3 video camera connected to a Zeiss Axiophot optical microscope. Identification was achieved through an initial morphological study. To this aim, the authors used the most relevant identification keys and literature sources [12,15–18]. Parasitoid specimens were also identified by molecular analyses on mitochondrial DNA sequences as described below.

2.2. Molecular Identification

The molecular identification of the collected parasitoids was performed by sequencing the amplified mtCOI gene fragment as follows. DNA was isolated from a single specimen of *A. arizonensis* and *A. matritensis* from each site using the E.Z.N.A.® Tissue DNA Kit (Omega Bio-Tek, Inc., Norcross, GA, USA). A non-destructive DNA extraction protocol was followed to allow for the morphological identification of the processed specimens. Samples of *Anagyrus vladimiri* Triapitsyn (Hymenoptera: Encyrtidae) as a positive DNA control and a negative control without DNA were also included in the extraction. Universal primer pairs LCO1490 (5′-GGTCAACAAATCATAAAGATATTGG-3′), HCO2198 (5′-TAAACTTCAGGGTGACCAAAAAATCA-3′), C1-J-2195 (alias Jerry, 5′-TTGATTTTTGGTCATCCAGAAG-3′) and TL2-N-3014 (alias Pat, 5′-TCCAATGCACTAAT-CTGCCATATTA-3′) were used to amplify the expected ≈630 bp and ≈800 bp of the mt-COI target region for *A. arizonensis* [13] and *A. matritensis* [19], respectively. A PCR was performed according to the protocol proposed by [4]. Reactions were performed in 20 μL volumes with 0.85X FailSafe™ PCR 2X PreMix F (Lucigen Corporation, Middleton, WI, USA), 0.5 μM of each primer 10 μM, 1.5 U Taq DNA Polymerase 5U (Invitrogen, Thermo Fisher Scientific, Waltham, MA, USA) and 2 μL DNA template. Cycling conditions were as follows: 96 °C for 5 min, 35 cycles at 96 °C for 45 s, 45 °C for 1 min and 72 °C for 1 min, followed by a final cycle at 72 °C for 10 min. Reactions and cycling conditions were carried out in an Applied Biosystems™ MiniAmp™ Plus Thermal Cycler (Waltham, MA, USA). PCR products were first checked by electrophoresis using a 1% agarose gel, and those of the expected size were shipped to BMR Genomics (Padova, Italy) for purification and sequencing using Sanger's method. The coding regions were manually checked for errors and trimmed for low quality in Unipro UGENE version 1.26.1. The resulting FASTA files were aligned to reference sequences from the National Center for Biotechnology Information (NCBI) GenBank® using the Basic Local Alignment Search Tool (BLAST) sequence analysis tool for species identification. All sequences were deposited in GenBank under the following accession numbers: PQ736881 for *A. arizonensis* and PQ736880 for *A. matritensis*.

3. Results

3.1. Natural Enemies

Among the surveyed sites in Sicily (Table 1), we recorded the exotic parasitoid *Aenasius arizonensis* on *P. solenopsis*-infested ornamental plants located in Catania and Siracusa (Figure 1A,B), while the native *Anagyrus matritensis* (Figure 1C) was only found in Catania. In tomato crops and urban ornamentals located in Marina di Ragusa and Catania, the generalist predators *Cryptolaemus montrouzieri* (Mulsant), *Hippodamia variegata* (Goeze) and *Parexochomus nigripennis* (Erichson) (Coleoptera: Coccinellidae) were often reported in association with *P. solenopsis* infestation (Table 1).

Figure 1. *Phenacoccus solenopsis* parasitized cocoons with *Aenasius arizonensis* pupae visible in transparency (**A**), *Phenacoccus solenopsis* cocoons with *Aenasius arizonensis* emerging holes (**B**), *Anagyrus matritensis* adult emerging from *Phenacoccus solenopsis* cocoon (**C**).

3.2. Materials Examined

Specimens were collected in the samples taken from the field; the further materials examined came from a laboratory rearing.

Depository: the examined specimens were deposited at the Department of Agriculture, Food and Environment of the University of Catania (Italy). Other specimens were also deposited as follows: *A. arizonensis* (Girault) (Hymenoptera: Encyrtidae), 10 ♀♀7 ♂♂Catania, Italy 2 October 2024 ex *P. solenopsis* Tinsley (Hemiptera: Pseudococcidae) on *H. rosa-sinensis*

L. at The Smithsonian Institution, National Museum of Natural History Washington, DC, USA.

Abbreviations:

AOL—the minimum distance between the posterior and anterior ocellus.
OOL—the minimum distance between the eye margin and the nearest posterior ocellus.
POL—the minimum distance between the posterior ocelli.
F_{1-6}—the antennal articles.

Aenasius arizonensis **(Girault)**

Aenasius bambawalei Hayat 2009 syn.nov.

The species was originally described in detail by Hayat [17], and additional descriptions were provided later by other authors [20–23]. Some featured morphological characters are reported below.

Distribution: China, India, Iran, Iraq, Israel, Pakistan, Turkey and USA [6,10,12,24,25]. This is the first report of the species and the genus in Italy and Europe.

Female. Length: 1.50–2.30 mm. Body shiny black with metallic reflections bluish green to blackish on thorax and bronzy-violet to bluish green on abdomen (Figure 2A).

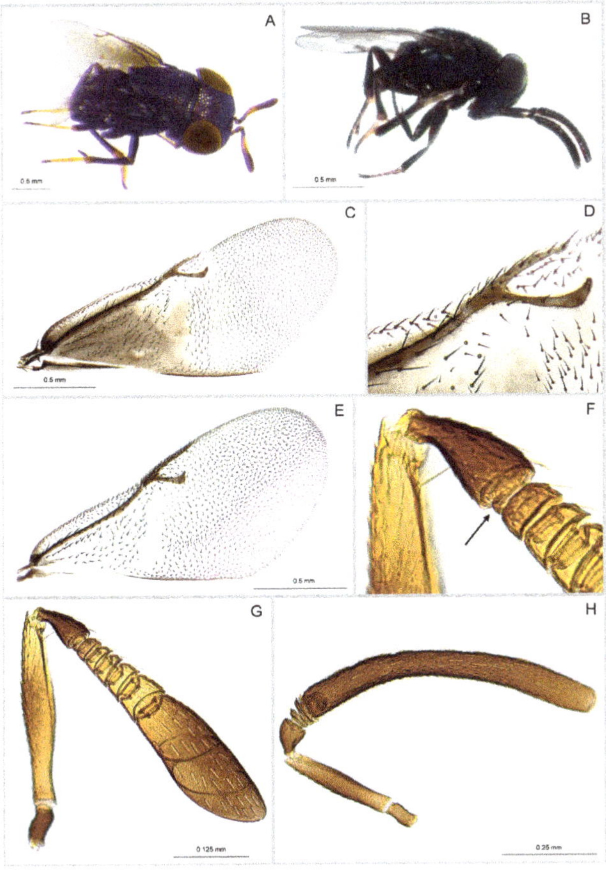

Figure 2. *Aenasius arizonensis* (Girault), female, habitus in dorso-lateral view (**A**); male, habitus in lateral view (**B**); female, fore wing (**C**); female, fore wing distal venations (**D**); male, fore wing (**E**); female antenna, particular of pedicel and first funicle segments, arrow indicates ring-shaped first segment (**F**); female antenna (**G**); male antenna (**H**).

Antenna (Figure 2G) with scape cylindrical, yellowish brown, with a brownish patch in the middle; pedicel black, segments F_1, F_2, F_3 brownish and F_4, F_5 and F_6 yellowish brown; clava dark brown, basally brownish yellow, strongly truncate at apex. Antennal funicle 6-segmented; segment F_1 is very narrow and poorly visible in the specimens that were collected in Sicily (Figure 2F). Fore wing (Figure 2C) brownish basally with distal half and costal cell hyaline; hind wing hyaline; the postmarginal vein slightly shorter than stigmal one. Legs, including coxae, black with apical portion of tibiae yellowish brown; all tarsi yellowish to yellowish brown and mid-tibial spur brown. Gaster smooth, very finely reticulated, shorter than thorax (thorax is 1.25× the abdomen length).

Male. Length: 1.10–1.50 mm. It is generally smaller than female (Figure 2B), and the sculpture and the colour of mesothoracic dorsum are similar to those of the female. Fore wing hyaline with the postmarginal vein longer than stigmal one (Figure 2E); hind wings hyaline. Antennal funicle with only two anelli and clava with a single, very long article (Figure 2H).

Anagyrus matritensis **(Mercet)**

Gyranusa matritensis Mercet, 1921.

Blastothrix orbitalis Ruschka, 1923. Synonymy by Koponen and Askew [26].

Anagyrus orbitalis (Ruschka): Trjapitzin, 1972: 255. Synonymy by Koponen and Askew, 2002 [24].

Anagyrus matritensis (Mercet): Noyes, 1981 [27].

The species was originally described by Mercet [28] on the basis of four females and five males collected in 1917 in the province of Madrid on a *Pinus halepensis* plant infested by an unidentified armoured scale; however, the scale has not been reported as a natural host. In this paper, only a small drawing of the female antenna is reported.

Distribution: Palaearctic with presence recorded in several European and Asian countries [25]. This is the first report of *A. matritensis* in Italy and its first record on *P. solenopsis*.

Blastothrix orbitalis was described by Ruschka [29] from four females collected in Austria in 1915 and two males from Mayr's collection; no information is given on the natural host.

The latest morphological description of *A. matritensis* was published in 2014 on two females caught in a yellow panel trap in Shandong Province, China [30]. The natural host and the male are unknown to the authors.

Other references on *A. matritensis* include morphological comparisons with other species, as follows:

(i) Compared with *A. descriptus* sp. n.; distinguishing characters for *A. matritensis* are not fully consistent with the original description; no details on males or host [31].
(ii) Compared with *A. kilinceri* sp. n.; distinguishing characters are not fully consistent with the original descriptions; no details on males or host [32].
(iii) The morphological details of *A. matritensis* are given in an identification key of the *Anagyrus* parasitoids of *Planococcus* and *Pseudococcus* mealybugs in eastern Spain; female habitus and fore wing are illustrated. Although the parasitoid is considered to be common in Spanish citrus orchards, information on the specific natural host was not provided [33].

Pending the complete review of this large genus-group of parasitoids, the identification of this species was confirmed by Dr. Emilio Guerrieri CNR-IPSP, Italy. The emergence of *A. matritensis* from *P. solenopsis* represents, to the best of our knowledge, a new association and opens new opportunities for the biological control of this pest in crop systems. The biology of this species is largely unknown, and studies of its host preference and parasitism behaviour are underway. The typical series of this species has not been studied, but the specimens collected are consistent with the original description [28] and other recent morphological descriptions of this species [30,33]. However, some additional characters and relevant variations found in our samples are included in the following description.

Female. Length 1.6–2.4 mm. Body slightly slender and not dorsoventrally flattened (Figure 3A); head testaceous brown with eye borders and occiput yellowish; a small yellowish spot is visible at the lower margin of the eye; antenna blackish; the base and the apical half of the scape and the basal part of F1 are whitish; the white proportion in the first antennal segment varies from one-fifths to three-fifths of the whole segment even though in the specimens examined, the white portion never exceed one-quarter; pronotum with lateral posterior margin and small light yellow spot on each side, prepectus and mesopleuron yellow; tegulae whitish with brown spots basally and apically; fore wing hyaline with a pattern of dark and pale setae forming a broad longitudinal hyaline band from the basal extreme to the apical one; hind wing entirely hyaline; legs whitish, with brown coxae and lines running through front and back on femurs and tibiae, particularly noticeable in the fore and hind legs (Figure 3F–H).

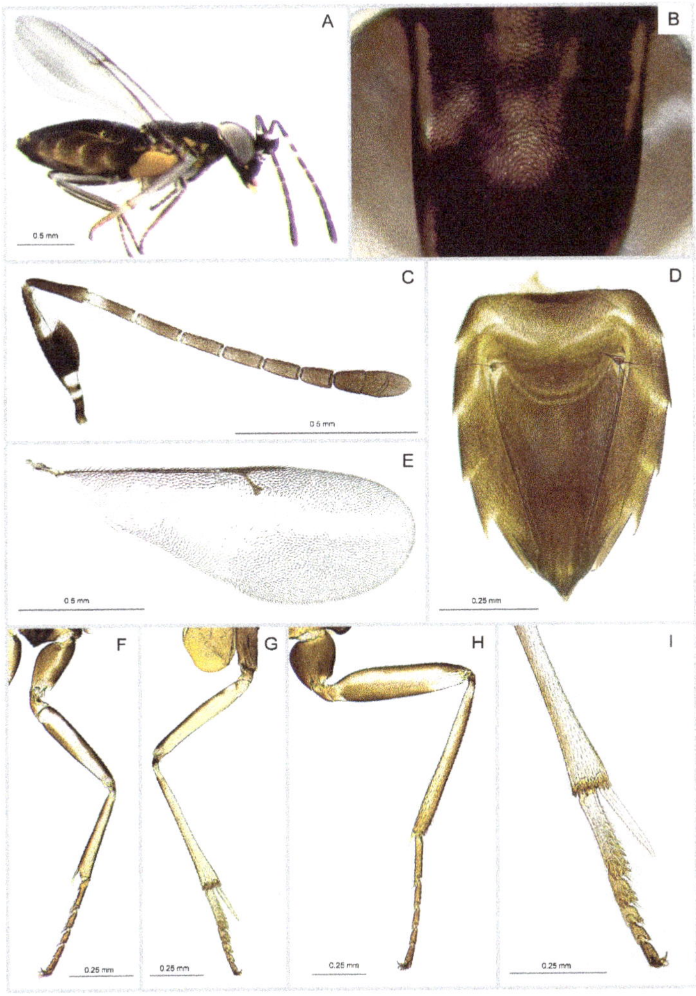

Figure 3. *Anagyrus matritensis* (Mercet) female. (**A**) Habitus in lateral view; (**B**) frontovertex with detail of sculpture; (**C**) antenna; (**D**) gaster; (**E**) fore wing; (**F**) fore leg; (**G**) mid-leg; (**H**) hind leg; (**I**) mid-tibial spur and tarsus.

Frontovertex, pronotum, mesoscutum, scutellum, propodeum laterally and gaster, with elongate to striate sculpture (Figure 3B,D); scape mostly reticulate; setae on frontovertex as long as diameter of anterior ocellus; head, mesosoma and gaster with dense white bristles.

Head in frontal view slightly (1.04×) wider than high; frontovertex 0.4× as broad as head; gena as long as the distance from antennal toruli; OOL larger than the diameter of one ocellus; ocelli forming an acute angle of about 85°; AOL = POL; eyes more than 2.5× as long as malar space; inner eye margins slightly diverging ventrally; dorsal margin of antennal torulus slightly above ventral eye margin; torulus separated from mouth margin by a little less than half its height; antennal scrobes moderately deep, not meeting apically; malar sulcus distinct. Mandible with two acute teeth; palp formula 4:3.

Antenna with scape elongate 3.7× longer than broad and about 2× as long as the pedicel (Figure 3C); the latter slightly longer than F_1; succeeding articles gradually slightly smaller; clava somewhat longer than previous two articles joined. Relative measurements: scape 986; pedicel 472; F_1 383; F_2 357; F_3 313; F_4 292; F_5 263; F_6 221; clava 607.

Mesoscutum barely as long as scutellum; posterior margin of pronotum concave; propodeum greatly irregular longitudinally with middle part very narrow; pleura reticulate-squamous; scutellum 0.9× as long as wide.

Fore wings with a hyaline longitudinal band running through the centre of the disc from the base to the apex; costal cell visible only in proximal third; marginal setae very short; fore wing 2.8× as long as broad (Figure 3E); linea calva partially visible and interrupted posteriorly; marginal and postmarginal veins combined are 2.2 × the stigma vein; postmarginal vein slightly longer than marginal vein and 1.4× as long as stigmal vein; hind wing completely hyaline and 4.3× as long as broad.

Distal end of tibia with a row of 11 thick and blunt spines; mid-tibial spur nearly as long as metatarsal (Figure 3I); apical half of metatarsus with a double row of five thick, blunt spines; second article with two apical spines; third article with only one blunt apical spine; fourth article without spines.

Gaster 1.7× longer than mesosoma (Figure 3D), narrowed towards the apex; dorsal surface of segments finely reticulated. Cercal plate is positioned at the upper quarter of the gaster, cercal bristles long and blackish in colour. Ovipositor short and not exserted, about one-third as long as the gaster.

Male. Length: 0.9–1.4 mm. General colouration as in female (Figure 4A), mostly testaceous and yellowish except for the head, the dorsal part of the mesosoma and partly the gaster, which are brownish. A small yellow spot is visible on each side of the base of the antennae; a yellow band is visible around the eye; legs are whitish, with brownish middle and hind coxae and lines running anteriorly and posteriorly through the femurs and tibiae, particularly pronounced on the fore and hind legs; the base of the middle tibiae has a whitish ring.

Antenna (Figure 4B) with the scape brownish and white at the base and at apex; clear-brownish funicle; scape elongated 4.4× long as broad and 4× the pedicel; all funicle segments are much longer than broad with F_1 the longest and F_6 the shortest; clava 1.7 longer than F_6 with scale-like sensilla in the basal half (Figure 4C); scattered long setae on funicle, the longest being more than three time as long as the segment width. Relative measurements: scape 612, pedicel 153, F_1 484, F_2 411, F_3 401, F_4 366, F_5 377, F_6 330, clava 584.

Wings entirely hyaline with blurred hyaline fringe; fore wing 2.7× as long as broad (Figure 4D); hind wing 4.1× as long as broad; linea calva well defined and interrupted by five rows of setae; postmarginal vein 2× longer than marginal vein and 1.5× the stigma vein (Figure 4E); marginal and postmarginal veins combined are 2.3× the stigma vein.

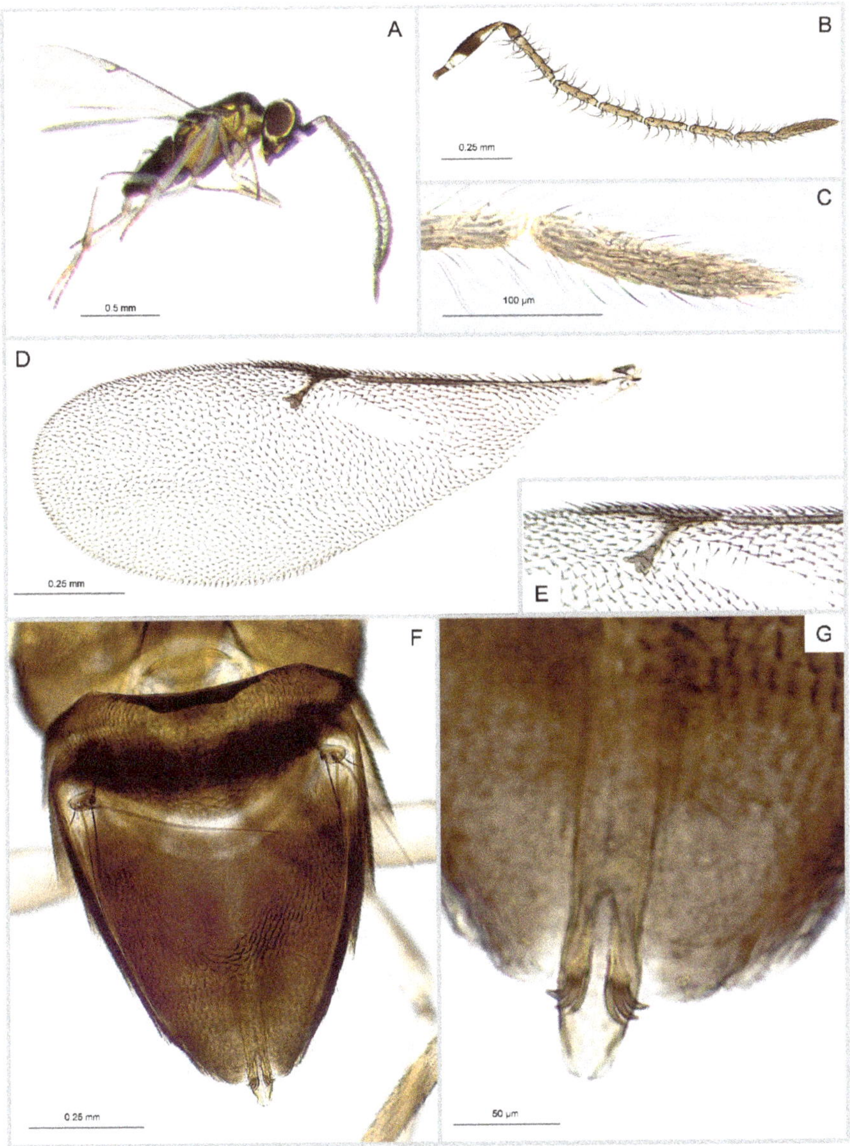

Figure 4. *Anagyrus matritensis* (Mercet) male. Habitus in lateral view (**A**); antenna (**B**); club showing scale-like sensilla (**C**); fore wing (**D**); fore wing distal venations (**E**); gaster (**F**); genitalia (**G**).

Gaster almost triangular (Figure 4F) about 1.2× as long as broad (at its base in dorsal view); cercal plate positioned at the upper quarter of the gaster, cercal bristles long (the longest about half the gaster length) and blackish; genitalia about 2× as long as mid-tibia; phallobase with well-developed digit, each with three short hooks (Figure 4G).

3.3. Molecular Identification

Based on mtCOI sequences, each parasitoid species that emerged from the *P. solenopsis* individuals sampled in geographically separated areas in Sicily shared the same haplotype. Subsequently, we confirmed the morphological identification of *A. arizonensis* through mtCOI fragment amplification from genomic DNA, followed by direct sequencing and BLAST searches. The resulting sequences were aligned to reference sequences from NCBI compared with publicly available data on GenBank and yielded an identity score of 100% and E-value = 0.0 with an *A. arizonensis* isolate from Southern China (accession number MT775615.1). On the other hand, no publicly available sequences for *A. matritensis* targeting the ≈800 mtCOI region were previously deposited on GenBank; therefore, our submission can be considered the first one for this taxon.

4. Discussion

Among alien invasive species, the solenopsis mealybug can cause a series of negative implications for the sustainability of cultivation systems in the Mediterranean because of its cryptic behaviour and the number of hosts [3]. For these reasons, the exploitation of sustainable control tools such as biological control are needed for mealybug pest control, and the quick identification of indigenous natural enemies is necessary to provide effective measures [10]. Following the introduction of *P. solenopsis* in Sicily [4], we reported for the first time in Europe the exotic parasitoid *A. arizonensis*, and we also recorded in Italy the presence of *A. matritensis* on *P. solenopsis* using morphological and molecular identification tools. In addition, three coccinellid predators (Coleoptera: Coccinellidae), namely *C. montrouzieri*, *H. variegata* and *P. nigripennis*, were consistently reported attacking *P. solenopsis* in different sites in the southern coast of the island.

The solitary endoparasitoid *A. arizonensis* was reported attacking *P. solenopsis* in India, and it has been documented as a very effective biological control agent with a parasitization rate up to 72% on cotton fields [20]. However, a hyperparasitoid that attacks *A. arizonensis*, *Promuscidea unfasciativentris*, may potentially reduce the efficiency of this biological control agent [10]. The discovery that the *A. arizonensis* haplotype from Sicily matches with the haplotype from individuals isolated in China suggests a possible introduction from Asia as we previously documented for *P. solenopsis* [4]. The introduction of exotic natural enemies by their hosts is likely to occur in the case of cryptic species such as mealybugs. Moreover, since we found *A. arizonensis* in the urban centre, this suggests a clear phenotypic plasticity of the species to adapt to a warmer and polluted environment with potential positive consequences for its biocontrol activity [34]. Future works should be promoted on the distribution, field parasitism rates and biology of *A. arizonensis* in regions of recent introduction, following the approach used for other invasive exotic pests [35,36].

The genus *Anagyrus* includes over 350 described species of primary endoparasitoids of mealybugs. Many of these species have been used in biological control programmes against mealybug pests [19]. Various recent revisions [19,37,38] pointed out and separated entities misidentified for a long time (e.g., the *Anagyrus pseudococci* species complex). Several species within this complex have been included in biocontrol programmes [19], and in this context, the correct identification of pests and released biological control agents is even more crucial for potential environmental and applied consequences. For this purpose, the combination, as performed in this study, of morphological and genetic approaches is essential [39]. To the best of our knowledge, this is the first record of *A. matritensis* in Italy, and more interestingly, this is the first confirmed report of a natural host for this parasitoid. The species is considered to be native to the Mediterranean area, and most likely, it switched to *P. solenopsis* from native mealybugs hosts which are still unknown and should be further investigated. Species identification is crucial when biological control strategies are implemented [40]. The accurate identification of natural enemies is crucial in rearing facilities, with the objective of targeting the cotton mealybug and thereby preventing potential economic losses.

More than 50 natural enemies, including both predators and parasitoids, have been recorded attacking *P. solenopsis* across its wide distribution range, with coccinellid predators (Coleoptera: Coccinellidae) being the most important taxon [10]. We consistently found the coccinellids *C. montrouzieri*, *H. variegata* and *P. nigripennis* to be in association with *P. solenopsis*. *Cryptolaemus montrouzieri* was often reported preying on *P. solenopsis* in China, India, Israel, Pakistan and Australia [6,10], and its role can be exploited in integrated pest management programmes against mealybug pests [41]. Meanwhile, *H. variegata* and *P. nigripennis* have been reported to feed on *P. solenopsis* in India [10] and Israel [6], respectively, and their function as promising biological control agents should be studied in laboratory and field conditions. Moreover, considering that the abundance of generalist natural enemy trophic groups leads to a significant reduction in pest abundance, understanding how plant diversity associated with tomato field structure arthropod food webs may improve the ecological management of *P. solenopsis* [42,43].

The integrated (morphological and molecular) taxonomic approach used in this study to investigate the natural enemy community associated with *P. solenopsis* should improve the accurate identification of these organisms, which is of great importance for the development of an effective biological control programme for this invasive pest in the Mediterranean. In addition, this study contributes to a better understanding of some of the ecological interactions between the invasive *P. solenopsis* and the natural enemies associated with it. Overall, continuously updating the morphological taxonomic keys associated with new molecular tools for the fast and precise identification of invasive pests is warranted.

Author Contributions: Conceptualization, M.R. and G.S.; methodology, M.R., E.P., A.R. and G.S.; formal analysis, M.R. and G.S.; investigation, M.R. and G.S.; resources, G.S. and L.Z.; writing—original draft preparation, G.S.; writing—review and editing, all authors; visualization and acquisition, L.Z. All authors have read and agreed to the published version of the manuscript.

Funding: This research was funded by the project "Agroecology-inspired Strategies and Tools to Enhance Resilience and ecosystem services in tomato crop" (ASTER), Prima Section 2 2021-MULTI-TOPIC, the Agritech National Research Center, and received funding from the European Union Next-Generation EU (Piano Nazionale di Ripresa e Resilienza (PNRR)–Missione 4 Componente 2, Investimento 1.4–D.D. 1032 17/06/2022, CN00000022) and the European Union's Horizon Europe research and innovation programme under grant agreement 101136611 (NextGenBioPest project). This manuscript reflects only the authors' views and opinions, neither the European Union nor the European Commission can be considered responsible for them.

Data Availability Statement: The datasets generated in the study are available from the corresponding author upon request.

Acknowledgments: We are very grateful to Emilio Guerrieri (CNR-IPSP, Italy) for confirming the initial identification of *A. matritensis* and valuable discussion on its taxonomical position; Luke Kresslein and Michael Gates (USDA, ARS, SEL, Smithsonian Institution, National Museum of Natural History Washington, DC) for confirming *A. arizonensis* and Karol Szawaryn (Museum and Institute of Zoology, Polish Academy of Sciences, Poland) for *P. nigripennis* identification.

Conflicts of Interest: The authors declare no conflicts of interest. The funders had no role in the design of the study; in the collection, analyses, or interpretation of data; in the writing of the manuscript; or in the decision to publish the results.

References

1. Hulme, P.E. Climate change and biological invasions: Evidence, expectations, and response options. *Biol. Rev.* **2017**, *92*, 1297–1313. [CrossRef]
2. Villegas-Luján, R.; Plowes, R.; Gilbert, L.E.; Rodríguez, J.C.; Canales-del-Castillo, R.; Gallegos-Morales, G.; España-Luna, M.P.; Fernández-Triana, J.; Sanchez-Peña, S.R. Redescription of *Apanteles mimoristae* (Hymenoptera, Braconidae), a parasitoid of the native pyralid cactus moth *Melitara* cf. *nephelepasa* in central Mexico. *J. Hymenopt. Res.* **2024**, *97*, 207–228. [CrossRef]

3. Waqas, M.S.; Shi, Z.; Yi, T.C.; Xiao, R.; Shoaib, A.A.; Elabasy, A.S.; Jin, D.C. Biology, ecology, and management of cotton mealybug *Phenacoccus solenopsis* Tinsley (Hemiptera: Pseudococcidae). *Pest Manag. Sci.* **2021**, *77*, 5321–5333. [CrossRef]
4. Ricupero, M.; Biondi, A.; Russo, A.; Zappalà, L.; Mazzeo, G. The cotton mealybug is spreading along the Mediterranean: First pest detection in Italian tomatoes. *Insects* **2021**, *12*, 675. [CrossRef] [PubMed]
5. Abbes, K.; Harbi, A.; Wanassi, T.; Ricupero, M.; Mazzeo, G.; Russo, A.; Biondi, A.; Zappalà, L.; Chermiti, B. The cotton mealybug *Phenacoccus solenopsis* Tinsley (Hemiptera: Pseudococcidae) is spreading in North Africa: First report in Tunisia. *Orient. Insects* **2024**, *58*, 172–186. [CrossRef]
6. Spodek, M.; Ben-Dov, Y.; Mondaca, L.; Protasov, A.; Erel, E.; Mendel, Z. The cotton mealybug, *Phenacoccus solenopsis* Tinsley (Hemiptera: Pseudococcidae) in Israel: Pest status, host plants and natural enemies. *Phytoparasitica* **2018**, *46*, 45–55. [CrossRef]
7. Zhou, A.; Lu, Y.; Zeng, L.; Xu, Y.; Liang, G. *Solenopsis invicta* (Hymenoptera: Formicidae), defend *Phenacoccus solenopsis* (Hemiptera: Pseudococcidae) against its natural enemies. *Environ. Entomol.* **2013**, *42*, 247–252. [CrossRef] [PubMed]
8. Menchetti, M.; Schifani, E.; Alicata, A.; Cardador, L.; Sbrega, E.; Toro-Delgado, E.; Vila, R. The invasive ant *Solenopsis invicta* is established in Europe. *Curr. Biol.* **2023**, *33*, 896–897. [CrossRef]
9. Shankarganesh, K.; Ricupero, M.; Sabtharishi, S. Field evolved insecticide resistance in the cotton mealybug *Phenacoccus solenopsis* and its direct and indirect impacts on the endoparasitoid *Aenasius arizonensis*. *Sci. Rep.* **2022**, *12*, 16764. [CrossRef]
10. Tong, H.J.; Yan, A.O.; Li, Z.H.; Ying, W.A.N.G.; Jiang, M.X. Invasion biology of the cotton mealybug, *Phenacoccus solenopsis* Tinsley: Current knowledge and future directions. *J. Integr. Agric.* **2019**, *18*, 758–770. [CrossRef]
11. Chen, H.Y.; Li, H.L.; Pang, H.; Zhu, C.D.; Zhang, Y.Z. Investigating the Parasitoid community associated with the invasive mealybug *Phenacoccus solenopsis* in Southern China. *Insects* **2021**, *12*, 290. [CrossRef] [PubMed]
12. Noyes, J.S.; Hayat, M. *Oriental Mealybug Parasitoids of the Anagyrini (Hymenoptera: Encyrtidae)*; CAB International: Wallingford, UK, 1994; p. viii+554.
13. Folmer, O.; Black, M.; Hoeh, W.; Lutz, R.; Vrijenhoek, R. DNA primers for amplification of mitochondrial cytochrome c oxidase subunit I from diverse metazoan invertebrates. *Mol. Marine Biol. Biotechnol.* **1994**, *3*, 294–299.
14. Mazzeo, G.; Ricupero, M.; Campo, G.; Conti, F.; Russo, A. New records of *Ripersiella* (Tinsley) (Hemiptera: Coccomorpha: Rhizoecidae) species and identification key of Rhizoecidae in Italy. *Phytoparasitica* **2023**, *51*, 1047–1057. [CrossRef]
15. Trjapitzin, V.A. Review of the genera of *Palaearctic encyrtids* (Hymenoptera, Encyrtidae). *Tr. Vsesoyuznogo Entomol. Obs.* **1971**, *5*, 68–155.
16. Guerrieri, E.; Noyes, J.S. Revision of european species of genus Metaphycus Mercet (Hymenoptera: Chalcidoidea: Encyrtidae), parasitoids of scale insects (Homoptera: Coccoidea). *Syst. Entomol.* **2000**, *25*, 147–222. [CrossRef]
17. Hayat, M. Description of a new species of *Aenasius* Walker (Hymenoptera: Encyrtidae), parasitoid of the mealybug, *Phenacoccus solenopsis* Tinsley (Homoptera: Pseudococcidae) in İndia. *Biosystematica* **2009**, *3*, 21–25.
18. Poorani, J. An illustrated guide to the lady beetles (Coleoptera: Coccinellidae) of the Indian Subcontinent. Part II. Tribe Chilocorini. *Zootaxa* **2023**, *5378*, 1–108. [CrossRef] [PubMed]
19. Andreason, S.A.; Triapitsyn, S.V.; Perring, T.M. Untangling the *Anagyrus pseudococci* species complex (Hymenoptera: Encyrtidae), parasitoids of worldwide importance for biological control of mealybugs (Hemiptera: Pseudococcidae): Genetic data corroborates separation of two new, previously misidentified species. *Biol. Control* **2019**, *129*, 65–82. [CrossRef]
20. Poorani, J.; Rajeshwaris, S.K.; Ankita, G. Notes on diagnosis and biology of *Aenasius bambawalei* Hayat (Hymenoptera: Encyrtidae), a parasitoid of the invasive mealybug, *Phenacoccus solenopsis* Tinsley (Hemiptera: Sternorrhyncha: Pseudococcidae). *J. Biol. Control* **2009**, *23*, 463–466.
21. Çalışkan, A.F.; Hayat, M.; Ulusoy, M.R.; Kaydan, M.B. Parasitoids (Hymenoptera: Encyrtidae) of an invasive mealybug *Phenacoccus solenopsis* Tinsley (Hemiptera: Pseudococcidae) in Turkey. *Turk. J. Entomol.* **2016**, *40*, 133–148. [CrossRef]
22. KeÇe, A.F.Ç.; Kahya, D.; Hayat, M.; Ulusoy, M.R. A new record of a parasitoid (Hymenoptera: Encyrtidae) of an invasive mealybug *Phenacoccus solenopsis* Tinsley (Hemiptera: Pseudococcidae) from Turkey. *Turk. J. Entomol.* **2018**, *9*, 31–37.
23. Bodlah, I.; Ahmad, M.; Nasir, M.F.; Naeem, M. Record of *Aenasius bambawalei* Hayat, 2009 (Hymenoptera: Encyrtidae), a parasitoid of *Phenacoccus solenopsis* (Sternorrhyncha: Pseudococcidae) from Punjab, Pakistan. *Pak. J. Zool.* **2010**, *42*, 533–536.
24. Khudhair, A.H.; Kareem, A.A.; Watson, G.W.; Kresslein, R.L.; Beasley, J.; Topakcı, N.; Polaszek, A. Parasitoids of the invasive cotton mealybug, *Phenacoccus solenopsis* Tinsley (Hemiptera, Pseudococcidae), in Iraq. *BioControl* **2024**, 1–9. Available online: https://link.springer.com/article/10.1007/s10526-024-10299-6 (accessed on 20 December 2024). [CrossRef]
25. Noyes, J.S. Universal Chalcidoidea Database. World Wide Web Electronic Publication, 2024. Available online: https://www.nhm.ac.uk/our-science/data/chalcidoids/database/ (accessed on 20 November 2024).
26. Koponen, M.; Askew, R.R. Chalcids from Madeira, Canary Islands and Azores (Hymenoptera, Chalcidoidea). *Vieraea (Folia Sci. Biol. Canar.)* **2002**, *30*, 115–145.
27. Noyes, J.S. On the types of the species of Encyrtidae described by R. García Mercet (Hymenoptera: Chalcidoidea). *EOS. Rev. Española Entomol.* **1981**, *55/56*, 165–189.
28. Mercet, R.G. *Fauna Iberica. Himenopteros Fam. Encírtidos*; Museo Nacional de Ciencas Naturales: Madrid, Spain, 1921; p. 733.

29. Ruschka, F. Beiträge zur Kenntnis einiger Encyrtiden-gattungen (Hym., Chalc.). *Verhandlungen Zool. Bot. Ges. Wien.* **1923**, *72*, 1–13.
30. Zu, G.H.; Li, C.D. Description of three new species and new distributional data for four species of *Anagyrus* from China. *Zootaxa* **2015**, *4028*, 257–273. [CrossRef]
31. Japoshvili, G. New data on encyrtid (Hymenoptera: Chalcidoidea: Encyrtidae) parasitoids of Coccoids (Hemiptera: Coccoidea) from Turkey, with description of five new species. *Entomol. Fenn.* **2012**, *23*, 72–82. [CrossRef]
32. Kaydan, M.B.; Japoshvili, G. The aphelinid and encyrtid parasitoids (Hymenoptera: Chalcidoidea) of Pseudococcidae (Hemiptera: Coccoidea) in the Van Lake basin of Turkey. *Turk. J. Entomol.* **2010**, *34*, 465–476.
33. Soler, J.M.; Falcó-Garí, J.V.; Herrero, B.; Aquino, D.A. The Anagyrus (Hymenoptera: Encyrtidae) parasitoids of the obscure mealybug *Pseudococcus viburni* (Hemiptera: Pseudococcidae) in Spain, with description of a new species. *Zootaxa* **2021**, *4980*, 83–98. [CrossRef]
34. Le Lann, C.; Van Baaren, J.; Visser, B. Dealing with predictable and unpredictable temperatures in a climate change context: The case of parasitoids and their hosts. *J. Exp. Biol.* **2021**, *224*, jeb238626. [CrossRef]
35. Zapponi, L.; Tortorici, F.; Anfora, G.; Bardella, S.; Bariselli, M.; Benvenuto, L.; Bernardinelli, I.; Butturini, A.; Caruso, S.; Colla, R.; et al. Assessing the Distribution of Exotic Egg Parasitoids of *Halyomorpha halys* in Europe with a Large-Scale Monitoring Program. *Insects* **2021**, *12*, 316. [CrossRef] [PubMed]
36. Zappalà, L. Citrus integrated pest management in Italy. In *Integrated Management of Arthropod Pests and Insect Borne Diseases, Integrated Management of Plant Pests and Diseases*; Ciancio, A., Mukerji, K.G., Eds.; Springer: Berlin/Heidelberg, Germany, 2010; Volume 5. [CrossRef]
37. Triapitsyn, S.V.; González, D.; Vickerman, D.B.; Noyes, J.S.; White, E.B. Morphological, biological, and molecular comparisons among the different geographical populations of *Anagyrus pseudococci* (Hymenoptera: Encyrtidae), parasitoids of *Planococcus* spp. (Hemiptera: Pseudococcidae), with notes on *Anagyrus dactylopii*. *Biol. Control* **2007**, *41*, 14–24. [CrossRef]
38. Andreason, S.A.; Triapitsyn, S.V.; Mendel, Z.; Protasov, A.; Levi, B.; Perring, T.M. Expanding the *Anagyrus pseudococci* species complex (Hymenoptera: Encyrtidae): A new species emerges from *Trabutina serpentina* (Hemiptera: Pseudococcidae) on tamarisk on the Dead Sea shore. *Zool. Middle East* **2019**, *65*, 347–360. [CrossRef]
39. Guerrieri, E.; Cascone, P. *Anagyrus* Howard (Hymenoptera: Encyrtidae) parasitoids of the invasive *Delottococcus aberiae* (De Lotto) (Hemiptera: Pseudococcidae) from South Africa, with description of two new species. *Zootaxa* **2018**, *4531*, 374–382. [CrossRef]
40. Sugawara, Y.; Mita, T.; Tabata, J.; Ueno, T. Genetic and morphological approach to reappraising species validity in two different Anagyrus wasps (Hymenoptera: Encyrtidae) attracted by cyclolavandulyl butyrate. *Entomol. Sci.* **2020**, *23*, 152–164. [CrossRef]
41. Modafferi, A.; Ricupero, M.; Mostacchio, G.; Latella, I.; Zappalà, L.; Palmeri, V.; Garzoli, S.; Giunti, G.; Campolo, O. Bioactivity of *Allium sativum* essential oil-based nano-emulsion against *Planococcus citri* and its predator *Cryptolaemus montrouzieri*. *Ind. Crops Prod.* **2024**, *208*, 117837. [CrossRef]
42. Hatt, S.; Xu, Q.; Francis, F.; Osawa, N. Aromatic plants of East Asia to enhance natural enemies towards biological control of insect pests. A review. *Entomol. Gen.* **2019**, *38*, 275–315. [CrossRef]
43. Dassou, A.G.; Vodouhé, S.D.; Bokonon-Ganta, A.; Goergen, G.; Chailleux, A.; Dansi, A.; Carval, D.; Tixier, P. Associated cultivated plants in tomato cropping systems structure arthropod communities and increase the *Helicoverpa armigera* regulation. *Bull. Entomol. Res.* **2019**, *109*, 733–740. [CrossRef]

MDPI AG
Grosspeteranlage 5
4052 Basel
Switzerland
Tel.: +41 61 683 77 34

Insects Editorial Office
E-mail: insects@mdpi.com
www.mdpi.com/journal/insects

www.ingramcontent.com/pod-product-compliance
Lightning Source LLC
Chambersburg PA
CBHW061542220926
51850CB00005B/232
* 9 7 8 3 7 2 5 8 5 4 1 1 0 *